D0215120

Algebra

Pure & Applied

Algebra

Pure & Applied

Aigli Papantonopoulou
The College of New Jersey

PRENTICE HALL, Upper Saddle River, NJ 07458

Library of Congress Cataloging-in-Publication Data

Papantonopoulou, Aigli
Algebra : pure & applied / Aigli Papantonopoulou
p. cm.
Includes bibliographical references and index.
ISBN 0-13-088254-2
1. Algebra, Abstract. I. Title.

QA162 .P36 2002
515'.02--dc21

2001021899

Acquisitions Editor: George Lobell
Vice President/Director of Production and Manufacturing: David W. Riccardi
Executive Managing Editor: Kathleen Schiaparelli
Senior Managing Editor: Linda Mihatov Behrens
Production Editor: Steven S. Pawlowski
Manufacturing Buyer: Alan Fischer
Manufacturing Manager: Trudy Pisciotti
Marketing Manager: Angela Battle
Marketing Assistant: Vince Jansen
Director of Marketing: John Tweeddale
Editorial Assistant: Melanie Van Benthuysen
Art Director: Jayne Conte
Cover Designer: Bruce Kenselaar
Cover Image: Paul Dance/Stone

Upper Saddle River, NJ 07458

Printed in the United States of America

10 9 8 7 6 5 4 3 2 1

ISBN 0-13-088254-2

Prentice-Hall International (UK) Limited, London
Prentice-Hall of Australia Pty. Limited, Sydney
Prentice-Hall Canada, Inc., Toronto
Prentice-Hall Hispanoamericana, (S.A.) Mexico
Prentice-Hall of India Private Limited, New Delhi
Pearson Education Asia Pte. Ltd.
Editora Prentice-Hall do Brasil, Ltda., Rio de Janeiro

Αφιερωμένο
στ' αξιοθαύμαστα παιδιά μου

Αλέξη και Φωκίων

Contents

Preface

This book aims to provide thorough coverage of the main topics of abstract algebra while remaining accessible to students with little or no previous exposure to abstract mathematics. It can be used either for a one-semester introductory course on groups and rings or for a full-year course. More specifics on possible course plans using the book are given in this preface.

Style of Presentation

Over many years of teaching abstract algebra to mixed groups of undergraduates, including mathematics majors, mathematics education majors, and computer science majors, I have become increasingly aware of the difficulties students encounter making their first acquaintance with abstract mathematics through the study of algebra. This book, based on my lecture notes, incorporates the ideas I have developed over years of teaching experience on how best to introduce students to mathematical rigor and abstraction while at the same time teaching them the basic notions and results of modern algebra.

Two features of the teaching style I have found effective are *repetition* and especially an *examples first, definitions later* order of presentation. In this book, as in my lecturing, the hard conceptual steps are always prepared for by working out concrete examples first, before taking up rigorous definitions and abstract proofs. Absorption of abstract concepts and arguments is always facilitated by first building up the student's intuition through experience with specific cases.

Another principle that is adhered to consistently throughout the main body of the book (Parts A and B) is that every algebraic theorem mentioned is given either with a complete proof, or with a proof broken up into to steps that the student can easily fill in, without recourse to outside references. The book aims to provide a self-contained treatment of the main topics of algebra, introducing them in such a way that the student can follow the arguments of a proof without needing to turn to other works for help.

Throughout the book all the examples, definitions, and theorems are consecutively numbered in order to make locating any particular item easier for the reader.

Coverage of Topics

In order to accommodate students of varying mathematical backgrounds, an optional Chapter 0, at the beginning, collects basic material used in the development of the main theories of algebra. Included are, among other topics, equivalence relations, the binomial theorem, De Moivre's formula for complex numbers, and the fundamental theorem of arithmetic. This chapter can be included as part of an introductory course or simply referred to as needed in later chapters.

Special effort is made in Chapter 1 to introduce at the beginning all main types of groups the student will be working with in later chapters. The first section of the chapter emphasizes the fact that concrete examples of groups come from different sources, such as geometry, number theory, and the theory of equations.

Chapter 2 introduces the notion of group homomorphism first and then proceeds to the study of normal subgroups and quotient groups. Studying the properties of the kernel of a homomorphism before introducing the definition of a normal subgroup makes the latter notion less mysterious for the student and easier to absorb and appreciate. A similar order of exposition is adopted in connection with rings. After the basic notion of a ring is introduced in Chapter 6, Chapter 7 begins with ring homomorphisms, after which consideration of the properties of the kernels of such homomorphisms gives rise naturally to the notion of an ideal in a ring.

Each chapter is designed around some central unifying theme. For instance, in Chapter 4 the concept of group action is used to unify such results as Cayley's theorem, Burnside's counting formula, the simplicy of A_5, and the Sylow theorems and their applications.

The ring of polynomials over a field is the central topic of Part B, Rings and Fields, and is given a full chapter of its own, Chapter 8. The traditional main topic in algebra, the solution of polynomial equations, is emphasized. The solutions of cubics and quartics are introduced in Chapter 8. In Chapter 9 Euclidean domains and unique factorization domains are studied, with a section devoted to the Gaussian integers. The fundamental theorem of algebra is stated in Chapter 10. In Chapter 11 the connection among solutions of quadratic, cubic, and quartic polynomial equations and geometric constructions is explored.

In Chapter 12, after Galois theory is developed, it is applied to give a deeper understanding of all these topics. For instance, the possible Galois groups of cubic and quartic polynomials are fully worked out, and Artin's Galois-theoretic proof of the fundamental theorem of algebra, using nothing from analysis but the intermediate value theorem, is presented. The chapter, and with it the main body of the book, culminates in the proof of the insolubility of the general quintic and the construction of specific examples of quintics that are not solvable by radicals.

A brief history of algebra is given in Chapter 13, after Galois theory (which was the main historical source of the group concept) has been treated, thus making a more meaningful discussion of the evolution of the subject possible.

A collection of additional topics, several of them computational, is provided in Part C. In contrast to the main body of the book (Parts A and B), where completeness is the goal, the aim in Part C is to give the student an introduction to — and some taste of — a topic, after which a list of further references is provided for those who wish to learn more. Instructors may include as much or as little of the material on a given topic as time and inclination indicate.

Each chapter in the book is divided into sections, and each section provided with a set of exercises, beginning with the more computational and proceeding to the more theoretical. Some of the theoretical exercises give a first introduction to topics that will be treated in more detail later in the book, while others introduce supplementary topics not otherwise covered, such as Cayley digraphs, formal power series, and the existence of transcendental numbers.

Suggestions for Use

A one-semester introductory course on groups and rings might include Chapter 0 (optional); Chapters 1, 2, and 3 on groups; and Chapters 6, 7, and 8 on rings.

For a full-year course, Parts A and B, Chapters 1 through 12, offer a comprehensive treatment of the subject. Chapter 9, on Euclidean domains, and Chapter 11, on geometric constructions, can be treated as optional supplementary topics, depending on time and the interest of the students and the instructor.

An instructor's manual, with solutions to all exercises plus further comments and suggestions, is available. Instructors can obtain it by directly contacting the publisher, Prentice Hall.

Acknowledgments

It is a pleasure to acknowledge various contributors to the development of this book. First I should thank the students of The College of New Jersey who have taken courses based on a first draft. I am grateful also to my colleagues Andrew Clifford, Tom Hagedorn, and Dave Reimer for useful suggestions.
Special thanks are due to my colleague Ed Conjura, who taught from a draft of the book and made invaluable suggestions for improvement that have been incorporated into the final version.
I am also most appreciative of the efforts of the anonymous referees engaged by the publisher, who provided many helpful and encouraging comments.
My final word of gratitude goes to my family — to my husband, John Burgess, and to our sons, Alexi and Fokion — for their continuous understanding and support throughout the preparation of the manuscript.

Aigli Papantonopoulou
The College of New Jersey
aigli@tcnj.edu

Algebra

Pure & Applied

Chapter 0

Background

In this preliminary chapter we review some fundamental mathematical notions that we refer to throughout later chapters.

Maps, *one to one* maps, *onto* maps, and *composition* of maps are basic concepts that appear and reappear, often in different forms. *Equivalence relations* on sets, and the *partitions* they determine, are also used frequently, especially in constructing new algebraic structures from old. The set of integers with the usual operations of addition and multiplication and all their various major properties repeatedly provide the root examples and models for general algebraic concepts and constructions. In connection with the integers, mathematical *induction* is an extremely useful method of proof with which it is important to become comfortable. Calculations to find *binomial coefficients* and algorithms to find *greatest common divisors* are among the methods of computation of which it is convenient to have background knowledge. The set of *complex numbers* with the usual operations also plays a significant role. *Matrices*, too, provide a number of examples to illustrate new algebraic notions, and knowledge of their most basic properties will be useful.

You probably have met most of these topics in one form or another in your previous study of mathematics. The essential information about them reviewed in this chapter is just what is needed for the main body of this book and is collected here in one place for easy reference.

0.1 Sets and Maps

In this section we introduce the basic notations for sets and operations on sets, as well as symbols for some specific sets of central importance. In addition we introduce the terminology for different kinds of maps between sets, and with it the notion of the cardinality of a set.

0.1.1 EXAMPLE Some sets that are especially important in algebra and have special names and symbols are the following:

$\mathbb{Z} = \{0, \pm 1, \pm 2, \pm 3, \ldots\}$ = the set of all integers

$\mathbb{Q} = \{a/b \mid a, b \in \mathbb{Z}, b \neq 0\}$ = the set of all rational numbers

$\mathbb{R} = \{\pm x_1 x_2 \ldots x_n . y_1 y_2 y_3 \ldots \mid 0 \leq x_i, y_j \leq 9\}$ = the set of all real numbers (as decimals)
$\mathbb{C} = \{a + bi \mid a, b \in \mathbb{R}, i^2 = -1\}$ = the set of all complex numbers ◇

0.1.2 EXAMPLE Other sets with special names and symbols include
$2\mathbb{Z} = \{2n \mid n \in \mathbb{Z}\}$ = the set of even integers
$\mathbb{R}^+ = \{x \in \mathbb{R} \mid x > 0\}$ = the set of positive real numbers ◇

0.1.3 DEFINITION Given two sets A and B, A is a **subset** of B, written $A \subseteq B$, if every element of A is an element of B. Two sets are equal, $A = B$, if and only if $A \subseteq B$ and $B \subseteq A$. The **union** of A and B is the set

$$A \cup B = \{x \mid x \in A \text{ or } x \in B\}$$

and the **intersection** of A and B is the set

$$A \cap B = \{x \mid x \in A \text{ and } x \in B\}$$

The **Cartesian product** of A and B is the set

$$A \times B = \{(a, b) \mid a \in A, b \in B\}$$

which is to say the set of all ordered pairs with the first component from A and the second component from B. ○

0.1.4 DEFINITION Given two sets A and B, a **function** or **map** from A to B assigns to each element of A exactly one element of B. We write $\phi: A \to B$ to indicate that ϕ is a map from A to B. A map must be well defined, which means that if ϕ is specified by a rule assigning to each element of A an element of B, the rule must unambiguously assign to each element of A *one* and *only one* element of B. If $\phi: A \to B$ is a map from A to B, we write $\phi(a)$ for the element of B assigned to the element a of A, called the **image** of a under ϕ. For A' a subset of A, we write $\phi(A') = \{\phi(a) \mid a \in A'\}$, also called the **image** of A' under ϕ. ○

0.1.5 EXAMPLE $\phi: \mathbb{Z} \to \{0, 1\}$ defined by the rule

$$\phi(n) = \begin{cases} 0 \text{ if } n \text{ is even} \\ 1 \text{ if } n \text{ is odd} \end{cases}$$

is well defined, but $\psi: \mathbb{Z} \to \{0, 1\}$ defined by the rule

$$\psi(n) = \begin{cases} 0 \text{ if } n \text{ is even} \\ 1 \text{ if } n \text{ is a multiple of } 3 \end{cases}$$

is not well defined, since the rule does not tell us what $\psi(1)$ is, and tells us both that $\psi(6) = 0$ and that $\psi(6) = 1$. ◇

0.1.6 EXAMPLE Let $\phi: \mathbb{Z} \to \mathbb{Z}$ be given by $\phi(n) = 2n$. Then for any two integers n and m, if $\phi(n) = \phi(m)$, then $2n = 2m$, which implies $n = m$. ◇

0.1.7 EXAMPLE Let $\chi: \mathbb{Z} \to \mathbb{Z}$ be given by $\phi(n) = n^2$. Then for any two integers n and m, if $\chi(n) = \chi(m)$, then $n^2 = m^2$, which implies $n = \pm m$. ◇

0.1.8 DEFINITION A map $\phi: A \to B$ is said to be **one to one** if $a_1 \neq a_2$ in A always implies $\phi(a_1) \neq \phi(a_2)$ in B. ○

Thus the map in Example 0.1.6 is one to one, while the map in Example 0.1.7 is not.

0.1.9 EXAMPLE Let ϕ: $\mathbb{Z} \to 2\mathbb{Z}$ be given by $\phi(n) = 2n$. Then for any $y \in 2\mathbb{Z}$, since y is even, $x = y/2$ is an integer and $\phi(x) = y$. So in this case $\phi(\mathbb{Z})$ is all of $2\mathbb{Z}$. ◇

0.1.10 EXAMPLE Let ϕ: $\mathbb{R} \to \mathbb{R}^+$ be defined by $\phi(x) = e^x$. Then for every $y \in \mathbb{R}^+$, since y is positive, $\ln y$ is a real number, and $\phi(\ln y) = e^{\ln y} = y$. So in this case $\phi(\mathbb{R})$ is all of $\mathbb{R}^+$. ◇

0.1.11 DEFINITION A map ϕ: $A \to B$ is said to be **onto** if for every $y \in B$ there is an $x \in A$ such that $\phi(x) = y$. In this case, $\phi(A) = B$. ○

Thus in Examples 0.1.10 and 0.1.11, the maps are onto, while in Examples 0.1.6 and 0.1.7 they are not.

0.1.12 EXAMPLE Let ϕ: $\mathbb{Z} \to 2\mathbb{Z}$ be given by $\phi(n) = 2n$ for all $n \in \mathbb{Z}$, and let χ: $2\mathbb{Z} \to 10\mathbb{Z}$ be given by $\chi(m) = 5m$ for all $m \in 2\mathbb{Z}$. Let $\chi \circ \phi$: $\mathbb{Z} \to 10\mathbb{Z}$ be given by $\chi \circ \phi(n) = \chi(\phi(n))$ for all $n \in \mathbb{Z}$. Then $\chi \circ \phi(n) = \chi(2n) = 5 \cdot 2n = 10n$. Note that each of ϕ, χ, and $\chi \circ \phi$ is both one to one and onto. ◇

0.1.13 EXAMPLE Let ϕ: $\mathbb{R} \to \mathbb{R}$ and χ: $\mathbb{R} \to \mathbb{R}$ be given by $\phi(x) = 2x$ and $\chi(x) = x^2$ for all $x \in \mathbb{R}$. Let $\chi \circ \phi$: $\mathbb{R} \to \mathbb{R}$ and $\phi \circ \chi$: $\mathbb{R} \to \mathbb{R}$ be given by $\chi \circ \phi(x) = \chi(\phi(x))$ and $\phi \circ \chi(x) = \phi(\chi(x))$ for all $x \in \mathbb{R}$. In this case, $\chi \circ \phi(x) = \chi(2x) = (2x)^2 = 4x^2$, while $\phi \circ \chi(x) = \phi(x^2) = 2x^2$. So $\chi \circ \phi \neq \phi \circ \chi$. Note also that though ϕ is one to one and onto, none of χ, $\chi \circ \phi$, or $\phi \circ \chi$ is either. ◇

0.1.14 DEFINITION Let ϕ: $A \to B$ and χ: $B \to C$ be two maps. Then we define the **composite** map $\chi \circ \phi$: $A \to C$ to be $\chi \circ \phi(a) = \chi(\phi(a))$ for all $a \in A$. ○

0.1.15 THEOREM Let ϕ: $A \to B$ and χ: $B \to C$ and ψ: $C \to D$ be three maps. Then

(1) (**Associativity**) $\psi \circ (\chi \circ \phi) = (\psi \circ \chi) \circ \phi$.

(2) If ϕ and χ are both one to one, then so is $\chi \circ \phi$.

(3) If ϕ and χ are both onto, then so is $\chi \circ \phi$.

Proof (1) For any $x \in A$ we have $\psi \circ (\chi \circ \phi)(x) = \psi((\chi \circ \phi)(x)) = \psi(\chi(\phi(x))) = (\psi \circ \chi)(\phi(x)) = (\psi \circ \chi) \circ \phi(x)$.

(2) Suppose that ϕ and χ are both one to one and consider any $x, y \in A$ for which we have $(\chi \circ \phi)(x) = (\chi \circ \phi)(y)$. Then $\chi(\phi(x)) = \chi(\phi(y))$, and since χ is one to one, we must have $\phi(x) = \phi(y)$. But then since ϕ is one to one, we must have $x = y$.

(3) Suppose that ϕ and χ are both onto and consider any $z \in C$. Since χ is onto, there must be some $y \in B$ such that $\chi(y) = z$. And since ϕ is onto, there must be some $x \in A$ such that $\phi(x) = y$. But then $\chi(\phi(x)) = \chi(y) = z$, and since $(\chi \circ \phi)(x) = \chi(\phi(x))$, we have found an element $x \in A$ with $(\chi \circ \phi)(x) = z$. □

0.1.16 DEFINITION For any set A we define the **identity** map ρ_0: $A \to A$ by $\rho_0(x) = x$ for all $x \in A$. ○

0.1.17 PROPOSITION Let A be any set and ρ_0: $A \to A$ the identity map. Then

(1) ρ_0 is one to one and onto.

(2) For any set B and any map ϕ: $A \to B$, we have $\phi \circ \rho_0 = \phi$.

(3) For any set B and any map ϕ: $B \to A$, we have $\rho_0 \circ \phi = \phi$.

Proof Left to the reader (as Exercise 11 at the end of this section). □

0.1.18 EXAMPLE Let ϕ: $\mathbb{Z} \to 3\mathbb{Z}$ be defined by $\phi(n) = 3n$ for all $n \in \mathbb{Z}$. Now consider the map χ: $3\mathbb{Z} \to \mathbb{Z}$ defined by $\chi(m) = m/3$ for all $m \in 3\mathbb{Z}$. Then $\chi \circ \phi(n) = \chi(\phi(n)) = 3n/3 = n$ for all $n \in \mathbb{Z}$. Hence $\chi \circ \phi$ is the identity map on $\mathbb{Z}$. Also, $\phi \circ \chi(m) = \phi(\chi(m)) = \phi(m/3) = 3(m/3) = m$ for all $m \in 3\mathbb{Z}$, so $\phi \circ \chi$ is the identity map on $3\mathbb{Z}$. ◇

0.1.19 DEFINITION Let ϕ: $A \to B$. Then ϕ is said to be **invertible** if there exists a map ϕ^{-1}: $B \to A$ such that $\phi^{-1} \circ \phi$ is the identity map on A and $\phi \circ \phi^{-1}$ is the identity map on B. ϕ^{-1} is said to be an **inverse** to ϕ. ○

0.1.20 THEOREM Let ϕ: $A \to B$ be invertible. Then

(1) There is a *unique* inverse ϕ^{-1} to ϕ.

(2) $(\phi^{-1})^{-1} = \phi$, that is, ϕ is the inverse to ϕ^{-1}.

Proof (1) Suppose there are two maps χ: $B \to A$ and ψ: $B \to A$ with $\chi \circ \phi = \psi \circ \phi = \rho_0{}^A$, the identity map on A, and $\phi \circ \chi = \phi \circ \psi = \rho_0{}^B$, the identity map on B. Then $\chi = \chi \circ \rho_0{}^B = \chi \circ (\phi \circ \psi) = (\chi \circ \phi) \circ \psi = \rho_0{}^A \circ \psi = \psi$.

(2) This is immediate from the definition of *inverse*. □

0.1.21 EXAMPLE Let ϕ: $\{1, 2, 3, 4\} \to \{1, 2, 3, 4\}$ be defined by $\phi(1) = 2$, $\phi(2) = 4$, $\phi(3) = 3$, $\phi(4) = 1$, or as on the left in Figure 1.

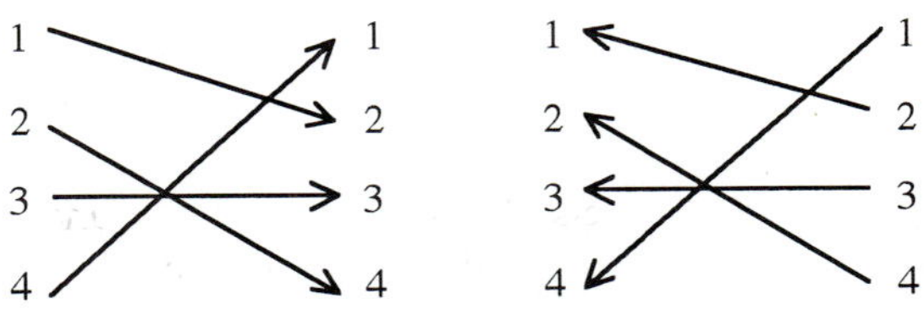

FIGURE 1

Then ϕ^{-1} is defined by $\phi^{-1}(1) = 4$, $\phi^{-1}(2) = 1$, $\phi^{-1}(3) = 3$, $\phi^{-1}(4) = 2$, or as on the right in Figure 1. ◇

0.1.22 EXAMPLE Let $\phi: \mathbb{R} \to \mathbb{R}^{\geq}$, where $\mathbb{R}^{\geq}$ is the set of nonnegative real numbers, be defined by $\phi(x) = |x|$ for all $x \in \mathbb{R}$. Note that ϕ is onto but not one to one, since for instance $\phi(-3) = \phi(3) = 3$.
And ϕ is not invertible, since to specify that $\phi^{-1}(y)$ should be some x such that $\phi(x) = y$ leaves it undetermined whether we should have $\phi^{-1}(3) = 3$ or $\phi^{-1}(3) = -3$. In other words, we do not get a well-defined map from this specification. ◇

0.1.23 EXAMPLE Let $\phi: \mathbb{Z} \to \mathbb{Z}$ be defined by $\phi(n) = 5n$ for all $n \in \mathbb{Z}$. Note that ϕ is one to one but not onto, since, for example, $6 \neq 5n$ for any n. And ϕ is not invertible, since to specify that $\phi^{-1}(m)$ should be some n such that $\phi(n) = m$ leaves it undetermined what $\phi^{-1}(6)$ should be. In other words, we do not get a well-defined map from this specification. ◇

0.1.24 THEOREM Let $\phi: A \to B$ and $\chi: B \to C$ be two maps. Then

(1) ϕ is invertible if and only if ϕ is one to one and onto.

(2) If ϕ and χ are invertible, then $\chi \circ \phi$ is invertible, and $(\chi \circ \phi)^{-1} = \phi^{-1} \circ \chi^{-1}$.

Proof (1) ($\Rightarrow$) Suppose $\phi^{-1}: B \to A$ exists. Then for any x and y in A, $\phi(x) = \phi(y)$ implies $x = \phi^{-1}(\phi(x)) = \phi^{-1}(\phi(y)) = y$. So ϕ is one to one. And given any $u \in B$, $\phi(\phi^{-1}(u)) = u$, so letting $x = \phi^{-1}(u)$ we have found an element $x \in A$ with $\phi(x) = u$. So ϕ is onto. ($\Leftarrow$) Suppose ϕ is one to one and onto. Define $\tau: B \to A$ as follows. For any $u \in B$, let $\tau(u)$ be the $x \in A$ such that $\phi(x) = u$. Since ϕ is onto, there will be some such x. And since ϕ is one to one, there will be only one, since if $\phi(x) = \phi(y) = u$, then $x = y$. So the specification just gives a well-defined function τ. Furthermore, $\phi(\tau(u)) = u$ for any $u \in B$ by definition of τ, and also $\tau(\phi(x)) = x$. Hence $\tau = \phi^{-1}$ and ϕ is invertible.

(2) Assume that both ϕ and χ are invertible. Then by (1) they are both one to one and onto. So by Theorem 0.1.15 $\chi \circ \phi$ is one to one and onto, and so by (1) again $\chi \circ \phi$ is invertible. Also, we have

$$(\phi^{-1} \circ \chi^{-1}) \circ (\chi \circ \phi) = \phi^{-1} \circ (\chi^{-1} \circ \chi) \circ \phi = \phi^{-1} \circ \rho_0^B \circ \phi = \phi^{-1} \circ \phi = \rho_0^A$$

and therefore $\phi^{-1} \circ \chi^{-1} = (\chi \circ \phi)^{-1}$. □

0.1.25 DEFINITION Given two sets A and B, we say A and B have the **same cardinality**, and we write $|A| = |B|$ if there exists a one-to-one and onto map $\phi: A \to B$. ○

0.1.26 EXAMPLE Two finite sets have the same cardinality if and only if they have the same number of elements. $|\mathbb{Z}| = |2\mathbb{Z}| = |n\mathbb{Z}|$ for any integer $n \geq 1$, since the map $\phi: \mathbb{Z} \to n\mathbb{Z}$ defined by $\phi(x) = nx$ for all $x \in \mathbb{Z}$ is one to one and onto. ◇

Exercises 0.1

In Exercises 1 through 7 determine whether or not the indicated maps are one to one.

1. $\phi: \mathbb{R} \to \mathbb{R}$, where $\phi(x) = 5x + 3$

2. $\phi: \mathbb{R} \to \mathbb{R}$, where $\phi(x) = e^x$

3. $\phi: \mathbb{R} \to \mathbb{R}$, where $\phi(x) = x^3$

4. $\phi: \mathbb{Z} \to \mathbb{Z}$, where $\phi(x) = x^2$

5. $\phi: \mathbb{Q}^* \to \mathbb{Q}^*$, where $\mathbb{Q}^*$ is the set of nonzero rational numbers, and $\phi(n/m) = m/n$

6. $\phi: \mathbb{R}^+ \to \mathbb{R}^+$, where $\mathbb{R}^+$ is the set of positive real numbers, and $\phi(x) = x^4$

7. $\phi: \mathbb{Z} \times \mathbb{Z}^* \to \mathbb{Q}$, where $\mathbb{Z}^*$ is the set of nonzero integers, and $\phi(m,n) = m/n$

In Exercises 8 through 10 determine whether or not the indicated maps are onto.

8. $\phi: \mathbb{R}^+ \to \mathbb{R}$, where $\phi(x) = \ln x$

9. $\phi: \mathbb{R} \to \mathbb{R}$, where $\phi(x) = x^2 - 4$

10. $\phi: \{1, 2, 3, 4\} \to \{1, 2, 3, 4\}$, where ϕ is any one-to-one map

11. Prove Proposition 0.1.17.

In Exercises 12 through 14 determine whether or not the indicated maps are invertible.

12. $\phi: \mathbb{R} \to \mathbb{R}$, where $\phi(x) = |x + 1|$

13. $\phi: \mathbb{R} \to \mathbb{R}$, where $\phi(x) = (5x + 3)/2$

14. $\phi: \{1, 2, 3, \ldots , n\} \to \{1, 2, 3, \ldots , n\}$, where $\phi(i) = i + 2$ for $1 \le i \le n - 2$ and $\phi(n - 1) = 1$ and $\phi(n) = 2$

15. Let $\phi: A \to B$ and $\chi: B \to C$ be two maps. Show that

(a) If $\chi \circ \phi$ is onto, then χ must be onto.

(b) If $\chi \circ \phi$ is one to one, then ϕ must be one to one.

16. Show that $|\mathbb{Z} \times \mathbb{Z}| = |2\mathbb{Z} \times 2\mathbb{Z}|$ (where $|\;|$ indicates cardinality as in Definition 0.1.25).

17. Show that if A is a finite set with $|A| = n$, then $|A \times A| = n^2$.

18. Show that if $|A| = |B|$ and $|C| = |D|$, then $|A \times C| = |B \times D|$.

0.2 Equivalence Relations and Partitions

The notion of an equivalence relation on a set plays an important role in many constructions in algebra. As we see in this section, an equivalence relation on a set determines a partition of the set into non-overlapping pieces and, conversely, any such partition determines an equivalence relation on the set.

0.2.1 EXAMPLE On the set $\mathbb{Z}$ of all integers, consider the relation $\sim$ defined by the condition $a \sim b$ if and only if $a - b$ is divisible by 5, for any $a, b \in \mathbb{Z}$. Note the following properties of $\sim$:

(1) For any integer a we have $a - a = 0$, which is divisible by 5, so $a \sim a$.

(2) For any integers a and b, $a - b = -(b - a)$, so if $a \sim b$, meaning that $a - b$ is divisible by 5, then so is $b - a$, and we have $b \sim a$.

(3) For any integers a, b, and c, if $a \sim b$ and $b \sim c$, then $a - b = 5n$ and $b - c = 5m$ for some integers n and m. But then $a - c = (a - b) + (b - c) = 5n + 5m = 5(n + m)$, and so we have $a \sim c$.

Now let us take, say, the integer 7, and find the subset $[7] = \{x \in \mathbb{Z} \mid x \sim 7\}$ of $\mathbb{Z}$ consisting of all integers x such that $x \sim 7$. Note that $7 \sim 2$, and therefore if $x \sim 7$, then $x \sim 2$ by property (3). Likewise, since $2 \sim 7$, if $x \sim 2$, then $x \sim 7$. So $x \sim 7$ if and only if $x \sim 2$, which is to say if and only if $x - 2 = 5k$, or, equivalently, $x = 2 + 5k$ for some integer k. Thus $[7] = 2 + 5\mathbb{Z}$, the set of all integers that can be written as the sum of 2 plus a multiple of 5. ◇

0.2.2 EXAMPLE Let $P(\mathbb{Z})$ be the set of all subsets of $\mathbb{Z}$, and consider the relation $\sim$ on $P(\mathbb{Z})$ defined by letting $S \sim T$ if and only if $|S| = |T|$, that is, if and only if S and T have the same cardinality. So $S \sim T$ if and only if there is a one-to-one, onto map $\phi: S \to T$. (See Definition 0.1.25.) Note the following properties of $\sim$:

(1) For any $S \in P(\mathbb{Z})$, the identity map on S is a one-to-one, onto map from S to itself. Therefore, $S \sim S$.

(2) For any $S, T \in P(\mathbb{Z})$, if $S \sim T$, then there is a one-to-one, onto map $\phi: S \to T$. Then $\phi^{-1}: T \to S$ is one to one and onto by Theorems 0.1.20 and 0.1.24, so $T \sim S$.

(3) For any $S, T, U \in P(\mathbb{Z})$, if $S \sim T$ and $T \sim U$, then there are one-to-one, onto maps $\phi: S \to T$ and $\chi: T \to U$. Then $\chi \circ \phi: S \to U$ is one to one and onto by Theorem 0.1.24, so $S \sim U$.

In this example, if S is finite, then $[S] = \{T \in P(\mathbb{Z}) \mid S \sim T\}$ consists of all subsets of $\mathbb{Z}$ that have the same number of elements as S. If S is infinite, $[S]$ consists of all infinite subsets of $\mathbb{Z}$. ◇

0.2.3 DEFINITION A **relation** on a nonempty set S is a subset R of $S \times S$. Let R be a relation on S and write aRb to mean that $(a, b) \in R$. Then R is an **equivalence relation** on S if it satisfies the following three conditions for all $a, b, c \in S$:

(1) **Reflexivity** aRa
(2) **Symmetry** If aRb, then bRa.
(3) **Transitivity** If aRb and bRc, then aRc.

If R is an equivalence relation on S, then for any $a \in S$, the **equivalence class** of a is the set $[a] = \{b \in S \mid aRb\}$. ○

In Examples 0.2.1 and 0.2.2 the relations $\sim$ were equivalence relations.

We prove some important properties of equivalence classes that are used frequently in algebraic constructions.

0.2.4 THEOREM Let $\sim$ be an equivalence relation on a set S, and let $a, b \in S$ be any elements of S. Then

(1) $a \in [a]$.
(2) If $a \in [b]$, then $[a] = [b]$.
(3) $[a] = [b]$ if and only if $a \sim b$.
(4) Either $[a] = [b]$ or $[a] \cap [b] = \varnothing$.

Proof (1) Reflexivity tells us $a \sim a$ and therefore $a \in [a]$.

(2) If $a \in [b]$, then by definition of equivalence classes we have $b \sim a$, and symmetry tells us we have $a \sim b$. Now if $x \in [a]$, then $a \sim x$, and transitivity tells us $b \sim x$, so $x \in [b]$. Thus $[a] \subseteq [b]$. Similarly, if $y \in [b]$ then $b \sim y$, and transitivity tells us $a \sim y$, so $y \in [a]$. Thus $[b] \subseteq [a]$ and $[a] = [b]$.

(3) ($\Rightarrow$) Suppose $[a] = [b]$. Since by (1) we have $b \in [b]$ it follows that $b \in [a]$, which by definition means $a \sim b$. ($\Leftarrow$) Suppose $a \sim b$. By definition, this means $b \in [a]$, and by (2) it follows that $[a] = [b]$.

(4) Suppose that $[a] \cap [b] \neq \varnothing$. This means there is some c such that $c \in [a]$ and $c \in [b]$. By (2) it follows that $[c] = [a]$ and $[c] = [b]$, so $[a] = [b]$. □

The fact that an equivalence relation divides a set into disjoint or non-overlapping pieces, the equivalence classes, is what makes equivalence relations so useful in algebraic constructions. In the next example, instead of starting with an equivalence relation and using it to divide up a set, we start by dividing a set and use the division to define an equivalence relation.

0.2.5 EXAMPLE Starting with the set $\mathbb{R}$ of all real numbers, let $[1] = \{x \in \mathbb{R} \mid 0 \leq x - 1 < 1\}$. In other words, $[1]$ is the half-closed, half-open interval $[1, 2)$ in $\mathbb{R}$. Similarly, for any integer n let $[n] = \{x \in \mathbb{R} \mid 0 \leq x - n < 1\} = [n, n + 1)$. Note that for any distinct integers $i \neq j$ we have $[i] \cap [j] = \varnothing$, and for any real number $x \in \mathbb{R}$, $x \in [n]$ where n is the greatest integer such that $n \leq x$. So we have divided $\mathbb{R}$ into disjoint pieces. If now we define a relation $\sim$ on $\mathbb{R}$ by letting $x \sim y$ if and only if $x \in [n]$ and $y \in [n]$ for the same integer n, then it can be checked that $\sim$ is an equivalence relation on $\mathbb{R}$. (See Exercise 9 at the end of this section.) ◇

It is convenient to name such a division of a set into disjoint pieces.

0.2.6 DEFINITION Let S be a nonempty set. A **partition** of S consists of a collection $\{P_i\}$ of nonempty subsets of S such that

(1) $S = \bigcup_i P_i$

(2) For any P_i, P_j in the collection, either $P_i = P_j$ or $P_i \cap P_j = \varnothing$.

The subsets P_i in the collection are called the **cells** of the partition. ○

We now come to the main theorem connecting equivalence relations and partitions, generalizing what we observed in Example 0.2.5.

0.2.7 THEOREM Let S be a nonempty set.

(1) Given an equivalence relation $\sim$ on S, the collection of equivalence classes under $\sim$ is a partition of S.

(2). Given a partition $\{P_i\}$ of S, there is an equivalence relation on S whose equivalence classes are precisely the cells of the partition.

Proof (1) Given an equivalence relation $\sim$, by Theorem 0.2.4, part (1), $a \in [a]$ for each $a \in S$, and therefore $S = \bigcup_a [a]$, which is the condition (1) for being a partition in Definition 0.2.6. Theorem 0.2.4, part (4), is precisely the condition (2) for being a partition in Definition 0.2.6.

(2) Given a partition $\{P_i\}$, define a relation $\sim$ by letting $a \sim b$ if and only if $a \in P_i$ and $b \in P_i$ for the same cell P_i. By condition (1) of Definition 0.2.6, any $a \in S$ does belong to some cell in the partition, and of course a then belongs to the same cell as itself, so we have $a \sim a$. If $a \sim b$, then a and b belong to the same cell of the partition, which is the same as saying b and a belong to the same cell, and we have $b \sim a$. If $a \sim b$ and $b \sim c$, then a belongs to the same cell P_i in the partition as b and b belongs to the same cell P_j in the partition as c. Since $b \in P_i \cap P_j$, by condition (2) of Definition 0.2.6 we must have $P_i = P_j$, and a and c belong to the same cell of the partition, and $a \sim c$. Finally, given $a \in S$, let $a \in P_i$. Then $x \in [a]$ if and only if $a \sim x$, hence if and only if a and x belong to the same cell of the partition or, in other words, if and only if $x \in P_i$. So the equivalence class of a is $[a] = P_i$. □

Exercises 0.2

In Exercises 1 through 8 determine whether the indicated relation is an equivalence relation on the indicated set and, if so, describe the equivalence classes.

1. In $\mathbb{R}$ $a \sim b$ if and only if $|a| = |b|$

2. In $\mathbb{R}$ $a \sim b$ if and only if $a \leq b$

3. In $\mathbb{Z}$ a $\sim$ b if and only if $a - b$ is even

4. In $\mathbb{R}$ $a \sim b$ if and only if $|a - b| \leq 1$

5. In $\mathbb{Z}$ $a \sim b$ if and only if $a = b$ + some multiple of 3

6. In $\mathbb{R} \times \mathbb{R}$ - {(0, 0)}, the real plane with the origin removed, $(x_1, y_1) \sim (x_2, y_2)$ if and only if $x_1y_2 = x_2y_1$

7. In $\mathbb{R} \times \mathbb{R}$ $(x_1, y_1) \sim (x_2, y_2)$ if and only if $x_1^2 + y_1^2 = x_2^2 + y_2^2$

8. In $\mathbb{R} \times \mathbb{R}$ $(x_1, y_1) \sim (x_2, y_2)$ if and only if $3y_1 - 5x_1 = 3y_2 - 5x_2$

9. Show that the relation on $\mathbb{R}$ defined in Example 0.2.5 is an equivalence relation.

10. In $\mathbb{R}$ consider the half-open, half-closed intervals $(n, n + 2]$, where n is any even integer. Show that the collection of these intervals is a partition of $\mathbb{R}$, and describe the equivalence relation this partition determines.

11. In the plane $\mathbb{R} \times \mathbb{R}$, explain why defining $(x_1, y_1) \sim (x_2, y_2)$ if and only if $x_1y_2 = x_2y_1$ does not give an equivalence relation.

12. Fix an integer n and define on $\mathbb{Z}$ the relation $a \sim b$ if and only if $a - b$ is divisible by n. Show that this is an equivalence relation on $\mathbb{Z}$ and describe the equivalence classes.

13. Let $\phi: S \rightarrow T$ be any map and define a relation $\sim$ on S by letting $a \sim b$ if and only if $\phi(a) = \phi(b)$. Show that $\sim$ is an equivalence relation on S.

0.3 Properties of $\mathbb{Z}$

In this section we establish some basic properties of the integers, many of which will be important later in identifying examples of various kinds of algebraic structures, where $\mathbb{Z}$ will play the role of a basic model. We begin with properties of the usual order relation on $\mathbb{Z}$ and then turn to properties involving the familiar operations of addition, subtraction, multiplication, and division. Finally, we introduce new algebraic structures closely related to the integers, called the *integers mod n*, for any integer $n > 1$.

Induction

Many proofs and constructions are based on the following fundamental property of the positive integers.

0.3.1 AXIOM (Well-ordering principle) Every nonempty set of positive integers contains a least element. ☆

Frequently in proving some theorem or constructing some structure we will want to pick the least element from some given nonempty set of positive integers. The well-ordering principle tells us such a least element always exists.
Closely related to the well-ordering principle is another principle, mathematical induction, that is equally important in proofs and constructions. We use the well-ordering principle to prove the principle of mathematical induction.

0.3.2 THEOREM (Principle of mathematical induction) Let $P(n)$ be a statement about a positive integer n such that

(1) $P(1)$ is true.

(2) If $P(k)$ is true, then $P(k + 1)$ is true.

Then $P(n)$ is true for all positive integers n.

Proof The proof will be by contradiction. Suppose there exists a positive integer n for which $P(n)$ is not true. Then the set S of all positive integers n for which $P(n)$ is not true is nonempty. By the well-ordering principle, S must have a least element m. By assumption (1), this least element cannot be 1, since $P(1)$ is true. So $m - 1$ is still positive. Since $m - 1 < m$, and m was the least positive integer for which the statement was not true, $P(m - 1)$ is true. So by assumption (2), $P((m - 1) + 1)$ is true, which is to say $P(m)$ is true. This is a contradiction, which shows that our original supposition that $P(n)$ was not true for some positive integers n is false. □

It is impossible to overemphasize the usefulness of mathematical induction. We next use it in a variety of examples to show something of the diversity of problems to which it can be applied.

0.3.3 EXAMPLE Let us prove that for any positive integer n the sum of the first n odd numbers is the square of n, or $1 + 3 + 5 + \dots + (2n - 1) = n^2$. As in all proofs by mathematical induction, we begin with the base step, proving the statement for 1. In this case the statement for 1 is just $1 = 1^2$, which is clear. Next we do the induction step, making the assumption, called the *induction hypothesis*, that the statement holds for k, and using this assumption to prove that the statement holds for $k + 1$. So what we are assuming is that $1 + 3 + 5 + \dots + (2k - 1) = k^2$, and what we want to prove is that $1 + 3 + 5 + \dots + (2k - 1) + (2(k + 1) - 1) = (k+1)^2$. What our assumption tells us is that $1 + 3 + 5 + \dots + (2k - 1) + (2(k + 1) - 1) = k^2 + (2(k + 1) - 1)$, and we easily see that $k^2 + (2(k + 1) - 1) = k^2 + 2k + 1 = (k+1)^2$, as required to complete the proof. ◇

0.3.4 EXAMPLE Let us prove that for any $n \geq 0$, a set S with n elements has exactly 2^n subsets. First let us dispose of the case $n = 0$. In this case the only set with n elements is the empty set $\varnothing$, and this has only one subset, itself, while $2^0 = 1$, so the statement is true in this case. Now consider the case $n = 1$. In this case a set $S = \{a\}$ with one element has just two subsets, $S = \{a\}$ and $\varnothing$, while $2^1 = 2$, so the statement is true in this case also. Now for the induction step. We assume that every set with k elements has 2^k subsets and want to prove that every set with $k + 1$ elements has 2^{k+1}

subsets. So let S be any set with $k + 1$ elements. Let a be any element of S, and consider the set $T = S - \{a\}$ consisting of S with a removed. T is a set with k elements, and so has 2^k subsets by assumption. The subsets of S that do not contain a are just the subsets of T. Therefore S has 2^k such subsets. Any subset of S that *does* contain a consists of a subset that does not, with a added to it. So there are again 2^k such subsets. Then, S has $2^k + 2^k = 2^{k+1}$ subsets, as required to complete the proof. ◇

0.3.5 EXAMPLE Let us prove that for any real numbers a, b and any integer $n \geq 1$ we have $a^{n+1} - b^{n+1} = (a - b)(a^n + a^{n-1}b + \dots + ab^{n-1} + b^n)$. For the base step $n = 1$ the statement is just $a^2 - b^2 = (a - b)(a + b)$, which is clear. For the induction step, assume $a^k - b^k = (a - b)(a^{k-1} + a^{k-2}b + \dots + ab^{k-2} + b^{k-1})$. We then have

$$(a - b)(a^k + a^{k-1}b + \dots + ab^{k-1} + b^k) =$$
$$(a - b)[a(a^{k-1} + a^{k-2}b + \dots + ab^{k-2} + b^{k-1}) + b^k) =$$
$$a(a - b)(a^{k-1} + a^{k-2}b + \dots + ab^{k-2} + b^{k-1}) + (a - b)b^k =$$
$$a(a^k - b^k) + (a - b)b^k = a^{k+1} - ab^k + ab^k - b^{k+1} = a^{k+1} - b^{k+1}$$

as required to complete the proof. ◇

Often it is convenient to use a slightly different strong version of mathematical induction, which is actually equivalent to the ordinary version used in the preceding examples. (See Exercise 11 at the end of this section.) The difference between this version and the original version of the principle is that the induction hypothesis is stronger. We do not just assume the statement we are interested in holds for $m - 1$ in order to prove it for m. Instead, we assume the statement holds for *all* $1 \leq k < m$.

0.3.6 THEOREM (Strong induction) Let $P(n)$ be a statement about a positive integer n such that

(1) $P(1)$ is true.

(2) If $P(k)$ is true for all k, $1 \leq k < m$, then $P(m)$ is true.

Then $P(n)$ is true for all positive integers n.

Proof Let $Q(n)$ be the statement that $P(k)$ holds for all $1 \leq k \leq n$. We prove that $Q(n)$ holds for all positive integers n by ordinary mathematical induction (as in Theorem 0.3.2). Since $Q(n)$ implies P(n), this implies that $P(n)$ holds for all positive integers n. For the base step, $Q(1)$ just amounts to $P(1)$ and is true by assumption (1). For the induction step, we assume $Q(m)$ holds and prove that $Q(m + 1)$ holds. Here $Q(m + 1)$ is the statement that $P(k)$ holds for all $1 \leq k \leq m + 1$. For $1 \leq k \leq m$, $P(k)$ follows from $Q(m)$. For $k = m + 1$, $P(m + 1)$ then follows by assumption (2), as required to complete the proof. □

Suppose that in Theorem 0.3.6 instead of (1) and (2), we assume for some positive integer n_0:

(1') $P(n_0)$ is true.

(2') If $P(k)$ is true for all k, $n_0 \leq k < m$, then $P(m)$ is true.

Then it follows that $P(n)$ is true for all $n \geq n_0$. (See Exercise 10 at the end of this section.)

Our next result is important not only as an illustration of proof by induction, but for its own sake.

0.3.7 EXAMPLE Given two real numbers a, b, by multiplying out we can obtain the following formulas for the first few powers of $a + b$:

$(a + b)^2 = a^2 + 2ab + b^2$

$(a + b)^3 = a^3 + 3a^2b + 3ab^2 + b^3$

$(a + b)^4 = a^4 + 4a^3b + 6a^2b^2 + 4ab^3 + b^4$

It does become cumbersome after a while to calculate all these powers! ◇

0.3.8 THEOREM (Binomial theorem) Given any real numbers a, b, then for any integer $n \geq 1$ we have

$$(a+b)^n = a^n + \binom{n}{1}a^{n-1}b + \binom{n}{2}a^{n-2}b^2 + \ldots + \binom{n}{n-2}a^2b^{n-2} + \binom{n}{n-1}ab^{n-1} + b^n$$

where the **binomial coefficients** are given by

$$\binom{n}{r} = \frac{n!}{r!(n-r)!}$$

for $0 \leq r \leq n$.

Proof For $n = 1$ the statement is just $(a + b)^1 = a^1 + b^1$, which is true. Assume the statement holds for k. We then have

$$(a + b)^{k+1} = (a + b)(a + b)^k =$$

$$(a+b)[\, a^k + \binom{k}{1}a^{k-1}b + \binom{k}{2}a^{k-2}b^2 + \ldots + \binom{k}{k-2}a^2b^{k-2} + \binom{k}{k-1}ab^{k-1} + b^k\,] =$$

$$a^{k+1} + \binom{k}{1}a^kb + \binom{k}{2}a^{k-1}b^2 + \ldots + \binom{k}{k-2}a^3b^{k-2} + \binom{k}{k-1}a^2b^{k-1} + ab^k +$$

$$+ a^kb + \binom{k}{1}a^{k-1}b^2 + \binom{k}{2}a^{k-2}b^3 + \ldots + \binom{k}{k-2}a^2b^{k-1} + \binom{k}{k-1}ab^k + b^{k+1} =$$

$$a^{k+1} + [\binom{k}{1} + 1]a^kb + [\binom{k}{2} + \binom{k}{1}]a^{k-1}b^2 + \ldots + [\binom{k}{r} + \binom{k}{r-1}]a^{k-r+1}b^r + \ldots + b^{k+1}$$

To complete the proof that $(a + b)^{k+1} =$

$$a^{k+1} + \binom{k+1}{1}a^kb + \binom{k+1}{2}a^{k-1}b^2 + \ldots + \binom{k+1}{k-1}a^2b^{k-1} + \binom{k+1}{k}ab^k + b^{k+1}$$

it will suffice to prove the following claim.

Claim (Pascal's identity)

$$\binom{k}{r} + \binom{k}{r-1} = \binom{k+1}{r}$$

Proof of claim We have

$$\binom{k}{r} + \binom{k}{r-1} = \frac{k!}{r!(k-r)!} + \frac{k!}{(r-1)!(k-r+1)!} =$$

$$\frac{k!(k-r+1) + rk!}{r!(k-r+1)!} = \frac{(k+1)!}{r!(k-r+1)!} = \binom{k+1}{r}$$

to complete the proof. □

0.3.9 EXAMPLE (Pascal's triangle)
Pascal's identity (the claim in the preceding proof) underlies the construction of the well-known **Pascal's triangle**:

```
                    1                              0th row

                 1     1                           1st row

              1     2     1                        2nd row

           1     3     3     1                     3rd row

        1     4     6     4     1                  4th row
         \   /  \  /  \  /  \  /
     1     5    10    10    5     1                5th row

   .    .    .    .    .    .    .    .
```

$$1 \quad k \quad \binom{k}{2} \quad \binom{k}{3} \quad \cdots \quad \binom{k}{k-3} \quad \binom{k}{k-1} \quad k \quad 1 \qquad k\text{th row}$$

As the lines indicate in the case of the 5th row, each entry in a row, except for the two 1s on the outside, is the sum of the two adjacent entries above it in the preceding row. ◇

Division Algorithm

We have all been familiar since elementary school with the process of division, by which a given integer a can always be represented as the sum of a multiple of another given integer $b \geq 1$ plus a remainder that is less than b. We see in the next theorem that the well-ordering principle guarantees the possibility of carrying out this process.

0.3.10 EXAMPLE We indicate the result of the process of division in the case of a few pairs of integers.

For 84 and 60 we have $84 = 1 \cdot 60 + 24$

For 924 and 105 we have $924 = 8 \cdot 105 + 84$

For -10 and 3 we have $-10 = (-4) \cdot 3 + 2$

In each case the first number is the sum of a multiple of the second number plus a nonnegative integer that is less than the second number. ◇

0.3.11 THEOREM (Division algorithm) Let a be any integer and b any integer with $b \geq 1$. Then there exist unique integers q and r such that

(1) $a = qb + r$

(2) $0 \leq r < b$

Proof Let $S = \{a - kb \mid k \in \mathbb{Z} \text{ and } a - kb \geq 0\}$. If $a \geq 0$, then $a - 0 \cdot b \in S$. If $a < 0$, then $a - 2ab \in S$. So in either case S is nonempty. By the well-ordering principle S has a least element r. Since $r \in S$, we have $r = a - qb$ or, equivalently, $a = qb + r$ for some integer q, and we have $r \geq 0$. It remains to show that $r < b$. But if we had $r \geq b$, then we would have $0 \leq r - b = a - (q + 1)b$ and therefore $a - (q + 1)b \in S$ while $a - (q + 1)b < r$, contrary to the choice of r as the *least* element of S. Thus the existence of a pair of integers q, r satisfying conditions (1) and (2) is proved.

Now let us prove uniqueness. Suppose we have another pair of integers q', r' also satisfying conditions (1) and (2). Let us assume $r \leq r'$. (The proof is similar on the opposite assumption.) We have $a = qb + r = q'b + r'$, from which we derive $0 \leq (q - q')b = r' - r < b$. Thus $(q - q')b$ is a nonnegative integer, a multiple of b, and less than b, which is only possible if $q - q' = 0$. Thus we have $q = q'$ and hence also $r = a - qb = a - q'b = r'$. □

The numbers q and r in Theorem 0.3.11 are called the **quotient** and **remainder** on dividing a by b.

0.3.12 DEFINITION Given two integers a and d, we say d is a **divisor** of a, written $d \mid a$, if $a = qd$ for some integer q. Note that we may have $d \leq 0$ in this definition. ○

0.3.13 DEFINITION Given two integers a and b, an integer d such that $d \mid a$ and $d \mid b$ is called a **common divisor** of a and b. ○

0.3.14 EXAMPLE 252 and 180 have the common positive divisors 1, 2, 3, 4, 6, 9, 12, 18, and 36. There is no common positive divisor larger than 36. ◇

We will often be interested in finding the largest among the common divisors of two integers.

0.3.15 DEFINITION Given two integers a and b, not both zero, the **greatest common divisor** of a and b is an integer $d \geq 1$ such that

(1) $d \mid a$ and $d \mid b$

(2) For any integer k, if $k \mid a$ and $k \mid b$, then $k \mid d$.

In this case we write $d = \gcd(a, b)$. ○

The division algorithm, with another application of the well-ordering principle, guarantees the existence of gcd(a, b), as shown in the next theorem.

0.3.16 THEOREM Let a and b be integers not both zero. Then

(1) $d = \gcd(a, b)$ exists.

(2) There exist integers u and v such that $d = ua + vb$.

Proof Let $S = \{xa + yb \mid x, y \in \mathbb{Z} \text{ and } xa + yb \geq 1\}$. S is nonempty since $aa + bb \in S$. The well-ordering principle implies S has a least element d. Since $d \in S$ we have $d = sa + tb$ for some $s, t \in \mathbb{Z}$, and $d \geq 1$. If k is a common divisor of a and b, we have $a = uk$ and $b = vk$ for some $u, v \in \mathbb{Z}$, and so $d = sa + tb = (su + tv)k$ and k is a divisor of d. Finally, to show $d = \gcd(a, b)$ it remains to show that d is a common divisor of a and b. Applying the division algorithm, we have $a = qd + r$, where $0 \leq r < d$. Hence $0 \leq r = a - qd = a - q(sa + tb) = (1 - qs)a + (-qt)b$. Since d was the least element of S, we cannot have $r \geq 1$, and so must have $r = 0$, so that $a = qd$ and d is a divisor of a. The proof that d is a divisor of b is exactly the same. □

Theorem 0.3.16, part (1), guarantees the existence of gcd(a,b). Theorem 0.3.16, part (2), which says that gcd(a,b) can be expressed as a **linear combination** $ua + vb$ of a and b, may at first glance appear less interesting, but in fact turns out to be extremely useful.

0.3.17 EXAMPLE Let us illustrate how to calculate the greatest common divisor in a simple case. We find gcd(84, 60) by repeatedly applying the division algorithm, as follows:

$$84 = 1 \cdot 60 + 24$$
$$60 = 2 \cdot 24 + 12$$
$$24 = 2 \cdot 12 + 0$$

From the third equation we know that $12 \mid 12$ and $12 \mid 24$. Hence the second equation implies that $12 \mid 60$. And since $12 \mid 24$ and $12 \mid 60$, the first equation implies $12 \mid 84$. So 12 is a common divisor of 84 and 60. If d' is any other common divisor of 84 and 60, we see from the first equation that $d' \mid 24$. Since $d' \mid 60$ and $d' \mid 24$, we see from the second equation that $d' \mid 12$. Therefore, $12 = \gcd(84, 60)$. ◇

0.3.18 PROPOSITION For any pair of integers a and $b \geq 1$ we can calculate gcd(a,b) by repeated application of the division algorithm, as follows:

$$a = q_1 b + r_1 \qquad \text{where } 0 \leq r_1 < b$$
$$b = q_2 r_1 + r_2 \qquad \text{where } 0 \leq r_2 < r_1$$
$$r_1 = q_3 r_2 + r_3 \qquad \text{where } 0 \leq r_3 < r_2$$
$$\vdots$$

stopping when we obtain a remainder of zero:

$$r_{n-3} = q_{n-1} r_{n-2} + r_{n-1} \qquad \text{where } 0 \leq r_{n-1} < r_{n-2}$$
$$r_{n-2} = q_n r_{n-1} + r_n \qquad \text{where } 0 \leq r_n < r_{n-1}$$
$$r_{n-1} = q_{n+1} r_n + 0$$

The last nonzero remainder $r_n = \gcd(a, b)$. This method of calculating the $\gcd(a, b)$ is called the **Euclidean algorithm**.

Proof That we must eventually come to a remainder of zero follows from the well-ordering principle, which implies that the set of positive remainders must have a least element. Since the sequence of remainders $b > r_1 > r_2 > r_3 > \ldots$ is strictly decreasing, the least positive remainder must be the *last* positive remainder, after which the next remainder must be zero. It follows as in the preceding example that if r_n is this last positive remainder, then $\gcd(a, b) = \gcd(b, r_1) = \gcd(r_1, r_2) = \ldots = \gcd(r_{i-1}, r_i) = \ldots = \gcd(r_{n-2}, r_{n-1}) = \gcd(r_{n-1}, r_n) = r_n$. □

0.3.19 EXAMPLE Let us calculate the gcd(924,105).

$$924 = 8 \cdot 105 + 84$$
$$105 = 1 \cdot 84 + 21$$
$$84 = 4 \cdot 21 + 0$$

So gcd(924, 105) = 21. From the second equation we obtain 21 = 105 - 84. From the first equation we obtain $84 = 924 - 8 \cdot 105$. Combining, we get $21 = 105 - (924 - 8 \cdot 105) = -924 + 9 \cdot 105$, and we have found a way of expressing the gcd(924, 105) as a linear combination $u924 + v105$. ◇

Fundamental Theorem of Arithmetic

Another important consequence of the division algorithm and the well-ordering principle is that every integer $n > 1$ can be written as a product of primes, numbers that cannot themselves be written as products in a nontrivial way.

0.3.20 EXAMPLE Let us calculate the gcd(385, 48).

$$385 = 8 \cdot 48 + 1$$
$$48 = 48 \cdot 1 + 0$$

So gcd(385, 48) = 1, or in other words, 1 is the largest, and therefore the only, positive integer that divides both 385 and 48. ◇

0.3.21 DEFINITION Two integers a and b are said to be **relatively prime** if $\gcd(a, b) = 1$, or in other words, if their only positive common divisor is 1. An integer $p > 1$ is said to be **prime** if its only positive divisors are 1 and p itself. ○

The next proposition is one of the important consequences of Theorem 0.3.16, part (2), as we will see in the proof of Theorem 0.3.26.

0.3.22 PROPOSITION Let a and b be relatively prime and c an integer. Then

(1) Any common divisor of a and bc is a common divisor of a and c.
(2) If a divides bc, then a divides c.
(3) If a and c are relatively prime, then a and bc are relatively prime.

Proof (1) Let $\gcd(a, b) = 1$ and let $d|a$ and $d|bc$. Then $1 = sa + tb$ for some integers s, t, by Theorem 0.3.16, and $a = dx$ and $bc = dy$ for some integers x and y. We then have

$$c = c \cdot 1 = c(sa + tb) = acs + bct = dxcs + dyt = d(xcs + yt)$$

and so $d|c$ as required to prove (1). Then (2) and (3) are immediate from (1). □

An immediate consequence is the following important corollary.

0.3.23 COROLLARY (Euclid's lemma) Let b and c be integers. If p is prime and $p|bc$, then $p|b$ or $p|c$.

Proof If we have $p|b$, there is nothing to prove. If we do not have $p|b$, then we must have $\gcd(p, b) = 1$, because 1 is the only positive divisor of p besides p itself. Therefore, we have $p|c$ by the preceding proposition. □

0.3.24 EXAMPLE It is very important in Euclid's lemma that p is a prime. Consider 6, which divides $3 \cdot 4 = 12$. We have neither $6|3$ nor $6|4$.

For the proof of the fundamental theorem of arithmetic we need Euclid's lemma in a more general form, provided by the next corollary.

0.3.25 COROLLARY Let $b_1, b_2, \ldots, b_r$ be integers. If p is a prime and $p|b_1b_2\ldots b_r$, then $p|b_i$ for some i with $1 \leq i \leq r$.

Proof We use induction on r. For $r = 1$ there is nothing to prove. The case $r = 2$ is Corollary 0.3.23. So assume the statement is true for $r = k$ and suppose $p|(b_1b_2\ldots b_k)b_{k+1}$. By Corollary 0.3.23, either $p|b_{k+1}$, in which case we are done, or $p|b_1b_2\ldots b_k$, in which case by our induction hypothesis $p|b_i$ for some i, $1 \leq i \leq k$. □

0.3.26 THEOREM (Fundamental theorem of arithmetic) Let n be an integer, $n > 1$. Then

(1) n is either a prime or a product of primes.

(2) The factorization of n into a product of primes is unique except for the order of the primes. That is, if

$$n = p_1p_2\ldots p_r \text{ and } n = q_1q_2\ldots q_s$$

where the p_i and q_j are primes, then $r = s$ and by reordering the q_j we can obtain $p_i = q_i$ for all i.

Proof For both statements (1) and (2) we use strong induction (Theorem 0.3.6 and Exercise 10) to prove that the statement holds for all $n \geq 2$.

(1) (Existence of the prime factorization of n) For $n = 2$, statement (1) is immediate, since 2 is itself a prime. So assume (1) holds for any integer k, $2 \leq k < n$, to prove (1) holds for n. If n is a prime, we are done. If n is not a prime, there are integers u and v, $1 < u, v < n$ such that $u \cdot v = n$. By our induction hypothesis, each of

u, v is either a prime or can be written as a product of primes, and it follows that $n = uv$ can be written as a product of primes.

(2) (Uniqueness of the prime factorization of n) For $n = 2$, statement (2) is immediate, since the prime 2 cannot be written as a product of other primes because any other prime is greater than 2. So assume (2) holds for any integer k, $2 \le k < n$, to prove (2) holds for n. Suppose

$$n = p_1 p_2 \cdots p_r = q_1 q_2 \cdots q_s$$

Then since $p_1 | n$ we have $p_1 | q_1 q_2 \ldots q_s$. By Corollary 0.3.25 $p_1 | q_i$ for some i, $1 \le i \le s$. We may reorder the q_j so that this q_i becomes q_1, and we have $p_1 | q_1$ and since q_1 is prime, we obtain $p_1 = q_1$. Now consider $k = n/p_1 = n/q_1 < n$. We have

$$k = p_2 \cdots p_r = q_2 \cdots q_s$$

By our induction hypothesis, the number of primes in each factorization must be the same. That is, $r - 1 = s - 1$, implying $r = s$. Also, the p_i, $2 \le i \le r$ and q_j, $2 \le j \le r$ must be the same except for order, completing the proof. □

The fundamental theorem of arithmetic we have just proved implies that given any integer $n > 1$, we can write n as a product $p_1{}^{a_1} p_2{}^{a_2} \ldots p_k{}^{a_k}$, where the p_i are distinct primes, and that these primes p_i, and their exponents a_i, are unique. When numbers are written this way, it is easy to find their greatest common divisor, as the next example illustrates.

0.3.27 EXAMPLE Consider the integers 924 and 105 from Example 0.3.19. In that example we found gcd(924, 105) = 21. We have

$$924 = 2^2 \cdot 3^1 \cdot 7^1 \cdot 11^1 \qquad 105 = 3^1 \cdot 5^1 \cdot 7^1 \qquad 21 = 3^1 \cdot 7^1$$

For purposes of comparison, we can write

$$924 = 2^2 \cdot 3^1 \cdot 5^0 \cdot 7^1 \cdot 11^1 \qquad 105 = 2^0 \cdot 3^1 \cdot 5^1 \cdot 7^1 \cdot 11^0$$

$$21 = 2^0 \cdot 3^1 \cdot 5^0 \cdot 7^1 \cdot 11^0$$

We see that the exponent on a prime in the factorization of 21 is whichever is less of the exponents on that same prime in the factorizations of 924 and 105. Suppose that instead we take whichever exponent is greater. Then we get the number

$$2^2 \cdot 3^1 \cdot 5^1 \cdot 7^1 \cdot 11^1 = 4620$$

Whereas 21 is the greatest integer that divides both the given numbers, 4620 is the least positive integer that both the given numbers divide. ◇.

0.3.28 DEFINITION Given two integers n and m, not both 0, the **least common multiple** of a and b is an integer $l \ge 1$such that

(1) $n|l$ and $m|l$

(2) For any integer k, if $n|k$ and $m|k$, then $l|k$.

In this case we write $l = \text{lcm}(n, m)$. ○

0.3.29 PROPOSITION Given

$n = p_1{}^{a_1} p_2{}^{a_2} \ldots p_k{}^{a_k}$ and $m = p_1{}^{b_1} p_2{}^{b_2} \ldots p_k{}^{b_k}$

where the p_i are distinct primes and a_i, $b_i \ge 0$, we have

(1) $\gcd(n, m) = p_1^{c_1} p_2^{c_2} \ldots p_k^{c_k}$, where $c_i = \min(a_i, b_i)$

(2) $\operatorname{lcm}(n, m) = p_1^{d_1} p_2^{d_2} \ldots p_k^{d_k}$, where $d_i = \max(a_i, b_i)$

(3) $\operatorname{lcm}(n, m) \cdot \gcd(n, m) = nm$

Proof See Exercises 19 through 21 at the end of this section. □

Integers mod n

We end this section by returning to the topic of equivalence relations. We examine one particular equivalence relation on $\mathbb{Z}$ that was mentioned in the preceding section.

0.3.30 DEFINITION Let $n > 0$ be a fixed positive integer. For any two integers a and b we say a and b are **congruent mod** n, and write $a \equiv b \bmod n$ if $n|(a - b)$. ○

0.3.31 PROPOSITION

(1) The relation of congruence mod n is an equivalence relation on $\mathbb{Z}$.

(2) This equivalence relation has exactly n equivalence classes:

$$n\mathbb{Z},\ 1 + n\mathbb{Z},\ 2 + n\mathbb{Z},\ \ldots,\ (n - 1) + n\mathbb{Z}$$

(3) If $a \equiv b \bmod n$ and $c \equiv d \bmod n$, then

$$a + c \equiv b + d \bmod n \qquad ac \equiv bd \bmod n$$

(4) If a and n are relatively prime, then

$$ab \equiv ac \bmod n \text{ implies } b \equiv c \bmod n$$

Proof (1) For all $a, b, c \in \mathbb{Z}$ we have the following. First, $a \equiv a$ since $n|(a - a) = 0$. Second, if $a \equiv b$, then $n|(a - b) = -(b - a)$, so $n|(b - a)$ and $b \equiv a$. Third, if $a \equiv b \bmod n$ and $b \equiv c \bmod n$, then $n|(a - b)$ and $n|(a - c)$ and so $n|((a - b) + (b - c)) = (a - c)$, and $a \equiv c \bmod n$.

(2) For any $a \in \mathbb{Z}$, let $[a]$ denote the equivalence class of $a \in \mathbb{Z}$. By the division algorithm we have $a = qn + r$ for some $q, r \in \mathbb{Z}$ with $0 \le r < n$. It follows that $a \equiv r$ and so $[a] = [r]$. So there are only the n equivalence classes

$$[0] = n\mathbb{Z},\ [1] = 1 + n\mathbb{Z},\ \ldots,\ [n - 1] = (n - 1) + n\mathbb{Z}$$

These are all distinct since for r, s with $0 \le r, s < n$ we have $n|(r - s)$ if and only if $r = s$.

(3) If $a \equiv b \bmod n$ and $c \equiv d \bmod n$, then $n|(a - b)$ and $n|(c - d)$, hence

$$n|((a - b) + (c - d)) = (a + c) - (b + d)$$

and $a + c \equiv b + d \bmod n$. Likewise

$$n|((a - b)c + (c - d)b = ac - bd$$

and $ac \equiv bd \bmod n$.

(4) If a and n are relatively prime and $ab \equiv ac \bmod n$, then

$$n|(ab - ac) = a(b - c)$$

and by Proposition 0.3.22 we have $n|(b - c)$ and $b \equiv c \bmod n$. □

0.3.32 EXAMPLE Let $\mathbb{Z}_5$ be the set consisting of the five equivalence classes of congruence mod 5. So $\mathbb{Z}_5 = \{[0], [1], [2], [3], [4]\}$. Proposition 0.3.31 guarantees that if $[r_1] = [r_2]$ and $[s_1] = [s_2]$, then $[r_1 + s_1] = [r_2 + s_2]$. So we can define unambiguously an addition operation on equivalence classes by setting $[r] + [s] = [r + s]$. Similarly, $[r_1 s_1] = [r_2 s_2]$ and we can define unambiguously a multiplication operation on equivalence classes by setting $[r]\cdot[s] = [rs]$. The tables for these operations are as follows.

TABLE 1 Addition mod 5

+	0	1	2	3	4
0	0	1	2	3	4
1	1	2	3	4	0
2	2	3	4	0	1
3	3	4	0	1	2
4	4	0	1	2	3

TABLE 2 Multiplication mod 5

·	0	1	2	3	4
0	0	0	0	0	0
1	0	1	2	3	4
2	0	2	4	1	3
3	0	3	1	4	2
4	0	4	3	2	1

In the tables we have omitted the brackets, as is often done. ◇

0.3.33 DEFINITION For any $n > 0$, let $\mathbb{Z}_n = \{[0], [1], \ldots, [n - 1]\}$, the set of equivalence classes of congruence mod n. Just as in the preceding example, Proposition 0.3.31 guarantees that the operations $[r] + [s] = [r + s]$ and $[r][s] = [rs]$ of **addition and multiplication mod** n are well defined in $\mathbb{Z}_n$, since if $[r_1] = [r_2]$ and $[s_1] = [s_2]$ in $\mathbb{Z}_n$, then

$$[r_1] + [s_1] = [r_1 + s_1] = [r_2 + s_2] = [r_2] + [s_2]$$
$$[r_1][s_1] = [r_1 s_1] = [r_2 s_2] = [r_2][s_2]$$

$\mathbb{Z}_n$ with these operations is called the **integers mod** n. ○.

The next proposition collects some basic properties of addition and multiplication in the integers mod n.

0.3.34 PROPOSITION For any $[r]$, $[s]$, and $[t]$ in $\mathbb{Z}_n$ we have

(1) **Commutative laws**

$[r] + [s] = [s] + [r]$ $\quad$ $[r][s] = [s][r]$

(2) **Associative laws**

$[r] + ([s] + [t]) = ([r] + [s]) + [t]$ $\quad$ $[r]([s][t]) = ([r][s])[t]$

(3) **Distributitivite law**

$[r]([s] + [t]) = [r][s] + [r][[t]$

(4) **Identity laws**

$[0] + [r] = [r] = [r] + [0]$ $\quad$ $[1][r] = [r] = [r][1]$

Proof Left to the reader (as Exercises 27 through 30 at the end of this section) □

0.3.35 EXAMPLE In $\mathbb{Z}_{10} = \{[0], [1], \ldots, [9]\}$ consider the subset $U(10) = \{[1], [3], [7], [9]\}$ that consists of those equivalence classes [s] mod 10, $1 \le s \le 9$, such that $\gcd(10, s) = 1$. Let us work out the multiplication table mod 10 for $U(10)$.

TABLE 3 Multiplication in $U(10)$

·	1	3	7	9
1	1	3	7	9
3	3	9	1	7
7	7	1	9	3
9	9	7	3	1

Calculating the table, we find that if $[r], [s] \in U(10)$, then $[r][s] \in U(10)$. The reason why will be seen in the next proposition. We also find that for any $[r] \in U(10)$ there is an $[s] \in U(10)$ such that $[r][s] = 1$. The reason is that if $\gcd(r,10) = 1$, then it follows from Theorem 0.3.16 that $rs + 10t = 1$ for some integers s and t, from which it follows that $rs \equiv 1 \bmod 10$ and $[r][s] = [1]$. ◇

0.3.36 DEFINITION In $\mathbb{Z}_n$ let $U(n)$ be the set of all equivalence classes $[s]$ mod n with $1 \le s < n$ and $\gcd(n,s) = 1$. The elements of $U(n)$ are called the **units mod** n. ○

0.3.37 PROPOSITION For any $[r], [s] \in U(n)$ we have $[r][s] = [rs] \in U(n)$

Proof If $[r], [s] \in U(n)$ then we have $\gcd(n, s) = \gcd(n, r) = 1$. By Proposition 0.3.22, part (3), it follows that $\gcd(n, rs) = 1$, and hence $[rs] \in U(n)$. □

0.3.38 PROPOSITION For any $[r] \in U(n)$ there is an $[s] \in U(n)$ such that $[r][s] = [1]$.

Proof Left to the reader (as Exercise 33 at the end of this section). □

Exercises 0.3

In Exercises 1 through 5 prove the statements by induction on n.

1. $1 + 2 + 3 + \ldots + n = n(n + 1)/2$ $\quad n \ge 1$

2. $1^2 + 2^2 + 3^2 + \ldots + n^2 = n(n + 1)(2n + 1)/6$ $\quad n \ge 1$

3. $1^3 + 2^3 + 3^3 + \ldots + n^3 = n^2(n + 1)^2/4$ $\quad n \ge 1$

4. If $0 \le x \le y$, then $x^n \le y^n$ $\quad n \ge 0$

5. $n < 2^n$ $\quad n \ge 0$

In Exercises 6 through 8 the **Fibonacci sequence** 1, 1, 2, 3, 5, 8, 13, ... is defined as follows: $F_1 = F_2 = 1$, $F_{n+2} = F_{n+1} + F_n$ for $n \geq 1$.

6. Show that $(F_{n+1})^2 - F_nF_{n+2} = (-1)^n$.

7. Show that $F_{n+1}F_{n+2} - F_nF_{n+3} = (-1)^n$.

8. Show that $F_n < 2^n$ for $n \geq 1$.

9. If $a, r \in \mathbb{R}$ and $r \neq 1$, show that $a + ar + ar^2 + \dots + ar^n = a(1 - r^{n+1})/(1 - r)$ for $n \geq 1$.

10. Let $P(n)$ be a statement about positive integers and $n_0 > 0$ any positive integer. Assume

(1) $P(n_0)$ is true.

(2) If $P(k)$ is true for all k, $n_0 \leq k < m$, then $P(m)$ is true.

Using the well-ordering principle, show that $P(n)$ is true for all $n \geq n_0$.

11. Show that the following principles are equivalent, so that assuming any one of them we can prove the others.

(a) The well-ordering principle

(b) The principle of mathematical induction

(c) The principle of strong induction

12. Write out the 10th row of Pascal's triangle.

13. Show that for $0 \leq r \leq n$ we have

$$\binom{n}{r} = \binom{n}{n-r}$$

14. Let p be a prime and suppose that we have

$$(1 + a)^p = 1 + c_1a + c_2a^2 + \dots + c_{p-1}a^{p-1} + a^p$$

for $a \in \mathbb{R}$. Show that $p | c_i$ for all i, $1 \leq i \leq p - 1$.

15. In the preceding exercise, take $p = 11$ and find $c_1, c_{10}, c_2, c_9, c_4, c_6$.

16. Use the Euclidean algorithm to calculate gcd(52,135) and write it as a linear combination of 52 and 135.

17. Let a and b be relatively prime. Show that for any integer n there exist integers x and y such that $n = xa + yb$.

18. Show that if $a = a'd$ and $b = b'd$, where $d = \gcd(a, b)$, then $\gcd(a', b') = 1$.

19. Prove Proposition 0.3.29, part (1).

20. Prove Proposition 0.3.29, part (2).

21. Prove Proposition 0.3.29, part (3).

22. Show that $\operatorname{lcm}(a, b) = ab$ if and only if a and b are relatively prime.

23. Find gcd(9750, 59400) and lcm(9750, 59400).

24. Show that for any integer n we have $n^3 \equiv n \bmod 6$.

25. Show that for any integer n if do not have $n \equiv 0 \bmod 5$, then we have $n^4 \equiv 1 \bmod 5$.

26. Show that for any integer n we have $n^5 \equiv n \bmod 5$.

27. Prove Proposition 0.3.34, part (1).

28. Prove Proposition 0.3.34, part (2).

29. Prove Proposition 0.3.34, part (3).

30. Prove Proposition 0.3.34, part (4).

31. Write the addition and multiplication tables for $\mathbb{Z}_n$, the integers mod n, for $n = 6$ and $n = 7$.

32. Show that n is prime if and only if in $\mathbb{Z}_n$, $[r][s] = [0]$ always implies $[r] = [0]$ or $[s] = [0]$.

33. Show that if $\gcd(n, r) = 1$, then there exists an integer s such that $\gcd(n, s) = 1$ and $rs \equiv 1 \bmod n$.

34. Write the multiplication table mod 7 for $U(7)$, and write the multiplication table mod 8 for $U(8)$.

35. If $\gcd(m, n) = 1$, show that for any pair of integers a and b there exists an integer x such that $x \equiv a \bmod m$ and $x \equiv b \bmod n$.

36. (The Chinese remainder theorem)
(a) If $m_1, m_2, \ldots, m_s$ are integers >1 such that any two of them are relatively prime to each other, and if $a_1, a_2, \ldots, a_s$ are any integers, show that there exists an integer x such that $x \equiv a_i \bmod m_i$ for all i, $1 \le i \le s$. (Use the preceding exercise.)
(b) Show that if x and x' both satisfy the congruences in part (a), then $x \equiv x' \bmod M$, where $M = m_1 m_2 \ldots m_s$.

0.4 Complex Numbers

Understanding complex numbers will be essential. As we see in this introductory section, we need complex numbers to find all the solutions of polynomial equations. When we look at the equation $x^2 - 2 = 0$, without any hesitation we write down the solutions as $\sqrt{2}$ and $-\sqrt{2}$. This is correct, except that often we forget that "$\sqrt{2}$" is just a symbol abbreviating the description "a number such that, when you square it and subtract 2, you get 0." We treat the equation $x^2 + 1 = 0$ in the same way. We write down its solutions as $\sqrt{-1}$ and $-\sqrt{-1}$, and we give these solutions names: $i = \sqrt{\sqrt{-1}}$ and $-i = -\sqrt{-1}$. In other words, i is assumed to be a number with the property $i^2 = -1$.

0.4.1 EXAMPLE Treating i as a number means allowing ourselves to combine it with other numbers, so that we get numbers such as $3i$, $i/2$, $1 + i$, $(-1 + \sqrt{3}i)/2$, and so on. The next definition states this more precisely. ◇

0.4.2 DEFINITION The set $\mathbb{C}$ of **complex numbers** is defined by
$\mathbb{C} = \{a + bi \mid a, b \in \mathbb{R} \text{ and } i^2 = -1\}$
If $z = a + bi$ is a complex number, then a is called the **real part** of z and b is called the **imaginary part** of z. ○

Every real number a counts as a complex number with imaginary part zero: $a = a + 0i$. Thus $\mathbb{Z} \subseteq \mathbb{Q} \subseteq \mathbb{R} \subseteq \mathbb{C}$.

0.4.3 EXAMPLE The real part of $3i$ is 0 and the imaginary part is 3. The real part of $i/2$ is 0 and the imaginary part is 1/2. The real part of $1 + i$ is 1 and the imaginary part is 1. The real part of $(-1 + \sqrt{3}i)/2$ is -1/2 and the imaginary part is $\sqrt{3}/2$. ◇

0.4.4 EXAMPLE Treating $a + bi$ as a number means allowing ourselves to apply the operations of addition and multiplication, assuming the usual laws for these operations, and remembering that $i^2 = -1$. For instance,

$$(4 + 3i) - (1 - 2i) = (4 - 1) + (3 - (-2))i = 3 + 5i$$
$$(4 + 3i)(1 - 2i) = 4 - 8i + 3i - 6i^2 = 10 - 5i$$

The next definition states this more precisely. ◇

0.4.5 DEFINITION Let $z = a + bi$ and $w = c + di$ be two complex numbers. We define their sum and product as follows:

$$z + w = (a + bi) + (c + di) = (a + c) + (b + d)i \in \mathbb{C}$$
$$zw = (a + bi)(c + di) = (ac - bd) + (ad + bc)i \in \mathbb{C}$$

The operations on a, b, c, d here are, of course, the usual ones on real numbers. ○

If we apply these definitions to real numbers, considered as complex numbers with imaginary part zero, $z = a + 0i$ and $w = c + 0i$, we get $z + w = a + b$ and $zw = ab$, so the sum and product as complex numbers is the same as the sum and product as real numbers. In other words, the operations of addition and multiplication on complex numbers are extensions of the corresponding operations on real numbers.
Subtraction can be done in the same way:

$$z - w = (a + bi) - (c + di) = (a - c) + (b - d)i \quad \in \mathbb{C}$$

Note that $z - w = 0 = 0 + 0i$ if and only if $a - c = 0$ and $b - d = 0$, hence if and only if $a = c$ and $b = d$, which is to say if and only if $z = w$.
We next need to consider how to perform division in $\mathbb{C}$ in a way compatible with division in $\mathbb{R}$. Recall that for any pair of real numbers $a, b \neq 0$ the division a/b can be seen as multiplication by the reciprocal of b or, in other words, $a/b = a \cdot (1/b)$. Thus we determine first the reciprocal of a nonzero complex number and show that this is also a complex number.

0.4.6 EXAMPLE Is $w = 1/(2+3i)$ a complex number? To count as a complex number, we have to be able to write it in the form $a + bi$. Here is how this can be done:

$$w = \frac{1}{2+3i} = \frac{1}{(2+3i)} \cdot \frac{(2-3i)}{(2-3i)} = \frac{2-3i}{4+9} = \frac{2}{13} - \frac{3}{13}i$$

Since 2/13, -3/13 are real numbers, w is a complex number. ◇

This example suggests how to do division in other cases.

0.4.7 EXAMPLE Let us calculate $w = (3-2i)/(5+3i)$.

$$w = \frac{3-2i}{5+3i} = \frac{(3-2i)}{(5+3i)} \cdot \frac{(5-3i)}{(5-3i)} = \frac{9-19i}{34} = \frac{9}{34} - \frac{19}{34}i \ \diamond$$

In these last two examples we had to calculate $(2-3i)(2+3i) = 2^2 + 3^2$ and $(5+3i)(5-3i) = 5^2 + 3^2$. We now see that these expressions $a^2 + b^2$ have a geometrical meaning. ◇

0.4.8 DEFINITION Just as a real number can be represented as a point on a line, so a complex number can be represented as a point in a plane, which we call the **complex plane**. The complex number $z = a + bi$ is represented by the point with coordinates (a, b).

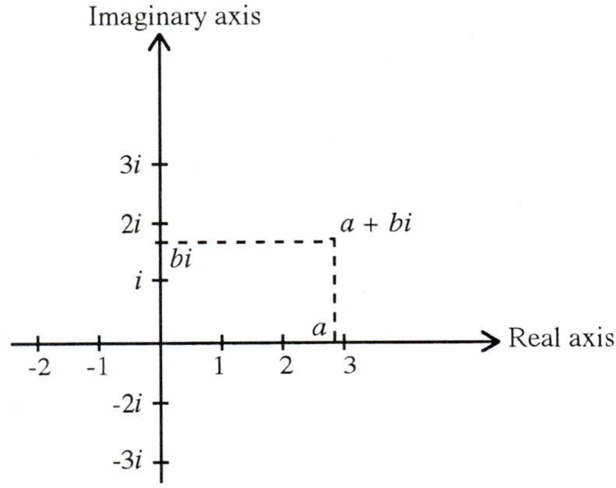

FIGURE 2

We call the x-axis the **real axis** and the y-axis the **imaginary axis**. ○

The first thing we clarify now using this geometric representation is the notion of the absolute value of a complex number. For a real number, say -3, the absolute value $|-3|$ means the distance of the point -3 from the origin 0 on the real line. We extend this definition and take the absolute value $|a + bi|$ of a complex number $a + bi$ to be the distance of the point (a, b) from the origin $(0, 0)$ in the complex plane.

0.4.9 DEFINITION For any complex number $z = a + bi$ define the **absolute value** of z by: $|z| = |a + bi| = \sqrt{a^2+b^2}$. Note that this is a nonnegative real number. ○

Applying this definition to a real number $a = a + 0i$ gives the positive square root of a^2, which is the usual absolute value, a if $a > 0$ and $-a$ if $a < 0$.

0.4.10 EXAMPLE The definition gives $|i| = |(1 + i)/\sqrt{2}| = |(-1 + \sqrt{3}\,i)/2)| = 1$. The points $(0, 1)$, $(1/\sqrt{2}, 1/\sqrt{2})$, and $(-1/2, \sqrt{3}/2)$ are all on the unit circle in the plane. ◇

0.4.11 DEFINITION For any complex number $z = a + bi$ define the **complex conjugate** or simply **conjugate** of z by $\bar{z} = \overline{a + bi} = a - bi$. ○

Applying this definition to a real number $a = a + 0i$ gives $a - 0i = a$. A real number is its own conjugate.

0.4.12 PROPOSITION

(1) If $z = a + bi$ is any complex number, then

$$z\bar{z} = |z|^2 = |\bar{z}|^2$$

(2) If $w = c + di$ is any other complex number, then

$$\frac{z}{w} = \frac{z\bar{w}}{|w|^2}$$

Proof (1) $(a + bi)(a - bi) = a^2 + b^2 = a^2 + (-b)^2$

(2) We have:

$$\frac{a+bi}{c+di} = \frac{(a+bi)}{(c+di)} \cdot \frac{(c-di)}{(c-di)} = \frac{(a+bi)(c-di)}{c^2+d^2} = \frac{ac+bd}{c^2+d^2} + \frac{bc-ad}{c^2+d^2}\,i$$

Note that here we have a general formula for division. □

We next use polar coordinates of points in the plane to represent complex numbers in a way that facilitates calculations in solving polynomial equations. Going back to the complex plane, draw the line from the origin $(0, 0)$ to the point (a, b) that represents the complex number $z = a + bi$, as in Figure 3.

If we let r be the length of this line segment, then we have $r^2 = a^2 + b^2 = |z|^2$ and $r = |z|$. If we let θ be the angle from the positive real axis to this line, then we have

$$a = r\cos\theta \quad \text{and} \quad b = r\sin\theta$$

$$z = r\cos\theta + i\,r\sin\theta = r(\cos\theta + i\sin\theta)$$

0.4.13 DEFINITION Let z be a complex number. Then the representation $z = a + bi$ is called the **Cartesian representation** of z while the representation $z = r(\cos\theta + i\sin\theta)$ is called the **polar representation** of z. ○

Note that r is always a nonnegative real number. Thus the polar representation of -2 is $2(\cos\pi + i\sin\pi)$.

0.4.14 EXAMPLE We have

$$i = 1(\cos {}^{\pi}/_{2} + i \sin {}^{\pi}/_{2}) \qquad -i = 1(\cos {}^{3\pi}/_{2} + i \sin {}^{3\pi}/_{2})$$

$$\sqrt{2}/_{2} + \sqrt{2}/_{2}\, i = 1(\cos {}^{\pi}/_{4} + i \sin {}^{\pi}/_{4}) \qquad 1 + i = \sqrt{2}(\cos {}^{\pi}/_{4} + i \sin {}^{\pi}/_{4})$$

Cartesian representations are on the left and polar representations are on the right. ◇

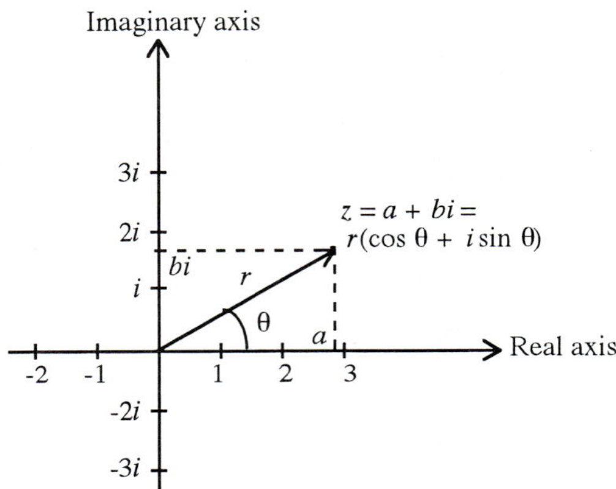

FIGURE 3

Writing complex numbers in polar form simplifies many calculations, as will be seen in the next several propositions.

0.4.15 PROPOSITION Given two complex numbers $z_1 = r_1(\cos \theta_1 + i\sin \theta_1)$ and $z_2 = r_2(\cos \theta_2 + i \sin \theta_2)$ in polar form, their product is given by

$$z_1 z_2 = r_1 r_2(\cos(\theta_1 + \theta_2) + i \sin(\theta_1 + \theta_2))$$

Proof This formula follows from the trigonometric identities:

$$\cos(\theta_1 + \theta_2) = \cos \theta_1 \cos \theta_2 - \sin \theta_1 \sin \theta_2$$
$$\sin(\theta_1 + \theta_2) = \sin \theta_1 \cos \theta_2 + \cos \theta_1 \sin \theta_2$$

and the definition of multiplication in $\mathbb{C}$. The proof is left to the reader (as Exercise 20 at the end of this section). □

0.4.16 COROLLARY Given a complex number $z = r(\cos \theta + i \sin \theta)$, we have

$$z^2 = r^2(\cos 2\theta + i \sin 2\theta)$$

Proof Left to the reader (as Exercise 21). □

0.4.17 COROLLARY (De Moivre's formula) Given any complex number $z = r(\cos \theta + i \sin \theta)$, for any positive integer n we have

$$z^n = r^n(\cos n\theta + i \sin n\theta)$$

Proof The proof uses induction on n and the preceding proposition and corollary. It is left to the reader (as Exercise 22). □

0.4.18 EXAMPLE Let $z = 1 - \sqrt{3}i$. Let us calculate z^8.

(1) We first put z in polar form. We have $r^2 = |z|^2 = 1^2 + (\sqrt{3})^2 = 4$, or $r = 2$. So we can rewrite z as $z = 2(^1/_2 - {}^{\sqrt{3}}/_2 i)$. We look for an angle θ, $0 \leq \theta < 2\pi$ with $\cos\theta = {}^1/_2$, $\sin\theta = -{}^{\sqrt{3}}/_2$. Such an angle is $\theta = {}^{5\pi}/_3$. So $z = 2(\cos {}^{5\pi}/_3 + i \sin {}^{5\pi}/_3)$.

(2) We next apply De Moivre's formula to get $z^8 = 2^8(\cos {}^{40\pi}/_3 + i \sin {}^{40\pi}/_3)$. Since ${}^{40\pi}/_3 = 12\pi + {}^{4\pi}/_3$ we have $\cos {}^{40\pi}/_3 = \cos {}^{4\pi}/_3$ and $\sin {}^{40\pi}/_3 = \sin {}^{4\pi}/_3$, and therefore $z^8 = 2^8(\cos {}^{4\pi}/_3 + i \sin {}^{4\pi}/_3)$.

(3) We finally put z^8 back in Cartesian form: $z^8 = 256(-{}^1/_2 - {}^{\sqrt{3}}/_2 i) = -128 - 128\sqrt{3}i$. ◇

0.4.19 EXAMPLE Let us calculate $\sqrt{i}$. What this means is that we want to find a complex number z such that $z^2 = i$. Let $z = r(\cos\theta + i \sin\theta)$. Then what we want is to have $z^2 = r^2(\cos 2\theta + i \sin 2\theta) = i = 1(\cos {}^{\pi}/_2 + i \sin {}^{\pi}/_2)$. Hence $r^2 = 1$ and so $r = 1$, while $2\theta = {}^{\pi}/_2 + k2\pi$ or $\theta = {}^{\pi}/_4 + k\pi$, for some integer k. Since we want $0 \leq \theta < 2\pi$, there are two possibilities, $\theta = {}^{\pi}/_4$ and $\theta = {}^{5\pi}/_4$. We have found the two solutions to the equation $z^2 = i$. They are

$$z_1 = 1(\cos {}^{\pi}/_4 + i \sin {}^{\pi}/_4) \qquad = {}^{\sqrt{2}}/_2(1 + i)$$
$$z_2 = 1(\cos {}^{5\pi}/_4 + i \sin {}^{5\pi}/_4) \qquad = -{}^{\sqrt{2}}/_2(1 + i)$$

These are the square roots of i. ◇

0.4.20 EXAMPLE Let us find all solutions of the equation $z^3 + 8i = 0$. We need to find the complex numbers $z = r(\cos\theta + i \sin\theta)$ such that $z^3 = r^3(\cos 3\theta + i \sin 3\theta) = -8i = 8(\cos {}^{3\pi}/_2 + i \sin {}^{3\pi}/_2)$. Hence $r^3 = 8$, while $3\theta = {}^{3\pi}/_2 + k2\pi$. Thus $r = 2$ and $\theta = {}^{\pi}/_2 + k({}^{2\pi}/_3)$ for $k = 0, 1$, or 2. Thus $\theta = {}^{\pi}/_2$ or ${}^{\pi}/_2 + {}^{2\pi}/_3 = {}^{7\pi}/_6$ or ${}^{\pi}/_2 + {}^{4\pi}/_3 = {}^{11\pi}/6$. We obtain three solutions as shown in Figure 4.

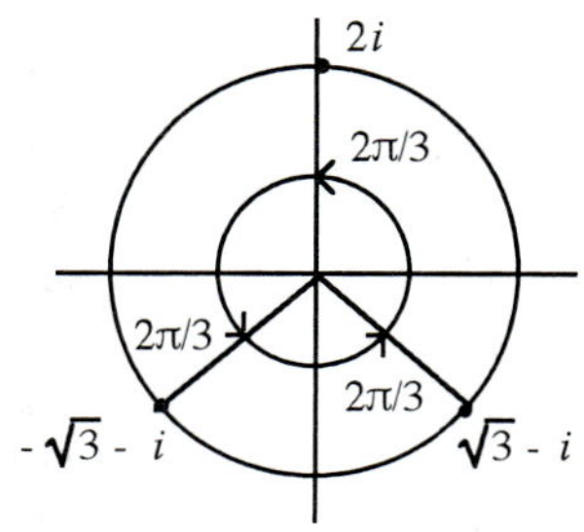

FIGURE 4

The three solutions are

$$z_1 = 2(\cos {}^{\pi}/_2 + i \sin {}^{\pi}/_2) \qquad = 2i$$

$$z_2 = 2(\cos {}^{7\pi}/_6 + i \sin {}^{7\pi}/_6) \qquad = 2(-{}^{\sqrt{3}}/_2 - {}^{1}/_2 i) = -\sqrt{3} - i$$

$$z_3 = 2(\cos {}^{11\pi}/_6 + i \sin {}^{11\pi}/_6) \qquad = 2({}^{\sqrt{3}}/_2 - {}^{1}/_2 i) = \sqrt{3} - i$$

Notice the position of the three points on the complex plane as illustrated in Figure 4. All three lie on a circle of radius $r = 2$, and the angle formed between any two of them is ${}^{2\pi}/_3$. ◇

Exercises 0.4

In Exercises 1 through 15 calculate the value of the given expression and express your answer in the form $a + bi$, where $a, b \in \mathbb{R}$.

1. $(3 + 2i) + (7 - 5i)$ **2.** $3i - (2 + i)$ **3.** $(4 - i) - (3 - 2i)$

4. i^5 **5.** i^6 **6.** i^7

7. i^{32} **8.** i^{38} **9.** $(-i)^5$

10. $(3 + 2i)(2 + 5i)$ **11.** $(5 - 2i)(3+ 4i)$

12. $(1 + i)^7$ (*Hint:* Use the binomial theorem, Theorem 0.3.8.)

13. $(1 + i)/i$ **14.** $(2 + i)/(1 + i)$ **15.** $i/(1 - i)$

16. Calculate $|2 - 3i|$, $|1 + i|$, $|\sqrt{2} - \sqrt{3}i|$.

In Exercises 17 through 19 express the given complex number in the polar form $r(\cos \theta + i \sin \theta)$.

17. $1 - i$ **18.** $-1 - i$ **19.** $-1 + \sqrt{3}i$

20. Prove Proposition 0.4.15. **21.** Prove Corollary 0.4.16.

22. Prove Corollary 0.4.17.

In Exercises 23 through 28 find all the solutions of the given equations.

23. $z^3 = 1$ **24.** $z^4 = 1$ **25.** $z^4 = -1$

26. $z^3 = -8$ **27.** $z^3 = -i$ **28.** $z^3 = -125i$

29. (Requires calculus) Using the power series expansion of e^x, $\cos x$, and $\sin x$, prove **Euler's formula** $e^{i\theta} = \cos \theta + i \sin \theta$.

30. Using Euler's formula in the preceding exercise, prove De Moivre's formula.

0.5 Matrices

We end this introductory chapter with a review of matrices. Some types of matrices provide us with important examples of the notions we will be studying in the next chapters.

0.5.1 EXAMPLE A 2×2 matrix with integer entries is a square array of four integers, such as

$$A = \begin{bmatrix} 3 & -1 \\ -2 & 5 \end{bmatrix} \text{ or } B = \begin{bmatrix} -1 & 2 \\ 0 & -4 \end{bmatrix}$$

A 2×3 matrix with integer entries is given by a rectangular array of six integers, with two rows and three columns, such as

$$C = \begin{bmatrix} 1 & -1 & 0 \\ 0 & 4 & 6 \end{bmatrix} \text{ or } D = \begin{bmatrix} 2 & 1 & 1 \\ -1 & 0 & 3 \end{bmatrix}$$

We can add matrices of the same shape, that is, with the same number of rows and the same number of columns, by adding corresponding entries. Thus

$$A + B = \begin{bmatrix} 3 & -1 \\ -2 & 5 \end{bmatrix} + \begin{bmatrix} -1 & 2 \\ 0 & -4 \end{bmatrix} = \begin{bmatrix} 3-1 & -1+2 \\ -2+0 & 5-4 \end{bmatrix} = \begin{bmatrix} 2 & 1 \\ -2 & 1 \end{bmatrix}$$

and

$$C + D = \begin{bmatrix} 1 & -1 & 0 \\ 0 & 4 & 6 \end{bmatrix} + \begin{bmatrix} 2 & 1 & 1 \\ -1 & 0 & 3 \end{bmatrix} = \begin{bmatrix} 1+2 & -1+1 & 0+1 \\ 0-1 & 4+0 & 6+3 \end{bmatrix} = \begin{bmatrix} 3 & 0 & 1 \\ -1 & 4 & 9 \end{bmatrix}$$

It would not make sense to try to add matrices of different shapes. ◇

0.5.2 DEFINITION An $n \times m$ **matrix** A is an array of entries in n rows and m columns. We write $A = \{a_{ij}\}$, where a_{ij} is the entry in the ith row and jth column, $1 \leq i \leq n$ and $1 \leq j \leq m$. ○

0.5.3 DEFINITION $M_{n \times m}(R)$ is the set of all $n \times m$ matrices with entries belonging to R. $M_{n \times n}(R)$, the set of all square $n \times n$ matrices with entries belonging to R, is also denoted $M(n, R)$. Here the set R may be any of $\mathbb{Z}$, $\mathbb{Q}$, $\mathbb{R}$, $\mathbb{C}$, or any $\mathbb{Z}_p$, p a prime.○

0.5.4 DEFINITION Let $A = \{a_{ij}\} \in M_{n \times m}(R)$ and $B = \{b_{ij}\} \in M_{n \times m}(R)$ be two $n \times m$ matrices. Then their **sum** is the matrix $A + B = \{a_{ij} + b_{ij}\}$ or, in other words, the matrix $\{c_{ij}\} \in M_{n \times m}(R)$, where $c_{ij} = a_{ij} + b_{ij}$ for all i and j, $1 \leq i \leq n$, $1 \leq j \leq m$. The **product** of an element t of R times a matrix $A = \{a_{ij}\}$ is defined by $tA = \{ta_{ij}\}$. ○

While addition of matrices is given by adding corresponding entries, and multiplication of a matrix by a number is given by multiplying each entry, multiplication of two matrices with each other is more complicated.

0.5.5 DEFINITION Let $A = \{a_{ij}\} \in M_{n\times m}(R)$ be an $n\times$m matrix and $B = \{b_{jk}\} \in M_{m\times r}(R)$ an $m \times r$ matrix, so a_{ij} is the entry in the ith row and jth column of A, and b_{jk} is the entry in the jth row and kth column of B. Then their **product** is the $n \times r$ matrix $AB = \{c_{ik}\} \in M_{n\times r}(R)$ where $c_{ik} = a_{i1}b_{1k} + a_{i2}b_{2k} + a_{i3}b_{3k} + \dots + a_{im}b_{mk}$, for all i and k, $1 \le i \le n$ and $1 \le k \le r$. ○

0.5.6 EXAMPLE Consider the 1×3 matrix $A = [-1\ \ 2\ \ 3]$ and the 3×1 matrix

$$B = \begin{bmatrix} 2 \\ 4 \\ -5 \end{bmatrix}$$

Then AB is the 1×1 matrix $[c]$, where $c = (-1)(2) + (2)(4) + (3)(-5) = -2 + 8 - 15 = -9$. That is, $AB = [-9]$. ◇

0.5.7 EXAMPLE Let us calculate

$$AB = \begin{bmatrix} 2 & 3 & 1 \\ 0 & -1 & 5 \end{bmatrix}\begin{bmatrix} 0 & -2 \\ 1 & 0 \\ 3 & 4 \end{bmatrix}$$

Note that AB will be a 2×2 matrix:

$$AB = \begin{bmatrix} c_{11} & c_{12} \\ c_{21} & c_{22} \end{bmatrix}$$

where we have

$c_{11} = (2)(0) + (3)(1) + (1)(3) = 6$ 1st row of A times 1st column of B

$c_{12} = (2)(-2) + (3)(0) + (1)(4) = 0$ 1st row of A times 2nd column of B

$c_{21} = (0)(0) + (-1)(1) + (5)(3) = 14$ 2nd row of A times 1st column of B

$c_{22} = (0)(-2) + (-1)(0) + (5)(4) = 20$ 2nd row of A times 2nd column of B

Hence

$$AB = \begin{bmatrix} 6 & 0 \\ 14 & 20 \end{bmatrix}$$

is the final answer. ◇

Note that matrix multiplication is not commutative, as the next example shows.

0.5.8 EXAMPLE Let

$$A = \begin{bmatrix} 1 & 1 \\ 1 & 0 \end{bmatrix} \text{ and } B = \begin{bmatrix} 0 & 1 \\ 1 & 1 \end{bmatrix}$$

Then we have

$$AB = \begin{bmatrix} 1 & 1 \\ 1 & 0 \end{bmatrix}\begin{bmatrix} 0 & 1 \\ 1 & 1 \end{bmatrix} = \begin{bmatrix} 1 & 2 \\ 0 & 1 \end{bmatrix} \text{ but } BA = \begin{bmatrix} 0 & 1 \\ 1 & 1 \end{bmatrix}\begin{bmatrix} 1 & 1 \\ 1 & 0 \end{bmatrix} = \begin{bmatrix} 1 & 0 \\ 2 & 1 \end{bmatrix}$$

Thus $AB \neq BA$. ◇

0.5.9 EXAMPLE In $M(n, R)$, where R is any of of $\mathbb{Z}$, $\mathbb{Q}$, $\mathbb{R}$, $\mathbb{C}$, or $\mathbb{Z}_p$, we have an **identity matrix** I_n with 1 on the diagonal ($a_{ii} = 1$ for all i, $1 \leq i \leq n$), and 0 off the diagonal ($a_{ij} = 0$ if $i \neq j$). Consider, for instance, the following:

$$I_3 = \begin{bmatrix} 1 & 0 & 0 \\ 0 & 1 & 0 \\ 0 & 0 & 1 \end{bmatrix}$$

It is easy to check that given any 3×3 matrix $A = \{a_{ij}\}$ we have $AI_3 = A = I_3A$. ◇

0.5.10 EXAMPLE Let

$$A = \begin{bmatrix} 2 & 1 \\ 1 & 1 \end{bmatrix} \text{ and } B = \begin{bmatrix} 1 & -1 \\ -1 & 2 \end{bmatrix}$$

Then we have

$$AB = \begin{bmatrix} 2 & 1 \\ 1 & 1 \end{bmatrix} \begin{bmatrix} 1 & -1 \\ -1 & 2 \end{bmatrix} = \begin{bmatrix} 1 & 0 \\ 0 & 1 \end{bmatrix} \text{ and } BA = \begin{bmatrix} 1 & -1 \\ -1 & 2 \end{bmatrix} \begin{bmatrix} 2 & 1 \\ 1 & 1 \end{bmatrix} = \begin{bmatrix} 1 & 0 \\ 0 & 1 \end{bmatrix}$$

Thus $AB = I_2 = BA$. ◇

The preceding example illustrates the notion of an invertible matrix introduced in the next definition: The matrix A in the example is invertible, and B is its inverse.

0.5.11 DEFINITION A matrix $A \in M(n, R)$ is **invertible** if there exists a matrix $A^{-1} \in M(n, R)$ such that $AA^{-1} = I_n = A^{-1}A$. Such a matrix A^{-1} is called an **inverse** to A. ○

In order to understand invertible matrices better, we need one more notion, introduced in the next definition. As we will see, it can be used to characterize completely which matrices are invertible, and to find their inverses. For simplicity we limit ourselves to the 2×2 case.

0.5.12 DEFINITION Given a 2×2 matrix in $M(n, R)$

$$A = \begin{bmatrix} a & b \\ c & d \end{bmatrix}$$

the **determinant** of A is defined to be $\det A = ad - bc \in R$. ○

We illustrate an important property of determinants.

0.5.13 EXAMPLE Let

$$A = \begin{bmatrix} 3 & 1 \\ 4 & 2 \end{bmatrix} \text{ and } B = \begin{bmatrix} 4 & 5 \\ 1 & 2 \end{bmatrix}$$

Then $\det A = (3)(2) - (1)(4) = 2$, and $\det B = (4)(2) - (5)(1) = 3$.

Now let us calculate the product:

$$AB = \begin{bmatrix} 3 & 1 \\ 4 & 2 \end{bmatrix}\begin{bmatrix} 4 & 5 \\ 1 & 2 \end{bmatrix} = \begin{bmatrix} 13 & 17 \\ 18 & 24 \end{bmatrix}$$

Then det AB = (13)(24) - (17)(18) = 312 - 306 = 6. Thus det AB = det A det B. ◇

The next proposition says that the relationship between the determinant of a product of two matrices and the product of the determinants of those matrices that we saw in the preceding example holds generally.

0.5.14 PROPOSITION For any $A, B \in M(2, R)$ we have det AB = det A det B.

Proof Let

$$A = \begin{bmatrix} a & b \\ c & d \end{bmatrix} \text{ and } B = \begin{bmatrix} a' & b' \\ c' & d' \end{bmatrix}$$

Then det A det B = $(ad - bc)(a'd' - b'c') = (ada'd' + bcb'c') - (bca'd' + adb'c')$. If we now calculate the product, we obtain

$$AB = \begin{bmatrix} aa' + bc' & ab' + bd' \\ a'c + dc' & cb' + dd' \end{bmatrix}$$

We therefore have

$$\begin{aligned} \det AB &= (aa' + bc')(cb' + dd') - (ab' + bd')(a'c + dc') = \\ &= (aa'dd' + bc'cb') + (aa'cb' + bc'dd') - (ab'a'c + bd'dc') - (ab'dc' + bd'a'c) \\ &= (ada'd' + bcb'c') - (bca'd' + adb'c') \qquad = \det A \det B. \ \square \end{aligned}$$

The definition of determinant and the proof of the multiplicative property of determinants can be extended to $n \times n$ matrices for any n. This material is covered in books and courses on linear algebra, and we do not go into it in our brief review, which is limited to basic notions we will need later in this book. (See Chapter 14.)
Going back to the question of when a matrix is invertible, we now have the answer for 2×2 matrices. (This characterization of invertible matices holds also for $n \times n$ matrices for any n. See Chapter 14.)

0.5.15 PROPOSITION A 2×2 matrix

$$A = \begin{bmatrix} a & b \\ c & d \end{bmatrix}$$

is invertible if and only if det $A \neq 0$, and if det $A \neq 0$, then

$$A^{-1} = \frac{1}{\det A}\begin{bmatrix} d & -b \\ -c & a \end{bmatrix}$$

Proof If $\det A = 0$, then by the preceding proposition, $\det AB = \det A \det B = 0 \cdot \det B = 0$ for any matrix B, while $\det I_2 = 1$. Hence $AB \neq I_2$ for any matrix B, and A is not invertible. If $\det A \neq 0$, then we have

$$\begin{bmatrix} a & b \\ c & d \end{bmatrix}\begin{bmatrix} d & -b \\ -c & a \end{bmatrix} \cdot \frac{1}{\det A} = \begin{bmatrix} ad-bc & 0 \\ 0 & ad-bc \end{bmatrix} \cdot \frac{1}{ad-bc} = \begin{bmatrix} 1 & 0 \\ 0 & 1 \end{bmatrix} = I_2$$

Thus A is invertible and the formula for the inverse stated above is proved. □

Exercises 0.5

In Exercises 1 through 10 perform the indicated matrix operations.

1. $\begin{bmatrix} 4 & 5 \\ 0 & -1 \\ 3 & 2 \end{bmatrix} + \begin{bmatrix} -2 & 1 \\ 3 & 6 \\ -3 & 3 \end{bmatrix}$ in $M_{3\times 2}(\mathbb{Z})$

2 $\begin{bmatrix} i & 2i \\ 1 & 0 \end{bmatrix} + \begin{bmatrix} 1 & 3 \\ i & 1 \end{bmatrix}$ in $M(2, \mathbb{C})$

3. $\begin{bmatrix} 1-i & 3+i \\ 0 & 4 \end{bmatrix} + \begin{bmatrix} 3i & 1-i \\ i & 2i \end{bmatrix}$ in $M(2, \mathbb{C})$

4. $\begin{bmatrix} i & 2i \\ 1 & 0 \end{bmatrix} + \begin{bmatrix} 1 & 3 \\ i & 1 \end{bmatrix}$ in $M(2, \mathbb{C})$

5. $i\begin{bmatrix} -1 & 1-i \\ 1+i & i \end{bmatrix}$ in $M(2, \mathbb{C})$

6. $2\begin{bmatrix} 3 & 2 \\ 4 & 1 \end{bmatrix}$ in $M(2, \mathbb{Z}_5)$

7. $\begin{bmatrix} 3 & 4 \\ 4 & 1 \end{bmatrix}\begin{bmatrix} 4 & 2 \\ 3 & 4 \end{bmatrix}$ in $M(2, \mathbb{Z}_5)$

8. $\begin{bmatrix} 1 & i \\ i & -1 \end{bmatrix}\begin{bmatrix} 2i & i \\ -i & 1 \end{bmatrix}$ in $M(2, \mathbb{C})$

9. $\begin{bmatrix} i & 0 \\ 0 & i \end{bmatrix}^5$ in $M(2, \mathbb{C})$

10. $\begin{bmatrix} 1 & i \\ -i & 1 \end{bmatrix}^4$ in $M(2, \mathbb{C})$

In Exercises 11 through 14 calculate the determinant of the indicated matrix.

11. $\begin{bmatrix} i & 1+i \\ 2i & -i \end{bmatrix}$ in $\mathbb{C}$

12. $\begin{bmatrix} 4 & 2 \\ 5 & 3 \end{bmatrix}$ in $\mathbb{Z}_7$

13. $\begin{bmatrix} 5 & 1 \\ 2 & 2 \end{bmatrix}$ in $\mathbb{Z}_7$

14. $\begin{bmatrix} 1 & -1 \\ -1 & 1 \end{bmatrix}$ in $\mathbb{Z}$

In Exercises 15 through 18 determine whether the indicated matrix is invertible and, if so, calculate the inverse matrix.

15. $\begin{bmatrix} 1 & 1 \\ 1 & -1 \end{bmatrix}$ in $M(2, \mathbb{Q})$

16. $\begin{bmatrix} \cos\theta & -\sin\theta \\ \sin\theta & \cos\theta \end{bmatrix}$ in $M(2, \mathbb{C})$

17. $\begin{bmatrix} i & i \\ i & -i \end{bmatrix}$ in $M(2, \mathbb{C})$

18. $\begin{bmatrix} 4 & 1 \\ -3 & 2 \end{bmatrix}$ in $M(2, \mathbb{Z}_5)$

19. Show that if $A, B \in M(2, \mathbb{C})$ are invertible, then so is AB.

20. Find all invertible matrices in $M(2, \mathbb{Z}_2)$.

21. Find all matrices A with $\det A = 1$ in $M(2, \mathbb{Z}_3)$.

Part A

Group Theory

Chapter 1
Groups

We are now ready to embark on our study of abstract algebra, beginning with the concept of a *group*.

Several sources contributed to the emergence of the *abstract group* concept. First, understanding the different deep properties of the integers was one of the most ancient preoccupations of mathematicians. Further, finding solutions to polynomial equations was for many centuries another important source of mathematical problems. Finally, the study of transformations of geometrical objects gave rise to new ideas in the development of mathematics in modern times. These three mathematical disciplines — number theory, the theory of algebraic equations, and the theory of geometric transformations — all contributed to the development of what in present-day mathematics is called the concept of an abstract group, or simply of a group.

In this chapter you will see how the notion of a group appears in several completely different situations and then learn how to work with an abstract group. The constructions in this first chapter serve as the basic examples of groups throughout later chapters.

1.1 Examples and Basic Concepts

We first look at some examples that will help us visualize the new concepts we introduce in this chapter.

1.1.1 EXAMPLE Let us find the roots of the polynomial $f(x) = x^3 - 1$ in $\mathbb{C}$. First note that $f(x)$ can be factored as $f(x) = x^3 - 1 = (x - 1)(x^2 + x + 1)$. Then if we use the quadratic formula, we find that $\omega = (-1 + \sqrt{3}\, i)/2$ and $\omega^2 = (-1 - \sqrt{3}\, i)/2$ are the two complex roots of $x^2 + x + 1$. Hence the three roots of $f(x)$ are $\{1, \omega, \omega^2\}$. Note that since $f(\omega) = 0$, we have $\omega^3 = 1$ and $\omega^4 = \omega$.

If we use the usual multiplication of complex numbers, we obtain the multiplication table shown in Table 1. ◇

TABLE 1 Multiplication for $\{1, \omega, \omega^2\}$

	1	ω	ω^2
1	1	ω	ω^2
ω	ω	ω^2	1
ω^2	ω^2	1	ω

1.1.2 EXAMPLE Let us find all the roots of $f(x) = x^4 - 1$ in $\mathbb{C}$. Note that $f(x)$ can be factored as $(x^2 - 1)(x^2 + 1) = (x - 1)(x + 1)(x - i)(x + i)$. So the four roots of $f(x)$ are $\{1, -1, i, -i\}$. If we use the usual multiplication of complex numbers, we obtain the multiplication table shown in Table 2. ◇

TABLE 2 Multiplication for $\{1, i, -1, -i\}$

	1	i	-1	$-i$
1	1	i	-1	$-i$
i	i	-1	$-i$	1
-1	-1	$-i$	1	i
$-i$	$-i$	1	i	-1

1.1.3 EXAMPLE For $\mathbb{Z}_3 = \{0, 1, 2\}$ and addition mod 3 as the operation we have the addition table shown in Table 3. ◇

TABLE 3 Addition mod 3

	0	1	2
0	0	1	2
1	1	2	0
2	2	0	1

1.1.4 EXAMPLE For $\mathbb{Z}_4 = \{0, 1, 2, 3\}$ and addition mod 4 as the operation we have the addition table shown in Table 4. ◇

TABLE 4 Addition mod 4

	0	1	2	3
0	0	1	2	3
1	1	2	3	0
2	2	3	0	1
3	3	0	1	2

If you compare now the tables in Examples 1.1.1 and 1.1.3, you will notice that they have the same pattern or structure, except that the names of the elements are different (similarly for Examples 1.1.2 and 1.1.4). These examples are of structures of the kind that are called groups. Before giving the formal definition of the group concept, we look at a few more, rather different-looking examples.

1.1.5 EXAMPLE Starting with an equilateral triangle, let us consider all the symmetries of the triangle, or motions of the triangle that leave it occupying the same place. There are six of these: the identity ρ_0, which leaves the triangle unmoved; the rotation ρ through 120° counterclockwise; the rotation ρ' through 240° counterclockwise; and three reflections or flips μ_1, μ_2, μ_3 along lines bisecting the triangle.

These are most easily visualized if we think of the three vertices of the triangle being labeled, as in Figure 1.

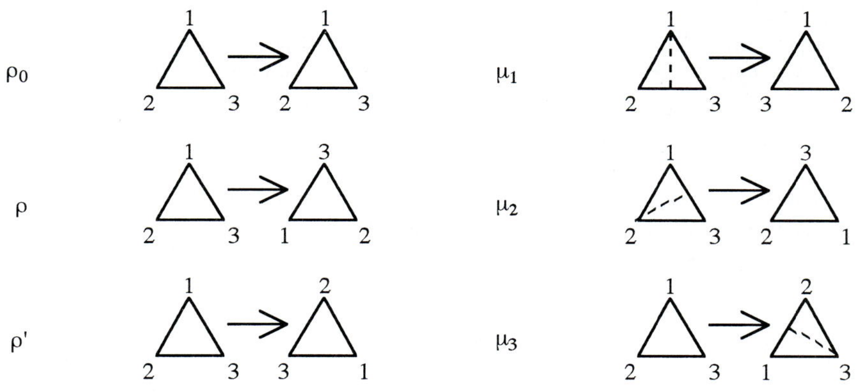

FIGURE 1

Let S_3 be the set of these six symmetries $\{\rho_0, \rho, \rho', \mu_1, \mu_2, \mu_3\}$. We can consider the elements of S_3 to be functions on $\{1, 2, 3\}$, and write, for instance, $\mu_1(2) = 3$ to indicate that μ_1 moves the vertex in position 2 to position 3, or $\rho(3) = 1$ to indicate that ρ moves the vertex in position 3 to position 1.

Let us now consider the operation of composition on this set S_3 of functions. For instance, performing ρ' is the same as performing ρ twice, so ρ' is the composition of ρ with itself, which we write $\rho' = \rho\rho = \rho^2$. The composition is computed almost equally easily in other cases.

We need to explain at this point a notation we use throughout the text. Since the operation is composition of functions, $\mu_1\mu_2$ stands for the composite function $\mu_1(\mu_2(x))$.

For example, let us first perform μ_1 and then perform ρ:

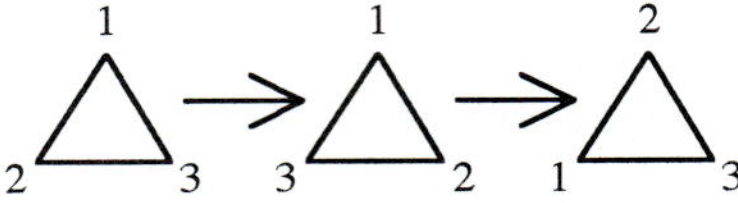

FIGURE 2

The result is the same as performing μ_3, and we write $\rho\mu_1 = \mu_3$. Note that the operation performed *first*, namely μ_1, is written on the *right*, as in the usual notation for composition of functions. For example, $\rho\mu_1(2) = \rho(\mu_1(2)) = \rho(3) = 1 = \mu_3(2)$. It can similarly be worked out that $\mu_1\rho = \mu_2$. So we have $\rho\mu_1 \neq \mu_1\rho$. ◇

1.1.6 EXAMPLE Let us consider similarly the symmetries of the square. Notice that if we remove a square with vertices labeled 1, 2, 3, 4 from a piece of paper, and then return the square to its original place, then no matter how we rotate or flip it, the two vertices next to vertex 1 must always be vertices 2 and 4. We have four choices where to put the vertex 1, but once we have placed vertex 1 we only have two choices where to put vertex 2, and once we have placed vertex 2, there will be only one choice where to put vertices 3 and 4. This means that there are $4 \cdot 2 = 8$ symmetries of the square. The four rotations may be named as shown in Figure 3.

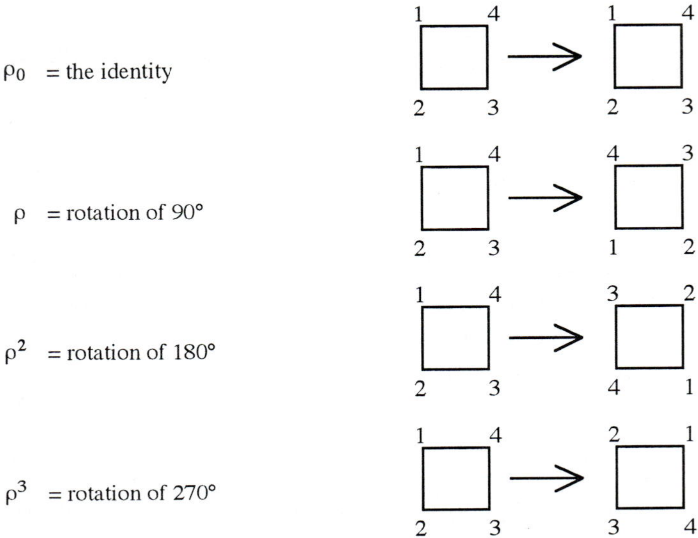

FIGURE 3

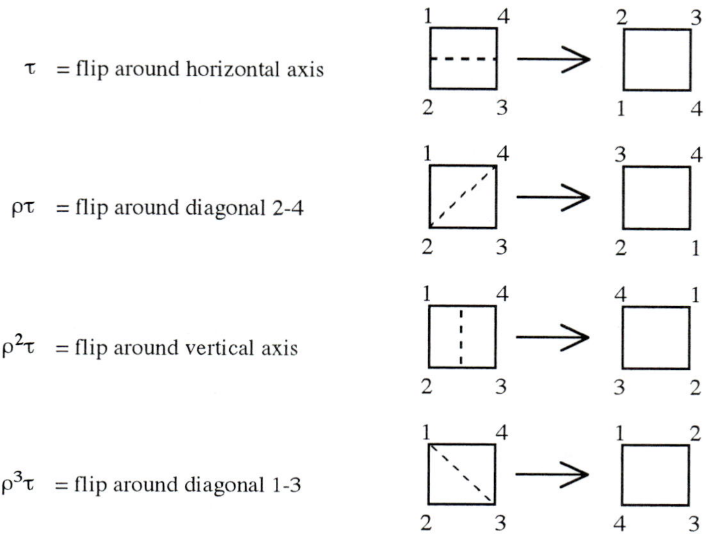

FIGURE 4

The four flips may be named as shown in Figure 4. Note again that when we write $\rho\tau$ we *first* perform the operation written on the *right*, namely, the flip τ, and then the operation written on the left, the rotation ρ. We name the set of symmetries of the square D_4. ◇

In all these examples we have a set and some kind of operation on it, and in all the examples the operation obeys certain laws, which we now define.

1.1.7 DEFINITION A nonempty set G equipped with an operation $*$ on it is said to form a **group** under that operation if the operation obeys the following laws, called the **group axioms**:

(1) **Closure** For any $a, b \in G$, we have $a * b \in G$.

(2) **Associativity** For any $a, b, c \in G$, we have $a * (b * c) = (a * b) * c$.

(3) **Identity** There exists an element $e \in G$ such that for all $a \in G$ we have $a * e = e * a = a$. Such an element $e \in G$ is called an **identity** in G.

(4) **Inverse** For each $a \in G$ there exists and element $a^{-1} \in G$ such that $a * a^{-1} = a^{-1} * a = e$. Such an element $a^{-1} \in G$ is called an **inverse** of a in G. ○

In simple terms, a group consists of two items, a set G and an operation $*$ on it, such that (1) for any elements a and b, the result $a * b$ of applying the operation to them is again an element of G; (2) when we perform a series of operations we can move parentheses around; (3) G has an element e that plays the role of an identity; and finally (4) every element a in G has an inverse in G. These conditions are met in all of Examples 1.1.1 through 1.1.6.

1.1.8 DEFINITION A group G with operation $*$ is said to be **Abelian** if the operation $*$ on G obeys the **commutative** law or, in other words, if for every $a, b \in G$ we have $a * b = b * a$. ○

Note that this condition is met in Examples 1.1.1 through 1.1.4, but not in Examples 1.1.5 and 1.1.6.

1.1.9 EXAMPLE $\langle\mathbb{Z}, +\rangle$, the set of all integers with or "under" the operation of addition forms a group, with $0 \in \mathbb{Z}$ as the identity element, and for every element $a \in \mathbb{Z}$, $-a$ as the inverse of a. ◇

1.1.10 EXAMPLE $\langle 2\mathbb{Z}, +\rangle$, the set of all even integers under the operation of addition, also is a group because the sum of two even integers is again even. More generally, $\langle n\mathbb{Z},+\rangle$ is a group for any n. ◇

1.1.11 EXAMPLE $\langle\mathbb{Q}, +\rangle$, $\langle\mathbb{R}, +\rangle$, and $\langle\mathbb{C}, +\rangle$, the rational, real, and complex numbers under addition, are all examples of groups. ◇

1.1.12 EXAMPLE $\langle\mathbb{Q}, \cdot\rangle$, the rational numbers under multiplication, is *not* a group. For though the rational numbers are closed under multiplication, multiplication is associative, there is a multiplicative identity 1, and every *nonzero* rational number a has the multiplicative inverse $1/a$, still the rational number 0 has no multiplicative inverse. ◇

This last example shows that a set may form a group under one operation and not another, which is why officially a group is a pair $\langle G, *\rangle$ consisting of two items. Often, however, when it is understood what operation we have in mind, we speak simply of the group G.

1.1.13 EXAMPLE If we consider the set $\mathbb{Q}^* = \mathbb{Q} - \{0\}$ of all nonzero rational numbers, then $\mathbb{Q}^*$ is closed under multiplication and $\langle\mathbb{Q}^*, \cdot\rangle$ is a group (similarly for $\mathbb{R}^* = \mathbb{R} - \{0\}$ and $\mathbb{C}^* = \mathbb{C} - \{0\}$). $\mathbb{Z}$ cannot be made a group under multiplication even if we remove 0, because for any nonzero integer $a \neq \pm 1$, the multiplicative inverse $1/a$ of a is not in $\mathbb{Z}$. ◇

1.1.14 EXAMPLE $\mathbb{Z}_n = \{0, 1, 2, \ldots, n\text{-}1\}$ under addition mod n is a group. 0 is the identity element, and for each $a \in \mathbb{Z}_n$, its inverse element is $n - a$. ◇

1.1.15 EXAMPLE Fix an integer $n > 1$, and let $U(n)$ be the set of all integers s, $1 \leq s \leq n$, such that $\gcd(n, s) = 1$ or, in other words, s and n are relatively prime. From Theorem 0.3.16 we know that there exist integers u and v such that $us + vn = 1$, and hence $us \equiv 1 \bmod n$. Moreover, if $1 \leq r \leq n$ and $u \equiv r \bmod n$, then also $\gcd(n, r) = 1$, $r \in U(n)$, and $rs \equiv 1 \bmod n$. Hence $U(n)$ under multiplication mod n forms a group. ◇

1.1.16 EXAMPLE Let us exhibit now the tables for the group operations in $U(10) = \{1, 3, 7, 9\}$ and $U(12) = \{1, 5, 7, 11\}$. ◇

TABLE 5 Multiplication mod 10

	1	3	7	9
1	1	3	7	9
3	3	9	1	7
7	7	1	9	3
9	9	7	3	1

TABLE 6 Multiplication mod 12

	1	5	7	11
1	1	5	7	11
5	5	1	11	7
7	7	11	1	5
11	11	7	5	1

1.1.17 EXAMPLE In Examples 1.1.1 and 1.1.2 we showed that the third and fourth roots of unity form groups under multiplication. This is actually true for the ***n*th roots of unity** for any n. These are the zeros of $x^n - 1$, namely, $\{1, \omega, \omega^2, \ldots, \omega^{n-1}\}$ where $\omega = \cos(2\pi/n) + i \sin(2\pi/n)$. ◇

1.1.18 EXAMPLE In Example 1.1.6 we described D_4, the group of symmetries of the square. Similarly, we can construct the group of symmetries of any regular n-gon for $n \geq 3$, the so-called **dihedral groups** D_n. Let us place a regular n-gon on the x,y-plane with its center at the origin. (See Figure 5.) Then we can describe the $2n$ elements of D_n as follows.

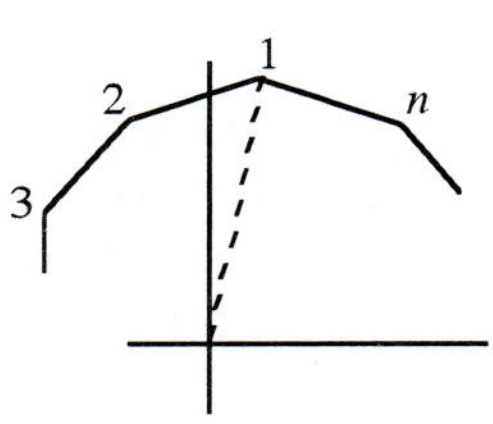

FIGURE 5

Let ρ be the rotation counterclockwise through $2\pi/n$ radians, and let τ be the flip around the axis through the origin and the vertex 1. Then $D_n = \{\rho_0, \rho, \rho^2, \ldots, \rho^{n-1}, \tau, \rho\tau, \rho^2\tau, \ldots, \rho^{n-1}\tau\}$, where ρ_0 is the identity and $\rho\tau = \tau\rho^{-1}$. (See Exercise 23.) Again, as in Example 1.1.6, we can think of the elements of D_n as functions on the set of n vertices and the operation used as composition of functions. ◇

In the next three examples we consider sets of 2×2 matrices with entries in R, where R may be $\mathbb{Q}$, $\mathbb{R}$, $\mathbb{C}$, or $\mathbb{Z}_n$.

1.1.19 EXAMPLE Let $M(2,R)$ be the set of all 2×2 matrices with entries in R. Then $M(2, R)$ is a group under addition of matrices. ◇

1.1.20 EXAMPLE As we have seen in Proposition 0.5.15, a matrix A is invertible if and only if $\det A \neq 0$. Note that since $\det AB = \det A \det B$, it follows that if A and B are invertible, then so is AB. Let $GL(2, R)$ be the set of all 2×2 matrices with nonzero determinant. Then $GL(2, R)$ is a group under matrix multiplication.

The identity and inverses are given by

$$I = \begin{bmatrix} 1 & 0 \\ 0 & 1 \end{bmatrix}, \begin{bmatrix} a & b \\ c & d \end{bmatrix}^{-1} = \frac{1}{ad - bc}\begin{bmatrix} d & -b \\ -c & a \end{bmatrix}$$

Since matrix multiplication does not in general obey the commutative law, this is another example of a non-Abelian group, called the **general linear group**. ◇

1.1.21 EXAMPLE The **special linear group**, $SL(2, R)$ consists of all 2×2 matrices A with entries from R having $\det A = 1$. This group also is non-Abelian.◇

Our next two examples of matrix groups will be finite groups.

1.1.22 EXAMPLE The **Klein four group** V can be represented as the group that consists of four matrices in $SL(2, R)$, namely

$$I = \begin{bmatrix} 1 & 0 \\ 0 & 1 \end{bmatrix}, \begin{bmatrix} -1 & 0 \\ 0 & 1 \end{bmatrix}, \begin{bmatrix} 1 & 0 \\ 0 & -1 \end{bmatrix}, \begin{bmatrix} -1 & 0 \\ 0 & -1 \end{bmatrix}$$

If you work out the group table for this group, you will find that it does *not* show the same pattern as was seen in Examples 1.1.2 and 1.1.4. ◇

1.1.23 EXAMPLE The **quaternion group** Q_8 can be represented as the group that consists of eight matrices in $SL(2, \mathbb{C})$, namely $Q_8 = \{\pm\mathbf{1}, \pm\mathbf{i}, \pm\mathbf{j}, \pm\mathbf{k}\}$, where

$$I = \begin{bmatrix} 1 & 0 \\ 0 & 1 \end{bmatrix}, \mathbf{i} = \begin{bmatrix} i & 0 \\ 0 & -i \end{bmatrix}, \mathbf{j} = \begin{bmatrix} 0 & 1 \\ -1 & 0 \end{bmatrix}, \mathbf{k} = \begin{bmatrix} 0 & i \\ i & 0 \end{bmatrix}$$

Since, for instance, $\mathbf{ij} = \mathbf{k}$ but $\mathbf{ji} = -\mathbf{k}$, this is yet another example of a non-Abelian group. ◇

There are a few basic properties of groups that may have become apparent from the various examples we have described. For instance, in every case we encountered only one identity element, and each element always had a unique inverse. (See the group tables in Examples 1.1.1 through 1.1.4, for instance.)
From now on, when we talk about an abstract group G we write the operation as a product, $a * b = ab$.

1.1.24 PROPOSITION (Basic group properties) For any group G

(1) The identity element of G is unique.

(2) For each $a \in G$, the inverse a^{-1} is unique.

(3) For any $a \in G$, $(a^{-1})^{-1} = a$.

(4) For any $a, b \in G$, $(ab)^{-1} = b^{-1}a^{-1}$.

(5) For any $a, b \in G$, the equations $ax = b$ and $ya = b$ have unique solutions or, in other words, the left and right cancellation laws hold.

Proof (1) If both e and e' are identities in G, then $ee' = e'$ because e is an identity, but also $ee' = e$ because e' is an identity. Hence $e = e'$.

(2) If both a' and a'' are inverses of $a \in G$, then $a' = a'e = a'(aa'') = (a'a)a'' = ea'' = a''$.

(3) Since $a^{-1}a = aa^{-1} = e$, and since [by (2)] the inverse of an element is unique, it follows that a is the inverse of a^{-1}, or $(a^{-1})^{-1} = a$.

(4) Since $(b^{-1}a^{-1})(ab) = b^{-1}(a^{-1}a)b = b^{-1}eb = b^{-1}b = e$ and similarly $(ab)(b^{-1}a^{-1}) = e$, and since the inverse of an element is unique, it follows that $b^{-1}a^{-1}$ is the inverse of ab, or $(ab)^{-1} = b^{-1}a^{-1}$.

(5) For $a, b \in G$, the equation $ax = b$ implies $a^{-1}(ax) = a^{-1}b$, and since $a^{-1}(ax) = (a^{-1}a)x = ex = x$, the equation implies $x = a^{-1}b$. Therefore, x is unique, since a^{-1} is unique. Similarly, the equation $ay = b$ implies $y = ba^{-1}$, and y is unique. □

One last note on the notation we will be using: For any positive integer n we abbreviate $a \cdot \ldots \cdot a$ (n times) as a^n and $a^{-1} \cdot \ldots \cdot a^{-1}$ (n times) as a^{-n}. We also let $a^0 = e$. When the group operation is written as addition, we write na instead of a^n for $n \in \mathbb{Z}$.

1.1.25 EXAMPLE In the group $U(9) = \{1, 2, 4, 5, 7, 8\}$ under multiplication mod 9 we have $2 \cdot 5 = 1$, $4 \cdot 7 = 1$, and $8 \cdot 8 = 1$. Hence $5 = 2^{-1}$, $7 = 4^{-1}$, and $8 = 8^{-1}$. Note $8^{-1} = (2 \cdot 4)^{-1} = 4^{-1} \cdot 2^{-1} = 7 \cdot 5 = 8$. ◇

1.1.26 EXAMPLE In the group S_3 of symmetries of the triangle (see Example 1.1.5), we have $\rho\mu_1 = \mu_3$, $\mu_1^{-1} = \mu_1$, $\rho^{-1} = \rho^2$, $\mu_3^{-1} = \mu_3$, and $(\rho\mu_1)^{-1} = \mu_3^{-1} = \mu_3 = \mu_1\rho^2 = \mu_1^{-1}\rho^{-1}$. ◇

We show now by one example how to construct an abstract group using a group table.

1.1.27 EXAMPLE Let us find all possible **group tables** for a group G with three elements. One of these elements must be the identity, call it e. So $G = \{e, a, b\}$, where a and b are two elements distinct from each other and from e. The fact that e is the identity enables us to fill in the top row and left column of the group table. Since G is a group, $ab, aa, bb \in G$. If $ab = a$, then $ab = ae$ and $b = e$, while if $ab = b$ then similarly $a = e$. Since a and b are distinct from e, we must have $ab = e$ and similarly $ba = e$. If $aa = e$, then $aa = ab$ and $a = b$, while if $aa = a$, then $aa = ae$ and $a = e$. Since these are impossible, we must have $aa = b$ and similarly $bb = a$. This determines the other four entries in the group table, which therefore is as in Table 7.

TABLE 7 Group of Order 3

	e	a	b
e	e	a	b
a	a	b	e
b	b	e	a

The group table for any group of order 3 must show this pattern, though the names of the elements may be different. (Compare with the tables in Examples 1.1.1 and 1.1.3.) ◇

1.1.28 DEFINITION The number of elements in a group G is called the **order** of G and is denoted $|G|$. G is a **finite** group if $|G|$ is finite. ○

Thus $\mathbb{Z}$ and $n\mathbb{Z}$ are groups of infinite order, while $|U(12)| = 4$, $|S_3| = 6$, $|D_4| = 8$, $|V| = 4$, $|Q_8| = 8$, and $|\mathbb{Z}_n| = n$ give examples of groups of finite order.

1.1.29 EXAMPLE The **Euler ϕ-function** $\phi(n)$ for an integer $n \geq 2$ is defined to be the number of positive integers s such that $1 \leq s \leq n$ and $\gcd(s, n) = 1$. Hence for any integer $n \geq 2$, $|U(n)| = \phi(n)$, and for a prime p, $|U(p)| = p - 1 = \phi(p)$. ◇

Exercises 1.1

In Exercises 1 through 5 show that the indicated set G with the specified operation forms a group by showing that the four axioms in the definition of a group are satisfied.

1. $G = 2\mathbb{Z}$ under addition

2. $G = \mathbb{Z}_5$ under addition mod 5

3. $G = U(10)$ under multiplication mod 10

4. $G = \mathbb{C}^* = \mathbb{C} - \{0\}$ under complex multiplication

5. $G = GL(2, \mathbb{Q})$ under matrix multiplication

In Exercises 6 through 9 construct the group table for the indicated group, and determine whether or not it is Abelian.

6. $G = S_3$ (see Example 1.1.5)

7. $G = D_4$ (see Example 1.1.6)

8. $G = V$ (see Example 1.1.22)

9. $G = Q_8$ (see Example 1.1.23)

10. Show that $GL(2,\mathbb{Q})$ is non-Abelian.

11. Show that if G is an Abelian group, then for all $a, b \in G$ and for all integers n, $(ab)^n = a^n b^n$.

12. In S_3 find two elements a, b such that $(ab)^2 \neq a^2 b^2$.

13. In S_3 find all elements a such that $a^2 = \rho_0 =$ the identity, and all elements b such that $b^3 = \rho_0$.

14. Find the inverse of each element of $U(10)$ and of $U(15)$.

15. Let G be the multiplicative group of all nth roots of unity. If $a \in G$, what is a^{-1}? (See Example 1.1.17.)

16. In D_4 find the inverse of ρ, of τ, and of $\rho\tau$. (See Example 1.1.6.)

17. In the Klein 4-group, show that every element is equal to its own inverse.

18. Show that if every element of a group G is equal to its own inverse, then G is Abelian.

19. Imitating Example 1.1.27, construct all possible group tables for a group G of order 4.

20. Construct all possible group tables for a group G of order 5. (*Hint:* Show first that for any $a \in G$, $a \neq e$, we have $a^k \neq e$ for all integers $1 \leq k < 5$.)

21. What is the order of $GL(2, \mathbb{Z}_2)$?

22. Show that if G is a finite group of even order, then G has an element $a \neq e$ such that $a^2 = e$.

23. In the dihedral groups D_n with $n \geq 3$, show that we have $\rho\tau = \tau\rho^{-1}$. (See Example 1.1.18.)

24. Prove that a finite group is Abelian if and only if its group table is a **symmetric matrix**, that is, a matrix $\{a_{ij}\}$ where $a_{ij} = a_{ji}$ for all i and j.

25. Let G be a group, $a \in G$, and m, n relatively prime integers. Show that if $a^m = e$, then there exists an element $b \in G$ such that $a = b^n$.

26. Let G be a finite Abelian group such that for all $a \in G$, $a \neq e$, we have $a^2 \neq e$. If $a_1, a_2, \ldots, a_n$ are all the elements of G with no repetitions, evaluate the product $a_1 a_2 \ldots a_n$.

27. Let G be a nonempty finite set closed under an associative operation such that both the left and the right cancellation laws hold. Show that G under this operation is a group.

28. Show that the nonzero elements of $\mathbb{Z}_p$, where p is a prime, form a group under multiplication mod p.

29. **(Wilson's theorem)** Prove that if p is prime, then $(p - 1)! \equiv -1 \bmod p$.

1.2 Subgroups

In several of the examples in the previous section, the set of elements of the group is a subset of the set of elements of another group, and the operation is the same.

1.2.1 EXAMPLE The set of even integers $2\mathbb{Z}$ is a subset of the set of integers $\mathbb{Z}$, and both are groups under addition. ◇

1.2.2 EXAMPLE The set of fourth roots of unity $\{\pm 1, \pm i\}$ is a subset of the set of nonzero complex numbers $\mathbb{C}^*$, and both are groups under multiplication. ◇

1.2.3 EXAMPLE Consider $\mathbb{Z}_8 = \{0, 1, 2, 3, 4, 5, 6, 7\}$ and the subset $H = \{0, 2, 4, 6\}$. Then H is also a group under addition mod 8, with the following group table shown in Table 8. ◇

TABLE 8 Group Table of H

	0	2	4	6
0	0	2	4	6
2	2	4	6	0
4	4	6	0	2
6	6	0	2	4

Subsets of a group G that are themselves groups under the same operation as G play an important role in identifying different groups.

1.2.4 DEFINITION A nonempty subset H of a group G is a **subgroup** of G if H is a group under the same operation as G. We use the notation $H \subseteq G$ to mean that H is a subset of G, and $H \leq G$ to mean that H is a subgroup of G. ○

1.2.5 EXAMPLE $\mathbb{Z} \leq \mathbb{Q} \leq \mathbb{R} \leq \mathbb{C}$ under addition. ◇

1.2.6 EXAMPLE $\mathbb{Q}^* \leq \mathbb{R}^* \leq \mathbb{C}^*$ under multiplication. ◇

1.2.7 EXAMPLE For any group G with identity element e, $\{e\}$ is a subgroup of G called the **trivial subgroup**, and G itself is a subgroup of G, called the **improper subgroup**. Any other subgroup of G besides these two is called a **nontrivial proper subgroup** of G.

1.2.8 EXAMPLE $\{\pm 1\} \leq \{\pm 1, \pm i\}$ under multiplication. ◇

1.2.9 EXAMPLE $\{\rho_0, \rho, \rho^2\} \leq S_3$ under composition of functions. ◇

To prove that a subset H of a group G forms a subgroup, we need to show that the four group axioms hold for H. This could be cumbersome in many cases, and before considering any more examples we prove a theorem that makes easier the verification that a subset is a subgroup.

1.2.10 THEOREM (Subgroup test) A nonempty subset H of a group G is a subgroup of G if and only if the following condition holds:

(∗) For every $a, b \in H$, $ab^{-1} \in H$

Proof (⇒) Assume H is a subgroup of G, and take any $a, b \in H$. Since H is a subgroup, it contains the inverse of any of its elements, and so $b^{-1} \in H$, since $b \in H$. Again, since H is a subgroup, it contains the product of any two of its elements, and so $ab^{-1} \in H$ since $a, b^{-1} \in H$.

(⇐) Assume condition (∗) holds. We verify the four group axioms:

(Identity) Since H is nonempty it has some element a. Apply (∗) to the pair a, a to conclude that $e = aa^{-1} \in H$.

(Inverse) For any $b \in H$, apply (∗) to the pair e, b to conclude that $b^{-1} = eb^{-1} \in H$.

(Closure) For any $a, b \in H$, we have just shown that $b^{-1} \in H$. Apply (∗) to the pair a, b^{-1} to conclude that $ab = a(b^{-1})^{-1} \in H$.

(Associativity) The operation on H is associative because it is the same as the operation on G, which is associative since G is a group. □

Note that if the group operation is written as addition, then (∗) reads as follows:

(∗) For every $a, b \in H$, $a - b \in H$

Using the subgroup test, we can now in many cases easily determine whether a subset of a group is a group.

1.2.11 EXAMPLE For any integer $n \geq 0$, $n\mathbb{Z}$ is a subgroup of $\mathbb{Z}$ under addition. For if $a, b \in n\mathbb{Z}$, then $a = nr$ for some integer r and $b = ns$ for some integer s. But then $a - b = nr - ns = n(r - s) \in n\mathbb{Z}$. ◇

1.2.12 EXAMPLE The set H of odd integers is not a subgroup of $\mathbb{Z}$ under addition, since $1, 3 \in H$, but $1 - 3 \notin H$. ◇

The following alternate subgroup test is easier to apply in some cases.

1.2.13 THEOREM A nonempty subset H of a group G is a subgroup of G if and only if the following conditions hold:

(1) (Closure) For any $a, b \in H$, $ab \in H$.

(2) (Inverse) For any $b \in H$, $b^{-1} \in H$.

Proof (⇒) If H is a subgroup of G, then the closure and inverse conditions hold, of course, because they are group axioms.

(⇐) Assuming the closure and inverse conditions hold, consider any $a, b \in H$. We have $b^{-1} \in H$ by the inverse condition, and so $ab^{-1} \in H$ by the closure condition, but by Theorem 1.2.10 this is enough to show that H is a subgroup. □

1.2.14 EXAMPLE $SL(2, \mathbb{Q})$ is a subgroup of $GL(2. \mathbb{Q})$. For if $A \in SL(2, \mathbb{Q})$, then $\det A = 1$, so $\det(A^{-1}) = 1/\det A = 1$, and $A^{-1} \in SL(2, \mathbb{Q})$. Further, if $A, B \in SL(2, \mathbb{Q})$, then $\det A = \det B = 1$, so $\det(AB) = \det A \cdot \det B = 1$, and $AB \in SL(2, \mathbb{Q})$. By Theorem 1.2.13, this suffices to show that $SL(2, \mathbb{Q})$ is a subgroup. ◇

1.2.15 THEOREM Let G be a group. A nonempty finite subset H of a group G is a subgroup of G if and only if the following condition holds:
(Closure) For any $a, b \in H$, $ab \in H$.

Proof By Theorem 1.2.13, we only need to show that the closure condition implies the inverse condition. So assume the closure condition and consider any $a \in H$. If $a = e$, then $a^{-1} = e^{-1} = e = a \in H$. If $a \neq e$, consider the powers $a = a^1, a^2, a^3, \ldots$. The closure condition implies that $a^i \in H$ for all i. Since H is finite, these powers cannot all be distinct. Therefore, there exist some i and j with $j < i$ and $a^i = a^j$. But then $a^{(i-j)} = a^i(a^j)^{-1} = a^i a^{-j} = e$, and so $aa^{(i-j-1)} = e$, and so $a^{-1} = a^{(i-j-1)} \in H$, and the inverse condition holds. □

1.2.16 DEFINITION Let G be a group and $a \in G$. Then we define

$$\langle a \rangle = \{a^n \mid n \in \mathbb{Z}\}$$

(If the group operation is written as addition, $\langle a \rangle = \{na \mid n \in \mathbb{Z}\}$.) ○

1.2.17 PROPOSITION Let G be a group and $a \in G$. Then $\langle a \rangle$ is a subgroup of G, called the **cyclic subgroup generated by** a.

Proof Let $x, y \in \langle a \rangle$. Then $x = a^m$, $y = a^n$ for some $m, n \in \mathbb{Z}$. Then $xy^{-1} = a^m(a^n)^{-1} = a^{m-n}$. Since $m - n \in \mathbb{Z}$, $xy^{-1} \in \langle a \rangle$. So by the subgroup test $\langle a \rangle$ is a subgroup of G. □

1.2.18 EXAMPLE In $\mathbb{Z}$ the subgroup generated by 3 is $\langle 3 \rangle = 3\mathbb{Z}$. ◇

1.2.19 EXAMPLE In $\mathbb{C}^*$ the subgroup generated by i is $\langle i \rangle = \{i, i^2, i^3, i^4\} = \{i, -1, -i, 1\}$. ◇

1.2.20 EXAMPLE In S_3 the subgroup generated by ρ is $\langle \rho \rangle = \{\rho_0, \rho, \rho^2\}$. ◇

1.2.21 EXAMPLE In D_4, $\langle \rho \rangle = \{\rho_0, \rho, \rho^2, \rho^3\}$ and $\langle \tau \rangle = \{\rho_0, \tau\}$. ◇

1.2.22 EXAMPLE Let us find all the subgroups of $\mathbb{Z}_6$. They are
$\{0\}$, the trivial subgroup
$\{0, 1, 2, 3, 4, 5, 6\} = \mathbb{Z}_6$, the improper subgroup
$\{0, 2, 4\} = \langle 2 \rangle = \langle 4 \rangle$
$\{0, 3\} = \langle 3 \rangle$
Note that if H is a subgroup and $5 \in H$, then $-5 = 1 \in H$, and that if $2, 3 \in H$, then $3 - 2 = 1 \in H$, and in either case $H = \mathbb{Z}_6$, so the subgroups listed are all there are. ◇

The next notion we introduce is crucial to the study of groups.

1.2.23 DEFINITION Let G be a group and $a \in G$. The **order** $|a|$ of a in G is the least positive integer n such that $a^n = e$, or infinite if there is no such n. (If the group operation is written as addition, the condition $a^n = e$ would be written $na = 0$.) ○

1.2.24 EXAMPLE In S_3, $|\mu_1| = 2$, $|\rho| = 3$. ◇

1.2.25 EXAMPLE In $\mathbb{Z}_6$, $|0| = 1$, $|1| = |5| = 6$, $|2| = |4| = 3$, $|3| = 2$. ◇

1.2.26 EXAMPLE In $\mathbb{Z}$, $|0| = 1$ and $|n|$ is infinite for all $n \neq 0$. ◇

1.2.27 EXAMPLE In $\mathbb{C}^*$, $|i| = 4$. ◇

Before closing this section, we introduce some very important subgroups of a group G.

1.2.28 DEFINITION Let G be any group. Then the **center** of G, denoted $Z(G)$, consists of the elements of G that commute with every element of G. In other words,

$$Z(G) = \{x \in G \mid xy = yx \text{ for all } y \in G\}$$

Note that $ey = y = ye$ for all $y \in G$, so $e \in Z(G)$, and the center is a nonempty subset of G. ○

We can actually say more.

1.2.29 THEOREM The center $Z(G)$ of a group G is a subgroup of G.

Proof We use Theorem 1.2.13, which means that we need to show that the product of any two elements of $Z(G)$ is in $Z(G)$ and that the inverse of any element of $Z(G)$ is in $Z(G)$. So first let $a, b \in Z(G)$, to show that $ab \in Z(G)$. By definition $ay = ya$ and $by = yb$ for any $y \in G$. It follows that $(ab)y = a(by) = a(yb) = (ay)b = (ya)b = y(ab)$ for any $y \in G$, which implies that $ab \in Z(G)$. Now let $a \in Z(G)$, to show that $a^{-1} \in G$. By definition, $ay = ya$ for all $y \in G$. It follows that $a^{-1}y = a^{-1}(y^{-1})^{-1} = (y^{-1}a)^{-1} = (ay^{-1})^{-1} = (y^{-1})^{-1}a^{-1} = ya^{-1}$, which implies $a^{-1} \in Z(G)$. □

Note that if G is an Abelian group, then $Z(G) = G$.

1.2.30 EXAMPLE Let us find the center of the non-Abelian group D_4. (See Example 1.1.6 and Exercise 7 in Section 1.1.) $D_4 = \{\rho_0, \rho, \rho^2, \rho^3, \tau, \rho\tau, \rho^2\tau, \rho^3\tau\}$. We have $\tau\rho = \rho^3\tau$, so $\tau \notin Z(D_4)$ and $\rho \notin Z(D_4)$ and $\rho^3 \notin Z(D_4)$. On the other hand, we have $\tau\rho^2 = (\tau\rho)\rho = (\rho^3\tau)\rho = \rho^3(\tau\rho) = \rho^3(\rho^3\tau) = (\rho^3\rho^3)\tau = \rho^6\tau = \rho^2\tau$, and it is then easy to see that ρ^2 commutes with all the other elements of D_4, so that $\rho^2 \in Z(D_4)$. Finally,

$$\begin{array}{lllll} (\rho\tau)\rho & = \rho(\rho^3\tau) & = \tau & \neq \rho^2\tau & = \rho(\rho\tau) \\ (\rho^2\tau)\rho & = \rho^2(\rho^3\tau) & = \rho\tau & \neq \rho^3\tau & = \rho(\rho^2\tau) \\ (\rho^3\tau)\rho & = \rho^3(\rho^3\tau) & = \rho^2\tau & \neq \tau & = \rho(\rho^3\tau) \end{array}$$

Hence $Z(D_4) = \{\rho_0, \rho^2\}$. ◇

The following definition introduces further important subgroups.

1.2.31 DEFINITION Let G be a group and $a \in G$. Then the **centralizer** of a in G, denoted $C_G(a)$, is the set of all elements of G that commute with a. In other words,

$$C_G(a) = \{y \in G \mid ay = ya\}$$

Note that for any $a \in G$ we have $Z(G) \subseteq C_G(a)$. In other words, the center is contained in the centralizer of any element. ○

We leave it as an exercise to show that $C_G(a)$ is a subgroup of G. (Exercise 32.) When it is understood what group G is involved, we just write $C(a)$ for $C_G(a)$.

1.2.32 EXAMPLE Let us find the centralizer of ρ in S_3. (See Example 1.1.5.) Obviously, $\rho_0, \rho, \rho^2 \in C(\rho)$. We know that $\rho\mu_1 \neq \mu_1\rho$. We can calculate that $\rho\mu_2 = \mu_1$ while $\mu_2\rho = \mu_3$, so that $\rho\mu_2 \neq \mu_2\rho$; while $\rho\mu_3 = \mu_2$ and $\mu_3\rho = \mu_1$, so that $\rho\mu_3 \neq \mu_3\rho$. Hence $C(\rho) = \{\rho_0, \rho, \rho^2\}$. ◇

Exercises 1.2

In Exercises 1 through 10 find the order of the indicated element in the indicated group.

1. $2 \in \mathbb{Z}_3$

2. $4 \in \mathbb{Z}_{10}$

3. $\mu_2 \in S_3$

4. $\rho \in D_4$

5. $\rho^2\tau \in D_4$

6. $(-1 + \sqrt{3}\, i)/2 \in \mathbb{C}^*$

7. $\mathbf{j} \in Q_8$

8. $-i \in \mathbb{C}^*$

9. $-1 + \sqrt{3}\, i \in \mathbb{C}^*$

10. $\cos(2\pi/7) + i \sin(2\pi/7) \in \mathbb{C}^*$

In Exercises 11 through 19 give at least two examples of a nontrivial proper subgroup of the indicated group.

11. $\mathbb{Z}$

12. $\mathbb{Q}$

13. $\mathbb{C}^*$

14. $\mathbb{Z}_8$

15 S_3

16. D_4

17. $8\mathbb{Z}$

18. $GL(2,\mathbb{Q})$

19. Q_8

20. Let G be a group and $a \in G$. Show that a and a^{-1} generate the same cyclic subgroup $\langle a \rangle = \langle a^{-1} \rangle$ and have the same order $|a| = |a^{-1}|$.

21. Let $G = \{a + b\sqrt{2} \mid a, b \in \mathbb{Q}\}$. Show that G is a subgroup of $\mathbb{R}$ under addition.

22. Let $G = \{n + mi \mid m, n \in \mathbb{Z}, i^2 = -1\}$. Show that G is a subgroup of $\mathbb{C}$ under addition.

23. Let $G = \{\cos(2k\pi/7) + i\sin(2k\pi/7) \mid k \in \mathbb{Z}\}$. Show that G is a subgroup of $\mathbb{C}^*$ under multiplication. What is the order of G?

24. Let $G = \{a + bi \mid a, b \in \mathbb{R},\ a^2 + b^2 = 1\}$. Determine whether or not G is a subgroup of $\mathbb{C}^*$ under multiplication.

25. For $\theta \in \mathbb{R}$, let $A(\theta) \in SL(2,\mathbb{R})$ be the matrix representing a rotation of θ radians:

$$A(\theta) = \begin{bmatrix} \cos\theta & -\sin\theta \\ \sin\theta & \cos\theta \end{bmatrix}$$

(a) Show that $H = \{A(\theta) \mid \theta \in \mathbb{R}\}$ is a subgroup of the special linear group $SL(2, \mathbb{R})$.

(b) Find the inverse of $A(2\pi/3)$. (c) Find the order of $A(2\pi/3)$.

26. In the special linear group $SL(2, \mathbb{Z}_{10})$, let $A = \begin{bmatrix} 1 & 2 \\ 0 & 1 \end{bmatrix}$

(a) Calculate A^3 and A^{11}. (b) Find the order of A.

27. In the special linear group $SL(3, \mathbb{R})$, for any $a, b, c \in \mathbb{R}$, let

$$D(a,b,c) = \begin{bmatrix} 1 & a & b \\ 0 & 1 & c \\ 0 & 0 & 1 \end{bmatrix}.$$

Show that $H = \{D(a, b, c) \mid a, b, c \in \mathbb{R}\}$ is a subgroup of $SL(3, \mathbb{R})$.

28. Show that in an Abelian group G the set consisting of all elements of G of finite order is a subgroup of G.

29. Show that if H and K are subgroups of G, then $H \cap K$ is also a subgroup of G.

30. Show that if G is a group and $a, b \in G$, then $|aba^{-1}| = |b|$.

31. Show that if G is a group and $a, b \in G$, then $|ab| = |ba|$.

32. Let G be a group and $a \in G$. Show that the centralizer $C(a)$ is a subgroup of G.

33. Find the centralizer $C(\mu_1)$ of μ_1 in S_3.

34. Find the centralizer $C(\rho^2)$ of ρ^2 in D_4.

35. Let G be a group, $a \in G$. Show that the centralizer $C(a) = G$ if and only if $a \in Z(G)$, the center of G.

36. Find the center $Z(S_3)$ of S_3.

1.3 Cyclic Groups

In this section we concentrate on the study of groups of a special kind, called *cyclic groups*. They will later be seen to be the building blocks out of which all finite Abelian groups can be constructed. We have already introduced (in Proposition 1.2.17) the notion of the cyclic subgroup $\langle a \rangle$ of a group G generated by an element a. We begin by reviewing some examples involving that notion.

1.3.1 EXAMPLE We have already observed that in $\mathbb{Z}_6$ the subgroup generated by 1 is the whole group, as is the subgroup generated by 5: $\mathbb{Z}_6 = \langle 1 \rangle = \langle 5 \rangle$. ◇

1.3.2 EXAMPLE Likewise, in $\mathbb{Z}$ the subgroup generated by 1 or by -1 is the whole group: $\mathbb{Z} = \langle 1 \rangle = \langle -1 \rangle$. ◇

1.3.3 EXAMPLE Likewise, in $G = \{1, i, -1, -i\}$, the subgroup generated by i is the whole group, since $i^2 = -1$, $i^3 = -i$, $i^4 = 1$. ◇

These are examples of the notion of a cyclic group introduced in the next definition.

1.3.4 DEFINITION A group G is called **cyclic** if there exists an element $a \in G$ such that $G = \langle a \rangle = \{a^n \mid n \in \mathbb{Z}\}$. Any such element is called a **generator** of G. ○

When the group operation is being written as addition, the condition $G = \{a^n \mid n \in \mathbb{Z}\}$ is written $G = \{na \mid n \in \mathbb{Z}\}$. When we calculate $\langle a \rangle$ for an element a of a group G, we calculate the successive powers of a, or when the group operation is written as addition, the successive multiples of a. If these give all the elements of G, then G is generated by a.

1.3.5 EXAMPLE For any $n > 1$, $\mathbb{Z}_n = \langle 1 \rangle = \langle n - 1 \rangle$. This is a cyclic group of order n. ◇

1.3.6 EXAMPLE $\mathbb{Z}_{10} = \langle 1 \rangle = \langle 3 \rangle = \langle 7 \rangle = \langle 9 \rangle$ or, in other words, 1, 3, 7, and 9 are all generators of $\mathbb{Z}_{10}$. We verify this for 7 by computing its successive multiples. (The case of 3 is similar and the cases of 1 and 9 were included already in the preceding example.) We have $2 \cdot 7 = 7 + 7 = 4$, since $14 \equiv 4 \bmod 10$. Likewise $3 \cdot 7 = 7 + 7 + 7 = 1$, since $21 \equiv 1 \bmod 10$. Similarly, $4 \cdot 7 = 8$, $5 \cdot 7 = 5$, $6 \cdot 7 = 2$, $8 \cdot 7 = 6$, $9 \cdot 7 = 3$, $10 \cdot 7 = 0$. ◇

1.3.7 EXAMPLE $U(10) = \{1, 3, 7, 9\} = \{3^0, 3^1, 3^2, 3^3\} = \langle 3 \rangle$. ◇

We have also encountered some examples of groups that are not cyclic.

1.3.8 EXAMPLE In S_3, $\langle \rho \rangle = \langle \rho^2 \rangle = \{\rho_0, \rho, \rho^2\}$, while $\langle \mu_i \rangle = \{\rho_0, \mu_i\}$ for $i = 1, 2, 3$. Hence none of the elements of S_3 generates the whole group, and S_3 is not cyclic. ◇

1.3.9 EXAMPLE $2\mathbb{Z} = \langle 2 \rangle$ and, in general, for any $n \geq 1$, $n\mathbb{Z} = \langle n \rangle$. These are all infinite cyclic groups. ◇

1.3.10 EXAMPLE $\mathbb{Z}_{10} \neq \{0, 2, 4\} = \langle 2 \rangle$. We verify this by computing successive multiples. We have $2{\cdot}2 = 2 + 2 = 4$, $3{\cdot}2 = 2 + 2 + 2 = 6$, $4{\cdot}2 = 8$, $5{\cdot}2 = 0$, $6{\cdot}2 = 2$, $7{\cdot}2 = 4$, $8{\cdot}2 = 6$, $9{\cdot}2 = 8$, $10{\cdot}2 = 0$, and the pattern repeats. ◇

The next theorem tells us that the kind of repetition seen in the preceding example happens generally.

1.3.11 THEOREM Let G be a group and $a \in G$. Then for all $i, j \in \mathbb{Z}$ we have

(1) If a has infinite order, then $a^i = a^j$ if and only if $i = j$.

(2) If a has finite order $|a| = n$, then $a^i = a^j$ if and only if n divides $i - j$.

Proof (1) Suppose a has infinite order. For one direction, if $i = j$, then clearly $a^i = a^j$. For the other direction, if $a^i = a^j$, then $a^{i-j} = a^i a^{-j} = e$. But since a has infinite order, $a^n = e$ if and only if $n = 0$, so we have $i - j = 0$ or $i = j$.

(2) Suppose a has finite order $|a| = n$. For one direction, if n divides $i - j$, so that $i = nk + j$ for some $k \in \mathbb{Z}$, then we have $a^i = a^{nk+j} = a^{nk}a^j = (a^n)^k a^j = e^k a^j = ea^j = a^j$. For the other direction, suppose $a^i = a^j$ or, equivalently, $a^{i-j} = e$. By the division algorithm we can write $i - j = qn + r$, where $0 \leq r < n$. Then $e = a^{i-j} = a^{nq+r} = a^{nq}a^r = (a^n)^q a^r = e^q a^r = ea^r = a^r$. Since $0 \leq r < n$ and n is by definition the *least* positive integer such that $a^n = e$, we must have $r = 0$, so $i - j = qn$ and $i - j$ is divisible by n. □

We leave the proofs of the following three corollaries to the reader. (See Exercises 13 through 15.)

1.3.12 COROLLARY Let G be a group and $a \in G$ an element of finite order $|a| = n$. Then for any $k \in \mathbb{Z}$, $a^k = e$ if and only if n divides k. □

1.3.13 COROLLARY Let G be a group and $a \in G$ an element of finite order $|a| = n$. Then $\langle a \rangle = \{e, a, a^2, \ldots, a^{n-1}\}$. □

1.3.14 COROLLARY Let G be a group and $a \in G$ an element of finite order. Then $|\langle a \rangle| = |a|$. □

1.3.15 EXAMPLE Let G be a cyclic group of order 6, $G = \langle a \rangle = \{e, a, a^2, a^3, a^4, a^5\}$. Let us find $|a^4|$. We have $(a^4)^2 = a^8 = a^{6+2} = a^6a^2 = ea^2 = a^2 \neq e$, while $(a^4)^3 = a^{12} = a^{6+6} = a^{6\cdot 2} = (a^6)^2 = e^2 = e$. Hence $|a^4| = 3$. ◇

The next theorem gives a formula enabling us to find the orders of elements of a cyclic group more easily.

1.3.16 THEOREM Let $G = \langle a \rangle$ be a cyclic group with generator a, of order $|G| = |a| = n$. Then for any element $a^s \in G$ we have $|a^s| = n / \gcd(n, s)$.

Proof By Corollary 1.3.12, $(a^s)^k = a^{sk} = e$ if and only if sk is a multiple of n. Since the order $|a^s|$ is by definition the least k such that $(a^s)^k = e$, it follows that the order $|a^s|$ is the least k such that the multiple sk of s is also a multiple of n or, equivalently, sk is the least positive integer that is a multiple of n as well as of s. This is to say that $sk = \operatorname{lcm}(n, s)$ and $k = \operatorname{lcm}(n, s)/s$. Since $\operatorname{lcm}(a, b) = ab/\gcd(a, b)$ by Proposition 0.3.29, part (3), we have $k = sn/s\gcd(n, s) = n/\gcd(n, s)$. □

1.3.17 EXAMPLE In a cyclic group $G = \langle a \rangle$ of order 210, the order of a^{80} is $|a^{80}| = 210/\gcd(210, 80) = 210/10 = 21$. ◇

1.3.18 EXAMPLE In $\mathbb{Z}_{105}$, the order of 84 is $|84| = 105/\gcd(105, 84) = 105/21 = 5$. ◇

The theorem can be used not only to simplify the calculations of orders of arbitrary elements of a cyclic group, as in the preceding examples, but also, as in the next example, to find all generators of the group.

1.3.19 EXAMPLE Consider $\mathbb{Z}_{12}$ and let us find all its generators, which is to say all elements $s \in \mathbb{Z}_{12}$ such that $|s| = 12$. By Theorem 1.3.16, $|s| = 12/\gcd(12, s)$. So $|s| = 12$ if and only if $\gcd(12, s) = 1$. Thus the generators of $\mathbb{Z}_{12}$ are the elements $s \in \mathbb{Z}_{12}$ such that $\gcd(12, s) = 1$, namely, the elements $s = 1, 5, 7$, or 11. ◇

1.3.20 COROLLARY Let $G = \langle a \rangle$ be a cyclic group with generator a, of order $|G| = |a| = n$. Then for any element $a^s \in G$, we have

$$a^s \text{ is a generator of } G \text{ if and only if } \gcd(n, s) = 1$$

Proof By definition, a^s is a generator of G if and only if $G = \langle a^s \rangle$, and hence if and only if $|\langle a^s \rangle| = n$. By Corollary 1.3.14, $|\langle a^s \rangle| = |a^s|$, and by Theorem 1.3.16, $|a^s| = n/\gcd(n, s)$. So a^s is a generator of G if and only if $n/\gcd(n, s) = n$ or, equivalently, $\gcd(n, s) = 1$. □

1.3.21 COROLLARY Let G be a cyclic group of order n. Then the number of generators of G is $\phi(n)$, where ϕ is the Euler function.

Proof This is immediate from the preceding corollary, since by definition $\phi(n)$ is the number of integers s, $1 \le s < n$, such that $\gcd(n, s) = 1$. □

The next example illustrates a property of cyclic groups that makes them especially easy to understand.

1.3.22 EXAMPLE Let us find all subgroups of $\mathbb{Z}_{15}$. To begin with we have the trivial subgroup $\langle 0 \rangle$. Let H be any nontrivial subgroup. By Corollary 1.3.20, if H contains any of the elements 1, 2, 4, 7, 8, 11, 13, 14, then H will be the improper subgroup $\mathbb{Z}_{15} = \langle 1 \rangle = \langle 2 \rangle = \langle 4 \rangle = \langle 7 \rangle = \langle 8 \rangle = \langle 11 \rangle = \langle 13 \rangle = \langle 14 \rangle$. Now let H be a nontrivial proper subgroup. First suppose $3 \in H$. Then for any $y \in H$, using the division algorithm we can write $y = q3 + r$ for some r with $0 \leq r < 3$. Since 3, $y \in H$, we have $r = y - q3 \in H$. But since H is proper, 1, 2 $\notin H$. So we must have $r = 0$, and y is a multiple of 3. Thus $H = \langle 3 \rangle = \{0, 3, 6, 9, 12\} = 3\mathbb{Z}_{15}$. Since 6 + 6 + 6 = 9 + 9 = 12 + 12 +12 + 12 = 3, any nontrivial proper subgroup containing any of 6, 9, or 12 will also contain 3 and be equal to $3\mathbb{Z}_{15} = \langle 3 \rangle = \langle 6 \rangle = \langle 9 \rangle = \langle 12 \rangle$. Now suppose H is a nontrivial proper subgroup and $5 \in H$. A similar argument shows that $H = \langle 5 \rangle = \{0, 5, 10\} = 5\mathbb{Z}_{15}$, and that any proper subgroup containing 10 will also be equal to $5\mathbb{Z}_{15} = \langle 5 \rangle = \langle 10 \rangle$. Since we have accounted for all possible elements, we have found all possible subgroups. ◇

1.3.23 THEOREM Every subgroup of a cyclic group is cyclic.

Proof Let $G = \langle a \rangle$ be a cyclic group and H a subgroup of G. If H is the trivial subgroup $\{e\}$, then $H = \langle e \rangle$ and is cyclic. Now assume that H is nontrivial, so there exists an element $b \in H$ with $b \neq e$. Since $b \in G = \langle a \rangle$, $b = a^s$ for some integer s, and since $b \neq e$, $s \neq 0$. Also, since $b \in H$, $a^{-s} = (a^s)^{-1} = b^{-1} \in H$. Since one or the other of s or $-s$ is positive, H contains some positive power of a. Now let m be the least positive integer such that $a^m \in H$. Consider any other element $y \in H$. Then $y = a^n$ for some integer n. Applying the division algorithm to m and n, we can write $n = qm + r$ for some integers q, r with $0 \leq r < m$. Then $y = a^n = a^{qm + r} = a^{qm}a^r = (a^m)^q a^r$, and $a^r = y(a^m)^{-q}$. Since y, $a^m \in H$, it follows that $a^r \in H$. But since $0 \leq r < m$ and m is the least positive integer with $a^m \in H$, we must have $r = 0$, and $y = (a^m)^q$. Thus every element of H is a power of a^m and $H = \langle a^m \rangle$ is cyclic. □

1.3.24 COROLLARY The subgroups of $\mathbb{Z}$ are $n\mathbb{Z} = \langle n \rangle$ for all $n \geq 0$.

Proof $\mathbb{Z} = \langle 1 \rangle$ is cyclic; hence every subgroup H of $\mathbb{Z}$ is also cyclic by Theorem 1.3.23, and hence is of the form $H = \langle m \rangle$ for some integer m. Since $\langle -m \rangle = \langle m \rangle$, and either $m \geq 0$ or $-m \geq 0$, $H = \langle n \rangle = n\mathbb{Z}$ for some $n \geq 0$. □

1.3.25 EXAMPLE Since $\mathbb{Z}_{12}$ is cyclic, all the subgroups of $\mathbb{Z}_{12}$ are cyclic, and if $H = \langle s \rangle$ is a subgroup, then $|H| = |s| = 12/\gcd(12, s)$ and is a divisor of 12. Let us consider a divisor of 12, say 4, and find all subgroups H with $|H| = 4$. We know that $|3| = 12/\gcd(12, 3) = 12/3 = 4$, and hence $H = \langle 3 \rangle = \{0, 3, 6, 9\}$ is one subgroup of order 4. It is in fact the only subgroup of order 4, since the only other element of order 4 in $\mathbb{Z}_{12}$ is 9, and $\langle 9 \rangle = \langle 3 \rangle$. ◇

1.3.26 THEOREM Let $G = \langle a \rangle$ be a cyclic group of order n. Then

(1) The order $|H|$ of any subgroup H of G is a divisor of $n = |G|$.

(2) For each positive integer d that divides n there exists a unique subgroup of order d, the subgroup $H = \langle a^{n/d} \rangle$.

Proof (1) Let H be a subgroup of $G = \langle a \rangle$. By Theorem 1.3.23, $H = \langle a^m \rangle$ for some integer $m \geq 0$, and by Theorem 1.3.16, $|H| = |a^m| = n/\gcd(n, m)$, which is a divisor of n.

(2) Since $e \in H$ for any subgroup H of G, the only subgroup of G of order 1 is the trivial subgroup $\{e\} = \langle e \rangle$. Let d be a divisor of n, $d > 1$. Then by Theorem 1.3.16, $|a^{n/d}| = n/\gcd(n, n/d) = d$. Hence $\langle a^{n/d} \rangle$ is a subgroup of G of order d. What remains to be shown is that this is the only subgroup of G of order d. So let H be a subgroup of G of order $|H| = d$. As in the proof of Theorem 1.3.23, $H = \langle a^s \rangle$, where s is the smallest positive integer such that $a^s \in H$. We know from Theorem 0.3.16 that there are integers u, v such that $\gcd(n, s) = un + vs$. Therefore, $a^{\gcd(n,s)} = a^{un + vs} = (a^n)^u(a^s)^v = e(a^s)^v \in H$. Since $1 \leq \gcd(n, s) \leq s$ and s was the least positive integer with $a^s \in H$, we must have $\gcd(n, s) = s$. Then by Theorem 1.3.16, $d = |H| = |a^s| = n/\gcd(n, s) = n/s$, so that $s = n/d$ and $H = \langle a^s \rangle = \langle a^{n/d} \rangle$ as desired. □

1.3.27 EXAMPLE Using Theorem 1.3.26, all the subgroups of $\mathbb{Z}_{12}$ and their orders are as follows:

$\langle 0 \rangle$ has order 1	$\langle 6 \rangle$ has order 2	$\langle 4 \rangle$ has order 3
$\langle 3 \rangle$ has order 4	$\langle 2 \rangle$ has order 6	$\langle 1 \rangle = \mathbb{Z}_{12}$ has order 12

We can display these in a diagram of the **subgroup lattice** of the group as in Figure 6. This is a diagram showing how the various subgroups of the group are related. Lines in the diagram represent inclusion. Thus the diagram indicates that $\langle 3 \rangle$ includes $\langle 6 \rangle$ and that $\langle 6 \rangle$ includes $\langle 0 \rangle$. It also indicates that the intersection of $\langle 3 \rangle$ and $\langle 2 \rangle$ is $\langle 6 \rangle$, and that the intersection of $\langle 6 \rangle$ and $\langle 4 \rangle$ is $\langle 0 \rangle$. ◇

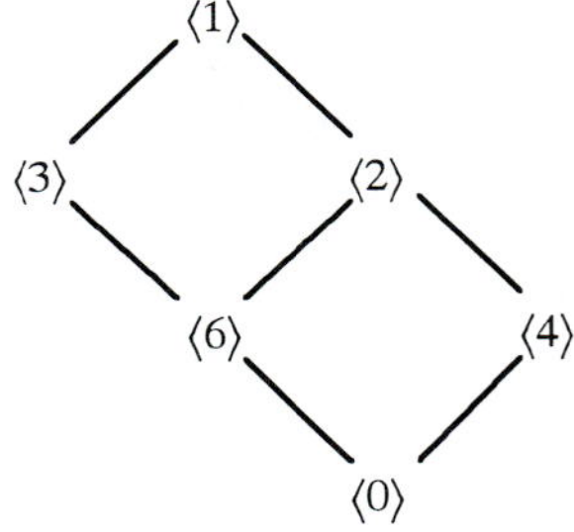

FIGURE 6

Exercises 1.3

1. Find the orders of the indicated elements in the indicated groups:

(a) $6 \in \mathbb{Z}_{10}$ (b) $6 \in \mathbb{Z}_{15}$ (c) $10 \in \mathbb{Z}_{42}$

(d) $77 \in \mathbb{Z}_{210}$ (e) $40 \in \mathbb{Z}_{210}$ (f) $70 \in \mathbb{Z}_{210}$

2. Let $G = \langle a \rangle$ be a cyclic group of order $|G| = 21$. Calculate the orders of a^2, a^6, a^8, a^9, a^{14}, a^{15}, a^{18}.

3. Let G be a group and $a \in G$ an element of order $|a| = 6$.
(a) Write all the elements of $\langle a \rangle$. (b) Find in $\langle a \rangle$ the elements a^{32}, a^{47}, a^{70}.

4. Find all the generators of $\mathbb{Z}_{10}$, $\mathbb{Z}_{12}$, and $\mathbb{Z}_{15}$.

5. Let $G = \langle a \rangle$ be a cyclic group of order 30. Find all the generators of G.

6. Draw the subgroup lattice diagram for $\mathbb{Z}_{18}$.

7. Find all the elements $b \in \mathbb{Z}_{15}$ of order $|b| = 5$.

8. Let $G = \langle a \rangle$ be a cyclic group of order 20. Find all the elements $b \in G$ of order $|b| = 10$.

9. List all the cyclic subgroups of S_3. Does S_3 have a noncyclic proper subgroup?

10. List all the cyclic subgroups of D_4. Does D_4 have a noncyclic proper subgroup?

11. Is it true that if every proper subgroup of a group G is cyclic, then G must also be cyclic? Answer with a proof if true or a counterexample if false.

12. Give examples of finite cyclic subgroups of $\mathbb{C}^*$.

13. Prove Corollary 1.3.12.

14. Prove Corollary 1.3.13.

15. Prove Corollary 1.3.14.

16. Show that every cyclic group is Abelian.

17. Give an example of a group with the indicated combination of properties:
(a) an infinite cyclic group
(b) an infinite Abelian group that is not cyclic
(c) a finite cyclic group with exactly six generators
(d) a finite Abelian group that is not cyclic

18. Let H and K be cyclic subgroups of an Abelian group G, with $|H| = 10$ and $|K| = 14$. Show that G contains a cyclic subgroup of order 70.

19. Let G be a group with no nontrivial proper subgroups.
(a) Show that G must be cyclic.
(b) What can you say about the order of G?

20. Let m and n be integers. By $m\mathbb{Z} + n\mathbb{Z}$ is meant the set $\{a + b \mid a \in m\mathbb{Z}, b \in n\mathbb{Z}\}$ of all sums of an element of $m\mathbb{Z}$ and an element of $n\mathbb{Z}$.
(a) Show that $m\mathbb{Z} + n\mathbb{Z}$ is a subgroup of $\mathbb{Z}$.
(b) Find a generator for the subgroup $12\mathbb{Z} + 21\mathbb{Z}$.
(c) Find a generator for the subgroup $m\mathbb{Z} + n\mathbb{Z}$.

21. Find a generator of the subgroup $6\mathbb{Z} \cap 15\mathbb{Z}$ of $\mathbb{Z}$.

22. Let m and n be integers. Find a generator for the subgroup $m\mathbb{Z} \cap n\mathbb{Z}$ of $\mathbb{Z}$.

23. Determine whether or not the following groups are cyclic:
(a) $U(10)$ (b) $U(12)$ (c) $U(20)$ (d) $U(24)$

24. Let a and b be elements of a group G with $|a| = 14$ and $|b| = 15$. Describe the subgroup $\langle a \rangle \cap \langle b \rangle$. Explain your answer.

25. Let $G = \langle a \rangle$ be a cyclic group of order 20, and H and K two distinct nontrivial proper subgroups of G such that $H \leq K$, and $a^4 \notin K$. Describe H and K.

26. Let $G = \langle a \rangle$ be a cyclic group of order n, and d a divisor of n. Show that the number of elements in G of order d is $\phi(d)$, where ϕ is the Euler ϕ-function.

1.4 Permutations

We now turn to the study of the most important examples of finite groups, the *groups of permutations*. The reason they are of central importance is that, as will be seen in Chapter 4, *any* finite group can be viewed as a subgroup of a permutation group. This means that studying finite groups amounts to studying permutation groups and their subgroups. When we want to construct a finite group with specific properties, we look for permutations that generate a subgroup with those properties.
We begin by looking at some examples of permutations and functions that are not permutations.

1.4.1 EXAMPLE The function f: $\mathbb{Z} \to \mathbb{Z}$ defined by $f(n) = n + 1$ is one to one, because if $f(n_1) = f(n_2)$, then $n_1 + 1 = n_2 + 1$ and hence $n_1 = n_2$. It is also onto, because for any $m \in \mathbb{Z}$, $f(m - 1) = m$. ◇

1.4.2 EXAMPLE The function g: $\mathbb{Z} \to \mathbb{Z}$ defined by $g(n) = 2n$ is one to one, because if $g(n_1) = g(n_2)$, then $2n_1 = 2n_2$ and hence $n_1 = n_2$. But it is not onto, because $g(\mathbb{Z}) = 2\mathbb{Z} \neq \mathbb{Z}$. ◇

1.4.3 EXAMPLE The function h: $\mathbb{Z} \to \mathbb{Z}$ defined by $h(2n) = n$ and $h(2n + 1) = n$ is not one to one, because, for instance, $h(0) = h(1) = 0$. But it is onto, because for any $m \in \mathbb{Z}$, $h(2m) = m$. ◇

1.4.4 EXAMPLE The function j: $\{1, 2, 3, 4\} \to \{1, 2, 3, 4\}$ defined by $j(1) = 3$, $j(2) = 4$, $j(3) = 1$, $j(4) = 2$ is both one to one and onto. ◇

1.4.5 EXAMPLE The function k: $\{1, 2, 3, 4\} \to \{1, 2, 3, 4\}$ defined by $k(1) = 2$, $k(2) = 2$, $k(3) = 4$, $k(4) = 3$ is neither one to one nor onto. ◇

1.4.6 DEFINITION A function ϕ: $A \to A$ is a **permutation of the set** A if ϕ is both one to one and onto. ○

Thus in Examples 1.4.1 and 1.4.4 we have examples of permutations, and in Examples 1.4.2, 1.4.3, and 1.4.5 we have examples of functions that are not permutations.

1.4.7 EXAMPLE Consider the set $A = \{1, 2, 3, 4, 5, 6\}$, and consider the two permutations ϕ and τ, which we represent as follows.

$$\phi = \begin{pmatrix} 1 & 2 & 3 & 4 & 5 & 6 \\ 4 & 6 & 1 & 2 & 3 & 5 \end{pmatrix} \quad \tau = \begin{pmatrix} 1 & 2 & 3 & 4 & 5 & 6 \\ 2 & 3 & 5 & 4 & 6 & 1 \end{pmatrix}$$

This is a concise way of writing $\phi(1) = 4$, $\phi(2) = 6$, and so on, and $\tau(1) = 2$, $\tau(2) = 3$, and so on. Since ϕ and τ are functions, we can construct the composite function $\phi \circ \tau$: $A \to A$ as follows: $\phi \circ \tau(1) = \phi(\tau(1)) = \phi(2) = 6$, $\phi \circ \tau(2) = \phi(\tau(2)) = \phi(3) = 1$, and so on. The resulting function can be written

$$\phi \circ \tau = \begin{pmatrix} 1 & 2 & 3 & 4 & 5 & 6 \\ 6 & 1 & 3 & 2 & 5 & 4 \end{pmatrix}$$

We leave it to you to check that

$$\tau \circ \phi = \begin{pmatrix} 1 & 2 & 3 & 4 & 5 & 6 \\ 4 & 1 & 2 & 3 & 5 & 6 \end{pmatrix}$$

Note that the composites are also one-to-one and onto functions, and hence permutations of the set A. ◇

1.4.8 DEFINITION Given two permutations ϕ and τ of a set A, we call the composite $\phi \circ \tau$ the **product permutation** of ϕ and τ, and we call the operation of composition **permutation multiplication**. ○

Permutation multiplication is the operation we use to make the set of all permutations on a set A into a group. Let us review one case in which this group is already familiar.

1.4.9 EXAMPLE Let us find all permutations of the set $A = \{1, 2, 3\}$. First notice that there will be exactly six such permutations. If we start with 1, we can send $1 \to 1$ or $1 \to 2$ or $1 \to 3$. Once we have chosen where 1 is going to be sent, we have two choices left for 2, and then only one choice remains for 3, so all together there are $3 \cdot 2 \cdot 1 = 6$ possibilities. The six permutations are as follows:

$$\rho_0 = \begin{pmatrix} 1 & 2 & 3 \\ 1 & 2 & 3 \end{pmatrix} \qquad \rho = \begin{pmatrix} 1 & 2 & 3 \\ 2 & 3 & 1 \end{pmatrix} \qquad \rho^2 = \begin{pmatrix} 1 & 2 & 3 \\ 3 & 1 & 2 \end{pmatrix}$$

$$\tau = \begin{pmatrix} 1 & 2 & 3 \\ 1 & 3 & 2 \end{pmatrix} \qquad \rho\tau = \begin{pmatrix} 1 & 2 & 3 \\ 2 & 1 & 3 \end{pmatrix} \qquad \rho^2\tau = \begin{pmatrix} 1 & 2 & 3 \\ 3 & 2 & 1 \end{pmatrix}$$

This is the familiar group S_3 of the symmetries of an equilateral triangle, which we examined in Example 1.1.5. ◇

1.4.10 THEOREM Let $A = \{1, 2, \ldots, n\}$ and let S_n be the collection of all permutations of A. Then S_n is a group under permutation multiplication.

Proof (Closure) Permutations are one to one, onto functions, and permutation multiplication is composition of functions. We know from Theorem 0.1.15 that a composition of two functions that are one to one and onto is one to one and onto.
(Associativity) We know from Theorem 0.1.15 that composition of functions is associative.

(Identity) The permutation defined by $\rho_0(i) = i$ for all $i \in A$ is the identity element for permutation multiplication.
(Inverse) For $\phi \in S_n$, since ϕ is one to one and onto, we know from Theorem 0.1.24 that the function given by $\phi^{-1}(i) = j$, where $\phi(j) = i$ is well defined, and it is the inverse of ϕ for permutation multiplication. □

Since the elements of any finite set with n elements could be labeled 1, 2, ..., n, Theorem 1.4.10 actually holds for any finite set A.

1.4.11 DEFINITION The group consisting of the set S_n of all permutations on $A = \{1, 2, \ldots, n\}$, under the operation of permutation multiplication is called the **symmetric group** of degree n. ○

1.4.12 PROPOSITION The symmetric group S_n has order $|S_n| = n!$.

Proof Let $\phi \in S_n$. Then ϕ can be written

$$\phi = \begin{pmatrix} 1 & 2 & 3 & \ldots & n-1 & n \\ \phi(1) & \phi(2) & \phi(3) & \ldots & \phi(n-1) & \phi(n) \end{pmatrix}$$

There are n choices for $\phi(1)$. Once we have chosen a value for $\phi(1)$, there are $n - 1$ choices for $\phi(2)$. Once $\phi(1)$ and $\phi(2)$ have been chosen, there are $n - 2$ choices for $\phi(3)$, and so on. Therefore, the total number of possibilities is $n \cdot (n - 1) \cdot (n - 2) \cdot \ldots \cdot 2 \cdot 1 = n!$ □

1.4.13 EXAMPLE Consider the set $A = \{1, 2, 3, \ldots, 9\}$ and the permutation

$$\phi = \begin{pmatrix} 1 & 2 & 3 & 4 & 5 & 6 & 7 & 8 & 9 \\ 7 & 9 & 1 & 8 & 6 & 4 & 3 & 5 & 2 \end{pmatrix} \in S_9$$

Let us look at where repeated application of ϕ takes various elements of the set A:

$1 \to \phi(1) = 7 \to \phi^2(1) = \phi(7) = 3 \to \phi^3(1) = \phi(3) = 1$
$2 \to \phi(2) = 9 \to \phi^2(2) = \phi(9) = 2$
$4 \to \phi(4) = 8 \to \phi^2(4) = \phi(8) = 5 \to \phi^3(4) = \phi(5) = 6 \to \phi^4(4) = \phi(6) = 4$

Let us call the permutation that takes 1 to 7, 7 to 3, 3 to 1, and leaves all other elements of the set alone (1 7 3). Similarly, let us write (2 9) and (4 8 5 6). These three permutations can be pictured as in Figure 7.

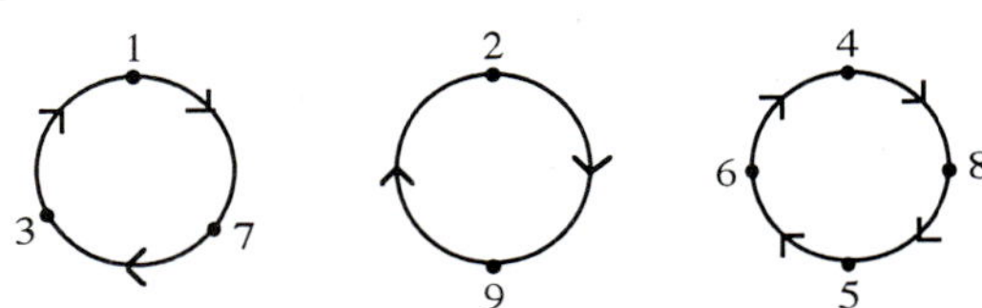

FIGURE 7

The original permutation ϕ is their product: $\phi = (1\ 7\ 3)(2\ 9)(4\ 8\ 5\ 6)$. In effect, ϕ partitions the set $\{1, 2, 3, \ldots, 9\}$ into three disjoint subsets and hence determines an equivalence relation on that set. Two elements $i, j \in A$ are equivalent, written $i \sim j$, if $\phi^n(i) = j$ for some $n \in \mathbb{Z}$. Thus $1 \sim 3$ because $\phi^2(1) = 3$, and $4 \sim 6$ because $\phi^3(4) = 6$. The equivalence classes are $\{1, 7, 3\}$, $\{2, 9\}$, and $\{4, 8, 5, 6\}$. ◇

1.4.14 EXAMPLE In Example 1.4.13, the permutation $\phi \in S_9$ in effect partitions the set $A = \{1, 2, 3, \ldots, 9\}$ into three disjoint pieces $\{1, 7, 3\}$, $\{2, 9\}$, $\{4, 8, 6\}$. ϕ moves around the elements in any one piece, but does not move elements between pieces. A partition determines an equivalence relation, which we may write $r \sim s$. Two elements r and s are equivalent, or belong to the same piece of the partition, if and only if we can get from one to the other by applying ϕ repeatedly. Thus $1 \sim 3$ because $\phi^2(1) = 3$, and $4 \sim 6$ because $\phi^3(4) = 6$. ◇

1.4.15 EXAMPLE In Example 1.4.13 we began with a two-row representation of a permutation and arrived at a more compact representation. Let us now consider the reverse process. Given a representation such as $\tau = (1\ 3\ 5)(2\ 6\ 8)(4\ 9)$, we have all the information about where τ sends each element of the set $A = \{1, 2, 3, \ldots, 9\}$. For example, this representation tells us that $\tau(3) = 5$ and $\tau(8) = 2$. If we combine all this information we obtain

$$\tau = \begin{pmatrix} 1 & 2 & 3 & 4 & 5 & 6 & 7 & 8 & 9 \\ 3 & 6 & 5 & 9 & 1 & 8 & 7 & 2 & 4 \end{pmatrix}$$

as the two-row representation of τ. ◇

1.4.16 LEMMA Each permutation $\phi \in S_n$ determines an equivalence relation on the set $A = \{1, 2, 3, \ldots, n\}$, defined by the condition that for all $r, s \in A$, $r \sim s$ if and only if $s = \phi^i(r)$ for some $i \in \mathbb{Z}$.

Proof There are three properties that must be checked to show that $\sim$ as just defined is an equivalence relation.
(Reflexivity) $r \sim r$ since $r = \phi^0(r)$
(Symmetry) If $r \sim s$, we have $s = \phi^i(r)$ for some $i \in \mathbb{Z}$. It follows that $r = \phi^{-i}(s)$, where $-i \in \mathbb{Z}$, and hence $s \sim r$.
(Transitivity) If $r \sim s$ and $s \sim t$, we have $s = \phi^i(r)$ and $t = \phi^j(s)$ for some $i, j \in \mathbb{Z}$. It follows that $t = \phi^j(s) = \phi^j(\phi^i(r)) = \phi^{j+i}(r)$, where $j + i \in \mathbb{Z}$, and hence $r \sim t$. □

1.4.17 DEFINITION For $\phi \in S_n$, the equivalence classes in $A = \{1, 2, 3, \ldots, n\}$ determined by ϕ are called the **orbits** of ϕ. ○

1.4.18 DEFINITION A permutation $\sigma \in S_n$ is called a **cycle** if it has at most one orbit with more than one element. The **length** of a cycle is the number of elements in its largest orbit. A cycle of length k is also called a k-**cycle** and may be written $(a_1\ a_2 \ldots\ a_k)$, where a_1 is an element of the largest orbit, $a_2 = \sigma(a_1)$, $a_3 = \sigma^2(a_1) =$

$\sigma(a_2)$, $a_3 = \sigma^3(a_1) = \sigma(a_2)$, and so on. Two cycles are called **disjoint** if their largest orbits are disjoint sets. ○

1.4.19 EXAMPLE Let us find all the 3-cycles in S_4. They are

(1 2 3)	(1 3 2)	(1 2 4)	(1 4 2)
(1 3 4)	(1 4 3)	(2 3 4)	(2 4 3)

Note that the same cycle can be written in more than one way. For instance, (1 2 3) = (2 3 1) = (3 1 2). ◇

In Examples 1.4.13 and 1.4.15 we saw two different permutations written as products of disjoint cycles. In fact, any permutation can be so written.

1.4.20 THEOREM Every permutation $\phi \in S_n$ can be written as a product of disjoint cycles.

Proof Let the orbits of ϕ be $O_1, \ldots, O_s$. For each orbit O_i we define the corresponding cycle σ_i as follows:

$$\sigma_i(a) = \begin{cases} \phi(a) \text{ if } a \in O_i \\ a \text{ if } a \notin O_i \end{cases}$$

The cycles σ_i are disjoint since the orbits O_i are equivalence classes and hence disjoint. It is now easy to see that $\phi = \sigma_1\sigma_2\ldots\sigma_s$. □

Let us now see why it is convenient to write a permutation as the product of disjoint cycles.

1.4.21 EXAMPLE In S_6 consider the cycles $\sigma = (1\ 3\ 5\ 4)$ and $\tau = (1\ 5\ 6)$. We have

$$\tau\sigma = (1\ 5\ 6)(1\ 3\ 5\ 4) = \begin{pmatrix} 1 & 2 & 3 & 4 & 5 & 6 \\ 3 & 2 & 6 & 5 & 4 & 1 \end{pmatrix} \qquad \sigma\tau = (1\ 3\ 5\ 4)(1\ 5\ 6) = \begin{pmatrix} 1 & 2 & 3 & 4 & 5 & 6 \\ 4 & 2 & 5 & 1 & 6 & 3 \end{pmatrix}$$

and hence $\tau\sigma \neq \sigma\tau$. ◇

We can show, however, that disjoint cycles do commute.

1.4.22 PROPOSITION Let σ_1 and σ_2 be two disjoint cycles in S_n. Then $\sigma_1\sigma_2 = \sigma_2\sigma_1$.

Proof Let O_1 and O_2 be the orbits of σ_1 and σ_2, which are disjoint. Note that for $i = 1$ or 2, $\sigma_i(a) \in O_i$ if $a \in O_i$, and $\sigma_i(a) = a$ if $a \notin O_i$.Therefore, if $b \in O_1$, then $\sigma_1\sigma_2(b) = \sigma_1(\sigma_2(b)) = \sigma_1(b)$ and $\sigma_2\sigma_1(b) = \sigma_2(\sigma_1(b)) = \sigma_1(b)$. Also, if $b \notin O_1$, then $\sigma_1\sigma_2(b) = \sigma_1(\sigma_2(b)) = \sigma_2(b)$ and $\sigma_2\sigma_1(b) = \sigma_2(\sigma_1(b)) = \sigma_2(b)$. In either case $\sigma_1\sigma_2(b) = \sigma_2\sigma_1(b)$, and $\sigma_1\sigma_2 = \sigma_2\sigma_1$. □

The preceding proposition can be used to compute the orders of permutations.

1.4.23 EXAMPLE Consider

$$\phi = \begin{pmatrix} 1 & 2 & 3 & 4 & 5 & 6 & 7 & 8 & 9 & 10 \\ 8 & 6 & 10 & 7 & 2 & 9 & 5 & 3 & 4 & 1 \end{pmatrix}$$

We first write ϕ as a product of disjoint cycles, $\phi = (1\ 8\ 3\ 10)(2\ 6\ 9\ 4\ 7\ 5)$. The order of $(1\ 8\ 3\ 10)$ is 4, and the order of $(2\ 6\ 9\ 4\ 7\ 5)$ is 6. (See Exercise 14.) Since the two cycles are disjoint and therefore commute, $|\phi| = \text{lcm}(4,6) = 12$. (See Exercise 15.) ◇

1.4.24 EXAMPLE The elements of D_4, the group of symmetries of the square, may be represented as permutations in S_4 by labeling the four vertices of the square 1, 2, 3, 4, as in Example 1.1.6. We then obtain

ρ_0	= identity	τ	$= (1\ 2)(3\ 4)$
ρ	$= (1\ 2\ 3\ 4)$	$\rho\tau$	$= (1\ 2\ 3\ 4)(1\ 2)(3\ 4) = (1\ 3)$
ρ^2	$= (1\ 3)(2\ 4)$	$\rho^2\tau$	$= (1\ 3)(2\ 4)(1\ 2)(3\ 4) = (1\ 4)(2\ 3)$
ρ^3	$= (1\ 4\ 3\ 2)$	$\rho^3\tau$	$= (1\ 4\ 3\ 2)(1\ 2)(3\ 4) = (2\ 4)$ ◇

1.4.25 EXAMPLE Let us compute the product $\phi = (1\ 4)(1\ 3)(1\ 2)$. First, $(1\ 2)$ sends 1 to 2, and neither $(1\ 3)$ nor $(1\ 4)$ moves 2, so $\phi(1) = 2$. Then $(1\ 2)$ sends 2 to 1, while $(1\ 3)$ sends 1 to 3, and $(1\ 4)$ does not move 3, so $\phi(2) = 3$. Then $(1\ 2)$ does not move 3, while $(1\ 3)$ sends 3 to 1 and $(1\ 4)$ sends 1 to 4, so $\phi(3) = 4$. Finally, neither $(1\ 2)$ nor $(1\ 3)$ moves 4, while $(1\ 4)$ sends 4 to 1, so $\phi(4) = 1$. Thus $\phi = (1\ 2\ 3\ 4)$. ◇

1.4.26 LEMMA Every cycle can be written as a product of 2-cycles.

Proof Exactly as in the preceding example, in general, $(a_1\ a_2\ a_3\ \ldots\ a_n) = (a_1\ a_n)\ldots(a_1\ a_3)(a_1\ a_2)$. □

1.4.27 PROPOSITION Every permutation can be written as a product of 2-cycles.

Proof To write a given permutation as a product of 2-cycles, first write it as a product of cycles by Theorem 1.4.20, and then write each of these cycles as a product of 2-cycles by Lemma 1.4.26. □

1.4.28 EXAMPLE In general, a permutation can be written as a product of 2-cycles in several different ways. For instance, in S_6 the identity can be written as $(1\ 2)(1\ 2)$, but also as $(1\ 2)(3\ 4)(1\ 2)(3\ 4)$, and so on. The permutation $(1\ 2\ 3)$ can be written as $(1\ 3)(1\ 2)$, but also as $(2\ 3)(1\ 3)$, or as $(4\ 5)(1\ 3)(4\ 5)(1\ 2)$. The permutation $(1\ 2\ 3\ 4)$ can be written as $(1\ 4)(1\ 3)(1\ 2)$, but also as $(5\ 6)(1\ 4)(1\ 3)(5\ 6)(1\ 2)$. Note, however, that in all these examples, the different ways of writing the same permutation as a product of 2-cycles either all involve an even number of 2-cycles, or all involve an odd number of 2-cycles. ◇

1.4.29 EXAMPLE Let n be a positive integer. For any finite sequence

$$s = (a_1, a_2, \ldots, a_n)$$

of integers define

$$p(s) = \prod_{1 \le i < j \le n} (a_i - a_j)$$

and for any $\tau \in S_n$, let $\tau s = (a_{\tau(1)}, a_{\tau(2)}, \dots , a_{\tau(n)})$. For instance, for $n = 6$ and $s = (1, 2, 3, 4, 5, 6)$, we have

$$p(s) = (1 - 2)(1 - 3)(1 - 4)(1 - 5)(1 - 6)(2 - 3)(2 - 4)(2 - 5)(2 - 6)$$
$$(3 - 4)(3 - 5)(3 - 6)(4 - 5)(4 - 6)(5 - 6)$$

and if $\tau = (2\ 5)$, then $\tau s = (1, 5, 3, 4, 2, 6)$. Note that we then have:

$$p(\tau s) = (1 - 5)(1 - 3)(1 - 4)(1 - 2)(1 - 6)\underline{(5 - 3)}\underline{(5 - 4)}\underline{(5 - 2)}(5 - 6)$$
$$(3 - 4)\underline{(3 - 2)}(3 - 6)\underline{(4 - 2)}(4 - 6)(2 - 6)$$

where we have underlined the terms that have changed sign. These are the term $(2 - 5)$, the terms $(2 - j)$ for $2 < j < 5$, and the terms $(i - 5)$ for $2 < i < 5$. Notice that there are five such terms, and so, without actually having to multiply out the long product, we see that $p(\tau s) = (-1)^5 p(s) = -p(s)$. ◇

1.4.30 LEMMA With the notation as before, for any positive integer n, any sequence of integers $s = (a_1, a_2, \dots , a_n)$, and any 2-cycle $\tau \in S_n$, we have $p(\tau s) = -p(s)$.

Proof Let $\tau = (k\ l)$, where $k < l$, and let us compare the two products $p(s)$ and $p(\tau s)$. We distinguish five cases:

(1) If $i < k$, then for any j with $i < j$, we have $i = \tau(i)$ and $i < \tau(j)$, and the term $(a_i - a_{\tau(j)})$ appears as a factor in both products.

(2) If $l < j$, then for any i with $i < j$, the term $(a_{\tau(i)} - a_j)$ similarly appears as a factor in both products.

(3) If $i = k$, then for any j with $k < j < l$, we have $j = \tau(j)$ and $(a_{\tau(k)} - a_{\tau(j)}) = (a_l - a_j) = -(a_j - a_l)$, so the term $(a_j - a_l)$ in the product $p(s)$ changes sign in $p(\tau s)$. There are $(l - k - 1)$ such terms.

(4) If $j = l$, then for any i with $k < i < l$, the term $(a_i - a_l)$ in the product $p(s)$ similarly changes sign in $p(\tau s)$. And there are again $(l - k - 1)$ such terms.

(5) Finally, $(a_{\tau(k)} - a_{\tau(l)}) = (a_l - a_k)$, and the term $(a_l - a_k)$ in the product $p(s)$ changes sign in $p(\tau s)$.

Thus the total number of terms that change sign is $2(l - k - 1) + 1$, which is odd, and therefore $p(\tau s) = -p(s)$. □

1.4.31 THEOREM No permutation in S_n can be written both as a product of an even number of 2-cycles and as a product of an odd number of 2-cycles.

Proof Using the same notations as in the preceding lemma, let τ and ρ be two 2-cycles in S_n. Then for any sequence $s = (a_1, a_2, \dots , a_n)$ we have

$$(\rho\tau)s = (a_{\rho\tau(1)}, a_{\rho\tau(2)}, \dots , a_{\rho\tau(n)}) = (a_{\rho(\tau(1))}, a_{\rho(\tau(2))}, \dots , a_{\rho(\tau(n))}) =$$
$$\rho(a_{\tau(1)}, a_{\tau(2)}, \dots , a_{\tau(n)}) = \rho(\tau s)$$

and hence $p((\rho\tau)s) = p(\rho(\tau s)) = -p(\tau s) = (-1)^2 p(s)$, applying the lemma twice. Similarly, given any permutation $\phi \in S_n$, if ϕ can be written as a product of k

2-cycles, then $p(\phi s) = (-1)^k p(s)$. Hence for a given ϕ, k is either always even or always odd for every way of writing ϕ as a product of 2-cycles. □

1.4.32 DEFINITION A permutation $\phi \in S_n$ is called an **even** permutation if it can be written as a product of an even number of 2-cycles, and it is called an **odd** permutation if it can be written as a product of an odd number of 2-cycles. ○

By Proposition 1.4.27, every permutation is either even or odd, and by Theorem 1.4.31 no permutation is both even and odd.

1.4.33 EXAMPLE Consider the permutation

$$\phi = \begin{pmatrix} 1 & 2 & 3 & 4 & 5 & 6 & 7 & 8 & 9 \\ 9 & 5 & 1 & 7 & 8 & 2 & 6 & 4 & 3 \end{pmatrix}$$

We can write $\phi = (1\ 9\ 3)(2\ 5\ 8\ 4\ 7\ 6) = (1\ 3)(1\ 9)(2\ 6)(2\ 7)(2\ 4)(2\ 8)(2\ 5)$, so ϕ is an odd permutation. Using the fact that in any group $(ab)^{-1} = b^{-1}a^{-1}$, we get $\phi^{-1} = (2\ 5)(2\ 8)(2\ 7)(2\ 4)(2\ 6)(1\ 9)(1\ 3) = (2\ 6\ 7\ 4\ 8\ 5)(1\ 3\ 9) = (1\ 3\ 9)(2\ 6\ 7\ 4\ 8\ 5)$. Note that ϕ^{-1} is also an odd permutation. ◇

1.4.34 THEOREM Let A_n be the set of all even permutations in S_n. Then A_n is a subgroup of S_n, called the **alternating group** of degree n.

Proof By Theorem 1.2.13 we need only show closure and the existence of inverses.
(Closure) If $\phi, \rho \in A_n$, then ϕ can be written as the product of some even number $2r$ of 2-cycles, and ρ as the product of some even number $2s$ of 2-cycles. But then $\phi\rho$ can be written as the product of the even number $2r + 2s = 2(r + s)$ of 2-cycles, and $\phi\rho \in A_n$.
(Inverses) If $\phi \in A_n$, then ϕ can be written as the product $\sigma_1\sigma_2\ldots\sigma_{2r}$ of some even number $2r$ of 2-cycles. But then $\phi^{-1} = (\sigma_1\sigma_2\ldots\sigma_{2r})^{-1} = \sigma_{2r}^{-1}\ldots\sigma_2^{-1}\sigma_1^{-1}$ can also be written as a product of the even number $2r$ of 2-cycles, and $\phi^{-1} \in A_n$. □

1.4.35 EXAMPLE In S_3, where $|S_3| = 3! = 6$, the even permutations are $A_3 = \{\rho_0 =$ identity, $\rho = (1\ 2\ 3)$, $\rho^2 = (1\ 3\ 2)\}$. So there are exactly three even permutations and three odd permutations. Let us find all the even permutations in S_4, where $|S_4| = 4! = 24$. If we write the permutations in A_4 as the product of disjoint cycles, we find three permutations that do not fix any element, namely the following:

$\sigma_1 = (1\ 2)(3\ 4)$ $\quad$ $\sigma_2 = (1\ 3)(2\ 4)$ $\quad$ $\sigma_3 = (1\ 4)(2\ 3)$

Since in A_3 there are two nontrivial permutations, for each i with $1 \leq i \leq 4$ there are two permutations in A_4 that fix i, making in all eight permutations in A_4 that fix exactly one element, namely the following:

$\rho_1 = (2\ 3\ 4)$	$\rho_1^2 = (2\ 4\ 3)$	which fix 1
$\rho_2 = (1\ 3\ 4)$	$\rho_2^2 = (1\ 4\ 3)$	which fix 2

$\rho_3 = (1\ 2\ 4)$ $\rho_3^2 = (1\ 4\ 2)$ which fix 3
$\rho_4 = (1\ 2\ 3)$ $\rho_4^2 = (1\ 3\ 2)$ which fix 4

In A_4 there are no permutations that fix exactly two elements, since all such permutations are 2-cycles. There is also the identity, making in all 12 elements: $A_4 = \{\rho_0 = \text{identity}, \rho_1, \rho_1^2, \rho_2, \rho_2^2, \rho_3, \rho_3^2, \rho_4, \rho_4^2, \sigma_1, \sigma_2, \sigma_3\}$. ◇

Having studied A_3 and A_4 and found that in S_3 and S_4 exactly half the permutations are even, it should come as no surprise that the same holds for every S_n, $n \geq 3$.

1.4.36 THEOREM The order of the alternating group of degree n is

$$|A_n| = |S_n|/2 = n!/2$$

Proof If we let O_n be the set of all odd permutations in S_n, then since every permutation in S_n is either in A_n or in O_n, but not both, we have $|S_n| = |A_n| + |O_n|$. We will show that $|A_n| = |O_n|$, which will prove the theorem. To show that two sets have the same number of elements, we need to construct a onetoone and onto map between them. Let $\gamma: A_n \to O_n$ be the map that sends an even permutation $\phi \in A_n$ to the odd permutation $\phi(1\ 2) \in O_n$. Then γ is one to one, since if $\gamma(\phi) = \gamma(\psi)$, then $\phi(1\ 2) = \psi(1\ 2)$, which implies that $\phi = \phi(1\ 2)(1\ 2) = \psi(1\ 2)(1\ 2) = \psi$. And γ is onto, since if $\tau \in O_n$ is an odd permutation, then $\tau(1\ 2) \in A_n$ is an even permutation with $\gamma(\tau(1\ 2)) = \tau(1\ 2)(1\ 2) = \tau$. □

TABLE 9 The Elements of D_5

ρ_0	= identity	τ	$= (2\ 5)(3\ 4)$
ρ	$= (1\ 2\ 3\ 4\ 5)$	$\rho\tau$	$= (1\ 2\ 3\ 4\ 5)(2\ 5)(3\ 4) = (1\ 2)(3\ 5)$
ρ^2	$= (1\ 3\ 5\ 2\ 4)$	$\rho^2\tau$	$= (1\ 3)(4\ 5)$
ρ^3	$= (1\ 4\ 2\ 5\ 3)$	$\rho^3\tau$	$= (1\ 4)(2\ 3)$
ρ^4	$= (1\ 5\ 4\ 3\ 2)$	$\rho^4\tau$	$= (1\ 5)(2\ 4)$ ◇

1.4.37 EXAMPLE As we saw in Example 1.4.24, D_4 is a subgroup of S_4. Similarly, D_5 is a subgroup of S_5. The elements of D_5 are as shown in Table 9. But while half the permutations in D_4 are even and half are odd, all the permutations in D_5 are even, and D_5 is a subgroup of A_5. (See Exercise 26.)

Exercises 1.4

In Exercises 1 through 5 determine which of the indicated functions is a permutation of the indicated set.

1. $f: \mathbb{R} \to \mathbb{R}$, where $f(x) = 3x + \sqrt{2}$

2. $f: \mathbb{R} \to \mathbb{R}$, where $f(x) = 3x^2 + 2$

3. $f: \mathbb{Z} \to \mathbb{Z}$, where $f(x) = |x|$

4. $f: U(5) \to U(5)$, where $f(x) = x^{-1}$

5. $f: \mathbb{Z}_6 \to \mathbb{Z}_6$, where $f(x) = x + 3$

In Exercises 6 through 9 find all the orbits of the indicated permutation.

6. $\phi = \begin{pmatrix} 1 & 2 & 3 & 4 & 5 & 6 & 7 \\ 7 & 6 & 3 & 2 & 1 & 4 & 5 \end{pmatrix}$

7. $\phi = \begin{pmatrix} 1 & 2 & 3 & 4 & 5 & 6 & 7 & 8 & 9 \\ 6 & 5 & 9 & 4 & 1 & 8 & 7 & 2 & 3 \end{pmatrix}$

8. $\tau: \mathbb{Z} \to \mathbb{Z}$, where $\tau(x) = x + 5$

9. $\tau: \mathbb{Z} \to \mathbb{Z}$, where $\tau(x) = x - 3$

10. Let

$$\phi = \begin{pmatrix} 1 & 2 & 3 & 4 & 5 & 6 & 7 & 8 \\ 3 & 8 & 4 & 1 & 6 & 7 & 2 & 5 \end{pmatrix}, \qquad \tau = \begin{pmatrix} 1 & 2 & 3 & 4 & 5 & 6 & 7 & 8 \\ 6 & 4 & 1 & 2 & 7 & 8 & 5 & 3 \end{pmatrix}$$

Calculate:

(a) $\phi\tau$ and $\tau\phi$

(b) $\phi^2\tau$ and $\phi\tau^2$

(c) the inverses ϕ^{-1} and τ^{-1}

(d) the orders $|\phi|$ and $|\tau|$

In Exercises 11 through 13 express each permutation as a product of disjoint cycles, and then calculate its order.

11. $\phi = \begin{pmatrix} 1 & 2 & 3 & 4 & 5 & 6 & 7 & 8 & 9 & 10 \\ 1 & 8 & 2 & 9 & 7 & 5 & 4 & 3 & 10 & 6 \end{pmatrix}$

12. $\phi = \begin{pmatrix} 1 & 2 & 3 & 4 & 5 & 6 & 7 \\ 5 & 6 & 2 & 4 & 3 & 7 & 1 \end{pmatrix}\begin{pmatrix} 1 & 2 & 3 & 4 & 5 & 6 & 7 \\ 7 & 6 & 4 & 5 & 1 & 3 & 2 \end{pmatrix}$

13. $\phi = \begin{pmatrix} 1 & 2 & 3 & 4 & 5 \\ 3 & 1 & 5 & 2 & 4 \end{pmatrix}\begin{pmatrix} 1 & 2 & 3 & 4 & 5 \\ 2 & 4 & 1 & 5 & 3 \end{pmatrix}\begin{pmatrix} 1 & 2 & 3 & 4 & 5 \\ 4 & 5 & 3 & 1 & 2 \end{pmatrix}$

14. Show that an n-cycle has order n.

15. Show that if ρ and σ in S_n are disjoint cycles, and $\phi = \rho\sigma$, then $|\phi| = \text{lcm}(|\rho|,|\sigma|)$.

16. Show that an m-cycle is an even permutation if and only if m is odd.

17. Determine whether the permutations in Exercises 11 through 13 are even or odd permutations.

18. Explain why the set of odd permutations in S_n is not a subgroup of S_n.

In Exercises 19 through 25 find the maximum possible order of an element in the indicated group.

19. S_4 **20.** S_5 **21.** S_6 **22.** S_7

23. A_5 **24.** A_6 **25.** A_7

26. Show that for any subgroup H of S_n either every element of H is an even permutation, or else exactly half of the elements of H are even permutations.

27. Construct the group table of A_4.

28. Let $H = \{\sigma \in S_4 \mid \sigma(2) = 2\}$.
(a) Show that H is a subgroup of S_4.
(b) What is the order of H?
(c) Find all the even permutations in H.

29. Let $n \geq 3$, $i \leq n$, and let $H = \{\sigma \in S_n \mid \sigma(i) = i\}$.
(a) Show that H is a subgroup of S_n.
(b) What is the order of H?
(c) Find all the even permutations in H.

30. Show that for any $n \geq 3$, S_n is a non-Abelian group.

31. Find all the elements in S_4 of order 2.

32. Show that if $\sigma \in S_n$ and $|\sigma| = 2$, then σ is a product of disjoint 2-cycles.

33. Show that if $\sigma \in A_n$, then σ can be written as a product of 3-cycles.

34. Show that if $\sigma \in A_n$, then σ can be written as a product of 3-cycles $(1\ 2\ s)$ where $s = 3, 4, \ldots, n$.

35. Show that every permutation $\rho \in S_n$ can be written as a product of 2-cycles of the form $(i\ \ i+1)$, where $1 \leq i \leq n$.

36. Show that every permutation $\phi \in S_n$ can be written as a product of powers of $\rho = (1\ 2\ 3\ \ldots\ n)$ and $\phi = (1\ 2)$.

37. Show that if $m \leq n$, then the number of m-cycles $(a_1, a_2, \ldots, a_m)$ in S_n is

$$n(n-1)(n-2)\ldots(n-m+1)/m$$

38. Consider the regular tetrahedron as in Figure 8.
(a) Find all possible rotations of the regular tetrahedron.
(b) The rotations of the regular tetrahedron correspond to elements of which known group?

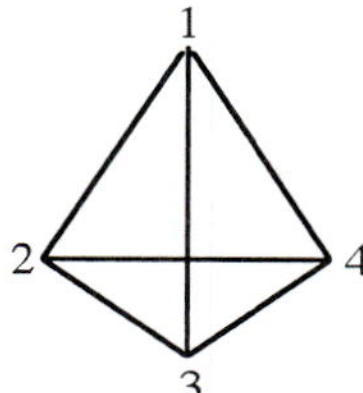

FIGURE 8

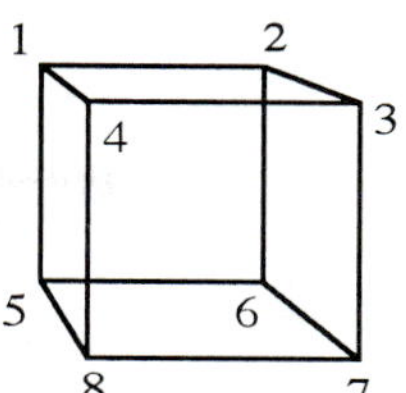

FIGURE 9

39. Consider the cube as in Figure 9. Find a group whose elements correspond to the rotations of the cube.

40. (The fifteen puzzle) The bottom right corner of the fifteen puzzle is empty, which allows us to slide the numbered squares without lifting them up. Show that every possible rearrangement that leaves the bottom right corner empty corresponds to an element of A_{15}. (The converse is also true, but harder to prove.)

1	2	3	4
5	6	7	8
9	10	11	12
13	14	15	

In Exercises 41 through 43 determine which of the indicated rearrangements of the fifteen puzzle are possible. (Assume Exercise 40 and its converse.)

41.

3	1	2	6
4	5	9	7
8	12	10	11
15	13	14	

42.

6	14	11	12
3	5	4	1
9	15	8	2
13	10	7	

43.

3	15	11	
14	12	1	4
2	6	9	5
13	8	7	10

44. A "perfect" shuffle of a deck of $2n$ cards is given by the permutation

$$\begin{pmatrix} 1 & 2 & 3 & \dots & n & n+1 & n+2 & n+3 & \dots & 2n \\ 2 & 4 & 6 & \dots & 2n & 1 & 3 & 5 & \dots & 2n-1 \end{pmatrix}$$

Find the least number of perfect shuffles that need to be performed on a deck of 52 cards before the cards are back in the original order. Answer the same question for a deck of 50 cards.

Chapter 2

Group Homomorphisms

We have established what groups are, and we have studied different types of groups and their subgroups. In this chapter we introduce four new basic notions, all closely related. First we show how a subgroup determines an equivalence relation on the elements of a group, and therefore a partition of the group into equivalence classes, which are called the *cosets* of the subgroup. We use this partition to give a simple and elegant proof of Lagrange's theorem, a very basic result in group theory. Second, we introduce certain maps we can use between groups, which are called *homomorphisms*. Third, we show how the notion of a homomorphism gives rise to the notion of a special kind of subgroup of a group, called a *normal subgroup*. Fourth, the image of a given group under a homomorphism will also turn out to be a group, and one with a special structure. We show that it can be regarded as a group whose elements are the cosets of a normal subgroup of the given group. Such groups are called *quotient groups*. The four new concepts of cosets, homomorphisms, normal subgroups, and quotient groups will come to seem natural once you realize how they are interrelated.

2.1 Cosets and Lagrange's Theorem

We have shown in Theorem 1.3.26 that if G is a finite cyclic group and H is a subgroup *of* G, then the order of H divides the order of G. In this section we prove Lagrange's theorem, which says that this conclusion holds for any finite group G. To prove this theorem we first show how a subgroup H determines an equivalence relation on G, and then show that each of the equivalence classes into which this equivalence relation partitions G has the same number of elements as H. It follows that the number of elements in G is a multiple of the number of elements in H. We begin by describing how any subgroup H of any group G determines an equivalence relation on G, even if G is not necessarily of finite order.

2.1.1 EXAMPLE In the group $\mathbb{Z}$ consider the subgroup $3\mathbb{Z} = \langle 3 \rangle$. For integers $a, b \in \mathbb{Z}$, consider the equivalence relation $a \sim b$ that holds if $b - a \in 3\mathbb{Z}$ or, in other words, if $a \equiv b \bmod 3$. Then the three equivalences classes are

$3\mathbb{Z}=\{\ldots,-6,-3,0,3,6,\ldots\}$ $1+3\mathbb{Z}=\{\ldots,-5,-2,1,4,7,\ldots\}$ $2+3\mathbb{Z}=\{\ldots,-4,-1,2,5,8,\ldots\}$ ◇

2.1.2 EXAMPLE In the group $\mathbb{Z}_{15}$, consider the subgroup $H = \langle 3 \rangle$. Then H partitions $\mathbb{Z}_{15}$ as follows: $H = \{0, 3, 6, 9, 12\}$, $1 + H = \{1, 4, 7, 10, 13\}$, and $2 + H = \{2, 5, 8, 11, 14\}$. The implicit equivalence relation can again be expressed as $a \sim b$ if $b - a \in H$. ◇

We now establish an equivalence relation determined by a subgroup H of any group G.

2.1.3 THEOREM Let G be any group and H a subgroup of G. Then

(1) For $a, b \in G$, the relation defined by $a \sim b$ if and only if $a^{-1}b \in H$ is an equivalence relation on G.

(2) For any $a \in G$, the equivalence class of a is $aH = \{ah \mid h \in H\}$.

Proof (1) There are three properties to be established to show that $\sim$ is an equivalence relation.
(Reflexivity) For any $a \in G$ we have $a^{-1}a = e \in H$ since H is a subgroup. Hence $a \sim a$.
(Symmetry) For any $a, b \in G$, if $a \sim b$, then $a^{-1}b \in H$. Thus $b^{-1}a = (a^{-1}b)^{-1} \in H$ since H is a subgroup, and hence $b \sim a$.
(Transitivity) For any $a, b, c \in G$, if $a \sim b$ and $b \sim c$, then $a^{-1}b \in H$ and $b^{-1}c \in H$. Then $a^{-1}c = (a^{-1}b)(b^{-1}c) \in H$ since H is a subgroup, and hence $a \sim c$.

(2) For any $a \in G$ the equivalence class of a consists of all $x \in G$ such that $a^{-1}x \in H$. If $x \sim a$, then $x = a(a^{-1}x) = ah$ where $h = a^{-1}x \in H$. Conversely, if $x = ah$ for some $h \in H$, then $a^{-1}x = a^{-1}ah = h \in H$, and $x \sim a$. □

We leave it to the reader (as Exercise 9 at the end of this section) to show that the relation defined by $a \sim b$ if and only if $ab^{-1} \in H$ is also an equivalence relation on G, with the equivalence class of a being $Ha = \{ha \mid h \in H\}$.
We now give a name to the equivalence classes aH and Ha.

2.1.4 DEFINITION Let G be a group, H a subgroup of G, and $a \in G$. Then the set $aH = \{ah \mid h \in H\}$ is called a **left coset** of H in G, and the set $Ha = \{ha \mid h \in H\}$ is called a **right coset** of H in G. ○

2.1.5 EXAMPLE In $S_3 = \{\rho_0, \rho, \rho^2, \mu_1, \mu_2, \mu_3\}$ let us find the left cosets and right cosets of $H = \langle \mu_1 \rangle$. The left cosets are

$H = \{\rho_0, \mu_1\}$ $\quad$ $\rho H = \{\rho, \rho\,\mu_1 = \mu_3\}\rho^2$ $\quad$ $H = \{\rho^2, \rho^2\mu_1 = \mu_2\}$

The right cosets are

$H = \{\rho_0, \mu_1\}$ $\quad$ $H\rho = \{\rho, \mu_1\rho = \mu_2\}$ $\quad$ $H\rho^2 = \{\rho^2, \mu_1\rho^2 = \mu_3\}$

Note that the two equivalence relations determined by $H = \langle \mu_1 \rangle$ give us different partitions of S_3. ◇

In an Abelian group, however, since all the elements commute, the left and right cosets coincide.

2.1.6 EXAMPLE In $\mathbb{Z}$ consider the subgroups $3\mathbb{Z} \supseteq 6\mathbb{Z}$. The cosets of $6\mathbb{Z}$ in $\mathbb{Z}$ are $6\mathbb{Z}$, $1 + 6\mathbb{Z}$, $2 + 6\mathbb{Z}$, $3 + 6\mathbb{Z}$, $4 + 6\mathbb{Z}$, $5 + 6\mathbb{Z}$. The cosets of $6\mathbb{Z}$ in $3\mathbb{Z}$ are $6\mathbb{Z}$ and $3 + 6\mathbb{Z}$. ◇

Since the cosets of a subgroup H in a group G are equivalence classes of an equivalence relation, we can obtain a number of facts about cosets as consequences of known properties of equivalence classes. (See Theorem 0.2.4.) The proofs of the following facts are left to the reader (as Exercises 11 and 12).

2.1.7 LEMMA Let G be a group, H a subgroup of G, and $a, b \in G$. Then

(1) Either $aH = bH$ or $aH \cap bH = \varnothing$ and either $Ha = Hb$ or $Ha \cap Hb = \varnothing$.

(2) $aH = bH$ if and only if $a^{-1}b \in H$ and $Ha = Hb$ if and only if $b^{-1}a \in H$. □

In all the examples of cosets we have seen so far, the size of any one coset has been the same as the size of any other coset. The next lemma says that this is always the case.

2.1.8 LEMMA Let G be a group and H a subgroup of G. Then for any $a \in G$, $|H| = |aH| = |Ha|$. □

Proof To show that the two sets H and aH have the same number of elements, we need to construct a map from H to aH that is one to one and onto. Let $\phi: H \to aH$ be the map defined by $\phi(h) = ah$. Then ϕ is one to one because if $\phi(h_1) = \phi(h_2)$, then $ah_1 = ah_2$, which implies $h_1 = h_2$. And ϕ is onto because for any $y \in aH$, we have $y = ah$ for some $h \in H$, hence $y = \phi(h)$. It is left to the reader (as Exercise 10) to show similarly that any right coset Ha of H has the same number of elements as H. □

Using Theorem 2.1.3 and Lemma 2.1.8, we can now prove Lagrange's theorem.

2.1.9 THEOREM (Lagrange's theorem) Let G be a finite group and H a subgroup of G. Then

(1) $|H|$ divides $|G|$.

(2) $|G| / |H|$ is equal to the number of distinct cosets of H.

Proof By Theorem 2.1.3 the left cosets of H are equivalence classes, and hence by Theorem 0.2.7 they partition G. Let $a_1H, a_2H, \ldots, a_sH$ be the distinct left cosets of H. Then $|G| = |a_1H| + |a_2H| + \ldots + |a_sH|$. By Lemma 2.1.8, $|a_iH| = |H|$ for all $a_i \in G$. Hence $|G| = s \cdot |H|$, where $s = |G|/|H|$ is the number of distinct cosets of H. □

Note that we could have proved Lagrange's theorem using right cosets of H instead of left cosets. Therefore, $|G|/|H|$ is also equal to the number of right cosets of H, and the numbers of left and of right cosets are the same.

We next introduce a name for this number of cosets.

2.1.10 DEFINITION Let G be a group and H a subgroup of G. Then the number of distinct left cosets of H in G is called the **index** of H in G and denoted by $\text{index}_G(H)$ or $[G:H]$. ○

Thus the two statements in Theorem 2.1.9 can be combined in the formula $|G| = |H|\cdot[G:H]$.

2.1.11 EXAMPLE Let H be a subgroup of a group G, where $|G| = 10$. Then by Lagrange's theorem, $|H| = 1, 2, 5$, or 10. For example, if $G = D_5 = \{\rho_0, \rho, \rho^2, \rho^3, \rho^4, \tau, \rho\tau, \rho^2\tau, \rho^3\tau, \rho^4\tau\}$ (as in Example 1.4.37), then $|\langle\rho_0\rangle| = 1$, $|\langle\rho^i\rangle| = 5$, $|\langle\tau\rangle| = |\langle\rho^i\tau\rangle| = 2$ for $1 \le i \le 4$, and, of course, $|D_5| = 10$. ◇

2.1.12 EXAMPLE S_3 can be viewed as a subgroup of S_4, where $S_3 = \{\phi \in S_4 \mid \phi(4) = 4\}$. Since $|S_4| = 4! = 24$ and $|S_3| = 3! = 6$, the index $[S_4:S_3] = 24/6 = 4$, and S_3 has four cosets in S_4. In order to find the four distinct cosets of S_3 we need to find first an element $\phi \in S_4$ such that $\phi \notin S_3$ or, in other words, $\phi(4) \neq 4$. One such element is $\phi = (1\ 2\ 3\ 4) \in S_4$. Since $\phi \notin S_3$, using Lemma 2.1.7, part (2), we have $\phi S_3 \neq S_3$. We have $\phi^2 = (1\ 3)(2\ 4) \notin S_3$, and also $\phi^{-1}\phi^2 \notin S_3$, hence again using Lemma 2.1.7, part (2), $\phi^2 S_3 \neq S_3$ and $\phi^2 S_3 \neq \phi S_3$. Finally, $\phi^3 = (1\ 4\ 3\ 2) \notin S_3$, $\phi^{-1}\phi^3 \notin S_3$, and $\phi^{-2}\phi^3 \notin S_3$, so again using Lemma 2.1.7, part (2), we see that S_3, ϕS_3, $\phi^2 S_3$, and $\phi^3 S_3$ are the four distinct left cosets of S_3 in S_4 ◇

The following are immediate consequences of Lagrange's theorem.

2.1.13 COROLLARY In a finite group G, the order $|a|$ of any element $a \in G$ divides the order $|G|$ of the group.

Proof For any element a of the finite group G, let $H = \langle a\rangle$. Then H is a subgroup of G, and $|H| = |a|$ by Corollary 1.3.14, while $|H|$ divides $|G|$ by Lagrange's theorem. □

2.1.14 COROLLARY Let G be a finite group and $a \in G$. Then $a^{|G|} = e$.

Proof This is immediate from Corollaries 2.1.13 and 1.3.12. □

2.1.15 EXAMPLE Let G be a group of order 7. Then by Lagrange's theorem G has no nontrivial proper subgroups. If we let $a \in G$, $a \neq e$, then $|a| \neq 1$, hence $|a| = 7$, and $G = \langle a\rangle$ is cyclic. ◇

2.1.16 THEOREM A group of prime order is cyclic.

Proof Let G be a group with $|G| = p$, where p is a prime, and let $a \in G$, $a \neq e$. Then $|a| \neq 1$, hence $|a| = p = |G|$, and $G = \langle a\rangle$. □

2.1.17 EXAMPLE The only noncyclic groups G of order $|G| \leq 7$ that we have encountered have been of two kinds

(1) Groups of order 4 such as $U(8)$, $U(12)$, and the Klein 4-group V (as in Example 1.1.22)

(2) A group of order 6, namely S_3 ◇

Finally, let us point out a famous result in number theory, Euler's theorem, that can be derived from Lagrange's theorem.

2.1.18 THEOREM (Euler's theorem) Let $n \geq 2$ and a be two relatively prime integers. Then $a^{\phi(n)} \equiv 1 \bmod n$.

Proof Using the division algorithm, $a = qn + r$, where $0 \leq r < n$. Since $\gcd(a, n) = 1$, we have $\gcd(r, n) = 1$, and hence $r \in U(n)$. Since $|U(n)| = \phi(n)$, by Corollary 2.1.14 we have $r^{\phi(n)} = 1$ in $U(n)$, and $a^{\phi(n)} \equiv r^{\phi(n)} \equiv 1 \bmod n$. □

2.1.19 THEOREM (Fermat's little theorem) If p is prime, then for any integer a we have $a^p \equiv a \bmod p$.

Proof If p divides a then $a \equiv 0 \bmod p$ and obviously $a^p \equiv 0 \bmod p$. If p does not divide a, then $\gcd(a, p) = 1$, and by Theorem 2.1.18, $a^{\phi(p)} \equiv 1 \bmod p$. But $\phi(p) = p - 1$, and so we have $a^{p-1} \equiv 1 \bmod p$ and $a^p \equiv a \bmod p$. □

There are a number of interesting applications of Fermat's little theorem. We illustrate just two of them.

2.1.20 EXAMPLE Let us compute the remainder of 5^{148} when divided by 7. Since $5^6 \equiv 1 \bmod 7$, and $148 = 24 \cdot 6 + 4$, we obtain $5^{148} \equiv (5^6)^{24} \cdot 5^4 \equiv 1 \cdot 5^4 \equiv (-2)^4 \equiv 2 \bmod 7$. ◇

2.1.21 EXAMPLE Let us show for any integer n that $n^{13} - n$ is divisible by 15. To show that $15 = 3 \cdot 5$ divides $n^{13} - n = n(n^{12} - 1)$, it is enough to show that both 3 and 5 divide $n(n^{12} - 1)$. Clearly, if 3 divides n, then 3 divides $n(n^{12} - 1)$. If 3 does not divide n, then $n^2 \equiv 1 \bmod 3$, hence $n^{12} - 1 \equiv (n^2)^6 - 1 \equiv 0 \bmod 3$, and again 3 divides $n(n^{12} - 1)$. Similarly, if 5 divides n, then 5 divides $n(n^{12} - 1)$. And if 5 does not divide n, then $n^4 \equiv 1 \bmod 5$, hence $n^{12} - 1 \equiv (n^4)^3 - 1 \equiv 0 \bmod 5$, and 5 divides $n(n^{12} - 1)$. ◇

Exercises 2.1

1. Find all the cosets of the subgroup $5\mathbb{Z}$ in $\mathbb{Z}$.

2. Find all the cosets of $9\mathbb{Z}$ in $\mathbb{Z}$ and of $9\mathbb{Z}$ in $3\mathbb{Z}$.

3. Find all the cosets of $\langle 6 \rangle$ in $\mathbb{Z}_{12}$ and all the cosets of $\langle 6 \rangle$ in the subgroup $\langle 2 \rangle$ of $\mathbb{Z}_{12}$.

4. In D_4 (see Example 1.4.24) find all the left and right cosets of $\langle \tau \rangle$.

5. Find the index of $\langle 10 \rangle$ in $\mathbb{Z}_{12}$.

6. Find the index of $\langle \mu_2 \rangle$ in S_3.

7. Find the index of $\langle \rho^2\tau \rangle$ in D_4. (See Example 1.4.24.)

8. Let $H = \{\phi \in S_n \mid \phi(n) = n\}$. Find the index of H in S_n.

9. Let H be a subgroup of a group G. For any $a, b \in G$, let $a \sim b$ if and only if $ab^{-1} \in H$. Show that the relation $\sim$ so defined is an equivalence relation on G, with equivalence classes the right cosets Ha of H.

10. Let H be a subgroup of a group G. Show that for any $a \in G$ we have $|Ha| = |H|$.

11. Prove part (1) of Lemma 2.1.7.

12. Prove part (2) of Lemma 2.1.7.

13. Let H be a subgroup of a group G. Show for any $a \in G$ that $aH = H$ if and only if $a \in H$.

14. Let $H = 5\mathbb{Z}$ in $\mathbb{Z}$. Determine whether the following cosets of H are the same:
(a) $12 + H$ and $27 + H$
(b) $13 + H$ and $-2 + H$
(c) $126 + H$ and $-1 + H$

15. Let G be a group of order 42. Find all possible orders $|H|$ for a subgroup H of G, and in each case determine the number of left cosets of H.

16. Let $G = \langle a \rangle$ be a cyclic group of order 60, and $H = \langle a^{35} \rangle$. List all the left cosets of H in G.

17. Let G be a group of order 36. If G has an element $a \in G$ such that $a^{12} \neq e$ and $a^{18} \neq e$, show that G is cyclic.

18. Let G be a group with $|G| < 300$. If G has a subgroup H of order 24 and a subgroup K of order 54, what is the order $|G|$ of G?

19. Let H and K be subgroups of a group G, where $|H| = 9$, $|K| = 12$ and where the index $[G:H \cap K] \neq |G|$. Find $|H \cap K|$.

20. Let G be a group with $|G| = p^2$, where p is prime. Show that every proper subgroup of G is cyclic.

21. Let G be a group with $|G| = pq$, where p and q are primes. Show that every proper subgroup of G is cyclic.

22. Let H and K be subgroups of a group G, with $|H| = n$ and $|K| = m$, where $\gcd(n, m) = 1$. Show that $H \cap K = \{e\}$.

23. Let G be a group and suppose $a, b \in G$ are elements such that $|a| = n$ and $|b| = m$, where $\gcd(n, m) = 1$. If for some integer k we have $a^k = b^k$, show that nm divides k. (*Hint:* Use the preceding exercise.)

24. Show that $n^{19} - n$ is divisible by 21 for any integer n.

25. Find the remainder of 9^{1573} when divided by 11.

26. Compute $\phi(p^2)$, where p is prime.

27. Compute $\phi(pq)$, where p and q are distinct primes.

28. Find the remainder of 5^{1258} when divided by 12.

29. Let G be a non-Abelian group with $|G| = 2p$, where p is prime. Show that there exists a $g \in G$ such that $|g| = p$.

30. Let G be a non-Abelian group with $|G| = 2p$, where p is prime. Show that G has p elements of order 2.

31. Let G be a group with $|G| > 1$ such that G has no nontrivial proper subgroups. Show that G is a finite cyclic group of prime order.

32. Let G be a group of order 15. Show that G contains an element of order 3.

33. Let H be a subgroup of a finite group G and K a subgroup of H. Suppose that the index $[G{:}H] = n$ and the index $[H{:}K] = m$. Show that the index $[G{:}K] = nm$. (*Hint:* Let x_iH be the distinct left cosets of H in G and y_jK the distinct left cosets of K in H. Show that x_iy_jK are the distinct left cosets of K in G.)

34. Let H and K be subgroups of a group G and for all $a, b \in G$ let $a \sim b$ if and only if $a = hbk$ for some $h \in H$ and $k \in K$. Show that the relation $\sim$ so defined is an equivalence relation. Describe the equivalence classes (which are called **double cosets**).

35. Let H and K be subgroups of a finite group G with index $[G{:}H] = n$ and index $[G{:}K] = m$. Show that $\operatorname{lcm}(n, m) \le [G{:}H \cap K] \le nm$.

36. Let H and K be finite subgroups of a group G. Let $HK = \{hk \mid h \in H, k \in K\}$. Show that $|HK| = |H||K| / |H \cap K|$. (Hint: $HK = \bigcup_{h \in H} hK$.)

37. Let H be a subgroup of the group G. Show that the map $a \to a^{-1}$ determines a one-to-one, onto map between the left cosets of H and the right cosets of H.

38. For any positive integer n show that $n = \sum_{d \mid n} \phi(d)$, where the sum is taken over all positive divisors d of n and ϕ is the Euler ϕ-function.

39. Show that the converse of Lagrange's theorem is false. (*Hint:* Show that A_4 has no subgroup of order 6.)

2.2 Homomorphisms

We have learned what groups and subgroups are and encountered several different kinds of groups: cyclic and noncyclic, Abelian and non-Abelian, finite and infinite. We have not, however, studied maps between groups. Since groups are not just plain sets, but rather are sets equipped with an operation that satisfies certain axioms, the maps we are interested in are ones that "respect" or "preserve" the operations on the groups involved.

2.2.1 EXAMPLE We look at three different functions from $\mathbb{Z}$ to $\mathbb{Z}$ and identify some properties of these functions. Let the functions $f,\ g,\ h : \mathbb{Z} \to \mathbb{Z}$ be given by

(1) $f(x) = x^2$
(2) $g(x) = x + 1$
(3) $h(x) = 2x$

In case (1), the image of f is not a subgroup of $\mathbb{Z}$. Also, if we take two elements $x, y \in \mathbb{Z}$, add them first, and then apply f, the result is not the same as if we first applied f to them and then added: $f(x) + f(y) = x^2 + y^2 \neq (x + y)^2 = f(x + y)$.
In case (2), the image of g is a subgroup of $\mathbb{Z}$, and in fact is $\mathbb{Z}$ itself, but again $g(x + y) = (x + y + 1) \neq (x + 1) + (y + 1) = g(x) + g(y)$.
In case (3), finally, the image of h is $2\mathbb{Z}$, which is a subgroup of $\mathbb{Z}$, and also we have $h(x + y) = 2(x + y) = 2x + 2y = h(x) + h(y)$. This is the only case where the function respects or preserves the group structure. ◇

2.2.2 DEFINITION A map $\phi\colon G \to G'$ from a group G to a group G' is called a **homomorphism** if

$$\phi(ab) = \phi(a)\phi(b) \text{ for all } a, b \in G$$

Note that in $\phi(ab)$ the product is being taken in G, while in $\phi(a)\phi(b)$ the product is being taken in G'. ○

2.2.3 EXAMPLE In Example 2.2.1, h is a homomorphism, while f and g are not. Note also that besides $h(x + y) = h(x) + h(y)$ we have $h(0) = 0$, and $h(-x) = -h(x)$. ◇

2.2.4 EXAMPLE The map $\phi\colon \mathbb{Z} \to \mathbb{Z}$ given by $\phi(x) = 5x$ is a homomorphism since $\phi(x + y) = 5(x + y) = 5x + 5y = \phi(x) + \phi(y)$. ◇

2.2.5 EXAMPLE The map $\phi\colon \mathbb{R}^* \to \mathbb{Z}_2$ given by

$$\phi(x) = \begin{cases} 0 \text{ if } x > 0 \\ 1 \text{ if } x < 0 \end{cases}$$

is a homomorphism. To check this, note that if x and y are both positive, then xy is positive and $\phi(xy) = 0 = 0 + 0 = \phi(x) + \phi(y)$. Also, if x and y are both negative, then xy is positive and $\phi(xy) = 0 = 1 + 1 = \phi(x) + \phi(y)$. Also, if x is positive and y is negative, then xy is negative and $\phi(xy) = 1 = 0 + 1 = \phi(x) + \phi(y)$, and similarly in the opposite case where x is negative and y is positive. ◇

2.2.6 EXAMPLE For any group G, the **identity** map is always a homomorphism, since if $\phi: G \to G$ is the identity $\phi(x) = x$, then $\phi(xy) = xy = \phi(x)\phi(y)$. ◇

2.2.7 EXAMPLE For any groups G and G', the map $\phi: G \to G'$ given by $\phi(x) = e'$, where e' is the identity element of G', is a homomorphism, called the **trivial** homomorphism between G and G'. For we have $\phi(xy) = e' = e' \cdot e' = \phi(x)\phi(y)$. ◇

2.2.8 EXAMPLE For any group G and any $a \in G$, consider the map $\phi: \mathbb{Z} \to \langle a \rangle$, called the **exponential map**, given by $\phi(n) = a^n$. Then ϕ is a homomorphism, since $\phi(n + m) = a^{n+m} = a^n a^m = \phi(n)\phi(m)$. ◇

2.2.9 EXAMPLE Let $\phi: \mathbb{Z} \to \mathbb{Z}_5$ be defined by $\phi(n) =$ the remainder of n mod 5. So, for instance, we have $\phi(7) = 2$, $\phi(8) = 3$, $\phi(7 + 8) = \phi(15) = 0$, and $\phi(7) + \phi(8) = 2 + 3 = 0$ in $\mathbb{Z}_5$. For any n, m in $\mathbb{Z}$ we can apply the division algorithm to write $n = q5 + \phi(n)$ and $m = p5 + \phi(m)$. We then have $n + m = (q + p)5 + (\phi(n) + \phi(m))$, and $\phi(n + m)$ is the sum of $\phi(n)$ and $\phi(m)$ in $\mathbb{Z}_5$, so ϕ is a homomorphism. ◇

2.2.10 PROPOSITION For any groups G, G', and G'', suppose $\phi: G \to G'$ and $\psi: G' \to G''$ are both homomorphisms. Then the composite map $\psi \circ \phi(x) = \psi(\phi(x))$ is a homomorphism from G to G''.

Proof Consider any $x, y \in G$. We have $\psi \circ \phi(xy) = \psi(\phi(xy)) = \psi(\phi(x)\phi(y)) = \psi(\phi(x))\psi(\phi(y)) = \psi \circ \phi(x) \psi \circ \phi(y)$. □

A homomorphism $\phi: G \to G'$ determines a special subgroup of G that plays a very important role in understanding the homomorphism.

2.2.11 DEFINITION Let $\phi: G \to G'$ be a homomorphism and let e' be the identity in G'. Then the **kernel** of ϕ is the set $\{x \in G \mid \phi(x) = e'\}$, denoted Kern ϕ ○

2.2.12 EXAMPLE The kernel of $\phi: \mathbb{Z} \to \mathbb{Z}_5$ in Example 2.2.9 is Kern $\phi = 5\mathbb{Z}$. ◇

2.2.13 EXAMPLE The kernel of $\phi: \mathbb{Z} \to \langle a \rangle$ in Example 2.2.8 is Kern $\phi = \{n \mid |a|$ divides $n\}$. ◇

2.2.14 EXAMPLE The kernel of $\phi: \mathbb{R}^* \to \mathbb{Z}_5$ in Example 2.2.5 is Kern $\phi = \{x \in \mathbb{R}^* \mid x$ is positive$\}$. ◇

After the examples we have already seen, the following list of properties of homomorphisms should not be surprising.

2.2.15 PROPOSITION (Basic group homomorphism properties) Let $\phi: G \to G'$ be a homomorphism. Then

(1) $\phi(e) = e'$, where e is the identity of G and e' the identity of G'.

(2) $\phi(a^{-1}) = (\phi(a))^{-1}$ for any $a \in G$.

(3) $\phi(a^n) = \phi(a)^n$ for any $n \in \mathbb{Z}$.

(4) If $|a|$ is finite, then $|\phi(a)|$ divides $|a|$.

(5) If H is a subgroup of G, then $\phi(H) = \{\phi(x) \mid x \in H\}$ is a subgroup of G'.

(6) If K is a subgroup of G', then $\phi^{-1}(K) = \{x \in G \mid \phi(x) \in K\}$ is a subgroup of G.

Proof (1) Since $\phi(e)\phi(e) = \phi(ee) = \phi(e) = e'\phi(e)$, we have $\phi(e) = e'$ by cancellation.

(2) Since $\phi(a)\phi(a^{-1}) = \phi(aa^{-1}) = \phi(e) = e' = \phi(a)(\phi(a))^{-1}$, we have $\phi(a^{-1}) = (\phi(a))^{-1}$ by cancellation.

(3) $\phi(a^n) = \phi(a)^n$ can be proved for $n > 0$ by induction, and then the statement for $n = 0$ follows by (1) and for $n < 0$ follows by (2). (See Execise 19.)

(4) Let $|a| = n$. Then by (3) we have $\phi(a)^n = \phi(a^n) = \phi(e) = e'$. Hence by Corollary 1.3.12, $|\phi(a)|$ divides n.

(5) Let $u, v \in \phi(H) = \{u \in G' \mid u = \phi(x)$ for some $x \in H\}$, and let $x, y \in H$ be such that $u = \phi(x)$ and $v = \phi(y)$. Then $xy^{-1} \in H$ since H is a subgroup, and $uv^{-1} = \phi(x)\phi(y)^{-1} = \phi(xy^{-1}) \in \phi(H)$. Hence by the subgroup test $\phi(H)$ is a subgroup of G'.

(6) We again use the subgroup test. Let $x, y \in \phi^{-1}(K)$. We have $\phi(xy^{-1}) = \phi(x)\phi(y)^{-1} \in K$. Hence $xy^{-1} \in \phi^{-1}(K)$ and $\phi^{-1}(K)$ is a subgroup of G. □

2.2.16 PROPOSITION Let $\phi: G \to G'$ be a homomorphism. Then Kern ϕ is a subgroup of G.

Proof This is immediate from Proposition 2.2.15, part (6). (See Exercise 20.) □

2.2.17 PROPOSITION Let $\phi: G \to G'$ be a homomorphism. Then ϕ is one to one if and only if the kernel is trivial, Kern $\phi = \{e\}$.

Proof ($\Rightarrow$) Suppose ϕ is one to one and suppose $x \in$ Kern ϕ. Then $\phi(x) = e' = \phi(e)$, hence $x = e$ and Kern $\phi = \{e\}$.
($\Leftarrow$) Suppose Kern $\phi = \{e\}$ and suppose for some $x, y \in G$ we have $\phi(x) = \phi(y)$. Then $\phi(xy^{-1}) = \phi(x)\phi(y)^{-1} = \phi(y)\phi(y)^{-1} = e'$. It follows that $xy^{-1} \in$ Kern $\phi = \{e\}$, hence $xy^{-1} = e$, so $x = y$ and ϕ is one to one. □

2.2.18 DEFINITION A homomorphism $\phi: G \to G'$ that is one to one and onto is called an **isomorphism**. Two groups G and G' are called **isomorphic**, written $G \cong G'$, if there exists some isomorphism $\phi: G \to G'$. ○

To show that two groups G and G' are isomorphic, we need to do four things:

(1) Define a map $\phi: G \to G'$.

(2) Show that ϕ is a homomorphism.

(3) Show that ϕ is one to one.

(4) Show that ϕ is onto.

The following example illustrates these four steps.

2.2.19 EXAMPLE $\mathbb{Z}$ and $3\mathbb{Z}$ are isomorphic groups. We carry out the four steps just indicated:

(1) Define $\phi: \mathbb{Z} \to 3\mathbb{Z}$ by $\phi(x) = 3x$.

(2) We have $\phi(x + y) = 3(x + y) = 3x + 3y = \phi(x) + \phi(y)$, so ϕ is a homomorphism.

(3) $\phi(x) = 0$ if and only if $3x = 0$, hence if and only if $x = 0$. So Kern $\phi = \{0\}$ and by Proposition 2.2.17 ϕ is one to one.

(4) Given $u \in 3\mathbb{Z}$, $u = 3x$ for some $x \in \mathbb{Z}$, so $u = \phi(x)$, and ϕ is onto. ◇

2.2.20 EXAMPLE $\mathbb{R}$, the real numbers under addition, is isomorphic to $\mathbb{R}^+$, the positive real numbers under multiplication. Again we carry out four steps to verify this claim:

(1) Let $\phi: \mathbb{R} \to \mathbb{R}^+$ be the exponential function $\phi(x) = \exp x = e^x$.

(2) $\phi(x + y) = e^{x+y} = e^x e^y = \phi(x)\phi(y)$, so ϕ is a homomorphism.

(3) The identity element in $\mathbb{R}^+$ is 1. Hence if $x \in$ Kern ϕ, then $\phi(x) = 1$, which is to say $e^x = 1$, which implies $x = 0$. So Kern $\phi = \{0\}$ and ϕ is one to one.

(4) For $u \in \mathbb{R}^+$, let $x = \ln u$, the natural logarithm of u. Then $\phi(x) = e^x = e^{\ln u} = u$, and ϕ is onto. ◇

2.2.21 EXAMPLE In Example 2.2.19, the map $\phi^{-1}: 3\mathbb{Z} \to \mathbb{Z}$, with $\phi^{-1}(u) = u/3$, is an isomorphism between $3\mathbb{Z}$ to $\mathbb{Z}$. In Example 2.2.20, the map $\phi^{-1}: \mathbb{R}^+ \to \mathbb{R}$, with $\phi^{-1}(u) = \ln u$, is an isomorphism between $\mathbb{R}^+$ and $\mathbb{R}$. ◇

2.2.22 PROPOSITION Let $\phi: G \to G'$ and $\psi: G' \to G''$ be isomorphisms. Then

(1) The composition $\psi \circ \phi: G \to G''$ is an isomorphism.

(2) The identity map $\phi: G \to G$ is an isomorphism.

(3) The inverse $\phi^{-1}: G' \to G$ is an isomorphism.

Proof (1) $\psi \circ \phi$ is a homomorphism by Proposition 2.2.10, and we know from Theorem 0.1.15 that a composition of one-to-one maps is one to one, and a composition of onto maps is onto.

(2) and (3) are left to the reader (as Exercise 21). □

2.2.23 PROPOSITION Let $G \cong G'$. Then

(1) $|G| = |G'|$.

(2) G is Abelian if and only if G' is Abelian.

(3) G is cyclic if and only if G' is cyclic.

(4) G has k elements of order n if and only if G' has k elements of order n.

Proof Let $\phi: G \to G'$ be an isomorphism.

(1) Since ϕ is a one-to-one and onto map between G and G', $|G| = |G'|$.

(2) Suppose G is Abelian and let $u, v \in G'$. Since ϕ is onto, there are $x, y \in G$ with $\phi(x) = u$, $\phi(y) = v$.

Then $uv = \phi(x)\phi(y) = \phi(xy)$ and $\phi(xy) = \phi(yx) = \phi(y)\phi(x) = vu$ since G is Abelian. So $uv = vu$ and G' is Abelian. If G' is Abelian, then, $\phi(xy) = \phi(x)\phi(y) = \phi(y)\phi(x) = \phi(yx)$, and since ϕ is one to one, $xy = yx$ and G is Abelian.
Before proving (3) and (4), we note the following:
Claim For any $a \in G$, $|a| = |\phi(a)|$.
Proof of Claim Suppose a is an element of order n in G, and let m be the order of $\phi(a)$ in G'. By Proposition 2.2.15, part (4), m divides $|a| = n$. But since $\phi(a^m) = \phi(a)^m = e'$, the identity of G', and since ϕ is one to one, we must have $a^m = e$, and so $n = |a|$ divides m. Thus a and $\phi(a)$ have the same order $n = m$.

(3) If $G = \langle a \rangle$ is cyclic, then by the claim $|\phi(a)| = |a| = |G| = |G'|$, so $G' = \langle \phi(a) \rangle$ and is cyclic. If $G' = \langle b \rangle$ and is cyclic, then since ϕ is onto there is an $a \in G$ with $\phi(a) = b$. But then $|a| = |\phi(a)| = |b| = |G'| = |G|$, and $G = \langle a \rangle$ and is cyclic.

(4) Suppose $a_1, a_2, \ldots, a_k$ are k distinct elements of order n in G. Then since ϕ is one to one, $\phi(a_1), \phi(a_2), \ldots, \phi(a_k)$ are all distinct, and by the claim, are all of order n. If $a_1, a_2, \ldots, a_k$ are *all* the elements of G of order n, then $\phi(a_1), \phi(a_2), \ldots, \phi(a_k)$ will be *all* the elements of G' of order n. For consider any other element u of G'. Since ϕ is onto, $u = \phi(x)$ for some $x \in G$, and u is distinct from all the $\phi(a_i)$, x is distinct from all the a_i. Since the a_i were all the elements of G of order n, we have $|u| = |\phi(x)| = |x| \neq n$. □

2.2.24 LEMMA Let G and H be cyclic groups of the same finite order n, and let a be any generator of G and b any generator of H. Then there is an isomorphism $\phi: G \to H$ with $\phi(a) = b$.

Proof We have $G = \langle a \rangle$, where $|G| = n$, so by Corollary 1.3.13

$$G = \{e, a, a^2, \ldots, a^{n-1}\}$$

where these elements are all distinct. Define a map $\phi: G \to H$ by $\phi(a^i) = b^i$ for $0 \leq i < n$.
Claim ϕ is an isomorphism.
Poof of Claim Left to the reader (as Exercise 39). □

2.2.25 PROPOSITION Let $G = \langle a \rangle$ be a cyclic group. Then

(1) If $|G| = \infty$, then $G \cong \mathbb{Z}$.
(2) If $|G| = n$, then $G \cong \mathbb{Z}_n$.

Proof (1)If $G = \langle a \rangle$, where $|a| = \infty$, let $\phi: \mathbb{Z} \to G$ be the exponential homomorphism as in Example 2.2.8, defined by $\phi(k) = a^k$. Since $|a| = \infty$, $a^k = e$ if and only if $k = 0$. Hence Kern $\phi = \{0\}$ and ϕ is one to one. Since G is cyclic, every $u \in G$ is of the form $u = a^k$ for some $k \in \mathbb{Z}$. Hence $u = \phi(k)$ and ϕ is onto. So ϕ is an isomorphism from $\mathbb{Z}$ to G, and by Proposition 2.2.22, ϕ^{-1} is an isomorphism from G to $\mathbb{Z}$.

(2) This is immediate from the preceding lemma. □

2.2.26 EXAMPLE D_4 and $\mathbb{Z}_8$ are not isomorphic, because D_4 is non-Abelian, and $\mathbb{Z}_8$ is Abelian. ◇

2.2.27 EXAMPLE $U(10) = \{1, 3, 7, 9\}$ and $U(12) = \{1, 5, 7, 11\}$ are not isomorphic, because $U(10)$ is cyclic and $U(12)$ is noncyclic. ◇

2.2.28 EXAMPLE $\mathbb{Q}$ under addition is not isomorphic to $\mathbb{Q}^*$ under multiplication. For suppose there is an isomorphism $\phi\colon \mathbb{Q} \to \mathbb{Q}^*$. Since ϕ is onto, there exists some $a \in \mathbb{Q}$ such that $\phi(a) = 2$. Consider the rational number $r = \phi(a/2)$. We have $r^2 = \phi(a/2)\phi(a/2) = \phi(a/2 + a/2) = \phi(a) = 2$, which is impossible. ◇

Exercises 2.2

In Exercises 1 through 10, determine whether or not the indicated map ϕ is a homomorphism, and in the cases where ϕ is a homomorphism, determine Kern ϕ.

1. $\phi\colon \mathbb{Z} \to \mathbb{Z}$, where $\phi(n) = n - 1$ **2.** $\phi\colon \mathbb{Z} \to \mathbb{Z}$, where $\phi(n) = 3n$

3. $\phi\colon \mathbb{R}^* \to \mathbb{R}^*$ (under multiplication), where $\phi(x) = |x|$

4. $\phi\colon GL(2,\mathbb{R}) \to \mathbb{R}^*$, where $GL(2, \mathbb{R})$ is the general linear group of 2×2 invertible matrices and $\phi(A) = \det A$

5. $\phi\colon S_3 \to \mathbb{Z}_2$, where

$$\phi(\sigma) = \begin{cases} 0 \text{ if } \sigma \text{ is an even permutation} \\ 1 \text{ if } \sigma \text{ is an odd permutation} \end{cases}$$

6. $\phi\colon D_4 \to \mathbb{Z}_2$, where $D_4 = \{\rho_0, \rho, \rho^2, \rho^3, \tau, \rho\tau, \rho^2\tau, \rho^3\tau\}$ is the dihedral group, and $\phi(\rho^i) = 0$, $\phi(\rho^i\tau) = 1$, for all i, $0 \le i \le 3$

7. $\phi\colon \mathbb{R} \to GL(2, \mathbb{R})$, where $\mathbb{R}$ is the group of real numbers under addition, $GL(2, \mathbb{R})$ is as in Exercise 4, and

$$\phi(x) = \begin{bmatrix} 1 & x \\ 0 & 1 \end{bmatrix}$$

8. $\phi\colon G \to G$, where G is any group, and $\phi(x) = x^{-1}$

9. $\phi\colon \mathbb{Z}_6 \to \mathbb{Z}_2$, where $\phi(x) =$ the remainder of x mod 2

10. $\phi\colon \mathbb{Z}_7 \to \mathbb{Z}_2$, where $\phi(x) =$ the remainder of x mod 2

In Exercises 11 through 15, compute the indicated values for the indicated homomorphisms.

11. $\phi(27)$, where $\phi\colon \mathbb{Z} \to \mathbb{Z}_5$ is as in Example 2.2.9

12. $\phi(27)$, where $\phi: \mathbb{Z} \to \mathbb{Z}_3$ is as in Example 2.2.9 with mod 3 in place of mod 5

13. $\phi((1\ 2)(2\ 3\ 1)(2\ 3))$, where $\phi: S_3 \to \mathbb{Z}_2$ is as in Exercise 5

14. $\phi(5)$ and $\phi(10)$, where $\phi: \mathbb{Z}_{15} \to \mathbb{Z}_3$ is a homomorphism with $\phi(1) = 2$

15. $\phi(10)$, where $\phi: \mathbb{Z} \to D_4$ is the exponential map with $\phi(1) = \rho$ (See Exercise 6.)

16. Find all possible homomorphisms from $\mathbb{Z}$ to $\mathbb{Z}$.

17. Find all possible homomorphisms from $\mathbb{Z}$ onto $\mathbb{Z}$.

18. Find all possible homomorphisms from $\mathbb{Z}_3$ to $\mathbb{Z}_6$. [Use Proposition 2.2.15, part (4).]

19. Prove (3) of Proposition 2.2.15.

20. Prove Proposition 2.2.16.

21. Prove (2) and (3) of Proposition 2.2.22.

22. Consider the relation R on the class of all groups defined by the condition that GRG' if and only if G and G' are isomorphic. Show that R has the properties of an equivalence relation (reflexivity, symmetry, transitivity).

In Exercises 23 through 30 construct an example of a nontrivial homomorphism between the two indicated groups, if this is possible, or explain why this is not possible.

23. $\phi: S_3 \to S_5$ **24.** $\phi: \mathbb{Z}_3 \to \mathbb{Z}_5$ **25.** $\phi: \mathbb{Z}_4 \to \mathbb{Z}_{12}$

26. $\phi: \mathbb{Z}_5 \to \mathbb{Z}_{12}$ **27.** $\phi: D_4 \to S_5$ **28.** $\phi: \mathbb{Z} \to \mathbb{Z}_7$

29. $\phi: \mathbb{Z}_{10} \to \mathbb{Z}_8$ **30.** $\phi: S_5 \to \mathbb{Z}_2$

31. Let $\phi: G \to G'$ be a homomorphism where $|G| = 9$. Find $|\text{Kern } \phi|$ if ϕ is
(a) trivial (b) one to one (c) neither

32. Let $\phi: \mathbb{Z}_{12} \to \mathbb{Z}_3$ be a homomorphism with Kern $\phi = \{0, 3, 6, 9\}$ and $\phi(4) = 2$. Find all the elements $x \in \mathbb{Z}_{12}$ such that $\phi(x) = 1$, and show that they form a coset of Kern ϕ in $\mathbb{Z}_{12}$.

33. Let G be a group, $a \in G$, and $\phi: \mathbb{Z} \to G$ the exponential homomorphism $\phi(n) = a^n$ as in Example 2.2.8. Describe all the possibilities for Kern ϕ.

34. Let $\phi: G \to G'$ be a homomorphism, $K =$ Kern ϕ, and $a \in G$. Show that $\{x \in G \mid \phi(x) = \phi(a)\} = aK$, the left coset of K to which the element a belongs.

In Exercises 35 through 38 determine whether the indicated map ϕ is an isomorphism. Justify your answer.

35. $\phi: 2\mathbb{Z} \to 3\mathbb{Z}$, where $\phi(2n) = 3n$ **36.** $\phi: U(10) \to \mathbb{Z}_4$, where $\phi(3) = 3$

37. ϕ: $U(10) \to \mathbb{Z}_4$, where $\phi(3) = 2$ **38.** ϕ: $\mathbb{Z} \to \mathbb{Z}$, where $\phi(n) = 3n$

39. Prove the claim in the proof of Lemma 2.2.24.

40. Show that $U(8)$ and $U(12)$ are isomorphic.

41. Show that in $\mathbb{C}^*$ the subgroup $\langle i \rangle$ generated by i is isomorphic to $\mathbb{Z}_4$.

42. Show that $\mathbb{Z}_4$ and the Klein 4-group V of Example 1.1.22 are not isomorphic.

43. Show that $U(14) \cong U(18)$.

44. Show that the dihedral group D_4 and the quaternion group Q_8 of Example 1.1.23 are not isomorphic.

45. Find four different subgroups of S_4 that are isomorphic to S_3.

46. Show that the alternating group A_4 contains a subgroup isomorphic to the Klein 4-group V.

47. Show that the dihedral group D_4 contains a subgroup isomorphic to the Klein 4-group V.

48. Let $G = \mathrm{GL}(2,\mathbb{Z}_2)$, the general linear group of 2×2 invertible matrices with coefficients in $\mathbb{Z}_2$. Show that $G \cong S_3$.

2.3 Normal Subgroups

We have seen that for any homomorphism ϕ: $G \to G'$, the kernel Kern ϕ is a subgroup of G. In this section we show that it is a subgroup with a special property concerning its cosets, and we begin the study of such special subgroups, called *normal* subgroups.

2.3.1 EXAMPLE Consider ϕ: $S_3 \to \mathbb{Z}_2$, where

$$\phi(\sigma) = \begin{cases} 0 \text{ if } \sigma \text{ is an even permutation} \\ 1 \text{ if } \sigma \text{ is an odd permutation} \end{cases}$$

as in Exercise 5 of the preceding section. Then Kern $\phi = A_3 = \{\rho_0, \rho, \rho^2\}$. The left cosets of A_3 are

$$A_3 = \{\rho_0, \rho, \rho^2\} \qquad \mu_1 A_3 = \{\mu_1, \mu_1\rho, \mu_1\rho^2\} = \{\mu_1, \mu_2, \mu_3\}$$

and the right cosets of A_3 are

$$A_3 = \{\rho_0, \rho, \rho^2\} \qquad A_3\mu_1 = \{\mu_1, \rho\mu_1, \rho^2\mu_1\} = \{\mu_1, \mu_3, \mu_2\}$$

Thus the left and right cosets are the same. *Warning*: The equation $\mu_1 A_3 = A_3\mu_1$ does *not* mean that μ_1 commutes with every element of A_3. On the contrary, we have $\mu_1\rho = \mu_2 \neq \mu_3 = \rho\mu_1$. ◇

2.3.2 EXAMPLE Let $Q_8 = \{\pm 1, \pm\mathbf{i}, \pm\mathbf{j}, \pm\mathbf{k}\}$ be the quaternion group (as in Example 1.1.23). Let $\phi: Q_8 \to \mathbb{Z}_2$ be defined by $\phi(\pm 1) = \phi(\pm\mathbf{i}) = 0$ and $\phi(\pm\mathbf{j}) = \phi(\pm\mathbf{k}) = 1$. Then ϕ is a homomorphism with Kern $\phi = \{\pm 1, \pm\mathbf{i}\}$. (See Exercise 9 at the end of this section.) The left cosets of K = Kern ϕ are

$$K = \{1, -1, \mathbf{i}, -\mathbf{i}\} \qquad \mathbf{j}K = \{\mathbf{j}, -\mathbf{j}, \mathbf{ji} = -\mathbf{k}, \mathbf{j}(-\mathbf{i}) = \mathbf{k}\}$$

The right cosets of K are

$$K = \{1, -1, \mathbf{i}, -\mathbf{i}\} \qquad K\mathbf{j} = \{\mathbf{j}, -\mathbf{j}, \mathbf{ij} = \mathbf{k}, -\mathbf{ij} = -\mathbf{k}\}$$

Again the left and right cosets are the same. And again the equation $\mathbf{j}K = K\mathbf{j}$ does not imply that $\mathbf{j}$ commutes with every element of K, since we have $\mathbf{ji} = -\mathbf{k} \neq \mathbf{k} = \mathbf{ij}$. ◇

These two examples illustrate an important property of the kernel of a homomorphism.

2.3.3 PROPOSITION Let $\phi: G \to G'$ be a homomorphism, and let K = Kern ϕ. Then for all $g \in G$ we have $gK = Kg$.

Proof Let $x \in gK$, so $x = gk_1$ for some $k_1 \in K$ = Kern ϕ. Then $\phi(x) = \phi(gk_1) = \phi(g)\phi(k_1) = \phi(g)e' = \phi(g)$. It follows that $e' = \phi(x)\phi(g)^{-1} = \phi(xg^{-1})$ and $xg^{-1} \in K$. Letting $k_2 = xg^{-1} \in K$, we have $x = k_2 g \in Kg$. Thus $gK \subseteq Kg$. The proof that $gK \subseteq Kg$ is similar. (See Exercise 10 at the end of this section.) □

Subgroups having the special property proved for kernels in the preceding proposition play an important role in understanding the structure of groups.

2.3.4 DEFINITION Let G be a group and H a subgroup of G. If for all $g \in G$ we have $gH = Hg$, then we say H is a **normal** subgroup of G and write $H \lhd G$. ○

2.3.5 COROLLARY Let $\phi: G \to G'$ be a homomorphism, and let K = Kern ϕ. Then $K \lhd G$.

Proof This is immediate from Proposition 2.3.3 and Definition 2.3.4. □

2.3.6 EXAMPLE $A_3 \lhd S_3$ and $\{\pm 1, \pm\mathbf{i}\} \lhd Q_8$ by Examples 2.3.1 and 2.3.2. ◇

2.3.7 PROPOSITION If G is an Abelian group, then every subgroup of G is a normal subgroup of G.

Proof This is immediate from Definition 2.3.4. □

2.3.8 EXAMPLE $n\mathbb{Z} \lhd \mathbb{Z}$ for all $n \geq 1$, and every subgroup of $\mathbb{Z}_n$ is normal in $\mathbb{Z}_n$. ◇

2.3.9 EXAMPLE In A_4 consider the subgroup

$$H = \{\rho_0, (1\ 2)(3\ 4), (1\ 3)(2\ 4), (1\ 4)(2\ 3)\}$$

We have $|A_4| = 12$ and $|H| = 4$, so the index $[A_4{:}H] = 3$. The three left cosets of H are H itself and

$$(1\ 2\ 3)H = \{(1\ 2\ 3), (1\ 3\ 4), (2\ 4\ 3), (1\ 4\ 2)\}$$
$$(1\ 3\ 2)H = \{(1\ 3\ 2), (2\ 3\ 4), (1\ 2\ 4), (1\ 4\ 3)\}$$

The three right cosets of H are H itself and

$$H(1\ 2\ 3) = \{(1\ 2\ 3), (2\ 4\ 3), (1\ 4\ 2), (1\ 3\ 4)\}$$
$$H(1\ 3\ 2) = \{(1\ 3\ 2), (1\ 4\ 3), (2\ 3\ 4), (1\ 2\ 4)\}$$

Therefore, $(1\ 2\ 3)H = H(1\ 2\ 3)$ and $(1\ 3\ 2)H = H(1\ 3\ 2)$, hence $H \lhd A_4$. ◇

The following theorem provides us with another tool for constructing examples of normal subgroups.

2.3.10 THEOREM Let G be a group and H a subgroup of G with index $[G{:}H] = 2$. Then H is a normal subgroup of G.

Proof Since $[G{:}H] = 2$, H has just two left cosets and just two right cosets. Since $H = eH = He$ is itself both a left and a right coset, for any $g \in G$ with $g \notin H$, the two distinct left cosets are H and gH and the two distinct right cosets are H and Hg. Since the cosets determine a partition of G, we must have $gH = \{k \in G \mid k \notin H\} = Hg$, and hence $H \lhd G$. □

2.3.11 EXAMPLE $A_n \lhd S_n$ for all $n \geq 3$, since $[S_n{:}A_n] = 2$. ◇

2.3.12 EXAMPLE The subgroup $H = \langle \rho \rangle$ in the dihedral group D_4 generated by the rotation ρ has order 4 and hence has index $[D_4{:}H] = 2$, so $H \lhd D_4$. ◇

The next theorem will provide an important test, making it easier to tell whether or not a given subgroup is normal.

2.3.13 THEOREM (Normal subgroup test) Let G be a group, H a subgroup of G. Then the following conditions are equivalent:

(1) $H \lhd G$
(2) $gHg^{-1} \subseteq H$ for all $g \in G$
(3) $gHg^{-1} = H$ for all $g \in G$

Proof (1) $\Rightarrow$ (2) Let $g \in G$ and let $x \in gHg^{-1}$, so $x = ghg^{-1}$ for some $h \in H$. By (1) we have $H \lhd G$ and hence $gH = Hg$. Then since $gh \in gH$ we have $gh \in Hg$, and $gh = h'g$ for some $h' \in H$. Hence $x = ghg^{-1} = h'gg^{-1} = h' \in H$. Therefore, $gHg^{-1} \subseteq H$.

(2) $\Rightarrow$ (3) By (2) we have $yHy^{-1} \subseteq H$ and hence $H \subseteq y^{-1}Hy$ for all $y \in G$. Given any $g \in G$, let $y = g^{-1}$. Then we have $gHg^{-1} \subseteq H \subseteq y^{-1}Hy = gHg^{-1}$ and $gHg^{-1} = H$.

(3) $\Rightarrow$ (1) Let $g \in G$ and let $x \in gH$, so $x = gh$ for some $h \in H$. It follows that $xg^{-1} = ghg^{-1} \in gHg^{-1}$. But by (3) we have $gHg^{-1} = H$, so $xg^{-1} \in H$ and $x \in Hg$. Thus $gH \subseteq Hg$. Similarly, if we start with $y \in Hg$ we can show $g^{-1}y \in H$ and hence $y \in gH$. □

2.3.14 EXAMPLE Let $G = \mathrm{GL}(2,\mathbb{R})$ and $H = \mathrm{SL}(2,\mathbb{R})$ be the general and special linear groups of 2×2 matrices with entries from $\mathbb{R}$ (as in Examples 1.1.20 and 1.1.21). Then $H \lhd G$ because if $A \in G$ and $B \in H$ we have $\det(ABA^{-1}) = \det A \cdot \det B \cdot (\det A)^{-1} = \det A^{-1} \cdot (\det A)^{-1} = 1$, hence $ABA^{-1} \in H$. ◇

2.3.15 EXAMPLE $Z(G)$, the center of a group G, is a normal subgroup of G, since the elements of $Z(G)$ commute with every element of G. ◇

2.3.16 EXAMPLE In S_3 consider the subgroup $H = \langle\mu_1\rangle$. If we calculate $\rho H\rho^{-1}$ we obtain the subgroup $\langle\mu_2\rangle$. If we calculate $\rho^2 H\rho^{-2}$ we obtain the subgroup $\langle\mu_3\rangle$. In this case, H, $\rho H\rho^{-1}$, and $\rho^2 H\rho^{-2}$ are three distinct subgroups of S_3 of order 2, and none of them is normal. ◇

2.3.17 THEOREM Let H be a subgroup of a group G. Then for any $g \in G$

(1) gHg^{-1} is a subgroup of G

(2) $|gHg^{-1}| = |H|$

Proof (1) If $x, y \in gHg^{-1}$, then $x = gh_1g^{-1}$ and $y = gh_2g^{-1}$ for some $h_1, h_2 \in H$. Therefore, $xy^{-1} = gh_1g^{-1}(gh_2g^{-1})^{-1} = gh_1g^{-1}gh_2^{-1}g^{-1} = gh_1h_2^{-1}g^{-1}$. Since H is a subgroup, $h = h_1h_2^{-1} \in H$, and so $xy^{-1} = ghg^{-1} \in gHg^{-1}$, and gHg^{-1} is a subgroup by the subgroup test.

(2) This is left to the reader (as Exercise 19). □

2.3.18 COROLLARY Let H be a subgroup of a group G. If H is the only subgroup of G of order $|H|$, then $H \lhd G$.

Proof This is immediate from Theorem 2.3.17. □

2.3.19 EXAMPLE In D_4 consider the subgroups $H = \langle\tau\rangle = \{\rho_0, \tau\}$ and $N = \{\rho_0, \rho^2, \tau, \rho^2\tau\}$. Then $H \lhd N$ and $N \lhd D_4$, since $[N{:}H] = [D_4{:}N] = 2$, but H is not a normal subgroup of D_4. Actually, N is the largest subgroup of D_4 that contains H as a normal subgroup. ◇

2.3.20 DEFINITION Let H be a subgroup of a group G. Then $N_G(H) = \{g \in G \mid gHg^{-1} = H\}$ is called the **normalizer** of H in G. ○

2.3.21 THEOREM Let H be a subgroup of a group G. Then

(1) $N_G(H)$ is a subgroup of G.

(2) $H \lhd N_G(H)$.

(3) If K is a subgroup of G and $H \lhd K$, then K is a subgroup of $N_G(H)$.

(4) $H \lhd G$ if and only if $N_G(H) = G$.

Proof Left to the reader. (See Exercises 20 and 21.) □

Given two subgroups H and K of a group G, the set $HK = \{hk \mid h \in H, k \in K\}$ turns out to be sometimes a subgroup of G and sometimes not. A partial answer to the question of when it is a subgroup is provided by the next proposition. (For a complete answer, see Exercise 26 at the end of this section.)

2.3.22 PROPOSITION Let H and K be subgroups of a group G, and assume $H \lhd G$. Then HK is a subgroup of G.

Proof Let $x, y \in HK$, so $x = h_1k_1$ and $y = h_2k_2$ for some $h_1, h_2 \in H$ and $k_1, k_2 \in K$. Then $xy^{-1} = h_1k_1(h_2k_2)^{-1} = h_1k_1k_2^{-1}h_2^{-1} = h_1kh_2^{-1}$, where $k = k_1k_2^{-1} \in K$, since K is a subgroup. We have $kh_2^{-1} \in kH$, and since $H \lhd G$, $kH = Hk$ and $kh_2^{-1} \in Hk$. So $kh_2^{-1} = h_3k$ for some $h_3 \in H$, and $xy^{-1} = h_1kh_2^{-1} = h_1h_3k = hk$, where $h = h_1h_3 \in H$. Thus $xy^{-1} \in HK$, and HK is a subgroup by the subgroup test. □

2.3.23 COROLLARY If H and K are subgroups of an Abelian group G, then HK is a subgroup of G.

Proof This is immediate from the preceding proposition. □

We end this section with a very useful counting principle.

2.3.24 THEOREM If H and K are finite subgroups of a group G, then

$$|HK| = |H||K| / |H \cap K|$$

Proof Let s be the index $[K:H \cap K]$, and let $(H \cap K)k_1, \dots, (H \cap K)k_s$ be the s distinct cosets of $H \cap K$ in K. These cosets form a partition of K and every element of K belongs to exactly one of them. Since $s = [K:H \cap K] = |K| / |H \cap K|$, it only remains to prove the following:

Claim $|HK| = |H|s$

Proof of Claim Consider $Hk_1, \dots, Hk_s$. They are distinct cosets of H, because if $Hk_i = Hk_j$ then $k_ik_j^{-1} \in H \cap K$, contrary to the choice of $k_1, \dots, k_s$. Also, every element of HK belongs to one of them, since given any $hk \in HK$, we know $k \in (H \cap K)k_i$ for some $1 \le i \le s$, and it follows that $hk \in Hk_i$. Thus the Hk_i partition HK, and each has $|H|$ elements, so the claim is proved. ◇

Exercises 2.3

In Exercises 1 through 8, determine whether the indicated subgroup is normal in the indicated group.

1. A_3 in S_3 **2.** A_3 in S_4 **3.** $3\mathbb{Z}$ in $\mathbb{Z}$

4. $\langle\rho\rangle$ in D_4 **5.** $\langle\rho\tau\rangle$ in D_4 **6.** $\{\pm 1, \pm\mathbf{j}\}$ in Q_8

7. $K = \{\rho_0, (1\ 2)(3\ 4), (1\ 3)(2\ 4), (1\ 4)(2\ 3)\}$ in S_4

8. $\langle(1\ 2\ 3)\rangle$ in S_4

9. Show that the map $\phi: Q_8 \to \mathbb{Z}_2$ defined by $\phi(\pm 1) = \phi(\pm \mathbf{i}) = 0$ and $\phi(\pm \mathbf{j}) = \phi(\pm \mathbf{k}) = 1$, as in Example 2.3.2, is a homomorphism.

10. Let $\phi: G \to G'$ be a homomorphism, and let $K = \text{Kern}\ \phi$ and $g \in G$. Complete the proof of Proposition 2.3.3 by showing that $gK \supseteq Kg$.

11. Find all the normal subgroups in $\text{GL}(2, \mathbb{Z}_2)$, the general linear group of 2×2 matrices with entries from $\mathbb{Z}_2$.

12. Find all the normal subgroups in D_4.

13. For $r \in \mathbb{R}^*$ let $rI = \begin{bmatrix} r & 0 \\ 0 & r \end{bmatrix}$. Show that $H = \{rI \mid r \in \mathbb{R}^*\}$ is a normal subgroup of $\text{GL}(2, \mathbb{R})$.

14. Let $\phi: G \to G'$ be a homomorphism and $H' \lhd G'$. Show that $H = \phi^{-1}(H') \lhd G$.

15. Give an example of a group G and two subgroups $H \leq K \leq G$ such that $H \lhd K$ and $K \lhd G$, but H is *not* normal in G.

16. Show that if $H \lhd G$ and $K \lhd G$, then $H \cap K \lhd G$.

17. Let H be a subgroup of a group G, and suppose that for every $x \in G$ there is a $y \in G$ such that $xH = Hy$. Show that $H \lhd G$.

18. Show that if $H \lhd G$ and $K \lhd G$, then $HK \lhd G$.

19. Prove part (2) of Theorem 2.3.17.

20. Prove parts (1) and (2) of Theorem 2.3.21.

21. Prove parts (3) and (4) of Theorem 2.3.21.

In Exercises 22 through 25, find the normalizer of the indicated subgroup in the indicated group.

22. A_3 in S_3 **23.** $\langle\mu_1\rangle$ in S_3 **24.** $\langle\tau\rangle$ in D_4 **25.** $\langle\mathbf{j}\rangle$ in Q_8

26. Let H and K be subgroups of a group G. Show that HK is a subgroup of G if and only if $HK = KH$.

27. Let G be a group of order pq, where p and q are distinct primes. Suppose G has a unique subgroup of order p and a unique subgroup of order q. Show that G is cyclic.

28. Let G be a group with a unique subgroup of order n and a unique subgroup of order m, where the positive integers n and m are relatively prime. Show that G has a normal subgroup of order nm.

29. Let $K \lhd G$ and let H be a subgroup of G. Show that $K \cap H \lhd H$.

2.4 Quotient Groups

In the preceding section we showed that if K is the kernel Kern $\phi = \{g \in G \mid \phi(g) = e\}$ of a homomorphism ϕ: $G \to G'$, then $gK = Kg$ for all $g \in G$, and we called subgroups with this property normal subgroups. In this section we study the images of homomorphisms, and among other things we show that, conversely, if K is a normal subgroup of a group G, then K is the kernel of a some homomorphism ϕ from G onto another group G'. We actually show how to construct the group G' and the homomorphism ϕ starting from G and K. This construction, called the quotient group construction, is very important in understanding the structure of groups.

2.4.1 EXAMPLE In $\mathbb{Z}$ consider the subgroup $5\mathbb{Z}$ and the set whose elements are the cosets of $5\mathbb{Z}$ in $\mathbb{Z}$. Since $m + 5\mathbb{Z} = n + 5\mathbb{Z}$ if and only if $m \equiv n \bmod 5$, the set of cosets is $\{5\mathbb{Z} = 0 + 5\mathbb{Z}, 1 + 5\mathbb{Z}, 2 + 5\mathbb{Z}, 3 + 5\mathbb{Z}, 4 + 5\mathbb{Z}\}$. There is a natural way to define an operation on this set, letting $(m + 5\mathbb{Z}) + (n + 5\mathbb{Z}) = (m + n) + 5\mathbb{Z} = (m +_5 n) + 5\mathbb{Z}$, where $+_5$ is addition mod 5. With this operation the set becomes a group isomorphic to $\mathbb{Z}_5$, with the coset $m + 5\mathbb{Z}$ corresponding to the element $\phi(m)$ of $\mathbb{Z}_5$, where $\phi(m)$ is the remainder of m mod 5. $5\mathbb{Z} = 5\mathbb{Z} + 0$ is the identity element. $1 + 5\mathbb{Z}$ and $4 + 5\mathbb{Z}$ are inverses of each other, as are $2 + 5\mathbb{Z}$ and $3 + 5\mathbb{Z}$. ◇

Given a subgroup H of a group G, we would like to imitate the construction of the preceding example and define an operation on the cosets of H in G by letting $aHbH = abH$. But for this definition to make sense, or as we say, for this operation to be well defined, we need to know that if a_1H and a_2H are the same coset and if b_1H and b_2H are the same coset, then a_1b_1H and a_2b_2H will be the same coset. In other words, we need to know that the result of the operation does not depend on which element, a_1 or a_2, we choose to represent the first coset, or on which element, b_1 or b_2, we choose to represent the second coset. The following lemma tells us when this condition is met.

2.4.2 LEMMA Let H be a subgroup of a group G. Then $H \lhd G$ if and only if $(aH)(bH) = abH$ is a well-defined operation on the left cosets of H.

Proof ($\Rightarrow$) Assume $H \lhd G$, and suppose $a_1H = a_2H$ and $b_1H = b_2H$. This means $a_1 = a_2h$ and $b_1 = b_2h'$ for some $h, h' \in H$. Then $a_1b_1 = (a_2h)(b_2h') = a_2(hb_2)h'$. Since $H \lhd G$, we know $b_2H = Hb_2$, and hence $hb_2 = b_2h''$ for some $h'' \in H$. So $a_1b_1 = a_2(hb_2)h' = a_2(b_2h'')h' = a_2b_2(h''h')$, where $h''h' \in H$, and therefore $a_1b_1H = a_2b_2H$ as required to show the operation is well defined.
($\Leftarrow$) Assume the indicated operation is well defined and let $g \in G$ and $h \in H$. Since $gH = (gh)H$ and the operation is well defined, we must have $(ghg^{-1})H = (gh)Hg^{-1}H = gHg^{-1}H = (gg^{-1})H = eH = H$. Therefore, $ghg^{-1} \in H$, and by the normal subgroup test (Theorem 2.3.13) this is enough to show $H \lhd G$. □

2.4.3 THEOREM Let H be a normal subgroup of G. Then the set of cosets of H in G forms a group under the operation $(aH)(bH) = (ab)H$.

Proof The indicated operation is well defined by the preceding lemma. We prove the four group axioms for this operation.
(Closure) It is immediate from the definition that the product $(aH)(bH)$ of two cosets is again a coset $(ab)H$.
(Associativity) $aH(bHcH) = aH(bc)H = (a(bc))H = ((ab)c)H = (ab)HcH = (aHbH)cH$.
(Identity) For all cosets aH we have $aHH = aHeH = (ae)H = aH$, and similarly $HaH = aH$, so $eH = H$ is the identity element.
(Inverses) For all cosets aH we have $aHa^{-1}H = (aa^{-1})H = eH = H$, and similarly $a^{-1}HaH = H$, so $a^{-1}H$ is the inverse of aH. □

2.4.4 DEFINITION Let H be a normal subgroup of G. Then the group consisting of the cosets of H in G under the operation $(aH)(bH) = (ab)H$ is called the **quotient group** of G by H, written G/H. ○

2.4.5 EXAMPLE $\mathbb{Z}/5\mathbb{Z} \cong \mathbb{Z}_5$, as we saw in Example 2.4.1. Similarly, in general, $\mathbb{Z}/n\mathbb{Z} \cong \mathbb{Z}_n$. ◇

2.4.6 EXAMPLE In $\mathbb{Z}_6$ consider the subgroup $\langle 3\rangle$. Then the cosets are $\{\langle 3\rangle, 1 + \langle 3\rangle, 2 + \langle 3\rangle\}$, and $\mathbb{Z}_6/\langle 3\rangle$ is a group of order 3. Any such group is isomorphic to $\mathbb{Z}_3$. (See Example 1.1.27.) So we obtain $\mathbb{Z}_6/\langle 3\rangle \cong \mathbb{Z}_3$. Note that $1 + \langle 3\rangle$ generates the quotient group. ◇

2.4.7 EXAMPLE In $\mathbb{Z}_{12}$ consider the subgroup $\langle 8\rangle = \{0, 4, 8\}$. The order of the quotient group $|\mathbb{Z}_{12}/\langle 8\rangle|$ is the number of cosets of $\langle 8\rangle$ or the index of $\langle 8\rangle$ in $\mathbb{Z}_{12}$, which is $[\mathbb{Z}_{12}:\langle 8\rangle] = |\mathbb{Z}_{12}|/|\langle 8\rangle| = 12/3 = 4$. Every group of order 4 is isomorphic either to $\mathbb{Z}_4$ or to the Klein 4-group V. (See Exercise 19 in Section 1.1.) Let us find the order of the element $1 + \langle 8\rangle$ in $\mathbb{Z}_{12}/\langle 8\rangle$. We have $2(1 + \langle 8\rangle) = 2 + \langle 8\rangle$, $3(1 + \langle 8\rangle) = 3 + \langle 8\rangle$, and $4(1 + \langle 8\rangle) = 4 + \langle 8\rangle = \langle 8\rangle$, so the order $|1 + \langle 8\rangle| = 4$. So $|1 + \langle 8\rangle| = |\mathbb{Z}_{12}/\langle 8\rangle|$, and $\mathbb{Z}_{12}/\langle 8\rangle$ is a cyclic group of order 4, and hence is isomorphic to $\mathbb{Z}_4$. ◇

2.4.8 EXAMPLE In D_4 consider the subgroup $\langle \rho^2\rangle = \{\rho_0, \rho^2\}$. Since $\rho^2 \in Z(D_4)$, the center of D_4, $\langle \rho^2\rangle$ is a normal subgroup. The index $[D_4:\langle \rho^2\rangle] = 8/2 = 4$, so the quotient group $D_4/\langle \rho^2\rangle$ has order $[D_4:\langle \rho^2\rangle] = 4$. Let us see to which of the two groups of order 4, $\mathbb{Z}_4$ or the Klein 4-group V, the quotient group is isomorphic. $D_4/\langle \rho^2\rangle = \{\langle \rho^2\rangle, \rho\langle \rho^2\rangle, \tau\langle \rho^2\rangle, \rho\tau\langle \rho^2\rangle\}$, and all the nonidentity elements have order 2. So $D_4/\langle \rho^2\rangle \cong V$. ◇

These last two examples point to several important facts about quotient groups, which we state in the next proposition.

2.4.9 PROPOSITION Let H be a normal subgroup of a group G. Then

(1) The order of an element aH in the quotient group G/H is the least positive integer k such that $a^k \in H$.

(2) If G is finite, then $|G/H| = |G|/|H|$.

(3) If G is an Abelian group, then G/H is an Abelian group.

(4) If G is a cyclic group, then G/H is a cyclic group.

Proofs (1) Let $aH \in G/H$. Since H is the identity element in G/H, the order $|aH|$ of aH is the least positive integer k such that $(aH)^k = H$. But $(aH)^k = a^kH$, so the order is the least positive integer such that $a^kH = H$. And $a^kH = H$ if and only if $a^k \in H$, completing the proof.

(2) through (4) are left as exercises. □

The converses of (3) and (4) of the preceding proposition are far from true, as can be seen from the next example.

2.4.10 EXAMPLE Consider the subgroup A_3 in S_3. We saw in Example 2.3.1 that the index $[S_3:A_3] = 2$. Hence S_3/A_3 is a group of order 2 and therefore is isomorphic to $\mathbb{Z}_2$, which is Abelian and even cyclic. But S_3 is neither. ◇

Let ϕ: $\mathbb{Z} \rightarrow \mathbb{Z}_5$ be the homomorphism with $\phi(n)$ = the remainder of n mod 5. Then Kern $\phi = 5\mathbb{Z}$. We constructed the quotient group $\mathbb{Z}/5\mathbb{Z}$ in Example 2.4.1 and saw that it was isomorphic to $\mathbb{Z}_5$. It is this isomorphism we would like to understand better. It is a link showing the relationships among a homomorphism, its image, its kernel, and the quotient group.

If ϕ: $G \rightarrow G'$ is any map, then for any subset $X \subseteq G$, we denote by $\phi(X)$ the image of X under ϕ. That is, $\phi(X) = \{x' \in G' \mid x' = \phi(x)$ for some $x \in X\}$. Similarly, for any subset $Y \subseteq G'$, we denote by $\phi^{-1}(Y)$ the preimage of Y. That is, $\phi^{-1}(Y) = \{x \in G \mid \phi(x) \in Y\}$.

2.4.11 EXAMPLE We pick an element of $\mathbb{Z}$, let us say 12, and apply the homomorphism ϕ: $\mathbb{Z} \rightarrow \mathbb{Z}_5$ as in Example 2.4.1, and then calculate ϕ^{-1}. We have $\phi(12) = 2$ and $\phi^{-1}(\phi(2)) = 2 + 5\mathbb{Z}$. This is the coset of the kernel $5\mathbb{Z}$ of ϕ to which the original element 12 belongs: $12 + 5\mathbb{Z} = 2 + 5\mathbb{Z}$. This relationship holds for other elements of $\mathbb{Z}$. Note in particular that $\phi^{-1}(\phi(0)) = 5\mathbb{Z}$ = Kern ϕ. ◇

2.4.12 PROPOSITION Let ϕ: $G \rightarrow G'$ be a homomorphism with Kern $\phi = K$. Then for any $g \in G$ we have $\phi^{-1}(\phi(g)) = gK$.

Proof Since for any $y \in G'$ we have $\phi^{-1}(y) = \{x \in G \mid \phi(x) = y\}$, it follows that $x \in \phi^{-1}(\phi(g))$ if and only if $\phi(x) = \phi(g)$. This condition is equivalent to the condition that $\phi(g)^{-1}\phi(x) = e'$, the identity of G'. Since $\phi(g)^{-1}\phi(x) = \phi(g^{-1}x)$, it follows that $x \in \phi^{-1}(\phi(g))$ if and only if $\phi(g^{-1}x) = e'$ or, in other words, if and only if $g^{-1}x \in$ Kern ϕ $= K$. This condition is equivalent to the condition that $x \in gK$. □

2.4.13 COROLLARY Let $\phi: G \to G'$ be a homomorphism. Then $\phi^{-1}(\phi(e)) = \text{Kern } \phi$.

Proof: This is immediate from the preceding proposition, since $eK = K$. □

2.4.14 DEFINITION Let $\phi: G \to G'$ be a homomorphism. The image $\phi(G) = \{\phi(x) \mid x \in G\}$ is always a subgroup of G' by Proposition 2.2.15, part (5), and will be equal to G' itself if and only if ϕ is onto. In this case, G' is said to be a **homomorphic image** of G. ○

The next theorem shows that there is a correspondence between normal subgroups and homomorphic images of a group.

2.4.15 THEOREM (First isomorphism theorem) For any groups G and G', let $\phi: G \to G'$ be a homomorphism with Kern $\phi = K$. Then

$$G/K \cong \phi(G)$$

Proof We follow our four basic steps for proving two groups are isomorphic, constructing a map and then showing it is a homomorphism, is one to one, and is onto.

(1) We construct a map $\chi: G/K \to \phi(G)$ as follows. Any element of G/K is a coset gK for some $g \in G$, and every element y of $\phi(G)$ is of the form $\phi(g)$ for some $g \in G$. So the natural way to define a map would be to let $\chi(gK) = \phi(g)$. We need to check that this map is well defined or, in other words, that if $g_1K = g_2K$, then $\phi(g_1) = \phi(g_2)$. To see this, note that if $g_1K = g_2K$, then $g_2^{-1}g_1 \in K$, which implies $\phi(g_2)^{-1}\phi(g_1) = \phi(g_2^{-1}g_1) = e'$, and so $\phi(g_1) = \phi(g_2)$.

(2) Let g_1K, g_2K be two elements of G/K. Then $\chi(g_1Kg_2K) = \chi(g_1g_2K) = \phi(g_1g_2) = \phi(g_1)\phi(g_2) = \chi(g_1K)\chi(g_2K)$. This shows χ is a homomorphism.

(3) Let g_1K, g_2K be two elements of G/K and suppose $\chi(g_1K) = \chi(g_2K)$. Since $\chi(g_1K) = \phi(g_1)$ and $\chi(g_2K) = \phi(g_2)$, we have $\phi(g_1) = \phi(g_2)$, and hence $g_1K = \phi^{-1}(\phi(g_1)) = \phi^{-1}(\phi(g_2)) = g_2K$, using Proposition 2.4.12. This shows χ is one to one.

(4) Let y be any element of $\phi(G)$. Then $y = \phi(x)$ for some $x \in G$, and since by definition $\chi(xK) = \phi(x)$, we have $y = \chi(xK)$. This shows χ is onto. □

The implications of this theorem are manifold. For instance, if we want to look for possible homomorphisms from a group G to another group G', we can look at the normal subgroups K of G and decide for each whether G/K is isomorphic to a subgroup of G'.

2.4.16 EXAMPLE Let us find all possible nontrivial homomorphisms $\phi: S_3 \to \mathbb{Z}_4$. The condition that ϕ be nontrivial means that we want Kern $\phi = K$ to be a proper subgroup of S_3. The only proper normal subgroups of S_3 are $\{e\}$ and A_3. But $K = \{e\}$ is impossible since then we would have $S_3/K \cong S_3$ and the fact that $S_3/K \cong \phi(S_3)$ would imply that $|\phi(S_3)| = 6$, which cannot be, since $\phi(S_3) \subseteq \mathbb{Z}_4$. So the only possibility is $K = A_3$. In that case, $\phi(S_3) \cong S_3/A_3$, which is a group of order 2. $\mathbb{Z}_4$ is cyclic, so it has a unique subgroup of order 2, namely, $\langle 2\rangle = \{0, 2\}$, and the map ϕ given by

$$\phi(\sigma) = \begin{cases} 0 \text{ if } \sigma \in A_3 \\ 2 \text{ if } \sigma \notin A_3 \end{cases}$$

is a homomorphism.◇

2.4.17 EXAMPLE Let us look for nontrivial homomorphisms $\phi: S_3 \to \mathbb{Z}_3$. The argument of the previous example shows that the only possibility would be to have $\phi(S_3)$ a subgroup of $\mathbb{Z}_3$ of order 2. But $\mathbb{Z}_3$ has no such subgroups, so there can be no such homomorphism. ◇

This last example illustrates how in the case of finite groups we can obtain useful consequences from the first isomorphism theorem.

2.4.18 PROPOSITION Let G and G' be finite groups, and let $\phi: G \to G'$ be a homomorphism. Then $|\phi(G)|$ divides both $|G|$ and $|G'|$.

Proof $\phi(G)$ is a subgroup of G', hence $|\phi(G)|$ divides $|G'|$ by Lagrange's theorem. $|\phi(G)| = |G/\text{Kern }\phi|$ by the first isomorphism theorem, and so $|\phi(G)|$ is equal to the index $[G:\text{Kern }\phi]$ and also divides $|G|$. □

Given a group G and a homomorphism $\phi: G \to G'$ to another group G' we obtain a normal subgroup K of G, namely $K = \text{Kern }\phi$. The next theorem shows that in a sense the converse is true.

2.4.19 THEOREM Given a group G and a normal subgroup K, there exists an onto homomorphism $\phi: G \to G/K$ with $\text{Kern }\phi = K$.

Proof We define ϕ by letting $\phi(g) = gK \in G/K$ for any $g \in G$. Since $K \lhd G$, by Theorem 2.4.3 G/K is a group under the operation $(g_1K)(g_2K) = g_1g_2K$. Then ϕ is a homomorphism because $\phi(g_1g_2) = g_1g_2K = (g_1K)(g_2K) = \phi(g_1)\phi(g_2)$. In the group G/K the identity element is K. So we have $x \in \text{Kern }\phi$ if and only if $\phi(x) = K$, and since $\phi(x) = xK$, we have $x \in \text{Kern }\phi$ if and only if $xK = K$, which is equivalent to $x \in K$. Thus $\text{Kern }\phi = K$. Finally, ϕ is onto since every element of G/K is of the form gK for some $g \in G$. □

This last theorem enables us to factor any homomorphism into two steps. Given a homomorphism $\phi: G \to G'$ with $\text{Kern }\phi = K$, let $\tau: G \to G/K$ be the homomorphism of Theorem 2.4.19, and let $\chi: G/K \to G'$ be the homomorphism in the proof of Theorem 2.4.15. Then since for any $g \in G$ we have $\chi(\tau(g)) = \chi(gK) = \phi(g)$, we have $\phi = \chi \circ \tau$, or as we say, the diagram in Figure 1 **commutes**.

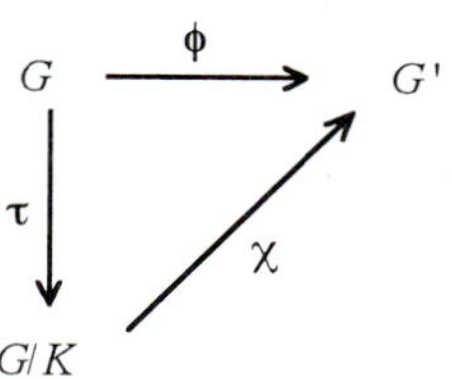

FIGURE 1

We end this chapter with an application to Abelian groups. As we will see in Chapter 4, if G is a finite group and p is a prime dividing the order of G, then G has an element of order p (Cauchy's theorem). We are now in a position to give a simple proof of this theorem for Abelian groups.

2.4.20 THEOREM (Cauchy's theorem for Abelian groups) Let G be a finite Abelian group and p a prime dividing the order of G. Then G has an element of order p.

Proof We will use induction on the order $|G|$ of G. If $|G| = 1$ there is nothing to prove, and if $|G| = 2$ or 3, then G is cyclic, and the statement of the theorem is true. So assume the statement is true for all Abelian groups of order less than $|G|$. If G has no nontrivial proper subgroups, then G is cyclic and the statement of the theorem again is true, by Theorem 1.3.26. So let H be a nontrivial proper subgroup of G. If p divides $|H|$, then since H is an Abelian group with order $|H| < |G|$, by our induction hypothesis there exists an element $a \in H \subseteq G$ of order p, and we are done. So assume p does not divide $|H|$. Since G is Abelian, $H \lhd G$, and G/H is an Abelian group with order $|G/H| = |G|/|H|$. Now $|G|/|H| < |G|$ since H is nontrivial, and p divides $|G|/|H|$ since p divides $|G|$ but does not divide $|H|$. So by our induction hypothesis there exists an element $X \in G/H$ of order p. X is of the form bH for some $b \in G$, where $bH \neq H$ but $(bH)^p = H$. Thus $b \notin H$, but since $b^pH = (bH)^p = H$, we have $b^p \in H$. Let $c = b^{|H|} \in G$. Then $c^p = b^{|H|p} = (b^p)^{|H|} = e$ since $b^p \in H$. So it only remains to show that $c \neq e$. But if $c = e$, then $e = b^{|H|}$ and hence $H = b^{|H|}H = (bH)^{|H|}$ and p must divide $|H|$ by Corollary 1.3.12, contradicting our assumption. Hence $c \neq e$ and $|c| = p$, completing the proof. ◇

Exercises 2.4

In Exercises 1 through 6, all the quotient groups are cyclic and therefore isomorphic to $\mathbb{Z}_n$ for some n. In each case, find this n.

1. $\mathbb{Z}_6/\langle 2\rangle$ **2.** $\mathbb{Z}_{12}/\langle 8\rangle$ **3.** $\mathbb{Z}_{15}/\langle 6\rangle$ **4.** S_4/A_4 **5.** $D_4/\langle \rho\rangle$ **6.** $Q_8/\langle \mathbf{j}\rangle$

In Exercises 7 through 11, find the order of the indicated element in the indicated quotient group.

7. $3+\langle 8\rangle$ in $\mathbb{Z}_{12}/\langle 8\rangle$ **8.** $3+\langle 6\rangle$ in $\mathbb{Z}_{15}/\langle 6\rangle$ **9.** $2+\langle 6\rangle$ in $\mathbb{Z}_{15}/\langle 6\rangle$

10. $\mathbf{i}\langle \mathbf{j}\rangle$ in $Q_8/\langle \mathbf{j}\rangle$ **11.** $\rho\langle \rho^2\rangle$ in $D_4/\langle \rho^2\rangle$

12. Prove part (2) of Proposition 2.4.9. **13.** Prove part (3) of Proposition 2.4.9.

14. Prove part (4) of Proposition 2.4.9.

In Exercises 15 through 18, find all possible nontrivial homomorphisms between the indicated groups.

15. $\phi\colon S_3 \to \mathbb{Z}_6$ **16.** $\phi\colon D_4 \to \mathbb{Z}_4$ **17.** $\phi\colon \mathbb{Z}_5 \to \mathbb{Z}_{10}$ **18.** $\phi\colon \mathbb{Z}_{10} \to \mathbb{Z}_5$

19. Let $\phi\colon G \to G'$ be an onto homomorphism with Kern $\phi = K$, and let H' be a subgroup of G'. Show that there exists a subgroup H of G such that $K \subseteq H$ and $H/K \cong H'$.

20. Let H and K be subgroups of a group G such that $H \lhd G$, $K \lhd G$, $[G{:}H] = 5$, and $[G{:}K] = 3$. Show that for all $g \in G$ we have $g^{15} \in H \cap K$.

21. Consider the dihedral group $D_6 = \{\rho^i\tau^j \mid 0 \le i < 6,\ 0 \le j < 2\}$, where $\rho^6 = \tau^2 =$ identity, and $\rho\tau = \tau\rho^{-1}$. Show that

(a) $\langle \rho^3\rangle \lhd D_6$ (b) $D_6/\langle \rho^3\rangle \cong S_3$

22. Consider the dihedral group $D_n = \{\rho^i\tau^j \mid 0 \le i < n,\ 0 \le j < 2\}$, where $\rho^n = \tau^2 =$ identity, and $\rho\tau = \tau\rho^{-1}$. For any divisor k of n show that

(a) $\langle \rho^k\rangle \lhd D_n$ (b) $D_n/\langle \rho^k\rangle \cong D_k$

23. Let $Z(G)$ be the center of a group G. Show that

(a) $Z(G) \lhd G$ (b) If $G/Z(G)$ is cyclic, then G is Abelian.

24. Let $Z(G)$ be the center of a group G. Show that if the index $[G{:}Z(G)] = p$, a prime, then G is Abelian.

25. Let G be a group and S a subset of G. Define $\langle S\rangle$ to be the smallest subgroup of G containing S, called the subgroup of G **generated** by S. Show that $\langle S\rangle$ exists.

26. Show that in S_4 the subgroup generated by $\{(1\ 2), (1\ 2\ 3\ 4)\}$ (in the sense of the preceding Exercise 25) is the whole group: $\langle (1\ 2), (1\ 2\ 3\ 4)\rangle = S_4$.

27. Let G be a group and let $S = \{xyx^{-1}y^{-1} \mid x, y \in G\}$. Let N be the subgroup $\langle S \rangle$ generated by S (in the sense of Exercise 25), called the **commutator** subgroup of G. Show that
(a) $N \triangleleft G$ (b) G/N is Abelian.
(c) If H is a normal subgroup of G and G/H is Abelian, then $N \subseteq H$.
(d) If H is a subgroup of G with $N \subseteq H$, then $H \triangleleft G$.

28. Find the commutator subgroup (in the sense of Exercise 27) of S_3.

29. Find the commutator subgroup (in the sense of Exercise 27) of D_4.

Cayley Digraphs

Let G be a finite group and S a subset of G that generates G. A set of equations satisfied by the generators that determine completely the operation table of G is called a set of **defining relations**.
For example, D_4 is generated by $S = \{\rho, \tau\}$ with defining relations $\rho^4 = \tau^2 = \rho_0$ and $\rho\tau = \tau\rho^{-1}$ and the quaternion group $Q_8 = \{\pm\mathbf{1}, \pm\mathbf{i}, \pm\mathbf{j}, \pm\mathbf{ij}\}$ is generated by $S = \{\mathbf{i}, \mathbf{j}\}$ with definitng relations $\mathbf{i}^4 = 1$, $\mathbf{i}^2 = \mathbf{j}^2$, and $\mathbf{ij} = -\mathbf{ji}$.
Given a set S that generates a finite group G, we can construct a **directed graph** or **Cayley digraph** of G with respect to S as follows:

(1) Each element of G is represented by a dot.
(2) Each element of S is represented by a directed edge.
(3) If $c \in S$ is represented by the directed edge $\rightarrow$, then for $a, b \in G$, $a \bullet \rightarrow \bullet\, b$ means that $ac = b$ in G.
(4) If $c \in S$ with $c^{-1} = c$, then the arrow is omitted from the edge that represents c.

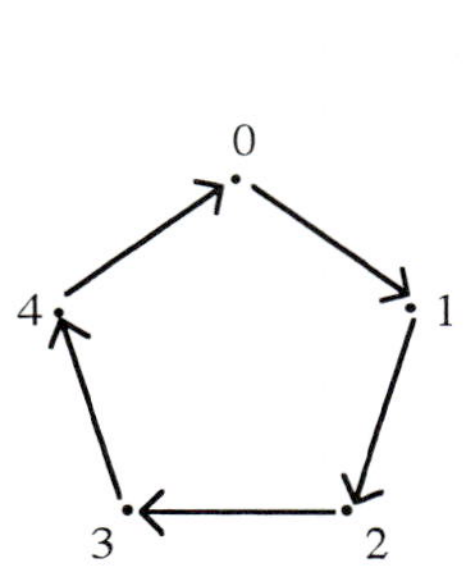

FIGURE 2

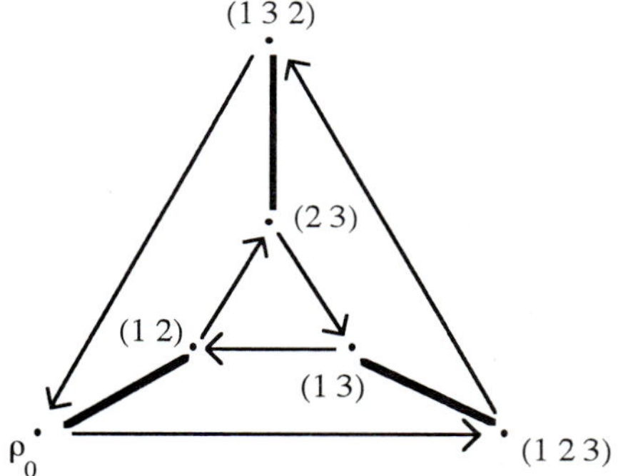

FIGURE 3

For example, a Cayley digraph of $\mathbb{Z}_5$ with $S = \{1\}$ can be drawn as in Figure 2, with $\rightarrow$ representing 1. A Cayley digraph of S_3 with $S = \{(1\ 2), (1\ 2\ 3)\}$ can be drawn as in Figure 3, with $\rightarrow$ representing $(1\ 2\ 3)$ and **—** representing $(1\ 2)$. Observe that from the diagram we can determine that S_3 is not Abelian.

30. Show that every Cayley digraph of a group must satisfy the following four conditions:

(1) For every pair of vertices x and y, there is a **path**, that is, a sequence of consecutive edges, that starts from x and ends at y.

(2) At most one edge goes from a vertex x to a vertex y.

(3) At each vertex x there is exactly one edge of each type that starts at x and exactly one edge of each type that ends at x.

(4) If two different paths start from a vertex x and both end at a vertex y, then the same two paths starting from any vertex z will end at the same vertex w.

In Exercises 31 through 35, construct Cayley digraphs of the indicated group G with the indicated generating set S, and specify the defining relations.

31. $G = \mathbb{Z}_6$, $S = \{1\}$

32. $G = \mathbb{Z}_6$ $S = \{2, 3\}$

33. $G = S_3$ $S = \{(1\ 2), (2\ 3)\}$

34. $G = D_4$ $S = \{\rho, \tau\}$

35. $G = A_4$ $S = \{(1\ 2\ 3), (1\ 2)(3\ 4)\}$

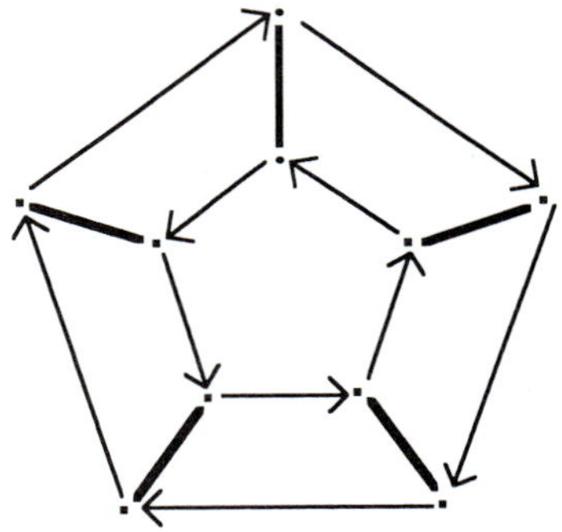

FIGURE 4

36. Identify the group, and its generating set and defining relations, that is represented by the Cayley digraph in Figure 4.

2.5 Automorphisms

We have studied isomorphisms between one group and another. In this section we consider isomorphisms between a group and itself. One of the first things we see is that the set of such isomorphisms itself forms a group in a natural way.

2.5.1 EXAMPLE Let us find all possible isomorphisms $\phi: \mathbb{Z}_6 \to \mathbb{Z}_6$. $\mathbb{Z}_6$ is cyclic and 1 is a generator, that is, $\mathbb{Z}_6 = \langle 1 \rangle$. By the claim in the proof of Proposition 2.2.22, if ϕ is an isomorphism, then $|\phi(a)| = |a|$ for all a. So if ϕ is an isomorphism, we must have $|\phi(1)| = |1| = 6$, so $\phi(1)$ must be a generator of $\mathbb{Z}_6$, and therefore $\phi(1) = 1$ or $\phi(1) = 5$. Also, once $\phi(1)$ is known, ϕ is completely determined, since we must have $\phi(2) = 2\phi(1)$, $\phi(3) = 3\phi(1)$, and so on, and these maps are isomorphisms by the claim in the proof of Lemma 2.2.23. So there exactly two isomorphisms. Let ϕ_0 be the identity mapping, with $\phi_0(n) = n$ for all $n \in \mathbb{Z}_6$, and let ϕ_1 be the isomorphism with $\phi_1(n) = 5n \bmod 6$ for all $n \in \mathbb{Z}_6$. Consider now the operation of composition of functions on the set $\{\phi_0, \phi_1\}$. Clearly $\phi_0 \circ \phi_0 = \phi_0$ and $\phi_0 \circ \phi_1 = \phi_1 \circ \phi_0 = \phi_1$. What about $\phi_1 \circ \phi_1$? We have $\phi_1 \circ \phi_1(1) = \phi_1(\phi_1(1)) = \phi_1(5) = 5 \cdot 5 = 25 = 1 \bmod 6$. So $\phi_1 \circ \phi_1 = \phi_0$. The set $\{\phi_0, \phi_1\}$ under the operation of composition thus forms a cyclic group of order 2. ◇

2.5.2 DEFINITION Let G be a group. An isomorphism $\phi: G \to G$ is called an **automorphism** of G, and the set of all automorphisms of G is denoted by Aut(G). ○

We show that what we found in Example 2.5.1 is always the case: The automorphisms of a group always form a group.

2.5.3 THEOREM Let G be a group. Then Aut(G) forms a group under composition of functions.

Proof We prove that the four group axioms hold.
(Closure) Let $\phi_1, \phi_2 \in \text{Aut}(G)$ and consider $\phi_1 \circ \phi_2$. We already know from Theorem 0.1.15 that a composition of one-to-one functions is one to one and a composition of onto functions is onto, so to show $\phi_1 \circ \phi_2 \in \text{Aut}(G)$ it only remains to check that it is a homomorphism, but this follows from Proposition 2.2.10.
(Associativity) We know from Theorem 0.1.15 that composition of functions is associative.
(Identity) Let ϕ_0 be the identity function on G, the mapping defined by letting $\phi_0(a) = a$ for all $a \in G$. Then ϕ_0 acts as an identity element in Aut(G).
(Inverses) For $\phi \in \text{Aut}(G)$, we know from Theorem 0.1.24 that the inverse function $\phi^{-1}: G \to G$ exists and is one to one and onto and satisfies $\phi \circ \phi^{-1} = \phi_0$. So to show that ϕ^{-1} acts as an inverse element to ϕ in Aut(G) it only remains to check that ϕ^{-1} is a homomorphism. So let $a, b \in G$, and let $c = \phi^{-1}(a)$, $d = \phi^{-1}(b)$. We have $\phi(c) = a$, $\phi(d) = b$, and since ϕ is a homomorphism, $\phi(cd) = \phi(c)\phi(d) = ab$, from which it

follows that $\phi^{-1}(ab) = cd = \phi^{-1}(a)\phi^{-1}(b)$, as required to show ϕ^{-1} is a homomorphism and complete the proof. □

2.5.4 EXAMPLE Let us determine Aut($\mathbb{Z}_8$). Since $\mathbb{Z}_8$ is cyclic with generator 1, if ϕ is any automorphism we must have $|\phi(1)| = |1| = 8$, and $\phi(1)$ must also be a generator of $\mathbb{Z}_8$. That is, we must have $\phi(1) = 1$ or 3 or 5 or 7. Also, once $\phi(1)$ is known, ϕ is completely determined, since in general $\phi(n) = n\phi(1)$, and these maps are isomorphisms. So there are exactly four automorphisms: the identity $\phi_1(n) = n$, $\phi_3(n) \equiv 3n \bmod 8$, $\phi_5(n) \equiv 5n \bmod 8$, and $\phi_7(n) \equiv 7n \bmod 8$. We have, for instance, $\phi_3 \circ \phi_5(n) = \phi_3(\phi_5(n)) = \phi_3(5n) \equiv 15n \equiv 7n \bmod 8$, so $\phi_3 \circ \phi_5 = \phi_7$ in Aut($\mathbb{Z}_8$). It can be checked that, in general, if $ij = k \bmod 8$, then $\phi_i \circ \phi_j = \phi_k$ in Aut($\mathbb{Z}_8$). (See Exercise 1 at the end of this section.) It follows that the mapping $T(\phi_i) = i$ gives an isomorphism between Aut($\mathbb{Z}_8$) and the multiplicative group $U(8)$. (See Exercise 2.) ◇

The last example can be generalized for any cyclic group G, as we show in the next theorem.

2.5.5 THEOREM Let G be a cyclic group of order n. Then Aut(G) $\cong U(n)$.

Proof We follow our four basic steps for proving two groups are isomorphic.

(1) First we define a map T: Aut(G) $\to U(n)$ as follows. Let $G = \langle a \rangle$, where $|a| = n$. Consider $\phi \in$ Aut(G). For any $g \in G$, we have $g = a^i$ for some integer i, $0 \leq i < n$, and then $\phi(g) = \phi(a^i) = \phi(a)^i$, so ϕ is completely determined once $\phi(a)$ is fixed. We must have $|\phi(a)| = |a| = n$, hence $\phi(a)$ is also a generator of G, and therefore $\phi(a) = a^r$ for some r with gcd(n, r) = 1, by Corollary 1.3.20, and by Theorem 1.3.11, $a^r = a^s$, where s is the remainder of r mod n. Thus there is an $s \in \{q \mid \gcd(n, q) = 1, 0 \leq q < n\} = U(n)$ such that $\phi(a) = a^s$. We let $T(\phi) = s$.

(2) Next we show that T is a homomorphism. If ϕ, $\psi \in$ Aut(G) with $\phi(a) = a^s$, $\psi(a) = a^r$, where r, $s \in U(n)$, then $\psi \circ \phi(a) = \psi(\phi(a)) = \psi(a^s) = a^{st} = a^u$, where $u = st \bmod n$. It follows that $T(\psi \circ \phi) = u = T(\psi)T(\phi) \bmod n$.

(3) T is one to one, because if $T(\phi) = T(\psi)$, then $\phi(a) = a^{T(\phi)} = a^{T(\psi)} = \psi(a)$, and hence $\phi = \psi$.

(4) T is onto, because if $s \in U(n)$, then a^s is a generator of G, and the map ϕ defined by $\phi(a^i) = a^{si}$ is an isomorphism by the claim in the proof of Lemma 2.2.24. □

For a cyclic group G, we now know what Aut(G) looks like. For noncyclic Abelian groups the situation is more complicated, though we do in this case have one nontrivial example of an automorphism, the mapping $\phi(x) = x^{-1}$. (See Exercise 4 at the end of this section.) For non-Abelian groups, the next proposition shows how to construct many examples of automorphisms.

2.5.6 PROPOSITION Let G be any group, $g \in G$, and T_g: $G \to G$ the map defined by letting $T_g(x) = gxg^{-1}$ for any $x \in G$. Then $T_g \in \text{Aut}(G)$.

Proof T_g is a homomorphism since for all x, $y \in G$ we have $T_g(xy) = gxyg^{-1} = (gxg^{-1})(gyg^{-1}) = T_g(x)T_g(y)$. T_g is one to one because $T_g(x) = T_g(y)$ means $gxg^{-1} = gyg^{-1}$, which implies $x = y$ by cancellation. T_g is onto because for any $x \in G$, setting $y = g^{-1}xg$ we have $T_g(y) = gyg^{-1} = gg^{-1}xgg^{-1} = x$. □

2.5.7 DEFINITION Let G be a group and $g \in G$. Then the automorphism T_g defined by $T_g(x) = gxg^{-1}$ is called an **inner automorphism**. The set of all inner automorphisms of G is denoted Inn(G). ○

2.5.8 PROPOSITION Let G be a group. Then Inn(G) forms a subgroup of Aut(G).

Proof The inner automorphism T_e is the identity element, since $T_e(x) = exe = x$. Given two inner automorphism T_h and T_g, their composition is the inner automorphism T_{hg}, since we have $T_{hg}(x) = hgx(hg)^{-1} = hgxg^{-1}h^{-1} = T_h(gxg^{-1}) = T_h(T_g(x)) = T_h \circ T_g(x)$. Given an inner automorphism T_g, its inverse is the inner automorphism $T_{g^{-1}}$ since $T_{g^{-1}} \circ T_g(x) = T_{g^{-1}}(T_g(x)) = T_{g^{-1}}(gxg^{-1}) = g^{-1}gxg^{-1}g = x$. □

2.5.9 EXAMPLE Let us determine Inn(D_4). First notice that if $g \in Z(D_4)$, the center of D_4, then T_g is the identity, since we have $T_g(x) = gxg^{-1} = xgg^{-1} = x$ for all $x \in D_4$. We know that the center is $Z(D_4) = \{\rho_0, \rho^2\}$. If g is any element of D_4, we have $T_g = T_{g\rho^2}$, since we have $T_{g\rho^2} = T_g \circ T_{\rho^2}$ and we have just established that T_{ρ^2} is the identity. We can calculate

$$T_\rho(\rho^i\tau) = \rho(\rho^i\tau)\rho^{-1} = \rho\rho^i(\tau\rho^{-1}) = \rho\rho^i(\rho\tau) = \rho^{i+2}\tau$$

$$T_\tau(\rho^i\tau) = \tau(\rho^i\tau)\tau^{-1} = \tau\rho^i = \rho^{-i}\tau$$

$$T_{\rho\tau}(\rho^i\tau) = \rho\tau(\rho^i\tau)\tau^{-1}\rho^{-1} = \rho\tau\rho^{i-1} = \rho^{-i+2}\tau = T_\rho(\rho^{-i}\tau) = T_\rho(T_\tau(\rho^i\tau))$$

If we call the identity mapping T_0, then T_0, T_ρ, T_τ, $T_{\rho\tau}$ are all the inner automorphisms of D_4. Note that $D_4/Z(D_4) = \{Z(D_4), \rho Z(D_4), \tau Z(D_4), \rho\tau Z(D_4)\}$ and the inner automorphisms correspond to the cosets of the center. The correspondence is in fact an isomorphism. ◇

The relationship between Inn(G) and $Z(G)$ noted in the preceding example holds generally, for any group.

2.5.10 THEOREM For any group G we have $\text{Inn}(G) \cong G/Z(G)$, where $Z(G)$ is the center of G.

Proof Let χ: $G \to \text{Inn}(G)$ be the map defined by letting $\chi(g) = T_g \in \text{Inn}(G)$ for all $g \in G$. By the first isomorphism theorem (Theorem 2.4.15), it suffices to show that χ is an onto homomorphism with Kern $\chi = Z(G)$. χ is a homomorphism because, as we saw in the proof of Proposition 2.5.8, $\chi(hg) = T_{hg} = T_h \circ T_g = \chi(h)\chi(g)$ for any $g, h \in G$. Furthermore, χ is onto by the definition of inner automorphism. And finally $g \in \text{Kern}\,\chi$ if and only if $\chi(g) = T_g$ is the identity map, and this condition is equivalent to the condition that $gxg^{-1} = x$ or $gx = xg$ for all $x \in G$. Thus $g \in \text{Kern}\,\phi$ if and only if g commutes with every element $x \in G$ or, in other words, if and only if $g \in Z(G)$, the center of G. □

Exercises 2.5

1. Let $\phi_1, \phi_3, \phi_5, \phi_7$ be the automorphisms of $\mathbb{Z}_8$, as in Example 2.5.4. Show that if $ij = k \bmod 8$, then $\phi_i \circ \phi_j = \phi_k$.

2. With the notation as in the preceding exercise, show that the mapping $T\colon \text{Aut}(\mathbb{Z}_8) \to U(8)$ defined by $T(\phi_i) = i$ is an isomorphism.

3. Let $G = \langle a \rangle$ be a cyclic group of order 10. Describe explicitly the elements of Aut(G).

4. Let G be an Abelian group. Show that the mapping $\phi\colon G \to G$ defined by letting $\phi(x) = x^{-1}$ for all $x \in G$ is an automorphism of G.

5. Determine Aut($\mathbb{Z}$).

6. Show that the mapping $\phi\colon S_3 \to S_3$ defined by letting $\phi(x) = x^{-1}$ for all $x \in S_3$ is not an automorphism of S_3.

7. Let G be a group, $H \lhd G$, $\phi \in \text{Aut}(G)$. Show that $\phi(H) \lhd G$.

8. For any group G show that $\text{Inn}(G) \lhd \text{Aut}(G)$.

9. Show that $\text{Inn}(S_3) \cong S_3$.

10. For p a prime show that $\text{Aut}(\mathbb{Z}_p) \cong \mathbb{Z}_{p-1}$.

11. Let Q_8 be the quarternion group. Show that $\text{Inn}(Q_8) \cong V$, the Klein 4-group. [*Hint:* First determine $Z(Q_8)$.]

12. Show that $\text{Inn}(D_4) \cong V$, the Klein 4-group.

13. Show that $|\text{Aut}(D_4)| \leq 8$.

14. For V the Klein 4-group show that $\text{Aut}(V) \cong \text{GL}(2, \mathbb{Z}_2)$, the general linear group of 2×2 matrices with entries from $\mathbb{Z}_2$.

For Exercises 15 through 19, a subgroup H of a group G is called a **characteristic** subgroup of G if for all $\phi \in \text{Aut}(G)$ we have $\phi(H) = H$.

15. Show that if H is a characteristic subgroup of G, then $H \lhd G$.

16. Show that if H is the only subgroup of G of order n, then H is a characteristic subgroup of G.

17. Let G be a group, H a normal subgroup of G, and K a characteristic subgroup of H. Show that K is a normal subgroup of G.

18. Let G be a group, H a characteristic subgroup of G, and K a characteristic subgroup of H. Show that K is a characteristic subgroup of G.

19. For any group G, show that the commutator subgroup of G (as in Exercise 27 in Section 2.4) is a characteristic subgroup of G.

20. Show that $\text{Aut}(D_4) \cong D_4$.

21. Show that $\text{Aut}(Q_8) \cong S_4$.

22. Show that $\text{Aut}(S_3) \cong S_3$.

Chapter 3

Direct Products and Abelian Groups

In Chapter 1 we learned what groups are and studied specific examples of very important groups such as $\mathbb{Z}_n$, $U(n)$, S_n, A_n, D_n, V, and Q_8 among the finite groups, as well as the matrix groups GL(2, R) and SL(2, R). In Chapter 2 we studied the maps we can use between groups, the so-called group homomorphisms, after establishing Lagrange's theorem. We studied the role normal subgroups play and their relation with group homomorphisms and quotient groups. We have already established the basic notions we need for the study of groups. In this chapter we learn how to use the groups we already know in order to construct new groups. We learn to identify Abelian groups and cyclic groups among these new groups, and we apply to them the theory we learned in Chapter 2. A very important theorem is derived that will allow us to find all Abelian groups of a given finite order.

3.1 Examples and Definitions

We use the groups we have already studied as building blocks to construct new groups, and we study the properties the new groups may have inherited from the original groups. We begin with some examples.

3.1.1 EXAMPLE Consider the set $\mathbb{Z}_2 \times \mathbb{Z}_3 = \{(a, b) \mid a \in \mathbb{Z}_2,\ b \in \mathbb{Z}_3\}$. The elements (a, b) of this set are pairs with the first component $a = 0$ or 1 and the second component $b = 0$, 1, or 2. Hence $\mathbb{Z}_2 \times \mathbb{Z}_3$ has exactly six elements, namely $\{(0, 0), (0, 1), (0, 2), (1, 0), (1, 1), (1, 2)\}$. We make this set into a group by choosing wisely an operation. For any two elements (a, b) and (c, d) in $\mathbb{Z}_2 \times \mathbb{Z}_3$ we let $(a, b) + (c, d) = (a + c \bmod 2, b + d \bmod 3)$. For example, $(0, 2) + (1, 1) = (1, 0)$, and $(1, 1) + (1, 2) = (0, 0)$. Obviously, we have closure. $(0, 0)$ is the identity element, and the inverse element of (a, b) is $-(a, b) = (2 - a, 3 - b)$. ◇

3.1.2 EXAMPLE Consider $\mathbb{Z} \times \mathbb{Z} = \{(a, b) \mid a, b \in \mathbb{Z}\}$. In this case the set is infinite. We define again a componentwise operation, namely for (a, b) and (c, d) in

$\mathbb{Z} \times \mathbb{Z}$, let $(a, b) + (c, d) = (a + c, b + d)$. Again $(0, 0)$ is the identity and $-(a, b) = (-a, -b)$. ◇

To emphasize that the two groups we put together can be totally unrelated, we look at another example.

3.1.3 EXAMPLE Consider the set $\mathbb{Z}_2 \times S_3 = \{(a, \sigma) \mid a \in \mathbb{Z}_2, \sigma \in S_3\}$. The first components are $a = 0$ or 1, and the second components are $\sigma =$ any permutation in S_3. In this case the operation we use is described as follows: For any (a, σ) and (b, τ) in $\mathbb{Z}_2 \times S_3$, let $(a, \sigma) * (b, \tau) = (a + b \bmod 2, \sigma \circ \tau)$, where $\sigma \circ \tau$ is the composition of the permutations. For example, $(1, \rho) * (1, \mu_1) = (0, \rho\mu_1) = (0, \mu_3)$. The identity element is $(0, \rho_0)$, $(1, \rho)^{-1} = (1, \rho^2)$, $(1, \mu_i)^{-1} = (1, \mu_i)$. ◇

For any two groups G_1 and G_2 we show that the set of pairs of elements from G_1 and G_2 can be made into a group. It then follows that the examples we have just looked at are indeed groups, and we can imitate them to construct new groups.

3.1.4 THEOREM Let $\langle G_1, \circ\rangle$ and $\langle G_2, \lozenge\rangle$ be two groups, and let

$$G_1 \times G_2 = \{(a_1, a_2) \mid a_1 \in G_1, a_2 \in G_2\}$$

Define the componentwise operation $*$ on $G_1 \times G_2$ by

$$(a_1, a_2) * (b_1, b_2) = (a_1 \circ b_1, a_2 \lozenge b_2)$$

Then $\langle G, *\rangle$ is a group.

Proof (Closure) Given (a_1, a_2) and (b_1, b_2) in $G_1 \times G_2$, by closure for G_1 and for G_2 we have $a_1 \circ b_1 \in G_1$ and $a_2 \lozenge b_2 \in G_2$, so $(a_1 \circ b_1, a_2 \lozenge b_2) \in G_1 \times G_2$.
(Associativity) Given (a_1, a_2), (b_1, b_2) and (c_1, c_2) in $G_1 \times G_2$, using associativity for G_1 and for G_2 we have

$$[(a_1, a_2) * (b_1, b_2)] * (c_1, c_2) = (a_1 \circ b_1, a_2 \lozenge b_2) * (c_1, c_2) =$$
$$((a_1 \circ b_1) \circ c_1, (a_2 \lozenge b_2) \lozenge c_2) = (a_1 \circ (b_1 \circ c_1), a_2 \lozenge (b_2 \lozenge c_2)) =$$
$$(a_1, a_2) * ((b_1 \circ c_1), (b_2 \lozenge c_2)) = (a_1, a_2) * [(b_1, b_2) * (c_1, c_2)]$$

(Identity) Let e_1 and e_2 be the identity elements in G_1 and G_2, respectively. Then given any (a_1, a_2) in $G_1 \times G_2$, we easily obtain

$$(a_1, a_2) * (e_1, e_2) = (a_1 \circ e_1, a_2 \lozenge e_2) = (a_1, a_2)$$

and similarly $(e_1, e_2) * (a_1, a_2) = (a_1, a_2)$, so (e_1, e_2) is the identity in $G_1 \times G_2$.
(Inverses) Given any (a_1, a_2) in $G_1 \times G_2$, let a_1^{-1} be the inverse of a_1 in G_1, and let a_2^{-1} be the inverse of a_2 in G_2. Then we easily obtain

$$(a_1, a_2) * (a_1^{-1}, a_2^{-1}) = (a_1 \circ a_1^{-1}, a_2 \lozenge a_2^{-1}) = (e_1, e_2)$$

and similarly $(a_1^{-1}, a_2^{-1}) * (a_1, a_2) = (e_1, e_2)$. □.

3.1.5 DEFINITION Given two groups G_1 and G_2, the group $G_1 \times G_2$ with the operation defined in the preceding theorem is called the **direct product** of G_1 and G_2. ○

We can construct the direct product of more than two groups. If G_1, G_2, and G_3 are groups, then we know by Theorem 3.1.4 that $G_1 \times G_2$ is a group under the componentwise operation. Now we can use Theorem 3.1.4 again, applied to the two groups $G_1 \times G_2$ and G_3, to conclude that $(G_1 \times G_2) \times G_3$ is a group with the componentwise operation. We can continue the process for any finite number of groups, as the next proposition states.

3.1.6 PROPOSITION Let $G_1, G_2, \dots, G_n$ be groups. Then

$$G_1 \times G_2 \times \dots \times G_n = \{(a_1, a_2, \dots, a_n) \mid a_i \in G_i\}$$

is a group under the componentwise operation

$$(a_1, a_2, \dots, a_n) * (b_1, b_2, \dots, b_n) = (a_1b_1, a_2b_2, \dots, a_nb_n)$$

where a_ib_i is the operation performed in G_i.

Proof Left to the reader (as Exercise 11). □

Note that we are only dealing with the direct product of a finite number of groups. We can define a similar construction for an infinite number of groups, which is of course more complicated.

We now establish some basic properties of the direct product $G_1 \times G_2$.

3.1.7 PROPOSITION Let G_1 and G_2 be groups. Then $G_1 \times G_2 \cong G_2 \times G_1$.

Proof (1) Let ϕ: $G_1 \times G_2 \to G_2 \times G_1$ be defined by $\phi((a, b)) = (b, a)$.

(2) ϕ is a homomorphism since for any (a, b) and (c, d) in $G_1 \times G_2$, we have $\phi((a, b)(c, d)) = \phi((ac, bd)) = (bd, ac) = (b, a)(d, c) = \phi((a, b))\phi((c, d))$.

(3) ϕ is one to one since $(a, b) \in$ Kern ϕ if and only if $(b, a) = (e_2, e_1)$, hence Kern $\phi = \{(e_1, e_2)\}$.

(4) That ϕ is onto is immediate from the definition of ϕ. □.

This last proposition tells us that the order in which we take the direct product does not matter. This is also true when we construct the direct product of more than two groups. (See Exercise 12.)

Among the first questions we ask when we encounter a new group is whether or not the group is Abelian, and what are the subgroups of the group. We answer the first question right away.

3.1.8 PROPOSITION Let G_1 and G_2 be groups. Then $G_1 \times G_2$ is Abelian if and only if G_1 and G_2 are both Abelian.

Proof Given (a, b) and (c, d) in $G_1 \times G_2$, $(a, b)(c, d) = (ac, bd)$ and $(c, d)(a, b) = (ca, db)$. Thus $(a, b)(c, d) = (c, d)(a, b)$ for all pairs of elements in $G_1 \times G_2$ if and only if $ac = ca$ for all pairs of elements in G_1 and $bd = db$ for all pairs of elements in G_2. □

Now we look at subgroups of the direct product $G_1 \times G_2$.

3.1.9 EXAMPLE Let us again consider the group $\mathbb{Z}_2 \times S_3$ from Example 3.1.3. Let $H = \mathbb{Z}_2 \times \{\rho_0\}$, where ρ_0 is the identity element of S_3. So $H = \{(0, \rho_0), (1, \rho_0)\}$, where $(0, \rho_0)$ is the identity element in $\mathbb{Z}_2 \times S_3$. Since $(1, \rho_0)(1, \rho_0) = (0, \rho_0)$, H is a subgroup of $\mathbb{Z}_2 \times S_3$. It is actually a normal subgroup, because if $(a, \sigma) \in \mathbb{Z}_2 \times S_3$, then

$$(a, \sigma)(0, \rho_0)(-a, \sigma^{-1}) = (0, \rho_0) \in H$$
$$(a, \sigma)(1, \rho_0)(-a, \sigma^{-1}) = (1, \rho_0) \in H$$

We can also construct the subgroup

$$K = \{0\} \times S_3 = \{(0, \rho_0), (0, \rho), (0, \rho^2), (0, \mu_1), (0, \mu_2), (0, \mu_3)\}$$

For any $(0, \sigma_1)$ and $(0, \sigma_2)$ in K we have

$$(0, \sigma_1)(0, \sigma_2)^{-1} = (0, \sigma_1\sigma_2^{-1}) \in K$$

and for any $(0, \sigma)$ in K and (a, τ) in $\mathbb{Z}_2 \times S_3$ we have

$$(a, \tau)(0, \sigma)(a, \tau)^{-1} = (a, \tau)(0, \sigma)(-a, \tau^{-1}) = (0, \tau\sigma\tau^{-1}) \in K$$

So K is a normal subgroup of $\mathbb{Z}_2 \times S_3$. ◇

The special subgroups of the direct product constructed as in this example will always be normal subgroups, as we now show.

3.1.10 PROPOSITION Let G_1 and G_2 be groups with e_i the identity element in G_i. Then

(1) $G_1 \times \{e_2\} \lhd G_1 \times G_2$ and $\{e_1\} \times G_2 \lhd G_1 \times G_2$

(2) $(G_1 \times G_2)/(G_1 \times \{e_2\}) \cong G_2$ and $(G_1 \times G_2)/(\{e_1\} \times G_2) \cong G_1$

Proof (1) Let $H = G_1 \times \{e_2\}$ and let $(a, e_2), (b, e_2) \in H$. Then

$$(a, e_2)(b, e_2)^{-1} = (a, e_2)(b^{-1}, e_2) = (ab^{-1}, e_2) \in H$$

Therefore, by the subgroup test H is a subgroup of $G_1 \times G_2$. Now let $(a_1, a_2) \in G_1 \times G_2$ and $(b, e_2) \in H$. Then

$$(a_1, a_2)(b, e_2)(a_1, a_2)^{-1} = (a_1, a_2)(b, e_2)(a_1^{-1}, a_2^{-1}) = (a_1ba_1^{-1}, e_2) \in H$$

So by the normal subgroup test $H \lhd G_1 \times G_2$.
Similar arguments prove the second half of the statement.

(2) Consider the map ϕ: $G_1 \times G_2 \to G_2$ defined by $\phi((a_1, a_2)) = a_2$. Then ϕ is a homomorphism since we have

$$\phi((a_1, a_2)(b_1, b_2)) = \phi((a_1b_1, a_2b_2)) = a_2b_2 = \phi((a_1, a_2))\phi((b_1, b_2))$$

ϕ is onto since given any $y \in G_2$, we have $y = \phi((x,y))$ for any $x \in G_1$. Finally, $(a_1, a_2) \in \text{Kern } \phi$ if and only if $a_2 = e_2$, hence if and only if $(a_1, a_2) \in G_1 \times \{e_2\} = H$. Therefore, by the first isomorphism theorem (Theorem 2.4.15), we have the required isomorphism. The second half of the statement can be proved by similar arguments. □

Exercises 3.1

1. For any two finite groups G_1 and G_2, determine the order of $G_1 \times G_2$.

2. Let V be the Klein 4-group. Show that $V \cong \mathbb{Z}_2 \times \mathbb{Z}_2$.

3. Show that D_4 and $\mathbb{Z}_2 \times \mathbb{Z}_4$ are not isomorphic.

4. Show that A_4 and $\mathbb{Z}_2 \times S_3$ are not isomorphic.

5. Show that $\mathbb{Z} \times \mathbb{Z} / \langle(1, 1)\rangle \cong \mathbb{Z}$. (Note that $\langle(1,1)\rangle = \{(a, a) \mid a \in \mathbb{Z}\}$.)

6. Show that $\mathbb{Z} \times \mathbb{Z} / \langle(1, 2)\rangle \cong \mathbb{Z}$.

7. In $\mathbb{Z}_2 \times \mathbb{Z}_4$ find a subgroup H such that $H \cong \mathbb{Z}_2 \times \mathbb{Z}_2$.

8. In D_4 find a subgroup H such that $H \cong \mathbb{Z}_2 \times \mathbb{Z}_2$.

9. In $\mathbb{Z}_4 \times \mathbb{Z}_4$ find two subgroups H and K of order 4 such that H is not isomorphic to K, but $(\mathbb{Z}_4 \times \mathbb{Z}_4) / H \cong (\mathbb{Z}_4 \times \mathbb{Z}_4) / K$.

10. Show that $(\mathbb{Z} \times \mathbb{Z} \times \mathbb{Z}) / \langle(1,1,1)\rangle \cong \mathbb{Z} \times \mathbb{Z}$.

11. Prove Proposition 3.1.6.

12. Let $G_1, G_2, \ldots, G_n$ be groups, and ϕ a permutation in S_n. Show that:

$$G_1 \times G_2 \times \ldots \times G_n \cong G_{\phi(1)} \times G_{\phi(2)} \times \ldots \times G_{\phi(n)}$$

13. Let G_1 and G_2 be groups. Show that $Z(G_1 \times G_2) \cong Z(G_1) \times Z(G_2)$.

14. Let $H \lhd G_1$ and $K \lhd G_2$. Show that

(a) $H \times K$ is a subgroup of $G_1 \times G_2$

(b) $H \times K \lhd G_1 \times G_2$

(c) $(G_1 \times G_2) / (H \times K) \cong G_1 / H \times G_2 / K$

15. Find a subgroup of $\mathbb{Z}_4 \times Q_8$ that is not normal, where Q_8 is the quaternion group.

16. Find the center and the commutator subgroups of $\mathbb{Z}_2 \times S_3$. (See Exercise 27 in Section 2.4.)

17. Find the center and the commutator subgroups of $S_3 \times D_4$. (See Exercise 27 in Section 2.4.)

18. Let r and s be two relatively prime positive integers. Show that $U(rs) \cong U(r) \times U(s)$.

19. Show that $U(105) \cong \mathbb{Z}_2 \times \mathbb{Z}_4 \times \mathbb{Z}_6$.

20. (a) Show that $U(8) \cong \mathbb{Z}_2 \times \mathbb{Z}_2$

(b) Find integers r, s, t, u such that $U(360) \cong \mathbb{Z}_r \times \mathbb{Z}_s \times \mathbb{Z}_t \times \mathbb{Z}_u$.

3.2 Computing orders

As we have seen in Chapter 2, if two groups are isomorphic, then they have the same number of elements with a given order. It is therefore important to know the orders of elements of a group. In this section we learn how to calculate the order of an element in the direct product of several groups in terms of the orders of the components.

3.2.1 EXAMPLE In $\mathbb{Z}_4 \times \mathbb{Z}_6$ consider the element (2, 5). Its order $|(2, 5)|$ is the least positive integer such that $n(2, 5) = (n2, n5) = (0, 0)$. Hence $n2 = 0$ in $\mathbb{Z}_4$ and $n5 = 0$ in $\mathbb{Z}_6$. In $\mathbb{Z}_4$ we have $|2| = 2$, so by Corollary 1.3.12 the condition that $n2 = 0$ in $\mathbb{Z}_4$ implies that 2 divides n. In $\mathbb{Z}_6$ we have $|5| = 6$, which similarly implies that 6 divides n. Therefore, we have $6 = \text{lcm}(2, 6) \leq |(2, 5)|$. On the other hand, since $|2| = 2$ in $\mathbb{Z}_4$ and 2 divides 6, we have $6{\cdot}2 = 0$ in $\mathbb{Z}_4$. Similarly, since $|5| = 6$ in $\mathbb{Z}_6$, we have $6{\cdot}5 = 0$ in $\mathbb{Z}_6$. Thus $6(2, 5) = (6{\cdot}2, 6{\cdot}5) = (0, 0)$ in $\mathbb{Z}_4 \times \mathbb{Z}_6$ and hence $|(2, 5)| \leq 6$. Thus $|(2,5)| = 6$ in $\mathbb{Z}_4 \times \mathbb{Z}_6$. ◇

3.2.2 EXAMPLE In $S_3 \times S_5$ consider the element (ρ, σ), where $\rho = (1\ 2\ 3) \in S_3$ and $\sigma = (1\ 2\ 4)(3\ 5) \in S_5$. Then $|\rho| = 3$ in S_3 and $|\sigma| = 6$ in S_5. As in the previous example, $|(\rho, \sigma)| = \text{lcm}(3, 6) = 6$. ◇

3.2.3 THEOREM Let G_1 and G_2 be groups and $(a_1, a_2) \in G_1 \times G_2$. Then

$$|(a_1, a_2)| = \text{lcm}(|a_1|, |a_2|)$$

Proof Let $n = |(a_1, a_2)|$ and $r = \text{lcm}(|a_1|, |a_2|)$. Then $(a_1, a_2)^r = (a_1{}^r, a_2{}^r) = (e_1, e_2)$ since $|a_1|$ divides r and $|a_2|$ divides r. Hence by Corollary 1.3.12, n divides r, hence $n \leq r$. On the other hand, $(a_1{}^n, a_2{}^n) = (a_1, a_2)^n = (e_1, e_2)$, which implies that $a_1{}^n = e_1$ and $a_2{}^n = e_2$, from which it follows by Corollary 1.3.12 that n is a common multiple of $|a_1|$ and $|a_2|$, and therefore $r \leq n$, completing the proof. □

3.2.4 COROLLARY Let $G = G_1 \times \ldots \times G_n$ be the direct product of a finite number of groups. Then the order of an element in G is given by

$$|(a_1, \ldots, a_n)| = \text{lcm}(|a_1|, \ldots, |a_n|)$$

Proof We use induction on n. For the case $n = 1$ there is nothing to prove. The case $n = 2$ is the preceding theorem. So suppose the Corollary holds for products of k groups and consider a product

$$G_1 \times \ldots \times G_k \times G_{k+1} = (G_1 \times \ldots \times G_k) \times G_{k+1}$$

of $k + 1$ groups. To complete the proof, by the preceding theorem we have

$$|(a_1, \ldots, a_k, a_{k+1})| = \text{lcm}(|(a_1, \ldots, a_k)|, |(a_{k+1})|) =$$
$$\text{lcm}(\text{lcm}((|a_1|, \ldots, |a_k|), |(a_{k+1})|) = \text{lcm}(|a_1|, \ldots, |a_k|, |(a_{k+1})|)$$ □

3.2.5 EXAMPLE Let us find $|(10, 10)|$ in $\mathbb{Z}_{12} \times \mathbb{Z}_{18}$. First we need to find the orders of the components separately. By Theorem 1.3.16 we have $|10| = 12/\gcd(12, 10) = 6$ in $\mathbb{Z}_{12}$ and $|10| = 18/\gcd(18, 10) = 9$ in $\mathbb{Z}_{18}$. Hence $|(10, 10)| = \text{lcm}(6, 9) = 18$ in $\mathbb{Z}_{12} \times \mathbb{Z}_{18}$ by Theorem 3.2.3. ◇

3.2.6 COROLLARY Let (r, s) be an element of $\mathbb{Z}_n \times \mathbb{Z}_m$. Then

$$|(r, s)| = \text{lcm}(n/\gcd(n, r), m/\gcd(m, s))$$

Proof This is immediate from Theorem 1.3.16 and Theorem 3.2.3. □

3.2.7 COROLLARY Let (r, s) be an element of $\mathbb{Z}_n \times \mathbb{Z}_m$. Then

$$|(r, s)| \leq |(1, 1)| = \text{lcm}(n, m)$$

Proof In $\mathbb{Z}_n$, $|r|$ divides n, and in $\mathbb{Z}_m$, $|s|$ = divides m, so any common multiple of n and m is a common multiple of $|r|$ and $|s|$, and $|(r, s)| = \text{lcm}(|r|, |s|) \leq \text{lcm}(n, m)$. On the other hand, we have $|1| = n$ in $\mathbb{Z}_n$ and $|1| = m$ in $\mathbb{Z}_m$, so $|(1, 1)| = \text{lcm}(n, m)$ by Theorem 3.2.3. □

We are now ready to state and prove as our next theorem a fact that follows from the corollaries we have just proved and that plays an important role in the classification of finite Abelian groups.

3.2.8 THEOREM The group $\mathbb{Z}_n \times \mathbb{Z}_m$ is isomorphic to the cyclic group $\mathbb{Z}_{nm}$ if and only if n and m are relatively prime.

Proof ($\Rightarrow$) If $\mathbb{Z}_n \times \mathbb{Z}_m$ is isomorphic to $\mathbb{Z}_{nm}$ and therefore cyclic, then $(1, 1)$ must be a generator by Corollary 3.2.7, and we must have $|(1, 1)| = nm$. But again by Corollary 3.2.7, $|(1, 1)| = \text{lcm}(n, m) = nm/\gcd(n, m)$, so we must have $\gcd(n, m) = 1$.
($\Leftarrow$) If $\gcd(n, m) = 1$, then

$$|(1, 1)| = \text{lcm}(n, m) = nm/\gcd(n, m) = nm$$

and (1,1) generates $\mathbb{Z}_n \times \mathbb{Z}_m$, which is cyclic and therefore is isomorphic to the group $\mathbb{Z}_{nm}$. □

3.2.9 COROLLARY $\mathbb{Z}_{n_1} \times \mathbb{Z}_{n_2} \times \ldots \times \mathbb{Z}_{n_s} \cong \mathbb{Z}_{n_1 n_2 \ldots n_s}$ if and only if for all $1 \leq i < j \leq s$, $\gcd(n_i, n_j) = 1$.

Proof Left to the reader (as Exercise 19). □

3.2.10 EXAMPLE Consider $(\mathbb{Z}_4 \times \mathbb{Z}_4)/\langle(1, 1)\rangle$. Since $|\mathbb{Z}_4 \times \mathbb{Z}_4| = 4 \cdot 4 = 16$ and $|\langle(1, 1)\rangle| = |(1, 1)| = \text{lcm}(4, 4) = 4$, it follows that $|(\mathbb{Z}_4 \times \mathbb{Z}_4)/\langle(1, 1)\rangle| = 16/4 = 4$ and so it must be isomorphic either to $\mathbb{Z}_4$ or to the Klein 4-group $V \cong \mathbb{Z}_2 \times \mathbb{Z}_2$. But if we calculate we find that $(1, 0) + \langle(1, 1)\rangle$ is an element of order 4, and hence the group is cyclic and isomorphic to $\mathbb{Z}_4$. ◇

3.2.11 EXAMPLE $U(10) = \{1, 3, 7, 9\}$ and $U(12) = \{1, 5, 7, 11\}$; hence $U(10) \times U(12)$ is a group of order 16. Let H be the subgroup generated by $(7, 7)$. So $H = \{(1, 1), (7, 7), (9, 1), (3, 7)\}$ and $(U(10) \times U(12))/H$ is a group of order 4. If we calculate, we find that H, $(3, 1)H$, $(3, 5)H$, and $(3, 11)H$ are four distinct cosets of H and hence all the elements of the quotient group, and that $(3, 1)(3, 1) = (3, 5)(3, 5) = (3, 11)(3, 11) = (9, 1) \in H$, hence $|(3,1)H| = |(3, 5)H| = |(3, 11)H| = 2$, and conclude that $(U(10) \times U(12))/\langle(7, 7)\rangle \cong \mathbb{Z}_2 \times \mathbb{Z}_2$. ◇

Exercises 3.2

In Exercises 1 through 6 find the order of the indicated element in the indicated group.

1. $(4, 6)$ in $\mathbb{Z}_6 \times \mathbb{Z}_8$ **2.** $(15, 15)$ in $\mathbb{Z}_{20} \times \mathbb{Z}_{27}$ **3.** $(\rho, 7)$ in $S_3 \times U(12)$

4. $(\rho, 7)$ in $D_4 \times U(12)$ **5.** $(\rho, \mathbf{i})$ in $S_3 \times Q_8$ **6.** $((2\,3\,4), 15)$ in $A_4 \times \mathbb{Z}_{18}$

In Exercises 7 through 10 find the distinct cosets of the indicated subgroup in the indicated group.

7. $H = \langle(8, 2)\rangle$ in $\mathbb{Z}_{10} \times \mathbb{Z}_4$ **8.** $H = \langle(3, 5)\rangle$ in $U(10) \times U(12)$

9. $H = \langle(6, 8)\rangle$ in $3\mathbb{Z} \times 2\mathbb{Z}$ **10.** $H = \langle(\rho, \tau)\rangle$ in $D_4 \times D_4$

In Exercises 11 through 14 find the order of the indicated element in the indicated quotient group.

11. $(1, 1) + \langle(8, 2)\rangle$ in $(\mathbb{Z}_{10} \times \mathbb{Z}_4)/\langle(8, 2)\rangle$

12. $(7, 7)\langle(3, 5)\rangle$ in $(U(10) \times U(12))/\langle(3, 5)\rangle$

13. $(3, 2) + \langle(6, 8)\rangle$ in $(3\mathbb{Z} \times 2\mathbb{Z})/\langle(6, 8)\rangle$ **14.** $(\rho^3, \tau)\langle(\rho, \tau)\rangle$ in $(D_4 \times D_4)/\langle(\rho, \tau)\rangle$

15. Show that $\mathbb{Z}_9 \times \mathbb{Z}_9$ and $\mathbb{Z}_{27} \times \mathbb{Z}_3$ are not isomorphic.

16. Show that $\mathbb{C} \cong \mathbb{R} \times \mathbb{R}$ as groups under addition.

17. Find the largest order of any element in $\mathbb{Z}_{21} \times \mathbb{Z}_{35}$.

18. Find all the elements of order 4 in $\mathbb{Z}_4 \times \mathbb{Z}_4$. **19.** Prove Corollary 3.2.9.

20. Find all the group homomorphisms from $\mathbb{Z}_6$ to $\mathbb{Z}_2 \times \mathbb{Z}_3$, and determine which ones are isomorphisms.

21. Show that if G is a finite group such that for all $g \in G$ we have $g^2 = e$, then

(a) G is Abelian. (b) $|G| = 2^n$ for some positive integer n.
(c) $G \cong \mathbb{Z}_2 \times \ldots \times \mathbb{Z}_2$ (n times).

3.3 Direct Sums

We have used direct products to construct new groups. We next use a similar notion, that of *direct sum*, and decompose some groups and write them as direct sums of certain normal subgroups. This will greatly improve our understanding of certain of these groups.
As we will see in the next section, such a decomposition can be used to characterize completely all finite Abelian groups.
We first illustrate the notion of direct sum with an example.

3.3.1 EXAMPLE In $\mathbb{Z}_{12}$ let $H = \langle 3 \rangle = \{0, 3, 6, 9\}$ be the subgroup of order 4, and $K = \langle 4 \rangle = \{0, 4, 8\}$ the subgroup of order 3. Since $\mathbb{Z}_{12}$ is Abelian, H and K are both normal subgroups and $H + K = \{h + k \mid h \in H, k \in K\}$ is a subgroup by Proposition 2.3.22. Note that $H \cap K = \{0\}$; hence $H + K$ is of order $|H + K| = |H||K| / |H \cap K| = 4 \cdot 3/1 = 12$ by Theorem 2.3.24. Hence $\mathbb{Z}_{12} = H + K$. Moreover, the fact that $H \cap K = \{0\}$ implies that every element $a \in \mathbb{Z}_{12}$ can be written as $a = h + k$ with $h \in H$ and $k \in K$ in a unique way. For if we have $a = h_1 + k_1 = h_2 + k_2$, then $h_1 - h_2 = k_2 - k_1 \in H \cap K = \{0\}$; hence $h_1 = h_2$ and $k_1 = k_2$. We can actually show that $\mathbb{Z}_{12} \cong H \times K \cong \mathbb{Z}_4 \times \mathbb{Z}_3$. Suppose that now instead we consider the subgroups H as before and $L = \langle 2 \rangle = \{0, 2, 4, 6, 8, 10\}$. You can easily check that $\mathbb{Z}_{12} = H + L$, so every element $a \in \mathbb{Z}_{12}$ can be written as $a = h + l$ with $h \in H$ and $l \in L$. But in this case we do not have uniqueness, since for instance $7 = 3 + 4 = 9 + 10$ in $\mathbb{Z}_{12}$, where $3, 9 \in H$ and $4, 10 \in L$. What "goes wrong" in this case is that $H \cap L = \{0, 6\} \neq \{0\}$. ◇

3.3.2 EXAMPLE In S_3 the subgroup A_3 of even permutations is normal. So if we let $H = A_3$ and $K = \langle \mu_1 \rangle$, then HK is a subgroup (again by Proposition 2.3.22) and $|HK| = |H||K| / |H \cap K| = 3 \cdot 2/1 = 6$ (again by Theorem 2.3.24). So $S_3 = HK$. But obviously S_3 is not isomorphic to $H \times K \cong \mathbb{Z}_3 \times \mathbb{Z}_2$, since S_3 is non-Abelian. What "goes wrong" in this case is that $K = \langle \mu_1 \rangle$ is not a normal subgroup. ◇

The next theorem best characterizes the notion we are introducing. We first need a simple lemma.

3.3.3 LEMMA Let G be a group and let H and K be subgroups of G such that

(1) $H \triangleleft G$ and $K \triangleleft G$
(2) $H \cap K = \{e\}$

Then for all $h \in H$ and $k \in K$ we have $hk = kh$.

Proof We want to show $hk = kh$, so consider $y = hkh^{-1}k^{-1}$. Since $H \triangleleft G$, $kh^{-1}k^{-1} \in H$ and hence $y = h(kh^{-1}k^{-1}) \in H$. Since $K \triangleleft G$, $hkh^{-1} \in K$ and $y = (hkh^{-1})k^{-1} \in K$. So $hkh^{-1}k^{-1} = y \in H \cap K = \{e\}$, and $hkh^{-1}k^{-1} = e$, implying $hk = kh$. □

3.3.4 THEOREM Let G be a finite group and H and K subgroups of G such that

(1) $H \lhd G$ and $K \lhd G$

(2) $H \cap K = \{e\}$

(3) $|HK| = |G|$

Then $G \cong H \times K$.

Proof (1) We define a map ϕ: $H \times K \to G$ by $\phi((h, k)) = hk \in G$.

(2) ϕ is a homomorphism since $\phi((h_1, k_1)(h_2, k_2)) = \phi((h_1h_2, k_1k_2)) = h_1h_2k_1k_2 = h_1k_1h_2k_2$ by the preceding lemma, and $h_1k_1h_2k_2 = \phi((h_1, k_1))\phi((h_2, k_2))$.

(3) ϕ is one to one because if $\phi((h_1, k_1)) = \phi((h_2, k_2))$, then $h_1k_1 = h_2k_2$, which implies $h_2^{-1}h_1 = k_2k_1^{-1} \in H \cap K = \{e\}$, and $h_1 = h_2$, $k_1 = k_2$.

(4) ϕ is onto since by the first isomorphism theorem $|\phi(H \times K)| = |H \times K| / |\text{Kern } \phi| = |H||K|/1 = |HK| = |G|$. □

For the rest of this section we concentrate on Abelian groups, for which condition (1) in the preceding theorem always holds. Let H_1 and H_2 be subgroups of an Abelian group G. Then they are normal subgroups of G and $H_1 + H_2$ is a subgroup of G. If, in addition, $H_1 \cap H_2 = \{0\}$, then every element x of $H_1 + H_2$ can be written uniquely as $x = h_1 + h_2$ with $h_1 \in H_1$ and $h_2 \in H_2$. In this case we call $H_1 + H_2$ the direct sum of H_1 and H_2 and write it $H_1 \oplus H_2$. We state the definition for sums of any finite number of subgroups.

3.3.5 DEFINITION Let $H_1, \ldots, H_n$ be subgroups of an Abelian group G. Then $H_1 + \ldots + H_n$ is called a **direct sum** and is written $H_1 \oplus \ldots \oplus H_n$ if for every $x \in H_1 + \ldots + H_n$ we have $x = h_1 + \ldots + h_n = h_1' + \ldots + h_n'$ with $h_i, h_i' \in H_i$ if and only if $h_i = h_i'$ for all i, $1 \le i \le n$. ○

Thus in a direct sum $H \oplus K$, any element x is *uniquely* represented as $x = h + k$, where $h \in H$ and $k \in K$. In Example 3.3.1, $\mathbb{Z}_{12} = H \oplus K$, but $\mathbb{Z}_{12} = H + L$ was not a direct sum.

We give some equivalent definitions of the direct sum, which will make the notion clearer.

3.3.6 THEOREM Let G be an Abelian group and $H_1, \ldots, H_n$ subgroups of G. Then the following statements are equivalent:

(1) $H_1 + \ldots + H_n$ is a direct sum.

(2) $(H_1 + H_2 + \ldots + H_{i-1} + H_{i+1} + \ldots + H_n) \cap H_i = \{0\}$, for all i, $1 \le i \le n$

(3) $(H_1 + H_2 + \ldots + H_{i-1}) \cap H_i = \{0\}$, for all i, $2 \le i \le n$

(4) $h_1 + \ldots + h_n = 0$, where $h_i \in H_i$, for all i, $1 \le i \le n$, implies $h_i = 0$, for all i, $1 \le i \le n$

Proof (1⇒2) Let $x \in (H_1 + H_2 + \ldots + H_{i-1} + H_{i+1} + \ldots + H_n) \cap H_i$. Then

$x = h_1 + h_2 + \ldots + h_{i-1} + h_{i+1} + \ldots + h_n$ for some $h_j \in H_j$, $j \ne i$

and since $x \in H_i$, we have two representations of $0 \in H_1 + \ldots + H_n$:

$$0 = h_1 + h_2 + \ldots + h_{i-1} + (-x) + h_{i+1} + \ldots + h_n = 0 + 0 + \ldots + 0$$

By the definition of direct sum it follows that each summand in the first representation is zero, and in particular $x = 0$.

(2⇒3) This follows immediately from the fact that

$$(H_1 + H_2 + \ldots + H_{i-1}) \cap H_i \subseteq H_1 + H_2 + \ldots + H_{i-1} + H_{i+1} + \ldots + H_n) \cap H_i$$

(3⇒4) If $h_1 + \ldots + h_n = 0$, where $h_i \in H_i$, we want to show $h_i = 0$ for all i. Let us assume the contrary (in other words, that some h_i are not zero) and consider the largest integer k, $1 \leq k \leq n$, such that $h_k \neq 0$. So $h_{k+1} = \ldots = h_n = 0$, and $0 = h_1 + \ldots + h_k$. We thus have

$$h_k = -h_1 - \ldots - h_{k-1} \in (H_1 + H_2 + \ldots + H_{k-1}) \cap H_k = \{0\}$$

So $h_k = 0$, which is a contradiction. Hence such a k does not exist, and $h_i = 0$ for all i, $1 \leq i \leq n$.

(4⇒1) We want to show that every x in $H_1 + \ldots + H_n$ can be written uniquely as a sum of elements of the subgroups. So suppose $x = h_1 + \ldots + h_n = h_1' + \ldots + h_n'$, where $h_i, h_i' \in H_i$. Then $(h_1 - h_1') + \ldots + (h_n - h_n') = 0$, where $(h_i - h_i') \in H_i$. So $(h_i - h_i') = 0$ and $h_i = h_i'$ for all i, $1 \leq i \leq n$. □

We now rephrase Theorem 3.3.4 in terms of direct sums of finite Abelian groups.

3.3.7 COROLLARY Let G be a finite Abelian group, and let H and K be subgroups of G such that

(1) $H \cap K = \{0\}$

(2) $|H + K| = |G|$

Then $G = H \oplus K \cong H \times K$.

Proof Since G is Abelian, $H \lhd G$ and $K \lhd G$; hence by Theorem 3.3.4, $G \cong H \times K$, and by Theorem 3.3.6, $H + K$ is a direct sum $H \oplus K$. Also, $H \oplus K$ is a subgroup of G with $|H \oplus K| = |H + K| = |G|$; hence $H \oplus K = G$. □

We can now actually show that direct sums and direct products, for any finite number of subgroups, are isomorphic.

3.3.8 THEOREM Let G be an Abelian group such that $G = H_1 \oplus \ldots \oplus H_n$. Then $G \cong H_1 \times \ldots \times H_n$.

Proof (1) We define a map $\phi: H_1 \oplus \ldots \oplus H_n, \rightarrow H_1 \times \ldots \times H_n$ by letting

$$\phi(h_1 + \ldots + h_n) = (h_1, \ldots, h_n)$$

Note that this map is well defined because if

$$h_1 + \ldots + h_n = h_1' + \ldots + h_n'$$

then $h_i = h_i'$ for all i, and $(h_1, \ldots, h_n) = (h_1', \ldots, h_n')$.

(2) ϕ is a homomorphism since
$\phi((h_1 + \ldots + h_n) + (h_1' + \ldots + h_n')) = \phi((h_1 + h_1') + \ldots + (h_n + h_n')) =$
$(h_1 + h_1', \ldots, h_n + h_n') = (h_1, \ldots, h_n) + (h_1', \ldots, h_n') =$
$\phi(h_1 + \ldots + h_n) + \phi(h_1' + \ldots + h_n')$
(3), (4) ϕ is clearly one to one and onto. □

3.3.9 COROLLARY Let G be a finite Abelian group such that $G = H_1 \oplus \ldots \oplus H_n$. Then $|G| = |H_1|\cdot\ldots\cdot|H_n|$.

Proof This is immediate from the preceding theorem. □

In the next corollaries we point out an important consequence of the representation $x = h_1 + \ldots + h_n$ for $x \in H_1 \oplus \ldots \oplus H_n$.

3.3.10 COROLLARY Let G be a finite Abelian group such that $G = H_1 \oplus \ldots \oplus H_n$, and $x = h_1 + \ldots + h_n$ an element of G, where $h_i \in H_i$ for $1 \le i \le n$. Then

$$|x| = \text{lcm}(|h_1|, \ldots, |h_n|)$$

Proof This is immediate from Theorem 3.3.8 and Corollary 3.2.4. (See Exercise 16.) □

3.3.11 COROLLARY If $G = G_1 \oplus G_2$, where G_1 is cyclic of order n and G_2 is cyclic of order m, then $G \cong \mathbb{Z}_{nm}$ if and only if $\gcd(n, m) = 1$.

Proof This is immediate from Proposition 3.3.8 and Theorem 3.2.8. (See Exercise 17.) □

3.3.12 LEMMA Let G be an Abelian group such that $G = H_1 \oplus \ldots \oplus H_n$, and let K_i be a subgroup of H_i for all i, $1 \le i \le n$, and let $K = K_1 + \ldots + K_n$. Then $K = K_1 \oplus \ldots \oplus K_n$.

Proof Consider any $x \in K$. If $x = k_1 + \ldots + k_n = k_1' + \ldots + k_n'$, where $k_i, k_i' \in K_i$, for all i, $1 \le i \le n$, then since K_i is a subgroup of H_i and $G = H_1 \oplus \ldots \oplus H_n$, we must have $k_i, = k_i'$ for all i, $1 \le i \le n$. □

3.3.13 PROPOSITION Let G be an Abelian group such that $G = H_1 \oplus \ldots \oplus H_n$, and let K_i be a subgroup of H_i for all i, $1 \le i \le n$, and let $K = K_1 + \ldots + K_n$. Then

$$G/K = G/(K_1 \oplus \ldots \oplus K_n) \cong H_1/K_1 \times \ldots \times H_n/K_n$$

Proof (1) Note that $K = K_1 \oplus \ldots \oplus K_n$ by the preceding lemma. We define a map $\phi: G \to H_1/K_1 \times \ldots \times H_n/K_n$ by

$$\phi(x) = \phi(h_1 + \ldots + h_n) = (h_1 + K_1, \ldots, h_n + K_n)$$

where $x = h_1 + \ldots + h_n$ with $h_i \in H_i$ for all i, $1 \le i \le n$. Since this representation of any $x \in G$ is unique, the map is well defined.

(2) ϕ is a homomorphism, since for $x = h_1 + \ldots + h_n$ and $x' = h_1' + \ldots + h_n'$ in G we have

$$\phi((h_1 + \ldots + h_n) + (h_1' + \ldots + h_n')) =$$
$$\phi((h_1 + h_1') + \ldots + (h_n + h_n')) =$$
$$((h_1 + h_1') + K_1, \ldots , (h_n + h_n') + K_n) =$$
$$(h_1 + K_1, \ldots , h_n + K_n) + (h_1' + K_1, \ldots , h_n' + K_n) =$$
$$\phi((h_1 + \ldots + h_n)) + \phi((h_1' + \ldots + h_n'))$$

(3) The identity element of $H_1/K_1 \times \ldots \times H_n/K_n$ is $(K_1, \ldots , K_n)$, so that if $x = h_1 + \ldots + h_n$, then $x \in$ Kern ϕ if and only if $h_i + K_i = K_i$ or, in other words, $h_i \in K_i$ for all i, $1 \le i \le n$. But this condition is equivalent to $x \in K_1 \oplus \ldots \oplus K_n$. So Kern $\phi = K_1 \oplus \ldots \oplus K_n$.

(4) Finally, ϕ is onto since for any $y = (h_1 + K_1, \ldots , h_n + K_n)$ in $H_1/K_1 \times \ldots \times H_n/K_n$, the element $x = h_1 + \ldots + h_n \in G$ satisfies $\phi(x) = y$. The proposition now follows by the first isomorphism theorem. □

The construction of direct sums we have been studying in this section is essential for the study of finite Abelian groups in the next section.

Exercises 3.3

In Exercises 1 through 5 find two nontrivial proper subgroups H and K of the indicated group G such that $G \cong H \oplus K$.

1. $\mathbb{Z}_{10}$ **2.** $\mathbb{Z}_{15}$ **3.** $\mathbb{Z}_{18}$ **4.** $\mathbb{Z}_{20}$ **5.** $\mathbb{Z}_{36}$

6. Explain why there are no nontrivial proper subgroups H and K in $\mathbb{Z}_9$ such that $\mathbb{Z}_9 = H \oplus K$.

7. Explain why there are no nontrivial proper subgroups H and K in $\mathbb{Z}_8$ such that $\mathbb{Z}_8 = H \oplus K$.

8. Find if possible proper nontrivial subgroups H_1, H_2, and H_3 in $\mathbb{Z}_{60}$ such that $\mathbb{Z}_{60} = H_1 \oplus H_2 \oplus H_3$.

9. Explain why there are no nontrivial proper subgroups H_1, H_2, and H_3 in $\mathbb{Z}_{36}$ such that $\mathbb{Z}_{36} = H_1 \oplus H_2 \oplus H_3$.

10. Find nontrivial proper subgroups H and K in $U(12)$ such that $HK = U(12)$.

11. Find nontrivial proper subgroups H and K in $U(15)$ such that $HK = U(15)$.

12. Show that there are no nontrivial proper subgroups H and K in $U(10)$ such that $HK = U(10)$.

13. Let H and K be subgroups of a group G such that $G = H \oplus K$, H is cyclic of order 4, and K is cyclic of order 35. Show that $G \cong \mathbb{Z}_{140}$.

14. Let H and K be subgroups of a group G such that $G = H \oplus K$, H is cyclic of order 6, and K is cyclic of order 15. Show that G is an Abelian group of order 90 that is not cyclic.

15. Let G be a finite group and H_i, $1 \le i \le n$, subgroups of G, such that

(1) $H_i \lhd G$ for all i, $1 \le i \le n$

(2) $(H_1 H_2 \ldots H_{i-1}) \cap H_i = \{e\}$ for all i, $1 \le i \le n$

(3) $|G| = |H_1||H_2| \ldots |H_n|$

Prove that $G = H_1 \oplus H_2 \oplus \ldots \oplus H_n$.

16. Let $G = H_1 \oplus \ldots \oplus H_n$, and let $x = h_1 + \ldots + h_n \in G$. Show that $|x| = \text{lcm}(|h_1|, \ldots, |h_n|)$.

17. Let $G = H \oplus K$, where H is cyclic of order n and K is cyclic of order m. Show that $G \cong \mathbb{Z}_{nm}$ if and only if n and m are relatively prime.

18. Let $G = \mathbb{Z}_6 \times \mathbb{Z}_8$ and define a map ϕ: $G \to \mathbb{Z}_3 \times \mathbb{Z}_4$ by $\phi((h_1,h_2)) = (h_1 \bmod 3, h_2 \bmod 4)$, for any $h_1 \in \mathbb{Z}_6$ and $h_2 \in \mathbb{Z}_8$.

(a) Show that ϕ is a homomorphism.

(b) Find Kern ϕ.

(c) Find $\text{Im}(\phi) = \phi(G)$.

19. Let H and K be subgroups of an Abelian group G and let ϕ: $G \to H$ be a homomorphism such that

(1) $\phi(h) = h$ for all $h \in H$

(2) Kern $\phi = K$

Show that $G = H \oplus K$.

20. Let H and K be subgroups of an Abelian group G and ϕ: $G \to H$ a homomorphism such that

(1) $\phi(h) = h$ for all $h \in H$

(2) Kern $\phi = K$

Show that there exists a homomorphism ψ: $G \to K$ such that

(1) $\psi(k) = k$ for all $k \in K$

(2) Kern $\psi = H$

(See Exercise 19.)

21. Let H and K be subgroups of an Abelian group G. Show that $G = H \oplus K$ if and only if there exists a homomorphism ϕ: $G \to H$ such that

(1) $\phi(h) = h$ for all $h \in H$

(2) Kern $\phi = K$

(See Exercise 19.)

22. Let H and K be normal subgroups of a group G and ϕ: $G \to H$ a homomorphism such that

(1) $\phi(h) = h$ for all $h \in H$

(2) Kern $\phi = K$

Show that $G \cong H \times K$.

23. Let G be an Abelian group and ϕ: $G \to G$ a homomorphism such that $\phi(\phi(g)) = g$ for all $g \in G$. (Such a homomorphism is called a **projection** homomorphism.) Show that $G \cong \phi(G) \times$ Kern ϕ.

24. Let G be a group with $|G| = nm$, where n and m are relatively prime. Assume that G has exactly one subgroup H of order $|H| = n$ and exactly one subgroup K of order $|K| = m$. Show that $G \cong H \times K$.

25. Show that every group of order 9 is Abelian.

26. Show that every group of order p^2 is Abelian, for any prime p.

3.4 Fundamental Theorem of Finite Abelian Groups

In this section we show how finite Abelian groups can be completely described in terms of direct products of some cyclic groups. We will be able to list all Abelian groups of a given finite order. Since we know how to construct subgroups of cyclic groups and how to calculate the order of an element in cyclic groups, we will be able to do the same for any finite Abelian group.

We know that any cyclic group G with $|G| = n$ is isomorphic to $\mathbb{Z}_n$ and that the direct product of such groups $\mathbb{Z}_n \times \mathbb{Z}_m$ is an Abelian group. The theorem we are going to prove tells us that indeed any Abelian group is isomorphic to a direct product of cyclic groups.

3.4.1 THEOREM (Fundamental theorem of finite Abelian groups) Let G be an Abelian group of finite order. Then

(1) $G \cong \mathbb{Z}_{p_1{}^{a_1}} \times \ldots \times \mathbb{Z}_{p_s{}^{a_s}}$

where the p_i are primes not necessarily distinct.

(2) The direct product is unique except for the order of factors. □

We show that any finite Abelian group G is the direct sum of some cyclic subgroups, $G = C_1 \oplus \ldots \oplus C_s$, where $|C_i| = p_i{}^{a_i}$ and hence $C_i \cong \mathbb{Z}_{p_i{}^{a_i}}$. It then follows by Theorem 3.3.8 that

$$G = C_1 \oplus \ldots \oplus C_s \cong \mathbb{Z}_{p_1{}^{a_1}} \times \ldots \times \mathbb{Z}_{p_s{}^{a_s}}$$

This is a difficult theorem to prove, so we break up the proof into four steps. Each is important enough to be called a proposition rather than a lemma.

The following example illustrates the first step.

3.4.2 EXAMPLE Let G be an Abelian group with $|G| = 24$. Let $H = \{x \in G \mid |x| = 1, 2, 4, \text{ or } 8\}$, and let $K = \{x \in G \mid |x| = 1 \text{ or } 3\}$. We use additive notation since G is Abelian. First note that H and K are both subgroups of G. We have $x \in H$ only if $8x = 0$; hence if $x, y \in H$, then $8(x - y) = 8x - 8y = 0 - 0 = 0$, and $x - y \in H$, so H is a subgroup by the subgroup test, and similarly for K. For any $g \in G$, since $1 = 2{\cdot}8 - 5{\cdot}3$ we have $g = (2{\cdot}8 - 5{\cdot}3)g = 2{\cdot}8g - 5{\cdot}3g$. Since $|G| = 24$, by Lagrange's theorem $3(16g) = 0$ and $16g \in K$, while $8(15g) = 0$ and $15g \in H$. So $g \in H + K$ and $H + K = G$. Since $H \cap K = \{0\}$, $G = H \oplus K$ and $24 = |G| = |H||K|$. Also, 3 cannot divide $|H|$, else H would have an element of order 3 by Cauchy's theorem for Abelian groups (Theorem 2.4.20), and similarly 2 cannot divide $|K|$. So $|H| = 8$ and $|K| = 3$ ◇

3.4.3 PROPOSITION Let G be an Abelian group of order $p^r m$, where p is prime and does not divide m. Let $H = \{x \in G \mid |x| = p^s,\ 0 \le s \le r\}$ and let $K = \{x \in G \mid |x| \text{ divides } m\}$. Then

(1) $G = H \oplus K$

(2) $|H| = p^r$ and $|K| = m$

Proof (1) First we show that H and K are subgroups of G. For $x \in H$ if and only if $p^r x = 0$, so if $x, y \in H$, then $p^r(x - y) = p^r x - p^r y = 0$, and $x - y \in H$, so H is a subgroup by the subgroup test, and similarly for K.

Next we show $G = H + K$. For since p^r and m are relatively prime, by Theorem 0.3.16 there exist integers u, v such that $up^r + vm = 1$. Therefore, for any $g \in G$ we have $g = 1g = (up^r + vm)g = (up^r)g + (vm)g$. Since $|G| = p^r m$, we have $p^r(vm)g = 0$, which implies $(vm)g \in H$, and $m(up^r)g = 0$, which implies $(up^r)g \in K$. Thus we have expressed g as a sum of an element of H and an element of K.

Finally, if $x \in H \cap K$, then the order of x must divide both p^r and m and so must divide $\gcd(p^r, m) = 1$. Thus $H \cap K = \{0\}$, and (1) follows by Corollary 3.3.7.

(2) K is a subgroup of an Abelian group, and therefore K is Abelian and by Cauchy's theorem (Theorem 2.4.20), if p divided $|K|$, K would contain an element of order p, which by definition it does not. Similarly, $|H|$ is not divisible by any prime other than p. Since $|G| = |H||K|$, the only possibility is $|H| = p^r$ and $|K| = m$. □

3.4.4 EXAMPLE Let G be an Abelian group of order $900 = 4{\cdot}9{\cdot}25 = 2^2{\cdot}3^2{\cdot}5^2$, and let $H = \{x \in G \mid |x| = 1, 2, \text{ or } 4\}$, and $K = \{x \in G \mid |x| \text{ divides } 3^2{\cdot}5^2 = 225\}$. Then by the preceding proposition, $G = H \oplus K$. Let now $L = \{x \in G \mid |x| = 1, 3, \text{ or } 9\}$ and $M = \{x \in G \mid |x| = 1, 5, \text{ or } 25\}$. Then L and M are subgroups of K, and again by the preceding proposition, $K = L \oplus M$. So $G = H \oplus L \oplus M$. ◇

Given an Abelian group G with $|G| = p_1^{a_1} p_2^{a_2} \ldots p_k^{a_k}$, where the p_i are distinct primes, let $G(p_i^{a_i}) = \{x \in G \mid |x| = p_i^s,\ 0 \le s \le a_i\}$. An immediate consequence of the preceding proposition is the following.

3.4.5 COROLLARY Let G be an Abelian group with $|G| = p_1^{a_1}p_2^{a_2}...p_k^{a_k}$, where the p_i are distinct primes. Then

(1) $G = G(p_1^{a_1}) \oplus \ldots G(p_k^{a_k})$

(2) $|G(p_i^{a_i})| = p_i^{a_i}$

Proof This is immediate from Proposition 3.4.3. □

3.4.6 DEFINITION Let G be a finite Abelian group and p a prime. Then G is called a ***p*-group** if $|G| = p^r$ for some positive integer r. ○

Corollary 3.4.5 states that any finite Abelian group can be decomposed as a direct sum of p-groups. We show that every finite Abelian group is a direct sum of cyclic groups by concentrating on p-groups and showing that every p-group is a direct sum of cyclic groups.

The next proposition, which is the second step in the proof of the fundamental theorem, is the hardest step in the proof. The next example illustrates the ideas involved in a fairly simple case.

3.4.7 EXAMPLE Let G be an Abelian group of order 8, and $a \in G$ an element of maximal order; in other words, an element such that $|x| \leq |a|$ for all other $x \in G$. We have $|a| = 8$ or 4 or 2.

Case 1 $|a| = 8$. Then $G = \langle a \rangle \cong \mathbb{Z}_8$.

Case 2 $|a| = 4$. Then $H = \langle a \rangle$ is a proper subgroup of G. Let $b \in G$, $b \notin \langle a \rangle$ be an element of minimal order or, in other words, an element such that $|y| \geq |b|$ for any other $y \in G$, $y \notin \langle a \rangle$. We have $|b| \leq |a| = 4$, since a was chosen to be of maximal order. If $|b| = 4$, then H and $K = \langle b \rangle$ will have a nontrivial intersection of order 2. In that case $2a = 2b$, which would imply that $|a + b| = 2$. Since $a + b \notin \langle a \rangle$, this is impossible, since b was chosen to be of minimal order. So we must have $|b| =2$ and $H \cap K = \{0\}$. In this case, $G = H \oplus K \cong \mathbb{Z}_4 \times \mathbb{Z}_2$.

Case 3 $|a| = 2$. Then, taking any $b \in G$ with $b \notin \langle a \rangle$, we have $|b| = 2$. Let $H = \langle a \rangle$ and $K = \langle b \rangle$. Then $H \cap K = \{0\}$, and $|H \oplus K| = 4$. Then, taking any $c \in G$ with $c \notin H \oplus K$, we have $|c| = 2$, and if $L = \langle c \rangle$, then $(H \oplus K) \cap L = \{0\}$ and $G = H \oplus K \oplus L \cong \mathbb{Z}_2 \times \mathbb{Z}_2 \times \mathbb{Z}_2$. ◇

With this example we demonstrated how to find all the Abelian groups of order 8, and we found that there are exactly three of them, up to isomorphism. At the same time, the example illustrates the two main ideas in the proof of the next proposition: the choice of an element a of maximal order and an element $b \notin \langle a \rangle$ of minimal order.

3.4.8 PROPOSITION Let p be a prime and G a finite Abelian p-group. Let a be an element of maximal order in G. Then $G = \langle a \rangle \oplus H$ for some subgroup H of G.

Proof Let $|G| = p^n$. We use induction on n. If $n = 1$, then G is cyclic and $G = \langle a \rangle \oplus \langle 0 \rangle$. So assume the proposition is true for all Abelian groups of order p^k, where $k < n$. Let $a \in G$ be of maximal order, say $|a| = p^r$, where $r \leq n$. So for all $x \in G$, $|x| \leq p^r$. Note that if $r = n$, then $G = \langle a \rangle$ and we are done. So suppose $r < n$. Let $b \in G$ be of minimal order with $b \notin \langle a \rangle$. So if $x \in G$ and $|x| < |b|$, then $x \in \langle a \rangle$.
Claim $\langle a \rangle \cap \langle b \rangle = \{0\}$
Proof of Claim We prove the claim by showing that $|b| = p$. Note that $|pb| = |b|/p$, so $pb \in \langle a \rangle$, and therefore $pb = ma$ for some integer m. Since $|a| = p^r$ and a was chosen to be of maximal order, we have $0 = p^r b = p^{r-1}(pb) = p^{r-1}(ma)$. Hence $|ma| \leq p^{r-1}$. Hence by Corollary 1.3.20, $\gcd(p^r, m) \neq 1$, and p divides m. So let $m = ps$. Then we have $pb = ma = psa$. Now consider the element $-sa + b$. Obviously, $p(-sa + b) = 0$, and $-sa + b \notin \langle a \rangle$ since $b \notin \langle a \rangle$. Therefore, $-sa + b$ is an element of G not in $\langle a \rangle$ of order p. Since b was chosen to be of minimal order for an element of G not in $\langle a \rangle$, we must have $|b| = p$, which implies the claim.
Having proved the claim, we now return to the proof of the proposition and proceed by considering the quotient group $G^* = G/\langle b \rangle$. Since $|G^*| = p^{n-1}$, by our induction hypothesis the proposition holds for G^*. Let $a^* = a + \langle b \rangle \in G^*$. The order $|a^*|$ is the least positive integer k such that $ka \in \langle b \rangle$. Since $\langle a \rangle \cap \langle b \rangle = \{0\}$, we have $|a^*| = |a| = p^r$. By Proposition 2.2.15, part (4), a^* must be an element of maximal order in G^*. Therefore, by our induction hypothesis, $G^* = \langle a^* \rangle \oplus H^*$ for some subgroup H^* of G^*. Now consider the homomorphism $\phi: G \to G/\langle b \rangle = G^*$ and let $H = \phi^{-1}(H^*)$. Then by Proposition 2.2.15, part (6), H is a subgroup of G with $H/\langle b \rangle \cong H^*$, and hence $|H| = |H^*|p$, and $|G| = |G^*||\langle b \rangle| = |G^*|p = |\langle a^* \rangle||H^*|p = p^r|H^*|p = p^r|H| = |\langle a \rangle||H|$. Finally, to show that $G = \langle a \rangle \oplus H$, by Corollary 3.3.7 it only remains to show that $\langle a \rangle \cap H = \{0\}$. So let $x \in \langle a \rangle \cap H$. Then $x + \langle b \rangle \in \langle a^* \rangle \cap H^* = \{0\}$, since $G^* = \langle a^* \rangle \oplus H^*$. Hence $x \in \langle b \rangle$, and since $\langle a \rangle \cap \langle b \rangle = \{0\}$, $x = 0$, and the proof that $G = \langle a \rangle \oplus H$ is complete. □

This was a long proof, but as you will see with the next examples, the proposition is a powerful result, and some of the steps in the proof are very useful.

3.4.9 EXAMPLE Let G be an Abelian group of order $|G| = 27 = 3^3$, and let $a \in G$ be of maximal order. Then $|a| = 3, 3^2$, or 3^3.

Case 1 $|a| = 3^3$. Then $G \cong \mathbb{Z}_{27}$.

Case 2 $|a| = 3^2$. Then $G = \langle a \rangle \oplus H$, where H is a subgroup of order 3. In this case $G \cong \mathbb{Z}_9 \times \mathbb{Z}_3$.

Case 3 $|a| = 3$. Then $G = \langle a \rangle \oplus H$, where H is a subgroup of order 9. Since a was of maximal order, H cannot have an element of order greater than 3. Applying Proposition 3.4.8 to H, we get $H \cong \mathbb{Z}_3 \times \mathbb{Z}_3$ and therefore $G \cong \mathbb{Z}_3 \times \mathbb{Z}_3 \times \mathbb{Z}_3$.

Note that no two of $\mathbb{Z}_{27}$, $\mathbb{Z}_9 \times \mathbb{Z}_3$, $\mathbb{Z}_3 \times \mathbb{Z}_3 \times \mathbb{Z}_3$ are isomorphic. ◇

In this example we were able to find all Abelian groups of order 27, and we found there are exactly three, up to isomorphism.
The next step in our proof, the third, is very brief and simple.

3.4.10 PROPOSITION Let p be a prime and G a finite Abelian p-group. Then $G = G_1 \oplus G_2 \oplus \ldots \oplus G_s$, where each G_i is cyclic and $|G_1| \geq |G_2| \geq \ldots \geq |G_s|$.

Proof We use induction on $|G|$. If $|G| = 1$, 2, or 3, there is nothing to prove. So assume the proposition is true for all Abelian groups of order less than $|G|$. By Proposition 3.4.8, $G = \langle a_1 \rangle \oplus H$, where a_1 is of maximal order. Therefore, if a_2 is of maximal order in H, then $|a_1| \geq |a_2|$. Let $G_1 = \langle a_1 \rangle$, and apply our induction hypothesis to H to get $H = G_2 \oplus \ldots \oplus G_s$, where each G_i is cyclic and $|G_2| \geq \ldots \geq |G_s|$, to complete the proof. □

We have now proved the main part of the fundamental theorem. By Corollary 3.4.5, if G is a finite Abelian group of order $|G| = p_1^{a_1} p_2^{a_2} \ldots p_k^{a_k}$, where the p_i are distinct primes, then G is a direct sum of subgroups of orders $p_i^{a_i}$. By Proposition 3.4.10, each p_i-subgroup can be further decomposed into the direct sum of cyclic groups. Putting these results together, $G = C_1 \oplus \ldots \oplus C_r$, where the C_i are cyclic and of order a power of a prime, and is therefore isomorphic to a direct product of groups $\mathbb{Z}_{n_i}$, where each n_i is a prime power.
What remains to be proved is that these n_i are unique except for the order in which they appear. It will suffice to prove this for any Abelian p-group.
In order to make the proof of the fourth and last step easier to follow, we again illustrate the main ideas of the proof with some examples.

3.4.11 EXAMPLE Let G and H be Abelian groups of order 2^4. Suppose $G = G_1 \oplus G_2$, where G_1 and G_2 are cyclic subgroups of order 4, so $G \cong \mathbb{Z}_4 \times \mathbb{Z}_4$. And suppose $H = H_1 \oplus H_2 \oplus H_3$, where H_1 is a cyclic subgroup of order 4, and H_2 and H_3 are cyclic subgroups of order 2, so $H \cong \mathbb{Z}_4 \times \mathbb{Z}_2 \times \mathbb{Z}_2$. Let us calculate the number of elements of order 1 or 2 in G and in H. Let $G^{(2)} = \{x \in G \mid |x| = 1 \text{ or } 2\}$. If $x, y \in G^{(2)}$, then $2(x - y) = 2x - 2y = 0 - 0 = 0$, and $(x - y) \in G^{(2)}$, so $G^{(2)}$ is a subgroup of G by the subgroup test. If $x = g_1 + g_2$, where $g_1 \in G_1$ and $g_2 \in G_2$, then by Corollary 3.3.10, $|x|$ divides 2 if and only if $|g_1|$ and $|g_2|$ both divide 2. This means $G^{(2)} = G_1^{(2)} + G_2^{(2)}$, where $G_i^{(2)} = G^{(2)} \cap G_i = \{x \in G_i \mid |x| = 1 \text{ or } 2\}$. By Lemma 3.3.12 this sum is direct, so by Corollary 3.3.9 $|G^{(2)}| = |G_1^{(2)}||G_2^{(2)}|$. But the set of elements of order 1 or 2 in a cyclic group of order divisible by 2 is just the unique cyclic subgroup of order 2, so $|G_1^{(2)}| = |G_2^{(2)}| = 2$, and $|G^{(2)}| = 2^2$. A similar calculation shows that if $H^{(2)} = \{x \in G \mid |x| = 1 \text{ or } 2\}$, then $|H^{(2)}| = |H_1^{(2)}||H_2^{(2)}||H_3^{(2)}| = 2^3$. Since they have different numbers of elements of order 2, G and H cannot be the group or be isomorphic groups, and $\mathbb{Z}_4 \times \mathbb{Z}_4$ is not isomorphic to $\mathbb{Z}_4 \times \mathbb{Z}_2 \times \mathbb{Z}_2$. ◇

3.4.12 EXAMPLE Let $G = G_1 \oplus G_2 \oplus G_3$ with $G_1 \cong \mathbb{Z}_8$, $G_2 \cong \mathbb{Z}_4$, and $G_3 \cong \mathbb{Z}_4$, and let $H = H_1 \oplus H_2 \oplus H_3$ with $H_1 \cong \mathbb{Z}_8$, $H_2 \cong \mathbb{Z}_8$, and $H_3 \cong \mathbb{Z}_2$. Let $G^{(2)} = \{x \in G \mid |x| = 1 \text{ or } 2\}$ and $H^{(2)} = \{x \in G \mid |x| = 1 \text{ or } 2\}$. Then a calculation like that in the preceding example shows $|G^{(2)}| = 2^3$ and $|H^{(2)}| = 2^3$. So we cannot distinguish G and H by their numbers of elements of order 2. Consider, however, the quotient group $G/G^{(2)}$. By Proposition 3.3.13, $G/G^{(2)} \cong G_1/G_1{}^{(2)} \oplus G_2/G_2{}^{(2)} \oplus G_3/G_3{}^{(2)}$, where, as in the preceding example, $G_i{}^{(2)} = G^{(2)} \cap G_i = \{x \in G_i \mid |x| = 1 \text{ or } 2\}$ is the cyclic subgroup of G_i of order 2, and hence $|G_i/G_i{}^{(2)}| = |G_i|/2$. So $G/G^{(2)} \cong \mathbb{Z}_4 \times \mathbb{Z}_2 \times \mathbb{Z}_2$. A similar argument shows $H/H^{(2)} \cong \mathbb{Z}_4 \times \mathbb{Z}_4 \times \{0\} \cong \mathbb{Z}_4 \times \mathbb{Z}_4$. Since we have seen in the preceding example that these quotient groups are not isomorphic, G and H cannot be the same group or be isomorphic groups, and $\mathbb{Z}_8 \times \mathbb{Z}_4 \times \mathbb{Z}_4$ is not isomorphic to $\mathbb{Z}_8 \times \mathbb{Z}_8 \times \mathbb{Z}_2$. ◇

3.4.13 PROPOSITION Let p be a prime and G a finite Abelian p-group. If

$$G = G_1 \oplus \ldots \oplus G_r \text{ and } G = H_1 \oplus \ldots \oplus H_s$$

where the G_i and H_i are cyclic with

$$|G_1| \geq \ldots \geq |G_r| \text{ and } |H_1| \geq \ldots \geq |H_r|$$

then

(1) $r = s$
(2) $G_i \cong H_i$ for all i

Proof (1) Let $G^{(p)} = \{x \in G \mid |x| = 1 \text{ or } p\}$, and for any subgroup K of G let $K^{(p)} = K \cap G^{(p)} = \{x \in K \mid |x| = 1 \text{ or } p\}$. If $x, y \in G^{(p)}$, then $p(x - y) = px - py = 0 - 0 = 0$, and $(x - y) \in G^{(p)}$, so $G^{(p)}$ is a subgroup of G by the subgroup test. If $x = g_1 + \ldots + g_r$, where $g_i \in G_i$ for all i, $1 \leq i \leq r$, then by Corollary 3.3.10, $|x|$ divides p if and only if $|g_i|$ divides p for all i. This means $G^{(p)} = G_1{}^{(p)} + \ldots + G_r{}^{(p)}$. By Lemma 3.3.12 this sum is direct, so by Corollary 3.3.9 $|G^{(p)}| = |G_1{}^{(p)}| \ldots |G_r{}^{(p)}|$. But the set of elements of order 1 or p in a cyclic group of order divisible by p is just the unique cyclic subgroup of order p. (See Exercise 21.) So $|G_i{}^{(p)}| = p$ for all i, and $|G^{(p)}| = p^r$. A similar calculation using the H_i in place of the G_i shows $|G^{(p)}| = p^s$. Hence $r = s$.

(2) To show $G_i \cong H_i$ for all i, $1 \leq i \leq r$, we use induction on n, where $|G| = p^n$. If $n = 1$, then G is cyclic and there is nothing to prove. So suppose the proposition is true for all Abelian p-groups of order p^r, where $r < n$. Consider the quotient $G/G^{(p)}$. By Proposition 3.3.13 we have

$$G/G^{(p)} \cong G_1/G_1{}^{(p)} \oplus \ldots \oplus G_r/G_r{}^{(p)}$$

$$G/G^{(p)} \cong H_1/H_1{}^{(p)} \oplus \ldots \oplus H_r/H_r{}^{(p)}$$

Thus we get two decompositions of the Abelian p-group $G/G^{(p)}$, whose order is less than p^n, and by our induction hypothesis $G_i/G_i{}^{(p)} \cong H_i/H_i{}^{(p)}$ for all i. Since as we noted in the proof of (1), $|G_i{}^{(p)}| = |H_i{}^{(p)}| = p$ for all i, it follows that $|G_i| = |H_i|$ for all i, and since G_i and H_i are cyclic, $G_i \cong H_i$ for all i, to complete the proof of the proposition. □

The proof of the Fundamental theorem now follows easily.

Proof (of Theorem 3.4.1, the fundamental theorem of finite Abelian groups) Let G be an Abelian group of finite order, $|G| = p_1^{a_1}p_2^{a_2}\ldots p_k^{a_k}$. Then by Corollary 3.4.5,

$$G \cong G(p_1^{a_1}) \oplus \ldots \oplus G(p_k^{a_k})$$

where $|G(p_i^{a_i})| = p_i^{a_i}$. By Proposition 3.4.10,

$$G(p_i^{a_i}) \cong \mathbb{Z}_{p_i^{t_1}} \times \ldots \times \mathbb{Z}_{p_i^{t_s}}$$

where $a_i = t_1 + \ldots + t_s$. Thus part (1) of Theorem 3.4.1 is proved. Finally, part (2) follows from Proposition 3.4.13. □

3.4.14 EXAMPLE Let us find up to isomorphism all Abelian groups of order 180. For such a group G we have $G \cong \mathbb{Z}_{p_1^{a_1}} \times \ldots \times \mathbb{Z}_{p_k^{a_k}}$, where $|G| = 180 = p_1^{a_1}\ldots p_k^{a_k}$ and the primes p_i are not necessarily distinct, and $180 = 2^2 \cdot 3^2 \cdot 5$. Therefore, G could be any of the following

$$G \cong \mathbb{Z}_4 \times \mathbb{Z}_9 \times \mathbb{Z}_5 \cong \mathbb{Z}_{180}$$
$$G \cong \mathbb{Z}_2 \times \mathbb{Z}_2 \times \mathbb{Z}_9 \times \mathbb{Z}_5 \cong \mathbb{Z}_2 \times \mathbb{Z}_{90}$$
$$G \cong \mathbb{Z}_4 \times \mathbb{Z}_3 \times \mathbb{Z}_3 \times \mathbb{Z}_5 \cong \mathbb{Z}_3 \times \mathbb{Z}_{60}$$
$$G \cong \mathbb{Z}_2 \times \mathbb{Z}_2 \times \mathbb{Z}_3 \times \mathbb{Z}_3 \times \mathbb{Z}_5 \cong \mathbb{Z}_6 \times \mathbb{Z}_{30}$$

Note that when we list all the nonisomorphic Abelian groups in a given order, we use the fundamental theorem and Theorem 3.2.8, which tells us that $\mathbb{Z}_m \times \mathbb{Z}_n \cong \mathbb{Z}_{mn}$ if and only if m and n are relatively prime. ◇

We end this section with some immediate consequences of the fundamental theorem.

3.4.15 DEFINITION A group G is said to be **decomposable** if it is the direct sum of two nontrivial proper subgroups. Otherwise it is said to be **indecomposable**. ○

3.4.16 THEOREM Let G be a finite Abelian group. Then G is indecomposable if and only if $G \cong \mathbb{Z}_{p^r}$ for some prime p and positive integer r.

Proof If G is indecomposable we obtain immediately by Theorem 3.4.1 that $G \cong \mathbb{Z}_{p^r}$. Conversely, if $G \cong \mathbb{Z}_{p^r}$, then G is indecomposable by Corollary 3.3.11. □

In the case of a finite cyclic group G we know that if m divides $|G|$, then G has a unique subgroup of order m. We are now able to prove something comparable for finite Abelian groups.

3.4.17 THEOREM If m divides the order of a finite Abelian group G, then G has a subgroup of order m.

Proof We have $G = G_1 \oplus \ldots \oplus G_k$, where G_i is a cyclic group of prime power order $p_i^{a_i}$ for all i, $1 \leq i \leq k$, and where $|G| = p_1^{a_1}\ldots p_k^{a_k}$. Since m divides $|G|$, m can be written as $m = p_1^{b_1}...p_k^{b_k}$, where $b_i \le a_i$ for all i. For each i, the cyclic group G_i of order $p_i^{a_i}$ has a subgroup H_i of order $p_i^{b_i}$. Consider $H = H_1 + \ldots + H_k$. By Lemma 3.3.12 this sum is direct, so by Corollary 3.3.9 $|H| = |H_1|\ldots|H_k| = m$. □

3.4.18 THEOREM Let m be a square-free integer or, in other words, an integer not divisible by the square of any prime. Then any Abelian group of order m is cyclic.

Proof Since m is square free, $m = p_1p_2\ldots p_k$ for distinct primes p_i. Hence by Theorem 3.4.1, for any group G of order m we have $G \cong \mathbb{Z}_{p_1} \times \mathbb{Z}_{p_2} \times \ldots \times \mathbb{Z}_{p_k}$, and by Corollary 3.3.11 this group is cyclic. □

Exercises 3.4

In Exercises 1 through 10 find up to isomorphism all Abelian groups of the indicated orders .

1. $n = 6$ **2.** $n = 9$ **3.** $n = 10$ **4.** $n = 12$ **5.** $n = 16$

6. $n = 20$ **7.** $n = 60$ **8.** $n = 60$ **9.** $n = 72$ **10.** $n = 108$

11. Find up to isomorphism all Abelian groups of order 32 that have exactly two subgroups of order 4.

12. Find up to isomorphism all Abelian groups of order 32 that have no elements of order 4.

13. Show that any Abelian group of order 6 contains an element of order 6.

14. Let p be a prime. Determine how many Abelian groups there are of order p^5.

15. Let p and q be distinct primes. Determine how many Abelian groups there are of the following orders:

(a) pq (b) pq^2 (c) p^2q^2

16. Let p be a prime. Determine up to isomorphism all Abelian groups of order p^n that contain an element of order p^{n-2}.

17. Determine whether the following pairs of Abelian groups are isomorphic.

(a) $\mathbb{Z}_{180} \times \mathbb{Z}_{42} \times \mathbb{Z}_{35}$ and $\mathbb{Z}_{315} \times \mathbb{Z}_{140} \times \mathbb{Z}_6$

(b) $\mathbb{Z}_{20} \times \mathbb{Z}_{70} \times \mathbb{Z}_{14}$ and $\mathbb{Z}_{28} \times \mathbb{Z}_{28} \times \mathbb{Z}_{25}$

18. Describe the positive integers n such that $\mathbb{Z}_n$ is up to isomorphism the only Abelian group of order n.

19. Let p and q be distinct primes and G and Abelian group of order $|G| = n$, where both p and q divide n. Show that G contains a cyclic subgroup of order pq.

20. Let G be a finite Abelian group and p a prime such that for all $x \in G$ the order of x is a power of p. Show that G is a p-group.

21. Let G be a finite Abelian p-group for some prime p. Consider $G^{(p)} = \{x \in G \mid |x| = 1 \text{ or } p\}$, and let H be a nontrivial cyclic subgroup of G. Show that $|G^{(p)} \cap H| = p$.

22. Determine up to isomorphism all Abelian groups of order 625. In each case

(a) Calculate $|G^{(5)}|$, where $G^{(5)} = \{x \in G \mid |x| = 1 \text{ or } 5\}$.

(b) Find $G/G^{(5)}$.

23. Let G be a finite Abelian p-group for some prime p. Let $G^{(p)}$ be as in Exercise 21, and let $pG = \{pg \mid g \in G\}$. Show that

(a) pG is a subgroup of G.

(b) $pG \cong G/G^{(p)}$

24. Let G_1 and G_2 be finite Abelian groups, and for every prime p that divides the order of G_i let $G_i(p) = \{g \in G_i \mid |g| \text{ is a power of } p\}$. (See Proposition 3.4.3.) If $\phi: G_1 \to G_2$ is a homomorphism, show that

(a) $\phi(G_1(p))$ is a subgroup of $G_2(p)$.

(b) $G_1 \cong G_2$ if and only if $G_1(p) \cong G_2(p)$ for all primes p.

25. Let G_1 and G_2 be finite Abelian groups. Show that G_1 and G_2 have the same number of elements of order n for all n, if and only if $G_1 \cong G_2$.

26. Let H and K be finite Abelian p-groups. Show that $H \times H \cong K \times K$ if and only if $H \cong K$.

Chapter 4

Group Actions

In this chapter we study the concept of a group *acting* on a set. The importance of this notion should become increasingly apparent over the course of this chapter as we consider a variety of applications. These include, among others, counting problems (applications of *Burnside's theorem*, in Section 4.3) and the analysis of the structure of groups of order pq or p^2q, where p and q are primes (applications of the *Sylow theorems*, in Section 4.7). Studying such topics using the concept of a *group action* helps bring unity to what otherwise would appear separate and unrelated developments.

4.1 Group Actions and Cayley's Theorem

The concept we introduce in this section involves two objects, a group G and a set X. Each element of the group G will determine a permutation of the elements of the set X. The operation on G will agree with the composition of the corresponding permutations of X. Our first example helps in visualizing this new concept.

4.1.1 EXAMPLE For any $\theta \in \mathbb{R}$ consider the matrix

$$A(\theta) = \begin{bmatrix} \cos\theta & -\sin\theta \\ \sin\theta & \cos\theta \end{bmatrix}$$

Let $G = \{A(\theta) \mid \theta \in \mathbb{R}\} \subseteq SL(2, \mathbb{R})$. With the operation of matrix multiplication, G is a subgroup of $SL(2, \mathbb{R})$. (See Exercise 1.) Let X be the plane $\mathbb{R}^2$. We represent a point as a column vector and use polar coordinates:

$$P(r, \phi) = \begin{bmatrix} x \\ y \end{bmatrix} = \begin{bmatrix} r\cos\phi \\ r\sin\phi \end{bmatrix}$$

So $X = \{P(r, \phi) \mid r, \phi \in \mathbb{R}, r \geq 0\}$. For any $A(\theta) \in G$ and any $P(r, \phi) \in X$ we have

$$A(\theta) \cdot P(r, \phi) = \begin{bmatrix} \cos\theta & -\sin\theta \\ \sin\theta & \cos\theta \end{bmatrix} \cdot \begin{bmatrix} r\cos\phi \\ r\sin\phi \end{bmatrix} =$$

$$\begin{bmatrix} r\cos\phi\cos\theta - r\sin\phi\sin\theta \\ r\cos\phi\sin\theta + r\sin\phi\cos\theta \end{bmatrix} = \begin{bmatrix} r\cos(\phi+\theta) \\ r\sin(\phi+\theta) \end{bmatrix} = P(r, \phi+\theta)$$

Here we have used the basic trigonometric identities for $\cos(\phi + \theta)$ and $\sin(\phi + \theta)$ as in the proof of Proposition 0.4.15. Note that the point $P(r, \phi + \theta)$ is obtained from the point $P(r, \phi)$ by rotating the plane around the origin by an angle of measure θ. Note also the following:

(1) The matrix

$$A(0) = \begin{bmatrix} \cos 0 & -\sin 0 \\ \sin 0 & \cos 0 \end{bmatrix} = \begin{bmatrix} 1 & 0 \\ 0 & 1 \end{bmatrix} = I$$

is the identity in G, and we have $A(0) \cdot P(r,\phi) = P(r,\phi + 0) = P(r,\phi)$ for any point $P(r,\phi) \in X$.

(2) For the product we have

$$A(\theta)A(\psi) = \begin{bmatrix} \cos\theta & -\sin\theta \\ \sin\theta & \cos\theta \end{bmatrix} \begin{bmatrix} \cos\psi & -\sin\psi \\ \sin\psi & \cos\psi \end{bmatrix} =$$

$$\begin{bmatrix} \cos\theta\cos\psi - \sin\theta\sin\psi & -\cos\theta\sin\psi - \sin\theta\cos\psi \\ \cos\theta\sin\psi + \sin\theta\cos\psi & \cos\theta\cos\psi - \sin\theta\sin\psi \end{bmatrix} =$$

$$\begin{bmatrix} \cos(\theta + \psi) & -\sin(\theta + \psi) \\ \sin(\theta + \psi) & \cos(\theta + \psi) \end{bmatrix} = A(\theta + \psi)$$

and $(A(\theta)A(\psi))P(r, \phi) = P(r, \phi + \theta + \psi) = A(\theta)(A(\psi)P(r, \phi))$. In other words, if we multiply the two matrices first and then rotate the point by the product of the two matrices, or if we first rotate by one matrix and then by the other, we get the same result in either case.

(3) Given two points $P(r, \phi)$ and $P(s, \psi)$ in X, then there exists a matrix $A(\theta)$ in G such that $A(\theta)P(r, \phi) = P(s, \psi)$ if and only if $s = r$, that is, if and only if the two points lie on the same circle $x^2 + y^2 = r^2$ centered at the origin. (See Exercise 2.) ◇

Having this example in mind, let us now give a formal definition. For any set X and group G we consider maps from $G \times X$ to X, written as $(g, x) \to g \cdot x$, where $g \in G$ and $x \in X$. We use, however, the information that G is actually a group and not just a set and concentrate on those maps that are in a certain sense compatible with the group structure of G.

4.1.2 DEFINITION A **group action** of a group G on a set X is a map from $G \times X$ to X with the following properties:

(1) $e \cdot x = x$ for all $x \in X$ and e the identity element of G.

(2) $g_1 \cdot (g_2 \cdot x) = (g_1 g_2) \cdot x$ for all $g_1, g_2 \in G$ and all $x \in X$.

When such an action exists, we say that the group G **acts** on X, and that X is a G-**set**. ○

In Example 4.1.1 we described a group action. The abstract definition can be illustrated by a variety of further concrete examples.

4.1.3 EXAMPLE The additive group $\mathbb{R}$ acts on the plane $\mathbb{R}^2$ by horizontal translation $(a, (x, y)) \to (x+a, y)$ and in a different way by vertical translation $(b,(x, y)) \to (x, y+b)$. ◇

4.1.4 EXAMPLE Let G be the cyclic group $\{e, a\}$ of order 2 and $X = \mathbb{C}$, the complex numbers. Then G acts on X by the action $(e, x + yi) \to x + yi$, $(a, x + yi) \to x - yi$, the complex conjugate of $x + yi$. ◇

4.1.5 EXAMPLE Every subgroup H of a group G (including the group G itself) acts on G by **left multiplication**, the action being $H \times G \to G$, where $(h, g) \to hg$. Properties (1) and (2) in this case are both obvious, being just the identity law and the associative law for the group operation. ◇

4.1.6 EXAMPLE Every subgroup H of a group G (again including the group G itself) acts on G by **conjugation**, the action being $H \times G \to G$, where $(h, g) \to hgh^{-1}$ for all $h \in H$ and $g \in G$. Property (1) is obvious, and property (2) follows from the equation $(h_1h_2)g(h_1h_2)^{-1} = h_1(h_2gh_2^{-1})h_1^{-1}$. ◇

Conjugation is a very important action that we will study closely in a later section. The following is also a fundamental example.

4.1.7 EXAMPLE Let $X = \{1, 2, \ldots, n\}$ and let S_n be the group of permutations of n elements. Then S_n acts on X as follows: $(\tau, i) \to \tau(i)$, where τ is a permutation in S_n and $i \in X$. Property (1) follows from the definition of the identity permutation, and property (2) from the definition of multiplication of permutations as composition of functions. ◇

The preceding example is fundamental because the action of any group G on a set X is closely related to the action of the group S_X of symmetries of X on X considered in the example. This relationship is spelled out in the next proposition.

4.1.8 PROPOSITION Let G be a group acting on X. Then

(1) for each $g \in G$ the map τ_g: $X \to X$ defined by $\tau_g(x) = g \cdot x$ is a permutation of X

(2) the map χ:$G \to S_X$ defined by $\chi(g) = \tau_g$ is a group homomorphism.

Proof (1) We need to show that τ_g:$X \to X$ is a permutation of X or, in other words, that it is one to one and onto. τ_g is one to one because if $\tau_g(x) = \tau_g(y)$, then $g \cdot x = g \cdot y$ and so $g^{-1} \cdot (g \cdot x) = g^{-1} \cdot (g \cdot y)$. Hence

$$x = e \cdot x = (g^{-1}g) \cdot x = g^{-1} \cdot (g \cdot x) = g^{-1} \cdot (g \cdot y) = (g^{-1}g) \cdot y = e \cdot y = y$$

And τ_g is onto because if $w \in X$, then

$$\tau_g(g^{-1} \cdot w) = g \cdot (g^{-1} \cdot w) = (gg^{-1}) \cdot w = e \cdot w = w$$

(2) We need to show that χ:$G \to S_X$ is a group homomorphism or, in other words, that for any $g_1,g_2 \in G$ we have $\chi(g_1g_2) = \chi(g_1) \circ \chi(g_2)$ in S_X. Recall that the operation on S_X is composition of functions, and that two permutations $\chi(g_1g_2)$ and

$\chi(g_1) \circ \chi(g_2)$ are equal if they agree on every element of X. So it is enough to show that for any $x \in X$ we have $\chi(g_1g_2)(x) = \chi(g_1) \circ \chi(g_2)(x)$. But we have the following identities:

$$\chi(g_1g_2)(x) = (g_1g_2) \cdot x = g_1 \cdot (g_2 \cdot x) = \tau_{g_1}(\tau_{g_2}(x))$$
$$= (\tau_{g_1} \circ \tau_{g_2})(x) = \chi(g_1) \circ \chi(g_2)(x)$$

which completes the proof. □

We can easily show that the converse of the preceding proposition is also true.

4.1.9 PROPOSITION Given a group homomorphism $\psi: G \to S_X$ from a group G to the group S_X of permutations of a set X, the map $G \times X \to X$ defined by $(g, x) \to g \cdot x = \psi(g)(x)$ is a group action of G on X.

Proof We need to show that the two conditions in Definition 4.1.2 are satisfied.

(1) $e \cdot x = \psi(e)(x) = \mathrm{id}(x) = x$, where $\mathrm{id} \in S_X$ is the identity permutation.

(2) $g_1 \cdot (g_2 \cdot x) = \psi(g_1)((\psi(g_2)(x)) = (\psi(g_1) \circ \psi(g_2))(x) = \psi(g_1g_2)(x) = (g_1g_2) \cdot x$. □

As will become apparent later, when we want to construct an example of a group that is not Abelian, we often look for subgroups of some S_n that satisfy appropriate conditions. The groups S_n seem to provide an inexhaustible supply of examples. The reason for this becomes obvious with the next result.

4.1.10 THEOREM (Cayley's theorem) Every group is isomorphic to a subgroup of a group of permutations.

Proof Let G be any group, and let it act on itself by left multiplication. So we have $G \times G \to G$, where $(g, x) \to gx$. By Proposition 4.1.8 there exists a group homomorphism $\chi: G \to S_G$ defined by $\chi(g) = \tau_g \in S_G$, where $\tau_g(x) = gx$ for all $x \in G$. The kernel of χ is the set of all $g \in G$ such that $\tau_g = \mathrm{id}$, the identity permutation. Hence $\mathrm{Kern}\,\chi = \{g \in G \mid gx = x \text{ for all } x \in G\}$ or, in other words, $\mathrm{Kern}\,\chi = \{e\}$, and χ is one to one and an isomorphism between G and $\mathrm{Im}(\chi)$, which is a subgroup of S_G. Therefore, G is isomorphic to a subgroup of a group of permutations. □

4.1.11 DEFINITION The homomorphism $\chi: G \to S_X$ associated with an action of a group G on a set X is called the **permutation representation** of that action. ○

4.1.12 EXAMPLE We have already given a permutation representation of the dihedral group D_4 in Example 1.4.24. More generally, the action of the dihedral group D_n on a regular n-gon gives a representation of D_n as a subgroup of the symmetric group S_n. ◇

4.1.13 DEFINITION The group G is said to act **faithfully** on the set X if $\mathrm{Kern}\,\chi = \{e\}$ or, in other words, if the only element of G that fixes every element of X is the identity $e \in G$. Thus G acts faithfully on X if and only if χ is one to one, in which case G is isomorphic to the subgroup $\mathrm{Im}(\chi)$ of S_X. ○

4.1.14 EXAMPLE The action of the dihedral group $D_n = \{e, \rho, \rho^2, , \rho^{n-1}, \tau, \rho\tau, \rho^2\tau, \ldots, \rho^{n-1}\tau\}$ on a regular n-gon is faithful. So is the action of any subgroup of D_n. For example, let $G = \{e, g, g^2\}$ be a cyclic group of order 3, let $X = \{1, 2, 3, 4, 5, 6\}$ be the six vertices of a regular hexagon, as in Figure 1.

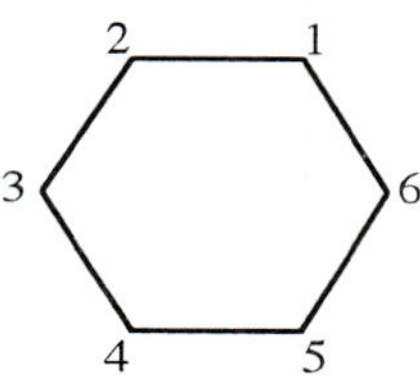

FIGURE 1

Let G act on X by having g rotate the hexagon counterclockwise by 120°. (So we are in effect identitifying G with the subgroup $\{e, \rho^2, \rho^4\}$ of the dihedral group D_6, where ρ is a counterclockwise rotation of 60°.) Then G acts faithfully on X and can be represented as the subgroup $\{e, (1\,3\,5)(2\,4\,6), (1\,5\,3)(2\,6\,4)\}$ of S_6, where the generator g is the permutation $g = (1\,3\,5)(2\,4\,6)$ in S_6. ◇

4.1.15 EXAMPLE Let $G = \{e, g, g^2\}$ be the cyclic group of order 3, let $X = \{1, 2, 3, 4, 5, 6, 7, 8\}$ be the eight vertices of a cube, and let G act on X by having g rotate the cube around the axis through vertices 2 and 8, so that $g \cdot 1 = 3$, $g \cdot 3 = 6$, and $g \cdot 6 = 1$, as in Figure 2.

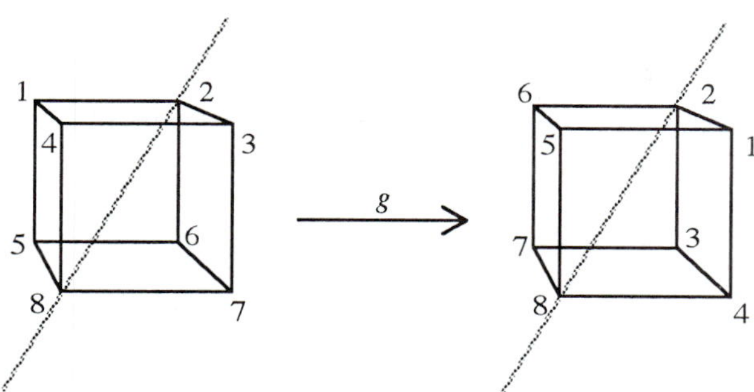

FIGURE 2

Then again the action is faithful and G may be represented as the subgroup $\{e, (1\,3\,6)(4\,7\,5), (1\,6\,3)(4\,5\,7)\}$ of S_8. In this case the generator g of G is represented by the permutation $g = (1\,3\,6)(4\,7\,5)$ ◇

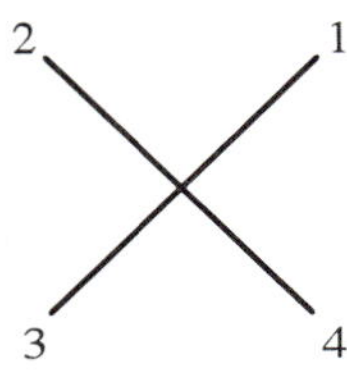

FIGURE 3

4.1.16 EXAMPLE As we have already said, the action of the dihedral group D_4 on the vertices $\{1, 2, 3, 4\}$ of the square is faithful. But we may consider D_4 instead acting on the set $\{d_1, d_2\}$ of the two diagonals of the square, where d_1 is the 1-3 diagonal and d_2 the 2-4 diagonal. (See Figure 3.) In this case the action is no longer faithful, since $\rho^2 \cdot d_1 = d_1$ and $\rho^2 \cdot d_2 = d_2$, where ρ is the group element $(1\ 2\ 3\ 4) \in D_4$ ◇

Exercises 4.1

1. Show that G as defined in Example 4.1.1 is a subgroup of $SL(2, \mathbb{R})$.

2. With G again as in Example 4.1.1, show that for any two points P and Q in the plane $\mathbb{R}^2$ there exists an $A \in G$ with $A \cdot P = Q$ if and only if both P and Q are points on the same circle $x^2 + y^2 = r^2$ for some $r > 0$.

In Exercises 3 through 7 (a) show how X may be regarded as a G-set in a natural way or, in other words, describe a natural group action of G on X; (b) show that the action has the two properties required by the definition of an action; and (c) give a permutation representation of the action.

3. $G = \{e, g\}$, the cyclic group of order 2. X = the vertices of an equilateral triangle.

4. $G = \{e, g, g^2\}$, the cyclic group of order 3. X = the vertices of an equilateral triangle.

5. $G \cong \mathbb{Z}_2 \times \mathbb{Z}_2$, a noncyclic group of order 4. X = the vertices of a square.

6. $G \cong \mathbb{Z}_2 \times \mathbb{Z}_2 \times \mathbb{Z}_2$. X = the vertices of a regular hexagon.

7. $G \cong \mathbb{Z}_2 \times \mathbb{Z}_2 \times \mathbb{Z}_2 \times \mathbb{Z}_2$. X = the vertices of a square.

8. Let $G = \mathbb{Z}$ and let X be the set of cosets of $5\mathbb{Z}$ in $\mathbb{Z}$. Give an example of an action of G on X, defined in a natural way, that is not faithful.

9. Show that the plane $\mathbb{R}^2$ is indeed a G-set with the horizontal translation action described in Example 4.1.3.

10. Let G be any group and X the set of all subgroups of G. Show that X is a G-set under conjugation: $(g, H) \to gHg^{-1}$.

11. Let $G = S_3$ and X the set of all subgroups of S_3. Write the table showing the action of G on X by conjugation (as in the preceding exercise). Is this a faithful action?

12. Let G and X again be as in Exercise 10, with G acting on X by conjugation. Describe the kernel of χ as in Proposition 4.1.8 in case G is an Abelian group.

13. Let H be a subgroup of a group G and X the set of all left cosets of H. Show how X can be made into a G-set in a natural way.

14.. Let X_1 and X_2 be G-sets for the same group G, and assume $X_1 \cap X_2 = \varnothing$. Show how $X_1 \cup X_2$ can be made into a G-set in a natural way.

In Exercises 15 through 18 let H be a subgroup of a group G and let X be the set $\{xH \mid x \in G\}$ of all left cosets of H in G. Let G act on H by left multiplication $(g, xH) \rightarrow gxH \in X$.

15. Show that this is indeed a group action.

16. Let $\chi: G \rightarrow S_X$ be the permutation representation of the action. Then

(a) Determine the kernel K of χ.

(b) Show that $K \subset H$.

(c) Show that if N is a normal subgroup of G and $N \subset H$, then $N \subset K$. In other words, show that K is the largest normal subgroup of G contained in H.

17. Show how Cayley's theorem follows from the preceding exercise.

18. Let $i = [G:H]$ be the index of H in G. Then

(a) Show that if χ is one to one, then $|G|$ divides $i!$

(b) Show that if $|G|$ does not divide $i!$, then the kernel K is nontrivial.

(c) Show that if $|G|$ does not divide $i!$, then G has a nontrivial proper normal subgroup.

4.2 Stabilizers and Orbits in a Group Action

In this section we show that a group action determines an equivalence relation on the G-set X, whose equivalence classes are called orbits, and we prove the main theorem that gives us the size of each orbit. This is a theorem that will be used several times in the rest of this chapter for various group actions.

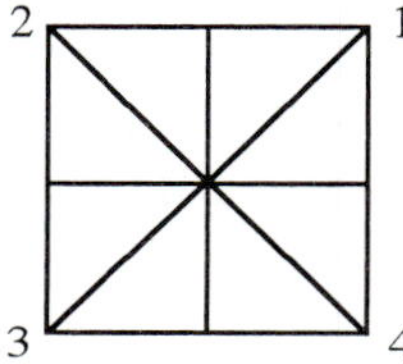

FIGURE 4

4.2.1 EXAMPLE Consider the dihedral group D_4. It acts in a natural way on the set consisting of the four vertices, 1, 2, 3, and 4 of the square, together with the 1-2, 2-3, 3-4, and 4-1 sides t_1, t_2, t_3, and t_4, and the 1-3 and 2-4 diagonals d_1 and d_2. Several of the new definitions we introduce in this section will be easier to visualize with the help of the adjoining table, which describes the action.

TABLE 1 Action of D_4

	1	2	3	4	t_1	t_2	t_3	t_4	d_1	d_2
ρ_0	1	2	3	4	t_1	t_2	t_3	t_4	d_1	d_2
ρ	2	3	4	1	t_2	t_3	t_4	t_1	d_2	d_1
ρ^2	3	4	1	2	t_3	t_4	t_1	t_2	d_1	d_2
ρ^3	4	1	2	3	t_4	t_1	t_2	t_3	d_2	d_1
τ	2	1	4	3	t_1	t_4	t_3	t_2	d_2	d_1
$\rho\tau$	3	2	1	4	t_2	t_1	t_4	t_3	d_1	d_2
$\rho^2\tau$	4	3	2	1	t_3	t_2	t_1	t_4	d_2	d_1
$\rho^3\tau$	1	4	3	2	t_4	t_3	t_2	t_1	d_1	d_2

4.2.2 PROPOSITION Let X be a G-set and $x \in X$, and let $G_x = \{ g \in G \mid g \cdot x = x\}$. Then G_x is a subgroup of G.

Proof It is enough to show (1) that if $g \in G_x$, then $g^{-1} \in G_x$; and (2) that if $g_1, g_2 \in G_x$, then $g_1g_2 \in G_x$. For (1), If $g \in G_x$, then $g \cdot x = x$ and hence $g^{-1} \cdot x = g^{-1} \cdot (g \cdot x) = (gg^{-1}) \cdot x = e \cdot x = x$ and $g^{-1} \in G_x$. For (2), If $g_1, g_2 \in G_x$, then $(g_1g_2) \cdot x = g_1 \cdot (g_2 \cdot x) = g_1 \cdot x = x$, and $g_1g_2 \in G_x$. □

4.2.3 DEFINITION The group G_x is called the **stabilizer** of x in G. ○

4.2.4 EXAMPLE Consider the preceding example of the action of D_4. From the table we can see, for example, that $g \cdot 2 = 2$ exactly when $g = \rho_0$ or $\rho\tau$, that $g \cdot d_1 = d_1$ exactly when $g = \rho_0, \rho^2, \rho\tau$, or $\rho^3\tau$ and that $g \cdot t_4 = t_4$ exactly when $g = \rho_0$ or $\rho^2\tau$. So $G_2 = \{\rho_0, \rho\tau\}$ and $G_{d_1} = \{\rho_0, \rho^2, \rho\tau, \rho^3\tau\}$ and $G_{t_4} = \{\rho_0, \rho^2\tau\}$. ◇

4.2.5 PROPOSITION Let G be a group acting on a set X. The relation on X defined by

$$a \sim b \text{ if and only if } a = g \cdot b \text{ for some } g \in G$$

for $a, b \in X$, is an equivalence relation.

Proof We verify the three properties of (1) reflexivity, (2) symmetry, and (3) transitivity. For (1), note that for any $a \in X$ we have $e \cdot a = a$ and hence $a \sim a$. For (2), note that for any $a, b \in X$, if $a \sim b$, then taking any $g \in G$ with $a = g \cdot b$ we have $g^{-1} \cdot a = g^{-1} \cdot (g \cdot b) = (g^{-1}g) \cdot b = e \cdot b = b$, and so $b \sim a$. For (3), note that for any $a, b, c \in X$, if $a \sim b$ and $b \sim c$, then taking any $g, h \in G$ with $a = g \cdot b$ and $b = h \cdot c$ we have $a = g \cdot b = g \cdot (h \cdot c) = (gh) \cdot c$, and so $a \sim c$. □

4.2.6 DEFINITION Let G be a group acting on a set X. Then the equivalence class $O_a = \{b \in X \mid a \sim b\}$ is called the **orbit** of G in X containing a. ○

4.2.7 EXAMPLE Consider the preceding example of the action of D_4. From the table we can see there are three orbits, the sets of vertices, of edges, and of diagonals. So, for instance, $O_2 = \{1, 2, 3, 4\}$, and $O_{d_1} = \{d_1, d_2\}$ and $O_{t_4} = \{t_1, t_2, t_3, t_4\}$. We have seen that G_2 is a group of order 2, while O_2 has 4 elements; also, G_{d_1} is a group of order 4, while O_{d_1} has 2 elements; again G_{t_4} is a group of order 2, while O_{t_4} has four elements. ◇

In the preceding example, we found $|G_a| \cdot |O_a| = |G|$ in every case. This is exactly what the next theorem tells us holds for *any* group action. ◇

4.2.8 THEOREM (The orbit-stabilizer relation) Let G be a group acting on a set X and a any element of the set X. Then the size of the orbit of a is equal to the index of the stabilizer of a; that is to say, $|O_a| = [G:G_a]$. If G is finite, then $|O_a| = |G| / |G_a|$.

Proof $[G:G_a]$ is the number of left cosets of G_a , so what we want to that show is that the set O_a and the set of left cosets of G_a have the same number of elements. To show that two sets have the same number of elements, we need to show that there is a one-to-one, onto map between the two sets. Now given $b \in O_a$, there exists a $g \in G$ such that $b = g \cdot a$. We define a map τ as follows:

$$b = g \cdot a \to gG_a$$

Now first, τ is well defined because if $b = g_1 \cdot a$ and $b = g_2 \cdot a$, then

$$(g_1{}^{-1}g_2) \cdot a = g_1{}^{-1} \cdot (g_2 \cdot a) = g_1{}^{-1} \cdot (g_1 \cdot a) = (g_1{}^{-1}g_1) \cdot a = e \cdot a = a$$

and so $g_1{}^{-1}g_2 \in G_a$ and $g_1G_a = g_2G_a$.

Further, τ is one to one since $\tau(b_1) = \tau(b_2)$ if and only if $g_1G_a = g_2G_a$, where $b_1 = g_1 \cdot a$ and $b_2 = g_2 \cdot a$, and $g_1G_a = g_2G_a$ if and only if $g_1 = g_2h$ for some $h \in G_a$, in which case

$$b_1 = g_1 \cdot a = (g_2h) \cdot a = g_2 \cdot (h \cdot a) = g_2 \cdot a = b_2$$

Finally, τ is onto because given any left coset $g'G_a$, we have $g' \cdot a \in O_a$, and $\tau(g' \cdot a) = g'G_a$. Thus $|O_a| = [G:G_a]$ as asserted.

Recall that if G is finite, then $[G:H] = |G|/|H|$ for any subgroup H. □

4.2.9 EXAMPLE Let $X = \{1, 2, \ldots, n\}$ be a set with n elements, and let $G = S_n$ the group of permutations of n elements. Given any two elements $i, j \in X$, there exists a permutation $\tau \in S_n$ such that $\tau(i) = j$. Hence under the action of S_n we have only one orbit, namely, all of X. If we pick any element of x, let's say $3 \in X$, the orbit O_3 of 3 is X, and the stabilizer G_3 of 3 is isomorphic to S_{n-1}, the group of permutations of the n-1 elements $\{1, 2, 4, 5, \ldots, n\}$. Hence we obtain $|S_n| = |O_3| \cdot |G_3| = n|S_{n-1}|$. ◇

4.2.10 DEFINITION Let G be a group acting on a set X. The action of G on X is called **transitive** if there is only one orbit in the action, that is, for any two elements $a, b \in X$, there exists some $g \in G$ such that $g \cdot a = b$. ○

4.2.11 EXAMPLE In Example 4.1.3 the action is not transitive, because the orbit of any point (a, b) in the plane under horizontal translation is just the horizontal line $y = b$. ◇

4.2.12 EXAMPLE Let X be the n vertices of a regular plane n-gon with $n \geq 3$. The action of the dihedral group D_n (illustrated in Example 1.1.18) on X is transitive. Hence for any $a \in X$, the orbit O_a has n elements. If we pick any vertex, say 1, and any $\sigma \in G_1$, the stabilizer of vertex 1, then either $\sigma(2) = 2$, in which case σ is the identity, or $\sigma(2) = n$, the vertex on the other side of 1 from 2, in which case σ is a flip through an axis passing through vertex 1, and $\sigma(i) = n - i + 2$ for all $i \in X$. Hence $|G_1| = 2$ and $|D_n| = |O_1| \cdot |G_1| = n \cdot 2$. ◇

4.2.13 EXAMPLE Let G be the group of proper rotations of a cube. The action of G on the 8 vertices is transitive. If we pick any vertex, say 3, and any $\tau \in G_2$, then either $\tau(3) = 3$, in which case τ is the identity; or $\tau(3) = 6$, in which case $\tau(6) = 1$ and $\tau(1) = 3$; or else $\tau(3) = 1$, in which case $\tau(1) = 6$ and $\tau(6) = 3$. Hence $|G_2| = 3$ and $|G| = |O_2| \cdot |G_2| = 8 \cdot 3 = 24$. (See the illustration in Example 4.1.15.) ◇

We end this section by pointing out an immediate consequence of Proposition 4.2.5 and Theorem 4.2.8 that will play a basic role later on in this chapter.

4.2.14 THEOREM Let X be a finite G-set. Let N be the number of orbits in the action, and let $a_1, a_2, \ldots, a_N$ be a complete set of representatives of the orbits (meaning that every element of X is in exactly one of the orbits O_{a_i}). Then

$$|X| = \sum_{i=1}^{N} [G{:}G_{a_i}]$$

where G_{a_i} is the stabilizer of a_i.

Proof By Proposition 4.2.5 the action of G on X determines an equivalence relation with the orbits O_{a_i} as the equivalence classes. By Theorem 0.2.7 we obtain a partition of the set X. Hence

$$|X| = \sum_{i=1}^{N} |O_{a_i}|$$

and the theorem follows by 4.2.8. □

Exercises 4.2

In Exercises 1 through 8 (a) find the stabilizer G_a for each a indicated, (b) find the orbit O_a, (c) check the orbit-stabilizer relation $|O_a| = [G{:}G_a]$, (d) determine whether the action is transitive, and (e) determine whether G acts faithfully on X.

1. $X = \{1, 2, 3\}$; $G = S_3$; $a = 1, 2, 3$

2. $X = \{1, 2, 3, 4\}$; $G = A_3 \subseteq S_3$; $a = 1, 2, 3$

3. $X = \{1, 2, 3, 4\}$; $G = S_4$; $a = 1, 3, 4$

4. $X = \{1, 2, 3, 4\}$; $G = A_4 \subseteq S_4$; $a = 1, 3, 4$

5. $X = \{1, 2, 3, 4, 5, 6, 7, 8\}$;
$G = \{\rho_0 = \text{identity}, (1\ 2\ 3\ 4)(5\ 7), (1\ 3)(2\ 4), (1\ 4\ 3\ 2)(5\ 7)\} \subseteq S_7$; $a = 1, 3, 6, 7$

6. X = the four vertices of a square $\{1, 2, 3, 4\}$; $G = D_4$; $a = 1, 2, 3$

7. $X = \{1, 2, 3, 4\}$, the four vertices of a square; $G = \langle\rho\rangle$, the subgroup generated by the rotation ρ of 90° in D_4; $a = 1, 3$

8. $X = \{1, 2, 3, 4\}$, the four vertices of a square; $G = \langle\rho^2, \tau\rangle$, the subgroup generated by the rotation ρ^2 of 180° and the flip τ in D_4; $a = 1, 3$

9. Let $X = \mathbb{C}\text{-}\{0,\text{-}1\}$, the complex plane with 0 and -1 deleted. For $z \in X$ let $T_0(z) = z$, $T_1(z) = -1/(1+z)$, $T_2 = (1+z)/-z$, and let $G = \{T_0, T_1, T_2\}$.

(a) Show that G is a group under composition of functions.
(b) Show how G acts on X in a natural way.
(c) Find all $a \in X$ such that $G_a = G$.

10. For $a, b \in \mathbb{R}$, let $g(a,b) \in M(2,\mathbb{R})$ be the matrix

$$g(a,b) = \begin{bmatrix} a & b \\ 0 & 1 \end{bmatrix}$$

Let $G = \{g(a, b) \mid a, b \in \mathbb{R}, a \neq 0\}$, with the operation of matrix multiplication, and let $X = \mathbb{R}$, the real line. For $g = g(a, b) \in G$ and $x \in X$, define $g \cdot x = ax + b$.

(a) Show that G is a subgroup of $GL(2, \mathbb{R})$.
(b) Show that $g \cdot x = ax + b$ defines an action of G on X.
(c) Find G_0 and O_0.
(d) Does G act faithfully on X?
(e) Is this action transitive?

11. Let H be a subgroup of a group G and let $X = \{gH \mid g \in G\}$, the set of all left cosets of H in G. Let G act on X by left multiplication (as in Exercises 15 through 18 of Section 4.1).

(a) Show that G_{aH}, the stabilizer of $aH \in X$, is the subgroup aHa^{-1} of G.
(b) Show that for any element $aH \in X$, $|O_{aH}| = [G:H]$.

12. Let G be a finite group with $|G| = n$, and let p be the smallest prime dividing n. Show that if H is a subgroup of G with $[G : H] = p$, then $H \lhd G$.

4.3 Burnside's Theorem and Applications

In this section we apply the orbit-stabilizer relation of the preceding section (Theorem 4.2.8) to prove Burnside's theorem, which gives a method of counting the number of orbits of a set under the action of a group of symmetries. We also illustrate how Burnside's theorem can be applied to various counting problems, in which we attempt to determine the number of "essentially different" designs or patterns of some kind.

4.3.1 EXAMPLE Consider all possible ways of coloring black two of the vertices of a square. If the question is how many pairs of vertices there are that we might choose to color, then the answer would be the binomial coefficient $4!/2!(4\text{-}2)! = 6$, the "designs" being illustrated in the adjoining figure. But if the problem is to find how many kinds of square tiles there are with two corners painted black, then the answer is just two. For the designs in each row of the figure are "essentially the same." Each row represents a single tile rotated in four different ways.

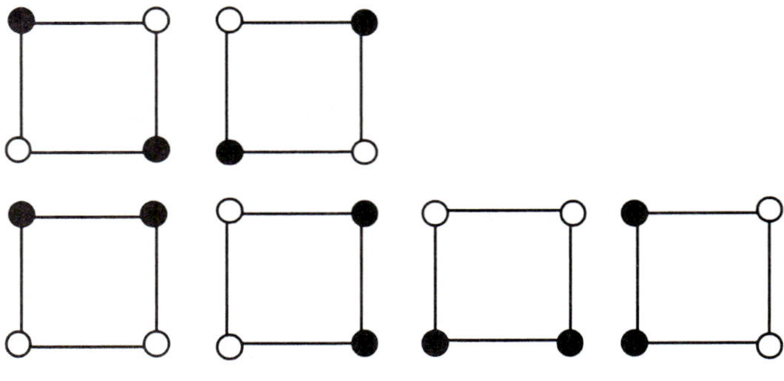

FIGURE 5

We can analyze this problem in terms of group actions as follows. If X is the set of the six different designs represented in Figure 5, and $G = <\rho>$ is the subgroup of the group of symmetries of the square, that is, of the dihedral group D_4, generated by the counterclockwise rotation ρ of 90°, then we see that when G acts on X in the natural way, X consists of just two orbits, illustrated by the two rows of the figure. The two items in the top row are equivalent under this action, since the second can be obtained from the first by applying ρ; and so we consider them essentially the same. (Applying ρ^2 takes us back to the first, and applying ρ^3 takes us back to the second again.) Likewise the four tiles in the bottom row are equivalent, since the second, third, and fourth can be obtained from the first by applying ρ, ρ^2, and ρ^3, respectively; and so they also are essentially the same. ◇

In the problems we consider in this section, X consists of different designs, and two designs A and B in X are considered essentially the same if they are equivalent under the action on X of some appropriate group G of permutations or, in other words, if they belong to the same orbit in the action of G on X. So if we want to count the number of essentially different designs, what we need to count is the number of orbits. The next theorem gives us a tool for doing so, but first we need one more definition.

4.3.2 DEFINITION Recall that for G acting on a set X, and for $a \in X$, we have already introduced the notation G_a for the stabilizer of a, which is a subgroup of G, and O_a for the orbit of a, which is a subset of X. Now for $g \in G$ we introduce the notation X_g for the set $\{x \in X \mid g \cdot x = x\}$ of elements in X **fixed** by g. ○

It is helpful to remember that X_g is a subset of X for a given $g \in G$ and G_a is a subgroup of G for any $a \in X$.

4.3.3 THEOREM (Burnside's theorem) Let G be a finite group acting on a finite set X. If N is the number of orbits in X under the action of G, then

$$N = \frac{1}{|G|} \cdot \Sigma_{g \in G} |X_g|$$

Proof In $G \times X$ consider all pairs (g, a) where $g \cdot a = a$. Let n be the number of such pairs. We count them in two different ways. First, for a fixed $g \in G$, the number of such pairs that are of form $(g, \text{-})$ is exactly $|X_g|$. Hence we get the following expression for n:

$$n = \Sigma_{g \in G} |X_g|$$

Second, for a fixed $a \in X$, the number of such pairs that are of the form $(\text{-}, a)$ is exactly $|G_a|$. Hence $n = \Sigma_{g \in G} |G_a|$. By Theorem 4.2.8 we know that $|O_a| = [G{:}G_a]$, and since G is a finite group, $|G_a| = |G| / [G{:}G_a] = |G| / |O_a|$. Hence we get the following expression for n:

$$n = |G| \cdot \Sigma_{a \in X} \frac{1}{|O_a|}$$

Finally, we note that $\Sigma_{a \in X} 1 / |O_a|$ is precisely the number N of orbits in X under the action of G. This is because for any $b \in O_a$, we have $O_b = O_a$, and hence

$$\Sigma_{b \in O_a} 1 / |O_b| = |O_a| / |O_b| = 1$$

Therefore, since every element of G belongs to exactly one orbit, we have

$$\Sigma_{a \in X} 1 / |O_a| = N$$

Hence from our two expressions for n we get

$$\Sigma_{g \in G} |X_g| = n = |G| \cdot N$$

And from this the theorem follows. □

Applications

We illustrate how to apply Burnside's theorem to specific counting problems. In all these problems, we are counting the number of orbits in some action. We first need to specify the set X and the group G acting on it, and then determine $|X_g|$ for all $g \in G$.

4.3.4 EXAMPLE (A necklace problem) Three black and three white beads are strung together to form a necklace, which can be rotated and turned over. Assuming that beads of the same color are indistinguishable, how many different kinds of necklaces can be made?

To solve this problem, we position the 3 black and the 3 white beads at the vertices 1, 2, 3, 4, 5, 6 of a regular hexagon. The number of ways of doing this is the binomial coefficient $6!/3!(6\text{-}3)! = 6{\cdot}5{\cdot}4/1{\cdot}2{\cdot}3 = 20$. So in this problem the set X consists of these 20 designs. Since the necklace can be both rotated and turned over, the group G acting on X consists of the dihedral group $D_6 = \{\rho_0, \rho, \rho^2, \ldots, \rho^5, \tau, \rho\tau, \rho^2\tau, \ldots, \rho^5\tau\}$, where $\rho = (1\ 2\ 3\ 4\ 5\ 6)$ is counterclockwise rotation by 60°, and $\tau = (2\ 6)(3\ 5)$ is reflection with the diagonal from the top left to the bottom right vertex fixed. Note that $\rho^2\tau$ and $\rho^4\tau$ are also reflections with a diagonal fixed, while $\rho\tau$ and $\rho^3\tau$ and $\rho^5\tau$ are reflections with a side-bisector fixed.

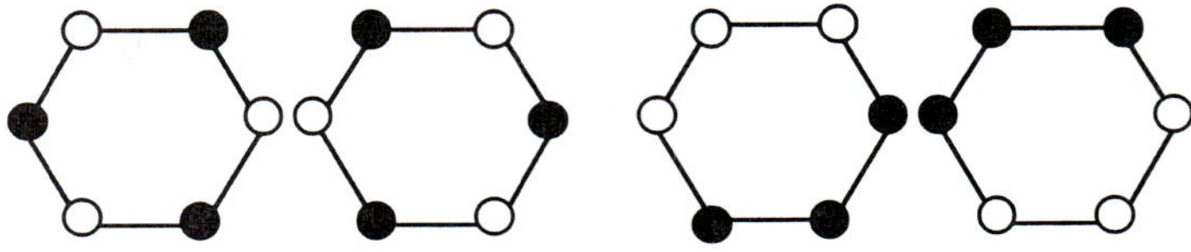

FIGURE 6

For example, the two designs on the left in Figure 6 are fixed by the rotations ρ^2 and ρ^4, while both these and the two designs on the right are fixed by the reflection τ.
If we count $|X_g|$ for every $g \in G$, the results are as in Table 2.

TABLE 2 Action of D_6 on X

g		$\lvert X_g\rvert$	g		$\lvert X_g\rvert$
ρ_0	= identity	20	τ	$= (2\ 6)(3\ 5)$	4
ρ	$= (1\ 2\ 3\ 4\ 5\ 6)$	0	$\rho\tau$	$= (1\ 2)(3\ 6)(4\ 5)$	0
ρ^2	$= (1\ 3\ 5)(2\ 4\ 6)$	2	$\rho^2\tau$	$= (1\ 3)(4\ 6)$	4
ρ^3	$= (1\ 4)(2\ 5)(3\ 6)$	0	$\rho^3\tau$	$= (1\ 4)(2\ 3)(5\ 6)$	0
ρ^4	$= (1\ 5\ 3)\ (2\ 6\ 4)$	2	$\rho^4\tau$	$= (1\ 5)(2\ 4)$	4
ρ^5	$= (1\ 6\ 5\ 4\ 3\ 2)$	0	$\rho^5\tau$	$= (1\ 6)(2\ 5)(3\ 4)$	0

Finally, using Burnside's theorem we obtain $N = {}^1/_{12}(20 + 2\cdot2 + 3\cdot4) = 3$. The three possible necklaces are illustrated in Figure 7.

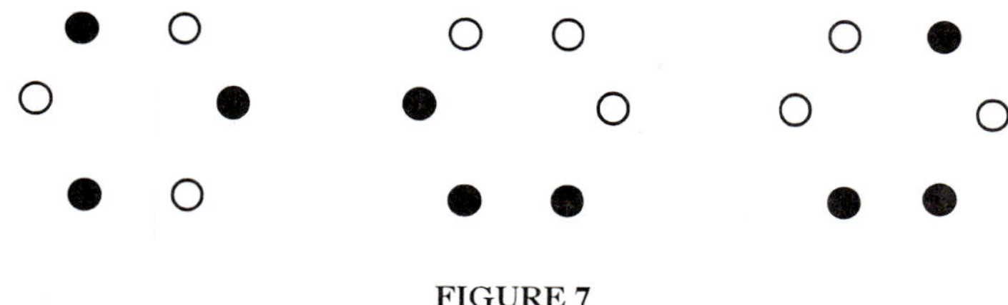

FIGURE 7

None of these three patterns can be carried into another by rotation orreflection.◇

4.3.5 EXAMPLE (A die problem) Let us position a cube on a table and name its six faces bottom, top, front, back, left, and right; and let us count the different ways we can mark the six faces to get a die. For the top we can choose any number from one to six of dots, for the bottom we can choose any of the remaining five numbers, and so on. So there are $6! = 720$ possibilities, and in this example X is the set consisting of these 720 marked cubes. To determine the group G acting on X in this example, consider all the possible ways a cube can be placed on a table. Any one of its six faces can be placed down as the bottom, and then the cube can be rotated so that any one of the four upright faces is placed to the front. Hence $|G| = 24$. If $g \in G$ and $g \neq e$, then $|X_g| = 0$, since no two sides are marked alike, and therefore any nontrivial rotation of the cube will change one of the markings. Hence the number of essentially different dice we obtain is $N = {}^1/_{24}\,(720) = 30$. ◇

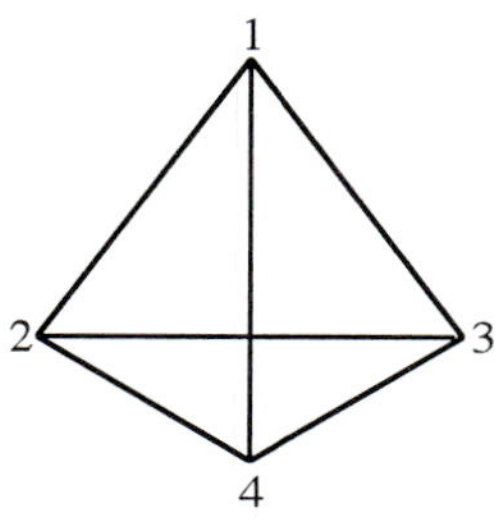

FIGURE 8

4.3.6 EXAMPLE Suppose we are coloring the edges of a regular tetrahedron, as in Figure 8, either white or black. First we note that there are 6 edges. The set X consists of all possible colorings, and has $2^6 = 64$ elements. Any one of the four faces of the tetrahedron can be placed on the bottom, and then any one of the three remaining faces can be rotated to be in the back. The group G therefore has $4 \cdot 3 = 12$ elements.

Let us now try to figure out the twelve elements of the group G before we attempt to calculate $|X_g|$ for each $g \in G$.

The 12 elements of G consist of e = the identity, plus three elements of order 2:

$\rho_1 = (1\ 2)(3\ 4)$ $\rho_2 = (1\ 3)(2\ 4)$ $\rho_3 = (1\ 4)(2\ 3)$

plus eight elements of order 3:

$\tau_1 = (2\ 3\ 4)$ $\tau_2 = (1\ 3\ 4)$

$\tau_3 = (1\ 2\ 4)$ $\tau_4 = (1\ 2\ 3)$

$\tau_1^{-1} = (2\ 4\ 3)$ $\tau_2^{-1} = (1\ 4\ 3)$

$\tau_3^{-1} = (1\ 4\ 2)$ $\tau_4^{-1} = (1\ 3\ 2)$

(G is in fact isomorphic to the alternating group A_4.) Now we must determine $|X_g|$ for each $g \in G$. Consider the rotation $\tau_1 = (2\ 3\ 4)$, and suppose that a specific coloring is in X_{τ_1}. Then the edges 2-3, 3-4, and 4-2 must all be the same color, and similarly the edges 1-2, 1-3, and 1-4. Hence $|X_{\tau_1}| = 2^2 = 4$. The same argument works for any of the 8 rotations of order 3.

Consider now the rotation $\rho_1 = (1\ 2)(3\ 4)$. Since the edges 1-2 and 3-4 remain fixed, either may be colored in either way, giving right away 2^2 choices. For the remaining four edges, 1-3 and 1-4 get exchanged, and so must have the same color, and similarly for 2-3 and 2-4. Hence $|X_{\rho_1}| = 2^2 \cdot 2 \cdot 2 = 16$. Again the same argument works for any of the 3 rotations of order 2.

Of course, for the identity $X_e = X$ and so $|X_e| = 64$. Hence we obtain

$$N = {}^1/_{12}(64 + 3 \cdot 16 + 8 \cdot 4) = 12 \ \diamond$$

4.3.7 EXAMPLE The molecules of methane CH_4 consist of a carbon atom at the center of a regular tetrahedron, bonded to each of four hydrogen atoms at the vertices. Suppose each hydrogen (H) is replaced either by fluorine (F), chlorine (Cl), or bromine (Br). How many essentially different compounds can be produced? Here there are four vertices and three possibilities for each vertex (F, Cl, Br), so that $|X| = 3^4 = 81$. The group G is isomorphic to A_4, as in the preceding example, whose notation is used in this example, too.

Consider $\tau_1 = (2\ 3\ 4)$. If a compound is in X_{τ_1}, then the same element must be at all the positions 2, 3 ,4, while any element may be at position 1. This gives $|X_{\tau_1}| = 3 \cdot 3 = 9$, and the same argument works for any permutation of order 3. Consider $\rho_1 = (1\ 2)(3\ 4)$. If a compound is in X_{ρ_1}, then positions 1 and 2 must have the same element at them, and likewise positions 3 and 4. This again gives $|X_{\rho_1}| = 3 \cdot 3 = 9$. Hence $N = {}^1/_{12}(81+3 \cdot 9+8 \cdot 9) = 15$. (Note that there are, for instance, two different compounds, mirror images of each other, having two F's and one Cl and one Br, and therefore having the same chemical formula CF_2ClBr.) $\diamond$

4.3.8 EXAMPLE In how many essentially different ways can we color the vertices of a cube if n colors are available? In this case, since there are 8 vertices and n possible colors for each, we have $|X| = n^8$. And as we have seen in Example 4.2.13, $|G| = 24$. What we need to do is determine $|X_g|$ for each $g \in G$. If g is a rotation ρ of 90° or 270° about an axis passing through the centers of two opposite faces, for instance $(1\ 2\ 3\ 4)(5\ 6\ 7\ 8)$, numbering the vertices as in Example 4.1.15, then for a design in X_ρ, all the vertices 1, 2, 3, 4 must be of the same color, and similarly all the vertices 5, 6, 7, 8, giving n^2 possibilities. If g is a rotation σ of 180° about such an axis, for instance (1 3)(2 4)(5 7)(6 8), then each of the pairs 1 and 3, 2 and 4, and so on must be colored the same, giving n^4 possibilities. If g is a rotation τ of 180° about an axis passing through the midpoints of two opposite edges, for instance (1 5)(2 8)(3 7)(4 6), then a similar analysis shows that again there are n^4 possibilities. If g is a rotation ϕ of 120° or 240° about an axis passing through two opposite vertices, for instance (2 4 5)(3 8 6), then 2, 4, and 5 must have the same color, as must 3, 8, and 6, while the color of the fixed point 1 may be chosen arbitrarily, as may that of the fixed point 7, giving again n^4 possibilities. The facts may be summarized as in Table 3.

TABLE 3 Rotations of the Cube with n Colors

Type of rotation	e	ρ	σ	τ	ϕ
Number of such rotations	1	6	3	6	8
$\lvert X_g \rvert$	n^8	n^2	n^4	n^4	n^4

From which we see $N = {}^1/_{24}(n^8 + 17n^4 + 6n^2)$. ◇

Exercises 4.3

In Exercises 1 through 11 find the number of essentially different ways in which we can do what is described.

1. Form a tetrahedral die by marking the four faces of a regular tetrahedron with one, two, three, or four dots, each number of dots appearing on exactly one face.

2. Color the edges of a square, if six colors are available and no color is to be used more than once.

3. Paint two faces of a regular tetrahedron red and the other two faces green.

4. Paint two faces of a cube red, two other faces blue, and the two remaining faces green.

5. Color the six faces of a cube with six different colors, if seven colors are available and no color is to be used more than once.

6. String three black and six white beads in a necklace, assuming the necklace can be turned over as well as rotated, and that beads of the same color are indistinguishable.

7. String one white, two red, and three black beads in a necklace, under the same assumptions as in the preceding exercise.

8. Replace each hydrogen atom in a molecule of methane (as in Example 4.3.7) with a fluorine, chlorine, bromine, or iodine atom.

9. Replace each of the hydrogen atom in a molecule of benzene with a fluorine, chlorine, or bromine atom. (A molecule of benzene consists of six carbon atoms in a regular hexagon, with one hydrogen atom bonded to each.)

10. Replace each of the hydrogen atom in a molecule of benzene with a fluorine, chlorine, bromine, or iodine atom.

11. Distribute nine balls equally among three children, if there are two white balls, three red balls, and four black balls.

12. Let X be a set with four elements. Find the number of equivalence relations on X that are not equivalent under any permutation in S_4.

4.4 Conjugacy Classes and the Class Equation

In this section we study one particular group action of special importance (already introduced in Example 4.1.6), the action of a group G on itself by conjugation or, in other words, the action $G \times G \to G$ taking (g, a) to gag^{-1}. In the preceding section we derived Burnside's counting formula concerning the orbits in a finite set X acted on by a finite group G and applied this formula to different types of problems. In this section we derive another important counting formula concerning the orbits in this particular action of a finite group G on itself. This formula will be important in understanding the structure of finite groups, and we apply it to the study of groups whose order is a power of a prime p.
We begin with a preliminary counting formula.

Conjugation

4.4.1 EXAMPLE Let $G = S_3 = \{\rho_0, \rho, \rho^2, \mu_1, \mu_2, \mu_3\}$ act on itself by conjugation, that is, the action $G \times G \to G$ is the defined by $(g,a) \to gag^{-1}$. (See Example 4.1.6.) Under this action the orbits are

$$O_{\rho_0} = \{\rho_0\}$$
$$O_\rho = O_{\rho 2} = \{\rho, \rho^2\}$$
$$O_{\mu_1} = O_{\mu_2} = O_{\mu_3} = \{\mu_1, \mu_2, \mu_3\}$$

The corresponding stabilizers are

$$G_\rho = G_{\rho 2} = \{\rho_0, \rho, \rho^2\} \qquad G_{\mu_i} = \{\rho_0, \mu_i\} \text{ for } i = 1, 2, 3$$

Note that

(1) $|G| = 6 = |O_{\rho_0}| + |O_{\rho}| + |O_{\mu_1}|$

(2) $|G| = 6 = |O_a| \cdot |G_a|$ for all $a \in G$

This should not come as a surprise, since these two facts illustrate theorems we have already proved (Proposition 4.2.5 and Theorem 4.2.8). ◇

Let us now define the terms we will be using.

4.4.2 DEFINITION If G is a group and $a, b \in G$, then a and b are said to be **conjugate** in G if there exists a $g \in G$ with $b = gag^{-1}$ or, in other words, if a and b are in the same orbit in the action of G on itself by conjugation. The **conjugacy class** $K_G(a) = \{gag^{-1} \mid g \in G\}$ of a in G is the set of all conjugates of a or, in other words, the orbit of a in this action. Recall that the **centralizer** $C_G(a) = \{g \in G \mid ga = ag\} = \{g \in G \mid gag^{-1} = a\}$ of a in G is the set of all elements that commute with a, or equivalently, the stabilizer of a in this action. When it is clear from context which group G is intended, the subscript is omitted and we just write $K(a)$ and $C(a)$. ○

4.4.3 PROPOSITION Let G be a group acting on itself by conjugation, and $a \in G$. Then the size of the conjugacy class of a is equal to the index of the centralizer of a, that is to say, $|K(a)| = [G:C(a)]$. If G is finite, then $|K(a)| = |G| / |C(a)|$.

Proof This is just a restatement of the orbit-stabilizer relation (Theorem 4.2.8) in terms of the notions introduced in the preceding definition. □

Now we are ready for the main theorem of this section.

4.4.4 THEOREM (The class equation) Let G be a finite group, $Z(G)$ the center of G, and let $a_1, a_2, \ldots, a_r$ be elements not in $Z(G)$ forming a complete set of representatives of the conjugacy classes not contained in $Z(G)$ (meaning that no two of the a_i are conjugate with each other, but every element not in the center is conjugate of one of them). Then

$$|G| = |Z(G)| + \sum_{i=1}^{r} [G:C(a_i)]$$

Proof If $b_1, b_2, \ldots, b_s$ are the elements of $Z(G)$, then since each b_i is conjugate only to itself, the b_j and a_i together are a complete set of representatives of *all* the conjugacy classes. Proposition 4.2.14 then tells us

$$|G| = \sum_{i=1}^{s} [G:C(b_i)] + \sum_{i=1}^{r} [G:C(a_i)]$$

where we have written $[G:C(g)]$ in place of $[G:G_g]$ in accordance with Proposition 4.4.3. But for each b_i in the center, $C(b_i)$ is simply G itself, and $[G:C(b_i)]$ is simply 1. From this the class equation follows. □

4.4.5 EXAMPLE Let us find the conjugacy classes and write out the class equation for the dihedral group D_4. It is not hard to work out that we have

$$\tau\rho\tau^{-1} = (\tau\rho)\tau = (\rho^3\tau)\tau = \rho^3\tau^2 = \rho^3$$
$$\rho\tau\rho^{-1} = \rho(\tau\rho^3) = \rho(\rho\tau) = \rho^2\tau$$
$$\rho(\rho\tau)\rho^{-1} = \rho(\rho\tau)\rho^3 = \rho^2(\tau\rho^3) = \rho^2(\rho\tau) = \rho^3\tau$$

From these we get the facts collected in Table 4.

TABLE 4 Conjugacy Classes in D_4

Centralizer	Conjugacy class	Index
$Z(D_4) = \{\rho_0, \rho^2\}$	$K(e) = \{e\}$, $K(\rho^2) = \{\rho^2\}$	
$C(\rho) = C(\rho^3) = \{\rho_0, \rho, \rho^2, \rho^3\}$	$K(\rho) = \{\rho, \rho^3\}$	$[G{:}C(\rho)] = 2$
$C(\tau) = C(\rho^2\tau) = \{\rho_0, \tau, \rho^2, \rho^2\tau\}$	$K(\tau) = \{\tau, \rho^2\tau\}$	$[G{:}C(\tau)] = 2$
$C(\rho\tau) = C(\rho^3\tau) = \{\rho_0, \rho\tau, \rho^2, \rho^3\tau\}$	$K(\rho\tau) = \{\rho\tau, \rho^3\tau\}$	$[G{:}C(\rho\tau)] = 2$

It follows that the class equation in this case is $8 = 2 + 2 + 2 + 2$. ◇

4.4.6 EXAMPLE Let us find the conjugacy classes and write out the class equation for the quaternion group $Q_8 = \{\pm 1, \pm\mathbf{i}, \pm\mathbf{j}, \pm\mathbf{k}\}$. Any element of any group commutes with its own powers, so the centralizer $C(\mathbf{i})$ of $\mathbf{i}$ contains $\mathbf{i}$, $\mathbf{i}^2 = -1$, $\mathbf{i}^3 = -i$, and $\mathbf{i}^4 = 1$. Since $C(\mathbf{i})$ is a subgroup, its order must divide that of Q_8, so if it were any larger than $\{1, \mathbf{i}\ {-1}, -\mathbf{i}\}$ it would be the whole group, which it is not, since $\mathbf{ij} = \mathbf{k} \neq -\mathbf{k} = \mathbf{ji}$. Hence $C(\mathbf{i}) = \{1, \mathbf{i}\ {-1}, -\mathbf{i}\}$ and the size of the conjugacy class of $\mathbf{i}$, which is equal to the index $[Q_8{:}C(\mathbf{i})]$, is 2. The elements of the conjugacy class are $\mathbf{i}$ itself and $\mathbf{jij}^{-1} = -\mathbf{i}$. A similar analysis applies to $\mathbf{j}$ and to $\mathbf{k}$, so that $Z(Q_8) = \{\pm 1\}$, the other conjugacy classes are $\{\pm\mathbf{i}\}$, $\{\pm\mathbf{j}\}$, and $\{\pm\mathbf{k}\}$, and the class equation is $8 = 2 + 2 + 2 + 2$. ◇

We now give two important consequences, one following immediately from the other, of the class equation.

4.4.7 THEOREM If G is a group of order p^n, where p is a prime and $n \geq 1$, then the center of G is nontrivial; in other words, $|Z(G)| > 1$ (and indeed $|Z(G)| = p^k$ for some k, $1 \leq k \leq n$).

Proof The class equation gives us

$$|G| = |Z(G)| + \sum_{i=1}^{r} [G{:}C(a_i)]$$

where $a_1, a_2, \ldots, a_r$ are a complete set of representatives of the conjugacy classes not contained in $Z(G)$. Since $C(a_i) \neq G$, the index $[G{:}C(a_i)]$ is not 1, and so p divides it.

Since p also divides $|G|$, it follows that p divides $|Z(G)|$, which is therefore not 1 (and since $Z(G)$ is a subgroup, $|Z(G)|$ divides $|G| = p^n$, and so must be p^k for some k, $1 \leq k \leq n$). □

4.4.8 COROLLARY If G is a group of order p^2, where p is a prime, then G is Abelian (and hence isomorphic to $\mathbb{Z}_{p^2}$ or $\mathbb{Z}_p \times \mathbb{Z}_p$).

Proof Suppose there is an element $a \in G$ with $a \notin Z(G)$, and consider the centralizer $C(a)$ of a. On the one hand, since $C(a)$ contains both $Z(G)$ and $a \notin Z(G)$, it is bigger than $Z(G)$. On the other hand, since a does not commute with every element of G, $C(a)$ is smaller than G. The order $|C(a)|$ of the centralizer must therefore be a number bigger than $|Z(G)|$ but smaller than $|G|=p^2$, and must divide p^2. But this is impossible, since by the preceding theorem, $|Z(G)| \geq p$. (See Exercises 20 through 22.) □

The notion of conjugacy can also be defined for subgroups as well as elements of a group, and this generalization will be useful later. (In fact, the notion can be defined for arbitrary subsets of the group, which need not be subgroups. See Exercises 10 through 12.)

4.4.9 DEFINITION Let G be a group and let X be the set of all subgroups of G. Consider the map $G \times X \to X$ sending (g, A) to $gAg^{-1} = \{gag^{-1} \mid a \in A\}$. This map is an action, called **conjugation**. Two subgroups $A, B \subseteq G$ are said to be **conjugate** in G if there is a $g \in G$ such that $B = gAg^{-1}$ or, in other words, if A and B are in the same orbit in the action of G on the set of its subgroups by conjugation. The **conjugacy class** $K_G(A) = \{gAg^{-1} \mid g \in G\}$ of A in G is the set of all conjugates of A or, in other words, the orbit of A in this action. Recall that the **centralizer** $C_G(A) = \{g \in G \mid ga = ag \text{ for all } a \in A\} = \{g \in G \mid gag^{-1} = a \text{ for all } a \in A\}$ of A in G is the set of all elements that commute with all elements of A. The **normalizer** $N_G(A) = \{g \in G \mid gA = Ag\} = \{g \in G \mid gAg^{-1} = A\}$ is the stabilizer of A in this action. When it is clear from context which group G is intended, the subscript is omitted and we just write $K(A)$ and $C(A)$ and $N(A)$. ○

4.4.10 PROPOSITION Let G be a group acting on the set of its subgroups by conjugation, and let A be a subgroup of G. Then the size of the conjugacy class of A is equal to the index of the normalizer of A, that is to say, $|K(A)| = [G:N(A)]$. If G is finite, then $|K(A)| = |G| / |N(A)|$.

Proof This is just a restatement of the orbit-stabilizer relation (Theorem 4.2.8) in terms of the notions introduced in the preceding definition. □

Exercises 4.4

1. Describe the conjugacy classes and the class equation of an Abelian group.

2. Let G_1 and G_2 be two groups. Show that in $G_1 \times G_2$, the elements (a, b) and (c, d) are conjugates if and only if a and c are conjugates in G_1, and b and d are conjugates in G_2.

In Exercises 3 through 8 describe the conjugacy classes and write the class equations of the indicated groups.

3. $\mathbb{Z}_2 \times S_3$ **4.** $\mathbb{Z}_2 \times D_4$ **5.** $\mathbb{Z}_3 \times S_3$

6. $S_3 \times S_3$ **7.** A_4 **8.** $\mathbb{Z}_3 \times A_4$

9. Let G be any group. Show that for any $a, b \in G$, if a and b are conjugates, then they have the same order.

In Exercises 10 through 12 let G be a group let $P(G)$ be the set of all subsets of G, and consider the map $G \times P(G) \to P(G)$ sending (g, A) to $gAg^{-1} = \{gag^{-1} \mid a \in A\}$.

10. Show that this map is a group action.

11. Show that the centralizer $C(A) = \{g \in G \mid gag^{-1}=a \text{ for all } a \in A\}$ is a subgroup of G.

12. Show that the normalizer $N(A) = \{g \in G \mid gAg^{-1} = A\}$ is a subgroup of G.

13. Let G be a group acting on the set of all its subsets by conjugation. Show that for any $S \subset G$ and $g \in G$ we have $gN(S)g^{-1} = N(gSg^{-1})$.

14. Let G be a group acting on the set of all its subsets by conjugation. Show that for any $S \subseteq G$ and $g \in G$ we have $gC(S)g^{-1} = C(gSg^{-1})$.

15. Let G be a group acting on itself by conjugation. Show that if a and b are conjugates in G, then $|C(a)| = |C(b)|$.

16. Let G be a group acting on itself by conjugation. Show that if a and b are conjugates in G, then the centralizers $C(a)$ and $C(b)$ are equal if and only if these centralizers are normal subgroups of G. (*Hint:* Use the preceding exercises.)

17. Explain why in Example 4.4.5 two elements a, b of D_4 are in the same orbit if and only if their centralizers $C(a)$, $C(b)$ are equal.

18. Let r be the index $[G:Z(G)]$ of the center $Z(G)$. Show that for any $g \in G$, the number of elements $|K(g)|$ of the conjugacy class $K(g)$ of the element g is less than or equal to r.

19. Explain why for any finite group G and any $g \in G$, the number of elements $|K(g)|$ of the conjugacy class $K(g)$ of the element g divides the order $|G|$ of the group G.

20. Explain why for any $g \in G$, the center $Z(G)$ of the group G is contained in the centralizer $C(g)$ of the element g.

21. Explain why for any $g \in G$, the center $Z(G)$ and the centralizer $C(g)$ are equal if and only if $g \in Z(G)$.

22. Show that for any non-Abelian group G, the index of the center $[G:Z(G)]$ cannot be a prime p.

23. Show that for any group G, the index of the center $[G:Z(G)]$ cannot be a prime p.

24. By definition, $a, b \in G$ are conjugate if there exists a $g \in G$ such that $b = gag^{-1}$. Give an example to show that this g need not be unique or, in other words, that there may be another $h \neq g$ such that $b = hah^{-1}$ also.

25. In the situation of the preceding exercise, show that the number of h such that $b = hah^{-1}$ is in fact equal to the order $|C(a)|$ of the centralizer.

26. Show that for any element $g \in G$ not equal to the identity e we have $|C(g)| \geq 2$.

4.5 Conjugacy in S_n and Simplicity of A_5

In this section we determine the conjugacy classes in the symmetric groups S_n and apply the results to give an elementary proof that A_5 has no nontrivial proper normal subgroups. The first two examples illustrate the main theorem of this section.

4.5.1 DEFINITION Given any $\sigma \in S_n$, it can be written as the product of disjoint cycles where shorter cycles are written before longer ones. Moreover, the lengths $n_1 \leq n_2 \leq \ldots \leq n_s$ of the cycles are uniquely determined and satisfy $n = n_1 + n_2 + \ldots + n_s$. We call $n_1, n_2, \ldots, n_s$ the **cycle type** of σ. ○

4.5.2 EXAMPLE In S_9, consider a permutation of, say, cycle type 2, 3, 4, such as

$$\sigma = (6\ 9)(1\ 4\ 5)(3\ 2\ 7\ 8)$$

Let us look at the conjugate $\tau\sigma\tau^{-1}$ of σ, where τ is, say, $(1\ 7\ 2\ 6)(9\ 3\ 5\ 4\ 8)$. Consider the permutation ϕ of the same cycle type 2, 3 ,4 as σ that is obtained by replacing each i in the cycle decomposition of σ by $\tau(i)$ or, in other words,

$$\phi = (\tau(6)\ \tau(9))\ (\tau(1)\ \tau(4)\ \tau(5))\ (\tau(3)\ \tau(2)\ \tau(7)\ \tau(8)) = (1\ 3)(7\ 8\ 4)(5\ 6\ 2\ 9)$$

We can easily calculate that we have

$$\tau\sigma\tau^{-1}(1) = \tau\sigma(6) = \tau(9) = 3 = \phi(1)$$
$$\tau\sigma\tau^{-1}(3) = \tau\sigma(9) = \tau(6) = 1 = \phi(3)$$
$$\tau\sigma\tau^{-1}(7) = \tau\sigma(1) = \tau(4) = 8 = \phi(7)$$

and so on, so that $\tau\sigma\tau^{-1}$, and the conjugate of σ is in this case is of the same cycle type as σ. ◇

4.5.3 EXAMPLE In S_6, consider two permutations of the same cycle type, say

$$\sigma = (2)(3\ 6)(4\ 1\ 5)$$
$$\rho = (4)(1\ 5)(6\ 3\ 2)$$

Let us look at the permutation τ that takes each i appearing in the decomposition of σ to the j that appears below it in the decomposition of ρ or, in other words,

$$\tau = (2\ 4\ 6\ 5)(3\ 1)$$

Consider $\tau\sigma\tau^{-1}$. We can easily calculate that $\rho = \tau\sigma\tau^{-1}$, and thus the permutation of the same cycle type as σ is in this case a conjugate of σ. ◇

4.5.4 THEOREM Two permutations σ and ρ in S_n are conjugate if and only if they have the same cycle type.

Proof ($\Rightarrow$) First suppose σ and ρ are conjugates, and take some τ such that $\rho = \tau\sigma\tau^{-1}$. Suppose that the cycle decomposition of σ is as follows:

$$(x_1\ \ldots\ x_p)(y_1\ \ldots\ y_q)(z_1\ \ldots\ z_r)\ldots$$

Consider the following permutation ϕ, which has the same cycle type as σ:

$$\phi = (\tau(x_1)\ \ldots\ \tau(x_p))\ (\tau(y_1)\ \ldots\ \tau(y_q))\ (\tau(z_1)\ \ldots\ \tau(z_r))\ldots$$

We claim $\rho = \phi$. For indeed, $\rho(\tau(x_i)) = \tau\sigma\tau^{-1}(\tau(x_i)) = \tau\sigma(x_i) = \tau(x_{i+1}) = \phi(\tau(x_i))$, and similarly $\rho(\tau(y_j)) = \phi(\tau(y_j))$, $\rho(\tau(z_k)) = \phi(\tau(z_k))$, and so on. Hence the conjugate $\rho = \phi$ as claimed, and ρ has the same cycle type as σ.

($\Leftarrow$) Now suppose we have two permutations with the same cycle type:

$$\sigma = (x_1\ \ldots\ x_p)(y_1\ \ldots\ y_q)(z_1\ \ldots\ z_r)\ldots$$
$$\rho = (u_1\ \ldots\ u_p)(v_1\ \ldots\ v_q)(w_1\ \ldots\ w_r)\ldots$$

Define a permutation τ by $\tau(x_i) = u_i$, $\tau(y_j) = v_j$, $\tau(z_k) = w_k$, and so on, and let ϕ be the conjugate $\tau\sigma\tau^{-1}$ of σ. We claim $\phi = \rho$. For indeed, $\phi(u_i) = \tau\sigma\tau^{-1}(u_i) = \tau\sigma(x_i) = \tau(x_{i+1}) = u_{i+1} = \rho(u_i)$, and similarly $\phi(v_j) = \rho(v_j)$, $\phi(w_k) = \rho(w_k)$, and so on. Hence $\phi = \rho$ as claimed, and ρ is a conjugate of σ. □

4.5.5 DEFINITION A **partition** of a positive integer n is any nondecreasing sequence $n_1 \leq n_2 \leq \ldots \leq n_s$ of positive integers whose sum $n_1+n_2+\ldots+n_s$ is n. ○

4.5.6 COROLLARY The number of conjugacy classes in S_n equals the number of partitions of n.

Proof This follows from the preceding theorem, since the cycle type of any permutation is a partition, and for every partition $n_1 \leq n_2 \leq \ldots \leq n_s$ there is a permutation having it as cycle type. For instance,

$$(1\ 2\ \ldots\ n_1)(n_1+1\ \ n_1+2\ \ldots\ n_1+n_2)\ldots(m+1\ \ m+2\ \ldots\ n)$$

where $m = n_1 + n_2 + \ldots + n_{s-1}$, so that $m + n_s = n$. □

In the proof of the theorem we showed how, given permutations σ, ρ of the same cycle type, to find a permutation τ such that $\rho = \tau\sigma\tau^{-1}$. The point of the next example is to illustrate that this τ is not unique (something we already know from Exercises 24 through 26 of the preceding section).

4.5.7 EXAMPLE In S_9 consider

$\sigma = (9)(4)(1\ 3)(5\ 8)(2\ 6\ 7)$ $\rho = (3)(8)(2\ 5)(9\ 7)(1\ 4\ 6)$

Using the construction in the theorem, we find that $\rho = \tau\sigma\tau^{-1}$, where

$\tau = (9\ 3\ 5)(4\ 8\ 7\ 6)(1\ 2)$

But we can equally well write the same two permutations in a different order:

$\sigma = (4)(9)(5\ 8)(1\ 3)(2\ 6\ 7)$ ρ = same as before

Using the construction in the theorem again, we also find that if

$\theta = (4\ 3\ 7\ 6)(9\ 8\ 5\ 2\ 1)$

then again $\rho = \theta\sigma\theta^{-1}$. ◇

4.5.8 EXAMPLE We can now find a complete set of representatives for all the conjugacy classes of S_n for, say, $n = 4$. According to the theorem and its corollary, we need only consider all partititons of 4. These are displayed in Table 5.

TABLE 5 Conjugacy Classes in S_4

Partition of 4	Conjugacy class representative	Number of conjugates
1, 1, 1, 1	(1)	1
1, 1, 2	(12)	6
1, 3	(123)	8
2, 2	(12)(34)	3
4	(1234)	6

Here we have also counted the number of permutations in each class. ◇

The next example illustrates how knowledge of the conjugacy classes enables us to determine certain other features.

4.5.9 EXAMPLE Let $\sigma = (1\ 2\ 3) \in S_4$. Let us determine the centralizer $C(\sigma)$ of σ in S_4. On the one hand, $C(\sigma)$ must contain the three powers e, σ, σ^2. On the other hand, by the preceding example any cycle $(x\ y\ z)$ is conjugate in S_4 to $\sigma = (1\ 2\ 3)$, and there are 8 such cycles. But by the orbit-stabilizer relation we know that the number of elements in the conjugacy class of σ is the index $[S_4{:}C(\sigma)]$, so $|S_4|/C(\sigma) = 8$. Since $|S_4| = 24$, we have $|C(\sigma)| = 3$, and $C(\sigma) = \{e, \sigma, \sigma^2\}$. ◇

The next example illustrates the fact that elements a and b that are conjugate in a group G and belong to a subgroup $H \subseteq G$ need not be conjugate in H.

4.5.10 EXAMPLE Let $\sigma = (1\ 2\ 3) \in A_4 \subseteq S_4$. According to the theorem, σ is conjugate to $\rho = (1\ 2\ 4)$ in S_4. But the permutation τ obtained by from the proof of the theorem is just $\tau = (3\ 4)$, an odd permutation, so that $\tau \notin A_4$. Moreover, there is no other permutation $\chi \in A_4$ such that $\rho = \chi\sigma\chi^{-1}$ either. For if there were, we would have $\tau\sigma\tau^{-1} = \chi\sigma\chi^{-1}$, and so (multiplying on the left by τ^{-1} and on the right by χ) we would have $\sigma\tau^{-1}\chi = \tau^{-1}\chi\sigma$, and so $\tau^{-1}\chi \in C(\sigma) \subseteq S_4$. Since τ is an odd permutation and χ would be an even permutation, $\tau^{-1}\chi \in C(\sigma)$ would be an odd permutation. But this is impossible since by the preceding example $C(\sigma)$ consists only of the even permutations $\{e, \sigma, \sigma^2\}$. Thus σ and ρ are conjugate in S_4 but not in A_4. ◇

The orbit-stablizer relation used in the preceding examples helped us calculate the order of the centralizer of a cycle. The next result gives a direct way of determining the centralizer.

4.5.11 PROPOSITION Let $\sigma = (x_1, \dots, x_m)$ be an m-cycle in S_n. Then

(1) The number of conjugates of σ in S_n is

$$|K(\sigma)| = n!\,/\,m\cdot(n - m)!$$

(2) The order of the centralizer of σ in S_n is

$$|C(\sigma)| = m\cdot(n\text{-}m)!$$

(3) Let $S_{n\text{-}m} \subseteq S_n$ be the subgroup of permutations that fix the m elements $x_1, \dots, x_m$. Then

$$C(\sigma) = \{\sigma^i\tau \mid 0 \le i \le m - 1, \tau \in S_{n\text{-}m}\}$$

Proof Parts (1) and (2) will be left to the reader (as Exercises 7 and 8). For part (3), note the following. First, all m powers σ^i of σ commute with σ and belong to its centralizer. Second, all $(n\text{-}m)!$ elements τ of $S_{n\text{-}m}$, being just the permutations whose cycles are disjoint from those of σ, commute with σ and hence belong to its centralizer. Therefore, all the products $\sigma^i\tau$ for $0 \le i \le m - 1$ and $\tau \in S_{n\text{-}m}$ belong to the centralizer. Moreover, these $m \cdot (n\text{-}m)!$ products are all distinct. For assume $\sigma^i\tau_1 = \sigma^j\tau_2$, where $0 \le i, j \le m$ and $\tau_1, \tau_2 \in S_{n\text{-}m}$. Then for any x_k with $1 \le k \le m$ we have $\tau_1(x_k) = \tau_2(x_k) = x_k$, and hence the fact that $\sigma^i\tau_1(x_k) = \sigma^j\tau_2(x_k)$ implies $\sigma^i(x_k) = \sigma^j(x_k)$. Since this holds for all k with $1 \le k \le m$, we have $\sigma^i = \sigma^j$, and this together with $\sigma^i\tau_1 = \sigma^j\tau_2$ implies $\tau_1 = \tau_2$. By part (2), then, these elements $\sigma^i\tau$ of the centralizer are all the elements that there are. □

4.5.12 EXAMPLE Let us determine the number of conjugates of $\rho = (1\ 2\ 3)$ in S_5 and in A_5. For S_5 we apply Proposition 4.5.11. This tells us that the number of conjugates of ρ in S_5 is $5 \cdot 4 \cdot 3 / 3 = 20$, that $|C(\sigma)|$ is $3\cdot 2! = 6$, and that $C(\sigma) = \{e, \rho, \rho^2 = (1\ 3\ 2), \theta = (4\ 5), \rho\theta = (1\ 2\ 3)(4\ 5), \rho^2\theta = (1\ 3\ 2)(4\ 5)\}$. The centralizer of ρ in A_5 consists of the even permutations among these, namely, e, ρ, ρ^2. Hence the size of the centralizer is 3, and the number of conjugates is $|A_5|/3 = 60/3 = 20$ again, and so ρ has the same conjugates in A_5 as in S_5, namely all 3-cycles, which thus form a single conjugacy class in A_5. ◇

4.5.13 EXAMPLE Let us determine the number of conjugates of $\sigma = (1\,2\,3\,4\,5)$ in S_5 and in A_5. For S_5 we apply Proposition 4.5.11. This tells us that the number of conjugates of σ in S_5 is $5 \cdot 4 \cdot 3 \cdot 2 \cdot 1 / 5 = 24$, that $|C(\sigma)| = |S_5|/24 = 120/24 = 5$, and that $C(\sigma) = \{e, \sigma, \sigma^2, \sigma^3, \sigma^4\}$. Since in this case we have $C(\sigma) \subseteq A_5$, the centralizer in A_5 is the same, and the number of conjugates is $|A_5|/5 = 60/5 = 12$, or half as many as for S_5. The same analysis applies to any 5-cycle and shows that while all 5-cycles form a single conjugacy class in S_5, they form two different conjugacy classes in A_5. ◇

4.5.14 EXAMPLE Let us determine the number of conjugates of $\tau = (1\,2)(3\,4)$ in S_5 and in A_5. (The calculation will apply to any permutation of the same cycle type.) By the main theorem of this section, the number of conjugates of τ is the number of permutations of form $(x\,y)(u\,v)$, which is 15. (See Exercise 9.) Hence $|C(\tau)| = 120/15 = 8$. To find the 8 permutations that commute with τ, first note that if $\phi = (1\,3\,2\,4)$, then $\tau = \phi^2$, so τ commutes with all the powers of ϕ, and then note that if $\psi = (1\,2)$, then $\tau\psi = \psi\tau = (3\,4)$, so τ commutes with ψ. From these facts it follows that $C(\tau) = \{e, \phi, \phi^2, \phi^3, \psi, \phi\psi, \phi^2\psi, \phi^3\psi\}$. The centralizer of τ in A_5 consists of the even permutations among these, namely, $\{e, \phi^2, \phi\psi, \phi^3\psi\}$. Hence the size of the centralizer is 4, and the number of conjugates is $60/4 = 15$ again, and so ρ has the same conjugates in A_4 as in S_5, namely, all products of two disjoint 2-cycles, which thus form a single conjugacy class in A_5. ◇

We introduce now a notion that is basic for the classification of finite groups.

4.5.15 DEFINITION A group G is called **simple** if it has no nontrivial proper normal subgroups. ○

From the definition we can see right away that the Abelian simple groups are exactly the cyclic groups of prime order. The groups A_n for $n \geq 5$ are examples of non-Abelian simple groups. We prove this now for $n = 5$. The case $n > 5$ is dealt with in the exercises of Section 5.3. The three examples we just studied give us a complete description of the conjugacy classes of A_5 and give us everything we need to prove the simplicity of A_5 once we prove the following lemma.

4.5.16 LEMMA Let G be a group and H a finite normal subgroup of G, and let $a_1, a_2, \ldots, a_r$ be elements of H forming a complete set of representatives of the conjugacy classes $K_G(b)$ in G of the elements b of H (meaning that no two of the a_i are conjugate in G with each other, but every element b of H is conjugate in G with one of them). Then

$$|H| = \sum_{i=1}^{r} |K_G(a_i)|$$

Proof Since no two of the a_i are conjugate in G, the $K_G(a_i)$ are disjoint. Since every element of H is conjugate to one of them, H is contained in their union. Moreover, since H is normal and $a_i \in H$, every conjugate ga_ig^{-1} belongs to H and the whole conjugacy class $K_G(a_i)$ is contained in H, so H is in fact equal to the disjoint union of the conjugacy classes, so the number of elements of H is equal to the sum of the number of elements in the conjugacy classes. □

4.5.17 THEOREM A_5 is a simple group.
Proof Examples 4.5.12, 4.5.13, and 4.5.14 cover all cycle types of even permutations and together show that A_5 consists of just 5 conjugacy classes: the identity, with 1 element; the class of all 3-cycles, with 20 elements; two classes of 5-cycles, each with 12 elements; and the class of all products of two disjoint 2-cycles, with 15 elements. By the preceding lemma, the order $|H|$ of any normal subgroup of A_5 would have to be $1 + A$, where A is the sum of zero to four numbers from among 12, 12, 15, 20. And, of course, $1 + A$ would have to divide $|A_5| = 60$. For a nontrivial *proper* normal subgroup, A would have to be a sum of from *one to three* of 12, 12, 15, 20; and $1 + A$ would have to be a *proper* divisor of 60. Simple arithmetic, trying all combinations, shows this to be impossible, so A_5 has no nontrivial proper normal subgroups and is therefore simple. □

Exercises 4.5

1. Let $\sigma = (1\ 2)(3\ 4\ 5)(6\ 7\ 8\ 9)$ in S_9.
(a) Write down two permutations ρ, τ that are conjugates of σ in S_9.
(b) Find permutations ϕ, χ, ψ such that $\rho = \phi\sigma\phi^{-1}$, $\tau = \chi\sigma\chi^{-1}$, $\tau = \psi\rho\psi^{-1}$.

2. In each case determine whether σ and ρ are conjugates in S_{10}. If so, find τ such that $\rho = \tau\sigma\tau^{-1}$.
(a) $\sigma = (1\ 3)(2\ 4\ 5)$ $\quad\rho = (3\ 1\ 5)(2\ 4)$
(b) $\sigma = (1\ 3)(2\ 4\ 5)$ $\quad\rho = (6\ 2)(8\ 9)(6\ 5)$
(c) $\sigma = (1\ 2\ 3)(4\ 5\ 6\ 7)$ $\quad\rho = (1\ 2)(3\ 4)(5\ 6\ 7)$
(d) $\sigma = (2\ 4\ 5)(3\ 5\ 6)$ $\quad\rho = (6\ 7\ 8\ 9\ 10)$
(e) $\sigma = (2\ 4\ 5)(4\ 5\ 6)$ $\quad\rho = (2\ 4\ 5\ 6\ 8)$
(f) $\sigma = (2\ 4)(8\ 3\ 6)(9\ 7\ 5\ 1)$, $\quad\rho = \sigma^2$

3. Prepare a table for the conjugacy classes of S_5 analogous to the one in Example 4.5.8 for S_4.

4. How many conjugacy classes are there in S_n for $n = 6$? $n = 7$? $n = 8$?

5. Find the center $Z(S_3)$ of S_3.

6. Find the center $Z(S_n)$ of S_n for $n > 3$.

7. Prove part (1) of Proposition 4.5.11.

8. Prove part (2) of Proposition 4.5.11.

9. Show that if $\tau = (x\ y)(u\ v)$ is the product of two disjoint 2-cycles in S_n, where $n \geq 4$, then the number of conjugates of τ in S_n is $n!/8 \cdot (n-4)!$

10. Show that if τ is as in the preceding exercise, then the order of the centralizer of τ in S_n is $8 \cdot (n\text{-}4)!$

11. For $\tau = (1\ 2)(3\ 4)$, determine all the elements of the centralizer $C(\tau)$ of τ in S_n for $n \geq 4$.

In Exercises 12 through 14 for the indicated permutation $\sigma \in S_5$, find (a) the number of conjugates of σ in S_5, (b) the centralizer of σ in S_5, (c) the centralizer of σ in A_5, (d) the number of conjugates of σ in A_5.

12. $\sigma = (5\ 2\ 4)$ **13.** $\sigma = (3\ 5\ 2\ 4\ 1)$ **14.** $\sigma = (4\ 1)(5\ 2)$

15. Find two permutations σ and ρ in S_5 that are conjugates in S_5 but not in A_5.

4.6 The Sylow Theorems

Lagrange's theorem tells us that the order of a subgroup divides the order of the group. In the case of Abelian groups, the converse is also true: If a number divides the order of the group, there is a subgroup of that order. This follows from the fundamental theorem for finite Abelian groups, which gives a complete understanding of the structure of finite Abelian groups. The structure of finite groups that are not Abelian is more complicated, but the converse of Lagrange's theorem does hold at least when the order of the group is a power of a prime. This is one of the consequences of the cluster of results known as the Sylow theorems.

The proof of these theorems is a direct application of the theory of group actions, and specifically of the action by conjugation of a group G on the set S of all its subgroups. The most basic result is the orbit-stabilizer relation, according to which the size of the conjugacy class $K(H)$ of a subgroup H in G is equal to the index of the normalizer $N(H)$ of H in G (Proposition 4.4.10).

4.6.1 DEFINITION Let p be a prime. Recall that a group whose order is p^n for some $n \geq 1$ is called a ***p*-group**. Let G be a finite group. Any subgroup of G that is a p-group is called a ***p*-subgroup** of G. If $|G| = p^n \cdot m$, where $n > 0$ and p does not divide m, then any p-subgroup of G of order p^n is called a **Sylow** p-subgroup of G. ○

4.6.2 EXAMPLE Let $G = S_3$, so $|G| = 3! = 2 \cdot 3$. For $p = 3$, there is the Sylow 3-subgroup $A_3 = \langle(1\ 2\ 3)\rangle$. Since $(1\ 2\ 3)$ and its inverse $(1\ 3\ 2)$ are the only elements of order 3, this is the only Sylow 3-subgroup. For $p = 2$, there are three Sylow p-subgroups in S_3, namely $\langle(1\ 2)\rangle$, $\langle(1\ 3)\rangle$, and $\langle(2\ 3)\rangle$. Since $(1\ 3)(1\ 2)(1\ 3) = (2\ 3)$ and $(2\ 3)(1\ 2)(2\ 3) = (1\ 3)$, the three Sylow 3-subgroups are all conjugates of each other. ◇

4.6.3 EXAMPLE Let $G = A_4$, so $|G| = 2^2 3$. For $p = 2$, G has the Sylow 2-subgroup $\{e, (1\,2)(3\,4), (1\,3)(2\,4), (1\,4)(2\,3)\}$. Since its three nonidentity elements are all the elements of order 2 in G, this is the only Sylow 2-subgroup. For $p = 3$, G also has at least four Sylow 3-subroups, namely, $P = \langle(1\,2\,3)\rangle$ and $\langle(1\,2\,4)\rangle$, $\langle(1\,3\,4)\rangle$, $\langle(2\,3\,4)\rangle$. Since the four elements displayed and their inverses are all the elements of order 3, these are all the Sylow 3-subgroups. We can verify that

$$(1\,2)(3\,4)\,P\,(1\,2)(3\,4) = \langle(1\,2\,4)\rangle$$
$$(1\,3)(2\,4)\,P\,(1\,3)(2\,4) = \langle(1\,3\,4)\rangle$$
$$(1\,4)(2\,3)\,P\,(1\,4)(2\,3) = \langle(2\,3\,4)\rangle$$

Thus they are all conjugates. ◇

4.6.4 EXAMPLE Let $G = D_6 = \{e, \rho, \rho^2, \dots, \rho^5, \tau, \rho, \rho^2\tau, \dots, \rho^5\tau\}$, where ρ is a rotation by 60°, and τ is a reflection with a diagonal fixed, so that $\rho\tau = \tau\rho^5$ (as in Example 4.3.4). $|G| = 2^2{\cdot}3$. For $p = 3$, since ρ^2 and ρ^4 are the only elements of order 3, there is just one Sylow 3-subgroup, $\{e,\rho^2,\rho^4\}$. For $p = 2$, G has at least three Sylow 2-subgroups, namely, $Q = \{e, \tau, \rho^3, \rho^3\tau\}$ and $\{e, \rho\tau, \rho^3, \rho^4\tau\}$, $\{e, \rho^2\tau, \rho^3, \rho^5\tau\}$. The seven nonidentity elements displayed are all the elements of order 2, and there are no elements of order 4, so any Sylow 2-subgroup would have to be made up of the identity and three of these seven elements. It follows that there are no other Sylow 2-subgroups than the three just displayed, because any group of order 4 is abelian, and no two of $\rho^3\tau$, $\rho^4\tau$, $\rho^5\tau$ commute with each other. We can verify that

$$\rho^2 Q \rho^4 = \{e, \rho^2\tau, \rho^3, \rho^5\tau\}$$
$$\rho Q \rho^5 = \{e, \rho\tau, \rho^3, \rho^4\tau\}$$

The three Sylow 2-subgroups are all conjugates. ◇

In these examples we have found Sylow p-subgroups for all primes p dividing the order of the group. The first of the Sylow theorems guarantees that Sylow p-subgroups indeed always exist.

4.6.5 THEOREM (First Sylow theorem) Let p be a prime, and let G be a finite group, and suppose $|G| = p^n m$, where $n \geq 1$ and p does not divide m. Then for all k with $1 \leq k \leq n$, G contains at least one subgroup of order p^k (and so in particular contains a Sylow p-subgroup).

Proof We use induction on $|G|$. If $|G| = 1$, there is nothing to prove, so we assume inductively that the theorem is true for all groups of order less than $|G|$. If G has a proper subgroup H such that p^k divides $|H|$, then by our inductive assumption H contains a subgroup of order p^k, and hence so does G. Recall the class equation (Theorem 4.4.4):

$$|G| = |Z(G)| + \sum_{i=1}^{r} [G{:}C(a_i)]$$

where the $a_1, a_2, \dots, a_r$ form a complete set of representatives of the conjugacy classes not contained in the center $Z(G)$. Since $a_i \notin A(G)$, the $C(a_i)$ are all proper subgroups of G. If p^k divides any $|C(a_i)|$, then by our induction hypothesis G contains a subgroup of order p^k. So assume p^k does not divide any $|C(a_i)|$. Then p must divide each index $[G{:}C(a_i)]$ and therefore must divide the order of the center $|Z(G)|$. But $Z(G)$ is an Abelian group, and therefore by Theorem 2.4.20 it contains an element of order p, generating a subgroup P of order p. If $k = 1$ we are done. If $k > 1$, then note that P is a normal subgroup of G since it is contained in the center of G, and therefore the quotient group G/P is defined and has order $p^{n-1}{\cdot}m < |G|$. By our inductive assumption, therefore, G/P has a subgroup K of order p^{k-1}. Consider the homomorphism $\phi\colon G \to G/P$ sending each $x \in G$ to its coset xP. By (6) of Proposition 2.2.15, $H = \phi^{-1}(K) = \{x \in G \mid xP \in K\}$ is a subgroup of G. If $x \in P$, then of course $xP = P =$ the identity element of $G/P \in K$, so $P \subseteq H$, and we have $H/P = \{xP \mid x \in H\} = K$. But then $|H| = |H/P|{\cdot}|P| = p^{k-1}{\cdot}p = p^k$, so G does have a subgroup of order p^k. □

The other Sylow theorems give information about the number of Sylow p-subgroups and the relationships among them. Before coming to these theorems, a couple of lemmas will be needed.

4.6.6 LEMMA Let p be a prime, G a finite group, P a Sylow p-subgroup of G, $N_G(P) = \{g \in G \mid gPg^{-1} = P\}$ the normalizer of P in G, and Q any p-subgroup of G. Then $Q \cap N_G(P) = Q \cap P$.

Proof Let $N = N_G(P)$. Since $P \subseteq N$, we have $Q \cap P \subseteq Q \cap N$, and only need to prove the opposite inclusion $Q \cap N \subseteq Q \cap P$. And since obviously $Q \cap N \subseteq Q$, we only need to prove $Q \cap N \subseteq P$. Write H for $Q \cap N$. Then P and H are both subgroups of N, and the orders of both are powers p^n and p^m of the same prime p. Moreover, P is a *normal* subgroup of N. By Proposition 2.3.22, PH is a subgroup of N, and by Theorem 2.3.24 we have $|PH| = |P||H|\,/\,|P \cap H|$ and so is also a power p^r of p. On the one hand, since the order p^n of P is the *highest* power of p that divides $|G|$, we must have $r \leq n$. But on the other hand, since $P \subseteq PH$, we must also have $n \leq r$. Therefore, $r = n$ and $PH = P$, from which it follows that $H \subseteq P$. □

4.6.7 LEMMA Let p be a prime, G a finite group, P a Sylow p-subgroup of G, and K the set of all conjugates of P in G. Let Q be any p-subgroup of G, and let Q act on K by conjugation. Let $P = P_1, P_2, \ldots, P_r$ be a complete set of representatives of the orbits in this action. Then

$$|K| = \sum_{i=1}^{r} \,[Q{:}Q \cap P_i]$$

Proof Since the P_i are a complete set of representatives, $|K|$ is equal to the sum of the sizes of the orbits of the P_i. By the orbit-stabilizer relation, the size of the orbit of P_i is the index of the stabilizer of P_i, which is to say, of the set of elements of Q that leave P_i fixed or, in other words, $Q \cap N_G(P_i)$, which by the preceding lemma is the same as $Q \cap P_i$. □

4.6.8 THEOREM (Second Sylow theorem) Let p be a prime, and let G be a finite group. Then if P is a Sylow p-subgroup of G and Q any p-subgroup of G, then Q is contained in some conjugate of P (and so in particular all Sylow p-subgroups are conjugate with each other).

4.6.9 THEOREM (Third Sylow theorem) Let p be a prime, and let G be a finite group, and suppose $|G| = p^n m$, where $n \geq 1$ and p does not divide m. Let n_p be the number of p-Sylow subgroups in G. Then

(1) $n_p \equiv 1 \bmod p$

(2) n_p divides m

Proof We prove the second and third Sylow theorems together.
With notation as in the statement of the theorems, let K be the set of all conjugates in G of a Sylow p-subgroup P of G.
Claim $|K| \equiv 1 \bmod p$.
Proof of Claim We apply Lemma 4.6.7 with $Q = P$. On the one hand, we have $P_1 = P$, so $[P:P \cap P_1] = 1$. On the other hand, for $i > 1$ we have $P_i \neq P$, so $[P:P \cap P_i] > 1$, and $[P:P \cap P_i]$ is then a power of p. Lemma 4.6.7 then tells us $|K| = 1 +$ (a sum of powers of p), and hence $|K| \equiv 1 \bmod p$ as claimed.
Now we are ready to prove Theorem 4.6.8, by supposing there is a p-subgroup Q that is *not* contained in any conjugate of P and deriving a contradiction. Again we apply Lemma 4.6.7. If Q is not contained in any P_i, then for each i we have $[Q:Q \cap P_i] > 1$, and $[Q:Q \cap P_i]$ is then a power of p. Lemma 4.6.7 then tells us $|K| =$ (a sum of powers of p), and hence $|K| \equiv 0 \bmod p$, contrary to the claim we have just proved. This contradiction proves Theorem 4.6.8.
Now (1) of the Theorem 4.6.9 follows immediately. For the claim tells us that the number $|K|$ of conjugates in G of a given Sylow p-subgroup P satisfies $|K| \equiv 1 \bmod p$, while Theorem 4.6.8 tells us that *all* Sylow p-subgroups are conjugates in G of P, so that $|K|$ is equal to the total number n_p of Sylow p-subgroups. It follows that $n_p \equiv 1 \bmod p$.
Finally, (2) of Theorem 4.6.9 also follows easily. For by Proposition 4.4.10 the number $|K|$ of conjugates of a given Sylow p-subgroup, which we now know to be the total number n_p of Sylow p-subgroups, is equal to the index $[G:N] = |G|/|N|$, where $N = N_G(P)$ is the normalizer of P in G. Now we have $|G| = p^n \cdot m$, with p not dividing m, and we also know $p^n = |P|$ divides $|N|$, since P is a subgroup of N. It follows that n_p divides m. □

The following are immediate consequences of the first and second Sylow theorems, respectively, and the proofs will be left to the reader.

4.6.10 COROLLARY (Cauchy's theorem) Let p be a prime and G a finite group. If p divides the order of G, then G has an element of order p. □

4.6.11 COROLLARY Let G be a finite group and p a prime that divides $|G|$. Then $n_p = 1$ if and only if the Sylow p-subgroup P is a normal subgroup of G. Hence, if the number of Sylow p-subgroups of G is one, then G is not simple. □

We conclude this section with a stronger version of the first Sylow theorem. For this we need one more lemma.

4.6.12 LEMMA Let p be a prime, G a finite group, and H a p-subgroup of G. If p divides $[G:H]$, then $H \neq N(H)$.

Proof By definition $[G:H] = |X|$, where X is the set of all left cosets of H in G. Let H act on X by left multiplication (as in Exercises 15 through 18 of Section 4.1 and Exercise 12 of Section 4.2). Then $|X|$ is the sum of the sizes of the different orbits in this action. The size of each orbit divides the order $|H|$ of the group by Theorem 4.2.8 and is therefore either 1 or a power of p. The size of the orbit of the coset gH will be 1 exactly in the case where $hgH = gH$ for all $h \in H$, which is equivalent to having $g^{-1}hg \in H$ for all $h \in H$, which by definition is equivalent to having $g \in N(H)$ the normalizer of H. If $H = N(H)$, then there is only *one* orbit, namely, the orbit of H itself, that has size 1, so in this case $[G:H] = |X| \equiv 1 \bmod p$ and is not divisible by p. □

4.6.13 THEOREM Let p be a prime, and let G be a finite group, and suppose $|G| = p^n m$, where $n \geq 1$ and p does not divide m. Then G contains for all k with $1 \leq k < n$, at least one subgroup of order p^k that is a normal subgroup of a subgroup of order p^{k+1}.

Proof By the first Sylow theorem there exists a subgroup H of G of order p^k, and by the second Sylow theorem it is contained in some Sylow p-subgroup P of G. Let N be the normalizer of H in P. Since $k < n$, p divides $[P:H]$, and therefore by the preceding lemma $N \neq H$. Since H is normal in N, the quotient group N/H is defined, and its order is a power of p. Hence by Corollary 4.6.10 it contains a subgroup $U \subseteq N/H$ of order p. Let $\phi: N \to N/H$ be the homomorphism, carrying each element $a \in N$ to its coset aH, and let $T = \phi^{-1}(U) = \{a \in N \mid \phi(a) \in U\}$. Then T is a subgroup of N containing H and $|T| = |U| \cdot |H| = p \cdot p^k = p^{k+1}$. Since H is normal in N it is also normal in the subgroup T. □

4.6.14 EXAMPLE Consider $G = S_4$, so $|G| = 4! = 24 = 2^3 \cdot 3$, and a Sylow 2-subgroup would have order $2^3 = 8$. According to the first Sylow theorem, S_4 should have a subgroup of order 8. And in agreement with the theorem, we have already seen that the dihedral group D_4 has a permutation representation as a subgroup of S_4. (See Examples 1.4.24 and 4.1.12.) ◇

4.6.15 EXAMPLE Consider $G = A_5$, so $|G| = 5!/2 = 60 = 2^2 \cdot 15$, and a Sylow 2-subgroup would have order $2^2 = 4$. The following are two Sylow 2-subgroups:

$$P = \{e, (1\ 2)(3\ 4), (1\ 3)(2\ 4), (1\ 4)(2\ 3)\}$$
$$Q = \{e, (1\ 2)(3\ 5), (1\ 3)(2\ 5), (1\ 5)(2\ 3)\}$$

According to the second Sylow theorem, these should be conjugate in A_5. And in agreement with the theorem, we can calculate that if $\tau = (2\ 3)(4\ 5)$, then $\tau(1\ 2)(3\ 4)\tau^{-1} = (1\ 3)(2\ 5)$ and $\tau(1\ 3)(2\ 4)\tau^{-1} = (1\ 2)(3\ 5)$ and $\tau(1\ 4)(2\ 3)\tau^{-1} = (1\ 5)(2\ 3)$, and so $\tau P\tau^{-1} = Q$. ◇

4.6.16 EXAMPLE Consider $G = A_5$, so $|G| = 5!/2 = 60 = 2^2 \cdot 3 \cdot 5$, and the Sylow 2-, 3-, and 5-subgroups would have orders $2^2 = 4$, 3, and 5, respectively. There are no elements of order 4 (since these would have to be 4-cycles, which are odd permutations), and the only elements of order 2 are products of disjoint 2-cycles, $(x\ y)(u\ v)$. Each such permutation belongs to a subgroup $\{e, (x\ y)(u\ v), (x\ u)(y\ v), (x\ v)(u\ y)\}$ as in the preceding example, and each such subgroup contains three such permutations. In Example 4.5.14 we saw that the number of such permutations is 15. Therefore, the number of Sylow 2-subgroups is $15/3 = 5$. As for Sylow 3- and 5-subgroups, these will be generated by 3-cycles and 5-cycles, respectively. In Examples 4.5.12 and 4.5.13 we saw that the number of 3-cycles is 20 and the number of 5-cycles is 24. Since each subgroup of order 3 contains two 3-cycles and each subgroup of order 5 contains four 5-cycles, the total numbers of such groups are $20/2 = 10$ and $24/4 = 6$, respectively. Note that we have

$$5 \equiv 1 \bmod 2 \text{ and } 5 \text{ divides } 3\cdot5$$
$$10 \equiv 1 \bmod 3 \text{ and } 10 \text{ divides } 2^2\cdot5$$
$$6 \equiv 1 \bmod 5 \text{ and } 6 \text{ divides } 2^2\cdot3$$

which agrees with the third Sylow theorem. ◇

4.6.17 EXAMPLE Let $P = \{e, \sigma=(1\ 2\ 3), \sigma^2=(1\ 3\ 2)\}$ be a Sylow 3-subgroup of A_5. Let us find the normalizer $N(P)$ of P in A_5. From the preceding example we know P has 10 conjugates, hence $[A_5:N(P)] = 10$, and so we should have $|N(P)| = |A_5|/10 = 60/10 = 6$. Since $P \subseteq N(P)$, we need only find an element τ of order 2 in $N(P)$. We can verify that if $\tau = (2\ 3)(4\ 5)$, then $\tau\sigma\tau = \sigma^2$ and $\tau\sigma^2\tau = \sigma$. So $\tau \in N(P)$ and $N(P) = \{e, \sigma, \sigma^2, \tau, \tau\sigma, \tau\sigma^2\}$. ◇

Exercises 4.6

1. Find the order of a Sylow p-subgroup of G for the indicated p and G:

(a) $p = 2$ $\quad G = S_5$
(b) $p = 2$ $\quad G = A_4$
(c) $p = 2$ $\quad G = D_6$
(d) $p = 3$ $\quad G =$ any group of order 270

2. Let P be a Sylow 2-subgroup of a group G of order 20. If P is not a normal subgroup of G, how many conjugates does P have in G?

In Exercises 3 through 5 find all Sylow p-subgroups of G for the indicated p and G, and show that they are conjugates.

3. $p = 3$, $G = S_4$ **4.** $p = 2$, $G = S_4$ **5.** $p = 2$, $G = A_5$

6. Find all Sylow 3-subgroups and Sylow 5-subgroups in S_5.

7. Show that the intersection of all the Sylow 2-subgroups in S_4 is a normal subgroup of S_4 isomorphic to the Klein 4-group.

8. Let P be as in Example 4.6.15. Find the normalizer of P in A_5.

9. Let P be as in Example 4.6.17. Find the normalizer of P in S_5.

10. In the dihedral group D_4, find the normalizer of $P = \{e, \tau\}$, where τ is a flip. Is P a normal subgroup of D_4?

11. In the dihedral group D_4, according to Theorem 4.6.13 there exists a subgroup Q of order 4 such that $P = \{e, \tau\}$ is a normal subgroup of Q. Find such a Q.

12. Let H be a normal subgroup of G and K a normal subgroup of H. Then K is a subgroup of G. Must K be a *normal* subgroup of G? If so, give a proof; if not, give a counterexample.

13. Prove Corollary 4.6.10, Cauchy's theorem. **14.** Prove Corollary 4.6.11.

15. Show that no group of order pq where p and q are distinct primes is simple. (*Hint:* Use the third Sylow theorem and Corollary 4.6.11.)

16. Let H be a normal p-subgroup of a finite group G. Show that H is contained in every Sylow p-subgroup of G.

17. Let G be a p-group and H a proper subgroup of G. Show that there exists a subgroup $K \leq G$ such that

(a) $H \leq K$ (b) $K \lhd G$ (c) $[G{:}K] = p$

18. Show that no group G of order $|G| = n$ is simple where

(a) $n = 45$ (b) $n = 16$

(c) $n = p^r$, p prime, $r > 1$ (d) $n = p^r m$, p prime, $r \geq 1, p > m$

4.7 Applications of the Sylow Theorems

In this section we give some applications of the Sylow theorems. In most cases, we will be using them to show that there are no simple groups of some particular order. In some cases, we will be able to determine *all* the groups of a particular order. We begin by looking at groups of order $2p$, where p is an odd prime, and first we look at two groups with which we are already familiar.

4.7.1 EXAMPLE Let G be a group of order 6. By Cauchy's theorem (Corollary 4.6.10), G has an element a of order 2 and an element b of order 3. Let $H = <a>$ and $K = <b>$. Since K is then a subgroup of G of index 2, it is normal, from which it follows that $G = HK$, and every element of G is of form $a^i b^j$, where $0 \leq i \leq 1$ and $0 \leq j \leq 2$. Moreover, also by normality, aba^{-1} is in K and so must equal either b or $b^2 = b^{-1}$. In case $aba^{-1} = b$ it follows $ba = ab$, and this relationship is enough to determine the whole multiplication table for G and show that G is Abelian, and therefore isomorphic to $\mathbb{Z}_6$. In case $aba^{-1} = b^{-1}$ it follows $ab = b^{-1}a = b^2a$, and again this relationship is enough to determine the whole multiplication table for G and show that G is isomorphic to S_3. ◇

The example we just calculated is an illustration of the following theorem, which holds for all groups of order $2p$, p an odd prime.

4.7.2 THEOREM Let G be a group of order $2p$, p an odd prime. Then G is isomorphic either to the cyclic group $\mathbb{Z}_{2p}$ or else to the dihedral group D_p.

Proof By Cauchy's theorem (Corollary 4.6.10), G has an element a of order 2, and an element b of order p. Let $H = <a>$ and $K = <b>$. Since K is then a subgroup of G of index 2, it is normal, from which it follows that $G = HK$, and every element of G is of form $a^i b^j$, where $0 \leq i \leq 1$ and $0 \leq j \leq p\text{-}1$. Moreover, also by normality, aba^{-1} is in K. Hence $aba^{-1} = b^j$ for some j with $1 \leq j \leq p\text{-}1$. Now we have $b^{j^2} = (b^j)^j = (aba^{-1})^j = ab^ja^{-1} = a(aba^{-1})a^{-1} = a^2ba^{-2} = b$, the last step following since a has order 2. So the order p of b must divide $j^2\text{-}1 = (j+1)(j\text{-}1)$, and since $1 \leq j \leq p\text{-}1$, the only possibilities are $j = 1$ and $j = p\text{-}1$. In case $j = 1$, we have $aba^{-1} = b$ or $ab = ba$. Much as in the preceding example, in case $j = 1$ we get $ba = ab$ and it follows by Theorem 3.2.8 that $G \cong \mathbb{Z}_{2p}$, while in $j = p\text{-}1$, we get $aba^{-1} = b^{p-1} = b^{-1}$ or $ab = b^{p-1}a = b^{-1}a$, and it follows that $G \cong D_p$. □

In the preceding theorem we only needed Cauchy's theorem, which follows from the first Sylow theorem. Because K was of index 2 we knew right away that it was therefore normal. In the case where the order of G is a product pq of distinct odd primes, we need to use the third Sylow theorem. Let us first note the following (a generalization of Exercise 15 of the preceding section).

4.7.3 PROPOSITION Let G be a group of order $q^n m$, where q is a prime and $m < q$. Then the Sylow q-subgroup of G is unique and hence normal in G, and hence G is not simple.

Proof By the third Sylow theorem, the number n_q of Sylow q-subgroups is of the form $1 + qk$ and divides m. Since $m < q$ the only possibility is $k = 0$ and $n_q = 1$, and the Sylow q-subgroup is unique. That it is normal, and hence that G is not simple, follows by Corollary 4.6.11. □

4.7.4 EXAMPLE Let G be a group of order $15 = 3 \cdot 5$. By Cauchy's theorem there is an element a of order 3 and an element b of order 5. Let H be the Sylow 3-subgroup $\langle a \rangle$, and K the Sylow 5-subgroup $\langle b \rangle$. By Proposition 4.7.3, the Sylow 5-subgroup K is unique and normal, and so $G = HK$ and every element of G is of form $a^i b^j$, where $0 \le i \le 2$ and $0 \le j \le 4$. By the third Sylow theorem, n_3 is of form $1+3k$ and divides 5. Clearly the only possibility is $k = 0$ and $n_3 = 1$, and the Sylow 3-subgroup H is also unique and normal. By the normality of H and K we have $aba^{-1} \in K$ and $bab^{-1} \in H$, and so $aba^{-1}b^{-1} = (aba^{-1})b = a(ba^{-1}b^{-1}) \in K \cap H = \{e\}$, so $aba^{-1}b^{-1}=e$ or $ab=ba$. It follows that G is Abelian and hence by Theorem 3.2.8 we obtain $G \cong \mathbb{Z}_{15}$. ◇

4.7.5 EXAMPLE Let G be a group of order $21 = 3 \cdot 7$. By Cauchy's theorem there is an element a of order 3 and an element b of order 7. Let H be the Sylow 3-subgroup $\langle a \rangle$, and K the Sylow 7-subgroup $\langle b \rangle$. By Proposition 4.7.3, the Sylow 7-subgroup K is unique and normal, and so $G = HK$ and every element of G is of form $a^i b^j$, where $0 \le i \le 2$ and $0 \le j \le 6$. By the third Sylow theorem, n_3 is of form $1+3k$ and divides 7. There are two possibilities, $k = 0$ and $k = 2$. In the case $k = 0$, the same argument as in the preceding example shows that $ab = ba$, and $G \cong \mathbb{Z}_{21}$. Let us consider the case $k = 2$. Since K is normal, $aba^{-1} \in K$ and $aba^{-1} = b^i$ for some i with $2 \le i < 6$. The same kind of computation as in the proof of Theorem 4.7.2 shows that that $b^{i^2} = a^2ba^{-2}$ and $b^{i^3} = a^3ba^{-3} = b$, where the last step follows since a has order 3. Thus $b^{i^3-1} = e$, and therefore the order 7 of b divides $i^3-1 = (i-1)(i^2+i+1)$. For $2 \le i \le 6$ this is only possible if $i = 2$ or $i = 4$. Actually the two possibilities turn out to be equivalent (according to Exercise 11), but we pursue here only the possibility $i=2$. From $aba^{-1} = b^2$ it follows that $ab^4a^{-1} = (aba^{-1})^4 = (b^2)^4 = b^8 = b$, since b has order 7, and hence that $ba = ab^4$. Actually the relations $a^3 = b^7 = e$ and $ba = ab^4$ completely determine the multiplication table for a non-Abelian group of order 21. (See Exercise 12.) To give a concrete example we look in the symmetric group S_7 for permutations satisfying these relations. It turns out that for $a = (2\ 3\ 5)(4\ 7\ 6)$ and $b = (1\ 2\ 3\ 4\ 5\ 6\ 7)$ it can be computed that $aba^{-1} = (1\ 5\ 2\ 6\ 3\ 7\ 4) = b^2$. Thus we obtain a non-Abelian group of order 21, namely $\{a^j b^k \mid 0 \le j \le 2, 0 \le k \le 6\}$. ◇

What makes the difference between 15 and 21 in the preceding examples is brought out in the following theorem.

4.7.6 THEOREM Let G be a group of order pq, where p and q are odd primes with $p < q$. If p does not divide q - 1, then G is cyclic.

Proof By Cauchy's theorem there are elements a, b of orders p, q generating subgroups H, K. By Proposition 4.7.3, the Sylow q-subgroup K is unique and normal, and so $G = HK$ and every element of G is of form $a^i b^j$, where $0 \le i \le p\text{-}1$ and $0 \le j \le q\text{-}1$. By the third Sylow theorem, the number n_p of Sylow p-subgroups divides q, which since q is prime means either $n_p = 1$ or $n_p = q$. But also by the same theorem

n_p is of form $1 + pk$, so that if p does not divide $q - 1$, then the possibility $1 + pk = np = q$ is excluded. So the Sylow p-subgroup H is also unique and normal. The argument of Example 4.7.4 then shows that the normality of H and K implies $aba^{-1}b^{-1} \in H \cap K = \{e\}$ and $ab = ba$, from which it then follows from Theorem 3.2.8, as in several earlier examples, that $G \cong \mathbb{Z}_{pq}$. □

4.7.7 EXAMPLE Any group of order $33 = 3 \cdot 11$ or $51 = 3 \cdot 17$ is cyclic, since 3 does not divide $11 - 1 = 10$ or $17 - 1 = 16$. ◇

It is natural to look next for information about groups of order p^2q, where p and q are distinct primes.

4.7.8 EXAMPLES A_4 and D_6 are examples of non-Abelian groups of orders $12 = 2^2 \cdot 3$ and, more generally, D_{2p} is an example of a non-Abelian group of order 2^2p. ◇

4.7.9 EXAMPLE In Example 4.7.5 we found a non-Abelian group G of order $21 = 3 \cdot 7$. Then $\mathbb{Z}_3 \times G$ would be an example of a non-Abelian group of order $63 = 3^2 \cdot 7$. More generally, if we have a non-Abelian group G of order pq, then $\mathbb{Z}_p \times G$ would be an example of a non-Abelian group of order p^2q. ◇

4.7.10 EXAMPLE By contrast to the preceding examples, every group G of order $45 = 3^2 \cdot 5$ is Abelian. For let H and K be a Sylow 5-subgroup and a Sylow 3-subgroup, respectively. These groups are of orders 5 and 9, respectively, and so we know them to be Abelian. By the third Sylow theorem, n_3 is of the form $1 + 3k$ and divides 5, and the only possibility is $n_3 = 1$, and the Sylow 3-subgroup is unique and normal. Similarly, n_5 is of the form $1 + 5k$ and divides 3^2. Again the only possibility is $n_5 = 1$, and the Sylow 5-subgroup is also unique and normal, and $G = HK$. It follows, as in earlier examples, that if $a \in H$ and $b \in K$, then $aba^{-1}b^{-1} \in H \cap K = \{e\}$, so that $ab = ba$. From this and the fact that H and K are themselves Abelian, it follows that $G = HK$ is Abelian (and by the analysis of finite Abelian groups isomorphic either to $\mathbb{Z}_{45}$ or to $\mathbb{Z}_3 \times \mathbb{Z}_3 \times \mathbb{Z}_5$). ◇

The preceding example suggests the following analogue of Theorem 4.7.6 for groups of order p^2q, whose proof will be left to the reader.

4.7.11 THEOREM Let G be a group of order p^2q, where p and q are distinct primes such that p does not divide $q - 1$ and q does not divide $p^2 - 1$. Then G is Abelian.

Proof Left to the reader (as Exercise 16). □

The next few examples illustrate how the Sylow theorems can be used to show that groups of certain orders cannot be simple. In some cases we can show more than that. For example, we can identify all possible groups of order 30.

4.7.12 EXAMPLE No group of order 30 is simple. For assume G is a simple group of order $30 = 2 \cdot 3 \cdot 5$ and consider the numbers n_5 and n_3 of Sylow 5-subgroups and 3-subgroups. We must have $n_5 > 1$ and $n_3 > 1$, else the corresponding Sylow subgroup would be normal and the group not simple. But we also know that n_5 divides 6 and is of form $1+5k$, so it must be 6. Similarly, we know n_3 divides 10 and is of form $1+3k$, so it must be 10. Now any two Sylow 5-subgroups or any two Sylow 3-subgroups must have intersection $\{e\}$, so the 6 Sylow 5-subgroups give us $6 \cdot (5 - 1) = 24$ elements of order 5, and the 10 Sylow 3-subgroups give us $10 \cdot (3 - 1) = 20$ elements of order 3, and that is too many elements. Hence $n_5 = 1$ or $n_3 = 1$, and in either case G is not simple. We return to the question of groups of order 30 shortly. ◇

4.7.13 EXAMPLE No group of order 56 is simple. Arguing as in the preceding example, we would conclude $n_7 > 1$, n_7 divides 8, and n_7 is of form $1+7k$, and hence $n_7 = 8$, and similarly $n_2 = 7$. The 8 Sylow 7-subgroups would give 48 elements of order 7, and even just 2 Sylow 2-subgroups would give more than 7 elements of even order. Since there is also the identity, this is too many elements. ◇

4.7.14 EXAMPLE No group of order 36 is simple. Here, assuming we had a simple group G of order $36 = 2^2 \cdot 3^2$, we would have $n_3 > 1$, n_3 divides 4, and n_3 of form $1+3k$, implying $n_3 = 4$. If P and Q are two Sylow 3-subgroups, we cannot have $P \cap Q = \{e\}$, since that would give 81 different elements in PQ, which is too many. So we must have $|P \cap Q| = 3$. Now consider the normalizer $N(P \cap Q)$. Since P and Q, having order 9, are both Abelian, for any $g \in P \cup Q$ and $a \in P \cap Q$ we have $gag^{-1} = a$, so $P \cup Q \subseteq N(P \cap Q)$ and $15 \leq |N(P \cap Q)|$, and since $|N(P \cap Q)|$ divides 36, $|N(P \cap Q)|$ must equal 18 or 36. If $N(P \cap Q)$ has 18 elements, then $N(P \cap Q)$ has index 2 and is normal. If $N(P \cap Q)$ has 36 elements, then it is the whole group, and $P \cap Q$ is normal. Either way, there is a normal subgroup and G is not simple. ◇

4.7.15 EXAMPLE Any group G of order $255 = 3 \cdot 5 \cdot 17$ is Abelian (and therefore isomorphic to the cyclic group $\mathbb{Z}_{255}$). For, on the one hand, we know $n_{17} = 1 + 17k$ divides $3 \cdot 5$, and so must be 1, and hence the Sylow 17-subgroup P of G is normal. Then the quotient G/P is a group of order 15, and so is Abelian. On the other hand, we know $n_3 = 1 + 3k$ divides $5 \cdot 17$, and so must be 1 or 85. Similarly, we know $n_5 = 1 + 5k$ divides $3 \cdot 17$, and so must be 1 or 51. Moreover, we cannot have both $n_3 = 85$ and $n_5 = 51$, since this would give 170 elements of order 3 and 204 elements of order 5. So there is a normal subgroup Q of order 3 or 5. Then the quotient G/P is a group of order $5 \cdot 17$ or $3 \cdot 17$ and so is Abelian by Theorem 4.7.6. By Exercise 27 of Section 2.4, if we let N be the commutator subgroup of G, then $N \subseteq P$ and $N \subseteq Q$, and since $|P| = 17$ and $|Q| = 3$ or 5, this implies $N = \{e\}$ and G is Abelian. ◇

4.7.16 EXAMPLE Any group of order 30 is isomorphic to one of the four groups $\mathbb{Z}_{30}$, D_{15}, $\mathbb{Z}_3 \times D_5$, or $\mathbb{Z}_5 \times S_3$. For from Example 4.7.12 we already know that if P is a Sylow 5-subgroup and Q a Sylow 3-subgroup, at least one of them is normal.

Hence by Proposition 2.3.22, PQ is a group. Moreover, since the order of PQ is 15, its index is 2 and it is normal, while by Example 4.7.4, it is cyclic. Let b be an element of order 15 generating PQ, and a any element of order 2, and consider aba^{-1}. Since PQ is normal, $aba^{-1} \in PQ$, and since PQ is cyclic, generated by b, we have $aba^{-1} = b^i$ for some i. Note that since a has order 2, $aba^{-1} = b^i$ gives $ba = ab^i$.As in several earlier examples, we can calculate $b^{i^2} = (b^i)^i = (aba^{-1})^i = ab^ia^{-1} = a(aba^{-1})a^{-1} = a^2ba^{-2} = b$, since a has order 2. It follows that the order 15 of b divides $i^2 - 1 = (i + 1)(i - 1)$. Simple arithmetic shows that the only possibilities are $i = 1$, 14, 4, or 11. In case $i = 1$ we have $ba = ab$ and the group is Abelian and G is isomorphic to $\mathbb{Z}_{30}$. In case $i = 14$ we have $ba = ab^{14} = ab^{-1}$. In case $i = 4$ we have $ba = ab^4$. And in case $i = 11$ we have $ba = ab^{11} = ab^{-4}$. In any of these three cases, the indicated relations are enough to determine the whole table for the group. We leave it to the reader (as Exercises 17 through 19) to work out that in these three cases G is isomorphic to D_{15} or to $\mathbb{Z}_3 \times D_5$ or to $\mathbb{Z}_5 \times D_3$. ◇

Exercises 4.7

In Exercises 1 through 5 determine how many nonisomorphic groups there are of the indicated order, and give examples of such groups.

1. Groups of order 9

2. Groups of order 10

3. Groups of order 14

4. Groups of order 49

5. Groups of order 99

In Exercises 6 through 10 show that no group of the indicated order is simple.

6. Groups of order 42

7. Groups of order 20

8. Groups of order 48

9. Groups of order 75

10. Groups of order 39

Exercises 11 and 12 pertain to Example 4.7.5, where we have a group G, an element b of order 7, and an element of order 3.

11. Show that if a_1 and a_2 are elements of G of order 3 that are inverses of each other, then $a_1ba_1^{-1} = b^2$ if and only if $a_2ba_2^{-1} = b^4$.

12. Assume that a is an element of G of order 3 and that $aba^{-1} = b^2$. Write out the multiplication table for all products of elements of G of form a^jb^k with $0 \le j \le 2$, $0 \le k \le 6$.

13. Construct a non-Abelian group of order 39. (*Hint:* Use Exercise 10 and the method of Example 4.7.5.)

14. Construct a non-Abelian group of order 117. (*Hint:* Use the preceding exercise and the method of Example 4.7.9.)

15. Construct a non-Abelian group of order 105. (*Hint:* See the preceding exercise.)

16. Prove Theorem 4.7.11. (*Hint:* Imitate the proof of Theorem 4.7.6 and Example 4.7.10.)

Exercises 17 through 19 pertain to Example 4.7.16, were we have a group G of order 30, an element a of order 2, an element b of order 15, and a relation of form $ba = ab^i$ that holds for some i.

17. Show that if $ba = ab^{-1}$, then G is isomorphic to D_{15}.

18. Show that if $ba = ab^4$, then G is isomorphic to $\mathbb{Z}_3 \times D_5$.

19. Show that if $ba = ab^{-4}$, then G is isomorphic to $\mathbb{Z}_5 \times D_3$.

20. Find four nonisomorphic groups of order 66.

21. Prove that there are *exactly* four nonisomorphic groups of order 66. (*Hint:* Use the previous exercise and the method of Example 4.7.16.)

22. Prove that every group of order 345 is Abelian and cyclic.

23. Let H be a normal subgroup of a group G. Show that if H is cyclic, then every subgroup of H is also normal in G.

24. Let G be a group of order 60. Suppose that G contains a normal subgroup of order 2. Show that G has normal subgroups of orders 6, 10, and 30.

25. Let G be a finite group, P a Sylow p-subgroup of G, and H a subgroup of G with $P \leq H \leq G$. Show that if P is normal in H and H is normal in G, then P is normal in G.

26. Let G be a finite group, P a Sylow p-subgroup of G, and H the normalizer of P in G. Show that the normalizer $N_G(H)$ of H in G is equal to H itself. (*Hint:* Use the preceding exercise.)

27. Show that a group of order 8 is isomorphic to one of the following groups:

$\mathbb{Z}_8$, $\mathbb{Z}_4 \times \mathbb{Z}_2$, $\mathbb{Z}_2 \times \mathbb{Z}_2 \times \mathbb{Z}_2$, D_4, or Q_8 the quaternion group

28. Classify all groups of order $n \leq 15$.

Chapter 5

Composition Series

In this chapter we learn one more tool that helps us understand and classify some groups. For a given group G we construct chains of subgroups of G, called *composition series*, such that the subgroups themselves and the quotient groups they determine satisfy certain special conditions. As we see from the *Jordan-Hölder theorem*, the lengths of these chains and the quotient groups are uniquely determined by the original group G and tell us something about its structure. At the end of Chapter 4 we found, using calculations based on the Sylow theorems, ways of determining whether certain groups are simple or not. The importance of these calculations becomes apparent when, in the last section of this chapter, we want to determine whether a given group is *solvable* or not. We begin this chapter with some fundamental theorems on isomorphisms that we use repeatedly in our constructions.

5.1 Isomorphism Theorems

We have already encountered the first isomorphism theorem, which expresses a homomorphic image of a group G as the quotient group of G by a normal subgroup of G, the kernel of the homomorphism. In this section we study two more isomorphism theorems that give us information on subgroups of the original group G and subgroups of the quotient group. The three isomorphism theorems allow us later in this chapter to study the composition series of a group and determine the solvability of a group.
Let us first review the first isomorphism theorem and draw one more important consequence from it.

5.1.1 THEOREM (First isomorphism theorem) Let $\phi: G \to G'$ be a homomorphism, with kernel Kern $\phi = K$. Then $G/K \cong \phi(G)$.

Proof This is Theorem 2.4.15. □

This theorem can be better understood if we have in mind the diagram in Figure 1. Here for any $g \in G$, $\tau(g) = gK \in G/K$, and $\chi(gK) = \phi(g) \in \phi(G)$. The diagram is commutative, meaning that we have $\chi(\tau(g)) = \phi(g)$ for all $g \in G$. Furthermore, χ is an isomorphism between the two groups G/K and $\phi(G)$.

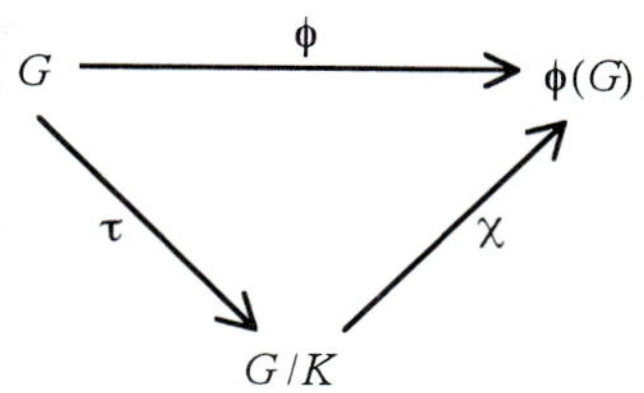

FIGURE 1

A consequence we derive from the first isomorphism theorem is illustrated by the next example.

5.1.2 EXAMPLE Let us now consider the commutative diagram in Figure 2, where ϕ: $D_4 \to \mathbb{Z}_2 \times \mathbb{Z}_2$ is defined as follows: For any element $\rho^i\tau^j \in D_4$, where $0 \le i \le 3$, $0 \le j \le 1$, we let $\phi(\rho^i\tau^j) = (i, j)$. So, for instance, $\phi(\rho) = (1, 0)$, $\phi(\tau) = (0, 1)$.

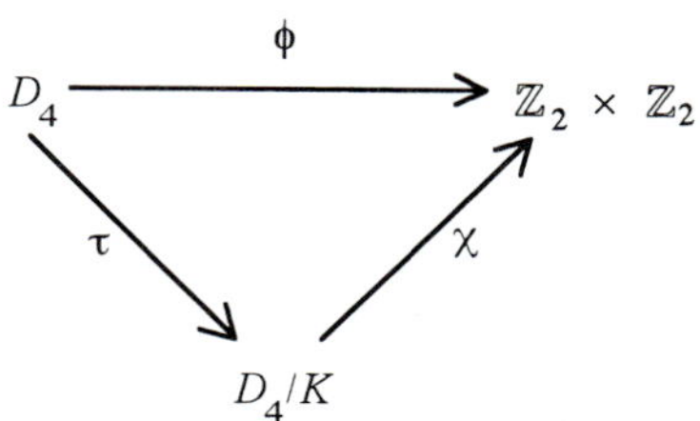

FIGURE 2

ϕ is an onto homomorphism, K = Kern $\phi = \{\rho_0, \rho^2\}$, and $D_4/K = \{K, \rho K, \tau K, \rho\tau K\}$. We look at the inverse images of subgroups of $\mathbb{Z}_2 \times \mathbb{Z}_2$.

(1) $\phi^{-1}(\mathbb{Z}_2 \times \{0\}) = \phi^{-1}(\langle(1,0)\rangle) = \{\rho_0, \rho, \rho^2, \rho^3\}$.

(2) $\phi^{-1}(\{0\} \times \mathbb{Z}_2) = \phi^{-1}(\langle(0,1)\rangle) = \{\rho_0, \rho^2, \tau, \rho^2\tau\}$

(3) $\phi^{-1}(\langle(1,1)\rangle) = \{\rho_0, \rho^2, \rho\tau, \rho^3\tau\}$

The three subgroups of $\mathbb{Z}_2 \times \mathbb{Z}_2$ (which are all normal, since $\mathbb{Z}_2 \times \mathbb{Z}_2$ is Abelian) give rise in a natural way to three subgroups of D_4 that contain K (which are also normal, since they have index 2). ◇

5.1.3 PROPOSITION Let G be a group and K a normal subgroup and $\tau: G \to G/K$ the homomorphism defined by letting $\tau(g) = gK$ for all $g \in G$. For any subgroup $H \leq G$ with $K \leq H$, let $\tau(H) = \{\tau(h) \mid h \in H\}$. Then

(1) This map is a one to one, onto map between the two sets of subgroups $\{H \mid H \leq G \text{ and } K \leq H\}$ and $\{L \mid L \leq G/K\}$.

(2) For any $H \leq G$ with $K \leq H$, if $\tau(H) = L \leq G/K$, then $H \lhd G$ if and only if $L \lhd G/K$.

Proof (1) First note that for any $H \leq G$, $\tau(H)$ is a subgroup of G/K by part (5) of Proposition 2.2.15. So we have a map between $\{H \mid H \leq G\}$ and $\{L \mid L \leq G/K\}$.
Next we want to show this map is one to one. So suppose $H_1 \leq G$ and $H_2 \leq G$ and $\tau(H_1) = \tau(H_2) = L$. Then given any $x \in H_1$ there exists a $y \in H_2$ such that $\tau(x) = \tau(y) \in G/K$, hence $xK = yK$. But $x \in xK = yK \subseteq H_2$ since H_2 contains K. So $x \in H_2$ and $H_1 \subseteq H_2$. Similarly, $H_2 \subseteq H_1$. Thus $H_1 = H_2$ and the map is one to one.
Finally, we want to show that the map is onto. Given $L \leq G/K$ we know $\tau^{-1}(L) = \{x \in G \mid xK \in L\}$ is a subgroup of G by part (6) of Proposition 2.2.15. For any $k \in K$, $kK = K$, which is the identity element of G/K and therefore an element of L, since L is a subgroup. Hence $k \in \phi^{-1}(L)$ and K is contained in $\phi^{-1}(L)$. So we have found an element $\tau^{-1}(L) \in \{H \mid H \leq G \text{ and } K \leq H\}$ such that $\tau(\tau^{-1}(L)) = L$, showing our map to be onto.

(2) Now suppose $H \lhd G$ and $K \leq H$. Then for any $gK \in G/K$ and any $\tau(h) \in \tau(H)$ we have $(gK)(\tau(h))(gK)^{-1} = \tau(g)\tau(h)\tau(g)^{-1} = \tau(ghg^{-1}) \in \tau(H)$ because $ghg^{-1} \in H$ since $H \lhd G$. So $\tau(H) \lhd G/K$.
Conversely, suppose $L \lhd G/K$. Then $\tau^{-1}(L) \lhd G$, since for any $g \in G$ and $x \in \tau^{-1}(L)$ we have $\tau(gxg^{-1}) = \tau(g)\tau(x)\tau(g)^{-1} \in L$ since $\tau(x) \in L$ and $L \lhd G/K$. □

5.1.4 DEFINITION Let G be a group and K a normal subgroup of G. The homomorphism $\tau: G \to G/K$ in the preceding proposition, with $\tau(g) = gK$, is called the **canonical homomorphism** from G onto G/K. The associated one-to-one, onto map from $\{H \mid H \leq G \text{ and } K \leq H\}$ to $\{L \mid L \leq G/K\}$ is called the **canonical correspondence**. ○

Given the canonical homomorphism $\tau: G \to G/K$ and a subgroup $H \leq G$ with $K \leq H$, we can restrict τ to H and obtain $\tau: H \to H/K \leq G/K$. Then $\tau(H) \cong H/K$ by the first isomorphism theorem. What we want to see now is what happens when we restrict τ to a subgroup $H \leq G$ where we do *not* have $K \leq H$.

5.1.5 EXAMPLE In the dihedral group $D_6 = \{\rho^i\tau^j \mid 0 \leq i \leq 5, 0 \leq j \leq 1, \tau\rho = \rho^{-1}\tau\}$ consider the subgroup $K = \langle\rho\rangle$. Since $|\rho| = 6$ and $|D_6| = 12$, K has index 2, hence $K \lhd D_6$, and $D_6/K = \{K, \tau K\} \cong \mathbb{Z}_2$. The canonical homomorphism $\phi: D_6 \to D_6/K$ is given by $\phi(\rho^i\tau^j) = \tau^j K$, which is K if j is even, τK if j is odd.
Now let $H = \{\rho_0, \rho^3, \tau, \rho^3\tau\}$. ϕ restricted to H gives us $\phi(\rho_0) = \phi(\rho^3) = K$, while $\phi(\tau) = \phi(\rho^3\tau) = \tau K$. The kernel of ϕ restricted to H is $K \cap H = \{\rho_0, \rho^3\}$, and

$H/\{\rho_0, \rho^3\} = \{K, \tau K\} \cong \mathbb{Z}_2$. Note that since $|HK| = |H||K|/|H \cap K| = 4{\cdot}6/2 = 12$, we have $HK = D_6$ and therefore $D_6/K = HK/K \cong \mathbb{Z}_2 \cong H/H \cap K$. ◇

We prove that this last isomorphism holds generally for any group G and subgroup H.

5.1.6 THEOREM (Second isomorphism theorem) Let G be a group, K a normal subgroup of G, and H any subgroup of G. Then

$$HK/K \cong H/(H \cap K)$$

Proof Let τ: $G \to G/K$ be the canonical homomorphism. Since $H \leq G$, $\tau(H) \leq G/K$ by part (5) of Proposition 2.2.15. Now restricting τ to H we have a homomorphism from H to $\tau(H)$. The kernel of this homomorphism consists of all elements of H that are in K = Kern τ. Therefore, by the first isomorphism theorem, $H/(H \cap K) \cong \tau(H)$. Now consider HK. Since $K \lhd H$, $HK \leq G$ by Proposition 2.3.22. Since the map τ is a homomorphism, $\tau(hk) = \tau(h)\tau(k) = \tau(h) \in \tau(H)$. Hence the map τ restricted to HK gives a homomorphism from HK onto $\tau(H)$, whose kernel is K. Therefore by the first isomorphism theorem, $HK/K \cong \tau(H)$. Therefore, $H/(H \cap K) \cong \tau(H) \cong HK/K$ and the theorem is proved. □

We are ready now to prove the third isomorphism theorem, but let us first look at a helpful example.

5.1.7 EXAMPLE In D_6 let us take two normal subgroups, $N = \langle\rho\rangle$ and $M = \langle\rho^3\rangle$. We have $M \leq N \leq D_6$. There are three quotient groups, D_6/N, N/M, and D_6/M. $D_6/N = \{N, \tau N\} \cong \mathbb{Z}_2$. $N/M = \{M, \rho M, \rho^2 M\} \cong \mathbb{Z}_3$, and $D_6/M = \{M, \rho M, \rho^2 M, \tau M, \rho\tau M, \rho^2\tau M\}$, a group of order 6. It is easily checked that $(\rho M)(\tau M) = \rho\tau M \neq \rho^2\tau M = (\tau M)(\rho M)$, hence D_6/M is non-Abelian. So $D_6/M \cong S_3$. Thus $(D_6/M)/(N/M) \cong S_3/\mathbb{Z}_3 \cong \mathbb{Z}_2 \cong (D_6/N)$. ◇

The relation among the three quotient groups exhibited in this last example is exactly what the third isomorphism theorem states generally for any group G.

5.1.8 THEOREM (Third isomorphism theorem) Let G be a group, and N and M normal subgroups of G with $M \leq N$. Then

$$G/N \cong (G/M)/(N/M)$$

Proof (1) $N \lhd G$, so let ϕ: $G \to G/N$ be the canonical homomorphism, $\phi(g) = gN$ for all $g \in G$, and $M \lhd G$, so let ψ: $G \to G/M$ be the canonical homomorphism, $\psi(g) = gM$ for all $g \in G$. Let ρ: $G/M \to G/N$ be the map defined by $\rho(gM) = \phi(g)$ for all $gM \in G/M$. Note that ρ is well defined since if $g_1M = g_2M$, then $g_1g_2^{-1} \in M \leq N$, so $\phi(g_1) = g_1N = g_2N = \phi(g_2)$.

(2) Now ρ is a homomorphism, since $\rho((g_1M)(g_2M)) = \rho((g_1g_2M)) = \phi(g_1g_2) = g_1g_2N = (g_1N)(g_2N) = \phi(g_1)\phi(g_2) = \rho(g_1M)\rho(g_2M)$.

(3) ρ is onto, since for any $gN \in G/N$, $gN = \phi(g) = \rho(gM)$.

(4) It remains to describe the kernel of ρ. Now gM is in the kernel of ρ if and only if $\rho(gM) = \phi(g) = gN$ is the identity element N of G/N. This is so if and only if $g \in N$. Hence Kern $\rho = \{gM \mid g \in N\} = N/M$.
The theorem now follows from the first isomorphism theorem.

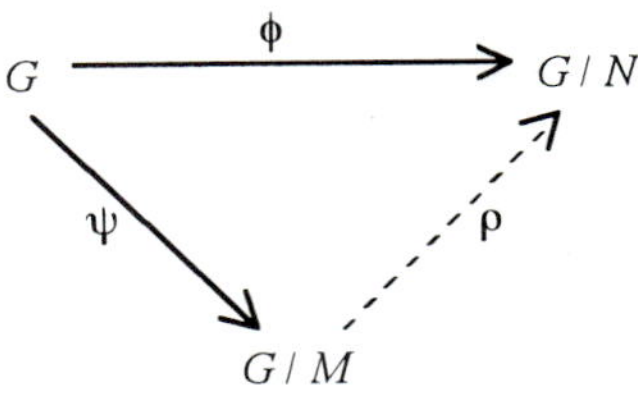

FIGURE 3

The relationships among ϕ, ψ, and ρ are represented in the commutative diagram of Figure 3. □

We this section with one more important illustration of the isomorphism theorems.

5.1.9 EXAMPLE Let us compare the subgroup lattices of D_4 and that of the quotient group D_4/K, where $K = \langle\rho^2\rangle$, so $D_4/K = \{K, \rho K, \tau K, \rho\tau K\}$.

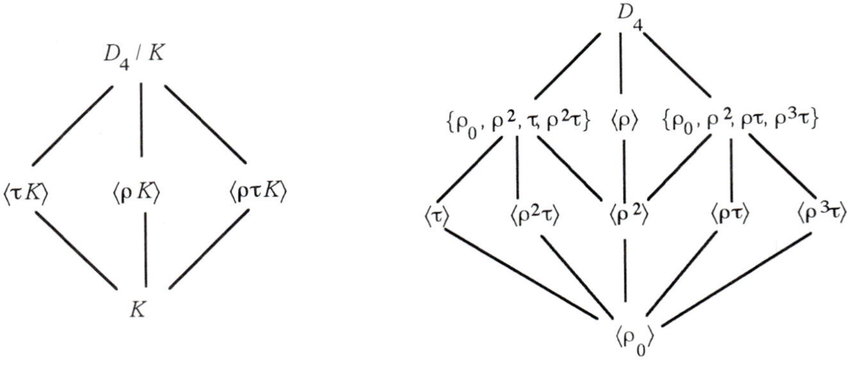

FIGURE 4 FIGURE 5

Where, in Figure 4

$\langle\tau K\rangle \cong \{\rho_0, \rho^2, \tau, \rho^2\tau\} / K$ $\langle\rho K\rangle \cong \langle\rho\rangle / K$ $\langle\rho\tau K\rangle \cong \{\rho_0, \rho^2, \rho\tau, \rho^3\tau\} / K$

Observe that we can obtain the subgroup lattice of the quotient group D_4/K by starting with the subgroup lattice of D_4, as in Figure 5, and removing from it the subgroups of D_4 that do not contain $K = \langle \rho^2 \rangle$. ◇

Exercises 5.1

1. Let ϕ: $\mathbb{Z}_{15} \to \mathbb{Z}_6$ be the homomorphism given by $\phi(1) = 4 \in \mathbb{Z}_6$.
(a) Find $K = \text{Kern } \phi$.
(b) List all the elements of the quotient group $\mathbb{Z}_{15} / \text{Kern } \phi$.
(c) List all the elements of the group $\phi(\mathbb{Z}_{15})$.
(d) Construct the isomorphism χ: $\mathbb{Z}_{15} / K \to \phi(\mathbb{Z}_{15})$.

2. Construct two different homomorphisms ϕ and ψ from $\mathbb{Z}_{10} \times \mathbb{Z}_{12}$ onto $\mathbb{Z}_6$.
(a) Find the kernel of each of the two homomorphisms.
(b) List the elements of each of the two quotient groups.
(c) Construct the isomorphism between each of the two quotient groups and $\mathbb{Z}_6$.

3. In the dihedral group D_6 consider the subgroup $M = \langle \rho^3 \rangle$.
(a) Show that $M \lhd D_6$.
(b) Construct the subgroup lattice of D_6.
(c) Construct the subgroup lattice of D_6/M and compare with (b).
(d) From (c) determine all the normal subgroups of D_6 that contain M.

4. In $\mathbb{Z}_{24}$ let $M = \langle 8 \rangle$ and $N = \langle 4 \rangle$.
(a) Construct the quotient groups $\mathbb{Z}_{24} / M$ and $\mathbb{Z}_{24} / N$.
(b) Construct the quotient group N / M.
(c) Construct the quotient group $(\mathbb{Z}_{24} / M) / (N / M)$
(d) Show that the quotient group in (c) is isomorphic to $\mathbb{Z}_{24} / N$ by constructing an isomorphism.

5. Let $G = G_1 \times G_2 \times G_3$ for some groups G_1, G_2, G_3. Show that there exist subgroups $H_i \leq G$, for $i = 1, 2, 3$, such that
(a) $G = H_1 \oplus H_2 \oplus H_3$
(b) $H_i \lhd G$ for all $i = 1, 2, 3$
(c) $H_i \oplus H_j \lhd G$ for i and j distinct
(d) $H_i \lhd H_i \oplus H_j$ for i and j distinct
(e) $G/H_i \cong H_j \oplus H_k$ for i, j, and k distinct
(f) $(H_i \oplus H_j)/H_i \cong H_j$ for i, j distinct
(g) $(G/H_i) / ((H_i \oplus H_j)/H_i) \cong H_k$ for i, j, and k distinct

6. Let H and K be subgroups of a group G such that $[G:H] = [G:K] = 2$ and $H \cap K \neq H$. Show that $G / (H \cap K) \cong \mathbb{Z}_2 \times \mathbb{Z}_2$.

7. Let M and N be normal subgroups of a group G such that $MN = G$. Show that $G / (M \cap N) \cong (G / M) \times (G / N)$.

8. Let $M \lhd G$ be such that $[G{:}M] = 5$, and let N be any subgroup of G. Show that either $N \leq M$ or $G = MN$ and $[N{:}M \cap N] = 5$.

9. Let $H_1 \lhd G_1$ and $H_2 \lhd G_2$. Show that

(a) $H_1 \times H_2 \lhd G_1 \times G_2$

(b) $(G_1 \times G_2) / (H_1 \times H_2) \cong (G_1/H_1) \times (G_2/H_2)$

10. Consider the subgroups $n\mathbb{Z} = \langle n \rangle$ and $kn\mathbb{Z} = \langle kn \rangle$ in $\mathbb{Z}$, where $n, k > 0$. Show that $\langle n \rangle / \langle kn \rangle \cong \mathbb{Z}_k$.

11. Let $K \lhd G$ with $[G{:}K]$ finite. Let H be a subgroup of G such that $|H|$ is finite and $[G{:}K]$ and $|H|$ are relatively prime. Show that $H \leq K$.

12. Let $K \lhd G$ with $|K|$ finite. Let H be a subgroup of G such that $[G{:}H]$ is finite and $[G{:}H]$ and $|K|$ are relatively prime. Show that $K \leq H$.

5.2 The Jordan-Hölder Theorem

The fundamental theorem of finite Abelian groups gave us a complete characterization of all finite Abelian groups in terms of cyclic groups. For non-Abelian groups we do not have a comparable result. Characterizing non-Abelian groups is a very complicated problem. We do, however, have some techniques available that help us understand the structure of some non-Abelian groups. One of these methods is the construction of a *composition series* for a given group, which we study in this section.

Suppose G is a group with proper normal subgroups H and K such that $G = H \oplus K$. If H and K are groups whose structure we know, then we can find all the basic characteristics of G. Being able to write a group as the direct sum of two normal subgroups is, however, a very strong condition, which many groups fail to meet. Let us look instead at a weaker condition that is a consequence of being expressible as a direct sum. If $G = H \oplus K$, then (1) $K \lhd G$ and (2) $G/K \cong H$. A group may satisfy (1) and (2) even though it is not a direct sum of H and K. Conditions (1) and (2) do, however, give some insight into the structure of G.

The next example illustrates the importance of a concept we define shortly.

5.2.1 EXAMPLE Consider the dihedral group D_6 and the following three normal subgroups:

(1) $N = \langle \rho \rangle \lhd D_6$	$D_6 / N \cong \mathbb{Z}_2$
(2) $K = \{\rho_0, \rho^2, \rho^4, \tau, \rho^2\tau, \rho^4\tau\}$	$D_6 / K \cong \mathbb{Z}_2$
(3) $L = \{\rho_0, \rho^3, \tau, \rho^3\tau\}$	$D_6 / L \cong \mathbb{Z}_3$

First note that since $\mathbb{Z}_2$ and $\mathbb{Z}_3$ are cyclic groups of prime order they are simple and have no nontrivial proper subgroups. Therefore, there are no proper subgroups of D_6 that contain N or K or L. Now, starting with each of these three normal subgroups of D_6, we construct a series of subgroups beginning with it.

(1) $D_6 \geq N \geq M = \langle \rho^3 \rangle \geq \langle \rho_0 \rangle$

Note that $D_6/N \cong \mathbb{Z}_2$, $N/M \cong \mathbb{Z}_3$, $M/\langle \rho_0 \rangle \cong \mathbb{Z}_2$.

(2) $D_6 \geq K \geq \langle \rho^2 \rangle \geq \langle \rho_0 \rangle$

Note that $D_6/K \cong \mathbb{Z}_2$, $K/\langle \rho^2 \rangle \cong \mathbb{Z}_2$, $\langle \rho^2 \rangle/\langle \rho_0 \rangle \cong \mathbb{Z}_3$.

(3) $D_6 \geq L \geq \langle \rho^3 \rangle \geq \langle \rho_0 \rangle$

Note that $D_6/L \cong \mathbb{Z}_3$, $L/\langle \rho^3 \rangle \cong \mathbb{Z}_2$, $\langle \rho^3 \rangle/\langle \rho_0 \rangle \cong \mathbb{Z}_2$.

In this example we observe that each subgroup in any one of these three series is normal in the preceding subgroup in that series, that each quotient group is simple, and that in each of the three series the same groups appeared as quotient groups, though in different orders. ◇

Let us now give some definitions of features we observed in this example.

5.2.2 DEFINITION A normal subgroup K of a group G is called a **maximal normal subgroup** of G if there is no proper normal subgroup H of G that contains but is not equal to K. In other words, if $K \leq H \lhd G$, then either $K = H$ or $H = G$. ○

The relation of this notion to the notion of a simple group is made clear by the next proposition.

5.2.3 PROPOSITION Let K be a normal subgroup of a group G. Then K is a maximal normal subgroup of G if and only if the quotient group G/K is simple.

Proof This is immediate from the correspondence between normal subgroups of G containing K and normal subgroups of G/K in Proposition 5.1.3. (See Exercise 11.) □

5.2.4 DEFINITION Let G be a group.

(1) A **subnormal series** for G is a chain of subgroups

$G = G_0 \geq G_1 \geq G_2 \geq \ldots \geq G_n = \{e\}$

such that $G_{i+1} \lhd G_i$ for all i, $0 \leq i \leq n - 1$.

(2) A **composition series** for G is a subnormal series in which the quotient groups G_i/G_{i+1}, called the **composition factors**, are all simple. The **length** of the composition series is the number n of composition factors. ○

Thus Example 5.2.1 was an example of three composition series for the same group D_6. Some further examples follow.

5.2.5 EXAMPLE $S_3 \geq A_3 \geq \langle \rho_0 \rangle$ is a composition series for S_3, with composition factors $S_3/A_3 \cong \mathbb{Z}_2$ and $A_3/\langle \rho_0 \rangle \cong \mathbb{Z}_3$. ◇

5.2.6 EXAMPLE Every simple group has a composition series of length 1. In particular, every cyclic group of prime order has such a composition series, and the alternating group A_5 has such a series. ◇

5.2.7 EXAMPLE Every finite Abelian group has a composition series in which all the composition factors are cyclic. This can be proved by induction on $|G|$. If $|G| = 1$ or 2, G is simple and has a composition series by the preceding example. So assume the statement is true for Abelian groups of order less than $|G|$. Let $|G| = p_1^{a_1} \ldots p_k^{a_k}$. If $k = 1$ and $a_1 = 1$, then $|G| = p_1$ is a prime, and hence G is simple. Otherwise, by Theorem 3.4.17, G has a subgroup H such that $|H| = p_1^{a_1-1} \ldots p_k^{a_k}$. Since G is Abelian, H is a normal subgroup of G, and $G/H \cong \mathbb{Z}_{p_1}$, which is simple. Since $|H| < |G|$, by our induction hypothesis there is a composition series $\{H_i\}$ for H, from which it follows there is a composition series $\{G_i\}$ for G with $G_0 = G$, $G_1 = H = H_0$, and $G_{i+1} = H_i$. ◇

5.2.8 DEFINITION Given two composition series $\{G_i\}$ and $\{H_i\}$ for the same group G

$$G = G_0 \geq G_1 \geq G_2 \geq \ldots \geq G_n = \{e\}$$
$$G = H_0 \geq H_1 \geq H_2 \geq \ldots \geq H_m = \{e\}$$

the two series are said to be **isomorphic series** if (1) $n = m$ and (2) there is a permutation τ of $\{0, 1, 2, \ldots, n-1\}$ such that for all i we have $G_i/G_{i+1} \cong H_{\tau(i)}/H_{\tau(i+1)}$. In other words, up to isomorphism, the same simple groups appear as composition factors, each the same number of times, in the two composition series, the only difference being the order in which they appear. ○

Thus the three composition series in Example 5.2.1 are all isomorphic, with $\mathbb{Z}_2$ appearing twice and $\mathbb{Z}_3$ once as a composition factor in each series.

5.2.9 EXAMPLE Here are three isomorphic composition series for $\mathbb{Z}_{30}$.

(1) $\mathbb{Z}_{30} \geq \langle 2 \rangle \geq \langle 6 \rangle \geq \langle 0 \rangle$

where $\mathbb{Z}_{30}/\langle 2 \rangle \cong \mathbb{Z}_2$, $\langle 2 \rangle/\langle 6 \rangle \cong \mathbb{Z}_3$, $\langle 6 \rangle/\langle 0 \rangle \cong \mathbb{Z}_5$

(2) $\mathbb{Z}_{30} \geq \langle 3 \rangle \geq \langle 15 \rangle \geq \langle 0 \rangle$

where $\mathbb{Z}_{30}/\langle 3 \rangle \cong \mathbb{Z}_3$, $\langle 3 \rangle/\langle 15 \rangle \cong \mathbb{Z}_5$, $\langle 15 \rangle/\langle 0 \rangle \cong \mathbb{Z}_2$

(3) $\mathbb{Z}_{30} \geq \langle 5 \rangle \geq \langle 10 \rangle \geq \langle 0 \rangle$

where $\mathbb{Z}_{30}/\langle 5 \rangle \cong \mathbb{Z}_5$, $\langle 5 \rangle/\langle 10 \rangle \cong \mathbb{Z}_2$, $\langle 10 \rangle/\langle 0 \rangle \cong \mathbb{Z}_3$ ◇

In the particular examples we have considered, all the composition series for the same group have been isomorphic. This is always the case according to the next theorem.

5.2.10 THEOREM (The Jordan-Hölder theorem) Any two composition series for the same group G are isomorphic. □

The proof follows easily once we have established the following two propositions, which are themselves of interest.

5.2.11 PROPOSITION Let G_1 and H_1 be two distinct maximal normal subgroups of the same group G, and let $K = G_1 \cap H_1$. Then

(1) $G = G_1H_1$

(2) $G/G_1 \cong H_1/K$ and $G/H_1 \cong G_1/K$

(3) K is a maximal normal subgroup of both G_1 and H_1.

Proof (1) For any $g \in G$ and any $x \in G_1H_1$, let $x = g_1h_1$ where $g_1 \in G_1$ and $h_1 \in H_1$. Then $gxg^{-1} = gg_1h_1g^{-1} = (gg_1g^{-1})(gh_1g^{-1})$. Since $G_1 \lhd G$ we have $gg_1g^{-1} \in G_1$ and since $H_1 \lhd G$ we have $gh_1g^{-1} \in H_1$. So $gxg^{-1} \in G_1H_1$, and G_1H_1 is normal. Since $G_1 \leq G_1H_1$, and G_1 is a maximal normal subgroup, we must have either $G_1H_1 = G_1$ or $G_1H_1 = G$, but since $H_1 \leq G_1H_1$ and G_1 and H_1 are distinct, we cannot have $G_1H_1 = G_1$, so we have $G_1H_1 = G$.

(2) is immediate from (1) and the second isomorphism theorem (Theorem 5.1.6).

(3) Since G_1 is a maximal normal subgroup, G/G_1 is simple by Proposition 5.2.3. Hence by (2), $H_1/K \cong G/G_1$ is also simple, and K is a maximal normal subgroup in H_1, again using Proposition 5.2.3. Similarly, using $G/H_1 \cong G_1/K$ we prove that K is a maximal normal subgroup in G_1. □

5.2.12 PROPOSITION Let G be a group with a composition series of length n, and let $K \lhd G$. Then K has a composition series of length $\leq n$.

Proof Given the composition series

$$G = G_0 \geq G_1 \geq G_2 \geq \ldots \geq G_n = \{e\}$$

for G, consider the series

$$K = G_0 \cap K \geq G_1 \cap K \geq G_2 \cap K \geq \ldots \geq G_n \cap K = \{e\} \cap K = \{e\}.$$

Claim 1 $G_{i+1} \cap K \lhd G_i \cap K$ for all i, $0 \leq i \leq n - 1$.

For since $K \lhd G$, it follows that $G_i \cap K \lhd G_i$. Since also $G_{i+1} \lhd G_i$, we obtain $G_{i+1} \cap K = G_{i+1} \cap (G_i \cap K) \lhd G_i$ and hence $G_{i+1} \cap K \lhd G_i \cap K$, proving the claim.

Claim 2 For each i, either $G_{i+1} \cap K = G_i \cap K$ or $(G_i \cap K)/(G_{i+1} \cap K) \cong G_i/G_{i+1}$.

By the second isomorphism theorem we have

$$G_i \cap K / G_{i+1} \cap K \cong G_{i+1}(G_i \cap K) / G_{i+1} \leq G_i/G_{i+1}$$

Since $G_{i+1} \lhd G_i$ and $(G_i \cap K) \lhd G_i$, we have $G_{i+1}(G_i \cap K) \lhd G_i$ and $G_{i+1} \leq G_{i+1}(G_i \cap K)$. Since G_{i+1} is a maximal normal subgroup of G_i, either $G_{i+1}(G_i \cap K) = G_{i+1}$, in which case $G_{i+1}(G_i \cap K) / G_{i+1}$ and therefore $(G_i \cap K)/(G_{i+1} \cap K)$ is trivial, and $G_i \cap K = G_{i+1} \cap K$; or else $G_{i+1}(G_i \cap K) = G_i$, in which case we have $G_i \cap K / G_{i+1} \cap K \cong G_{i+1}(G_i \cap K) / G_{i+1} = G_i/G_{i+1}$, and the claim is proved.

The proposition follows on dropping from the series for K any $G_{i+1} \cap K$ that is equal to the preceding $G_i \cap K$. □

Proof (of the Jordan-Hölder theorem) Let G be a group having a composition series and let

(1) $G \geq G_1 \geq G_2 \geq \ldots \geq G_n = \{e\}$

be a composition series of minimal length n for G, and let

(2) $G \geq H_1 \geq H_2 \geq \ldots \geq H_m = \{e\}$

be any other composition series for G. We want to prove that $m = n$ and that (1) and (2) are isomorphic composition series. We proceed by induction on n. If $n = 1$, G is simple and the one and only composition series is $G = G_0 \geq G_1 = \{e\}$. So assume the theorem holds for groups with composition series of length less than n. If $G_1 = H_1$, then since (1) gives a composition series for G_1 of length $n - 1$, by our induction hypothesis any two composition series for G_1 must be isomorphic, so the composition series for $G_1 = H_1$ given by (2) is isomorphic to that given by (1), from which it follows that (1) and (2) are isomorphic composition series for G. So suppose G_1 and H_1 are distinct maximal normal subgroups of G, and let $K = G_1 \cap H_1$. By Proposition 5.2.12, K has a composition series $\{K_i\}$ of length $r \leq n - 1$, since G_1 does. By part (3) of Proposition 5.2.11, K is a maximal normal subgroup of both G_1 and H_1. This gives us two more composition series for G:

(3) $G \geq G_1 \geq K \geq K_1 \geq \ldots \geq K_r = \{e\}$

(4) $G \geq H_1 \geq K \geq K_1 \geq \ldots \geq K_r = \{e\}$

By part (2) of Proposition 5.2.11, $G/G_1 \cong H_1/K$ and $G/H_1 \cong G_1/K$, so (3) and (4) are isomorphic composition series for G. By our induction hypothesis, since G_1 has a composition series of length $n - 1$, any two composition series for G_1 are isomorphic, from which it follows that (1) and (3) are isomorphic composition series for G. So $r = n - 2$, and (4) shows that H_1 has a composition series of length $n - 1$, so our induction hypothesis implies that any two composition series for H_1 are isomorphic, and (2) and (4) are isomorphic composition series for G. It follows that (1) is isomorphic to (2), to complete the proof. □

The Jordan-Hölder theorem that we have just proved states that a given group G may have more than one composition series, but the lengths of all composition series and the factors are completely determined by G. The following example illustrates the theorem in a familiar case.

5.2.14 EXAMPLE Let $G = \langle a \rangle$ be a cyclic group of order n, and let $n = p_1 p_2 \cdots p_r$ where the p_i are primes not necessarily distinct. If we let $G_i = \langle a^{p_1 p_2 \cdots p_i} \rangle$, then $|G_i| = p_{i+1} \cdots p_r$ and $G_i / G_{i+1} \cong \mathbb{Z}_{p_{i+1}}$, a simple group. Hence $\{G_i\}$ is a composition series for G. The Jordan-Hölder theorem implies that the integer r and the primes p_i are uniquely determined by n. In other words, the factorization of n into primes is unique except for the order of the prime factors. ◇

The Jordan-Hölder theorem tells us that the composition factors and the length of the composition series are unique for a given group. What it does not tell us is whether from the given simple groups $C_1, C_2, \ldots, C_r$ that appear as composition factors we can recover the group. We can see right away that in fact we cannot, since, for instance, the non-Abelian group D_6 has the same composition factors $\mathbb{Z}_2, \mathbb{Z}_2, \mathbb{Z}_3$ as the noncyclic Abelian group $\mathbb{Z}_2 \times \mathbb{Z}_2 \times \mathbb{Z}_3$ and the cyclic group $\mathbb{Z}_{12}$. Thus non-isomorphic groups can have composition series with the same length and the same composition factors. ◇

Exercises 5.2

In Exercises 1-10 find a composition series for the indicated group. In each case find the composition factors.

1. D_4 **2.** D_5 **3.** S_3 **4.** S_4 **5.** A_4

6. D_n for any $n \geq 3$ **7.** G an Abelian group of order 42

8. G a noncyclic Abelian group of order 60 **9.** $S_3 \times D_4$

10. $G = \mathbb{Z}_{p_1} \times \ldots \times \mathbb{Z}_{p_s}$, where the p_i are primes not necessarily distinct

11. Prove Proposition 5.2.3.

12. Show that every finite group has a composition series.

13. Let G_1 and G_2 be groups having composition series of lengths n_1 and n_2, respectively. Show that $G_1 \times G_2$ has a composition series of length $n_1 + n_2$.

14. Show that $\mathbb{Z}$ does not have a composition series.

15. Show that a cyclic group G has a composition series if and only if G is finite.

16. Show that an infinite Abelian group cannot be a simple group.

17. Let $G = G_0 \geq G_1 \geq G_2 \geq \ldots \geq G_n = \{e\}$ be a subnormal series for a group G such that $|G_i/G_{i+1}| = s_i < \infty$ for $0 \leq i \leq n - 1$. Show that $|G| = s_0 s_1 \ldots s_{n-1}$.

18. Show that an infinite Abelian group can have no composition series. (*Hint:* use the previous two exercises.)

19. Show that if G is a finite Abelian group of order $p_1^{a_1} p_2^{a_2} \ldots p_k^{a_k}$, where the p_i are distinct primes, then any composition series for G has length $a_1 + a_2 + \ldots + a_k$.

20. Let G be a group and $K \lhd G$. Show that G has a composition series if and only if K and G/K have composition series.

5.3 Solvable Groups

In this section we continue our study of subnormal series of a group G, adding new conditions. We look at groups that have a subnormal series with Abelian groups as the factor groups. The groups with such a subnormal series turn out, as we will see in Chapter 12, to play an important role in determining whether a polynomial equation is solvable by radicals.

5.3.1 DEFINITION A group G is said to be **solvable** if it has a subnormal series

$$G = G_0 \geq G_1 \geq G_2 \geq \ldots \geq G_n = \{e\}$$

in which the factors G_i/G_{i+1} are Abelian for all i, $0 \leq i \leq n - 1$. ○

We saw in the preceding chapter many examples of groups with composition series whose composition factors were simple Abelian groups. For instance, D_6 has composition factors $\mathbb{Z}_2$, $\mathbb{Z}_2$, $\mathbb{Z}_3$. The Jordan-Hölder theorem implies that if G is a finite solvable group, then every composition series for G must have Abelian composition factors. (See Exercise 9.) Hence if we find a composition series with simple but non-Abelian factors, then we can immediately conclude the group is not solvable.

5.3.2 EXAMPLE S_5 is not solvable. For consider the series $S_5 \geq A_5 \geq \{e\}$. $S_5/A_5 \cong \mathbb{Z}_2$, and A_5 is simple, as shown in Theorem 4.5.17. Hence this is a composition series for S_5, and it contains a non-Abelian factor, A_5. Therefore, S_5 is an example of a group that is not solvable. ◇

From this example we observe that in order to construct examples of groups that are not solvable, we need examples of simple non-Abelian groups, such as A_5.

5.3.3 EXAMPLE Let G be a p-group; in other words, let $|G| = p^n$ for some prime p and positive integer n. Then it follows from Theorem 4.6.13 that G is solvable. (See Exercise 3.) ◇

5.3.4 EXAMPLE Let G be a group of order pq, where p and q are distinct primes. Then it follows from Proposition 4.7.3 that G is solvable. (See Exercise 5.) ◇

5.3.5 EXAMPLE Let G be a group of order $30 = 2 \cdot 3 \cdot 5$. Then, as we saw in Example 4.7.12, G is not simple, since either the 3-Sylow subgroup or the 5-Sylow subgroup is normal in G. We consider the two cases separately.

(1) If P_5 is a normal 5-Sylow subgroup, then $P_5 \cong \mathbb{Z}_5$ and G/P_5 is a group of order 6, hence $G/P_5 \cong \mathbb{Z}_6$ and we are done, or $G/P_5 \cong S_3$. In the latter case, then using Proposition 5.1.3, there is a normal subgroup $H \lhd G$ such that $P_5 \leq H$ and $H/P_5 \cong A_3 \lhd S_3$. By the third isomorphism theorem, $G/H \cong (G/P_5)/(H/P_5) \cong S_3/A_3 \cong \mathbb{Z}_2$. Thus we obtain $G \geq H \geq P_5 \cong \mathbb{Z}_5 \geq \{e\}$, and G is solvable.

(2) If P_3 is a normal 3-Sylow subgroup, then $P_3 \cong \mathbb{Z}_3$ and G/P_3 is a group of order 10. By Cauchy's theorem (Corollary 4.6.10) any group of order 10 has a cyclic

subgroup of order 5. Hence using Proposition 5.1.3, there is a normal subgroup $H \lhd G$ such that $P_3 \leq H$ and $H/P_3 \cong \mathbb{Z}_5$, and $[G:H] = 2$, so H is normal. Thus we obtain $G \geq H \geq P_3$ with $G/H \cong \mathbb{Z}_2$, $H/P_3 \cong \mathbb{Z}_5$, and $P_3 \cong \mathbb{Z}_3$, and G is solvable. ◇

As we saw in these last examples, the Sylow theorems give us powerful tools to determine whether a group is solvable or not. There is also another useful result that completely characterizes solvable groups, but first we need to construct a specific subnormal series, which we describe next.

5.3.6 DEFINITION For any group G and $a, b \in G$, the **commutator** of a and b is the element $aba^{-1}b^{-1} \in G$. The **commutator subgroup** of G is the subgroup G' generated by the set of all commutators in G, that is, the set of all $z \in G$ that can be written as a product

$$z = (a_1b_1a_1^{-1}b_1^{-1})(a_2b_2a_2^{-1}b_2^{-1})\ldots(a_nb_na_n^{-1}b_n^{-1})$$

of finitely many commutators. ○

5.3.7 THEOREM Let G be a group and G' its commutator subgroup. Then

(1) $G' \lhd G$

(2) G/G' is Abelian.

(3) If $N \lhd G$ and G/N is Abelian, then $G' \leq N$.

Proof (1) G' is a subgroup of G by the subgroup test, since if

$$z = a_1b_1a_1^{-1}b_1^{-1}\ldots a_nb_na_n^{-1}b_n^{-1}$$
$$w = c_1d_1c_1^{-1}d_1^{-1}\ldots c_md_mc_m^{-1}d_m^{-1}$$

then

$$zw^{-1} = a_1b_1a_1^{-1}b_1^{-1}\ldots a_nb_na_n^{-1}b_n^{-1}d_mc_md_m^{-1}c_m^{-1}\ldots d_1c_1d_1^{-1}c_1^{-1}$$

which is again a finite product of commutators. It is actually a normal subgroup of G since for any $g \in G$,

$$gzg^{-1} = ga_1b_1gg^{-1}a_1^{-1}b_1^{-1}g^{-1}\ldots ga_nb_ngg^{-1}a_n^{-1}b_n^{-1}g^{-1}$$

which is again a finite product of commutators.

(2) For any elements xG' and yG' in G/G', we have $xG'yG' = xyG'$ and $yG'xG' = yxG'$, and since $xy(yx)^{-1} = xyx^{-1}y^{-1} \in G'$, we have $xyG' = yxG'$, and G/G' is Abelian.

(3) Consider now any commutator $xyx^{-1}y^{-1} \in G$. If G/N is Abelian, then $xNyN = yNxN$, hence $xyx^{-1}y^{-1} = (xy)(yx)^{-1} \in N$, so N contains all commutators, and therefore all products of commutators, and $G' \leq N$. □

Now we construct the series we need for our main theorem. Given a group G, we let $G = G^{(0)}$, let $G^{(1)}$ be its commutator subgroup, and let $G^{(i+1)}$ be the commutator subgroup of $G^{(i)}$. Hence we obtain

(*) $\quad G = G^{(0)} \geq G^{(1)} \geq G^{(2)} \geq \ldots$

Note that by Theorem 5.3.7, $G^{(i+1)} \lhd G^{(i)}$ and $G^{(i)}/G^{(i+1)}$ is Abelian for all $i \geq 1$.

What is missing from (*) to make G a solvable group is the information that the series (*) ends. It turns out, as we prove in the next theorem, that this series ends exactly when G is solvable.

5.3.8 THEOREM Let G be a group and $G^{(i)}$ the subgroups of G defined previously. Then G is solvable if and only if $G^{(n)} = \{e\}$ for some positive integer n.

Proof ($\Rightarrow$) Suppose G is solvable and let

$$G = G_0 \geq H_1 \geq H_2 \geq \ldots \geq H_n = \{e\}$$

be a subnormal series with Abelian factors H_i/H_{i+1}. By Theorem 5.3.7, $G^{(1)} \leq H_1$. Now assume $G^{(k)} \leq H_k$. Since H_k/H_{k+1} is Abelian, again by Theorem 5.3.7 we obtain $H_k' \leq H_{k+1}$. Hence $G^{(k+1)} = (G^{(k)})' \leq H_k' \leq H_{k+1}$. Thus using induction we have shown that $G^{(i)} \leq H_i$ for all $i \geq 1$ and hence $G^{(n)} = \{e\}$ since $H_n = \{e\}$.

($\Leftarrow$) Suppose $G^{(n)} = \{e\}$ for some positive integer n. Then

$$G \geq G^{(1)} \geq G^{(2)} \geq \ldots \geq G^{(n)} = \{e\}$$

is a subnormal series since by Theorem 5.3.7, $G^{(i+1)} \lhd G^{(i)}$. Also $G^{(i)}/G^{(i+1)}$ is Abelian, so G is solvable. □

We give a consequence of the preceding theorem and because of its importance we call it a theorem instead of a corollary.

5.3.9 THEOREM S_n is not solvable for any $n \geq 5$.

Proof Let H be any subgroup of S_n such that H contains all the 3-cycles in S_n.
Claim The commutator subgroup H' of H also contains all 3-cycles in S_n.
Proof of Claim Let $\sigma = (i\ j\ k)$ be any 3-cycle in H. Now consider the following 3-cycles in H $\tau = (i\ l\ k)$ and $\rho = (k\ j\ m)$, where i, j, k, l, m are all distinct. Then $\tau\rho\tau^{-1}\rho^{-1} \in H'$. But we have

$$\tau\rho\tau^{-1}\rho^{-1} = (i\ l\ k)(k\ j\ m)(i\ k\ l)(k\ m\ j) = (i\ j\ k) = \sigma$$

So $\sigma \in H'$ and the claim is proved.
Thus S_n' must contain all 3-cycles, and $S_n^{(i)}$ must contain all 3-cycles for all i, and therefore *no* positive integer r exists such that $S_n^{(r)} = \{e\}$, and Theorem 5.3.8 implies that S_n is not solvable. □

Notice in the proof that we needed $n \geq 5$ to give us the five distinct integers i, j, k, l, m that we used.
We already knew that S_5 is not solvable (Example 5.3.2) in consequence of A_5 being simple (Theorem 4.5.17). Actually, A_n is simple for all $n \geq 5$ (see Exercises 15 through 18), a fact that can be used to give another proof of Theorem 5.3.9. By this fact, $S_n \geq A_n \geq \{e\}$ is a composition series, whereas if S_n were solvable, it would have a composition series with Abelian factors, which is impossible by the Jordan-Hölder theorem.

We end this section with two more consequences of Theorem 5.3.8. They turn out to be very useful in Chapter 12, where we characterize radical extensions in terms of solvable groups.

5.3.10 THEOREM Let G be a solvable group. Then
(1) Every subgroup of G is solvable.
(2) Every homomorphic image of G is solvable.

Proof (1) Let H be a subgroup of G and assume that $G^{(n)} = \{e\}$. We show by induction that $H^{(i)} \leq G^{(i)}$ for all $0 \leq i \leq n$. For $i = 0$ there is nothing to prove. So assume $H^{(i)} \leq G^{(i)}$ for some $0 \leq i \leq n - 1$. Then $H^{(i+1)} = (H^{(i)})' \leq (G^{(i)})' = G^{(i+1)}$.

(2) Let ϕ: $G \to M$ be an onto homomorphism. We show by induction that $M^{(i)} \leq \phi(G^{(i)})$ for all all $0 \leq i \leq n$. For $i = 0$ we have $M^{(0)} = M = \phi(G) = \phi(G^{(0)})$. So assume $M^{(i)} \leq \phi(G^{(i)})$ for some $0 \leq i \leq n - 1$. Then given any u and v in $M^{(i)}$, there exist x and y in $G^{(i)}$ such that $u = \phi(x)$ and $v = \phi(y)$. Therefore,

$$uvu^{-1}v^{-1} = \phi(x)\phi(y)\phi(x)^{-1}\phi(y)^{-1} = \phi(xyx^{-1}y^{-1}) \in \phi(G^{(i+1)})$$

Hence $M^{(i+1)} \leq \phi(G^{(i+1)})$. □

5.3.11 THEOREM Let $H \lhd G$. Then G is a solvable group if and only if H and G/H are both solvable groups.

Proof ($\Rightarrow$) Assume G is solvable. Then part (1) of the preceding theorem implies that H is solvable, and part (2), using the canonical homomorphism ϕ: $G \to G/H$, implies that G/H is solvable.

($\Leftarrow$) Assume that H and G/H are solvable and let

$$H = H_0 \geq H_1 \geq H_2 \geq \ldots \geq H_m = \{e\}$$
$$G/H = G_0/H \geq G_1/H \geq G_2/H \geq \ldots \geq G_n/H = \{e\}$$

be subnormal series with H_i/H_{i+1} and $(G_i/H)/(G_{i+1}/H)$ Abelian factors. By the third isomorphism theorem (Theorem 5.1.8), G_i/G_{i+1} are Abelian groups, hence

$$G = G_0 \geq G_1 \geq G_2 \geq \ldots \geq G_n = H = H_0 \geq H_1 \geq H_2 \geq \ldots \geq H_m = \{e\}$$

is a subnormal series for G with Abelian factors. Hence G is solvable.□

Exercises 5.3

1. Show that the dihedral groups D_n, $n \geq 3$, are solvable.

2. Show that every Abelian group is solvable.

3. Show that every p-group is solvable.

4. Show that every group of order $2p$, where p is an odd prime, is solvable.

5. Show that every group of order pq, where p and q are distinct primes, is solvable.

6. Construct an example of a group that is solvable but has no composition series.

7. Show that every group of order 42 is solvable. (Use Exercise 6 in Section 4.7.)

8. Show that every group of order 20 is solvable. (Use Exercise 7 in Section 4.7.)

9. Let G be a finite solvable group. Show that G has a composition series in which every composition factor is Abelian.

10. Let G_1 and G_2 be solvable groups. Show that $G_1 \times G_2$ is also solvable.

11. Let G be a solvable group. Show that for every nontrivial subgroup $H \leq G$, $H' \neq H$, where H' is the commutator subgroup of H.

12. For any group G, show that G is solvable if and only if $G/Z(G)$ is solvable.

13. For any group G, show that $G^{(i)} \lhd G$ for all $i \geq 1$.

14. Show that if $N \lhd G$, then $N' \lhd G$.

15. Let H be a nontrivial normal subgroup of A_n, $n \geq 6$. Show that there exists $\tau \in H$ and $1 \leq i \leq n$ such that $\tau(i) = i$ and τ is not the identity permutation.

16. Let H be a nontrivial normal subgroup of A_n, $n \geq 6$. Assume that there exists a nontrivial element $\tau \in H$ such that $\tau(i) = i$. Show that $G_i \leq H$, where $G_i = \{\rho \in A_n \mid \rho(i) = i\} \cong A_{n-1}$.

17. Let H be a nontrivial normal subgroup of A_n, $n \geq 6$. Assume that for some i, $1 \leq i \leq n$, we have $G_i = \{\rho \in A_n \mid \rho(i) = i\} \leq H$. Show that $G_i \leq H$ for *all* i, $1 \leq i \leq n$.

18. Use induction on n and Exercises 15 through 17 to show that A_n is simple for all $n \geq 6$.

19. Show that if G is a solvable group, then it has a subnormal series

$$G = G_0 \geq G_1 \geq G_2 \geq \ldots \geq G_n = \{e\}$$

in which the factors G_i / G_{i+1} are cyclic groups of prime order for all $0 \leq i \leq n - 1$.

20. Show that a solvable group having a composition series must be a finite group.

21. Let H and K be solvable subgroups of a group G with $H \lhd G$. Show that HK is solvable.

22. For any group G we define the following subgroups: $Z_0(G) = \{e\}$, $Z_1(G) = Z(G)$ the center of G, and $Z_{i+1}(G) \geq Z_i(G)$ the subgroup such that $Z_{i+1}(G)/Z_i(G) \cong Z(G/Z_i(G))$. Consider the series

(†) $\{e\} = Z_0(G) \leq Z_1(G) \leq Z_2(G) \leq \ldots$

G is called **nilpotent** if for some n, $Z_n(G) = G$. Show that if G is nilpotent, then (†) is a subnormal series for G.

23. Show that every nilpotent group is solvable.

24. Show that $Z_i(G)$ is a characteristic subgroup of G for all i. (See Exercises 15 through 19 in Section 2.5.)

25. Give an example of a solvable group that is not nilpotent.

26. Show that if G is nilpotent, then $G/Z(G)$ is also nilpotent.

27. Show that D_4 is nilpotent.

28. Show that if G is a non-Abelian group of order p^3, where p is prime, then $|Z(G)| = p$ and $G/Z(G) \cong \mathbb{Z}_p \times \mathbb{Z}_p$.

29. Show that for any prime p, every p-group is nilpotent. (*Hint:* Use Proposition 4.4.7.)

Part B

Rings and Fields

Chapter 6

Rings

Using our familiar number systems as our models, we now begin looking at algebraic structures with more than one operation. We identify important properties of these operations to give us a better understanding of the different algebraic structures they determine. The concepts of *rings*, *integral domains*, and *fields*, which are introduced step by step in this chapter, form the basis for the algebraic theories we study in later chapters.

6.1 Examples and Basic Concepts

In our study of groups we restricted ourselves to sets with one operation that satisfied four conditions: closure, associativity, existence of an identity, and existence of an inverse for each element. The number systems we studied, $\mathbb{Z}$, $\mathbb{Q}$, $\mathbb{R}$, $\mathbb{C}$, and $\mathbb{Z}_n$, are groups under addition, but they are also equipped with a second operation, multiplication. We identify some of the important properties of this second operation, as well as the relation between the two operations. The set $\mathbb{Z}$ of integers is our starting point.

6.1.1 EXAMPLE Let us single out some of the important properties that addition and multiplication in $\mathbb{Z}$ satisfy:

(1) $\mathbb{Z}$ is an Abelian group under addition.

(2) Given any two integers a and b, ab is also an integer.

(3) Given any three integers a, b, and c, we have $a(bc) = (ab)c$; in other words, when we multiply a row of integers, it does not matter where we put the parentheses.

(4) Given any three integers a, b, and c, we have $a(b + c) = ab + ac$ and $(b + c)a = ba + ca$.

Here the first property (1) refers only to addition, the properties (2) and (3) refer only to multiplication, while property (4) points out how the two operations "interact." As we will see, many of the basic facts about the familiar number systems follow from the properties just listed. ◇

We concentrate now on sets with two operations having these four properties.

6.1.2 DEFINITION A set R equipped with two operations, written as addition and multiplication, is said to be a **ring** if the following four **ring axioms** are satisfied, for any elements a, b, and c in R:

(1) R is an Abelian group under addition.
(2) **Closure** $ab \in R$
(3) **Associativity** $a(bc) = (ab)c$
(4) **Distributivity** $a(b + c) = ab + ac$ and $(b + c)a = ba + ca$ ○

It is important to note that (1) tells us that everything we have learned about groups in general and Abelian groups in particular will apply to addition in rings.

6.1.3 EXAMPLE $\mathbb{Z}$, $\mathbb{Q}$, $\mathbb{R}$, and $\mathbb{C}$ are rings under the usual operations. ◇

6.1.4 EXAMPLE By Propositions 0.3.31 and 0.3.34, for every positive integer n, $\mathbb{Z}_n$ is a ring under the operations of addition mod n and multiplication mod n. ◇

6.1.5 EXAMPLE Let R be any of the rings $\mathbb{Z}$, $\mathbb{Q}$, $\mathbb{R}$, $\mathbb{C}$, or $\mathbb{Z}_n$. Then the set $M(2, R)$ of 2×2 matrices with entries from R is a ring under the addition and multiplication operations defined in Definition 0.5.4. ◇

6.1.6 EXAMPLE Let $F(\mathbb{R})$ be the set of all functions $f: \mathbb{R} \to \mathbb{R}$. We define **pointwise** addition and multiplication as follows. For all f and g in $F(\mathbb{R})$ we let

(1) $(f + g)(x) = f(x) + g(x)$ for all $x \in \mathbb{R}$
(2) $(f \cdot g)(x) = f(x) \cdot g(x)$ for all $x \in \mathbb{R}$

We leave it to the reader to show that $F(\mathbb{R})$ forms a ring under these two operations. (See Exercise 8). ◇

6.1.7 EXAMPLE If $R_1, R_2, \ldots, R_n$ are rings, then as we know from Propositions 3.1.6 and 3.1.8, their **direct product** $R_1 \times R_2 \times \ldots \times R_n$ is an Abelian group under componentwise addition:

$$(a_1, a_2, \ldots, a_n) + (b_1, b_2, \ldots, b_n) = (a_1 + b_1, a_2 + b_2, \ldots, a_n + b_n)$$

If we now define componentwise multiplication:

$$(a_1, a_2, \ldots, a_n)(b_1, b_2, \ldots, b_n) = (a_1b_1, a_2b_2, \ldots, a_nb_n)$$

it is easy to see that we obtain a ring. (See Exercise 9.) ◇

Throughout our study of rings, 0 always denotes the additive identity of the ring. For the additive inverse of an element a we write $-a$. We abbreviate $a + a + \ldots + a$ (n times) as $n \cdot a$, where n is any positive integer.

Warning: $n \cdot a$ *is not the product* of n times a, since n may not belong to the same ring as a.

For n a negative integer we similarly abbreviate $(-a) + (-a) + \ldots + (-a)$ ($|n|$ times) as $-n \cdot a$.

The first basic and important properties of rings, which we prove next, involve relations between the two operations of the ring. Note the important role of the distributive laws in the proofs.

6.1.8 PROPOSITION Let a and b be elements of a ring R. Then

(1) $a0 = 0a = 0$

(2) $(-a)b = a(-b) = -ab$

(3) $(-a)(-b) = ab$

(4) $(n \cdot a)(m \cdot b) = nm \cdot (ab)$ for all integers n and m

Proof (1) $a0 + a0 = a(0 + 0)$ by the distributive law. Hence $a0 + a0 = a0$, and $a0 = 0$ follows by the cancellation law for the additive group. The proof that $0a = 0$ is similar.

(2) We want to show that $(-a)b = -ab$ or, in other words, that $(-a)b$ is the additive inverse of ab. So consider $(-a)b + ab = (-a + a)b$ by the distributive law. Hence $(-a)b + ab = 0b = 0$ by (1). The proof that $a(-b) = -ab$ is similar.

(3) From (2) we obtain $(-a)(-b) = a[-(-b)] = ab$, since in any group the inverse of the inverse of an element is the original element.

(4) We use induction twice. If $n = 0$ or $m = 0$, there is nothing to prove. Let $n = 1$. Then (4) holds for $m = 1$. Assume (4) holds for $n = 1$ and $m = k \geq 0$. Then, using the distributive law, we obtain $a[(k + 1) \cdot b] = a[k \cdot b + b] = a(k \cdot b) + ab = k \cdot (ab) + ab = (k + 1) \cdot ab$, and (4) holds for $n = 1$ and $m = k + 1$. This shows that (4) holds for $n = 1$ and for all $m \geq 0$. If $m < 0$, let $r = -m$. We have $a(m \cdot b) = a[(-r) \cdot b] = a[-(r \cdot b)] = -[a(r \cdot b)] = -r \cdot (ab) = m \cdot (ab)$. Hence (4) holds for $n = 1$ and all $m \in \mathbb{Z}$. Now assume (4) holds for $n = k \geq 0$ and all $m \in \mathbb{Z}$. Then $[(k + 1) \cdot a](m \cdot b) = (k \cdot a + a)(m \cdot b) = (k \cdot a)(m \cdot b) + a(m \cdot b) = km \cdot (ab) + m \cdot (ab) = (km + m) \cdot (ab) = [(k + 1)m] \cdot (ab)$, and (4) holds for $n = k + 1$ and all $m \in \mathbb{Z}$. This shows that (4) holds for all $n \geq 0$ and all $m \in \mathbb{Z}$. If $n < 0$, let $s = -n$. We have $(n \cdot a)(m \cdot b) = (-s \cdot a)(m \cdot b) = -(s \cdot a)(m \cdot b) = -sm \cdot (ab) = nm \cdot (ab)$, and the proof of (4) is complete. □

The ring of integers $\mathbb{Z}$, which served as a model for the definition of a ring in general, has some other properties that are not mentioned in Definition 6.1.2. For example, multiplication in $\mathbb{Z}$ is commutative, and there is a multiplicative identity $1 \in \mathbb{Z}$ such that for any $n \in \mathbb{Z}$ we have $1 \cdot n = n \cdot 1 = n$. Rings with these two extra properties are identified in the next definition.

6.1.9 DEFINITION Let R be a ring. Then

(1) R is said to be a **commutative ring** if for all $a, b \in R$ we have $ab = ba$.

(2) R is said to be a **ring with unity** if there is an element denoted $1 \in R$ such that for all $a \in R$ we have $1a = a1 = a$. ○.

6.1.10 EXAMPLE $\mathbb{Z}$, $\mathbb{Q}$, $\mathbb{R}$, $\mathbb{C}$, and $\mathbb{Z}_n$ are all commutative rings with unity. ◇

6.1.11 EXAMPLE $M(2, R)$, as in Example 6.1.5, is a ring with unity. The identity matrix I plays the role of the multiplicative identity in $M(2, R)$. Note, however, that this is not a commutative ring. (See Example 0.5.8.) ◇

If R is a ring with unity, $1 \in R$, then 1 plays the role of a multiplicative identity in R. Note that 1, if it exists in R, is unique in R. For suppose that 1 and 1' were both unities in R. Then we would have $1 = 11' = 1'$, just as in the proof the uniqueness of the identity element of a group.

In the case of groups we defined a subgroup of a group G to be a nonempty subset of G that is itself a group under the same operation as G. We then proved a very useful subgroup test (Theorem 1.2.10), which we restate here in additive notation:

Given a nonempty subset H of a group G under addition, then H is a subgroup of G if and only if for all $a, b \in H$ we have $a - b \in H$.

We imitate this definition and theorem for the case of rings and subrings.

6.1.12 DEFINITION A nonempty subset S of a ring R is a **subring** of R if S is a ring with the same operations as R.

6.1.13 THEOREM (Subring test) A nonempty subset S of a ring R is a subring of R if and only if for all $a, b \in S$ we have

(1) $a - b \in S$

(2) $ab \in S$

Proof ($\Leftarrow$) If (1) and (2) hold, then by the subgroup test, (1) implies that S is a subgroup of R under addition and therefore an Abelian group. (2) gives us closure under multiplication. Finally, the associative laws of multiplication and the distributive laws hold in R, since R is a ring, and therefore hold in S, since we are using the same operations.

($\Rightarrow$) If S is a subring of R, then S is a subgroup of R under addition, and the subgroup test implies (1), while (2) is just the ring axiom of closure under multiplication. □

The subring test becomes very useful when we want to check whether a set S is a ring under some operations in the case where S is a subset of a ring. We illustrate the use of the test in a few examples.

6.1.14 EXAMPLE $2\mathbb{Z}$ is a ring under the usual operations on integers. For $2\mathbb{Z} \subseteq \mathbb{Z}$, and for all $x, y \in 2\mathbb{Z}$ we have (1) $x - y \in 2\mathbb{Z}$ and (2) $xy \in 2\mathbb{Z}$, and thus $2\mathbb{Z}$ is a subring of $\mathbb{Z}$. The same can be said for $n\mathbb{Z}$ for all $n \geq 1$. Note that these are examples of commutative rings without unity, if $n \geq 2$.

6.1.15 EXAMPLE The set $\mathbb{Z}[i] = \{a + bi \mid a, b \in \mathbb{Z}, i^2 = -1\}$ is a ring, called the ring of **Gaussian integers**, under the usual operations on complex numbers. To see this, we note that $\mathbb{Z}[i] \subseteq \mathbb{C}$, so we can apply the subring test. If $x, y \in \mathbb{Z}[i]$, then $x = a + bi$ and $y = c + di$ for some $a, b, c, d \in \mathbb{Z}$. We then have

(1) $x - y = (a + bi) - (c + di) = (a - c) + (b - d)i \quad \in \mathbb{Z}[i]$

(2) $xy \quad = (a + bi)(c + di) \quad = (ac - bd) + (ad + bc)i \quad \in \mathbb{Z}[i]$

since $a - c$, $b - d$, $ac - bd$, $ad + bc \in \mathbb{Z}$. It follows by the subring test that $\mathbb{Z}[i]$ is a subring of $\mathbb{C}$. ◇

6.1.16 EXAMPLE The set $\mathbb{Q}(\sqrt{2}) = \{a + b\sqrt{2} \mid a, b \in \mathbb{Q}\}$ is a ring under the usual operations on real numbers. To see this, we note that $\mathbb{Q}(\sqrt{2}) \subseteq \mathbb{R}$, so we can apply the subring test. If $x, y \in \mathbb{Q}(\sqrt{2})$, then $x = a + b\sqrt{2}$ and $y = c + d\sqrt{2}$ for some a, b, c, and $d \in \mathbb{Q}$. We then have

$$(1)\ x - y = (a + b\sqrt{2}) - (c + d\sqrt{2}) = (a - c) + (b - d)\sqrt{2} \in \mathbb{Q}[\sqrt{2}]$$
$$(2)\ xy = (a + b\sqrt{2})(c + d\sqrt{2}) = (ac + 2bd) + (ad + bc)\sqrt{2} \in \mathbb{Q}[\sqrt{2}]$$

since $a - c$, $b - d$, $ac + 2bd$, $ad + bc \in \mathbb{Q}$. It follows by the subring test that $\mathbb{Q}(\sqrt{2})$ is a subring of $\mathbb{R}$. ◇

Note that this last example can be generalized to $\mathbb{Q}(\sqrt{p})$ for any prime p, obtaining in this way an infinite number of subrings of $\mathbb{R}$ with $\mathbb{Q} \subseteq \mathbb{Q}(\sqrt{p}) \subseteq \mathbb{R}$. For $\mathbb{Q}$ is a subring of $\mathbb{Q}(\sqrt{p})$ since any rational number a can be written as $a = a + 0\sqrt{p}$.

We end this section with one more example of a ring. In Chapter 1 we defined Q_8, the quaternion group in terms of matrices in $SL(2, \mathbb{C})$. (See Example 1.1.23.) In the next example we use Q_8 to construct a very important example of a noncommutative ring.

6.1.17 EXAMPLE For any complex numbers u and v, consider the matrix

$$h(u, v) = \begin{bmatrix} u & v \\ -\overline{v} & \overline{u} \end{bmatrix}$$

and let $\mathbb{H} = \{h(u, v) \mid u, v \in \mathbb{C}\} \subseteq M(2, \mathbb{C})$. We leave it as an exercise to show that $\mathbb{H}$ is a subring of $M(2, \mathbb{C})$. (See Exercise 10.) $\mathbb{H}$ is called the ring of **quaternions**. If we let $Q_8 = \{\pm 1, \pm\mathbf{i}, \pm\mathbf{j}, \pm\mathbf{k}\}$, as in Example 1.1.23, note that if $u = a + bi$ and $v = c + di$, where a, b, c, d are real numbers, then $h(u,v)$ can be expressed as

$$a\begin{bmatrix} 1 & 0 \\ 0 & 1 \end{bmatrix} + b\begin{bmatrix} i & 0 \\ 0 & -i \end{bmatrix} + c\begin{bmatrix} 0 & 1 \\ -1 & 0 \end{bmatrix} + d\begin{bmatrix} 0 & i \\ i & 0 \end{bmatrix}$$

or $a\mathbf{1} + b\mathbf{i} + c\mathbf{j} + d\mathbf{k}$. In other words, the elements of $\mathbb{H}$, the quarternions, are linear combinations of elements of Q_8 with real coefficients. ◇

Exercises 6.1

In Exercises 1 through 6 determine whether the indicated sets form a ring under the indicated operations.

1. $S = \{a + b\sqrt{3} \mid a, b \in \mathbb{Z}\}$, under the usual operations on real numbers

2. $S = \{a + bi \mid a, b \in \mathbb{Q}\}$, under the usual operations on complex numbers

3. $S = \{\begin{bmatrix} a & b \\ 0 & a \end{bmatrix} \mid a, b \in \mathbb{R}\}$, under matrix addition and multiplication

4. $S = \{\begin{bmatrix} a & b \\ -b & a \end{bmatrix} \mid a, b \in \mathbb{R}\}$, under matrix addition and multiplication

5. $S = \{A \in M(2, \mathbb{R}) \mid \det A = 0\}$, under matrix addition and multiplication

6. $S = \{m/n \in \mathbb{Q} \mid n \text{ odd}\}$, under the usual operations on real numbers

7. $S = \{ri \mid r \in \mathbb{R}, i^2 = -1\}$, under the usual operations on complex numbers

8. Show that the set $F(\mathbb{R})$ of all functions $f: \mathbb{R} \to \mathbb{R}$ forms a ring under the operations defined in Example 6.1.6.

9. Let $R_1, R_2, \ldots, R_n$ be any rings and $S = R_1 \times R_2 \times \ldots \times R_n$ their direct product as defined in Example 6.1.7. Show that

(a) S is a ring.

(b) S is commutative if and only if R_i is commutative for all i, $1 \le i \le n$.

(c) S is a ring with unity if and only if R_i is a ring with unity for all i, $1 \le i \le n$.

10. Show that $\mathbb{H}$ as defined in Example 6.1.17 is a ring.

11. If S and T are subrings of a ring R, show that $S \cap T$ is a subring of R.

12. Determine all the subrings of $\mathbb{Z}$.

13. Let R be a ring. The **center** of R is defined as follows:

$$Z(R) = \{x \in R \mid xy = yx \text{ for all } y \in R\}$$

Show that $Z(R)$ is a subring of R.

14. Find the center $Z(\mathbb{H})$ of $\mathbb{H}$, the ring of quaternions. (See Example 6.1.17 and Exercise 10.)

15. Give an example of a ring R and elements a, b, and c in R such that $a \neq 0$, $ab = ac$, but $b \neq c$.

16. Let R be a ring. Show that $(a + b)(a - b) = a^2 - b^2$ for all a and b in R if and only if R is a commutative ring.

17. Let R be a ring. Show that $(a + b)^2 = a^2 + 2ab + b^2$ for all a and b in R if and only if R is a commutative ring.

18. Show that the binomial theorem (Theorem 0.3.8) holds for all elements a and b of a commutative ring R.

19. A **Boolean ring** is a ring with the property that $a^2 = a$ for all $a \in R$. Show that a Boolean ring is a commutative ring with $2a = 0$ for all $a \in R$.

20. For any set X, let $P(X) = \{A \mid A \subseteq X\}$ be the **power set** of X. For any A and B in $P(X)$ define

$$A + B = \{x \mid x \in A \cup B, x \notin A \cap B\} \qquad A \cdot B = A \cap B$$

Show that under these two operations

(a) $P(X)$ is a ring with unity. (b) $P(X)$ is a Boolean ring.

21. Let R be a ring with unity 1 and let $S = \{n \cdot 1 \mid n \in \mathbb{Z}\}$. Show that S is a subring of R.

6.2 Integral Domains

In this section we identify a property of some rings that will play an important role in our study. Again the ring of integers, $\mathbb{Z}$, is our model. In the number systems in which we are used to making calculations, when we encounter an equation like $ab = ac$, where $a \neq 0$, we immediately conclude that $b = c$. As we will soon see, such a conclusion may be incorrect in some rings.

6.2.1 EXAMPLE $\mathbb{Z}_5$ and $\mathbb{Z}_6$ are both rings. We are familiar with their addition tables, under which they are Abelian groups. Let us now compare their multiplication tables, but let us remove 0, since multiplying by 0 simply gives a row or column of 0s.

TABLE 1 Multiplication mod 5

	1	2	3	4
1	1	2	3	4
2	2	4	1	3
3	3	1	4	2
4	4	3	2	1

TABLE 2 Multiplication mod 6

	1	2	3	4	5
1	1	2	3	4	5
2	2	4	0	2	4
3	3	0	3	0	3
4	4	2	0	4	2
5	5	4	3	2	1

There are many differences you can observe between these two tables. The one we want to point out here is that 0 does not appear in any of the rows in multiplication mod 5, while 0 does appear in three rows in multiplication mod 6, namely $2 \cdot 3 = 3 \cdot 2 = 3 \cdot 4 = 4 \cdot 3 = 0$. Note also that in $\mathbb{Z}_6$ we have $3 \cdot 2 = 3 \cdot 4$, even though $2 \neq 4$. ◇

6.2.2 DEFINITION If a and b are two nonzero elements of a ring R such that $ab = 0$, then a and b are said to be **zero divisors** in R. ○

6.2.3 EXAMPLE In $\mathbb{Z}_6$ the zero divisors are 2, 3, and 4. In $\mathbb{Z}_{12}$ the zero divisors are

2 since $2 \cdot 6 = 0$
3 since $3 \cdot 4 = 0$
4 since $4 \cdot 3 = 0$
6 since $6 \cdot 2 = 0$
8 since $8 \cdot 3 = 0$
9 since $9 \cdot 4 = 0$
10 since $10 \cdot 6 = 0$

Observe that the zero divisors in $\mathbb{Z}_{12}$ are exactly the elements of $\mathbb{Z}_{12}$ that are *not* relatively prime to 12. That this is always the case is what we show next. ◇

6.2.4 THEOREM A nonzero element r in $\mathbb{Z}_n$ is a zero divisor if and only if r and n are not relatively prime.

Proof ($\Rightarrow$) Suppose $r \in \mathbb{Z}_n$, $r \neq 0$, and for some $m \in \mathbb{Z}_n$, $m \neq 0$, we have $rm \equiv 0 \bmod n$. Since $m \neq 0$ in $\mathbb{Z}_n$, n does not divide m in $\mathbb{Z}$, and it follows from Proposition 0.3.22, part (2), that r and n are not relatively prime.
($\Leftarrow$) Suppose r and n are not relatively prime. Hence $\gcd(r,n) = \mathrm{d} > 1$, and $n/d < n$. We have $r\cdot(n/d) = (r/d)\cdot n = 0$ in $\mathbb{Z}_n$. Thus r is a zero divisor in $\mathbb{Z}_n$. □

6.2.5 COROLLARY $\mathbb{Z}_p$ has no zero divisors if and only if p is prime.

Proof Immediate from Theorem 6.2.4. □

6.2.6 EXAMPLE $\mathbb{Z}$, $\mathbb{Q}$, $\mathbb{R}$, $\mathbb{C}$, and $\mathbb{Z}_p$, where p is a prime, are all rings with no zero divisors.

The multiplicative **cancellation laws** hold in a ring R if for all a, b, and c in R, with $a \neq 0$, $ab = ac$ implies $b = c$ and $ba = ca$ implies $b = c$. We will see that the rings in which the cancellation laws hold are precisely the rings that have no zero divisors.

6.2.7 THEOREM In a ring R the cancellation laws hold if and only if R has no zero divisors.

Proof ($\Rightarrow$) Suppose the cancellation laws hold in a ring R, and suppose for some $a, b \in R$, $a \neq 0$, we have $ab = 0$. We want to show this can happen only if $b = 0$. Since $ab = 0$ and $a0 = 0$, the cancellation laws give us $b = 0$. Hence R has no zero divisors.
($\Leftarrow$) Suppose R has no zero divisors and for some $a, b, c \in R$, $a \neq 0$, we have $ab = ac$. Then $a(b - c) = ab - ac = 0$. Since a is not a zero divisor, we must have $b - c = 0$, and $b = c$. Similarly, $ba = ca$ implies $b = c$. □

6.2.8 DEFINITION A ring R is said to be an **integral domain** if

(1) R is commutative.
(2) R has unity, $1 \in R$.
(3) R has no zero divisors. ○

6.2.9 EXAMPLE $\mathbb{Z}$, $\mathbb{Q}$, $\mathbb{R}$, $\mathbb{C}$, and $\mathbb{Z}_p$, where p is a prime, are all integral domains. ◇

6.2.10 DEFINITION A nonempty subset S of an integral domain D is called a **subdomain** of D if S is an integral domain under the same operations as D. ○

We leave the proof of the following proposition as an exercise.

6.2.11 PROPOSITION (Subdomain test) A nonempty subset S of an integral domain D is a subdomain of D if and only if

(1) S is a subring of D.

(2) $1 \in S$, where 1 is unity in D.

Proof See Exercise 10. □

6.2.12 EXAMPLE $\mathbb{Z}[i]$ is an integral domain, since $\mathbb{Z}[i]$ is a subring of $\mathbb{C}$ and $1 \in \mathbb{Z}[i]$. ◇

6.2.13 EXAMPLE $\mathbb{Q}(\sqrt{2})$ is an integral domain, since $\mathbb{Q}(\sqrt{2})$ is a subring of $\mathbb{R}$ and $1 \in \mathbb{Q}(\sqrt{2})$. ◇

6.2.14 EXAMPLE The ring of 2×2 matrices $M(2, \mathbb{R})$ is *not* an integral domain. For $M(2, \mathbb{R})$ has zero divisors; for example,

$$\begin{bmatrix} 0 & 1 \\ 0 & 0 \end{bmatrix} \cdot \begin{bmatrix} 1 & 0 \\ 0 & 0 \end{bmatrix} = \begin{bmatrix} 0 & 0 \\ 0 & 0 \end{bmatrix}$$

Also, $M(2, \mathbb{R})$ is not commutative. ◇

6.2.15 EXAMPLE $\mathbb{Z} \times \mathbb{Z}$ is not an integral domain, since $(2, 0) \cdot (0, 3) = (0, 0)$. In general, for any two non-trivial rings R_1 and R_2, $R_1 \times R_2$ is not an integral domain, since all the elements of the forms $(r_1, 0)$ and $(0, r_2)$ with $r_1 \neq 0$ and $r_2 \neq 0$ are zero divisors in $R_1 \times R_2$. ◇

6.2.16 EXAMPLE We point out here another property of integral domains. In $\mathbb{Z}$ the equation $a^2 = a$ has exactly two solutions, $a = 0$ or $a = 1$. By contrast, in $\mathbb{Z}_6$, which we know is *not* an integral domain, $a = 3$ is also a solution of $a^2 = a$. In an integral domain D, $a^2 = a$ implies $a(a - 1) = a^2 - a = 0$, and since there are no zero divisors, $a = 0$ or $a = 1$ are the only solutions in D. ◇

We have not mentioned in this section a very important example of an integral domain, the ring of polynomials with coefficients in a given integral domain. We devote a whole chapter to the study of this ring of polynomials later, after we have established and clarified some basic concepts.

Exercises 6.2

In Exercises 1 through 7 find all the zero divisors in the indicated rings.

1. $\mathbb{Z}_4$ **2.** $\mathbb{Z}_8$ **3.** $\mathbb{Z}_{11}$ **4.** $\mathbb{Z}_2 \times \mathbb{Z}_2$

5. $\mathbb{Z}_4 \times \mathbb{Z}_6$ **6.** $\mathbb{Z} \times \mathbb{Q}$ **7.** $M(2, \mathbb{Z}_2)$

8. Give an example of a commutative ring with no zero divisors that is not an integral domain.

9. Give an example of a ring with unity and no zero divisors that is not an integral domain.

10. Prove Proposition 6.2.11.

11. Show that the intersection of two subdomains of an integral domain D is also a subdomain of D.

12. Let D be an integral domain and let $S = \{n \cdot 1 \mid n \in \mathbb{Z}\}$, where 1 is unity in D. Show that

(a) S is a subdomain of D.
(b) If R is any subdomain of D, then $S \subseteq R$.

In Exercises 13 through 15 show that the indicated rings are integral domains.

13. $\mathbb{Q}(i) = \{a + bi \mid a, b \in \mathbb{Q}, i^2 = -1\}$

14. $\mathbb{Z}[\sqrt{5}] = \{a + b\sqrt{5} \mid a, b \in \mathbb{Z}\}$

15. $\mathbb{Q}(\sqrt{2}, \sqrt{3}) = \{a + b\sqrt{2} + c\sqrt{3} + d\sqrt{2}\sqrt{3} \mid a, b, c, d \in \mathbb{Q}\}$

16. Find all the subdomains of $\mathbb{Z}$.

17. Show that the only subdomain of $\mathbb{Z}_p$, where p is prime, is $\mathbb{Z}_p$.

18. Show that $\mathbb{Z}_2$ is the only Boolean ring (see Exercise 19 in Section 6.1) that is an integral domain.

19. Let R be a ring with at least two elements such that for every nonzero element $a \in R$ there exists a unique element $b \in R$ with $aba = a$. Show that

(a) R has no zero divisors.
(b) $bab = b$
(c) R has unity.

20. Consider the ring $\mathbb{Z}_7$.

(a) Show that $\mathbb{Z}_7$ is a ring, as in the preceding exercise.
(b) For any nonzero $a \in \mathbb{Z}_7$ find the corresponding $b \in \mathbb{Z}_7$ with $aba = a$.

6.3 Fields

In the preceding section we introduced the notion of an integral domain, a commutative ring with unity and no zero divisors. In an integral domain the cancellation laws hold, as we showed by arguing that if $ab = ac$, $a \neq 0$, then $a(b - c) = ab - ac = 0$, and so since there are no zero divisors, $b - c = 0$ and $b = c$. Note that our proof did *not* involve "dividing" both sides by a, simply because we don't know whether a has a multiplicative inverse.

6.3.1 EXAMPLE In $\mathbb{Z}$, 1 and -1 are the only elements that have a multiplicative inverse: $1 \cdot 1 = 1$ and $(-1) \cdot (-1) = 1$. ◇

6.3.2 EXAMPLE In $\mathbb{Z}_5$ (see Example 6.2.1), $1 \cdot 1 = 1$, $2 \cdot 3 = 1$, $3 \cdot 2 = 1$, $4 \cdot 4 = 1$, and thus every nonzero element has a multiplicative inverse. ◇

6.3.3 DEFINITION In a ring R with unity 1, an element $a \in R$ is called a **unit** if a has a multiplicative inverse in R. ○

The multiplicative inverse of a, if it exists, is unique. (See Exercise 11). It is denoted by a^{-1}.

6.3.4 EXAMPLE The units in $\mathbb{Z}_{12}$ are 1, 5, 7, and 11, since $1 \cdot 1 = 5 \cdot 5 = 7 \cdot 7 = 11 \cdot 11 = 1$. Observe that in $\mathbb{Z}_{12}$ the units are precisely the nonzero elements that are not zero divisors. Also note that the set of units in $\mathbb{Z}_{12}$ form the multiplicative group $U(12)$. These observations are verified in general in the next two theorems. ◇

6.3.5 THEOREM In a ring R with unity 1, if an element $a \in R$ is a unit, then a is not a zero divisor.

Proof Let $a \in R$ be a unit in R, so that a^{-1} exists in R. If for some $b \in R$ we have $ab = 0$, then we have $b = 1b = (a^{-1}a)b = a^{-1}(ab) = a^{-1}0 = 0$, so a is not a zero divisor. □

6.3.6 THEOREM Let R be a commutative ring with unity 1, and let

$$U(R) = \{a \in R \mid a \text{ is a unit in } R\}$$

the set of all units in R. Then $U(R)$ is a group under the multiplication operation of R.

Proof We show that $U(R)$ satisfies the four group axioms.
(Closure) Let $a, b \in U(R)$, so a^{-1} and b^{-1} and hence $b^{-1}a^{-1}$ exist in R. We have $(ab)(b^{-1}a^{-1}) = a(bb^{-1})a^{-1} = aa^{-1} = 1$, and similarly $(b^{-1}a^{-1})(ab) = 1$, so $b^{-1}a^{-1}$ is the multiplicative inverse of ab, and ab is a unit, so $ab \in U(R)$.
(Associativity) The multiplication operation of $U(R)$ is just the multiplication operation of R, which is associative since R is a ring.
(Identity) $1 \cdot 1 = 1$, hence 1 is its own multiplicative inverse, 1 is a unit, and $1 \in U(R)$.
(Inverse) If $a \in U(R)$, then a has a multiplicative inverse a^{-1} in R. But then a is the multiplicative inverse of a^{-1}, so a^{-1} has a multiplicative inverse, and $a^{-1} \in U(R)$. □

6.3.7 THEOREM In $\mathbb{Z}_n$ the multiplicative group of units is $U(\mathbb{Z}_n) = U(n)$.

Proof If $a \in U(\mathbb{Z}_n)$ or, in other words, if a is a unit in $\mathbb{Z}_n$, then $a \neq 0$ and by Theorem 6.3.5 a is not a zero divisor in $\mathbb{Z}_n$. Therefore, from Theorem 6.2.4 it follows that a and n are relatively prime and $a \in U(n)$.

If $a \in U(n)$, then since $U(n)$ is a group under multiplication, there is a multiplicative inverse a^{-1} for a in $U(n) \subseteq \mathbb{Z}_n$, and thus $a \in U(\mathbb{Z}_n)$. □

6.3.8 EXAMPLE $U(\mathbb{Z}_6) = \{1,5\} = U(6)$. $U(\mathbb{Z}_5) = \mathbb{Z}_5 - \{0\} = U(5)$. ◇

6.3.9 EXAMPLE For the integers, $U(\mathbb{Z}) = \{+1, -1\}$, the multiplicative group of order 2. For the rational numbers, $U(\mathbb{Q}) = \mathbb{Q}^*$, the nonzero rational numbers under multiplication. ◇

6.3.10 EXAMPLE Let us calculate the group of units in $\mathbb{Z}_4 \times \mathbb{Z}_6$. Here (a, b) is a unit in $\mathbb{Z}_4 \times \mathbb{Z}_6$ if and only if there exists an element (c, d) in $\mathbb{Z}_4 \times \mathbb{Z}_6$ such that $(a, b)(c, d) = (ac, bd) = (1, 1)$. In other words, a must be a unit in $\mathbb{Z}_4$ and b must be a unit in $\mathbb{Z}_6$. Hence $U(\mathbb{Z}_4 \times \mathbb{Z}_6) = \{(1, 1), (1, 5), (3, 1), (3, 5)\} = U(\mathbb{Z}_4) \times U(\mathbb{Z}_6)$. (See Exercise 13.) ◇

We now define one of the most basic notions in the theory of rings.

6.3.11 DEFINITION A ring R is said to be a **field** if

(1) R is commutative.
(2) R has unity, $1 \in R$.
(3) Every nonzero element in R is a unit. ○

Observe that by Theorem 6.3.6, condition (3) can be replaced by

(3') The set of nonzero elements in R is a group under multiplication in R.

6.3.12 EXAMPLE $\mathbb{Q}$, $\mathbb{R}$, and $\mathbb{C}$ are fields, and $\mathbb{Z}$ is an integral domain that is not a field. ◇

The relation between integral domains and fields is explained in the following theorems.

6.3.13 THEOREM Every field is an integral domain.

Proof This is immediate from Definition 6.2.8, Definition 6.3.11, and Theorem 6.3.5. □

6.3.14 THEOREM Every finite integral domain is a field.

Proof Comparing Definition 6.2.8 with Definition 6.3.11, we note that all we need to show is that in a finite integral domain every nonzero element is a unit. Let D be a finite integral domain, $a \in D$, $a \neq 0$. Consider

$$D - \{0\} = \{1, a_1, a_2, \dots, a_{n-1}\}$$

the set of all nonzero elements in D. Since D is an integral domain and $a \neq 0$, $\{a, aa_1, aa_2, \dots, aa_{n-1}\}$ are all nonzero. Also, they are all distinct, since if $aa_i = aa_j$, then $a_i = a_j$ by the cancellation laws, which hold in any integral domain. Thus

$$D - \{0\} = \{a, aa_1, aa_2, \dots, aa_{n-1}\}$$

and either $1 = a$, in which case $a^{-1} = 1$, or else $1 = aa_i$ for some i, $1 \le i \le n - 1$, in which case $a^{-1} = a_i$. Therefore, a is a unit, and D is a field. □

6.3.15 COROLLARY $\mathbb{Z}_p$ is a field if and only if p is prime.

Proof This is immediate from the preceding theorem and Corollary 6.2.5. □

6.3.16 DEFINITION A nonempty subset S of a field F is a **subfield** of F if S is a field under the same operations as F. ○

6.3.17 THEOREM (Subfield test) A nonempty subset S of a field F is a field under the same two operations as F if and only if for all $a, b \in S$ we have

(1) $a - b \in S$

(2) For $b \neq 0$, $ab^{-1} \in S$

Proof Left as an exercise. (See Exercise 15.) □

6.3.18 EXAMPLE Let us use the subfield test to show that

$$\mathbb{Q}(\sqrt{2}) = \{a + b\sqrt{2} \mid a, b \in \mathbb{Q}\}$$

is a field under the usual operations. Since we have already shown in Example 6.1.16 that $\mathbb{Q}(\sqrt{2})$ is a subring of $\mathbb{R}$, in order to prove that it is a subfield of $\mathbb{R}$ it suffices to prove condition (2) in Theorem 6.3.17. So let $x, y \in \mathbb{Q}(\sqrt{2})$ with $y \neq 0$. Then $x = a + b\sqrt{2}$ and $y = c + d\sqrt{2}$ for some $a, b, c, d \in \mathbb{Q}$. Then

$$xy^{-1} = (a + b\sqrt{2})\cdot\frac{1}{c + d\sqrt{2}} = \frac{a + b\sqrt{2}}{c + d\sqrt{2}} = \frac{a + b\sqrt{2}}{c + d\sqrt{2}} \cdot \frac{c - d\sqrt{2}}{c - d\sqrt{2}} =$$

$$\frac{(ac - 2bd) + (bc - ad)\sqrt{2}}{c^2 - 2d^2} = \frac{ac - 2bd}{c^2 - 2d^2} + \frac{bc - ad}{c^2 - 2d^2}\sqrt{2} \in \mathbb{Q}(\sqrt{2})$$

Note that since $c + d\sqrt{2} = y \neq 0$, we cannot have $c^2 - 2d^2 = 0$, because this would imply that $\sqrt{2} = \pm(c/d)$, which is impossible for $c, d \in \mathbb{Q}$. Hence the expressions we obtained previously as coefficients are in $\mathbb{Q}$. ◇

6.3.19 EXAMPLE So far we have encountered infinite fields such as $\mathbb{Q}$, $\mathbb{R}$, $\mathbb{C}$, and $\mathbb{Q}(\sqrt{2})$, and finite fields such as $\mathbb{Z}_p$ with p elements, where p is a prime. We give here an example of a finite field with n elements where n is *not* a prime. Consider

$$\mathbb{Z}_3[i] = \{a + bi \mid a, b \in \mathbb{Z}_3,\ i^2 = -1\}$$

Since a and b are either 0, 1, or $2 = -1$, $\mathbb{Z}_3[i]$ has exactly nine elements:

$$\mathbb{Z}_3[i] = \{0, 1, 2, i, 1 + i, 2 + i, 2i, 1 + 2i, 2 + 2i\}$$

Addition and multiplication in $\mathbb{Z}_3[i]$ are defined as follows. For any $x = a + bi$ and $y = c + di$ with $a, b, c, d \in \mathbb{Z}_3$, we have

$$x + y = (a +_3 c) + (b +_3 d)i \qquad x \cdot y = (a \cdot_3 c -_3 b \cdot_3 d) + (a \cdot_3 d +_3 b \cdot_3 c)i$$

where $+_3$, $-_3$, $\cdot_3$ are addition, subtraction, and multiplication mod 3.With these two operations we obtain a commutative ring with unity. Every nonzero element in $\mathbb{Z}_3[i]$ is a unit, with $1^{-1} = 1$, $2^{-1} = 2$, $i^{-1} = 2i$, $(1 + i)^{-1} = 2 + i$, $(1 + 2i)^{-1} = 2 + 2i$. Therefore, $\mathbb{Z}_3[i]$ is a field. ◇

6.3.20 EXAMPLE If we let $Q_8 = \{\pm 1, \pm\mathbf{i}, \pm\mathbf{j}, \pm\mathbf{k}\}$ be the quaternion group, then $\mathbb{H}$ the ring of quaternions as described in Example 6.1.17 can be defined as

$$\mathbb{H} = \{a + b\mathbf{i} + c\mathbf{j} + d\mathbf{k} \mid a, b, c, d \in \mathbb{R} \text{ and } \mathbf{i}, \mathbf{j}, \mathbf{k} \in Q_8\}$$

$\mathbb{H}$ is a ring with unity. If $x \in \mathbb{H}$, $x \neq 0$, and $x = a + b\mathbf{i} + c\mathbf{j} + d\mathbf{k}$, then let $x^* = a - b\mathbf{i} - c\mathbf{j} - d\mathbf{k}$. We have $xx^* = a^2 + b^2 + c^2 + d^2 \neq 0$ and $x \cdot (x^*/xx^*) = 1$, hence x is a unit. Note, however, that since $\mathbf{ij} = \mathbf{k}$ and $\mathbf{ji} = -\mathbf{k}$, $\mathbb{H}$ is not commutative and so is not a field. ◇

6.3.21 DEFINITION A ring R with unity such that every nonzero element $a \in R$ is a unit in R is called a **division ring**. ○

6.3.22 EXAMPLE Every field is a division ring, and $\mathbb{H}$ is an example of a division ring that is not a field. ◇

We introduce one last concept in this section, one that helps differentiate between two classes of integral domains and fields.

6.3.23 EXAMPLE In $\mathbb{Z}_6$ we can find a least positive integer n such that $na = 0$ for all $a \in \mathbb{Z}_6$, namely $n = 6$. Similarly, in $\mathbb{Z}_4 \times \mathbb{Z}_6$ if we take $n = 12$, then $12(a, b) = (12a, 12b) = (0, 0)$ for all $(a, b) \in \mathbb{Z}_4 \times \mathbb{Z}_6$, and no smaller positive integer has this property. In the ring of integers $\mathbb{Z}$, there is no positive integer n such that $na = 0$ for all $a \in \mathbb{Z}$. ◇

6.3.24 DEFINITION In a ring R the **characteristic** of R, denoted char R, is the least positive integer n such that $n \cdot a = 0$ for all $a \in R$. If no such n exists, we say char $R = 0$. ○

6.3.25 EXAMPLE For every integer n there is a ring of characteristic n, since

(1) char $\mathbb{Z}_n = n$

(2) char $\mathbb{Z}$ = char $\mathbb{Q}$ = char $\mathbb{R}$ = char $\mathbb{C} = 0$ ◇

In the case where R is a ring with unity, we have an easy way to determine the characteristic of R, as we see next.

6.3.26 THEOREM Let R be a ring with unity 1. Then

(1) char $R = 0$ if 1 has infinite order under addition.

(2) char $R = n$ if 1 has order n under addition.

Proof (1) If $|1| = \infty$, then there is no finite integer n such that $n \cdot 1 = 0$, hence char $R = 0$.

(2) If $|1| = n$, then n is the least positive integer such that $n \cdot 1 = 0$, and for all $a \in R$ by Proposition 6.1.8 part (4) we have $n \cdot a = n(1 \cdot a) = (n \cdot 1)a = 0a = 0$. Hence char $R = n$. □

6.3.27 EXAMPLE In $R = \mathbb{Z}_4 \times \mathbb{Z}_6$, unity is the element $(1, 1)$ and char $R = |(1, 1)| = \text{lcm}(4, 6) = 12$, as we have seen in Example 6.3.23. Consider $S = \mathbb{Z}_4 \times \{0\} \subseteq R$. Then S is a subring of R with unity (1,0). Hence char $S = |(1, 0)| = 4 \neq$ char R. In other words, the characteristic of a subring S may be different from the characteristic of the ring R. ◇

As we see in the next example, this cannot occur in an integral domain.

6.3.28 EXAMPLE Let D be any integral domain and S a subdomain of D. Then by Proposition 6.2.11, if 1 is unity in D, then $1 \in S$. In other words, D and S have the same multiplicative identity. Thus char $D = |1| =$ char S. ◇

We end this section with a description of the characteristic of an integral domain.

6.3.29 THEOREM Let D be an integral domain. Then char $D = 0$ or char $D = p$, where p is a prime.

Proof Assume D is an integral domain with char $D \neq 0$. Therefore, there exists a least positive integer n such that $n \cdot 1 = 0$. If n is not prime, then $n = uv$ for some positive integers with $u < n$ and $v < n$. By Proposition 6.1.8, part (4) we have $0 = n \cdot 1 = (uv) \cdot 1 = (u \cdot 1)(v \cdot 1) \in D$. Since D is an integral domain, we must have either $u \cdot 1 = 0$ or $v \cdot 1 = 0$. In either case we get a contradiction with the choice of n as the *least* positive integer with $n \cdot 1 = 0$. Thus $n =$ char D must be prime. □

Figure 1 may help you visualize the different types of rings we have introduced in this chapter. In each region a representative example is given. You may want to construct more examples for each of the seven regions.

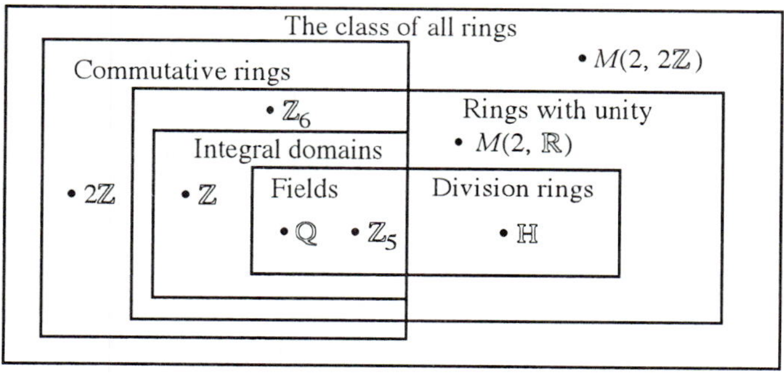

FIGURE 1

Exercises 6.3

In Exercises 1 through 10 find all the units in the indicated rings.

1. $\mathbb{Z}_{10}$ **2.** $\mathbb{Z}_2 \times \mathbb{Z}_4$ **3.** $\mathbb{Z}[i]$

4. $\mathbb{Z} \times \mathbb{Z}$ **5.** $\mathbb{H}$ **6.** $\mathbb{C}$

7. $\mathbb{Q}(\sqrt{3})$ **8.** $M(2, \mathbb{Z}_2)$ **9.** $M(2, \mathbb{Z})$

10. $M(2, \mathbb{R})$

11. Let a be a unit in a ring R with unity. Show that the multiplicative inverse of a in R is unique.

12. Let R be a ring with unity $1 \in R$ and S a subring of R with $1 \in S$. Show that if $a \in S$ is a unit in S, then a is a unit in R. Show by example that the converse is not necessarily true.

13. Let R_1 and R_2 be commutative rings with unity. Show that the group of units $U(R_1 \times R_2) \cong U(R_1) \times U(R_2)$.

14. Let R be a commutative ring with unity and $M(2, R)$ the ring of 2×2 matrices with entries from R. Show that $A \in M(2, R)$ is a unit if and only if $\det A$ is a unit in R.

15. Prove Theorem 6.3.17.

In Exercises 16 through 22 determine which of the indicated rings are fields.

16. $\mathbb{Z}[i]$ **17.** $\mathbb{Q} \times \mathbb{Q}$

18. $\mathbb{Z}_{13}$ **19.** $\mathbb{Q}(\sqrt{3})$

20. $\mathbb{Q}(i) = \{a + bi \mid a, b \in \mathbb{Q}\}$ **21.** $\mathbb{H}$

22. $\mathbb{Z}_2[i] = \{a + bi \mid a, b \in \mathbb{Z}_2\}$

23. Let S and T be two subfields of a field F. Show that $S \cap T$ is a subfield of F.

In Exercises 24 thorugh 30 determine the characteristic of the indicated rings.

24. $\mathbb{Z}_{10} \times \mathbb{Z}_8$ **25.** $\mathbb{C}$ **26.** $\mathbb{Z} \times \mathbb{Z}$

27. $\mathbb{H}$ **28.** $\mathbb{Q}(\sqrt{2})$ **29.** $\mathbb{Z}_3[i]$

30. $\mathbb{Z}_2 \times \mathbb{Q} \times \mathbb{Z}_3$

31. Let F be a field with char $F = p > 0$. Show that for any elements $a, b \in F$ we have $(a + b)^p = a^p + b^p$.

32. Show that if D is an integral domain with char $D = 0$, then D is infinite.

33. Let F be a field with q elements. Show that for all $a \in F$ we have $a^q = a$.

34. Let p be a prime and consider the equation $x^p - 1 = 0$. Show that

(a) In $\mathbb{C}$, $x^p - 1 = 0$ has p distinct solutions.

(b) In a field F with char $F = p$, $x^p - 1 = 0$ has only one solution. (*Hint:* Use Exercise 31.)

35. An element a in a ring R is called **nilpotent** if for some $k \geq 1$ we have $a^k = 0$. Show that the set of nilpotent elements in a commutative ring R form a subring of R.

36. Find all nilpotent elements in $\mathbb{Z}_{24}$.

37. Show that if D is an integral domain, then 0 is the only nilpotent element in D.

38. Let a be a nilpotent element in a commutative ring R with unity. Show that

(a) $a = 0$ or a is a zero divisor.

(b) ax is nilpotent for all $x \in R$.

(c) $1 + a$ is a unit in R.

(d) If u is a unit in R, then $u + a$ is also a unit in R.

39. In a ring R an element $a \in R$ is called **idempotent** if $a^2 = a$. Show that in an integral domain, 0 and 1 are the only idempotent elements.

40. Find all the idempotent elements in $\mathbb{Z}_6$, $\mathbb{Z}_{12}$, and $\mathbb{Z}_6 \times \mathbb{Z}_{12}$.

41. Show that a ring R is a division ring if and only if for any element $a \in R$ there exists a unique element $b \in R$ such that $aba = a$. (See Exercise 19 in Section 6.2.)

42. Show that the center of a division ring is a field.

43. Show that a finite ring R with unity and with no zero divisors is a division ring.

44. Define the ring of **integral quaternions** as follows:

$$I = \{a + b\mathbf{i} + c\mathbf{j} + d\mathbf{k} \mid a, b, c, d \in \mathbb{Z}\} \subseteq \mathbb{H}$$

(a) Show that I is a subring of $\mathbb{H}$.

(b) Let N: $I \to \mathbb{Z}$ be the function defined as follows, called the **norm**:

$$N(a + b\mathbf{i} + c\mathbf{j} + d\mathbf{k}) = a^2 + b^2 + c^2 + d^2 \in \mathbb{Z}$$

Show that for all $z = a + b\mathbf{i} + c\mathbf{j} + d\mathbf{k} \in I$, $N(z) = zz^*$ where $z^* = a - b\mathbf{i} - c\mathbf{j} - d\mathbf{k}$.

(c) Show that $N(zw) = N(z)N(w)$.

(d) Show that $z \in I$ is a unit if and only if $N(z) = 1$.

(e) Show that the group of units of I is $U(I) = Q_8$, the quaternion group.

Chapter 7

Ring Homomorphisms

In Chapter 2 we identified the maps we can use from a group G to another group G' and called them group homomorphisms. In this chapter we define *ring homomorphisms*. We show how ring homomorphisms give rise to the notion of a special kind of subring, called an *ideal*. The homomorphic image of a ring turns out to be isomorphic to a *quotient ring*. Ring homomorphisms, ideals, and quotient rings are interrelated in exactly the same way that group homomorphisms, normal subgroups, and quotient groups were interrelated in the case of groups. We end this chapter with the construction of the *field of quotients* of an integral domain.

7.1 Definitions and Basic Properties

As in the case of groups, the maps we use between rings must be maps that preserve the algebraic structures of the rings. We consider maps that respect both operations of the ring. Since rings are groups under addition to begin with, our maps are group homomorphisms under addition that also respect the second operation.

7.1.1 EXAMPLE Consider the two rings $\mathbb{Z}$ and $2\mathbb{Z}$ and the natural map $\phi(x) = 2x$ for all $x \in \mathbb{Z}$. We know that ϕ is a group homomorphism under addition, since $\phi(x + y) = 2x + 2y = \phi(x) + \phi(y)$. But now look what happens under multiplication: $2 = \phi(1 \cdot 1) \neq \phi(1) \cdot \phi(1) = 4$. In other words, this map does not respect multiplication in $\mathbb{Z}$. ◇

7.1.2 EXAMPLE Consider now the map $\phi\colon \mathbb{Z} \to \mathbb{Z}_3$ defined by $\phi(x) = x \bmod 3$ for all $x \in \mathbb{Z}$. By Proposition 0.3.31 part (3), for any x and y in $\mathbb{Z}$ we have

$$\phi(x+y) = (x+y) \bmod 3 = (x \bmod 3) + (y \bmod 3) = \phi(x) +_3 \phi(y)$$
$$\phi(x \cdot y) = (x \cdot y) \bmod 3 = (x \bmod 3) \cdot (y \bmod 3) = \phi(x) \cdot_3 \phi(y)$$

Therefore, in this example the map ϕ respects both operations. ◇

7.1.3 DEFINITION A map $\phi\colon R \to R'$ from a ring R to a ring R' is called a **ring homomorphism** if for all $a, b \in R$ we have

(1) $\phi(a + b) = \phi(a) + \phi(b)$
(2) $\phi(a \cdot b) = \phi(a) \cdot \phi(b)$

Note that on the left side we use the two operations in R, while on the right side we use the two operations in R'. ○

Observe that (1) in the definition of a ring homomorphism tells us that ϕ is a group homomorphism under addition. It follows that whatever we know from Chapter 2 about group homomorphisms under addition also applies to a ring homomorphism.

7.1.4 EXAMPLE Consider the map $\phi: \mathbb{Z}_4 \to \mathbb{Z}_6$ defined by $\phi(x) = 3x$ for all $x \in \mathbb{Z}_4$. Again using Proposition 0.3.31, part (3), we obtain

$$\phi(x + y) = 3(x + y) \bmod 6 = (3x \bmod 6) + (3y \bmod 6) = \phi(x) +_6 \phi(y)$$
$$\phi(x \cdot y) = 3(x \cdot y) \bmod 6 = 9(x \cdot y) \bmod 6 = (3x \bmod 6) \cdot (3y \bmod 6) = \phi(x) \cdot_6 \phi(y)$$

and ϕ is a ring homomorphism. Note that in our calculation we have used the fact that $3 \equiv 9 \bmod 6$. The jump from 3 to 9 mod 6 can be better seen in $3 \bmod 6 = \phi(1) = \phi(1 \cdot 1) = \phi(1) \cdot \phi(1) = 9 \bmod 6$. ◇

7.1.5 EXAMPLE Let us find all possible ring homomorphisms $\phi: \mathbb{Z} \to \mathbb{Z}$. The homomorphism ϕ must be a group homomorphism under addition, and since $\mathbb{Z}$ is a cyclic group generated by 1, the image $\phi(1)$ of 1 determines ϕ completely. So let $\phi(1) = a \in \mathbb{Z}$. Then we have

$$a = \phi(1) = \phi(1 \cdot 1) = \phi(1) \cdot \phi(1) = a^2$$

Hence $a^2 = a$ in $\mathbb{Z}$, which is equivalent to $a(a - 1) = 0$. Since $\mathbb{Z}$ has no zero divisors, we obtain $a = 0$ or $a = 1$. If $\phi(1) = a = 0$, then $\phi(x) = 0$ for all $x \in \mathbb{Z}$, and hence ϕ is the zero homomorphism. If $\phi(1) = a = 1$, then $\phi(x) = x$ for all $x \in \mathbb{Z}$, and ϕ is the identity homomorphism. These are the only possible ring homomorphisms from $\mathbb{Z}$ to $\mathbb{Z}$. ◇

7.1.6 EXAMPLE Let us find all possible ring homomorphisms $\phi: \mathbb{Z}_6 \to \mathbb{Z}_{12}$. Again using the fact that $\mathbb{Z}_6$ is a cyclic group under addition with 1 as a generator, $\phi(1) = a \in \mathbb{Z}_{12}$ completely determines ϕ. From Proposition 2.2.15, part (4), we know that the order $|\phi(1)|$ of $\phi(1)$ in $\mathbb{Z}_{12}$ divides the order $|1| = 6$ of 1 in $\mathbb{Z}_6$. Hence either

$|a| = 1$ and $a = 0$, or
$|a| = 2$ and $a = 6$, or
$|a| = 3$ and $a = 4$ or 8, or
$|a| = 6$ and $a = 2$ or 10

So far we have used only condition (1) in the definition of ring homomorphism. Let us now use condition (2), which implies

$$a = \phi(1) = \phi(1 \cdot 1) = \phi(1) \cdot \phi(1) = a^2$$

In other words, among the elements of $\mathbb{Z}_{12}$ we have found so far as candidates we need to pick those that satisfy $a^2 = a$. (Such elements in a ring are called idempotents; (see Exercise 39 in Section 6.3). Since in $\mathbb{Z}_{12}$ we have $6^2 = 0$, $8^2 = 2^2 = 10^2 = 4$, the only possibilities are $a = 0$ and $a = 4$. Thus either $\phi(x) = 0$ for all $x \in \mathbb{Z}_6$ or $\phi(x) = 4x$ for all $x \in \mathbb{Z}_6$, and these are the only ring homomorphisms from $\mathbb{Z}_6$ to $\mathbb{Z}_{12}$. ◇

In a ring homomorphism we define the kernel as the kernel of the group homomorphism under addition.

7.1.7 DEFINITION Let $\phi: R \to R'$ be a ring homomorphism. Then the **kernel** of ϕ is the set Kern $\phi = \{x \in R \mid \phi(x) = 0 \in R'\}$. ○

In the examples we have seen so far we have repeatedly used familiar properties of group homomorphisms. Since rings are groups under addition, the following properties of ring homomorphisms follow from Propositions 2.2.15 and 2.2.16.

7.1.8 PROPOSITION Let $\phi: R \to R'$ be a ring homomorphism. Then

(1) $\phi(0) = 0$

(2) $\phi(-a) = -\phi(a)$ for all $a \in R$

(3) $\phi(na) = n\phi(a)$ for all $a \in R$ and $n \in \mathbb{Z}$

(4) ϕ is one to one if and only if Kern $\phi = \{0\}$. □

Let us now show some properties of ring homomorphisms that involve both conditions (1) and (2) in the definition of ring homomorphism.

7.1.9 PROPOSITION Let $\phi: R \to R'$ be a ring homomorphism. Then

(1) $\phi(a^n) = \phi(a)^n$ for all $a \in R$ and all $n > 0$, $n \in \mathbb{Z}$.

(2) If A is a subring of R, then $\phi(A) = \{\phi(x) \in R' \mid x \in A\}$ is a subring of R'.

(3) If R is a ring with unity 1, then $\phi(1)$ is unity in $\phi(R)$, and $\phi(1)^2 = \phi(1)$.

(4) If R is a ring with unity 1 and $a \in U(R)$ is a unit in R, then $\phi(a^n) = \phi(a)^n$ in $\phi(R)$ for all $n \in \mathbb{Z}$.

(5) If B is a subring of R', then $\phi^{-1}(B) = \{x \in R \mid \phi(x) \in B\}$ is a subring of R, and Kern $\phi \subseteq \phi^{-1}(B)$.

(6) Kern ϕ is a subring of R.

(7) If R is a commutative ring, then $\phi(R)$ is a commutative ring.

Proof (1) For $n = 1$ we have $\phi(a^1) = \phi(a) = \phi(a)^1$. If $k > 0$ and $\phi(a^k) = \phi(a)^k$, then $\phi(a^{k+1}) = \phi(a^k \cdot a) = \phi(a^k)\phi(a) = \phi(a)^k \cdot \phi(a) = \phi(a)^{k+1}$, and (1) is proved by induction.

(2) If A is a subring of R, then by Proposition 2.2.15, part (5), $\phi(A)$ is a subgroup of R' under addition. For any $x, y \in \phi(A)$ we have $x = \phi(a)$ and $y = \phi(b)$ for some $a, b \in A$. Note that since A is a subring of R we have $ab \in A$. Then $xy = \phi(a)\phi(b) = \phi(ab) \in \phi(A)$, and therefore by the subring test $\phi(A)$ is subring of R'.

(3) Let $x \in \phi(R)$. Then $x = \phi(a)$ for some $a \in R$. Hence $x \cdot \phi(1) = \phi(a) \cdot \phi(1) = \phi(a \cdot 1) = \phi(a) = x$. Similarly, we can show $\phi(1) \cdot x = x$, and therefore $\phi(1)$ is unity in $\phi(R)$, and $\phi(1) \cdot \phi(1) = \phi(1)$.

(4) Let $a \in U(R)$. For $n > 0$, (4) follows from (1). For $n = 0$, $\phi(a^0) = \phi(1)$ is unity in $\phi(R)$ by (3), and hence $\phi(1) = \phi(a)^0$, and (4) follows. For $n = -1$ we have $\phi(a) \cdot \phi(a^{-1}) = \phi(a \cdot a^{-1}) = \phi(1)$, which is unity in $\phi(R)$ by (3), and similarly we can show $\phi(a^{-1}) \cdot \phi(a) = \phi(1)$, and hence $\phi(a^{-1}) = \phi(a)^{-1}$, the inverse of $\phi(a)$ in $\phi(R)$. For

any $n < 0$, let $n = -s$, where $s > 0$. Then using (1) and the case $n = -1$ of (4) we have $\phi(a^n) = \phi(a^{-s}) = \phi((a^{-1})^s) = (\phi(a^{-1}))^s = (\phi(a)^{-1}))^s = \phi(a)^{-s} = \phi(a)^n$.

(5) If B is a subring of R', then by Proposition 2.2.15, part (6), $\phi^{-1}(B)$ is a subgroup of R under addition. For any $a, b \in \phi^{-1}(B)$ we have $\phi(ab) = \phi(a)\cdot\phi(b) \in B$, since B is a subring of R'. Therefore, by the subring test $\phi^{-1}(B)$ is a subring of R. Furthermore, since $0 \in B$, Kern $\phi = \phi^{-1}(0) \subseteq \phi^{-1}(B)$.

(6) Let $B = \{0\} \subseteq R'$ be the trivial subring. Then by (5), Kern $\phi = \phi^{-1}(0)$ is a subring of R.

(7) Given $x, y \in \phi(R)$ we have $x = \phi(a)$ and $y = \phi(b)$ for some $a, b \in R$. If R is commutative, then $xy = \phi(a)\cdot\phi(b) = \phi(ab) = \phi(ba) = \phi(b)\cdot\phi(a) = yx$, and $\phi(R)$ is commutative. □

7.1.10 DEFINITION A ring homomorphism $\phi: R \to R'$ that is one to one and onto is called a **ring isomorphism**. Two rings R and R' are said to be **isomorphic**, written $R \cong R'$, if there exists a ring isomorphism $\phi: R \to R'$. ○

To show that two rings are isomorphic, we follow the four steps we described in Section 2.2 for groups. We leave it as an exercise to show that Proposition 2.2.22 holds also for ring isomorphisms. (See Exercise 12.) Proposition 2.2.23 gave us important tools to determine whether two groups are isomorphic. The next proposition gives something comparable for rings.

7.1.11 PROPOSITION Let $R \cong R'$ be isomorphic rings. Then

(1) R is a commutative ring with unity if and only if R' is a commutative ring with unity.

(2) R is an integral domain if and only if R' is an integral domain.

(3) R is a field if and only if R' is a field.

Proof See Exercise 13. □

7.1.12 EXAMPLE For $a, b \in \mathbb{R}$, let $A(a, b) \in M(2, \mathbb{R})$ be defined by

$$A(a,b) = \begin{bmatrix} a & b \\ -b & a \end{bmatrix}$$

Let $R = \{A(a, b) \mid a, b \in \mathbb{R}\} \subseteq M(2, \mathbb{R})$. We show that $R \cong \mathbb{C}$.

(1) Let $\phi: R \to \mathbb{C}$ be defined by $\phi(A(a, b)) = a + bi \in \mathbb{C}$.

(2) We show ϕ is a ring homomorphism. For addition we have

$$\phi(A(a, b + A(c, d)) =$$

$$\phi\left(\begin{bmatrix} a & b \\ -b & a \end{bmatrix} + \begin{bmatrix} c & d \\ -d & c \end{bmatrix}\right) = \phi\left(\begin{bmatrix} a+c & b+d \\ -(b+d) & a+c \end{bmatrix}\right) =$$

$$\phi(A(a + c, b + d)) = (a + c) + (b + d)i =$$
$$(a + bi) + (c + di) = \phi(A(a,b)) + \phi(A(c,d))$$

For multiplication we have

$$\phi(A(a, b)\cdot A(c, d)) =$$

$$\phi(\begin{bmatrix} a & b \\ -b & a \end{bmatrix} \cdot \begin{bmatrix} c & d \\ -d & c \end{bmatrix}) = \phi(\begin{bmatrix} ac - bd & ad + bc \\ -(ad + bc) & ac - bd \end{bmatrix}) =$$

$$\phi(A(ac - bd, ad + bc)) = (ac - bd) + (ad + bc)i =$$

$$(a + bi)\cdot(c + di) = \phi(A(a, b))\cdot\phi(A(c, d))$$

(3) ϕ is one to one since $\phi(A(a, b)) = a + bi = 0$ if and only if $a = b = 0$, and Kern $\phi = \{A(0, 0)\}$ is trivial.

(4) ϕ is obviously onto. ◇

We close this section with an interesting application of ring homomorphisms.

7.1.13 EXAMPLE We show that the equation $2x^3 - 5x^2 + 7x - 8 = 0$ has no solutions in $\mathbb{Z}$. Let $\phi: \mathbb{Z} \to \mathbb{Z}_3$ be the natural homomorphism $\phi(x) = x \bmod 3$. Suppose there exists an integer $a \in \mathbb{Z}$ such that $2a^3 - 5a^2 + 7a - 8 = 0$. Then

$$0 = \phi(0) = \phi(2a^3 - 5a^2 + 7a - 8) = 2\phi(a)^3 - 5\phi(a)^2 + 7\phi(a) - 8$$

using Propositions 7.1.8, part (3), and 7.1.9, part (1). Since $-5 \equiv 7 \equiv -8 \equiv 1 \bmod 3$, in $\mathbb{Z}_3$ we have

$$2\phi(a)^3 - 5\phi(a)^2 + 7\phi(a) - 8 = 2\phi(a)^3 + \phi(a)^2 + \phi(a) + 1$$

and so $2b^3 + b^2 + b + 1 = 0$, where $b = \phi(a) \in \mathbb{Z}_3$. But we can easily check to see that no element $b = 0$ or 1 or 2 in $\mathbb{Z}_3$ is a solution to this equation. Therefore, no integer $a \in \mathbb{Z}$ can be a solution to the original equation. ◇

Exercises 7.1

In Exercises 1 through 11 find all possible ring homomorphisms between the indicated rings.

1. $\phi: \mathbb{Z} \to \mathbb{Z}_3$ **2.** $\phi: 3\mathbb{Z} \to \mathbb{Z}$ **3.** $\phi: \mathbb{Z}_4 \to \mathbb{Z}_6$

4. $\phi: \mathbb{Z}_6 \to \mathbb{Z}_{10}$ **5.** $\phi: \mathbb{Z}_{12} \to \mathbb{Z}_6$ **6.** $\phi: \mathbb{Q} \to \mathbb{Q}$

7. $\phi: \mathbb{Q}(\sqrt{2}) \to \mathbb{Q}(\sqrt{2})$ **8.** $\phi: \mathbb{Q}(\sqrt{2}) \to \mathbb{Q}(\sqrt{3})$ **9.** $\phi: \mathbb{Z}[i] \to \mathbb{C}$

10. $\phi: \mathbb{Z}[i] \to \mathbb{H}$, the ring of quaternions **11.** $\phi: \mathbb{Z} \times \mathbb{Z} \to \mathbb{Z} \times \mathbb{Z}$

12. Prove Proposition 2.2.22 for ring isomorphisms. **13.** Prove Proposition 7.1.11.

14. Show that $\mathbb{Z}_n \times \mathbb{Z}_m$ and $\mathbb{Z}_{nm}$ are isomorphic rings if and only if n and m are relatively prime.

15. For $a, b \in \mathbb{Z}$, let $B(a, b) \in M(2, \mathbb{Z})$ be defined by $B(a, b) = \begin{bmatrix} a & 3b \\ b & a \end{bmatrix}$. Let $S = \{B(a, b) \mid a, b \in \mathbb{Z}\} \subseteq M(2, \mathbb{Z})$. Show that $S \cong \mathbb{Z}[\sqrt{3}] = \{a + b\sqrt{3} \mid a, b \in \mathbb{Z}\}$.

16. Show that $\mathbb{R}$ and $\mathbb{C}$ are not isomorphic rings.

17. Show that if $R_1 \cong R_2$ then char R_1 = char R_2.

18. Show that $\mathbb{Q}(\sqrt{3})$ and $\mathbb{Q}(\sqrt{5})$ are not isomorphic rings.

19. Let R be a commutative ring with char $R = p$, where p is prime. Show that $\phi: R \to R$ defined by $\phi(x) = x^p$ for all $x \in R$ is a ring homomorphism (ϕ is called the **Frobenius map**).

20. Let R_1 and R_2 be rings.

(a) Let ϕ: $R_1 \times R_2 \to R_1$ be defined by $\phi(a, b) = a$. Show that ϕ is a ring homomorphism.

(b) Show that $R_1 \times R_2 \cong R_2 \times R_1$.

21. Show that the isomorphism relation $\cong$ is an equivalence relation on the class of all rings.

22. Show that $x^3 + 10x^2 + 6x + 1 = 0$ has no solutions in $\mathbb{Z}$.

23. Determine whether $x^3 + 6x^2 + 22x + 1 = 0$ has a solution in $\mathbb{Z}$.

7.2 Ideals

In the case of a group homomorphism ϕ: $G \to G'$, the kernel Kern ϕ is a subgroup of G with the property that every left coset is a right coset. We called any such subgroup a normal subgroup. We then studied normal subgroups and showed that they give rise to quotient groups. Similarly, for the case of a ring homomorphism ϕ: $R \to R'$, Kern ϕ is a subring of R, as we have shown. Actually, Kern ϕ is a subring with an extra property that will give rise to the notion of a quotient ring.

7.2.1 EXAMPLE Let K = Kern ϕ, where ϕ is the homomorphism ϕ: $\mathbb{Z} \to \mathbb{Z}_2$ defined by $\phi(x) = x \bmod 2$. In this case $K = 2\mathbb{Z} \subseteq \mathbb{Z}$. Note now that for any $n \in \mathbb{Z}$ and any $k = 2x \in K$ we have $\phi(nk) = \phi(n2x) = \phi(n)\phi(2x) = 0$, and, similarly, $\phi(kn) = \phi(2xn) = \phi(2x)\phi(n) = 0$. In other words, for any $n \in \mathbb{Z}$ and $k \in K$ we have $nk \in K$ and $kn \in K$. We show next that this property holds for the kernel of any ring homomorphism. ◇

7.2.2 PROPOSITION Let ϕ: $R \to R'$ be a ring homomorphism, and let K = Kern ϕ. Then

(1) K is a subring of R.

(2) For all $r \in R$ and for all $k \in K$ we have $rk \in K$ and $kr \in K$.

Proof (1) has already been shown in Proposition 7.1.9, part (6).

(2) Let $r \in R$ and $k \in K$. Then $\phi(rk) = \phi(r)\phi(k) = \phi(r) \cdot 0 = 0$, and, similarly, $\phi(kr) = \phi(k)\phi(r) = 0 \cdot \phi(r) = 0$. Hence both rk and kr are in the kernel: $rk \in K$ and $kr \in K$. □

7.2.3 DEFINITION Let R be a ring and I a subset of R. Then I is called an **ideal** in R if

(1) I is a subring of R and
(2) For all $r \in R$ and for all $x \in I$ we have $rx \in I$ and $xr \in I$. ○

7.2.4 COROLLARY Let $\phi: R \to R'$ be a ring homomorphism with $K = \text{Kern}\ \phi$. Then K is an ideal of R.

Proof Immediate from Proposition 7.2.2 and the preceding definition. □

7.2.5 EXAMPLE Let us find all the ideals in $\mathbb{Z}$. If I is an ideal in $\mathbb{Z}$, then I is a subring of $\mathbb{Z}$ by (1) of the definition. Hence $I = n\mathbb{Z}$ for some $n \geq 1$. So let $I = n\mathbb{Z}$ and let $a \in \mathbb{Z}$, $b \in I$. Then $b = nk$ for some $k \in \mathbb{Z}$ and $ab = a(nk) = n(ak) \in n\mathbb{Z} = I$, so I is an ideal. Thus all the ideals in Z are precisely $n\mathbb{Z}$ for any $n \geq 1$. ◇

The next example is a warning *not* to assume that all the subrings of a given ring will be ideals.

7.2.6 EXAMPLE $\mathbb{Z}$ is a subring of $\mathbb{Q}$ but is *not* an ideal in $\mathbb{Q}$ since, for instance, $1/2 \in \mathbb{Q}$, $5 \in \mathbb{Z}$, but $(1/2) \cdot 5 \notin \mathbb{Z}$. ◇

7.2.7 EXAMPLE In any ring R, $\{0\}$ is an ideal in R, called the **trivial** ideal, and R itself is an ideal in R, called the **improper** ideal. ◇

7.2.8 EXAMPLE Let R be a commutative ring and let $a \in R$. Then we define the **principal ideal generated by** a, denoted by $\langle a \rangle$ as follows:

$$\langle a \rangle = \{ra \mid r \in R\}$$

In other words, $\langle a \rangle$ consists of all multiples of a by any element r of R. Note that $\langle a \rangle$ is an ideal in R because

(1) Given any x and y in $\langle a \rangle$ we have $x = ra$ and $y = sa$ for some $r, s \in R$, and

$$x - y = ra - sa = (r - s)a \in \langle a \rangle \qquad xy = (ra)(sa) = (ras)a \in \langle a \rangle$$

and thus $\langle a \rangle$ is a subring of R.

(2) For any $r \in R$ and $x = sa \in \langle a \rangle$ we have

$$rx = r(sa) = (rs)a \in \langle a \rangle \qquad xr = (sa)r = (rs)a \in \langle a \rangle$$

so both conditions in the definition of ideal are satisfied. ◇

7.2.9 EXAMPLE Every ideal in $\mathbb{Z}$ is a principal ideal. For in Example 7.2.5 we showed that any ideal I in $\mathbb{Z}$ is of the form $I = n\mathbb{Z}$ for some $n \geq 1$. Thus $I = \langle n \rangle$, the principal ideal generated by n. ◇

7.2.10 EXAMPLE Let us find all the ideals in $\mathbb{Q}$. Let $I \neq \{0\}$ be a nontrivial ideal in $\mathbb{Q}$. So there exists an element $a \in I$, $a \neq 0$. Since $a \in \mathbb{Q}^*$, and $\mathbb{Q}$ is a field, a is a unit. In other words, there exists a multiplicative inverse $a^{-1} \in \mathbb{Q}$ with $a^{-1}a = 1$. But since $a^{-1} \in \mathbb{Q}$, $a \in I$, and I is an ideal, we have

$$1 = a^{-1}a \in I$$

Now let us use again the fact that I is an ideal, and let us take any rational number $b \in \mathbb{Q}$. Then since $1 \in I$ we have

$$b = b \cdot 1 \in I$$

and this implies $I = \mathbb{Q}$. Therefore, $\{0\}$ and $\mathbb{Q}$ are the only ideals in $\mathbb{Q}$. ◇

This is a striking conclusion. If we look carefully at the proof, we notice that the key ingredient was the fact that $a \neq 0$ implied that a is a unit or, in other words, the fact that $\mathbb{Q}$ is a field. So the next proposition should not come as a surprise.

7.2.11 PROPOSITION Let R be a commutative ring with unity. Then R is a field if and only if $\{0\}$ and R are the only ideals in R.

Proof ($\Rightarrow$) If R is a field and $I \neq \{0\}$ is an ideal in R, then there exists $a \in I$, $a \neq 0$. Since R is a field, every nonzero element is a unit, and therefore $1 = a^{-1}a \in I$ since I is an ideal. Hence for all $r \in R$ we have $r = r \cdot 1 \in I$, and $I = R$.
($\Leftarrow$) R is a commutative ring with unity, so to show that R is a field we only need to show that all the nonzero elements in R are units. So let $0 \neq a \in R$ and consider $I = \langle a \rangle$, the principal ideal generated by a. $I \neq \{0\}$ since $0 \neq a \in I$. If $\{0\}$ and R are the only ideals in R, then we must have $I = R$. R is a ring with unity, hence $1 \in I = \langle a \rangle$. In other words, there exists $r \in R$ such that $1 = ra$. Hence $r = a^{-1}$ and a is a unit as desired. □

Thus, for instance, the only ideals in $\mathbb{Q}$, $\mathbb{R}$, $\mathbb{C}$, and $\mathbb{Z}_p$, where p is a prime, are the trivial and the improper ideals.

7.2.12 EXAMPLE Let us consider the ideal $I = 5\mathbb{Z}$ in $R = \mathbb{Z}$. $\mathbb{Z}$ is an Abelian group under addition and $5\mathbb{Z}$ is a normal subroup. Hence by Theorem 2.4.3,

$$\mathbb{Z}/5\mathbb{Z} = \{5\mathbb{Z}, 1 + 5\mathbb{Z}, 2 + 5\mathbb{Z}, 3 + 5\mathbb{Z}, 4 + 5\mathbb{Z}\}$$

is a group under the commutative operation of addition of cosets:

$$(a + 5\mathbb{Z}) + (b + 5\mathbb{Z}) = (a + b) + 5\mathbb{Z}$$

Now let us try to define a multiplication operation on $\mathbb{Z}/5\mathbb{Z}$. The natural way would be

$$(a + 5\mathbb{Z}) \cdot (b + 5\mathbb{Z}) = a \cdot b + 5\mathbb{Z}$$

Let us see whether this is well defined. If $x + 5\mathbb{Z} = a + 5\mathbb{Z}$ and $y + 5\mathbb{Z} = b + 5\mathbb{Z}$, then $x \in a + 5\mathbb{Z}$ and $y \in b + 5\mathbb{Z}$, and so $x = a + 5k$ and $y = b + 5j$ for some $k, j \in \mathbb{Z}$. Thus

$$xy = (a + 5k)(b + 5j) = ab + b(5k) + a(5j) + (5k)(5j)$$

Since $b(5k)$ and $a(5j)$ and $(5k)(5j)$ are elements of $5\mathbb{Z}$, we have $xy \in ab + 5\mathbb{Z}$, $xy + 5\mathbb{Z} = ab + 5\mathbb{Z}$, and the operation is well defined. Note where we used the fact that $5\mathbb{Z}$ is an ideal in $\mathbb{Z}$: We used this fact to conclude that $b(5k)$ and $a(5j)$ are elements of $5\mathbb{Z}$. ◇

A ring R is a group under addition and an ideal I in R is a subring of R, therefore a subgroup of R under addition. Since R is an Abelian group under addition, I is a normal subgroup of R under addition. Thus R/I is an Abelian group under the addition of cosets, by Theorems 2.4.3 and 2.4.9, part (3). In the case of groups we showed in Lemma 2.4.2 that a subgroup being a normal subgroup is equivalent to the group operation on cosets being well defined. We next show that a subring being an ideal is equivalent to the mutliplication operation on cosets being well defined. It is instructive to compare Lemma 2.4.2 with Lemma 7.2.13, and Theorem 2.4.3 with Theorem 7.2.14.

7.2.13 LEMMA Let I be a subring of a ring R. Then I is an ideal in R if and only if multiplication $(a + I)(b + I) = (ab + I)$ is a well-defined operation on the cosets of I in R.

Proof ($\Rightarrow$) Assume I is an ideal in R and suppose $a_1 + I = a_2 + I$ and $b_1 + I = b_2 + I$. This implies that $a_1 = a_2 + k$ and $b_1 = b_2 + j$ for some $k, j \in I$. Then

$$a_1b_1 = a_2b_2 + a_2j + kb_2 + kj$$

Since I is a subring of R, and therefore closed under multiplication as well as addition, $kj \in I$. Since I is an ideal, $a_2j \in I$ and $kb_2 \in I$, and so $a_2j + kb_2 + jk \in I$. Therefore, $a_1b_1 \in a_2b_2 + I$ and $a_1b_1 + I = a_2b_2 + I$ by Lemma 2.1.7. Thus multiplication on the set of cosets of I is well defined.

($\Leftarrow$) Assume that the indicated operation is well defined. We need to show that for all $r \in R$ and $x \in I$ we have $rx \in I$ and $xr \in I$. So let $r \in R$ and $x \in I$. Since $x \in I$ we have $x + I = 0 + I = I$ by Lemma 2.1.7. Hence

$$rx + I = (r + I)(x + I) = (r + I)(0 + I) = 0 + I = I$$

Again by Lemma 2.1.7 we obtain $rx \in I$. By a similar argument we obtain $xr \in I$. Thus I is an ideal in R. □

7.2.14 THEOREM Let I be an ideal in a ring R. Then R/I is a ring under the operations defined by

(1) $(a + I) + (b + I) = (a + b) + I$ (2) $(a + I)\cdot(b + I) = ab + I$

for all $a + I, b + I \in R/I$.

Proof (1) By Theorems 2.4.3 and 2.4.9 we already know that R/I is an Abelian group under addition.

(2) By Lemma 7.2.13 multiplication is well defined, and by definition it is closed.

(3) For any $a + I$, $b + I$, $c + I \in R/I$, using the fact that multiplication is associative in R we have

$$[(a + I)(b + I)](c + I) = (ab + I)(c + I) = (ab)c + I =$$
$$a(bc) + I = (a + I)(bc + I) = (a + I)[(b + I)(c + I)]$$

Hence multiplication is associative in R/I

(4) We leave it as an exercise for you to show that the distributive laws hold in R/I. (Exercise 14.) □

7.2.15 DEFINITION Let I be an ideal in a ring R. Then R/I with the operations on cosets specified in the preceding theorem is called the **quotient ring** of R by I. ○
An immediate consequence of the definition of multiplication in the quotient ring R/I is the next proposition, whose proof we leave as an exercise.

7.2.16 PROPOSITION Let I be an ideal in a commutative ring R with unity 1. Then R/I is a commutaive ring with unity $1 + I$.

Proof See Exercise 13. □

We can easily show that Theorems 2.4.15 and 2.4.19 can be restated for ideals in rings.

7.2.17 THEOREM (First isomorphism theorem for rings) Let $\phi: R \to R'$ be a ring homomorphism with kernel $K = \text{Kern } \phi$. Then

$$R/K \cong \phi(R)$$

Proof Since $\phi: R \to R'$ is a group homomorphism under addition, Theorem 2.4.15 tells us that the map $\chi: R/K \to \phi(R)$ defined by $\chi(a + K) = \phi(a)$ is a group isomorphism. Now to show that it is actually a ring isomorphism, we only need to check that it is a ring homomorphism or, in other words, that it respects multiplication. Let $a + K$ and $b + K$ be elements of R/K. Then we have

$$\chi[(a + K)(b + K)] = \chi(ab + K) = \phi(ab) = \phi(a)\phi(b) = \chi(a + K)\chi(b + K)$$

since ϕ is a ring homomorphism, and the proof is complete. □

7.2.18 THEOREM Given a ring R and an ideal K in R, there exists an onto ring homomorphism $\phi: R \to R/K$ with $\text{Kern } \phi = K$.

Proof By Theorem 2.4.19 the map $\phi: R \to R/K$ defined by $\phi(a) = a + K$ is a group homomorphism under addition, with $\text{Kern } \phi = K$. We need to check that it respects multiplication. So let $a, b \in R$. Then

$$\phi(ab) = ab + K = (a + K)(b + K) = \phi(a)\phi(b)$$

and ϕ is a ring homomorphism. Finally, ϕ is onto by Theorem 7.2.17. □

Now that we are familiar with the concept of an ideal, we can show, using Proposition 7.1.9, parts (2) and (5), that ring homomorphisms "preserve" ideals. Compare the next proposition with Proposition 2.2.15, parts (5) and (6).

7.2.19 PROPOSITION Let $\phi: R \to R'$ be a ring homomorphism. Then

(1) If I is an ideal in R, then $\phi(I)$ is an ideal in $\phi(R)$.

(2) If J is an ideal in R', then $\phi^{-1}(J)$ is an ideal in R with $\text{Kern } \phi \subseteq \phi^{-1}(J)$.

Proof See Exercise 15. □

If we apply Proposition 7.2.19 to the homomorphism $\phi: R \to R/K$ as defined in Theorem 7.2.18, we can establish the correspondence between ideals in R/K and ideals in R that contain K.

7.2.20 PROPOSITION Let K be an ideal in a ring R. Then

(1) Given an ideal I in R with $K \subseteq I$, then $I^* = I/K = \{a + K \mid a \in I\}$ is an ideal in R/K.

(2) Given an ideal J^* in R/K, then there exists an ideal J in R with $K \subseteq J$ such that $J^* = J/K = \{a + K \mid a \in J\}$.

(3) Given ideals I and J in R, both containing K, then $I \subseteq J$ if and only if $I/K \subseteq J/K$.

Proof (1) and (2) are immediate from Proposition 7.2.19 and Theorem 7.2.18.

(3) If $I \subseteq J$, then $I/K = \{a + K \mid a \in I\} \subseteq \{a + K \mid a \in J\} = J/K$.

If $I/K \subseteq J/K$, then for any element $a \in I$ we have $a + K \in I/K$ and hence $a + K \in J/K$. Therefore, $a + K = b + K$ for some $b \in J$. Since $a \in a + K$ we have $a \in b + K$, thus $a = b + k$ for some $k \in K$. Since $K \subseteq J$ we have $k \in J$ and $b + k \in J$. Hence $a \in J$, showing $I \subseteq J$. □

We now study two special types of ideals and the corresponding quotient rings.

7.2.21 EXAMPLE In $\mathbb{Z}$ consider the two ideals $5\mathbb{Z}$ and $6\mathbb{Z}$. By Euclid's lemma (Corollary 0.3.23), if b and c are integers such that $bc \in 5\mathbb{Z}$, then either $5 \mid b$, in which case $b \in 5\mathbb{Z}$, or else $5 \mid c$, in which case $c \in 5\mathbb{Z}$. On the other hand, we can have integers x and y such that $x \notin 6\mathbb{Z}$ and $y \notin 6\mathbb{Z}$, but $xy \in 6\mathbb{Z}$, for example, $x = 4$ and $y = 9$.

Notice now that using the ring homomorphism $\phi: \mathbb{Z} \to \mathbb{Z}_n$ defined by $\phi(x) = x \bmod n$, and using Theorem 7.2.17, we have $\mathbb{Z}/n\mathbb{Z} \cong \mathbb{Z}_n$. Hence $\mathbb{Z}/5\mathbb{Z}$ is an integral domain, while $\mathbb{Z}/6\mathbb{Z}$ is a ring with zero divisors.

Let us also point out one more difference between the two ideals $5\mathbb{Z}$ and $6\mathbb{Z}$. First note that $6\mathbb{Z} \subset 3\mathbb{Z} \subset \mathbb{Z}$, where $6\mathbb{Z} \neq 3\mathbb{Z}$ and $3\mathbb{Z} \neq \mathbb{Z}$. Now suppose that J is an ideal with $5\mathbb{Z} \subseteq J \subseteq \mathbb{Z}$. By Example 7.2.5 there exists an integer $n \in \mathbb{Z}$ such that $J = n\mathbb{Z}$. Since $5\mathbb{Z} \subseteq J = n\mathbb{Z}$ we have $5 \in n\mathbb{Z}$ and $5 = nk$ for some $k \in \mathbb{Z}$. But then either $n = 5$ and $k = 1$, in which case $5\mathbb{Z} = J$, or else $n = 1$ and $k = n$, in which case $J = \mathbb{Z}$. ◇

7.2.22 DEFINITION A nontrivial proper ideal $I \neq R$ in a commutative ring R is called a **prime** ideal if $ab \in I$ implies $a \in I$ or $b \in I$ for all a and b in R. ○

7.2.23 DEFINITION A nontrivial proper ideal $I \neq R$ in a ring R is called a **maximal** ideal if the only ideals J in R such that $I \subseteq J \subseteq R$ are $J = I$ and $J = R$. ○

Therefore, in Example 7.2.21, $5\mathbb{Z}$ is both a prime ideal and a maximal ideal, while $6\mathbb{Z}$ is neither.

7.2.24 EXAMPLE In $\mathbb{Z}$ an ideal I is a prime ideal if and only if $I = p\mathbb{Z}$, where p is a prime. For, on the one hand, if p is prime and $ab \in p\mathbb{Z}$, then by Euclid's lemma (Corollary 0.3.23), either $a \in p\mathbb{Z}$ or $b \in p\mathbb{Z}$ and $p\mathbb{Z}$ is a prime ideal. But on the other hand, if $I = n\mathbb{Z}$ where $n > 1$ is not a prime, then $n = uv$ for some positive integers $u, v < n$. Hence $uv \in n\mathbb{Z}$ while $u \notin n\mathbb{Z}$ and $v \notin n\mathbb{Z}$, and $n\mathbb{Z}$ is not a prime ideal. ◇

7.2.25 EXAMPLE In $\mathbb{Z}$ an ideal I is a maximal ideal if and only if I is a prime ideal. For, on the one hand, if $I = p\mathbb{Z}$ is a prime ideal and $p\mathbb{Z} = I \subseteq r\mathbb{Z} \subseteq \mathbb{Z}$, then $p = rk$ for some $k \in \mathbb{Z}$. But then either $r = p$ and $k = 1$, in which case $r\mathbb{Z} = p\mathbb{Z} = I$, or else $r = 1$ and $k = p$, in which case $r\mathbb{Z} = 1\mathbb{Z} = \mathbb{Z}$. This shows that $I = p\mathbb{Z}$ is a maximal ideal. On the other hand, if $I = n\mathbb{Z}$, $n > 1$, is a not a prime ideal, then n is not a prime and $n = uv$ for some u, v with $1 < u < n$ and $1 < v < n$. But then $I = n\mathbb{Z} \subseteq u\mathbb{Z} \subseteq \mathbb{Z}$, where $n\mathbb{Z} \neq u\mathbb{Z}$ since $u < n$, and $u\mathbb{Z} \neq \mathbb{Z}$ since $1 < u$, and $I = n\mathbb{Z}$ is not a maximal ideal. ◇

7.2.26 EXAMPLE With this example we point out that prime ideals and maximal ideals do not always coincide. In the ring $R = \mathbb{Z} \times \mathbb{Z}$ consider $I = \mathbb{Z} \times \{0\} = \{(a, 0) \mid a \in \mathbb{Z}\}$. I is an ideal in R. (See Exercise 11.) Let $x = (a_1, b_1)$ and $y = (a_2, b_2)$ be two elements of R such that $xy \in I$. Then $(a_1a_2, b_1b_2) = (a_1, b_1)(a_2, b_2) = xy \in I$ implies $b_1b_2 = 0$. Since $\mathbb{Z}$ has no zero divisors, either $b_1 = 0$ and $x \in I$ or $b_2 = 0$ and $y \in I$. Therefore, I is a prime ideal. But I is not a maximal ideal in R since we have

$$I = \mathbb{Z} \times \{0\} \subseteq \mathbb{Z} \times 2\mathbb{Z} \subseteq \mathbb{Z} \times \mathbb{Z} = R$$

where $\mathbb{Z} \times \{0\} \neq \mathbb{Z} \times 2\mathbb{Z}$ and $\mathbb{Z} \times 2\mathbb{Z} \neq \mathbb{Z} \times \mathbb{Z}$, and $\mathbb{Z} \times 2\mathbb{Z}$ is an ideal in R. (See Exercise 12.) ◇

The importance of the notions of prime ideal and maximal ideal becomes apparent with the next theorem.

7.2.27 THEOREM Let R be a commutative ring with unity, and let I be an ideal in R. Then

(1) I is a prime ideal in R if and only if R/I is an integral domain.
(2) I is a maximal ideal in R if and only if R/I is a field.

Proof Note that by Theorem 7.2.14 and Proposition 7.2.16, R/I is a commutative ring with unity.

(1) R/I will therefore be an integral domain if and only if it has no zero divisors. This condition is equivalent to the condition that $(a + I)(b + I) = I$ if and only if $a + I = I$ or $b + I = I$. Thus R/I is an integral domain if and only if $ab + I = I$ implies that $a + I = I$ or $b + I = I$ or, in other words, if and only if $ab \in I$ implies that $a \in I$ or $b \in I$, which is to say that I is a prime ideal in R.

(2) By Proposition 7.2.11, R/I will be a field if and only if its only ideals are $\{0\}$ and R/I. Suppose R/I is a field and consider any ideal J in R such that $I \subseteq J \subseteq R$. By Proposition 7.2.20, $\{0\} \subseteq J/I \subseteq R/I$ and J/I is an ideal in R/I. Thus either $J/I = \{0\}$, in which case $J = I$, or $J/I = R/I$, in which case $J = R$. Thus I is a maximal ideal in R. Conversely, suppose that I is a maximal ideal in R and consider any ideal

J^* in R/I. Again by Proposition 7.2.20, $J^* = J/I$ for some ideal J in R such that $I \subseteq J \subseteq R$. Since I is a maximal ideal, either $J = I$, in which case $J^* = \{0\}$, or $J = R$, in which case $J^* = R/I$. Thus R/I is a field. □

7.2.28 COROLLARY In a commutative ring R with unity, every maximal ideal is a prime ideal.

Proof If I is a maximal ideal in R, then R/I is a field. Every field is an integral domain, so R/I is an integral domain, and I is a prime ideal. □

7.2.29 EXAMPLE Let $\phi: \mathbb{Z} \times \mathbb{Z} \to \mathbb{Z}$ be the ring homomorphism defined by $\phi((a, b)) = b$. Then Kern $\phi = \mathbb{Z} \times \{0\}$, and by Theorem 7.2.17 we have $(\mathbb{Z} \times \mathbb{Z}) / (\mathbb{Z} \times \{0\}) \cong \mathbb{Z}$, which is an integral domain, but not a field. This agrees with our conclusion in Example 7.2.26 that $\mathbb{Z} \times \{0\}$ is a prime ideal but not a maximal ideal. ◇

7.2.30 EXAMPLE Let $\phi: \mathbb{Z} \times \mathbb{Z} \to \mathbb{Z}_3$ be the ring homomorphism defined by $\phi((a, b)) = b \bmod 3$. Then Kern ϕ = $\mathbb{Z} \times 3\mathbb{Z}$ and $(\mathbb{Z} \times \mathbb{Z}) / (\mathbb{Z} \times 3\mathbb{Z}) \cong \mathbb{Z}_3$, which is a field. Thus $\mathbb{Z} \times 3\mathbb{Z}$ is a maximal ideal in $\mathbb{Z} \times \mathbb{Z}$. ◇

Theorem 7.2.27 gives us a correspondence between prime ideals or maximal ideals, on the one hand, and quotient rings that are integral domains or fields, on the other hand. In the next chapter, this theorem is seen to be very important, as the basis for the construction of new examples of integral domains and fields.

Exercises 7.2

In Exercises 1 through 10 determine whether the indicated set I is an ideal in the indicated ring R.

1. $I = \mathbb{Q}$ in $R = \mathbb{R}$

2. $I = 2\mathbb{Z}$ in $R = \mathbb{Q}$

3. $I = 2\mathbb{Z} \times 2\mathbb{Z}$ in $R = \mathbb{Z} \times \mathbb{Z}$

4. $I = \{0, 2, 4, 6, 8\}$ in $\mathbb{Z}_{10}$

5. $I = \{(n, n) \mid n \in \mathbb{Z}\}$ in $R = \mathbb{Z} \times \mathbb{Z}$

6. $I = \{(2x, 2y) \mid x, y \in \mathbb{Z}\}$ in $R = \mathbb{Z} \times \mathbb{Z}$

7. $I = \{(x, -x) \mid x \in \mathbb{Z}\}$ in $R = \mathbb{Z} \times \mathbb{Z}$

8. $I = \{\begin{bmatrix} 0 & n \\ 0 & m \end{bmatrix} \mid n, m \in \mathbb{Z}\}$ in $R = \{\begin{bmatrix} a & b \\ 0 & c \end{bmatrix} \mid a, b, c \in \mathbb{Z}\}$.

9. $I = \{(2x, 3x) \mid x \in \mathbb{Z}_6\}$ in $R = \mathbb{Z}_6 \times \mathbb{Z}_6$

10. $I = \{n + ni \mid n \in \mathbb{Z}\}$ in $R = \mathbb{Z}[i]$.

11. Show that $\mathbb{Z} \times \{0\}$ is an ideal in $\mathbb{Z} \times \mathbb{Z}$.

12. Show that $\mathbb{Z} \times 2\mathbb{Z}$ is an ideal in $\mathbb{Z} \times \mathbb{Z}$.

13. Prove Proposition 7.2.16.

14. Prove Proposition 7.2.14, part (4).

15. Prove Proposition 7.2.19

16. Let I and J be ideals in a ring R. Show that $I \cap J$ is an ideal in R.

17. Find all the maximal ideals in $\mathbb{Z}_{12}$, and in each case describe the quotient ring.

18. Find all the maximal ideals in $\mathbb{Z}_6 \times \mathbb{Z}_{15}$, and in each case describe the quotient ring.

19. Show that I is a maximal ideal in $\mathbb{Z}_n$ if and only if $I = \langle p \rangle$, where p is a prime divisor of n.

20. For any two rings R_1 and R_2 with unity, show that

(a) If I_1 is and ideal in R_1 and I_2 is an ideal in R_2, then $I_1 \times I_2$ is an ideal in $R_1 \times R_2$.

(b) If M is an ideal in $R_1 \times R_2$, then there exist ideals M_1 in R_1 and M_2 in R_2 such that $M = M_1 \times M_2$.

(c) M is a maximal ideal in $R_1 \times R_2$ if and only if $M = M_1 \times R_2$, where M_1 is a maximal ideal in R_1, or $M = R_1 \times M_2$, where M_2 is a maximal ideal in R_2.

21. Let R be a finite commutative ring with unity. Show that every prime ideal in R is a maximal ideal in R.

22. Let R be a ring with unity 1 and I an ideal in R. Show that

(a) If $1 \in I$, then $I = R$.

(b) If I contains a unit, then $I = R$.

23. Let ϕ: $F \to R$ be a ring homomorphism where F is a field. Show that ϕ either is one to one or else is the 0-homomorphism.

24. Let I be an ideal in a ring R. Show that $M(2, I)$ is an ideal in $M(2, R)$.

25. Let I and J be ideals in a ring R.

(a) Show that $I + J = \{a + b \mid a \in I, b \in J\}$ is an ideal.

(b) Show that $IJ = \{a_1b_1 + a_2b_2 + \ldots + a_nb_n \mid a_i \in I, b_i \in J\}$ is an ideal.

(c) Show that $IJ \subseteq I \cap J$.

(d) If R is commutative and $I + J = R$, show that $IJ = I \cap J$.

26. Let $I = n\mathbb{Z}$ and $J = m\mathbb{Z}$ be two ideals in $\mathbb{Z}$. By the previous exercise, $I + J$ is an ideal in $\mathbb{Z}$ and therefore $I + J = k\mathbb{Z}$ for some k. Express k in terms of n and m.

27. Let D be an integral domain and a, $b \in D$. Show that $\langle a \rangle = \langle b \rangle$ if and only if $a = ub$ for some unit u in D.

28. Let ϕ: $R \to R'$ be a homomorphism of commutative rings. Show that

(a) If I is a prime ideal in R' and $\phi^{-1}(I) \neq R$, then $\phi^{-1}(I)$ is a prime ideal in R.

(b) If J is a maximal ideal in R' and $\phi^{-1}(J) \neq R$, then $\phi^{-1}(J)$ is a maximal ideal in R.

29. In the ring of Gaussian integers $\mathbb{Z}[i]$, describe the ideal $\langle i \rangle$.

30. In the ring of Gaussian integers $\mathbb{Z}[i]$, consider the ideal $J = \langle 1 + i \rangle$.
(a) Show that $2 \in J$.
(b) Find all the cosets of J in $\mathbb{Z}[i]$.
(c) Describe the quotient ring $\mathbb{Z}[i]/J$.

31. Determine whether the ideal $\langle 3 + i \rangle$ is a prime ideal in $\mathbb{Z}[i]$.

32. Let R be a commutative ring. The **annihilator** of R is defined as follows:

$$\text{Ann}(R) = \{a \in R \mid ax = 0 \text{ for all } x \in R\}$$

Show that $\text{Ann}(R)$ is an ideal in R.

33. Let R be a commutative ring and I an ideal in R. The **radical** of I is defined as follows:

$$\text{rad}(I) = \{a \in R \mid a^n \in I \text{ for some } n \in \mathbb{Z}\}$$

Show that $\text{rad}(I)$ is an ideal in R containing I.

34. (**The second isomorphism theorem for rings**) Let S be a subring of a ring R and let I be an ideal in R. Show that
(a) $S + I = \{a + b \mid a \in S, b \in I\}$ is a subring of R.
(b) $S \cap I$ is an ideal in S.
(c) $(S + I)/I \cong S/(S \cap I)$

35. (**The third isomorphism theorem for rings**) Let I and J be ideals in a ring R with $I \subseteq J$. Show that $(R/I)/(J/I) \cong R/J$.

7.3 The Field of Quotients

The ring of integers $\mathbb{Z}$ is our basic example of an integral domain which is not a field. Only 1 and -1 are units in $\mathbb{Z}$. We know, however, that multiplicative inverses of other nonzero integers exist outside $\mathbb{Z}$, namely in $\mathbb{Q}$, the field of rational numbers. In this section we study closely the relationship between $\mathbb{Z}$ and $\mathbb{Q}$. We observe that the construction of $\mathbb{Q}$ from $\mathbb{Z}$ is made possible by the fact that $\mathbb{Z}$ is an integral domain. This observation enables us to construct for any integral domain a field related to it in the same way $\mathbb{Q}$ is related to $\mathbb{Z}$.

7.3.1 EXAMPLE Let us describe the construction of the field of rational numbers $\mathbb{Q}$ from the ring of integers $\mathbb{Z}$. To begin, let us consider the set of all fractional expressions S with integer numerator and nonzero integer denominator:

$$S = \left\{\frac{a}{b} \mid a, b \in \mathbb{Z} \text{ and } b \neq 0\right\}$$

To obtain the set of all rational numbers from S we need to take into account the fact that a rational number can be represented by many different fractions. For instance,

$$\frac{1}{2} = \frac{2}{4} = \frac{3}{6} = \cdots$$

Note that the condition $a/b = c/d$ is equivalent to $ad = bc$.

(1) On the set S, define a relation $\sim$ by

$$\frac{a}{b} \sim \frac{c}{d} \text{ if and only if } ad = bc$$

Then $\sim$ is an equivalence relation. (See Exercise 17.) When we write the rational number 1/2, what we mean is the equivalence class [1/2], and similarly for other rational numbers. For instance,

$$[\tfrac{3}{5}] = \{ \frac{3k}{5k} \mid k \in \mathbb{Z}, k \neq 0\}$$

Hence the set of rational numbers can be described as the set of these equivalence classes

$$\mathbb{Q} = \{[\frac{a}{b}] \mid a, b \in \mathbb{Z}, b \neq 0\}$$

(2) The operations on $\mathbb{Q}$ are given by the following:

$$[\frac{a}{b}] + [\frac{c}{d}] = [\frac{ad + bc}{bd}] \qquad \text{and} \qquad [\frac{a}{b}] \cdot [\frac{c}{d}] = [\frac{ac}{bd}]$$

These operations are well defined. In other words, with either operation, the result is independent of the choice of representatives of the equivalence classes. (See Exercise 18.)

(3) The zero element in $\mathbb{Q}$ is $[0/b]$ and the additive inverse of $[a/b]$ is $[-a/b]$, as can be checked using the definition of addition in (2). The unity in $\mathbb{Q}$ is $[a/a]$ for $a \neq 0$, and for $a \neq 0$ the multiplicative inverse of $[a/b]$ is $[b/a]$. $\mathbb{Z}$ is isomorphic to the subring of $\mathbb{Q}$ consisting of the elements $[a/1]$ for $a \in \mathbb{Z}$. ◇

This description of $\mathbb{Q}$ serves us as a guide for the construction of such a field for any integral domain.

7.3.2 LEMMA Let D be an integral domain and let

$$S = \{(a, b) \mid a, b \in D, b \neq 0\}$$

Define a relation on S by

$$(a, b) \sim (c, d) \text{ if and only if } ad = bc$$

Then $\sim$ is an equivalence relation.

Proof (1) (Reflexivity) $(a, b) \sim (a, b)$ because $ab = ba$, since D is a commutative ring.

(2) (Symmetry) If $(a, b) \sim (c, d)$, then $ad = bc$, which implies $cb = da$, since D is a commutative ring, and thus $(c, d) \sim (a, b)$.

(3) (Transitivity) If $(a, b) \sim (c, d)$ and $(c, d) \sim (e, f)$, then $ad = bc$ and $cf = de$. Therefore, $adf = bcf = bde$, and since multiplication is commutative,

$afd = bed$. Since $(c, d) \in S$, $d \neq 0$. Since D has no zero divisors, the cancellation laws hold, and therefore we have $af = be$, and so $(a, b) \sim (e, f)$. □

Note how the facts that D is a commutative ring and that D has no zero divisors were used in the proof that $\sim$ is an equivalence relation. The equivalence relation $\sim$ partitions the set S into disjoint equivalence classes. (See Theorem 0.2.7.) To make our notation more intuitive, for any element $(a, b) \in S$ we denote its equivalence class $[(a,b)]$ by a/b. Thus we have:

$$\frac{a}{b} = \frac{c}{d} \text{ if and only if } ad = bc$$

We now consider the set

$$F = \left\{\frac{a}{b} \mid a, b \in D, b \neq 0\right\}$$

In order to make the set F into a field, we first need to define the two operations.

7.3.3 LEMMA For any elements a/b, $c/d \in F$, the following two operations are well defined:

(1) Addition:

$$\frac{a}{b} + \frac{c}{d} = \frac{ad + bc}{bd}$$

(2) Multiplication:

$$\frac{a}{b} \cdot \frac{c}{d} = \frac{ac}{bd}$$

Proof First note that $ad + bc$ and ac are both in D and that since D is an integral domain and $b \neq 0$, $d \neq 0$, we have $bd \neq 0$. Hence the right-hand side of each of the equations (1) and (2) is an element of F.

(1) Suppose $a/b = a'/b'$ and $c/d = c'/d'$. To show that addition is well defined we need to show that

$$\frac{ad + bc}{bd} = \frac{a'd' + b'c'}{b'd'}$$

or, in other words, that $b'd'(ad + bc) = bd(a'd' + b'c')$. Since $ab' = a'b$ and $cd' = c'd$ we obtain

$$b'd'(ad + bc) = (ab')d'd + (cd')b'b = (a'b)d'd + (c'd)b'b = bd(a'd' + b'c')$$

as desired. Note that we used the fact that D is a commutative ring repeatedly.

(2) Suppose $a/b = a'/b'$ and $c/d = c'/d'$. To show that mutliplication is well defined we need to show that

$$\frac{ac}{bd} = \frac{a'c'}{b'd'}$$

or, in other words, that $(b'd')(ac) = (bd)(a'c')$. Since $ab' = a'b$ and $cd' = c'd$ we obtain

$$(b'd')(ac) = (ab')(cd') = (a'b)(cd') = (a'b)(c'd) = (bd)(a'c')$$

as desired, again using the commutativity of multiplication in D. □

We have now constructed the set F and introduced two well-defined operations of addition and multiplication, under which F is closed. We next verify that with these operations F is a field.

7.3.4 LEMMA The set F as defined previously forms a field under the operations of addition and multiplication.

Proof There are nine field axioms to be checked.

(1) Addition in F is associative:

$$\frac{a}{b} + \left(\frac{c}{d} + \frac{e}{f}\right) = \frac{a}{b} + \frac{cf + ed}{df} = \frac{a(df) + (cf + ed)b}{b(df)} =$$
$$\frac{(ad + cb)f + e(db)}{(bd)f} = \frac{ad + bc}{bd} + \frac{e}{f} = \left(\frac{a}{b} + \frac{c}{d}\right) + \frac{e}{f}$$

(2) Addition in F is commutative:

$$\frac{a}{b} + \frac{c}{d} = \frac{ad + bc}{bd} = \frac{cb + da}{db} = \frac{c}{d} + \frac{a}{b}$$

(3) 0/1 is the additive identity element:

$$\frac{0}{1} + \frac{a}{b} = \frac{0 \cdot b + a \cdot 1}{1 \cdot b} = \frac{a}{b}$$

Note that $0/1 = 0/c$ for all $c \neq 0$.

(4) $-a/b$ is the additive inverse of a/b:

$$\frac{a}{b} + \frac{-a}{b} = \frac{ab - ba}{bb} = \frac{0}{bb} = \frac{0}{1}$$

Thus far we have shown that F is an Abelian group under addition and is closed under multiplication. We leave the remaining field axioms as exercises. (Exercises 19 and 20.).

(5) Multiplication in F is associative.

(6) The distributive laws hold in F.

(7) Multiplication in F is commutative.

(8) 1/1 is the multiplicative identity element. Note that $1/1 = c/c$ for all $c \neq 0$.

(9) If $a/b \neq 0$ in F, then $a \neq 0$ in D, and b/a is the multiplicative inverse of a/b. □

7.3.5 LEMMA $D' = \{a/1 \mid a \in D\}$ is a subring of F with $D' \cong D$.

Proof By the subring test, since we have $(a/1) - (b/1) = (a - b)/1 \in D'$ and since $(a/1)(b/1) = ab/1 \in D'$, it follows that D' is a subring of F. Let $\phi: D \to D'$ be defined by $\phi(a) = a/1$. Then

$$\phi(a + b) = \frac{a}{1} + \frac{b}{1} = \frac{a + b}{1} = \phi(a) + \phi(b)$$

$$\phi(ab) = \frac{a}{1} \cdot \frac{b}{1} = \frac{ab}{1} = \phi(a)\phi(b)$$

Hence ϕ is a ring homomorphism. ϕ is obviously onto. ϕ is also one to one, since $\phi(a) = \phi(b)$ means $a/1 = b/1$, hence $a \cdot 1 = 1 \cdot b$ and $a = b$. Thus ϕ is an isomorphism. □

We can now state the main theorem.

7.3.6 THEOREM (Existence of the field of quotients) Let D be an integral domain. Then there exists a field F consisting of quotients a/b, where $a, b \in D$ and $b \neq 0$. Moreover, we can identify every element $a \in D$ with the element $a/1 \in F$, and then D becomes a subring of F and each element of F is of the form $a/b = ab^{-1}$, where $a, b \in D$ and $b \neq 0$.

Proof The only thing we have not already shown is that for all $a/b \in F$ we have $a/b = ab^{-1}$. But this is clearly so, since

$$\frac{a}{b} = \frac{a}{1} \cdot \frac{1}{b} = \frac{a}{1} \cdot \left(\frac{b}{1}\right)^{-1}$$

So if we identify every element $x \in D$ with $x/1 \in F$, we obtain $a/b = ab^{-1}$. □

7.3.7 DEFINITION Any field F as in the preceding theorem is called a **field of quotients** of the integral domain D. ○

From the construction of the field of quotients F we can see that F consists of the elements of the original integral domain D, the multiplicative inverses of elements of D, and the product of elements of D with multiplicative inverses of elements of D. In other words, F is obtained by "adding" the absolute minimum needed to extend D to a field. The next theorem shows that F is indeed the smallest field that contains D and that any two fields of quotients of D must be isomorphic.

7.3.8 THEOREM Let F be a field of quotients of an integral domain D, and let E be any field containing D. Then there exists a ring homomorphism $\phi: F \to E$ such that ϕ is one to one and $\phi(a) = a$ for all $a \in D$.

Proof Define ϕ as follows. For any $a \in D$, let $\phi(a) = a$, and for any $ab^{-1} \in F$, let $\phi(ab^{-1}) = \phi(a)\phi(b)^{-1} \in E$. Note that since $b \neq 0$ in D, $\phi(b) = b \neq 0$ in E. The map ϕ is well defined, since if $ab^{-1} = cd^{-1}$ in F, then $ad = bc$ in D and since ϕ is the identity map on D, $\phi(a)\phi(d) = ad = bc = \phi(b)\phi(c)$ in E, so $\phi(a)\phi(b)^{-1} = \phi(c)\phi(d)^{-1}$ in E. It can be checked that ϕ is a ring homomorphism. (See Exercise 21.) Finally, ϕ is one to one since $\phi(ab^{-1}) = \phi(cd^{-1})$ implies $\phi(a)\phi(d) = \phi(b)\phi(c)$ in E, and $ad = bc$ in D, which implies $ab^{-1} = cd^{-1}$ in F. □

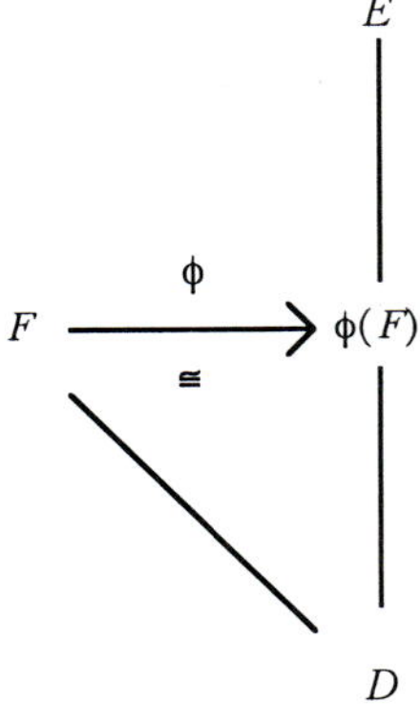

FIGURE 1

Note in the preceding proof that every element of the subfield $\phi(F) \subseteq E$ is of the form $\phi(a)\phi(b)^{-1}$, that is, is a quotient of elements of D in E.

7.3.9 COROLLARY Every field E containing an integral domain D contains a field of quotients of D.

Proof Immediate from the preceding theorem and remark. □

7.3.10 COROLLARY (Uniqueness of the field of quotients) Any two fields of quotients of an integral domain D are isomorphic.

Proof Suppose that F and E are two fields of quotients of the integral domain D. This means that every element of E is of the form $ab^{-1} \in E$ for $a, b \in D$, $b \neq 0$. Therefore, $E = \phi(F)$ where ϕ is as in the proof of the preceding theorem. Hence $E \cong F$. □

Let us put together everything we have proved about the field of quotients F of an integral domain D.
Let D be an integral domain. Then

(1) The field F of quotients of D exists.
(2) The field F of quotients of D is unique up to isomorphism.
(3) $F = \{a/b \mid a, b \in D, b \neq 0\}$.
(4) If E is a field such that $D \subseteq E$, then there is a subfield $F' \subseteq E$ with $F \cong F'$.

This last item says that F is the smallest field that contains D.

We close this section by going back to the concept of the characteristic of an integral domain, which we introduced in Section 6.3. We have developed some new tools that enable us to appreciate the importance of the characteristic. We show that $\mathbb{Q}$ and $\mathbb{Z}_p$, where p is a prime, are the smallest fields of their characteristic. As we will see in a later chapter, any other field is, in a sense, an *extension* of one of these fields.

7.3.11 THEOREM Let D be an integral domain. There there exists a subdomain $D' \subseteq D$ such that

(1) If char $D = 0$, then $\mathbb{Z} \cong D' \subseteq D$.
(2) If char D = p, then $\mathbb{Z}_p \cong D' \subseteq D$.

Proof Let $D' = \{m\cdot 1' \mid m \in \mathbb{Z}\}$, where $1'$ is unity in D, and consider the map $\phi\colon \mathbb{Z} \to D'$ defined by $\phi(m) = m\cdot 1'$ for all $m \in \mathbb{Z}$.
ϕ is a ring homomorphism because

$$\phi(n + m) = (n + m) \cdot 1' = n \cdot 1' + m \cdot 1' = \phi(m) + \phi(n)$$
$$\phi(nm) = (nm) \cdot 1' = (n \cdot 1')(m \cdot 1') = \phi(n)\phi(m)$$

by Proposition 6.1.8, part (4).

(1) If char $D = 0$, then by Theorem 6.3.26, for all positive $m \in \mathbb{Z}$, $m\cdot 1' \neq 0$. Hence in this case we have Kern $\phi = \{0\}$ and ϕ is one to one, and $\mathbb{Z} \cong \phi(\mathbb{Z}) = D' \subseteq D$.

(2) If char $D = p$, then again by Theorem 6.3.26, $|1'| = p$ and Kern $\phi = p\mathbb{Z} \subseteq \mathbb{Z}$. Hence in this case by Theorem 7.2.17 we have $\mathbb{Z}/p\mathbb{Z} \cong \phi(\mathbb{Z}) = D' \subseteq D$, and thus $\mathbb{Z}_p \cong D'$.
Note that in both cases D' is the image of a ring under a ring homomorphism and hence by Proposition 7.1.9, part (2), D' is a subring of D. Since $1' \in D'$, by Proposition 6.2.11 D' is a subdomain of D. □

Finally, an immediate consequence of the preceding theorem is the next result, which is actually just the restatement of the theorem for fields but which is so important that we call it a theorem rather than a corollary.

7.3.12 THEOREM Let F be a field. Then there exists a subfield $F' \subseteq F$ such that

(1) If char $F = 0$, then $\mathbb{Q} \cong F' \subseteq F$.

(2) If char $F = p$, then $\mathbb{Z}_p \cong F' \subseteq F$.

Proof (1) If char $F = 0$, since F is an integral domain, by Theorem 7.3.11 $\mathbb{Z} \cong D' = \{n \cdot 1' \mid n \in \mathbb{Z}\} \subseteq F$, where $1'$ is unity in F. D' is an integral domain contained in F and therefore by Corollary 7.3.9, F contains the field of quotients F' of D'. Since $\mathbb{Z} \cong D' \subseteq F$ we obtain $\mathbb{Q} \cong F' \subseteq F$.

(2) This is immediate from Theorem 7.3.11. □

We have just shown that $\mathbb{Q}$ is the smallest field of characteristic 0. In other words, every field of characteristic 0 contains a subfield isomorphic to $\mathbb{Q}$. In the case of characteristic p, $\mathbb{Z}_p$ is the smallest such field.

7.3.13 DEFINITION The fields $\mathbb{Q}$ and $\mathbb{Z}_p$ are called **prime fields**. ○

As we will see in a later chapter, the prime fields are the basis on which all other fields are built.

Exercises 7.3

In Exercises 1 through 15 determine the field of quotients of the indicated rings if it exists. If it does not exist, explain why.

1. $\mathbb{Z}$ **2.** $\mathbb{Z}[i]$ **3.** $\mathbb{Z} \times \mathbb{Z}$

4. $\mathbb{R}$ **5.** $\mathbb{R} \times \mathbb{R}$ **6.** $\mathbb{Z}_3[i]$

7. $\mathbb{Q}(\sqrt{2})$ **8.** $\mathbb{Z}_5$ **9.** $\mathbb{Z}_{12}$

10. $\mathbb{Z}[\sqrt{2}]$ **11.** $(\mathbb{Z} \times \mathbb{Z})/(\mathbb{Z} \times \{0\})$ **12.** $\mathbb{Z}/7\mathbb{Z}$

13. $\mathbb{Z}_{12}/\langle 4 \rangle$ **14.** $\mathbb{Z}_{15}/\langle 3 \rangle$

15. $\mathbb{Q}(i) = \{a + bi \mid a, b \in \mathbb{Q}, i^2 = -1\}$

16. Show that every subfield of $\mathbb{R}$ must contain $\mathbb{Q}$.

17. Show that the relation on the set S defined in Example 7.3.1 is an equivalence relation on S.

18. Show part (2) of Example 7.3.1.

19. Show parts (5), (6), and (7) of Lemma 7.3.4.

20. Show parts (8) and (9) of Lemma 7.3.4.

21. Show that the map ϕ as defined in the proof of Theorem 7.3.8 is a ring homomorphism.

22. Explain why any two integral domains with 17 elements must be isomorphic. Generalize.

23. Let D be an integral domain and D' a subdomain of D. Show that char D' = char D.

24. Give an example of a ring R and a subring R' with char $R' \neq$ char R.

25. Let D be an integral domain with a subdomain $D' \subseteq D$ such that $D' \cong \mathbb{Z}_p$. Show that char $D = p$.

26. Let D be a finite field. Show that there exists a unique prime integer p such that $\mathbb{Z}_p$ is isomorphic to a subfield of D.

27. Explain why there is no integral domain with exactly 10 elements.

28. Explain why there is no integral domain D with $|D| = pq$, where p and q are distinct primes.

29. Show that if D is a finite integral domain, then $|D| = p^n$ for some integer $n > 0$, where p = char D.

Chapter 8

Rings of Polynomials

In this chapter we study polynomials with coefficients in a given ring R, construct the *ring of polynomials* $R[x]$, and study its different properties. In the important special case where the ring R is also a field F, the ring of polynomials $F[x]$ turns out to have many properties in common with the ring of integers $\mathbb{Z}$. For instance, we find that they are both integral domains that are not fields, and both have a division algorithm. We also find that there are important similarities between the ideals in $F[x]$ and the ideals in $\mathbb{Z}$.

8.1 Basic Concepts and Notation

In this section we construct the ring of polynomials with coefficients in a fixed ring R and establish some basic definitions and properties. Our first example illustrates the importance of the ring R from which the coefficients of a polynomial come.

8.1.1 EXAMPLE Finding the zeros of a polynomial is one of the classical problems in algebra. Let us consider the polynomial $f(x) = x^2 + x + 1$. If we ask what are the zeros of $f(x)$, the question is incomplete, as we can show:

(1) If the coefficients are taken to be from the ring $\mathbb{R}$ of real numbers, then $f(x)$ has no zeros.

(2) If the coefficients are taken to be from the ring $\mathbb{Z}_3$ of integers mod 3, then $f(x) = x^2 + x + 1 = (x - 1)^2$ and $f(1) = 0$, hence $1 \in \mathbb{Z}_3$ is the only zero of $f(x)$.

(3) If the coefficients are taken to be from the ring $\mathbb{Z}_7$ of integers mod 7, then $f(x) = x^2 + x + 1 = (x - 2)(x - 4)$ and $f(2) = f(4) = 0$, hence $2, 4 \in \mathbb{Z}_7$ are the two zeros of $f(x)$. ◇

Thus when we work with polynomials, we must always indicate the ring from which we are taking the coefficients.

8.1.2 EXAMPLE Consider $f(x) = x^2 + 8$.

(1) $f(x)$ has no zeros in $\mathbb{R}$, as can easily be verified.

(2) In $\mathbb{Z}_{11}$, $f(x) = x^2 + 8 = x^2 - 3 = (x - 5)(x - 6)$, and $f(x)$ has exactly two zeros.

(3) In $\mathbb{Z}_{12}$, $f(x) = x^2 + 8 = x^2 - 4 = (x - 2)(x - 10) = (x - 4)(x - 8)$, and $f(x)$ has four zeros. ◇

Let us now give some formal definitions, which will facilitate verification of the ring axioms later.

8.1.3 DEFINITION Let R be a ring. A **polynomial** with coefficients in R and indeterminate x is a finite sum:

$$f(x) = \Sigma_{i=0}^{n} a_i x^i = a_n x^n + \ldots + a_2 x^2 + a_1 x^1 + a_0 x^0$$

where a_i belongs to R for each $i \in \mathbb{Z}$ with $0 \leq i \leq n$. The a_i are called the **coefficients** of the polynomial. In the special case where all the coefficients are zero, the polynomial is called the **zero polynomial**, written $f(x) = 0$. The set of all polynomials with coefficients in R and indeterminate x is denoted $R[x]$. ○

In writing out a polynomial, we write $a_1 x$ for $a_1 x^1$ and write a_0 for $a_0 x^0$; also, we generally omit any summand $a_i x^i$ with $a_i = 0$ and write any summand $a_i x^i$ with $a_i = 1$ as x^i and any summand $a_i x^i$ with $a_i = -1$ as $-x^i$. For example,

$$1x^3 + (-1)x^2 + 0x^1 + 1x^0 = x^3 - x^2 + 1$$

8.1.4 DEFINITION Let R be a ring, $f(x) = a_n x^n + \ldots + a_1 x^1 + a_0 x^0$ a polynomial in $R[x]$, and b an element of R. The **value** $f(b)$ of $f(x)$ for **argument** b is the element $a_n b^n + \ldots + a_1 b^1 + a_0 b^0$ of R. The **polynomial function** determined by $f(x)$ is the function from R to R, taking each $b \in R$ to $f(b)$. An argument b for which the value $f(b)$ of the polynomial $f(x)$ is 0 is called a **zero** of the polynomial $f(x)$ and a **solution** to the **polynomial equation** $f(x) = 0$. ○

Polynomials are not the same as polynomial functions. In particular, two polynomials are not the same unless all their coefficients are the same, even if they determine the same polynomial function, a point illustrated in the next example.

8.1.5 EXAMPLE Let $R = \mathbb{Z}_2$. The polynomials $f(x) = x + 1$ and $g(x) = x^2 + 1$ are distinct, since f has coefficients $a_1 = 1$, $a_0 = 1$ while g has coefficients $b_2 = 1$, $b_1 = 0$, $b_0 = 1$. But it is easily computed that $f(0) = 1 = g(0)$ and $f(1) = 0 = g(1)$, so that f and g determine the same polynomial function from $\mathbb{Z}_2$ to $\mathbb{Z}_2$. ◇

8.1.6 DEFINITION Let $f(x)$ be a polynomial in the indeterminate x with coefficients a_i in a ring R. The greatest n with $a_n \neq 0$ is called the **degree** of $f(x)$, written $n = \deg f(x)$, and the coefficient a_n is called the **leading** coefficient. Note that the notions of degree and leading coefficient are undefined in the special case of the zero polynomial. If the degree is 0, so that the polynomial is just $f(x) = a_0$, the polynomial is called **constant**. Every $a \in R$ can be identified with the polynomial $f(x) = a$, and in this way R can be viewed as a subset of $R[x]$. Polynomials of degrees 1, 2, 3, 4, and 5 are called **linear**, **quadratic**, **cubic**, **quartic**, and **quintic**, respectively. If the leading coefficient is 1, so the polynomial looks like

$$f(x) = x^n + \dots + a_2x^2 + a_1x^1 + a_0x^0$$

the polynomial is called **monic**. ○

We write $R[x]$ for the set of all polynomials in the indeterminate x with coefficients from the ring R. We make $R[x]$ into a ring by defining suitable operations of addition and multiplication. We generally do not distinguish between the constant polynomial a_0 and the element a_0 of the ring R that is its only nonzero coefficient, nor do we distinguish between the zero polynomial 0 and the zero element 0 of the ring R. The addition and multiplication operations on $R[x]$ are defined to agree with those on R, so that the ring R will be a subring of the ring $R[x]$.

8.1.7 DEFINITION If $f(x), g(x) \in R[x]$, where $f(x) = \Sigma a_i x_i$ and $g(x) = \Sigma b_i x_i$, then we define the sum and product as follows:

(1) $f(x) + g(x) = \Sigma c_i x^i$

where $c_i = a_i + b_i$

(2) $f(x) \cdot g(x) = \Sigma d_i x^i$

where $d_i = \Sigma_{k=0}^{i} a_k b_{i-k} = a_0 b_i + a_1 b_{i-1} + \dots + a_{i-1} b_1 + a_i b_0$. ◇

8.1.8 EXAMPLE Let's take $f(x) = x^3 + 2x^2 + 3x + 4$ and $g(x) = 5x^2 + 6x + 7$ in $\mathbb{Z}[x]$. Then $f(x) \cdot g(x) = \Sigma d_i x_i$, where

$d_0 = 4 \cdot 7$
$d_1 = 3 \cdot 7 + 4 \cdot 6$
$d_2 = 2 \cdot 7 + 3 \cdot 6 + 4 \cdot 5$
$d_3 = 1 \cdot 7 + 2 \cdot 6 + 3 \cdot 5$
$d_4 = 1 \cdot 6 + 2 \cdot 5$
$d_5 = 1 \cdot 5$

In other words, the product is $5x^5 + 16x^4 + 34x^3 + 52x^2 + 45x + 28$. ◇

8.1.9 EXAMPLE Let us take the same $f(x)$ and $g(x)$ as in Example 8.1.8, but let us consider them as polynomials in $\mathbb{Z}_{10}[x]$. Then

$$f(x) \cdot g(x) = 5x^5 + 6x^4 + 4x^3 + 2x^2 + 5x + 8$$

is their product. ◇

We are used to calculating the degree of a product of two polynomials with real or complex coefficients to be the sum of their degrees. But this is not always the case for polynomials with coefficients from an arbitrary ring, as the following examples show.

8.1.10 EXAMPLE Consider the polynomials $f(x) = 2x^3 + x + 1$ and $g(x) = 3x^5 + 1$ in $\mathbb{Z}_6[x]$. Then $f(x) \cdot g(x) = 3x^6 + 3x^5 + 2x^3 + x + 1$. Although the degree of $f(x)$ is 3 and the degree of $g(x)$ is 5, the degree of the product is only $6 \neq 3 + 5$. ◇

8.1.11 EXAMPLE Consider the polynomials $f(x) = 2x^3$ and $g(x) = 2x^2$ in $\mathbb{Z}_4[x]$. Then $f(x) \cdot g(x) = 0$, the zero polynomial! ◇

When *can* we count on the degree of a product of two given polynomials to be the sum of the degrees of the given polynomials? From the preceding two examples you may well have guessed the answer, given in the last part of the next theorem.

8.1.12 THEOREM Let R be a ring. Then under the operations introduced in Definition 8.1.7,

(1) $R[x]$ is a ring containing the ring R as a subring.

(2) If R is a commutative, then $R[x]$ is commutative.

(3) If R has a unity 1, then it is also unity for $R[x]$.

(4) If R is an integral domain, then $R[x]$ is an integral domain, and the product of any two nonzero polynomials $f(x)$, $g(x) \in R[x]$ such that $\deg f(x) = m$ and $\deg g(x) = n$, is a nonzero polynomial $f(x) \cdot g(x)$ of degree $m + n$.

Proof Parts (1) through (3) are left to the reader. (See Exercise 23.). For (4), let the leading coefficient of $f(x)$ be a_m, and let the leading coefficient of $g(x)$ be b_n, so $f(x) = a_m x^m + \ldots + a_0$ and $g(x) = b_n x^n + \ldots + b_0$, where $a_m \neq 0$ and $b_n \neq 0$ in R. Then $f(x) \cdot g(x) = a_m b_n x^{m+n} + \ldots + a_0 b_0$. If R is an integral domain, then $a_m b_n \neq 0$, and so it is the leading coefficient of $f(x) \cdot g(x)$. Hence $f(x) \cdot g(x) \neq 0$, and $R[x]$ is an integral domain. Furthermore, $\deg(f(x) \cdot g(x)) = m + n$. □

8.1.13 EXAMPLE Let us find all the units in $\mathbb{Z}_3[x]$. Suppose $0 \neq f(x) \in \mathbb{Z}_3[x]$ is a unit. This means that there exists a polynomial $g(x) \in \mathbb{Z}_3[x]$ such that $f(x) \cdot g(x) = 1 \in \mathbb{Z}_3[x]$. By Theorem 8.1.12, part (4), $\deg f(x) + \deg g(x) = \deg(1) = 0$, and $f(x)$ is a nonzero constant. Therefore, $f(x) = 1$ and $f(x) = 2$ are the only units in $\mathbb{Z}_3[x]$. ◇

8.1.14 PROPOSITION Let D be an integral domain. Then the units in $D[x]$ are precisely the units of D.

Proof If $c \in D$ is a unit in D and d is its multiplicative inverse in D, then considered as constant polynomials $c, d \in D[x]$ and their product in $D[x]$ is the same as their product in the subring D, namely $cd = 1$, so c is again a unit in $D[x]$ and d is its multiplicative inverse in $D[x]$.
Conversely, let $f(x)$ be a unit in $D[x]$, and let $g(x)$ be its multiplicative inverse in $D[x]$. Since the product $f(x) \cdot g(x) = 1$, and so is of degree 0, by the last part of the preceding theorem, $f(x)$ and $g(x)$ must also be of degree 0, which is to say they must both be constants $f(x) = c_0$ and $g(x) = d_0$. Moreover, c_0 must be a unit of D with multiplicative inverse d_0, since we have $c_0 \cdot d_0 = f(x) \cdot g(x) = 1$. □

8.1.15 COROLLARY If F is a field, then $F[x]$ is an integral domain but not a field.

Proof Since F is a field, F is an integral domain. By Theorem 8.1.12, part (4), $F[x]$ is an integral domain. By Proposition 8.1.14, all the nonconstant polynomials in $F[x]$ are nonzero elements that are not units. Hence $F[x]$ is not a field. □

8.1.16 PROPOSITION Let R be a ring. Then $R[x]$ has the same characteristic as R.

Proof For any integer $n > 0$ we write as usual na for $a + \ldots + a$ (n times), and $nf(x) = f(x) + \ldots + f(x)$ (n times). By Definition 6.3.24, the characteristic of $R[x]$ is the least integer $n > 0$ such that $nf(x) = 0$ for all $f(x)$ in $R[x]$, and is zero if no such n exists; similarly for R. Now since R is contained in $R[x]$, it is clear that if $nf(x) = 0$ for all $f(x)$ in $R[x]$, then also $na = 0$ for all a in R. What remains to be shown is that if $na = 0$ for all a in R, then $nf(x) = 0$ for all $f(x)$ in $R[x]$. To see this, let

$$f(x) = a_m x^m + \ldots + a_0 \in R[x]$$

where m is the degree of $f(x)$. Then

$$nf(x) = na_m x^m + \ldots + na_0 = 0x^m + \ldots + 0 = 0. \ \square$$

We now can construct integral domains of many different characteristics that are not fields.

8.1.17 EXAMPLES

(1) Characteristic 0: $\mathbb{Z}[x]$, $\mathbb{Q}[x]$, $\mathbb{R}[x]$, $\mathbb{C}[x]$ are integral domains of characteristic 0 that are not fields.

(2) Characteristic p: for any prime p, the ring of polynomials $\mathbb{Z}_p[x]$ is an integral domain characteristic p that is not a field. ◇

The construction of the ring of polynomials can be generalized to more than one indeterminate.

8.1.18 DEFINITION Starting with a ring R and taking one indeterminate x, we formed the ring $R[x]$. Taking now another indeterminate y, we could go on to form the ring $R[x][y]$. Its elements would be polynomials in y whose coefficients are taken from the ring of polynomials in x, for instance, the following:

(1) $(x+2)y^2 + (x^3+2x)y + (x^2+1)$

Instead, we could generalize our definition and define a ring $R[x,y]$. Its elements would be finite sums $\Sigma a_{ij}x^i y^j$, for instance, the following:

(2) $xy^2 + 2y^2 + x^3y + 2xy + x^2 + 1$

The results of these two constructions are equivalent by the next proposition, and we may think of the ring of **polynomials in two indeterminates** with coefficients in R in either way. ○

8.1.19 PROPOSITION Let R be a ring. Then $R[x][y]$ and $R[x,y]$ are isomorphic. Moreover, both are integral domains if R is.

Proof Let ϕ map the element of $R[x][y]$ to the element of $R[x, y]$ obtained by "multiplying out" in the way that turns (1) in the preceding definition into (2). More formally, consider an element h of $R[x][y]$,

$$h = f_n(x)y^n + \ldots + f_1(x)y + f_0(x)$$

where $f_i(x) = (\Sigma_{j=0}^{m_i} a_{ij}x^j)$, deg $f_i(x) = m_i$.

Then we have

$$\phi(h) = \Sigma_{i=0}^{n} \Sigma_{j=0}^{m_i} a_{ij}x^j y^i$$

The verification that this is an isomorphism is left to the reader. (See Exercise 25.) By Theorem 8.1.12, if R is an integral domain, then so is $R[x]$. Repeating this argument, so is $R[x][y]$. But then so is $R[x,y]$, since it is isomorphic to $R[x][y]$. □

More generally, in a similar way we can define the ring $R[x_1, \dots, x_n]$ of polynomials in any given finite number n of indeterminates $x_1, \dots, x_n$ It will be an integral domain if R is. There is one more construction to be mentioned. We have seen in Section 7.3 that for any integral domain we can construct the field of its quotients. We have seen that if D is an integral domain, then so is $D[x]$. We put the two ideas together in the next definition.

8.1.20 DEFINITION Let D be an integral domain and $D[x]$ the ring of polynomials with coefficients in D. The field of quotients of $D[x]$ is called the field of **rational functions** with coefficients in D and is written $D(x)$. Its elements are fractions of the form $f(x)/g(x)$, where $f(x)$ and $g(x)$ are polynomials in $D[x]$, with $g(x) \neq 0$, and where we consider $f_1(x)/g_1(x) = f_2(x)/g_2(x)$ if and only if $f_1(x)g_2(x) = f_2(x)g_1(x)$. ○

Exercises 8.1

In Exercises 1 through 6 compute the sum $f(x) + g(x)$ and product $f(x) \cdot g(x)$ for the given polynomials $f(x)$ and $g(x)$ in the given polynomial ring $R[x]$.

1. $2x^2 + x + 1$ and $3x^3 + 2$ in $\mathbb{Z}[x]$

2. $2x^2 + x + 1$ and $3x^2 + 2$ in $\mathbb{Z}_6[x]$

3. $2x^2 + x + 1$ and $3x^2 + 2$ in $\mathbb{Z}_7[x]$

4. $6x^2 + 2$ and $2x^3 + 6x$ in $\mathbb{Z}_4[x]$

5. $2x^2+2x+1$ and $2x^2 + 2x + 1$ in $\mathbb{Z}_4[x]$

6. $3x^3 + 4x^2 + 2x + 1$ and $4x^3 + 3x + 2$ in $\mathbb{Z}_5[x]$

7. Find all the polynomials of degree ≤ 2 in $\mathbb{Z}_2[x]$.

8. Find all the polynomials of degree ≤ 2 in $\mathbb{Z}_3[x]$.

9. Let F be a subfield of a field E and $\alpha \in E$. Define ϕ_α: $F[x] \to$ E as follows. For

$$f(x) = a_n x^n + \dots a_1 x + a_0 \in F[x]$$

let

$$\phi_\alpha(f(x)) = a_n \alpha^n + \dots a_1 \alpha + a_0 \in E$$

Show that ϕ_α is a ring homomorphism. It is called an **evaluation homomorphism**.

In Exercises 10 through 14 calculate $\phi_\alpha(f(x))$ for the given fields E and F.

10. $\phi_1(x^4 + 1)$ $F = E = \mathbb{Z}_2$

11. $\phi_1(x^2 + x + 1)$ $F = E = \mathbb{Z}_2$

12. $\phi_{\sqrt{2}}(x^2 - 2)$ $F = \mathbb{Q}, E = \mathbb{R}$

13. $\phi_i(x^3)$ $F = \mathbb{Q}, E = \mathbb{C}$

14. $\phi_2[(x^4 + 1)(x^3 + 4x + 2)]$ $E = F = \mathbb{Z}_5$

15. Let ϕ: $R \to S$ be a ring homomorphism, and define $\phi^*: R[x] \to S[x]$ as follows. For

$$f(x) = a_n x^n + \dots a_1 x + a_0 \in R[x]$$

let

$$\phi^*(f(x)) = \phi(a_n)x^n + \dots \phi(a_1)x + \phi(a_0) \in S[x]$$

Show that ϕ^* is a ring homomorphism. It is called the **induced homomorphism**.

16. Let ϕ: $\mathbb{Z} \to \mathbb{Z}_5$ be the natural homorphism from $\mathbb{Z}$ to $\mathbb{Z}_5$ that sends $n \in \mathbb{Z}$ to its remainder mod 5. Show that the induced homomorphism ϕ^*: $\mathbb{Z}[x] \to \mathbb{Z}_5[x]$ as defined in the preceding problem is also onto. Describe the kernel of ϕ^*.

17. Show that if the rings R and S are isomorphic, then $R[x]$ is isomorphic to $S[x]$.

18. Show that if S is a subring of R, then $S[x]$ is a subring of $R[x]$.

19. Find all the units in

(a) $\mathbb{Z}[x]$ (b) $\mathbb{Q}[x]$ (c) $\mathbb{Z}_5[x]$ (d) $\mathbb{Z}[x, y]$ (e) $\mathbb{Q}(x)$

20. Give an example of a natural number $n > 1$ and a polynomial $f(x) \in \mathbb{Z}_n[x]$ of degree > 0 that is a unit in $\mathbb{Z}_n[x]$.

21. Find a zero divisor, if possible, in each of the following. If not possible, explain why not.

(a) $\mathbb{Z}_4[x]$ (b) $\mathbb{Z}_5[x]$ (c) $\mathbb{Z}_6[x]$

22. Give an example, if possible, of two polynomials $f(x)$ and $g(x)$ in the indicated rings such that the degree of $f(x) \cdot g(x)$ is not equal to the sum of the degrees of $f(x)$ and $g(x)$. If not possible, explain why not.

(a) $\mathbb{Z}_8[x]$ (b) $\mathbb{Z}_7[x]$ (c) $\mathbb{Z}_9[x]$

23. Prove parts (1) through (3) of Theorem 8.1.12.

24. Let $f(x, y)$ be the following element of $\mathbb{Q}[x, y]$:

$$4x^2y^3 + 5xy^3 - 7x^4 + 3x^3y^2 - 2y^3 + 3x^4y + 2x^2 - x^3y + 10xy^2 + 3$$

(a) Rewrite $f(x, y)$ as an element of $(\mathbb{Q}[x])[y]$.

(b) Rewrite $f(x, y)$ as an element of $(\mathbb{Q}[y])[x]$.

25. Prove that the map ϕ defined in the proof of Proposition 8.1.19 is an isomorphism.

Formal Power Series

In Exercises 26 through 31 for any ring R define the set $R[[x]]$ of **formal power series** in the indeterminate x with coefficients from R to be the set of all infinite formal sums

$$\Sigma_{i=0}^{\infty} a_i x^i$$

with all a_i in R.

26. Define two operations on $R[[x]]$, addition and multiplication, such that
(a) $R[[x]]$ is a ring under these operations.
(b) $R[x]$ is a subring of $R[[x]]$.
(c) If R is an integral domain, then $R[[x]]$ is an integral domain.

27. Let $f(x) = \Sigma_{i=0}^{\infty} a_i x^i \in R[[x]]$. Show that $f(x)$ is a unit in $R[[x]]$ if and only if a_0 is a unit in R.

28. Given a ring homomorphism $\phi: R \to S$, define a map $\phi^\dagger: R[[x]] \to S[[x]]$ by setting

$$\phi^\dagger(\Sigma_{i=0}^{\infty} a_i x^i) = \Sigma_{i=0}^{\infty} \phi(a_i) x^i$$

Show that $\phi^\dagger$ is a ring homomorphism and that $\phi^\dagger(R[x]) = \phi^*(R[x]) \subseteq S[x]$.

29. For $\phi^\dagger$ as defined in the preceding exercise show that
(a) $\phi^\dagger$ is one to one if and only if ϕ is.
(b) $\phi^\dagger$ is onto if and only if ϕ is.

30. Describe the polynomials that are units in $\mathbb{Z}[[x]]$.

31. Find the multiplicative inverse of $x+1$ in $\mathbb{Z}[[x]]$.

8.2 The Division Algorithm in $F[x]$

We have already seen that if D is an integral domain, then $D[x]$ is an integral domain but not a field. In this section we see that in the case where the original integral domain is a field F, the ring $F[x]$ has several properties analogous to properties of $\mathbb{Z}$ that we already know.

A fundamental property of $\mathbb{Z}$ is the existence of a division algorithm, according to which for any two integers a and b, with $b \neq 0$, there exists a unique pair of integers q and r such that

(1) $a = b \cdot q + r$

(2) $0 \le r < |b|$

We have used this property extensively. The analogous property for $F[x]$, as we will see, plays a fundamental role.

8.2.1 EXAMPLE Consider the polynomials $f(x) = 3x$ and $g(x) = 2x$ in $\mathbb{Z}_7[x]$. Let us divide $f(x)$ by $g(x)$ or, in other words, let us find $q(x) \in \mathbb{Z}_7[x]$ such that $f(x) = q(x)g(x)$. Since $\mathbb{Z}_7[x]$ is an integral domain, $\deg f(x) = \deg q(x) + \deg g(x)$. Hence $\deg q(x) = 0$ and $q(x) = c \in \mathbb{Z}_7[x]$ is a non-zero constant. Hence $3x = c(2x)$ and $c = 3 \cdot 2^{-1} = 3 \cdot 4 = 5$ in $\mathbb{Z}_7[x]$. Note that c exists and is unique because the multiplicative inverse 2^{-1} exists and is unique in $\mathbb{Z}_7[x]$. ◇

To imitate Theorem 0.3.11 in the case of $F[x]$, where F is a field, we need to say something about the remainder of a division. In the case of $\mathbb{Z}$ the remainder r upon division by b satisfies $0 \leq r < b$: The remainder is less than the divisor. Now we have no way of comparing two polynomials as greater or lesser, except for comparing their degrees. Comparison of degrees is exactly what is used in the following analogue of Theorem 0.3.11.

8.2.2 THEOREM (Division algorithm) Let F be a field and $f(x)$ and $g(x)$ elements of $F[x]$, with $g(x) \neq 0$. Then there exist unique elements $q(x)$ and $r(x)$ of $F[x]$ such that

(1) $f(x) = q(x) \cdot g(x) + r(x)$

(2) $r(x) = 0$ or $\deg r(x) < \deg g(x)$

Proof First we prove the existence assertion. In case $f(x) = 0$, then $f(x) = 0g(x) + 0$. In case $\deg f(x) < \deg g(x)$, then $f(x) = 0g(x) + f(x)$. It remains to consider the case where $\deg f(x) \geq \deg g(x)$. For this case, let $n = \deg f(x)$ and $m = \deg g(x)$. We will prove the theorem by induction on n. To begin the induction, consider the case $n = m = 0$. Then both the given polynomials are constants, $f(x) = a_0$ and $g(x) = b_0 \neq 0$, where a_0 and b_0 are elements of the field F. Then $a_0 b_0^{-1}$ is also an element of the field F, and $f(x) = a_0 b_0^{-1} g(x) + 0$. (Note that $b_0 \neq 0$ is a unit in F.)
Now consider the case $n > 0$, and assume as our induction hypothesis that the theorem holds for polynomials of degree $< n$. We have

$$f(x) = a_n x^n + \ldots + a_2 x^2 + a_1 x + a_0$$
$$g(x) = b_m x^m + \ldots + b_2 x^2 + b_1 x + b_0$$

with $n \geq m$. (Note again that $b_m \neq 0$ is a unit in F.) Consider the element $a_n b_m^{-1} x^{n-m} g(x)$ of $F[x]$. It looks like

$$a_n b_m^{-1} x^{n-m} g(x) = a_n x^n + (\text{terms of lower degree})$$

So the polynomial

(1) $$h(x) = f(x) - a_n b_m^{-1} x^{n-m} g(x)$$

will either be the zero polynomial, in which case we are done, since then $f(x) = a_n b_m^{-1} x^{n-m} g(x) + 0$, or else will have $\deg h(x) < \deg f(x)$. In that case, by our induction hypothesis there will be elements $q'(x)$ and $r'(x)$ of $F[x]$ such that

(2) $$h(x) = q'(x)g(x) + r'(x)$$

and either $r'(x) = 0$ or $\deg r'(x) < \deg g(x)$. Now using the equations (1) and (2) we see that

$$f(x) = a_n b_m^{-1} x^{n-m} g(x) + q'(x)g(x) + r'(x) = (a_n b_m^{-1} x^{n-m} + q'(x))g(x) + r'(x)$$

which gives the required $q(x)$ and $r(x)$.
Now we prove the uniqueness assertion. Suppose we have

$$f(x) = q_1(x)g(x) + r_1(x)$$
$$f(x) = q_2(x)g(x) + r_2(x)$$

where $r_i(x) = 0$ or $\deg r_i(x) < \deg g(x)$. Subtracting the two equations, we get

$$0 = [q_1(x)-q_2(x)]g(x) + [r_1(x)-r_2(x)]$$

Hence

(3) $$[q_1(x)-q_2(x)]g(x) = r_2(x)-r_1(x)$$

We must have $r_2(x)-r_1(x)= 0$, for otherwise $\deg[r_2(x)-r_1(x)] < \deg g(x)$, which would make equation (3) impossible. So we must also have $q_1(x)-q_2(x) = 0$ or, in other words, $r_1(x) = r_2(x)$ and $q_1(x) = q_2(x)$. □

8.2.3 EXAMPLE In $\mathbb{Z}_5[x]$, let us divide $f(x)$ by $g(x)$, where

$$f(x) = 2x^4 + x^3 + 3x^2 + 3x + 1,\ g(x) = 2x^2 - x + 2$$

The calculations can be set out in the same format as long division for integers.

$$\begin{array}{r|l} & x^2 + x + 1 \\ \hline 2x^2 - x + 2 & 2x^4 + x^3 + 3x^2 + 3x + 1 \\ & -(2x^4 - x^3 + 2x^2) \\ \hline & 2x^3 + x^2 + 3x + 1 \\ & -(2x^3 - x^2 + 2x) \\ \hline & 2x^2 + x + 1 \\ & -(2x^2 - x + 2) \\ \hline & 2x - 1 = 2x + 4 \end{array}$$

Therefore, $q(x) = x^2+x+1$ and $r(x) = 2x+4$. ◇

An immediate application of the division algorithm for $\mathbb{Z}$ was the Euclidean algorithm for $\mathbb{Z}$, which allows us to calculate the gcd of two integers a and b and to write it as a linear combination of a and b. We next show that we can do the same kind of calculations in $F[x]$. First we need some definitions analogous to those used for $\mathbb{Z}$.

8.2.4 DEFINITION For any $f(x)$ and $g(x) \neq 0$ in $F[x]$, where F is a field, the polynomials $q(x)$ and $r(x)$ in the division algorithm are called the **quotient** and **remainder** on division of $f(x)$ by $g(x)$. If the remainder $r(x) = 0$ or, in other words, if there exists a $q(x)$ such that $f(x) = q(x)\cdot g(x)$, then $g(x)$ is called a **divisor** of $f(x)$ or is said to **divide** $f(x)$, written $g(x) \mid f(x)$; also, $f(x)$ is called a **multiple** of $g(x)$. Writing a polynomial $f(x)$ as a product $f(x) = g(x)\cdot h(x)$ of two other polynomials is called **factoring** the polynomial, and such a factorization is called **nontrivial** if the factors $g(x)$ and $h(x)$ both have degree > 0. ○

8.2.5 PROPOSITION Let F be a field and $f(x)$ and $g(x)$ elements of $F[x]$. Then

(1) If $g(x) \mid f(x)$, then $eg(x) \mid f(x)$ for any element $e \neq 0$ of F.

(2) If $g(x) \mid f(x)$ and $f(x) \mid g(x)$, then $f(x) = eg(x)$ for some element $e \neq 0$ of F.

Proof (1) Note that since F is a field, the assumption $e \neq 0$ implies e is a unit in F. If $f(x) = q(x) \cdot g(x)$, then also $f(x) = [e^{-1}q(x)] \cdot [eg(x)]$.

(2) If $f(x) = q(x) \cdot g(x)$ and $g(x) = p(x) \cdot f(x)$, then $f(x) = [q(x) \cdot p(x)] \cdot f(x)$, which since $F[x]$ is an integral domain implies $q(x) \cdot p(x) = 1$, so $q(x)$ and $p(x)$ must both be of degree zero, $q(x) = e$ and $p(x) = e^{-1}$ for some e in F. □

8.2.6 DEFINITION For $f(x)$ and $g(x)$ in $F[x]$, where F is a field, a **common divisor** of $f(x)$ and $g(x)$ is any polynomial $c(x)$ in $F[x]$ such that $c(x) \mid f(x)$ and $c(x) \mid g(x)$. A **greatest common divisor** of $f(x)$ and $g(x)$ is a common divisor $d(x)$ such that for any other common divisor $c(x)$ we have $c(x) \mid d(x)$. If the only common divisors, and therefore the only greatest common divisors, of $f(x)$ and $g(x)$ are constants, then $f(x)$ and $g(x)$ are called **relatively prime**. Note that by the preceding Proposition, if $d_1(x)$ and $d_2(x)$ are both greatest common divisors, then $d_1(x) = ed_2(x)$ for some $e \neq 0$ in F, from which it follows that there can be only one *monic* greatest common divisor, which we write $\mathbf{gcd}(f(x), g(x))$. ○

8.2.7 EXAMPLE Let us show that

$$f(x) = 2x^4 - x^3 + 4x^2 + 3 \text{ and } g(x) = x^3 + 2x + 1$$

in $\mathbb{C}[x]$ have no zeros in common. If we apply the division algorithm to $f(x)$ and $g(x)$ we obtain

$$f(x) = (2x - 1)g(x) + 4 \tag{4}$$

Hence for any α such that $g(\alpha) = 0$ we must have $f(\alpha) = 4 \neq 0$. Note that equation (4) implies that $\gcd(f(x), g(x)) = 1$ or, in other words, $f(x)$ and $g(x)$ are relatively prime. ◇

We now prove for $F[x]$ a theorem about the gcd analogous to the one that holds for $\mathbb{Z}$, Theorem 0.3.16.

8.2.8 THEOREM Let F be a field and $f(x)$ and $g(x)$ elements of $F[x]$, not both 0. Then there exists a greatest common divisor $d(x)$ of $f(x)$ and $g(x)$ that can be written as a linear combination of $f(x)$ and $g(x)$. That is, there exist elements $u(x)$ and $v(x)$ of $F[x]$ such that $d(x) = u(x) \cdot f(x) + v(x) \cdot g(x)$ is a greatest common divisor of $f(x)$ and $g(x)$.

Proof The proof proceeds in three steps.

(1) Consider the set

$$I = \{m(x)f(x) + n(x)g(x) \mid m(x), n(x) \in F[x]\}$$

Note that $f(x), g(x) \in I$, so I has elements other than 0. Let $d(x)$ be an element of I of minimum degree. Since it is easily seen that for any $e \in F$, $ed(x) \in I$ and has the same degree as $d(x)$, we may take $d(x)$ to be monic, for if it is not monic already, we can simply replace it by $a^{-1}d(x)$, where a is its leading coefficient. Since $d(x)$ is in I, we have $d(x) = u(x)f(x) + v(x)g(x)$ for some $u(x)$ and $v(x)$ in $F[x]$.

(2) We now show that $d(x)$ is a common divisor of $f(x)$ and $g(x)$. Using the division algorithm we obtain $f(x) = q(x) \cdot d(x) + r(x)$, where either $r(x) = 0$ or $\deg r(x) < \deg d(x)$. We want to show $r(x) = 0$. Solving for $r(x)$, we get

$$r(x) = f(x) - q(x)d(x) = [1 - q(x)u(x)]f(x) - [q(x)v(x)]g(x)$$

This shows $r(x)$ is in I, and since $d(x)$ was chosen to be of minimum degree, we cannot have $\deg r(x) < \deg d(x)$ but must have $r(x) = 0$ and so $d(x) \mid f(x)$, as required. A similar argument shows $d(x) \mid g(x)$.

(3) To complete the proof, we need to show that any common divisor $c(x)$ of $f(x)$ and $g(x)$ divides $d(x)$. But if $f(x) = q(x)c(x)$ and $g(x) = p(x)c(x)$, then

$$d(x) = u(x)f(x) + v(x)g(x) = [u(x)q(x) + v(x)p(x)]c(x)$$

This shows that $c(x) \mid d(x)$ as required. □

The division algorithm can be used to find the gcd in $F[x]$ much as in $\mathbb{Z}$ (Proposition 0.3.18).

8.2.9 THEOREM Let F be a field and $f(x)$ and $g(x)$ elements of $F[x]$. If we apply the division algorithm repeatedly,

(1) $f(x) = q_1(x)g(x) + r_1(x)$

(2) $g(x) = q_2(x)r_1(x) + r_2(x)$

(3) $r_1(x) = q_3(x)r_2(x) + r_3(x)$

⋮

then we must come to some finite n for which we have

(n) $r_{n-2}(x) = q_n(x)r_{n-1}(x) + r_n(x)$

(n+1) $r_{n-1}(x) = q_{n+1}(x)r_n(x) + r_{n+1}(x)$

with $r_{n+1}(x) = 0$. Then $r_n(x)$ is a gcd of $f(x)$ and $g(x)$.

Proof The sequence of divisions cannot go on forever, since $\deg g(x) > \deg r_1(x) > \deg r_2(x) > \ldots > 0$, hence we must come to an n as asserted. For such an n, from equation (n+1) we see $r_n(x) \mid r_{n-1}(x)$; then from equation (n) we see $r_n(x) \mid r_{n-2}(x)$; and so on back to equations (2) and (1), which show $r_n(x) \mid g(x)$ and $r_n(x) \mid f(x)$. Furthermore, if $c(x)$ is any common divisor of $f(x)$ and $g(x)$, from equation (1) we see $c(x) \mid r_1(x)$; and then from equation (2) we see $c(x) \mid r_2(x)$; and so on down to equation (n+1), which shows $c(x) \mid r_n(x)$. Thus $r_n(x)$ is a gcd of $f(x)$ and $g(x)$. □

8.2.10 DEFINITION The sequence of calculations (1), (2), (3), ... in the preceding theorem is called the **Euclidean algorithm**. ○

Just as for $\mathbb{Z}$, we can also use the Euclidean algorithm to write the gcd as a linear combination.

8.2.11 EXAMPLE Consider the polynomials

$$f(x) = x^4 + 2x^2 + 1 \text{ and } g(x) = x^2 + x + 2$$

in $\mathbb{Z}_3[x]$. Applying the Euclidean algorithm, we obtain

(1) $x^4 + 2x^2 + 1 = (x^2 - x + 1)(x^2 + x + 2) + (x + 2)$

(2) $x^2 + x + 2 = (x + 2)(x + 2) + 1$

Therefore, gcd($f(x)$, $g(x)$) = 1, and $f(x)$ and $g(x)$ are relatively prime. Also, we have

$$\begin{aligned} 1 &= (x^2 + x + 2) - (x + 2)(x + 2) \\ &= (x^2 + x + 2) - (x + 2)[(x^4 + 2x^2 + 1) - (x^2 - x + 1)(x^2 + x + 2)] \\ &= [-x - 2](x^4 + 2x^2 + 1) + [1 + (x + 2)(x^2 - x + 1)](x^2 + x + 2) \end{aligned}$$

These kinds of calculations will be useful later, when we are looking for multiplicative inverses in quotient rings of $F[x]$. ◇

Exercises 8.2

In Exercises 1 through 5 calculate the quotients and remainders on division of the indicated $f(x)$ by the indicated $g(x)$ in the indicated polynomial rings $F[x]$.

1. $f(x) = 4x^3 - 2x^2 + 5x - 3$ $\quad g(x) = x^2 + x + 1,$ $\quad \mathbb{Q}[x]$

2. $f(x) = 3x^4 + 2x + 3$ $\quad g(x) = x^2 + 2x + 1,$ $\quad \mathbb{Z}_5[x]$

3. $f(x) = x^4 + x^3 + x^2 + x + 1$ $\quad g(x) = x + 1,$ $\quad \mathbb{Z}_2[x]$

4. $f(x) = x^5 + 5x^3 + 3x^2 + 2$ $\quad g(x) = x^2 + 4x + 5$ $\quad \mathbb{Z}_7[x]$

5. $f(x) = x^4 + 9x^2 + 5$ $\quad g(x) = x^2 + 3x + 4$ $\quad \mathbb{Z}_{11}[x]$

In Exercises 6 through 9 calculate gcd($f(x)$, $g(x)$) for the indicated $f(x)$ and $g(x)$ in the indicated polynomial rings $F[x]$. Also, in each case find $u(x)$ and $v(x)$ such that gcd($f(x)$, $g(x)$) = $u(x)f(x) + v(x)g(x)$.

6. $f(x) = x^4 - x^2 - 2$ $\quad g(x) = x^3 + x^2 + x + 1$ $\quad \mathbb{Q}[x]$

7. $f(x) = x^4 + x^3 + x + 1$ $\quad g(x) = x + 1$ $\quad \mathbb{Z}_2[x]$

8. $f(x) = x^3 + 1$ $\quad g(x) = x + 2$ $\quad \mathbb{Z}_5[x]$

9. $f(x) = x^3 + 2x + 1$ $\quad g(x) = x + 2$ $\quad \mathbb{Z}_3[x]$

10. Let I be the set defined in the proof of Theorem 8.2.8. Show that

(a) I is an ideal in $F[x]$.

(b) $I = \langle d(x) \rangle$, the principal ideal generated by $d(x)$.

11. Show that for any polynomial $h(x)$ in $\mathbb{Z}_5[x]$ there exist $m(x)$ and $n(x)$ in $\mathbb{Z}_5[x]$ such that $h(x) = m(x)(x^2 + x + 1) + n(x)(x + 1)$.

12. Find $m(x)$ and $n(x)$ in $\mathbb{Z}_5[x]$ such that

$$x^4 + 4x^2 + 3 = m(x)(x^2 + x + 1) + n(x)(x + 1)$$

13. Find $m(x)$ and $n(x)$ in $\mathbb{Q}[x]$ such that

$$x = m(x)(x^3 + 1) + n(x)(x^2 + 1)$$

14. Let $f(x)$ and $g(x)$ be polynomials in $F[x]$ for some field F, with $\deg f(x)$, $\deg g(x) \geq 1$. Show that there exist unique polynomials

$$q_0(x), q_1(x), q_2(x), \ldots, q_s(x)$$

such that each $q_i(x)$ is either zero or has $\deg q_i(x) < \deg g(x)$, and

$$f(x) = q_s(x)(g(x))^s + \ldots + q_2(x)(g(x))^2 + q_1(x)g(x) + q_0(x)$$

In Exercises 15 thorugh 17 find the polynomials $q_i(x)$ as in Exercise 14 for the indicated $f(x)$ and $g(x)$ in the indicated $F[x]$.

15.	$f(x) = x^3 + x + 1$	$g(x) = x + 1$	$\mathbb{Z}_2[x]$
16.	$f(x) = x^3 + 2x + 3$	$g(x) = x + 1$	$\mathbb{Q}[x]$
17.	$f(x) = x^6 + x^4 + 2x^3 + x + 1$	$g(x) = x^2 + 1$	$\mathbb{Z}_5[x]$

8.3 More Applications of the Division Algorithm

We have already seen how to use the division algorithm to find the gcd of two polynomials. In this section we use it to establish some basic properties of polynomials and their zeros. (The notion of a zero of polynomial has been defined in Definition 8.1.4.)

8.3.1 EXAMPLE

(1) As we saw in Example 8.1.1, in $\mathbb{Z}_3[x]$, $f(x) = x^2 + x + 1$ has exactly one zero, $f(1) = 0$, and $f(x)$ factors as follows: $x^2 + x + 1 = (x - 1)^2$ in $\mathbb{Z}_3[x]$.

(2) By contrast, in $\mathbb{Z}_7[x]$, $f(x) = x^2 + x + 1$ has exactly two zeros, $f(2) = f(4) = 0$, and $f(x)$ factors as follows: $x^2 + x + 1 = (x - 2)(x - 4)$ in $\mathbb{Z}_7[x]$.

(3) In $\mathbb{Q}[x]$, $f(x) = x^3 + x^2 + x + 1$ has a zero in $\mathbb{Q}$, $f(-1) = 0$, and $f(x)$ factors as follows: $x^3 + x^2 + x + 1 = (x + 1)(x^2 + 1)$ in $\mathbb{Q}[x]$.

(4) By contrast, in $\mathbb{C}[x]$, $f(x) = x^3 + x^2 + x + 1 = (x + 1)(x - i)(x + i)$ and -1, $\pm i$ are the three zeros of $f(x)$ in $\mathbb{C}[x]$. ◇

This example illustrates a very important consequence of the division algorithm, which we now prove.

8.3.2 THEOREM (Factor theorem) Let F be a field, $f(x)$ a polynomial in $F[x]$, and a an element of F. Then a is a zero of $f(x)$ if and only if $(x - a)$ is a divisor of $f(x)$ in $F[x]$.

Proof ($\Rightarrow$) Assume first that a is a zero of $f(x)$. Applying the division algorithm, we can write $f(x) = q(x)(x - a) + r(x)$, where either $r(x) = 0$ or $\deg r(x) < \deg(x - a) = 1$.

Thus $r(x)$ is a constant $c \in F$ and $f(x) = q(x)(x - a) + c$. Hence $0 = f(a) = q(a)(a - a) + c = c$, so $f(x) = q(x)(x - a)$ and $(x - a)$ is a divisor of $f(x)$ as required.
($\Leftarrow$) Assume $(x - a)$ is a divisor of $f(x)$, so that $f(x) = q(x)(x - a)$ for some $q(x)$ in $F[x]$. Then we have $f(a) = q(a)(a - a) = 0$ □

8.3.3 EXAMPLE Working in $\mathbb{Z}_5[x]$, consider the polynomial $f(x) = x^4 + x^3 + x - 1$. Now $f(2) = 0$, so $(x - 2)$ must be a divisor of $f(x)$ in $\mathbb{Z}_5[x]$. Let us calculate the quotient.

$$
\begin{array}{r|l}
 & x^3 + 3x^2 + x + 3 \\
\hline
x - 2 & x^4 + x^3 + x - 1 \\
 & -(x^4 - 2x^3) \\
\hline
 & 3x^3 + x - 1 \\
 & -(3x^3 - x^2) \\
\hline
 & x^2 + x - 1 \\
 & -(x^2 - 2x) \\
\hline
 & 3x - 1 \\
 & -(3x - 1) \\
\hline
 & 0
\end{array}
$$

Thus $f(x) = (x^3 + 3x^2 + x + 3)(x - 2)$ in $\mathbb{Z}_5[x]$. Since 2 is also a zero of $g(x) = x^3 + 3x^2 + x + 3$, we see that $x - 2$ must also be a divisor of $g(x)$ in $\mathbb{Z}_5[x]$. Again let us calculate the quotient.

$$
\begin{array}{r|l}
 & x^2 + 1 \\
\hline
x - 2 & x^3 + 3x^2 + x + 3 \\
 & -(x^3 - 2x^2) \\
\hline
 & x + 3 \\
 & -(x - 2) \\
\hline
 & 0
\end{array}
$$

In $\mathbb{Z}_5[x]$ we then have

$$(x^2 + 1)(x - 2)^2 = (x^2 - 4)(x - 2)^2 = (x + 2)(x - 2)(x - 2)^2 = (x + 2)(x - 2)^3$$

giving the factorization $f(x) = (x + 2)(x - 2)^3$. ◇

8.3.4 THEOREM (Remainder theorem) Let F be a field, $f(x)$ a polynomial in $F[x]$, and a an element of F. Then $f(a)$ is the remainder on dividing $f(x)$ by $(x - a)$ in $F[x]$.

Proof Left to the reader. (See Exercise 8.) □

8.3.5 DEFINITION Let F be a field and let $f(x) \in F[x]$ and $\alpha \in F$ be such that $f(\alpha) = 0$. Then we say that α is a zero of $f(x)$ of **multiplicity** s if

$$f(x) = q(x)(x - \alpha)^s$$

where $q(x) \in F[x]$ and $q(\alpha) \neq 0$. ○

8.3.6 EXAMPLE

(1) As we saw in Example 8.3.3, $f(x) = x^4 + x^3 + x - 1 = (x + 2)(x - 2)^3$ in $\mathbb{Z}_5[x]$. Hence $2 \in \mathbb{Z}_5$ is a zero of $f(x)$ of multiplicity 3.

(2) $g(x) = x^5 - 1 = (x - 1)^5$ in $\mathbb{Z}_5[x]$, hence $1 \in \mathbb{Z}_5$ is a zero of $g(x)$ of multiplicity 5.

(3) $g(x) = x^5 - 1 = (x - 1)(x^4 + x^3 + x + 1)$ in $\mathbb{Q}[x]$, and $1 \in \mathbb{Q}$ is a zero of $g(x)$ of multiplicity 1. ◇

8.3.7 THEOREM Let F be a field, $f(x)$ a nonzero polynomial in $F[x]$, and n its degree. Then $f(x)$ has at most n zeros in F.

Proof We use induction on n. If $n = 0$, then $f(x)$ is a nonzero constant, and so has no zeros. Now assume the theorem is true for all polynomials in $F[x]$ of degree $n - 1$. If $f(x)$ has no zeros in F, then the theorem holds. If $f(x)$ has a zero a in F, then by the factor theorem $f(x) = q(x)(x - a)$, where $q(x)$ has degree $n - 1$. If $b \neq a$ is a zero of $f(x)$, then $0 = f(b) = q(b)(b - a)$, and since F, being a field, has no zero divisors, we must have $0 = q(b)$. This tells us that a zero of $f(x)$ is either a or a zero of $q(x)$, and by our induction hypothesis $q(x)$ has no more than $n - 1$ zeros. It follows that $f(x)$ has no more than n zeros, as required. □

The assumption that F is a field is very important for the preceding theorem, as we can see from the following example.

8.3.8 EXAMPLE Let $f(x) = x^2 - 4$ in $\mathbb{Z}_{12}[x]$. Then, as we saw in Example 8.1.2, $f(x)$ has two factorizations:

$$f(x) = (x - 2)(x + 2) = (x - 2)(x - 10)$$
$$f(x) = (x - 4)(x + 4) = (x - 4)(x - 8)$$

in $\mathbb{Z}_{12}$, and hence has four zeros $f(2) = f(4) = f(8) = f(10) = 0$. So with the ring $\mathbb{Z}_{12}$, which is not a field, we have a polynomial of degree 2 with four zeros. ◇

8.3.9 EXAMPLE Let $f(x) \in \mathbb{R}[x]$ be a polynomial of degree 3, and $g(x) \in \mathbb{R}[x]$ a polynomial of degree 1. Consider $f(x)$ and $g(x)$ as polynomial functions, and draw their graphs in the xy-plane $\mathbb{R}^2$. Then Theorem 8.3.7 tells us that the two graphs intersect at no more than three points. For $f(x) - g(x) \in \mathbb{R}[x]$ is a polynomial of degree 3, and hence there exist at most three $\alpha \in \mathbb{R}$ such that $f(\alpha) - g(\alpha) = 0$. ◇

In the next theorem we show, using Theorem 8.3.7, an important property of finite fields that will be very useful later.

8.3.10 THEOREM Let F be any finite field and G a subgroup of F^*, the multiplicative group of nonzero elements of F. Then G is cyclic.

Proof G is a finite Abelian multiplicative group. By the fundamental theorem for finite Abelian groups, Theorem 3.4.1, $G \cong G_{d_1} \times \dots \times G_{d_n}$, where the d_i are powers of primes and G_{d_i} is a multiplicative cyclic group of order d_i.

Now let

$$N = \prod_{i=1}^{n} d_i \text{ and let } M = \text{lcm}(d_1, d_2, \ldots, d_n)$$

So $M \leq N$. For every b_i in G_{d_i} we have $b_i^{d_i} = 1$ by Langrange's theorem, so $b_i^M = 1$. Therefore, for every a in G we have $a^M = 1$ or, in other words, every element of G is a zero of the polynomial $f(x) = x^M - 1$ in $F[x]$. It follows by Theorem 8.3.7 that $|G| \leq M$. But we also know $|G| = N$. It follows that $M = N$, which is only possible if distinct d_i and d_j are relatively prime, and then by Theorem 3.2.8 $G \cong G_N$, the cyclic group of order N. □

We end this section with some useful facts about the zeros of a polynomial in $\mathbb{R}[x]$.

8.3.11 EXAMPLE The zeros of the polynomial $x^n - 1$ are called the ***n*th roots of unity**. Working in $\mathbb{C}$, if we let $\omega = \cos(2\pi/n) + i\sin(2\pi/n)$, then by DeMoivre's theorem we have $\omega^n = 1$ and $\omega^k \neq 1$ for $1 \leq k < n$. Moreover, we can easily see that $1, \omega, \omega^2, \ldots, \omega^{n-1}$ are zeros of $f(x) = x^n - 1$ and distinct from each other. Since $f(x)$ can have at most n zeros, these are *all* the nth roots of unity. ◇

8.3.12 THEOREM Let $f(x)$ be a polynomial in $\mathbb{R}[x]$. If $z = a + bi \in \mathbb{C}$ is a zero of $f(x)$, then the complex conjugate $\bar{z} = a - bi$ of z is also a zero of $f(x)$.

Proof Consider the polynomial

$$(x - (a + bi))(x - (a - bi)) = x^2 - 2ax + (a^2 + b^2) = g(x)$$

in $\mathbb{R}[x]$. Applying the division algorithm, we obtain $q(x)$ and $r(x)$ in $\mathbb{R}[x]$ such that $f(x) = q(x)g(x) + r(x)$, where $r(x)$ is either 0 or of degree 0 or 1, or in other words, $r(x) = cx + d$ for some c and d in $\mathbb{R}$. Now $0 = f(a + bi) = q(x)\cdot 0 + r(a + bi)$. Hence $0 = r(a + bi) = c(a + bi) + d = (ca + d) + cbi$. Hence $ca + d = cb = 0$. But then $r(a - bi) = c(a - bi) + d = (ca + d) - cbi = 0$. So $a - bi$ is a zero of $r(x)$, as well as a zero of $g(x)$, and hence is a zero of $f(x)$ as required. □

8.3.13 THEOREM Let $f(x)$ be a polynomial in $\mathbb{Q}[x]$. If $u = a + b\sqrt{c}$ is a zero of $f(x)$, where $a, b \in \mathbb{Q}$ and $\sqrt{c} \notin Q$, then $\bar{u} = a - b\sqrt{c}$ is also a zero of $f(x)$.

Proof Left to the reader. (See Exercise 9.) □

Exercises 8.3

In Exercises 1 through 7 find all zeros of the indicated $f(x)$ in the indicated field.

1. $f(x) = x^2 + x + 1$ in $\mathbb{Z}_3$
2. $f(x) = x^3 + x^2 + x + 1$ in $\mathbb{R}$
3. $f(x) = x^3 + x^2 + x + 1$ in $\mathbb{C}$

4. $f(x) = x^8 - 1$ in $\mathbb{R}$

5. $f(x) = x^4 - 4x^2 + 4$ in $\mathbb{R}$

6. $f(x) = x^3 + 3x + 5$ in $\mathbb{Z}_7$

7. $f(x) = x^4 + 4$ in $\mathbb{Z}_5$

8. Prove Theorem 8.3.4.

9. Prove Theorem 8.3.13.

In Exercises 10 through 12 working in $\mathbb{C}$, find the nth roots of unity for the indicated n, and show that they form a cyclic subgroup of $\mathbb{C}^*$ of order n.

10. $n = 3$

11. $n = 4$

12. $n = 6$

13. Find all zeros of $x^2 - 1$ in $\mathbb{Z}_{15}$. Does your answer contradict Theorem 8.3.7?

14. Let F be an infinite field and $f(x)$ a polynomial in $F[x]$. Show that if $f(a) = 0$ for an infinite number of elements a of F, then $f(x)$ must be the zero polynomial.

15. Let F be a field and $f(x)$ and $g(x)$ polynomials in $F[x]$. Show that if $f(x) \neq g(x)$, then $f(a) = g(a)$ for at most finitely many a in F.

In Exercises 16 through 20 find the remainder on dividing the indicated $f(x)$ by $x - a$ for the indicated a in $F[x]$ for the indicated F.

16. $f(x) = x^3 + 3x^2 + 5x + 1$ $a = 1$ $F = \mathbb{Z}_7$

17. $f(x) = x^5 + 4x^3 + 2x + 3$ $a = 1$ $F = \mathbb{Z}_5$

18. $f(x) = x^3 + x^2 + 1$ $a = -1$ $F = \mathbb{Z}_3$

19. $f(x) = x^5 + x^3 + x^2 + 1$ $a = -1$ $F = \mathbb{Q}$

20. $f(x) = x^3 + x^2 - 1$ $a = 2$ $F = \mathbb{Z}_5$

21. Let F be a field, $f(x) = a_nx^n + \ldots + a_1x + a_0$ a polynomial in $F[x]$, and let a be a nonzero element of F. Show that if $f(a) = 0$, then a^{-1} is a zero of the polynomial $g(x) = a_0x^n + a_1x^{n-1} + \ldots + a_n$.

22. Let $a_1, \ldots, a_{n+1}$ be distinct elements of an integral domain D, and $b_1, \ldots, b_{n+1}$ some elements of D. Show that there is at most one polynomial $f(x)$ in $D[x]$ of degree $\leq n$ with $f(a_i) = b_i$ for all $i = 1, \ldots, n + 1$.

23. **(Lagrange interpolation)** Let F be a field and let $a_1, \ldots, a_{n+1}$ be distinct elements of F, and $b_1, \ldots, b_{n+1}$ some elements of F. Let $f(x)$ be the polynomial:

$$\sum_{i=1}^{n+1} b_i\,(x - a_1)\ldots(x - a_{i-1})(x - a_{i+1})\ldots(x - a_{n+1}) \,/\, (a_i - a_1)\ldots(a_i - a_{i-1})(a_i - a_{i+1})\ldots(a_i - a_{n+1})$$

Show that $f(x)$ is the unique polynomial of degree n in $F[x]$ such that $f(a_i) = b_i$ for all $i = 1, \ldots, n + 1$.

In Exercises 24 thorugh 26 use Lagrange interpolation to find the unique polynomial $f(x)$ in $\mathbb{R}[x]$ of the indicated degree n such that the graph of $f(x)$ goes through the indicated points in the plane $\mathbb{R}^2$.

24. $n = 2$ points (1, 2) (2, 4) (3, 2)

25. $n = 3$ points (1, 0) (2, -1) (3, 0) (4, 1)

26. $n = 2$ points (0, 2) (1, 0) (2, 0)

8.4 Irreducible Polynomials

Finding and studying the zeros of a polynomial $f(x)$, or solutions of a polynomial equation $f(x) = 0$, has always been a fundamental part of algebra. The factor theorem (Theorem 8.3.2) shows that factoring a polynomial and finding zeros of a polynomial are closely related problems. In this section we concentrate on the factorization of a polynomial with coefficients in a field F. We first define a notion for $F[x]$ comparable to the notion of a prime for $\mathbb{Z}$ and show that a unique factorization theorem holds in $F[x]$ analogous to the fundamental theorem of arithmetic for $\mathbb{Z}$.

8.4.1 DEFINTION Let F be a field and $f(x)$ a nonconstant polynomial in $F[x]$. Then $f(x)$ is **irreducible** over F if $f(x)$ cannot be expressed as a product $f(x) = g(x)h(x)$ of polynomials $g(x)$ and $h(x)$ in $F[x]$ both of lower degree than $f(x)$. Also, $f(x)$ is **reducible** over F if it is not irreducible. ○

Specifying over which field F a polynomial $f(x)$ is irreducible or reducible is extremely important, as we see from the next examples.

8.4.2 EXAMPLE $x^2 - 2$ is irreducible over $\mathbb{Q}$, but $x^2 - 2 = (x - \sqrt{2})(x + \sqrt{2})$ is reducible over $\mathbb{R}$. ◇

8.4.3 EXAMPLE $x^2 + 1$ is irreducible over $\mathbb{Q}$ and $\mathbb{R}$, but $x^2 + 1 = (x - i)(x + i)$ is reducible over $\mathbb{C}$. ◇

8.4.4 EXAMPLE x^2+1 is reducible over $\mathbb{Z}_2$, where $x^2 + 1 = (x + 1)^2$, but irreducible over $\mathbb{Z}_3$. ◇

The fundamental theorem of arithmetic (Theorem 0.3.23), as was shown in Section 0.3, is an important consequence of the division algorithm for $\mathbb{Z}$. Since we have proved a division algorithm for the ring of polynomials $F[x]$ where F is a field, it should not come as a surprise that an analogue to the fundamental theorem of arithmetic exists for $F[x]$.

8.4.5 THEOREM Let F be a field and let $f(x) \in F[x]$ be irreducible over F. If $g(x), h(x) \in F[x]$ and $f(x) \mid g(x)h(x)$, then $f(x) \mid g(x)$ or $f(x) \mid h(x)$.

Proof Let $d(x) = \gcd(f(x),g(x))$, which exists by Theorem 8.2.8. Since $d(x) \mid f(x)$ and $f(x)$ is irreducible, then either

(1) $\deg d(x) = \deg f(x)$ and $d(x) = af(x)$, where $a \in F$ and $a \neq 0$; or

(2) $\deg d(x) = 0$ and $d(x) = c \in F$, and $c \neq 0$.

In case (1), by Proposition 8.2.5, part (1), $f(x) \mid g(x)$. In case (2), by Theorem 8.2.8, there exist $u(x)$ and $v(x)$ in $F[x]$ such that

$$c = u(x)f(x) + v(x)g(x)$$

Hence

$$h(x) = c^{-1}u(x)f(x)h(x) + c^{-1}v(x)g(x)h(x)$$

and since $f(x) \mid g(x)h(x)$ it follows that $f(x) \mid h(x)$. □

8.4.6 THEOREM (Unique factorization theorem) Let F be a field and $f(x) \in F[x]$ a nonconstant polynomial. Then

(1) $f(x) = up_1(x)p_2(x)\dots p_s(x)$, where $u \in F$, $u \neq 0$, and each $p_i(x)$ is a monic irreducible polynomial over F.

(2) Except for the order of the irreducible factors, the factorization in (1) is unique.

Proof (1) We use induction on $\deg f(x) = n$. If $n = 1$, then $f(x) = a_1x + a_0$, where $a_1 \neq 0$ in F. Hence $f(x) = a_1(x + a_0a_1^{-1})$, and $x + a_0a_1^{-1} \in F[x]$ is monic and irreducible. So assume that the theorem holds for all polynomials of degree $< n = \deg f(x)$. If $f(x)$ is irreducible, there is nothing to prove. Otherwise, $f(x) = g(x)h(x)$, where $\deg g(x) < n$ and $\deg h(x) < n$. Therefore, by our induction hypothesis we have factorizations:

$$g(x) = up_1(x)p_2(x)\dots p_r(x) \qquad h(x) = vp_{r+1}(x)p_{r+2}(x)\dots p_s(x)$$

where u, $v \in F$ and each each $p_i(x)$ is a monic irreducible. This gives the factorization:

$$f(x) = (uv)p_1(x)p_2(x)\dots p_r(x)p_{r+1}(x)p_{r+2}(x)\dots p_s(x)$$

(2) Again we use induction on $\deg f(x) = n$. If $n = 1$, uniqueness holds. So assume uniqueness holds for all polynomials of degree $< n = \deg f(x)$. If

$$f(x) = up_1(x)p_2(x)\dots p_s(x) = vq_1(x)q_2(x)\dots q_t(x)$$

where u, $v \in F$ and the $p_i(x)$ and $q_j(x)$ are monic and irreducible, then $p_1 \mid q_1(x)\dots q_r(x)$. By Theorem 8.4.5, $p_1(x) \mid q_j(x)$ for some $1 \leq j \leq t$. Reordering the irreducible factors in the second factorization if necessary, we may assume $j = 1$ and $p_1(x) \mid q_1(x)$. Since $p_1(x)$ and $q_1(x)$ are monic and irreducible, we must have $p_1(x) = q_1(x)$. (See Exercise 32.) Therefore,

$$up_2(x)\dots p_s(x) = vq_2(x)\dots q_t(x)$$

is a polynomial of degree $< n$. Therefore, by our induction hypothesis, $s = t$, and for each $2 \leq i \leq s$, $p_i(x) = q_j(x)$ for some $2 \leq j \leq t$, and the proof is complete. □

Determining whether a given polynomial is or is not irreducible over a given field can be a very difficult task. In what follows we introduce various techniques that can be used to show that a polynomial is irreducible. To begin, the factor theorem gives a useful criterion for the irreducibility of polynomials of degree 2 or 3.

8.4.7 THEOREM Let F be a field, $f(x)$ a polynomial in $F[x]$ of degree 2 or 3. Then $f(x)$ is reducible over F if and only if $f(x)$ has a zero in F.

Proof ($\Rightarrow$) First suppose $f(x)$ is reducible over F, so we have $f(x) = g(x)h(x)$ for some polynomials $g(x)$ and $h(x)$ both of degree less than that of $f(x)$. Then at least one of $g(x)$ or $h(x)$ would have to be of degree 1 or, in other words, of the form $a_1x + a_0$ for some $a_0, a_1 \in F$ with $a_1 \neq 0$. Then $\alpha = -a_1^{-1}a_0 \in F$ is a zero of $f(x)$ in F.
($\Leftarrow$) Conversely, if α is a zero of $f(x)$ in F, then the factor theorem implies that $f(x)$ factors in $F[x]$ as $f(x) = (x - \alpha)q(x)$ for some polynomial $q(x)$ in $F[x]$ of degree less than deg $f(x)$, so that $f(x)$ is reducible. □

8.4.8 EXAMPLE We can show that $f(x) = 2x^3 + x + 1$ is irreducible over $\mathbb{Z}_5$, simply by calculating $f(a)$ for each element a of $\mathbb{Z}_5$ and showing that no such a is a zero: $f(0) = 1, f(1) = 4, f(2) = 4, f(3) = 3, f(4) = 3$. ◇

8.4.9 EXAMPLE $x^2 + x + 1$ is reducible over $\mathbb{Z}_3$, where $x^2 + x + 1 = (x + 2)^2$, but we can show that x^2+x+1 is irreducible over $\mathbb{Q}$ by showing it has no zeros as follows: Since $x^3 - 1 = (x - 1)(x^2 + x + 1)$, any zeros of $x^2 + x + 1$ would be third roots of unity, whereas we have seen that the only third roots of unity in $\mathbb{C}$ are 1, which is not a zero of $x^2 + x + 1$, and $-{}^1/_2 + {}^{\sqrt{3}}/_2\, i$ and $-{}^1/_2 - {}^{\sqrt{3}}/_2\, i$, which do not belong to $\mathbb{Q}$. ◇

8.4.10 EXAMPLE It should be emphasized that the criterion of Theorem 8.4.7 only applies to polynomials of degree 2 or 3. Consider $x^4 + 2x^2 + 1$ in $\mathbb{Q}[x]$. We can easily see that it is reducible over $\mathbb{Q}$, since $x^4 + 2x^2 + 1 = (x^2 + 1)^2$, but the only zeros are $\pm i$, which do not belong to $\mathbb{Q}$. ◇

All the theorems that follow give us tools to determine whether a given polynomial is or is not irreducible over $\mathbb{Q}$.

8.4.11 THEOREM (Rational roots theorem) Let

$$f(x) = a_nx^n + \ldots + a_1x + a_0$$

be a polynomial in $\mathbb{Z}[x]$. Let a be a zero of $f(x)$ in $\mathbb{Q}$, and write $a = r/s$, where r and s are relatively prime integers. Then r divides a_0 and s divides a_n in $\mathbb{Z}$.

Proof We have

$$0 = f(a) = a_n(r/s)^n + a_{n-1}(r/s)^{n-1} + \ldots + a_1(r/s) + a_0$$

and hence

$$a_nr^n + a_{n-1}r^{n-1}s + \ldots + a_1rs^{n-1} + a_0s^n = 0$$

This can be rewritten in two different ways:

(1) $a_n r^n = -s[a_{n-1}r^{n-1} + \dots + a_0 s^{n-1}]$

(2) $a_0 s^n = -r[a_n r^{n-1} + \dots + a_1 s^{n-1}]$

Since r and s are relatively prime, (1) shows that s divides a_n and (2) shows that r divides a_0. □

8.4.12 EXAMPLE Consider $f(x) = 2x^3 + x^2 - 1$. If $f(x)$ were reducible over $\mathbb{Q}$, by Theorem 8.4.7 it would have a zero a in $\mathbb{Q}$, and by Theorem 8.4.11 we would have $a = r/s$, where $r|-1$ and $s|2$. Hence we would have $r = \pm 1$, $s = \pm 1$ or ± 2, and $a = \pm 1$ or $\pm 1/2$. But we can calculate $f(1) = 2$, $f(-1) = -2$, $f(1/2) = -1/2$, $f(-1/2) = -1$. So $f(x)$ is irreducible over $\mathbb{Q}$. ◇

8.4.13 EXAMPLE We can show for any prime p and integer $n \geq 2$ that the nth root of p is irrational. For $\sqrt[n]{p}$ is a zero of $f(x) = x^n - p$. But Theorem 8.4.11 implies that if a in $\mathbb{Q}$ is a zero of $x^n - p$, then $a = \pm 1$ or $\pm p$, and we can calculate that $f(\pm 1) = \pm 1 - p \neq 0$ and $f(\pm p) = \pm p^n - p \neq 0$. ◇

We have had several examples of polynomials that are irreducible over one field but reducible over a larger field. The next theorem shows us that if we have a polynomial with integer coefficients that cannot be factored in $\mathbb{Z}[x]$, then it still cannot be factored in $\mathbb{Q}[x]$. This is because $\mathbb{Q}$ is not just any field containing $\mathbb{Z}$, since $\mathbb{Q}$ is the field of quotients of $\mathbb{Z}$. What we have said about the relationship between factoring over $\mathbb{Z}$ and factoring over $\mathbb{Q}$ in fact generalizes to the case of any integral domain D and its field of quotients F (see Theorem 9.2.22.) Here we prove it only in the case of $\mathbb{Z}$ and $\mathbb{Q}$. We first need some definitions and lemmas.

8.4.14 DEFINITION Let $f(x) = a_n x^n + \dots + a_0$ be a polynomial in $\mathbb{Z}[x]$. Then $c = \gcd(a_n, \dots, a_0)$ is called the **content** of $f(x)$, and if $c = 1$, then $f(x)$ is said to be a **primitive** polynomial. Note that if $g(x)$ has content c, then $g(x) = cg_1(x)$, where $g_1(x)$ is primitive. ○

8.4.15 LEMMA (Gauss's lemma) Let $f(x)$ and $g(x)$ be primitive polynomials in $\mathbb{Z}[x]$. Then their product $f(x)g(x)$ is also a primitive polynomial.

Proof We prove the lemma by contradiction. Suppose $f(x)g(x)$ is not primitive, and let p be a prime that divides the content of $f(x)g(x)$. We use the induced ring homomorphism $\phi: \mathbb{Z}[x] \to \mathbb{Z}_p[x]$ from Exercise 15 of Section 8.1, which sends $f(x) \in \mathbb{Z}[x]$ to $\overline{f(x)} \in \mathbb{Z}_p[x]$, obtained by reducing the coefficients of $f(x)$ mod p. We have $\overline{f(x)g(x)} = \overline{f(x)} \cdot \overline{g(x)}$ since ϕ is a ring homomorphism. Now since p divides every coefficient of $f(x)g(x)$, we obtain $\overline{f(x)g(x)} = \overline{f(x)} \cdot \overline{g(x)} = 0$ in $\mathbb{Z}_p[x]$. But since $\mathbb{Z}_p[x]$ is an integral domain, this means we have either $\overline{f(x)} = 0$ or $\overline{g(x)} = 0$ in $\mathbb{Z}_p[x]$, and either p divides all the coefficients of $\overline{f(x)}$ or all those of $\overline{g(x)}$ so that either $f(x)$ or $g(x)$ is not primitive, a contradiction. □

8.4.16 THEOREM Let $f(x)$ be a nonzero polynomial in $\mathbb{Z}[x]$. Then $f(x)$ factors into a product of two polynomials of degrees r and s in $\mathbb{Q}[x]$ if and only if it factors into a product of two polynomials of those same degrees r and s in $\mathbb{Z}[x]$.

Proof To prove the theorem, we need to show that if $f(x) = g(x)h(x)$, where $g(x)$ and $h(x)$ are polynomials in $\mathbb{Q}[x]$ with deg $g(x) = r$ and deg $h(x) = s$, then there exist polynomials $g_1(x)$ and $h_1(x)$ in $\mathbb{Z}[x]$ with deg $g_1(x) = r$ and deg $h_1(x) = s$ such that $f(x) = g_1(x)h_1(x)$. It suffices to prove the theorem for primitive polynomials. For then if $f(x)$ is any polynomial, we can write $f(x) = cf_1(x)$, where c is the content of $f(x)$ and $f_1(x)$ is primitive, and any factorization $f(x) = g(x)h(x)$ of $f(x)$ over $\mathbb{Q}$ will give us a factorization $f_1(x) = (c^{-1}g(x))h(x)$ of $f_1(x)$ over $\mathbb{Q}$, and applying the theorem to $f_1(x)$ to get a factorization $f_1(x) = g_1(x)h_1(x)$ of $f_1(x)$ over $\mathbb{Z}$, we then get a factorization $f(x) = (cg_1(x))h_1(x)$ of $f(x)$ over $\mathbb{Z}$.

So given the factorization $f(x) = g(x)h(x)$ in $\mathbb{Q}[x]$ with $f(x)$ a primitive polynomial in $\mathbb{Z}[x]$, let a be the least common multiple of the denominators of the coefficients of $g(x)$ when written in lowest terms, and let b be the same for $h(x)$. Then $ag(x)$ and $bh(x)$ are polynomials in $\mathbb{Z}[x]$, and $abf(x) = (ag(x))(bh(x))$. Now write $ag(x) = cg_1(x)$ and $bh(x) = dh_1(x)$, where c and d are the contents of $ag(x)$ and of $bh(x)$, respectively, and where $g_1(x)$ and $h_1(x)$ are primitive polynomials in $\mathbb{Z}[x]$. Since $f(x)$ is primitive, the content of $abf(x)$ is ab. By Gauss's lemma, $g_1(x)h_1(x)$ is primitive, and therefore cd is the content of $(cg_1(x))(dh_1(x)) = abf(x)$. It follows that $ab = cd$, and since $abf(x) = cdg_1(x)h_1(x)$ we have $f(x) = g_1(x)h_1(x)$, a factorization in $\mathbb{Z}[x]$, where moreover $g_1(x)$ has the same degree as $cg_1(x) = ag(x)$ and as $g(x)$, and $h_1(x)$ has the same degree as $h(x)$.

The other direction of the theorem is immediate. □

8.4.17 EXAMPLE Consider $f(x) = 6x^3 + 4x^2 - 3x - 2$ in $\mathbb{Z}[x]$. The gcd of 6, 4, 3, 2 is 1, so $f(x)$ is primitive. Now we have the factorization $f(x) = ({}^9/_2x + 3)({}^4/_3x^2 - {}^2/_3) = g(x)h(x)$. With the same notation as in Theorem 8.4.16 and its proof, we have $r = 1$, $s = 2$, $a = 2$, $b = 3$, $c = 3$, $d = 2$, so $ag(x) = 9x + 6 = 3(3x + 2) = cg_1(x)$ and $bh(x) = 4x^2 - 2 = 2(2x^2 - 1) = dh_1(x)$, and we have the factorization $f(x) = g_1(x)h_1(x) = (3x + 2)(2x^2 - 1)$ in $\mathbb{Z}[x]$. ◇

8.4.18 EXAMPLE We now apply Theorem 8.4.16 to show that $f(x) = x^4 - 5x^2 + 6x + 1$ is irreducible over $\mathbb{Q}$. If $f(x)$ had a divisor of degree 1, it would have a zero in $\mathbb{Q}$, and by Theorem 8.4.11 it would have to be ± 1. But calculations show that $f(1) = 3$ and $f(-1) = -9$. If $f(x)$ had two divisors of degree 2, we would have

$$x^4 - 5x^2 + 6x + 1 = (x^2 + ax + b)(x^2 + cx + d)$$

where a, b, c, d may be taken to be integers by Theorem 8.4.16. Now

$$(x^2 + ax + b)(x^2 + cx + d) = x^4 + (a + c)x^3 + (b + d + ac)x^2 + (bc + ad)x + bd$$

Equating coefficients on the same powers of x, we obtain $a + c = 0$, $b + d + ac = -5$, $bc + ad = 6$, and $bd = 1$. But since a, b, c, d are integers, from the condition $bd = 1$

we must have $b = d = \pm 1$, and from the condition $bc + ad = 6$ we get $a + c = \pm 6$, contrary to the condition $a + c = 0$. This contradiction shows that $f(x)$ cannot be factored over $\mathbb{Z}$ and therefore cannot be factored over $\mathbb{Q}$. ◇

The next theorem gives us another way to test whether polynomials with integer coefficients are irreducible over $\mathbb{Q}$.

8.4.19 THEOREM (Eisenstein's criterion) Let

$$f(x) = a_n x^n + \ldots + a_1 x + a_0$$

be a polynomial in $\mathbb{Z}[x]$, and suppose that there exists a prime p such that

(1) p does not divide a_n.

(2) p divides a_i for all $i < n$.

(3) p^2 does not divide a_0.

Then $f(x)$ is irreducible over $\mathbb{Q}$.

Proof Suppose $f(x)$ is reducible over $\mathbb{Q}$. Then by Theorem 8.4.16, $f(x)$ is reducible over $\mathbb{Z}$ or, in other words,

$$f(x) = (b_r x^r + \ldots + b_0)(c_s x^s + \ldots + c_0)$$

where the b_i and c_i are integers, r and s are $\neq 0$, and $r + s = n$. The constant term a_0 of $f(x)$ must be equal to $b_0 c_0$. From condition (3) we know that p does not divide both b_0 and c_0, while from condition (2) we know p divides $a_0 = b_0 c_0$ and therefore divides exactly one of b_0 and c_0. So let's say p divides b_0 but not c_0. The leading coefficient a_n of $f(x)$ must be equal to $b_r c_s$. From condition (1) we know p does not divide $a_n = b_r c_s$ and therefore does not divide either b_r or c_s. Let m be the least integer for which p does not divide b_m: In other words, p divides b_i for $i < m$, but p does not divide b_m. We know that $1 \leq m \leq r < n$. Consider the coefficient a_m in $f(x)$. It can be calculated to be

$$a_m = b_m c_0 + b_{m-1} c_1 + \ldots + b_1 c_{m-1} + b_0 c_m$$

Here p divides b_i for all $i < m$, but p does not divide b_m or c_0. It follows from this that p does not divide a_m, where $m < n$, contrary to condition (2). The contradiction completes the proof. □

8.4.20 EXAMPLE Let $f(x) = 10x^7 - 6x^3 + 27x + 12$. Taking $p = 3$, Eisenstein's criterion shows that $f(x)$ is irreducible over $\mathbb{Q}$. ◇

We apply Eisenstein's criterion to the so-called cyclotomic polynomials.

8.4.21 DEFINITION Consider the polynomial $x^n - 1$ whose zeros are the nth roots of unity. Since 1 is clearly an nth root of unity, $x^n - 1$ must be divisible by $x - 1$. In fact, the quotient can be calculated:

$$(x^n - 1) = (x - 1)(x^{n-1} + x^{n-2} + \ldots + x + 1)$$

If p is a prime, then the polynomial $\Phi_p(x)$ defined by

$$\Phi_p(x) = (x^p - 1)/(x - 1) = x^{p-1} + x^{p-2} + \ldots + x + 1$$

is called the **cyclotomic** polynomial for p. ○

8.4.22 COROLLARY The cyclotomic polynomial $\Phi_p(x)$ is irreducible over $\mathbb{Q}$ for every prime p.

Proof Eisenstein's criterion does not apply directly, of course, since the coefficients of $\Phi_p(x)$ are all 1. But consider the polynomial $f(x) = \Phi_p(x + 1)$. Any factorization $\Phi_p(x) = g(x)h(x)$ would give rise to a factorization $f(x) = g_1(x)h_1(x)$, where $g_1(x) = g(x + 1)$ and $h_1(x) = h(x + 1)$. So if $\Phi_p(x)$ were reducible over $\mathbb{Q}$, $f(x)$ would also be reducible over $\mathbb{Q}$. Now we can calculate using the binomial theorem (Theorem 0.3.8):

$$f(x) = [(x+1)^p-1]/[(x+1)-1] = x^{p-1} + px^{p-2} + \binom{p}{2}x^{p-3} + \ldots + \binom{p}{2}x + p$$

Hence by Exercise 14 in Section 0.3 Eisenstein's criterion applies and tells us that $f(x)$ is irreducible over $\mathbb{Q}$, so $\Phi_p(x)$ must be irreducible over $\mathbb{Q}$. □

8.4.23 EXAMPLE By the corollary, $f(x) = \Phi_5(x) = x^4 + x^3 + x^2 + x + 1$ is irreducible over $\mathbb{Q}$. But note that $g(x) = x^3 + x^2 + x + 1$ is reducible, since $g(-1) = 0$, and it factors as $g(x) = (x + 1)(x^2 + 1)$. ◇

We conclude this section with one more irreducibility test.

8.4.24 THEOREM Let $f(x)$ be a polynomial in $\mathbb{Z}[x]$ of degree at least 1, let p be a prime, and let $\overline{f(x)}$ be the polynomial in $\mathbb{Z}_p[x]$ obtained from $f(x)$ after reducing its coefficients mod p. Then if

(1) $\deg f(x) = \deg \overline{f(x)}$ and

(2) $\overline{f(x)}$ is irreducible over $\mathbb{Z}_p$

then $f(x)$ is irreducible over $\mathbb{Q}$.

Proof If $f(x)$ had a factorization in $\mathbb{Q}[x]$, by Theorem 8.4.16 it would have a factorization in $\mathbb{Z}[x]$, so $f(x) = g(x)h(x)$, where $g(x)$ and $h(x)$ are polynomials in $\mathbb{Z}[x]$ and $\deg g(x)$ and $\deg h(x)$ are both $< \deg f(x)$. Since reduction mod p is a ring homomorphism between $\mathbb{Z}[x]$ and $\mathbb{Z}_p[x]$, we would have $\overline{f(x)} = \overline{g(x)h(x)} = \overline{g(x)}\cdot\overline{h(x)}$ where

$$\deg \overline{g(x)} \leq \deg g(x) < \deg f(x) = \deg \overline{f(x)}$$

by condition (1), and, similarly, $\deg \overline{h(x)} < \deg \overline{f(x)}$. But this would be a factorization of $\overline{f(x)}$ in $\mathbb{Z}_p[x]$, contrary to assumption (2). □

8.4.25 EXAMPLE Consider $f(x) = 7x^3 - 6x^2 + 3x + 9$ in $\mathbb{Z}[x]$. This reduces mod 2 to $\overline{f(x)} = x^3 + x + 1$ in $\mathbb{Z}_2[x]$, which still has degree 3, and is irreducible over $\mathbb{Z}_2$ since it has no zeros in $\mathbb{Z}_2$, because $\overline{f(0)} = \overline{f(1)} = 1$. Thus $f(x)$ is irreducible over $\mathbb{Q}$ by Theorem 8.4.24. ◇

Note that to apply Theorem 8.4.24 we only need to find *one* prime p such that $f(x)$ is irreducible mod p. It may be that $f(x)$ is reducible mod q for some *other* prime q. For instance, it can be shown that $f(x)$ in the preceding example is reducible mod 3, since $f(x) = 7x^3 - 6x^2 + 3x + 9$ reduces to $\overline{f(x)} = x^3$ in $\mathbb{Z}_3[x]$.

8.4.26 EXAMPLE Consider $f(x) = 3x^4 - 6x^3 + 10x^2 - 5x + 9$ in $\mathbb{Z}[x]$. This reduces mod 2 to $\overline{f(x)} = x^4 + x + 1$ in $\mathbb{Z}_2[x]$, which still has degree 4. But in $\mathbb{Z}_2[x]$, this polynomial cannot have a divisor of degree 1 because it has no zeros in $\mathbb{Z}_2$, since $\overline{f(0)} = \overline{f(1)} = 1$. Moreover, if this polynomial had a divisor of degree 2 we would have

$$x^4 + x + 1 = (x^2 + ax + b)(x^2 + cx + d)$$

which would require $bd = 1$, $bc + ad = 1$, $b + d + ac = 0$, and $a + c = 0$, and these are conditions that cannot all be fulfilled, because $bd = 1$ implies $b = d = 1$, given which $bc + ad = 1$ implies $c + a = 1$, contrary to $a + c = 0$, so $\overline{f(x)}$ is irreducible over $\mathbb{Z}_2$, and hence $f(x)$ is irreducible over $\mathbb{Q}$. ◇

The next example shows that the converse of Theorem 8.4.24 does not hold: A polynomial in $\mathbb{Z}[x]$ can be reducible over $\mathbb{Z}_p$ for all primes p and yet irreducible over $\mathbb{Q}$.

8.4.27 EXAMPLE Consider $f(x) = x^4 + 1$. We first show that this polynomial is reducible over $\mathbb{Z}_p$ for all primes p. For $p = 2$ we have $x^4 + 1 = (x + 1)^4$. For an odd prime p, consider the subgroup $H = \{a^2 \mid a \in \mathbb{Z}_p^*\}$ of the multiplicative group $\mathbb{Z}_p^*$ of nonzero elements of $\mathbb{Z}_p$. Since p is odd, $|\mathbb{Z}_p^*| = p - 1$ is even, let's say $p - 1 = 2k$. In that case, $2k + 1 = 0$ in $\mathbb{Z}_p$, and $k + 1 = -k$ and, in general, $k + i = -(k - i + 1)$ in $\mathbb{Z}_p$. Hence

$$\mathbb{Z}_p^* = \{1, 2, 3, \ldots, k, k + 1, k + 2, \ldots, 2k\} = \{1, 2, 3, \ldots, k, -k, -(k - 1), \ldots, -2, -1\}$$

and

$$H = \{1^2 = (-1)^2, 2^2 = (-2)^2, \ldots, k^2 = (-k)^2\}$$

and therefore $|H| \leq k$. Furthermore, for all a and b in $\mathbb{Z}_p$, $a^2 = b^2$ if and only if $a^2 - b^2 = (a + b)(a - b) = 0$, and since p is prime, there are no zero divisors in $\mathbb{Z}_p$. Hence $a^2 = b^2$ if and only if $a + b = 0$ or $a - b = 0$ or, in other words, if and only if $b = \pm a$. Thus $|H| = k = {}^1/_2(p - 1) = {}^1/_2|\mathbb{Z}_p^*|$, and H is a subgroup of $\mathbb{Z}_p^*$ of index 2.
If $-1 \in H$, then taking an a such that $a^2 = -1$, we have the factorization

$$x^4 + 1 = x^4 - a^2 = (x^2 + a)(x^2 - a)$$

If $-1 \notin H$, we proceed as follows. We have seen that H has index 2 in $\mathbb{Z}_p^*$ or, in other words, H has just two cosets in $\mathbb{Z}_p^*$. Since $-1 \notin H$ these cosets are just H and $-H = \{-b \mid b \in H\}$. This means either $2 \in H$ or else $2 \in -H$ and $-2 \in H$. In the former case, take a such that $a^2 = 2$.

We then have the factorization

$$x^4 + 1 = (x^2 + ax + 1)(x^2 - ax + 1)$$

In the latter case, take a such that $a^2 = -2$. We then have the factorization

$$x^4 + 1 = (x^2 + ax - 1)(x^2 - ax - 1)$$

Thus in every case $f(x)$ is reducible over $\mathbb{Z}_p$. But $f(x)$ has no divisor of degree 1 in $\mathbb{Z}[x]$, since it has no zeros, because $f(\pm 1) = 2$. If it had a divisor of degree 2 in $\mathbb{Z}[x]$ we would have a factorization

$$x^4 + 1 = (x^2 + ax + b)(x^2 + cx + d)$$

which would require $bd = 1$, $bc + ad = 0$, $b + d + ac = 0$, and $a + c = 0$. The condition $bd = 1$ implies $b = d = \pm 1$; the condition $bc + ad = 0$ then implies $c = -a$; and the condition $b + d + ac = 0$ then implies $a^2 = \pm 2$, which is impossible for an integer a. This shows that $f(x)$ is irreducible over $\mathbb{Q}$. Thus we have an example of a polynomial irreducible over $\mathbb{Q}$ but reducible over $\mathbb{Z}_p$ for all primes p. ◇

8.4.28 EXAMPLE From the calculations in the preceding example we see that $x^4 + 1$ is reducible over $\mathbb{R}$, for we can take $b = d = 1$ and $a = \sqrt{2}$ and obtain a factorization into quadratic factors:

$$x^4 + 1 = (x^2 + \sqrt{2}\,x + 1)(x^2 - \sqrt{2}\,x + 1)$$

Applying the quadratic formula to this factorization, we see that $x^4 + 1$ factors over $\mathbb{C}$ as follows:

$$(x + \tfrac{\sqrt{2}}{2} + \tfrac{\sqrt{2}}{2}\,i\,)(x + \tfrac{\sqrt{2}}{2} - \tfrac{\sqrt{2}}{2}\,i\,)(x - \tfrac{\sqrt{2}}{2} + \tfrac{\sqrt{2}}{2}\,i\,)(x - \tfrac{\sqrt{2}}{2} - \tfrac{\sqrt{2}}{2}\,i\,)$$

into a product of linear factors. ◇

Exercises 8.4

In Exercises 1 through 12 factor the indicated polynomial $f(x)$ completely into irreducible factors in the polynomial ring $F[x]$ for the indicated field F.

1. $f(x) = x^4 - 1$ $\quad F = \mathbb{R}, F = \mathbb{C}$

2. $f(x) = x^4 + 1$ $\quad F = \mathbb{Z}_2$

3. $f(x) = x^4 + 4$ $\quad F = \mathbb{Z}_5$

4. $f(x) = x^3 + 4x^2 + 5x + 2$ $\quad F = \mathbb{Z}_7$

5. $f(x) = x^2 + 2x - 1$ $\quad F = \mathbb{Z}_7$

6. $f(x) = x^2 + x + 3$ $\quad F = \mathbb{Z}_5$

7. $f(x) = x^2 + x + 1$ $\quad F = \mathbb{Q}, F = \mathbb{C}$

8. $f(x) = x^3 + x^2 + x + 1$ $\quad F = \mathbb{Z}_5, F = \mathbb{Q}, F = \mathbb{C}$

9. Show that $f(x) = x^3 + 2x + 1$ is irreducible over $\mathbb{Z}_5$.

10. Show that $f(x) = x^3 + x + 1$ is irreducible over $\mathbb{Z}_7$.

11. Show that $f(x) = x^4 - 2$ is irreducible over $\mathbb{Q}$ but reducible over $\mathbb{R}$.

12. Show that $f(x) = x^4 - 2x^2 - 4$ is irreducible over $\mathbb{Q}$.

In Exercises 13 through 18 find all the zeros of the indicated polynomial $f(x)$ in the indicated field F.

13. $f(x) = x^3 + 2x^2 - 5x - 6$ $\quad F = \mathbb{Q}$

14. $f(x) = x^4 - 2x^2 - 4$ $\quad F = \mathbb{R}$

15. $f(x) = x^4 - 2x^2 - 4$ $\quad F = \mathbb{C}$

16. $f(x) = x^4 - x^3 - 8x^2 + 11x - 3$ $\quad F = \mathbb{R}$

17. $f(x) = x^4 - 3x^3 + 3x^2 - 6x + 2$ $\quad F = \mathbb{C}$

18. $f(x) = x^4 + x^3 + 3x^2 + 2x + 2$ $\quad F = \mathbb{C}$

In Exercises 19 through 25 determine if possible, using any of the criteria given by theorems in this section, whether the indicated polynomial $f(x)$ in $\mathbb{Z}[x]$ is reducible over $\mathbb{Q}$. Justify your answers.

19. $f(x) = 10x^7 - 6x^4 + 15x^2 + 18x - 6$

20. $f(x) = x^4 - 4x^2 + 4x - 1$

21. $f(x) = x^6 + x^5 + x^4 + x^3 + x^2 + x + 1$

22. $f(x) = 2x^4 + x^3 - 2x^2 + x + 1$

23. $f(x) = 3x^4 + 5x + 1$

24. $f(x) = x^6 + 2x^3 - 3x^2 + 1$

25. $f(x) = x^4 + 4$

26. Construct a new example of a polynomial $f(x)$ in $\mathbb{Z}[x]$ that is irreducible over $\mathbb{Q}$ but reducible over $\mathbb{R}$.

27. Construct a new example of a polynomial $f(x)$ in $\mathbb{Z}[x]$ that is irreducible over $\mathbb{Q}$ even though the polynomial $\overline{f(x)}$ in $\mathbb{Z}_p[x]$ is reducible over $\mathbb{Z}_p$ for every prime p.

28. Find all the polynomials of degree 2 that are irreducible over $\mathbb{Z}_3$.

29. For any prime p, show that the number of polynomials of form $x^2 + ax + b$ that are irreducible over $\mathbb{Z}_p$ is $p(p-1)/2$.

30. For any prime p and any element a of $\mathbb{Z}_p$, show that $x^p - a$ and $x^p + a$ are reducible over $\mathbb{Z}_p$.

31. Let $f(x) = x^{n-1} + x^{n-2} + \ldots + x + 1 \in \mathbb{Q}[x]$, where n is not prime. Show that $f(x)$ is not irreducible over $\mathbb{Q}$.

32. Let F be a field $f(x)$ and $g(x)$ in $F[x]$ be monic polynomials. Show that if $g(x)$ is irreducible over F and $f(x) \mid g(x)$, then either $f(x) = 1$ or $f(x) = g(x)$.

33. Let F be a field and $F(x)$ the field of rational functions over F. (See Definition 8.1.20.) Then $f(x)/g(x) \in F(x)$ is said to be in **reduced form** if $g(x)$ is monic and $\gcd(f(x), g(x)) = 1$. Show that every element $f(x)/g(x) \in F(x)$ can be put in reduced form in a unique way.

8.5 Cubic and Quartic Polynomials

We all learned in high school algebra the quadratic formula that allows us to find the zeros of quadratic (degree 2) polynomials with coefficients in $\mathbb{R}$ in terms of square roots. (The zeros may belong to $\mathbb{C}$ and not to $\mathbb{R}$, of course.) Algorithms also exist for finding the zeros in $\mathbb{C}$ of cubic (degree 3) and quartic (degree 4) polynomials with coefficients in $\mathbb{R}$, in terms of square and cube roots, though they are seldom mentioned in high school. Since finding zeros of a polynomial $f(x)$ or, what is the same thing, finding the solutions to a polynomial equation $f(x) = 0$ is such an old and fundamental part of algebra, the reader ought to be given an account of these algorithms, and we give one in this chapter. In Chapter 12 we study such polynomials more thoroughly and, using Galois theory, we show that there can be no similar algorithm for quintic (degree 5) polynomials.

Cubics

We start with an example that is already familiar, pertaining to cube or third roots.

8.5.1 EXAMPLE Let $f(x) = x^3 - 1 \in \mathbb{R}[x]$. Since $f(1) = 0$, $f(x)$ factors over $\mathbb{R}$

$$f(x) = (x - 1)(x^2 + x + 1)$$

and, using the quadratic formula, we can find the zeros of the second factor:

$$\omega = -\frac{1}{2} + \frac{\sqrt{3}}{2}i \text{ and } \omega^2 = -\frac{1}{2} - \frac{\sqrt{3}}{2}i$$

So 1, ω, and ω^2 are the third roots of unity. Note that in agreement with Theorem 8.3.12, ω^2 is the complex conjugate of ω.

8.5.2 EXAMPLE Let $f(x) = x^3 - a \in \mathbb{R}[x]$, where $a > 0$. Then $\sqrt[3]{a}$ is one of the zeros of $f(x)$. To obtain the other two zeros, note that if ω is a third root of unity as in the preceding example, then $(\sqrt[3]{a}\,\omega)^3 = a \cdot \omega^3 = a \cdot 1 = a$. Hence $\sqrt[3]{a}$, $\sqrt[3]{a}\,\omega$, and $\sqrt[3]{a}\,\omega^2$ are the three zeros of $f(x) = x^3 - a$. Again $\sqrt[3]{a}\,\omega^2$ is the complex conjugate of $\sqrt[3]{a}\,\omega$. ◇

8.5.3 EXAMPLE Let $f(x) = x^3 + 1 \in \mathbb{R}[x]$. Then $f(-1) = 0$ and $f(x)$ factors over $\mathbb{R}$:

$$f(x) = (x + 1)(x^2 - x + 1)$$

and, using the quadratic formula, we can find the two complex zeros of $f(x)$:

$$-\omega = \frac{1 - \sqrt{3}i}{2} \text{ and } -\omega^2 = \frac{1 + \sqrt{3}i}{2}$$

So 1, $-\omega$, $-\omega^2$ are the three zeros of $f(x)$. ◇

8.5.4 EXAMPLE Let $f(x) = x^3 + a \in \mathbb{R}[x]$, where $a > 0$. Then from the preceding two examples we obtain the result that $-\sqrt[3]{a}$, $-\sqrt[3]{a}\,\omega$, and $-\sqrt[3]{a}\,\omega^2$ are the three zeros of $f(x)$. ◇

These examples bring out the importance of the third roots of unity 1, ω, ω^2. In particular, Examples 8.5.2 and 8.5.4 show that once we know one real zero of $f(x)$ we can immediately write down the other two complex zeros. The next example is a harder one, and it illustrates the method used in the general case, but again in this example the third roots of unity play an important role.

8.5.5 EXAMPLE Let $f(x) = x^3 + 3x + 1 \in \mathbb{R}[x]$. The solution in this case uses a trick — and one it took thousands of years for anyone to think of! The idea is to show that we could find solutions for the equation we are interested in *if* we could find solutions for some *other* equation, and eventually reduce the problem we want to solve to one we know we can solve. First, we substitute $u + v$ for x. To solve the equation $f(x) = 0$, it would be enough to solve the equation

$$(u + v)^3 + 3(u + v) + 1 = 0, \text{ or}$$
$$u^3 + 3u^2v + 3uv^2 + v^3 + 3u + 3v + 1 = 0, \text{ or}$$
$$u^3 + v^3 + 1 = -3(u^2v + uv^2 + u + v), \text{ or}$$
$$u^3 + v^3 + 1 = -3(u + v)(uv + 1)$$

One way this equation could be satisfied would be if both sides were zero, which would be the case if we had

(1) $$u^3 + v^3 + 1 = 0$$

(2) $$uv + 1 = 0$$

To satisfy (2), we can let $v = -1/u = -u^{-1}$. Then (1) becomes

$$u^3 + v^3 + 1 = u^3 - u^{-3} + 1 = 0, \text{ or}$$
$$u^3 + 1 - u^{-3} = 0, \text{ or}$$
$$u^6 + u^3 - 1 = 0, \text{ or}$$
$$(u^3)^2 + u^3 - 1 = 0$$

This is a quadratic equation in u^3, so we solve it using the quadratic formula, which gives

$$u^3 = \frac{-1 \pm \sqrt{5}}{2} \in \mathbb{R}$$

Since $u^3 + v^3 = -1$ we can take

(3)
$$u^3 = \frac{-1+\sqrt{5}}{2} \text{ and } v^3 = \frac{-1-\sqrt{5}}{2}$$

(It doesn't really matter which of these we call u and which we call v, because the roles of u and v are completely symmetric.) Then, taking the real cube roots in (3), the real zero we are looking for can be written as

$$x = u + v = \sqrt[3]{\frac{-1+\sqrt{5}}{2}} + \sqrt[3]{\frac{-1-\sqrt{5}}{2}} \in \mathbb{R}$$

Now, to obtain the other two zeros we use the third roots of unity ω and ω^2. Recall that $1 + \omega + \omega^2 = 0$, $\omega^2 = \overline{\omega}$ and $\omega^3 = (\omega^2)^3 = 1$. Letting u and v be as before, consider now $u\omega + v\omega^2$ and its complex conjugate $u\omega^2 + v\omega$. We claim that these are the two complex zeros of $f(x) = x^3 + 3x + 1$. Since they are complex conjugates, it is enough to check one of them:

$$(u\omega + v\omega^2)^3 + 3(u\omega + v\omega^2) + 1 =$$
$$u^3\omega^3 + 3u^2\omega^2v\omega^2 + 3u\omega v^2\omega^4 + v^3\omega^6 + 3u\omega + 3v\omega^2 + 1 =$$
$$u^3 + 3u^2v\omega + 3uv^2\omega^2 + v^3 + 3u\omega + 3v\omega^2 + 1 =$$
$$(u^3 + v^3 + 1) + 3(u^2v\omega + uv^2\omega^2 + u\omega + v\omega^2) =$$
$$(u^3 + v^3 + 1) + 3(uv + 1)(u\omega + v\omega^2) = 0 + 3\cdot 0\cdot(u\omega + v\omega^2) = 0$$

Here we have used (1) and (2) at the next to last step. Thus

$$u + v = \sqrt[3]{\frac{-1+\sqrt{5}}{2}} + \sqrt[3]{\frac{-1-\sqrt{5}}{2}}$$

$$u\omega + v\omega^2 = \sqrt[3]{\frac{-1+\sqrt{5}}{2}} \cdot \frac{-1+\sqrt{3}i}{2} + \sqrt[3]{\frac{-1-\sqrt{5}}{2}} \cdot \frac{-1-\sqrt{3}i}{2}$$

$$u\omega^2 + v\omega = \sqrt[3]{\frac{-1+\sqrt{5}}{2}} \cdot \frac{-1-\sqrt{3}i}{2} + \sqrt[3]{\frac{-1-\sqrt{5}}{2}} \cdot \frac{-1+\sqrt{3}i}{2}$$

are the three zeros of $f(x)$. ◇

Observe that in the preceding example, u^3 and hence v^3 turned out to be real numbers. This will not always be the case.

Following exactly the same steps as in the example, we obtain the following more general solution.

8.5.6 PROPOSITION Let $f(x) = x^3 + px + q \in \mathbb{R}[x]$, and let u and v be given by

$$u^3 = \frac{-q}{2} + \sqrt{\left(\frac{q}{2}\right)^2 + \left(\frac{p}{3}\right)^3} \quad \text{and} \quad uv = -p/3$$

Then the zeros of $f(x)$ are $u + v$, $u\omega + v\omega^2$, and $u\omega^2 + v\omega$.

Proof Left to the reader. (See Exercise 35.) □

From here it is a comparatively short step to a completely general solution. First, since $a_3x^3 + a_2x^2 + a_1x + a_0$ and $x^3 + (a_2/a_3)x^2 + (a_1/a_3)x + (a_0/a_3)$ have the same zeros, we need only consider the case of a monic polynomial $f(x) = x^3 + ax^2 + bx + c$.

To find the zeros of $f(x)$, substitute $z - (a/3)$ for x and notice that the equation $f(x) = 0$ then reduces to an equation:

$$z^3 + pz + q = 0$$

where p and q are rational expressions in a, b, c. (See Exercise 17.) The solution to this latter equation can be obtained by the preceding proposition, and any such solution, z, gives a solution to the original equation by substituting $x = z - (a/3)$. Thus in principle the discussion to this point should equip the reader to find the zeros of any cubic polynomial in $\mathbb{R}[x]$, using square and cube roots.

Quartics

We end this section with some discussion of finding the zeros of a quartic polynomial $f(x)$ in $\mathbb{R}[x]$. Suppose we wish to solve an equation

(1) $$f(x) = x^4 + ax^3 + bx^2 + cx + d = 0$$

If we can find numbers h, k, u, v such that we have

(2) $$f(x) = (x^2 + hx + k)^2 - (ux + v)^2$$

then the original equation (1) will be equivalent to

(3) $$[(x^2 + hx + k) + (ux + v)][(x^2 + hx + k) - (ux + v)] = 0$$

which can be solved by applying the quadratic formula twice.

How can we find such numbers? Expanding the right side of (2) and comparing the coefficients with the coefficients in (1), we obtain the following:

$a = 2h$	or $h = a/2$
$b = h^2 + 2k - u^2$	or $u^2 = h^2 + 2k - b = (a/2)^2 + 2k - b$
$c = 2hk - 2uv$	or $2uv = 2hk - c = ak - c$
$d = k^2 - v^2$	or $v^2 = k^2 - d$

These equations suffice to determine h, u, v provided that k can be determined. To determine k, note that $4u^2v^2 - (2uv)^2 = 0$. Hence, using the preceding expressions for u^2, $2uv$, v^2 in terms of k, we have a condition that k must satisfy

$$4[(a/2)^2 + 2k - b][k^2 - d] - [ak - c]^2 = 0, \text{ or}$$

(4) $$8k^3 - 4bk^2 + (2ac - 8d)k + (4bd - a^2d - c^2) = 0$$

This is a cubic equation that in principle we know how to solve in terms of square and cube roots. Solving it and obtaining k, we then go back to determine h, u, v, and then solve the original equation $f(x) = 0$.

8.5.7 DEFINITION The cubic in (4) is called the **auxiliary cubic** of the original quartic in (1). ○

8.5.8 EXAMPLE Let $f(x) = x^4 + x^3 + 2x^2 + x + 1$. Then the auxiliary cubic is

$$g(k) = 8k^3 - 8k^2 - 6k + 6$$

To solve $4k^3 - 4k^2 - 3k + 3 = 0$, before going through the long process required for a solution to a cubic equation in general, we first check for rational zeros and in this case. Fortunately, we find that 1 is a zero of $g(k)$.

Hence we obtain

$$h = 1/2$$
$$u = \sqrt{h^2 + 2k - b} = 1/2$$
$$v = \sqrt{k^2 - d} = 0$$

and from (3) we obtain

$$[(x^2 + (x/2) + 1) + (x/2)][(x^2 + (x/2) + 1) - (x/2)] = 0$$

or, in other words, $(x^2 + x + 1)(x^2 + 1) = 0$, and by Example 8.5.1, ω, ω^2, i, $-i$ are the four zeros of $f(x)$. ◇

8.5.9 COROLLARY Let

$$f(x) = x^4 + px^2 + qx + r \in \mathbb{Q}[x]$$

Then its auxiliary cubic is

$$8x^3 - 4px^2 - 8rx + (4pr - q^2)$$

Proof Immediate from (4). □

Exercises 8.5

In Exercises 1 through 5 find all the complex numbers $z = a + bi \in \mathbb{C}$ that satisfy the indicated equation

1. $z^2 = i$ **2.** $z^3 = i$ **3.** $z^2 = 1 + i$ **4.** $z^4 = i$ **5.** $z^3 = 1 + i$

In Exercises 6 through 10 find the zeros in $\mathbb{C}$ of the indicated polynomials from $\mathbb{R}[x]$, and express them in the form $a + bi$, where $a, b \in \mathbb{R}$.

6. $x^2 + x + 1$ **7.** $x^2 + 2x + 2$ **8.** $x^2 + 6x + 3$

9. $x^2 + 6x + 9$ **10.** $3x^2 - 2x + 2$

Discriminants

11. Let $f(x)$ be a polynomial of degree 2 in $\mathbb{R}[x]$ with two distinct zeros r, s in $\mathbb{C}$. The **discriminant** D of $f(x)$ is defined as the product $D = (r - s)^2$. Show that

(a) If $f(x) = x^2 + bx + c$, then $D = b^2 - 4c$

(b) $D > 0$ if and only if r and s are distinct real numbers.

(c) $D < 0$ if and only if r and s are nonreal complex conjugates.

12. Let $f(x)$ be a polynomial of degree 3 in $\mathbb{R}[x]$ with three distinct zeros r, s, t in $\mathbb{C}$. The **discriminant** D of $f(x)$ is the product $D = [(r - s)(r - t)(s - t)]^2$. Show that:

(a) If $f(x) = x^3 + px + q$, then $D = -4p^3 - 27q^2$

(b) $D > 0$ if and only if r, s, t are distinct real numbers.

(c) $D < 0$ if and only if one of r, s, t is real and the others are nonreal complex conjugates.

In Exercises 13 through 16 find the discriminant of the indicated cubic polynomial, and in each case determine the number of distinct real zeros.

13. $x^3 + x - 1$

14. $x^3 + x + 1$

15. $x^3 - 3x - 1$

16. $x^3 - 5x + 3$

17. Let $f(x) = x^3 + ax^2 + bx + c$ and let $g(x) = f(x - (a/3))$. Show that $g(x)$ is of the form $x^3 + px + q$.

18. With the same notation as in the preceding exercise, show that $f(x)$ and $g(x)$ have the same discriminant.

In Exercises 19 through 22 use Exercises 12 [part (a)], 17, and 18 to find the discriminant of the indicated cubic polynomial, and in each case determine the number of distinct real zeros.

19. $x^3 - x^2 + 2x + 1$

20. $x^3 + x^2 + 2x + 1$

21. $x^3 - x^2/2 + x/2 + 1/2$

22. $x^3 - 6x^2 + 7x + 9$

In Exercises 23 through 30 find all the zeros of the indicated cubic polynomials in $\mathbb{C}$. (*Hint:* Use Proposition 8.5.6.)

23. $x^3 + 3x - 1$

24. $x^3 - 3x - 1$

25. $x^3 - 3x + 2$

26. $x^3 + x + 1$

27. $x^3 - 3x + 1$

28. $x^3 - x^2 + 2x + 1$

29. $x^3 + x^2 + 2x + 1$

30. $x^3 - 2x^2 - x + 2$

In Exercises 31 through 34 find all the zeros of the indicated quartic polynomials in $\mathbb{C}$. (*Hint:* Check first for rational zeros.)

31. $x^4 - 4x^3 + 6x^2 - 4x + 1$

32. $x^4 + 2x^2 + 1$

33. $x^4 - 2x^3 - x + 2$

34. $x^4 - 2x^3 + 3x^2 - 7x + 2$

35. Prove Proposition 8.5.6.

36. Show that solving $x^4 + ax^2 + bx + c = 0$ is equivalent to finding the intersection of the parabola $y = x^2$ with another parabola.

37. Let $f(x) = x^4 + x^3 - x^2 - x + 1 \in \mathbb{Q}[x]$.

(a) Show that if $r \in \mathbb{C}$ is a zero of $f(x)$, then $-(1/r)$ is also a zero of $f(x)$.

(b) Find all the zeros of $f(x)$ in $\mathbb{C}$.

8.6 Ideals in $F[x]$

In this section we study the ideals in the polynomial ring $F[x]$, where F is a field. We already know that the $n\mathbb{Z}$ are the only ideals in $\mathbb{Z}$, that $p\mathbb{Z}$ is a prime ideal in $\mathbb{Z}$ if and only if p is a prime integer, and that $p\mathbb{Z}$ is a maximal ideal in $\mathbb{Z}$ if and only if it is a prime ideal. We see that ideals in $F[x]$ have a similar structure and that irreducible polynomials in $F[x]$ behave much like primes in $\mathbb{Z}$.

8.6.1 EXAMPLE Recall from Example 7.2.8 the definition of a principal ideal in a commutative ring R: If $a \in R$, then the principal ideal $\langle a \rangle$ generated by a is the ideal $\{ra \mid r \in R\}$ consisting of all multiples of a. Thus if F is a field, then the principal ideal $\langle x \rangle$ in $F[x]$ generated by x is the set of all multiples of x, which is to say, the set of all polynomials in $F[x]$ with constant term 0. ◇

8.6.2 DEFINITION Let D be an integral domain. Then D is called a **principal ideal domain (PID)** if every ideal in D is a principal ideal. ○

8.6.3 EXAMPLE From Examples 7.2.8 and 7.2.9 we know that $\mathbb{Z}$ is a PID, since as we know every ideal I in $\mathbb{Z}$ is generated by a fixed element $n \in I$, so $I = n\mathbb{Z} = \langle n \rangle$. ◇

8.6.4 EXAMPLE Any field F is a PID, since the only ideals are $\{0\} = \langle 0 \rangle$ and $F = \langle 1 \rangle$. ◇

8.6.5 THEOREM Let F be a field. Then $F[x]$ is a PID.

Proof We know by Theorem 8.1.12 that $F[x]$ is an integral domain. We need to show that for any ideal I in $F[x]$ there exists an element $f(x)$ of I such that $I = \langle f(x) \rangle$. If I is the zero ideal $\{0\}$, then $I = \langle 0 \rangle$. If I is not the zero ideal, let $g(x)$ be a nonzero element of I of minimal degree. We show that $g(x)$ generates I or, in other words, that $I = \langle g(x) \rangle$. What we must show is that if $f(x)$ is any other element of I, then $g(x)$ is a divisor of $f(x)$. To show this we apply the division algorithm to write $f(x) = q(x)g(x) + r(x)$ with $r(x) = 0$ or $\deg r(x) < \deg g(x)$. Since $r(x) = f(x) - q(x)g(x)$ is in I and $g(x)$ was chosen to have minimal degree, we cannot have $\deg r(x) < \deg g(x)$, and so must have $r(x) = 0$, so that $f(x) = q(x)g(x)$ is a multiple of $g(x)$ as required. □

From the proof of the theorem we see that in order to find a generator for an ideal I in $F[x]$ we just need to choose an element of I of minimal degree. If $g(x)$ and $h(x)$ are both elements of I of minimal degree, then according to the proof of the theorem $I = \langle g(x) \rangle = \langle h(x) \rangle$ and $g(x) = ch(x)$, where c is a unit of $F[x]$, in other words, a nonzero constant.

8.6.6 THEOREM Let F be a field. A nontrivial ideal $I = \langle p(x) \rangle$ is a maximal ideal in $F[x]$ if and only if $p(x)$ is irreducible over F.

Proof ($\Rightarrow$) Suppose $I = \langle p(x) \rangle$ is a maximal ideal in $F[x]$. I is neither $\{0\} = \langle 0 \rangle$ nor $F[x] = \langle 1 \rangle$, so $p(x)$ is neither the zero polynomial nor a unit of $F[x]$, which is to say a

constant polynomial. If $p(x) = g(x)h(x)$, then $p(x) \in \langle g(x) \rangle$ and so $I = \langle p(x) \rangle \subseteq \langle g(x) \rangle \subseteq F[x]$. By our assumption that I is a maximal ideal, we must have either $\langle p(x) \rangle = \langle g(x) \rangle$ or $\langle g(x) \rangle = F[x]$. In the former case, $\deg g(x) = \deg p(x)$, while in the latter case $\deg g(x) = 0$ and $\deg h(x) = \deg p(x)$. This shows that $p(x)$ is irreducible over F.
($\Leftarrow$) Suppose $p(x)$ is irreducible over F and let $J = \langle f(x) \rangle$ be an ideal with $\langle p(x) \rangle \subseteq J = \langle f(x) \rangle \subseteq F[x]$. Then $p(x) \in \langle f(x) \rangle$, which implies $p(x) = q(x)f(x)$ for some $q(x)$ in $F[x]$. By our assumption that $p(x)$ is irreducible, we must either have $\deg f(x) = \deg p(x)$ and so $q(x)$ a nonzero constant, or $\deg q(x) = \deg p(x)$ and so $f(x)$ a nonzero constant. In the former case, $\langle p(x) \rangle = \langle f(x) \rangle$. In the latter case, $\langle f(x) \rangle = F[x]$. This shows that $I = \langle p(x) \rangle$ is a maximal ideal. □

8.6.7 COROLLARY Let F be a field and $p(x)$ a nonzero polynomial in $F[x]$. Then $\langle p(x) \rangle$ is a prime ideal in $F[x]$ if and only if $p(x)$ is irreducible over F.

Proof Suppose first that $p(x)$ is irreducible over F. Then by the theorem, $\langle p(x) \rangle$ is a maximal, and hence a prime, ideal in $F[x]$. Conversely, if $\langle p(x) \rangle$ is a prime ideal in $F[x]$ and $p(x) = g(x)h(x)$, either $g(x) \in I = \langle p(x) \rangle$ or $h(x) \in I = \langle p(x) \rangle$. In the former case $\deg g(x) = \deg p(x)$ and in the latter case $\deg h(x) = \deg p(x)$. This shows that $p(x)$ is irreducible over F. □

8.6.8 COROLLARY Let F be a field and I a nontrivial ideal in $F[x]$. Then I is a prime ideal in $F[x]$ if and only if I is a maximal ideal in $F[x]$.

Proof Left to the reader. (See Exercise 23.) □

The characterization of the maximal ideals in $F[x]$ provided by Theorem 8.6.6 gives us a very important tool for constructing new fields that contain the original field F.

8.6.9 COROLLARY Let F be a field and $p(x)$ a nonzero polynomial in $F[x]$. Then $p(x)$ is irreducible over F if and only if $F[x]/\langle p(x) \rangle$ is a field.

Proof Left to the reader. (See Exercise 24.) □

The assumption in the Theorem 8.6.5 that F is a field is crucial, as can be seen from the following examples, where F is not a field.

8.6.10 EXAMPLE If F is a field, then $F[x][y] = F[x, y]$, the ring of polynomials in two indeterminates x and y with coefficients from F, is not a principal ideal domain. For consider the ideal

$$I = \langle x, y \rangle = \{xf(x, y) + yg(x, y) \mid f(x, y), g(x, y) \in F[x, y]\}$$

$I = \langle x, y \rangle$ is not a principal ideal, because if $I = \langle h(x, y) \rangle$, then since $x, y \in I$, x and y would both have to be multiples of $h(x, y)$, which is impossible unless $h(x, y)$ is a constant, in which case $h(x, y) \notin \langle x, y \rangle$. ◇

8.6.11 EXAMPLE $\mathbb{Z}[x]$ is not a principal ideal domain. For consider the ideal

$$I = \langle 2, x \rangle = \{2f(x) + xg(x) \mid f(x), g(x) \in \mathbb{Z}[x]\}$$

$I = \langle 2, x\rangle$ is not a priniciple ideal, because if $I = \langle h(x)\rangle$, then since $2, x \in I$, 2 and x would both have to be multiples of $h(x)$, which is impossible unless $h(x) = \pm 1$, in which case $h(x) \notin \langle 2, x\rangle$. ◇

Exercises 8.6

In Exercises 1 through 7 (a) Show that the indicated set I is an ideal in $\mathbb{Q}[x]$. (b) Find a polynomial $g(x)$ in $\mathbb{Q}[x]$ such that $I = \langle g(x)\rangle$ (c) Determine whether I is a maximal ideal in $\mathbb{Q}[x]$.

1. $I = \{f(x) \in \mathbb{Q}[x] \mid f(2) = 0\}$

2. $I = \{f(x) \in \mathbb{Q}[x] \mid f(2) = f(3) = 0\}$

3. $I = \{f(x) \in \mathbb{Q}[x] \mid f(\sqrt{2}) = 0\}$

4. $I = \{f(x) \in \mathbb{Q}[x] \mid f(\sqrt{2}) = f(\sqrt{3}) = 0\}$

5. $I = \{f(x) \in \mathbb{Q}[x] \mid f(1-i) = f(1+i) = 0\}$

6. $I = \{f(x) \in \mathbb{Q}[x] \mid f(i) = f(-i) = 0\}$

7. $I = \{f(x) \in \mathbb{Q}[x] \mid f(1 - \sqrt{2}) = f(1 + \sqrt{2}) = 0\}$

In Exercises 8 through 14 determine whether the indicate ideals are maximal or not in $\mathbb{Q}[x]$. Justify your answers.

8. $I = \langle x^4 - 4\rangle$

9. $I = \langle x^2 - 5\rangle$

10. $I = \langle x^2 + x + 1\rangle$

11. $I = \langle x^4 - 1\rangle$

12. $I = \langle 6x^5 + 14x^3 - 21x + 42\rangle$

13. $I = \langle x^4 + 4\rangle$

14. $I = \langle 3x^4 + 5x + 1\rangle$

15. Let R be any ring and I any ideal in R. Show that $I[x]$ is an ideal in $R[x]$.

16. With $I[x]$ as in the preceding problem, describe all the ideals $I[x]$ in $R[x]$ for
(a) $R = \mathbb{Z}$ (b) $R = \mathbb{Q}$

17. Find all the maximal ideals $I = \langle g(x)\rangle$ in $\mathbb{Z}_3[x]$ with $g(x)$ of form $x^2 + ax + b$.

18. Find all the maximal ideals $I = \langle g(x)\rangle$ in $\mathbb{Z}_5[x]$ with $g(x)$ of form $x^2 + ax + 1$.

In Exercises 19 through 22 construct a ring homomorphism $\phi: \mathbb{Q}[x] \to \mathbb{C}$ having the indicated ideal K in $\mathbb{Q}[x]$ as its kernel.

19. $K = \langle x^2 - 3\rangle$

20. $K = \langle x^2 + 1\rangle$

21. $K = \langle x^2 + x + 1\rangle$

22. $K = \langle x^4 - 5\rangle$

23. Prove Corollary 8.6.8.

24. Prove Corollary 8.6.9.

25. Let $f(x)$ and $g(x)$ be two nonzero polynomials in $F[x]$, where F is a field. Show that there exists a unique monic polynomial $m(x) \in F[x]$ such that

(a) $f(x) \mid m(x)$ and $g(x) \mid m(x)$

(b) If $f(x) \mid q(x)$ and $g(x) \mid q(x)$ for some $q(x) \in F[x]$, then $m(x) \mid q(x)$.

Such a polynomial $m(x)$ is called the **least common multiple** of $f(x)$ and $g(x)$: $m(x) = \text{lcm}(f(x), g(x))$.

26. Let $f(x)$ and $g(x)$ be two nonzero polynomials in $F[x]$, where F is a field. Show that

(a) $\langle f(x)\rangle + \langle g(x)\rangle = \langle \gcd(f(x), g(x))\rangle$

(b) $\langle f(x)\rangle \cap \langle g(x)\rangle = \langle \text{lcm}(f(x), g(x))\rangle$

27. Let F be a field and $F[[x]]$ the ring of formal power series over F. (See Section 8.1, Exercises 26 through 31.) Show that

(a) For any $0 \neq f(x) \in F[[x]]$, $f(x) = x^n u$ for some positive integer n and some unit u in $F[[x]]$. (See Section 8.1, Exercise 27.)

(b) The only ideals in $F[[x]]$ are $\langle 0\rangle$, $F[[x]]$, and $\langle x^k\rangle$ for any positive integer k.

(c) $\langle x\rangle$ is the unique maximal ideal in $F[[x]]$.

8.7 Quotient Rings of $F[x]$

We will use what we know about the integers $\mathbb{Z}$ and quotient rings $\mathbb{Z}/n\mathbb{Z}$ as a guide to understanding the structure of the quotient rings $F[x]/I$ of the polynomial ring $F[x]$ over a field F.

Let's take an ideal $I \neq \{0\}$ in $F[x]$. By Theorem 8.6.5, $I = \langle g(x)\rangle$ for some polynomial $g(x) \in F[x]$. Let $\deg g(x) = n$. Consider any polynomial $f(x) \in F[x]$. Apply the division algorithm to write $f(x) = q(x)g(x) + r(x)$ with the remainder $r(x) = 0$ or $\deg r(x) < \deg g(x) = n$. Thus $f(x) \in r(x) + I$, where

$$r(x) = a_{n-1}x^{n-1} + \dots + a_1x + a_0$$

and the a_i are elements of F. Thus $f(x)$ belongs to the coset of I that can be written as

$$(a_{n-1}x^{n-1} + \dots + a_1x + a_0) + I$$

Now recall that I is the zero element, the additive identity, in $F[x]/I$. We make a change of notation to make the calculations less cumbersome. From now on we write the element $x+I \in F[x]/I$ as α. Then since I is an ideal, $\alpha^i = (x + I)^i = x^i + I$ and for any $c_i \in F$:

$$c_{n-1}\alpha^{n-1} + \dots + c_1\alpha + c_0 = (c_{n-1}x^{n-1} + \dots + c_1x + c_0) + I$$

In particular, $g(\alpha) = g(x) + I$, and since $I = \langle g(x)\rangle$ we have $g(\alpha) = 0$ in $F[x]/I$.

8.7.1 EXAMPLE We can now describe the elements of the quotient ring $\mathbb{Q}[x]/\langle x^2 + x + 1\rangle$ as follows. Since $I = \langle x^2 + x + 1\rangle$, any remainder $r(x)$ mod I will have degree at most 1, $r(x) = a_0 + a_1x$. So

$$\mathbb{Q}[x]/\langle x^2 + x + 1\rangle = \{a_0 + a_1\alpha \mid a_0, a_1 \in \mathbb{Q}\}$$

where $\alpha^2 + \alpha + 1 = 0$. ◇

8.7.2 EXAMPLE Since $x^2 + x + 1$ is irreducible over $\mathbb{Z}_2$, the quotient ring $\mathbb{Z}_2[x]/\langle x^2 + x + 1\rangle$ is a field. The elements of this quotient ring are

$$\{a_0 + a_1\alpha \mid a_0, a_1 \in \mathbb{Z}_2\} = \{0, 1, \alpha, \alpha + 1\}$$

where $\alpha^2 + \alpha + 1 = 0$. The multiplication is given in Table 1.

TABLE 1 Multiplication in the Quotient Ring

	1	α	$\alpha + 1$
1	1	α	$\alpha + 1$
α	α	$\alpha + 1$	1
$\alpha + 1$	$\alpha + 1$	1	α

Thus this construction gives us a field with four elements. ◇

8.7.3 EXAMPLE We can construct a field with $8 = 2^3$ elements as follows. We start with $\mathbb{Z}_2[x]$ and take a polynomial of degree 3 that is irreducible over $\mathbb{Z}_2$, for example, $g(x) = x^3 + x + 1$. Then $\mathbb{Z}_2[x]/\langle x^3 + x + 1\rangle$ is a field, and a remainder mod $I = \langle x^3 + x + 1\rangle$ is a polynomial of degree at most 2, $r(x) = a_2x^2 + a_1x + a_0$ where the a_i are elements of $\mathbb{Z}_2$. Therefore,

$$\mathbb{Z}_2[x]/\langle x^3 + x + 1\rangle = \{a_0 + a_1\alpha + a_2\alpha^2 \mid a_i \in \mathbb{Z}_2\}$$

where $g(\alpha) = \alpha^3 + \alpha + 1 = 0$. So the eight elements of this field are $0, 1, \alpha + 1, \alpha, \alpha^2, \alpha^2 + 1, \alpha^2 + \alpha$, and $\alpha^2 + \alpha + 1$. We can easily work out the multiplication table for this field using the fact that $\alpha^3 + \alpha + 1 = 0$. For instance,

$$\alpha^2(\alpha + 1) = \alpha^3 + \alpha^2 = (\alpha^3 + \alpha + 1) + \alpha^2 + \alpha + 1 = \alpha^2 + \alpha + 1$$

$$(\alpha^2)^2 = \alpha^4 = \alpha(\alpha^3 + \alpha + 1) + \alpha^2 + \alpha = \alpha^2 + \alpha$$

$$(\alpha^2 + 1)(\alpha^2 + \alpha) = \alpha^4 + \alpha^3 + \alpha^2 + \alpha = \alpha(\alpha^3 + \alpha + 1) + (\alpha^3 + \alpha + 1) + \alpha + 1 = \alpha + 1$$

and similarly for other products. ◇

From these last two examples we can easily see a method to construct finite fields with p^n elements, where p is any prime and n is any positive integer. Later we will see that all finite fields F have order p^n for $p = \text{char } F$ and some n.

8.7.4 EXAMPLE Let us see how to find the multiplicative inverse of a nonzero element of $F = \mathbb{Q}[x]/\langle x^2 - 2\rangle$, which is a field since $x^2 - 2$ is irreducible over $\mathbb{Q}$.

In fact, as in the preceding examples, we can work out that

$$\mathbb{Q}[x]/\langle x^2 - 2\rangle = \{a_0 + a_1\alpha \mid a_0, a_1 \in \mathbb{Q}\}$$

where $\alpha^2 = 2$ or, in other words, $F = \mathbb{Q}(\sqrt{2})$: We can write any element b of F as

$$b = a_0 + a_1\sqrt{2}$$

where $a_i \in \mathbb{Q}$. If $b \neq 0$, then a_0 and a_1 cannot both be zero. The inverse b^{-1} will also be an element of $\mathbb{Q}(\sqrt{2})$ and hence of the form

$$b^{-1} = c_0 + c_1\sqrt{2}$$

To find c_0 and c_1, consider in $\mathbb{R}$:

$$\frac{1}{a_0 + a_1\sqrt{2}} \cdot \frac{a_0 - a_1\sqrt{2}}{a_0 - a_1\sqrt{2}} = \frac{a_0 - a_1\sqrt{2}}{a_0^2 - 2a_1^2} = \frac{a_0}{a_0^2 - 2a_1^2} - \frac{a_1\sqrt{2}}{a_0^2 - 2a_1^2}$$

Note $a_0^2 - 2a_1^2 \neq 0$ for the following reason: Since a_0 and a_1 cannot both be zero, if $a_0 - 2a_1^2 = 0$ or, in other words, $a_0 = 2a_1^2$, neither can be zero, and we would have $a_0/a_1 = \pm\sqrt{2}$, which is impossible if the $a_i \in \mathbb{Q}$, since $\sqrt{2} \notin \mathbb{Q}$. Thus we have found the inverse:

$$b^{-1} = [a_0/(a_0^2 - 2a_1^2)] - [a_1/(a_0^2 - 2a_1^2)]\sqrt{2}$$

as required. ◇

This example can be generalized.

8.7.5 PROPOSITION Let $f(x)$ and $g(x)$ be polynomials in $F[x]$ both irreducible over F such that $f(x) \neq cg(x)$ for any unit c in $F[x]$. Then

(1) $\gcd(f(x), g(x)) = 1$

(2) There exist $u(x)$ and $v(x)$ in $F[x]$ such that $1 = u(x)f(x) + v(x)g(x)$.

(3) $u(x)$ is the multiplicative inverse of $f(x)$ in $F[x]/\langle g(x)\rangle$.

Proof Left to the reader (See Exercise 32.) □

8.7.6 EXAMPLE We can find the multiplicative inverse of x in $\mathbb{Q}[x]/\langle x^3 - 2\rangle$, which is a field since $x^3 - 2$ is irreducible over $\mathbb{Q}$, as follows. We use the Euclidean algorithm:

$$x^3 - 2 = x^2 \cdot x + (-2)$$
$$x = (-x/2)\cdot(-2) + 0$$

to find $1 = (x^2/2)x - (^1/_2)(x^3 - 2)$, and so $x^2/2$ is the multiplicative inverse of x in $\mathbb{Q}[x]/\langle x^3 - 2\rangle$. ◇

8.7.7 EXAMPLE We now find the multiplicative inverse of the element $x + 4$ in $\mathbb{Z}_5[x]/\langle x^3 + x + 1\rangle$ as follows. (Note that $x^3 + x + 1$ is irreducible over $\mathbb{Z}_5$, since it is of degree 3 and has no zeros in $\mathbb{Z}_5$.) We use the Euclidean algorithm:

$$x^3 + x + 1 = (x^2 + x + 2)(x + 4) + 3$$

Multiplying both sides by 2 mod 5, we get

$$2(x^3 + x + 1) \quad = 2(x^2 + x + 2)(x + 4) + 1$$

from which it follows that $(3x^2 + 3x + 1)(x + 4) = 1$ in $\mathbb{Z}_5[x]/\langle x^3 + x + 1\rangle$ and $3x^2 + 3x + 1$ is the multiplicative inverse of $x + 4$ in $\mathbb{Z}_5[x]/\langle x^3 + x + 1\rangle$. ◇

Exercises 8.7

In Exercises 1 through 4 construct a field with the indicated number n of elements.

1. $n = 9$

2. $n = 27$

3. $n = 16$

4. $n = 25$

In Exercises 5 through 14 determine whether the indicate quotient rings are fields. Justify your answers.

5. $\mathbb{Q}[x]/\langle x^2 - 5\rangle$

6. $\mathbb{Q}[x]/\langle x^2 + 3x + 2\rangle$

7. $\mathbb{Z}_3[x]/\langle x^2 + x + 1\rangle$

8. $\mathbb{C}[x]/\langle x^2 + x + 1\rangle$

9. $\mathbb{Z}_2[x]/\langle x^2 + x + 1\rangle$

10. $\mathbb{Q}[x]/\langle x^4 + 1\rangle$

11. $\mathbb{Q}[x]/\langle x^4 - 1\rangle$

12. $\mathbb{Q}[x]/\langle x - 2\rangle$

13. $\mathbb{C}[x]/\langle x - 2\rangle$

14. $\mathbb{C}[x]/\langle x^2 + 1\rangle$

15. Show that for any prime p there exists a field with p^2 elements.

16. Describe the elements of $\mathbb{Q}[x]/\langle x^2 - 3\rangle$, and show that this quotient ring of $\mathbb{Q}[x]$ is isomorphic to $\mathbb{Q}(\sqrt{3})$.

17. Show that $\mathbb{R}[x]/\langle x^2 + 1\rangle$ is isomorphic to $\mathbb{C}$.

18. Show that the ideal $I = \langle x^2 + 1\rangle$ is a prime ideal but not a maximal ideal in $\mathbb{Z}[x]$.

In Exercises 19 through 24 compute the product of the indicated polynomials in the indicated quotient rings.

19. $3x + 2$ and $5x - 3$ in $\mathbb{Q}[x]/\langle x - 2\rangle$

20. $5x + 1$ and $2x + 3$ in $\mathbb{Q}[x]/\langle x^2 - 2\rangle$

21. $x^2 + 2x - 3$ and $x^2 + 3x + 1$ in $\mathbb{Q}[x]/\langle x^3 + 2\rangle$

22. $x^2 + x + 1$ and $x^2 + 1$ in $\mathbb{Z}_2[x]/\langle x^4 + x + 1\rangle$

23. $x^2 + 2x + 2$ and $x^2 + 2$ in $\mathbb{Z}_3[x]/\langle x^3 + 2x + 1\rangle$

24. $ax + b$ and $cx + d$ in $\mathbb{Q}[x]/\langle x^2 + 1\rangle$

In Exercises 25 through 30 find a generator of the indicated ideals in the indicated rings.

25. $\langle 4\rangle \cap \langle 6\rangle$ in $\mathbb{Z}$

26. $\langle x - 1\rangle \cap \langle x^2 - 1\rangle$ in $\mathbb{Q}[x]$

27. $\langle x^2 + x + 1\rangle \cap \langle x^3 - 1\rangle$ in $\mathbb{Q}[x]$

28. $\langle x + 1\rangle \cap \langle x^2 + 1\rangle$ in $\mathbb{Z}_2[x]$

29. $\langle x^2 - 5x + 6\rangle \cap \langle x^2 - 3x + 2\rangle$ in $\mathbb{Q}[x]$

30. $\langle x^2 + x + 1\rangle \cap \langle x^2 + 2\rangle$ in $\mathbb{Z}_3[x]$

31. Let F be a field and let $I = \langle f(x)\rangle$ and $J = \langle g(x)\rangle$ be two ideals in $F[x]$. Prove a general result about the generator of the ideal $I \cap J$. (*Hint:* Look at the preceding exercises.)

32. Prove Proposition 8.7.5.

In Exercises 33 through 37 find the multiplicative inverse of the indicated element in the indicated field.

33. x in $\mathbb{R}[x]/\langle x^2 + 1\rangle$

34. $x + 1$ in $\mathbb{Z}_5[x]/\langle x^3 + x + 1\rangle$

35. $x^2 + 1$ in $\mathbb{Z}_3[x]/\langle x^3 + 2x + 1\rangle$

36. $x^2 - x + 1$ in $\mathbb{Q}[x]/\langle x^3 - 2\rangle$

37. $x + 3$ in $\mathbb{Q}[x]/\langle x^2 - 2\rangle$

8.8 The Chinese Remainder Theorem for $F[x]$

In all the examples of the previous section we considered quotient rings $F[x]/\langle p(x)\rangle$, where $p(x)$ is an irreducible polynomial over F, so that the quotient ring is a field. In this section we consider quotient rings where $p(x)$ is not necessarily irreducible over F so that the quotient rings may not be integral domains. As a key to understanding the structure of such rings we use a Chinese remainder theorem for $F[x]$. We try to understand at least some quotient rings $F[x]/\langle p(x)\rangle$, where $p(x)$ is not irreducible over F.

8.8.1 EXAMPLE Consider

$$\mathbb{Z}_2[x]/\langle x^2 + x\rangle = \{a_0 + a_1\alpha \mid a_i \in \mathbb{Z}_2,\ \alpha^2 + \alpha = 0\}$$

This is a ring with four elements 0, 1, α, $1 + \alpha$ with the addition and multiplication tables shown in Tables 2 and 3.

TABLE 2 Addition

+	0	1	α	$1+\alpha$
0	0	1	α	$1+\alpha$
1	1	0	$1+\alpha$	α
α	α	$1+\alpha$	0	1
$1+\alpha$	$1+\alpha$	α	1	0

TABLE 2 Multiplication

$\cdot$	1	α	$1+\alpha$
1	1	α	$1+\alpha$
α	α	α	0
$1+\alpha$	$1+\alpha$	0	$1+\alpha$

If we let $\phi: \mathbb{Z}_2[x]/\langle x^2 + x\rangle \to \mathbb{Z}_2 \times \mathbb{Z}_2$ be the ring homomorphism such that

$$\phi(0) = (0, 0),\ \phi(1) = (1, 1),\ \phi(\alpha) = (1, 0),\ \phi(1 + \alpha) = (0, 1)$$

We can easily verify that ϕ is a ring isomorphism, so $\mathbb{Z}_2[x]/\langle x^2 + x\rangle \cong \mathbb{Z}_2 \times \mathbb{Z}_2$. ◇

8.8.2 EXAMPLE Consider

$$\mathbb{Q}[x]/\langle x^3 - 2x\rangle = \{a_0 + a_1\alpha + a_2\alpha^2 \mid a_i \in \mathbb{Q} \text{ and } \alpha^3 - 2\alpha = 0\}$$

For any polynomial $f(x)$ in $\mathbb{Q}[x]$ we consider three divisions, dividing $f(x)$ by $x^3 - 2x$ and by its two irreducible factors, x and $x^2 - 2$:

$$f(x) = q(x)(x^3 - 2x) \quad + (a_0 + a_1x + a_2x^2)$$
$$f(x) = q(x)(x^2 - 2)x \quad + (a_1x + a_2x)x \quad + a_0$$
$$f(x) = q(x)x(x^2 - 2) \quad + a_2(x^2 - 2) \quad + (a_0 + 2a_2 + a_1x)$$

Thus we have

$$f(x) = a_0 + a_1x + a_2x^2 \quad \mathrm{mod}\ \langle x^3 - 2x\rangle$$
$$f(x) = a_0 \quad \mathrm{mod}\ \langle x\rangle$$
$$f(x) = a_0 + 2a_2 + a_1x \quad \mathrm{mod}\ \langle x^2 - 2\rangle$$

Consider the evaluation homomorphisms:

$$\phi_0: \mathbb{Q}[x] \to \mathbb{Q}[x]/\langle x\rangle \quad \cong \mathbb{Q}$$

where $\phi_0(f(x)) = f(0) = a_0$

$$\phi_{\sqrt{2}}: \mathbb{Q}[x] \to \mathbb{Q}[x]/\langle x^2 - 2\rangle \quad \cong \mathbb{Q}(\sqrt{2})$$

where $\phi_{\sqrt{2}}(f(x)) = f(\sqrt{2}) = (a_0 + 2a_2) + a_1\sqrt{2}$

They induce a ring homomorphism:

$$\phi: \mathbb{Q}[x] \to \mathbb{Q} \times \mathbb{Q}(\sqrt{2})$$

where $\phi(f(x)) = (\ \phi_0(f(x)),\ \phi_{\sqrt{2}}(f(x))\ )$. Note that $f(x)$ is in the kernel of ϕ exactly when $f(x)$ is in the kernels of both ϕ_0 and $\phi_{\sqrt{2}}$ or, in other words, when

$$f(x) \in \langle x\rangle \cap \langle x^2 - 2\rangle = \langle x^3 - 2x\rangle$$

Also, ϕ is onto because for any element $(a_0, b + a_1\sqrt{2})$ in $\mathbb{Q} \times \mathbb{Q}(\sqrt{2})$, we may let

$g(x) = a_0 + a_1x + ((b-a_0)/2)x^2 \in \mathbb{Q}[x]$

and we will have $\phi(g(x)) = (a_0, b + a_1\sqrt{2})$.

Thus ϕ gives rise to the isomorphism $\mathbb{Q}[x]/\langle x^3 - 2x\rangle \cong \mathbb{Q} \times \mathbb{Q}(\sqrt{2})$. ◇

In both examples we took the quotient mod an ideal with a reducible polynomial as a generator. In the first example,

$$I = \langle x^2 + x\rangle = \langle x(x + 1)\rangle$$

In the second example,

$$I = \langle x^3 - 2x\rangle = \langle x(x^2 - 2)\rangle$$

The generator in both cases factors into two factors that are relatively prime and the quotient ring turned out to be of the form $R_1 \times R_2$. In the first example,

$$\mathbb{Z}_2[x]/\langle x^2 + x\rangle \cong \mathbb{Z}_2 \times \mathbb{Z}_2$$

In the second example,

$$\mathbb{Q}[x]/\langle x^3 - 2x\rangle \cong \mathbb{Q} \times \mathbb{Q}(\sqrt{2})$$

This is not a coincidence, as we see in the following theorem.

8.8.3 THEOREM (Chinese remainder theorem). Let F be a field and let

$$I_1 = \langle g_1(x)\rangle, I_2 = \langle g_2(x)\rangle, \dots , I_n = \langle g_n(x)\rangle$$

be ideals in $F[x]$ such that for all i and j with $i \neq j$ we have

$$\gcd(g_i(x), g_j(x)) = 1$$

and let $f_1(x), \dots , f_n(x)$ be any polynomials in $F[x]$. Then

(1) There exists an $f(x) \in F[x]$ such that

$$f(x) - f_i(x) \in I_i \quad \text{for } i = 1, \dots , n$$

(2) This $f(x)$ is uniquely determined up to congruence modulo the ideal $J = \langle g_1(x)\cdot g_2(x)\cdot \dots \cdot g_n(x)\rangle$.

Proof (1) For any $s = 1, \dots , n$, let

$$J_s = \bigcap_{i\neq s} I_i = \langle g_1(x)\cdot\ldots\cdot g_{s-1}(x)\cdot g_{s+1}(x)\ldots g_n(x)\rangle$$

Then J_s and $I_s = \langle g_s(x)\rangle$ are two ideals in $F[x]$ whose generators are relatively prime. Hence $F[x] = I_s + J_s$. Consequently, for each $f_s(x)$ in $F[x]$ there exist $h_s(x)$ in I_s and $k_s(x)$ in J_s such that $f_s(x) = h_s(x) + k_s(x)$. Furthermore, $k_s(x) + I_s = f_s(x) + I_s$ and $k_s(x) + I_j = 0 + I_j$ for $j \neq s$. Let $f(x) = k_1(x) + \ldots + k_n(x)$; then $[f(x) - f_i(x)] + I_i = [f(x) - k_i(x)] + I_i = 0 + I_i$, and (1) follows.

(2) If $f'(x) \in F[x]$ is such that $f'(x) - f_i(x) \in I_i$ for $i = 1, \dots , n$, then $f(x) - f'(x) \in I_i$ for $i = 1, \dots , n$ and

$$f(x) - f'(x) \in \bigcap_{i=1}^{n} I_i = J$$

and (2) follows. □

The next corollary gives us an important tool to calculate quotient rings of $F[x]$.

8.8.4 COROLLARY Let F be a field and let $I_i = \langle g_i(x)\rangle$ for $i = 1, \dots , n$ be ideals in $F[x]$, with $\gcd(g_i(x),g_j(x)) = 1$ for $i \neq j$, and let $J = \langle g_1(x)\ldots g_n(x)\rangle$. Then

$$F[x]/J \cong F[x]/I_1 \times \ldots \times F[x]/I_n$$

Proof First we define a homomorphism. Consider the ring homomorphisms ϕ_i: $F[x] \to F[x]/I_i$, where ϕ_i takes every polynomial $f(x) \in F[x]$ to its remainder mod I_i. The ϕ_i induce a ring homomorphism:

$$\phi: F[x] \to F[x]/I_1 \times \ldots \times F[x]/I_n$$

where

$$\phi(f(x)) = (\phi_1(f(x)), \ldots, \phi_n(f(x)))$$

Next we show Kern $\phi = J$. Let K be the kernel of ϕ. Then $f(x) \in K$ if and only if $\phi(f(x)) = (0,\ldots,0)$, and this occurs if and only if $f(x) \in I_i$ for all $i = 1,\ldots,n$, and hence if and only if $f(x) \in J$.

Finally, we show that ϕ is onto. For given

$$(g_1(x), \ldots, g_n(x)) \in F[x]/I_1 \times \ldots \times F[x]/I_n$$

since each homomorphism is ϕ_i is onto, $g_i = \phi_i(f_i(x))$ for some $f_i(x) \in F[x]$. Hence by Theorem 8.8.3 there exists $f(x) \in F[x]$ such that $f(x) - f_i(x) \in I_i$ for all $i = 1, \ldots, n$. Hence $\phi_i(f(x) - f_i(x)) = 0$ and $\phi(f(x)) = (\phi_1(f_1(x)), \ldots, \phi_n(f_n(x))) = (f_1(x), \ldots, f_n(x))$. Thus the corollary follows by the first isomorphism theorem for rings (Theorem 7.2.17). □

8.8.5 EXAMPLE Let us describe the quotient ring $\mathbb{Q}[x]/\langle x^4 - 8x^2 + 15\rangle$. We have $x^4 - 8x^2 + 15 = (x^2 - 3)(x^2 - 5)$, where the factors $x^2 - 3$ and $x^2 - 5$ are relatively prime. Hence

$$\mathbb{Q}[x]/\langle x^4 - 8x^2 + 15\rangle \cong \mathbb{Q}[x]/\langle x^2 - 3\rangle \times \mathbb{Q}[x]/\langle x^2 - 5\rangle \cong \mathbb{Q}(\sqrt{3}) \times \mathbb{Q}(\sqrt{5})$$

by the preceding Corollary 8.8.4. ◇

8.8.6 EXAMPLE Let us describe the quotient ring $\mathbb{Q}[x]/\langle x^3 + 3x^2 + 2x\rangle$. We have $x^3 + 3x^2 + 2x = x(x + 1)(x + 2)$, where the three factors are relatively prime, and where $\mathbb{Q}[x]/\langle x+a\rangle \cong \mathbb{Q}$. Hence $\mathbb{Q}[x]/\langle x^3 + 3x^2 + 2x\rangle \cong \mathbb{Q} \times \mathbb{Q} \times \mathbb{Q}$. ◇

8.8.7 EXAMPLE Let us describe the quotient ring $\mathbb{R}[x]/\langle x^3 + x\rangle$. We have $x^3 + x = x(x^2 + 1)$, where the factors are relatively prime, $\mathbb{R}[x]/\langle x\rangle \cong \mathbb{R}$, and $\mathbb{R}[x]/\langle x^2 + 1\rangle \cong \mathbb{C}$. Hence $\mathbb{R}[x]/\langle x^3 + x\rangle \cong \mathbb{R} \times \mathbb{C}$. ◇

Finally, we look at some quotient rings $F[x]/\langle f(x)\rangle$ where the polynomial $f(x)$ has repeated factors.

8.8.8 EXAMPLE Let us describe the ring

$$\mathbb{Z}_2[x]/\langle x^2\rangle = \{a_0 + a_1\alpha \mid a_i \in \mathbb{Z}_2 \text{ and } \alpha^2 = 0\}\}$$

This is a ring with four elements 0, 1, α, $1+\alpha$ with the addition and multiplication tables as in Tables 4 and 5. Note that the addition table, Table 4, is the same as the addition table for $\mathbb{Z}_2 \times \mathbb{Z}_2$, but the multiplication table, Table 5, is the same as the multiplication table for $\mathbb{Z}_4$.

TABLE 4 Addition

+	0	1	α	$1+\alpha$
0	0	1	α	$1+\alpha$
1	1	0	$1+\alpha$	α
α	α	$1+\alpha$	0	1
$1+\alpha$	$1+\alpha$	α	1	0

TABLE 5 Multiplication

$\cdot$	1	α	$1+\alpha$
1	1	α	$1+\alpha$
α	α	0	α
$1+\alpha$	$1+\alpha$	α	1

Hence this is a ring with four elements that is not an integral domain and that is not isomorphic either to $\mathbb{Z}_2 \times \mathbb{Z}_2$ or to $\mathbb{Z}_4$. ◇

8.8.9 EXAMPLE Consider the quotient ring

$$\mathbb{Q}[x]/\langle x^2\rangle = \{a_0 + a_1\alpha \mid a_i \in \mathbb{Q} \text{ and } \alpha^2 = 0\}\}$$

The addition table of this quotient is the same as the addition table of $\mathbb{Q} \times \mathbb{Q}$. But the multiplication table is given by

$$(a_0 + a_1\alpha)(b_0 + b_1\alpha) = a_0b_0 + (a_1b_0 + a_0b_1)\alpha$$

To get a better picture of this quotient ring, for any $a_0, a_1 \in \mathbb{Q}$, consider the matrix $M(a_0,a_1)$ given by

$$M(a_0,a_1) = \begin{bmatrix} a_0 & 0 \\ a_1 & a_0 \end{bmatrix}$$

And let $M_2 = \{M(a_0,a_1) \mid a_i \in \mathbb{Q}\}$. Then M_2 is a ring and any element of M_2 can be written as $a_0\mathbf{1} + a_1\mathbf{u}$ where

$$\mathbf{1} = \begin{bmatrix} 1 & 0 \\ 0 & 1 \end{bmatrix}, \quad \mathbf{u} = \begin{bmatrix} 0 & 0 \\ 1 & 0 \end{bmatrix}$$

And finally, $\mathbb{Q}[x]/\langle x^2\rangle \cong M_2$. (See Exercise 14.) ◇

Exercises 8.8

In Exercises 1 through 6 describe the indicated quotient rings (as in Examples 8.8.5 through 8.8.7).

1. $\mathbb{Q}[x]/\langle x^2 + x\rangle$

2. $\mathbb{Q}[x]/\langle x^3 + x\rangle$

3. $\mathbb{R}[x]/\langle x^2 + x\rangle$

4. $\mathbb{C}[x]/\langle x^3 + x\rangle$

5. $\mathbb{Z}_2[x]/\langle x^3 + x^2 + x\rangle$

6. $\mathbb{Z}_3[x]/\langle x^3 + x\rangle$

7. Let F be a field and let $I = \langle f(x)\rangle$ and $J = \langle g(x)\rangle$ be two ideals in $F[x]$, where $\gcd(f(x), g(x)) = 1$. Show that $I + J = F[x]$.

8. Let R be a ring with unity 1, and let $I_1, \dots, I_n$ be ideals in R such that $I_i + I_j = R$ for all $i \neq j$. Show that for any $a_1, \dots, a_n$ in R there exists a b in R such that $b - a_i \in I_i$ for all $i = 1, \dots, n$. (*Hint:* First show that $I_s + \bigcap_{i \neq s} I_i = R$. Then imitate the proof of Theorem 8.8.3.)

9. Let R be a ring with unity 1, and let $I_1, \dots, I_n$ be ideals in R such that $I_i + I_j = R$ for all $i \neq j$. Show that there exists a ring isomorphism:

$$\theta: R/(I_1 \cap \dots \cap I_n) \cong R/I_1 \times \dots \times R/I_n$$

(*Hint:* Use Exercise 8.)

In Exercises 10 through 13 write out the addition and multiplication tables for the following quotient rings (as in Example 8.8.8).

10. $\mathbb{Z}_2[x]/\langle x^3 \rangle$

11. $\mathbb{Z}_3[x]/\langle x^2 \rangle$

12. $\mathbb{Z}_2[x]/\langle x^2 + 1 \rangle$

13. $\mathbb{Z}_3[x]/\langle x^2 + x + 1 \rangle$

14. Let $M_2 = \{\begin{bmatrix} a_0 & 0 \\ a_1 & a_0 \end{bmatrix} \mid a_0, a_1 \in \mathbb{Q}\}$. Show that

(a) M_2 is a ring under matrix addition and multiplication.
(b) As a group under addition, M_2 is isomorphic to $\mathbb{Q} \times \mathbb{Q}$.
(c) M_2 is isomorphic to $\mathbb{Q}[x]/\langle x^2 \rangle$.

15. Using the preceding exercise, write four matrices that correspond to the four elements of $\mathbb{Z}_2[x]/\langle x^2 \rangle$ (as in Example 8.8.8).

16. Using Example 8.8.9 and the preceding exercises as your model, construct a ring of matrices isomorphic to $\mathbb{Q}[x]/\langle x^3 \rangle$.

17. Generalize the result of the preceding exercise to $\mathbb{Q}[x]/\langle x^n \rangle$ for any $n > 1$.

18. Let $N_2 = \{\begin{bmatrix} a_0 - a_1 & 0 \\ a_1 & a_0 - a_1 \end{bmatrix} \mid a_0, a_1 \in \mathbb{Q}\}$. Show that

(a) N_2 is a ring under matrix addition and multiplication.
(b) N_2 is isomorphic to $\mathbb{Q}[x]/\langle (x + 1)^2 \rangle$.
(c) $\mathbb{Q}[x]/\langle (x + 1)^2 \rangle$ is isomorphic to $\mathbb{Q}[x]/\langle x^2 \rangle$.

Chapter 9

Euclidean Domains

We have encountered two integral domains, $\mathbb{Z}$ and $F[x]$, where F is a field, with striking similarities. We proved a unique factorization theorem for $\mathbb{Z}$ (fundamental theorem of arithmetic, 0.3.26) and for $F[x]$ (Theorem 8.4.6). We have shown that they are both principal ideal domains (Example 7.2.5 and Theorem 8.6.5). If we look back at the way we derived such fundamental properties shared by $\mathbb{Z}$ and $F[x]$, we cannot fail to observe that they are all consequences of one key theorem: the division algorithm, Theorem 0.3.11 for $\mathbb{Z}$ and Theorem 8.2.2 for $F[x]$. In this chapter we first study integral domains for which a division algorithm holds (*Euclidean domains*). We show that such integral domains are always principal ideal domains (PIDs), and that in PIDs a unique factorization theorem always holds. We then study further those integral domains that have this unique factorization property (UFDs, or unique factorization domains), and we prove the following inclusions of classes of integral domains:

$$\text{Integral domains} \supset \text{UFDs} \supset \text{PIDs} \supset \text{Euclidean domains} \supset \text{Fields}$$

We end this chapter with a study of the Gaussian integers $\mathbb{Z}[i]$ and an application to a well-known theorem in number theory due to Fermat.

9.1 Division Algorithms and Euclidean Domains

The division algorithm (Theorem 0.3.11) allows us to define such notions as the gcd and the lcm of two integers. Analogous definitions were introduced in Chapter 8 for polynomials with coefficients in a field, as a consequence of Theorem 8.2.2. In the case of $\mathbb{Z}$ the absolute value of an integer was used to express the remainder r upon division by an integer b, $|r| < b$. In the case of $F[x]$ we used the degree of a polynomial to express the remainder $r(x)$ upon division by a polynomial $g(x)$, $\deg r(x) < \deg g(x)$. Therefore, in order to look for other integral domains with a division algorithm, we must look for integral domains equipped with a function similar to that of the absolute value as in $\mathbb{Z}$ or that of the degree as in $F[x]$, which assigns to every element of the domain a nonnegative integer.

9.1.1 EXAMPLE If F is a field, then for any a, $b \neq 0$ in F, $a = qb + r$, where $r = 0$ and $q = ab^{-1}$. In particular, in the field of complex numbers $\mathbb{C}$, consider $z = 3 + 2i$ and $w = 1 + i$. Then

$$zw^{-1} = (3 + 2i)(1 + i)^{-1} = (3 + 2i)(1 - i)(1 - i)^{-1}(1 + i)^{-1} = {}^5/_2 - {}^i/_2$$

Hence in $\mathbb{C}$, $z = ({}^5/_2 - {}^i/_2)w = qw$.

Now let us consider $z = 3 + 2i$ and $w = 1 + i$ as elements of the Gaussian integers $\mathbb{Z}[i]$, which is a subdomain of $\mathbb{C}$. The division we just performed in (2) does not make sense in $\mathbb{Z}[i]$, since $q = {}^5/_2 - {}^i/_2 \notin \mathbb{Z}[i]$. Let us manipulate q to get rid of the denominators:

$$q = {}^5/_2 - {}^i/_2 = (2 - i) + ({}^1/_2 + {}^i/_2)$$

Therefore,

$$z = 3 + 2i = qw = (2 - i)w + ({}^1/_2 + {}^i/_2)w = (2 - i)(1 + i) + ({}^1/_2 + {}^i/_2)(1 + i)$$

Thus $z = (2 - i)w + i \in \mathbb{Z}[i]$. So we were able to write z in the form $z = q'q + r'$, where $q', r' \in \mathbb{Z}[i]$. Note that the remainder $r' = i$ that we found satisfies the following condition:

$$|i|^2 = 1 < 2 = |1 + i|^2$$

For any element $x = a + bi \in \mathbb{Z}[i]$, $|x|^2 = a^2 + b^2 = x \cdot \overline{x}$ and if $y \in \mathbb{Z}[i]$, then

$$|x|^2 \leq |xy|^2 = |x|^2|y|^2$$

As we see in Proposition 9.1.4, the function v: $\mathbb{Z}[i] - \{0\} \to \mathbb{Z}$ defined by $v(x) = |x|^2$ gives rise to a division algorithm for the Gaussian integers. ◇

We now introduce the basic definition for this chapter.

9.1.2 DEFINITION An integral domain D is called a **Euclidean domain** if there exists a function v: $D - \{0\} \to \mathbb{Z}^+ \cup \{0\}$ from the set of nonzero elements of D to the set of nonnegative integers such that

(1) For $x \neq 0$ and $y \neq 0$ in D

$$v(x) \leq v(xy)$$

(2) Given a and $b \neq 0$ in D, there exist q and r in D such that

$$a = qb + r, \text{ where } r = 0 \text{ or } v(r) < v(b)$$

The element q is called the **quotient** and the element r the **remainder** of the division. ○

9.1.3 EXAMPLES We know several examples of Euclidean domains already.

(1) $\mathbb{Z}$ is a Euclidean domain with $v(a) = |a|$ for all $0 \neq a \in \mathbb{Z}$, since for nonzero integers a, b, $1 \leq |b|$ implies $v(a) = |a| \leq |a||b| = |ab| = v(ab)$.

(2) Any field F is a Euclidean domain with $v(a) = 0$ for all $a \in F^*$.

(3) If F is a field, then $F[x]$ is a Euclidean domain with $v(f(x)) = \deg f(x)$. Note that condition (1) of the definition follows from Theorem 8.1.12, part (4), and condition (2) is given by Theorem 8.2.2. ◇

9.1.4 PROPOSITION The Gaussian integers $\mathbb{Z}[i]$ with the function $v(z) = a^2 + b^2$ for all $z = a + bi \neq 0$ in $\mathbb{Z}[i]$ is a Euclidean domain.

Proof (1) If $z = a + bi \neq 0$, then $v(z) = a^2 + b^2 \geq 1$. In addition, $a^2 + b^2 = z \cdot \overline{z}$, and hence for any $w \neq 0$ in $\mathbb{Z}[i]$, $v(zw) = zw\overline{z}\,\overline{w} = z\overline{z}\,w\overline{w} = v(z)v(w)$ and $1 \leq v(z)$, $1 \leq v(w)$. Therefore $v(z) \leq v(z)v(w) = v(zw)$.

(2) We show that a division algorithm exists in $\mathbb{Z}[i]$. Given z and $w \neq 0$ in $\mathbb{Z}[i]$, consider z and w first as elements of $\mathbb{C}$. Then

$$z(w)^{-1} = z\overline{w}(\overline{w})^{-1}w^{-1} = z\overline{w}(\overline{w}w)^{-1}$$

Therefore, $z(w)^{-1} \in \mathbb{Q}(i)$. Thus $z(w)^{-1} = u + vi$, where $u, v \in \mathbb{Q}$. Now let

$$u = a + u' \text{ and } v = b + v'$$

where a and b are the integers closest to u and v, respectively. Thus

$$|u'| \leq {}^1/_2 \text{ and } |v'| \leq {}^1/_2$$

Hence $z = (a + bi)w + (u' + v'i)w$. Observe that $(u' + v'i)w = z-(a + bi)w \in \mathbb{Z}[i]$, and therefore $z = qw + r$, where $q = a + bi \in \mathbb{Z}[i]$ and $r = (u' + v'i)w \in \mathbb{Z}[i]$. It remains to show that

$$r = 0 \text{ or } v(r) < v(w)$$

This holds, for if $r \neq 0$, then $v(r) = v((u' + v'i)w) = v(u' + v'i)v(w) = (|u'|^2 + |v'|^2)v(w) \leq ({}^1/_4 + {}^1/_4)v(w) = {}^1/_2 v(w)$. □

Observe that in the definition of a Euclidean domain the quotient q and remainder r are not required to be unique.

9.1.5 EXAMPLE In Example 9.1.1 we divided $z = 3 + 2i$ by $w = 1 + i$ and obtained

$$3 + 2i = (2 - i)(1 + i) + i$$

We actually had three more choices for the quotient q and remainder r:

$$3 + 2i = (3 - i)(1 + i) - 1$$
$$3 + 2i = (2)(1 + i) + 1$$
$$3 + 2i = (3)(1 + i) - i$$

In all four cases, $v(r) = 1 < 2 = v(1 + i)$. ◇

An immediate consequence of Definition 9.1.2 is that in a Euclidean domain every ideal is principal. Note that the proof we give next is almost the same as the proof of Theorem 8.6.5, where the function v is the degree function.

9.1.6 THEOREM Every Euclidean domain is a PID.

Proof Let I be an ideal in a Euclidean domain D. If $I = \{0\}$, then $I = \langle 0 \rangle$. If $I \neq \{0\}$, let $0 \neq a \in I$ be an element of I such that $v(a) \leq v(x)$ for all $0 \neq x \in I$. We show that $I = \langle a \rangle$. Let $b \in I$. Then there exist q and r in D such that $b = qa + r$ with $r = 0$ or $v(r) < v(a)$. Since $r = b - qa$ we have $r \in I$. By the minimality of $v(a)$ we obtain $r = 0$ and $b = qa \in \langle a \rangle$. Thus $b \in \langle a \rangle$ for all $b \in I$ and $I = \langle a \rangle$. □

Determining whether or not a given integral domain is a Euclidean domain can be a difficult problem. But the theorem we have just proved can be useful since it is usually easier to show that a given integral domain is not a PID. In that case, we can automatically conclude that it is not a Euclidean domain.

9.1.7 EXAMPLE Consider $\mathbb{Z}[\sqrt{5}\,i]$ and for any $z = a + b\sqrt{5}\,i$ define the function $v(z) = z\bar{z} = a^2 + 5b^2$. Let $I = \langle 3, 1 + 2\sqrt{5}i \rangle$, the ideal generated by 3 and $1 + 2\sqrt{5}i$.

(1) Note that I is a proper ideal because if we had $1 \in I$ we would have

$$1 = 3u + (1 + 2\sqrt{5}i)w \text{ where } u, w \in \mathbb{Z}[\sqrt{5}i]$$

Multiplying both sides by $(1 - 2\sqrt{5}i)$, we obtain

$$(1 - 2\sqrt{5}i) = 3(1 - 2\sqrt{5}i)u + 21w$$

This implies that $(1 - 2\sqrt{5})$ is a multiple of 3 in $\mathbb{Z}[\sqrt{5}i]$, which is impossible.

(2) Suppose I is principal, say $I = \langle x + y\sqrt{5}i \rangle$. Then

$$3 = \alpha(x + y\sqrt{5}i) \text{ and } 1 + 2\sqrt{5}i = \beta(x + y\sqrt{5}i)$$

But this implies that $v(x + y\sqrt{5}) = x^2 + 5y^2$ is a divisor of

$$v(3) = 9 \text{ and } v(1 + 2\sqrt{5}i) = 21$$

Hence $x^2 + 5y^2 = 1$ or 3. But $x^2 + 5y^2 = 3$ has no integer solutions, and $x^2 + 5y^2 = 1$ if and only if $x = \pm 1$ and $y = 0$, in which case $I = \mathbb{Z}[\sqrt{5}\,i]$, which contradicts (1). Therefore, I is not a principal ideal, and $\mathbb{Z}[\sqrt{5}\,i]$ is not a PID. and hence is not a Euclidean domain. ◇

One of the immediate consequences of the division algorithm in Chapter 0 and in Chapter 8 was the existence of a greatest common divisor and the Euclidean algorithm. A generalization of Theorem 0.3.16 and Theorem 8.2.8 is given now for any Euclidean domain, but first we need some definitions.

9.1.8 DEFINITION Let R be a commutative ring and let $a, b \in R$, with $b \neq 0$.

(1) b is said to be a **divisor** of a in R, written $b|a$, if there exists an $x \in R$ such that $a = xb$.

(2) $c \in R$ is said to be a **common divisor** of a and b if $c|a$ and $c|b$. ○

9.1.9 DEFINITION Let R be a commutative ring and let $a, b \in R$. A **greatest common divisor** of a and b is a nonzero element $d \in R$ such that

(1) d is a common divisor of a and b.

(2) If c is any other common divisor of a and b, then $c|d$. ○

9.1.10 THEOREM Let D be a Euclidean domain and $a, b \in D$ two nonzero elements of D. Then there exists an element $d \in D$ such that

(1) d is a greatest common divisor of a and b.

(2) There exist $u, v \in D$ such that $d = ua + vb$.

Proof Let $I = \{xa + yb \mid x, y \in D\}$. I is an ideal in D, the ideal generated by a and b. By Theorem 9.1.6, $I = \langle d \rangle$ for some $d \in D$. Since $d \in I$, $d = ua + vb$ for some $u, v \in D$. Since $I = \langle d \rangle$, every element of I is of the form xd for some $x \in D$. Since $a \in I$ we obtain $d|a$, and since $b \in I$ we obtain $d|b$. If $c|a$ and $c|b$, say $a = xc$ and $b = yc$, then $d = ua + vb = uxc + vyc = (ux + vy)c$ and $c|d$ and the proof is complete. □

In a Euclidean domain D we can use the Euclidean algorithm as in Chapter 0 and Chapter 8 to obtain a greatest common divisor. Let $a, b \in D$ with $b \neq 0$.

Then applying the Euclidean algorithm, we have

$$\begin{array}{llll} a = & q_1 b + r_1 & \text{where } v(r_1) < v(b) & \text{if } r_1 \neq 0 \\ b = & q_2 r_1 + r_2 & \text{where } v(r_2) < v(r_1) & \text{if } r_2 \neq 0 \\ r_1 = & q_3 r_2 + r_3 & \text{where } v(r_3) < v(r_2) & \text{if } r_3 \neq 0 \\ \vdots & & & \\ r_{k-1} = & q_{k+1} r_k + r_{k+1} & \text{where } v(r_{k+1}) < v(r_k) & \text{if } r_{k+1} \neq 0 \end{array}$$

Since $v(b) > v(r_1) > v(r_2) > \ldots > v(r_k) > v(r_{k+1}) > \ldots \geq 0$ is a decreasing sequence of nonnegative integers, for some integer n we must have $r_{n+1} = 0$. Hence $r_n | r_{n-1}$ and $r_n | r_i$ for all $i \leq n$, and therefore $r_n | b$ and $r_n | a$. In addition, if $c | a$ and $c | b$, then $c | r_i$ for all $i \leq n$, and hence $c | r_n$. Thus r_n is a greatest common divisor of a and b.

9.1.11 EXAMPLE In the case of polynomials $f(x)$ and $g(x)$ in $F[x]$, if $d(x)$ and $d'(x)$ are both greatest common divisors of $f(x)$ and $g(x)$, then $d(x) = c(x)d'(x)$, where $\deg c(x) = 0$. Hence $c(x)$ is a nonzero constant, and therefore a unit in $F[x]$. ◇

9.1.12 PROPOSITION Let D be an integral domain and let $a, b \in D$. Then if d and d' are greatest common divisors of a and b in D, then $d = ud'$ for some unit $u \in D$.

Proof Since both d and d' are greatest common divisors of a and b, $d | d'$ and $d' | d$. Hence $d = ud'$ and $d' = vd$ for some u and v in D. Therefore, $d = u(vd) = (uv)d$, and $(1 - uv)d = 0$. Since $d \neq 0$ and D is an integral domain we obtain $uv = 1$, and u is a unit with inverse v in D. □

9.1.13 DEFINITION Let R be a commutative ring with unity. Two elements a and b in R are said to be **associates** if $a = ub$ for some unit $u \in R$. ○

Proposition 9.1.12 says that two greatest common divisors of a and b are associates. This can be generalized to the following proposition, whose proof we leave as an exercise.

9.1.14 PROPOSITION Let D be an integral domain and I a principal ideal in D. Then $I = \langle d \rangle = \langle d' \rangle$ implies that d and d' are associates.

Proof See Exercise 18. □

9.1.15 THEOREM Let D be a Euclidean domain. Then

(1) $v(1) \leq v(a)$ for all $0 \neq a \in D$.

(2) $v(1) = v(a)$ if and only if a is a unit in D.

Proof (1) $v(1) \leq v(1 \cdot a) = v(a)$ for all $0 \neq a \in D$.

(2) If a is a unit in D, then $v(a) \leq v(a \cdot a^{-1}) = v(1) \leq v(a)$, hence $v(1) = v(a)$. Conversely, if $v(1) = v(a)$, by the division algorithm there exist q and r in D such that $1 = qa + r$ with $r = 0$ or $v(r) < v(a)$. But since $v(a) = v(1)$ and since by (1) $v(1) \leq v(r)$, we cannot have $v(r) < v(a)$, and must have $r = 0$. Thus $1 = qa$ and a is a unit. □

9.1.16 EXAMPLE In $\mathbb{Z}[i]$, $v(z) = a^2 + b^2 = z\overline{z}$, hence $v(1) = 1$ and z is a unit in $\mathbb{Z}[i]$ if and only if $v(z) = 1$. Hence 1, -1, i, $-i$ are the units in $\mathbb{Z}[i]$. ◇

Exercises 9.1

In Exercises 1 through 3 show that the indicated integral domains $\mathbb{Z}[\sqrt{d}\,]$ are Euclidean domains with the function $v(z) = |z\overline{z}|$, where for $z = a + b\sqrt{d}$, $\overline{z} = a - b\sqrt{d}$.

1. $\mathbb{Z}[\sqrt{2}]$ **2.** $\mathbb{Z}[\sqrt{2}i]$ **3.** $\mathbb{Z}[\sqrt{3}]$

In Exercises 4 through 7 find a quotient q and remainder r in the indicated Euclidean domain, where $a = qb + r$.

4. $a = 5 + 3i$ $b = 2 + i$ in $\mathbb{Z}[i]$

5. $a = 3 + 4i$ $b = 4 - 3i$ in $\mathbb{Z}[i]$

6. $a = 3 + 2\sqrt{2}$ $b = 1 + \sqrt{2}$ in $\mathbb{Z}[\sqrt{2}]$

7. $a = 5 + 2\sqrt{2}$ $b = 3 + \sqrt{2}$ in $\mathbb{Z}[\sqrt{2}]$

In Exercises 8 through 11 find a greatest common divisor d of a and b in the indicated Euclidean domain, and express $d = ua + vb$.

8. $a = 7 + 5\sqrt{2}$ $b = 1 + \sqrt{2}$ in $\mathbb{Z}[\sqrt{2}]$

9. $a = -3 + 7\sqrt{3}$ $b = 7 - \sqrt{3}$ in $\mathbb{Z}[\sqrt{3}]$

10. $a = 2 + 8i$ $b = 6 + 8i$ in $\mathbb{Z}[i]$

11. $a = 4 + 7i$ $b = 8 - i$ in $\mathbb{Z}[i]$

In Exercises 12 through 14 find a generator for the ideal I in the indicated Euclidean domain.

12. I = the ideal generated by $f(x) = x^3 + x^2 - 2x - 2$ and $g(x) = x^3 - x^2 - 2x + 1$ in $\mathbb{Q}[x]$

13. I = the ideal generated by $5 + 5i$ and $3 - i$ in $\mathbb{Z}[i]$

14. I = the ideal generated by 13 and $3 + 2i$ in $\mathbb{Z}[i]$

15. Prove or disprove that $\mathbb{Z}[x]$ is a Euclidean domain.

16. Prove or disprove that for any field F, $F[x, y]$ is a Euclidean domain.

17. Let D be a Euclidean domain and a and b elements of D. Show that

(a) If a and b are associates, then $v(a) = v(b)$.

(b) If $v(a) = v(b)$ and $a|b$, then a and b are associates.

18. Prove Proposition 9.1.14.

19. Let D be a Euclidean domain and a and b nonzero elements of D. Show that $v(a) < v(ab)$ if and only if b is not a unit in D.

20. Show that $\mathbb{Z}[\sqrt{3}i]$ is not a Euclidean domain. (Use Theorem 9.1.6.)

21. Let D be an integral domain. Show that the following three statements are equivalent:

(a) D is a field. (b) $D[x]$ is a Euclidean domain. (c) $D[x]$ is a PID.

22. Let D be a Euclidean domain, a, b, and c nonzero elements of D, and d a greatest common divisor of a and b. Show that if $a|bc$, then $(a/d)|c$.

23. Suppose x_0, y_0 in $\mathbb{Z}$ is a solution of the equation $ax + by = c$, where $a \neq 0$, $b \neq 0$, and c are in $\mathbb{Z}$. Show that the complete set of solutions x, y in $\mathbb{Z}$ is given by

$$x = x_0 + k\,[{}^{b}/_{\gcd(a,b)}],\ y = y_0 - k\,[{}^{a}/_{\gcd(a,b)}]$$

for all $k \in \mathbb{Z}$.

24. Use the preceding exercise to find the complete set of solutions in $\mathbb{Z}$ of

(a) $3x - 6y = 10$ (b) $3x - 6y = 9$ (c) $385x - 275y = 495$

25. Let R be a commutative ring with unity. For a, b in R, a **least common multiple** of a and b is an element $m \in R$ such that

(1) $a|m$ and $b|m$

(2) If $a|n$ and $b|n$ for $n \in R$, then $m|n$.

Show that if R is a Euclidean domain, then

(a) A least common multiple m of a and b exists.

(b) Any two least common multiples of a and b are associates.

(c) If d is a greatest common divisor of a and b, then ab/d is a least common multiple of a and b.

(d) The ideal $\langle a\rangle \cap \langle b\rangle$ is generated by any least common multiple of a and b.

26. Let $d \in \mathbb{Z}$ be such that $\sqrt{d} \notin \mathbb{Q}$. In $\mathbb{Z}[\sqrt{d}]$ let

$$v(z) = |z\bar{z}| = |a^2 - db^2|, \text{ where } z = a + b\sqrt{d}$$

(a) (**Pell's equation**) Show that $a + b\sqrt{d}$ is a unit in $\mathbb{Z}[\sqrt{d}]$ if and only if $a^2 - db^2 = \pm 1$.

(b) Assume that for any rational numbers x and y there exist integers n and m such that $|(x - n)^2 - d(y - m)^2| < 1$. Show that in this case $\mathbb{Z}[\sqrt{d}]$ is a Euclidean domain with $v(a + b\sqrt{d}) = |a^2 - db^2|$.

27. In $\mathbb{Z}[\sqrt{2}]$ show that

(a) $1 + \sqrt{2}$ is a unit.

(b) $\pm(1 + \sqrt{2})^n$ for $n \in \mathbb{Z}$ is a unit.

(c) $\pm(1 + \sqrt{2})^n$ for $n \in \mathbb{Z}$ are all the units.

9.2 Unique Factorization Domains

As was pointed out in the introduction to this chapter, the fundamental theorem of arithmetic (Theorem 0.3.26) and the unique factorization theorem for $F[x]$ (Theorem 8.4.6) were each a consequence of the division algorithm. In the preceding section we called an integral domain with a division algorithm a Euclidean domain and showed that every Euclidean domain is a PID. In this section we show that every PID. has the unique factorization property. An integral domain with this property is called a UFD. Thus in this section we complete the proof of the following inclusions of classes of integral domains:

$$\text{Integral domains} \supset \text{UFDs} \supset \text{PIDs} \supset \text{Euclidean domains} \supset \text{Fields}$$

We first look at two concepts that in our studies so far have seemed to coincide.

9.2.1 EXAMPLE Consider the element 3 in $\mathbb{Z}[\sqrt{5}i]$ and the function

$$v(z) = z\bar{z} = a^2 + 5b^2, \text{ where } z = a + b\sqrt{5}i$$

(1) Let us try to factor $3 = zw$ in $\mathbb{Z}[\sqrt{5}i]$. We must have

$$9 = v(3) = v(zw) = zw\overline{zw} = zw\bar{z}\,\bar{w} = z\bar{z}\,w\bar{w} = v(z)v(w)$$

If $z = a + b\sqrt{5}i$, then $v(z) = a^2 + 5b^2$ must divide 9, and since $a^2 + 5b^2 = 3$ has no integer solutions, $v(z)$ must equal 1 or 9. If $v(z) = 1$, then z is a unit, while if $v(z) = 9$, then $v(w) = 1$ and w is a unit. Thus if $3 = zw$, one of z or w must be a unit in $\mathbb{Z}[\sqrt{5}i]$.
(2) Now consider $z = 2 + \sqrt{5}i$ and $w = 2 - \sqrt{5}i$. Here 3 does not divide z and does not divide w, but 3 divides $zw = 9$. ◇

9.2.2 DEFINITION Let D be an integral domain and $a \in D$, $a \neq 0$, and a not a unit in D. Then

(1) a is called **irreducible** in D if whenever $a = xy$ with $x, y \in D$, then one of x or y must be a unit in D. Otherwise, a is called **reducible** in D.

(2) a is called **prime** in D if whenever $a|xy$ with $x, y \in D$, then either $a|x$ or $a|y$ in D.

9.2.3 EXAMPLE Note the contrast in the following examples.

(1) In $\mathbb{Z}$ and in $F[x]$, where F is a field, primes and irreducible elements coincide.

(2) In Example 9.2.1 we show that in $\mathbb{Z}[\sqrt{5}i]$, the element 3 is irreducible but not prime. ◇

9.2.4 PROPOSITION In an integral domain D, every prime is irreducible.

Proof Let D be an integral domain and let p be a prime in D. If $p = xy$ for some $x, y \in D$, then $p|x$ or $p|y$. Say $p|x$. Then $x = cp$ for some $c \in D$. We then have

$p = xy = (cp)y = p(cy)$. Therefore, $(1 - cy)p = 0$, and since D is an integral domain, $cy = 1$, and y is a unit in D. Similarly, if $p|y$, then x is a unit in D. So p is irreducible. □

In an integral domain D, an element p is prime if and only if the principal ideal $I = \langle p \rangle$ generated by p is a prime ideal in D. (See Exercise 17.) The next two propositions will show us that in a PID the two concepts irreducible and prime coincide.

9.2.5 PROPOSITION Let D be a PID. Then $p \in D$ is irreducible if and only if the principal ideal $I = \langle p \rangle$ generated by p is a maximal ideal in D.

Proof ($\Leftarrow$) Assume that D is a PID and that $I = \langle p \rangle$ is a maximal ideal in D. If $p = xy \in D$, then $I = \langle p \rangle \subseteq \langle x \rangle \subseteq D$. Since I is maximal, either $\langle p \rangle = \langle x \rangle$ or $\langle x \rangle = D = \langle 1 \rangle$. If $\langle p \rangle = \langle x \rangle$, then by Theorem 9.1.14, p and x are associates, and y is a unit. If $\langle x \rangle = \langle 1 \rangle$, then x and 1 are associates, and x is a unit. Hence if $p = xy \in D$, then either x or y is a unit, and p is irreducible.
($\Rightarrow$) Assume that D is a PID and that p is irreducible. If J is an ideal in D such that $I = \langle p \rangle \subseteq J \subseteq D$, then since D is a PID, $J = \langle q \rangle$ for some $q \in D$. Therefore, $p \in \langle q \rangle$ and $p = cq$ for some $c \in D$. Since p is irreducible, either c or q is a unit in D. If c is a unit, then p and q are associates and $I = \langle p \rangle = \langle q \rangle = J$. If q is a unit, then $1 \in J$ and $J = D$. Thus $I = \langle p \rangle$ is a maximal ideal. □

Note that we proved the previous proposition earlier in the special case where $D = F[x]$ as Theorem 8.6.6.

9.2.6 PROPOSITION Let D be a PID. Then a nonzero element p is prime in D if and only if p is irreducible in D.

Proof See Exercise 18. □

In Example 9.1.7 we show that $\mathbb{Z}[\sqrt{5}i]$ is not a Euclidean domain by showing that it is not a PID. As was shown in Example 9.2.1, $\mathbb{Z}[\sqrt{5}i]$ contains irreducible elements that are not prime. Therefore Proposition 9.2.6 gives us another proof of the fact that this integral domain is not a PID and therefore not a Euclidean domain.

9.2.7 EXAMPLE In $\mathbb{Z}[\sqrt{5}i]$ consider the two factorizations of 21:

$$21 = 3 \cdot 7$$
$$21 = (1 + 2\sqrt{5}i)(1 - 2\sqrt{5}i)$$

(1) In Example 9.2.1 we show that 3 is irreducible in $\mathbb{Z}[\sqrt{5}i]$. Similarly, 7 is irreducible, for if $7 = zw$, where $z = a + b\sqrt{5}\,i$, then $49 = v(7) = v(z)v(w)$ and $v(z) = a^2 + 5b^2$. Hence $v(z)$ divides 49, and since $a^2 + 5b^2 = 7$ has no integer solutions, either $v(z) = 1$ or $v(z) = 49$ and $v(w) = 1$, and so either z or w is a unit.

(2) We can also show that $p = 1 + 2\sqrt{5}i$ and $q = 1 - 2\sqrt{5}i$ are irreducible. For if $p = zw$, then $21 = v(p) = v(z)v(w)$ and $v(z) = a^2 + 5b^2$. So $v(z)$ divides 21, and since

$a^2 + 5b = 3$ and $a^2 + 5b = 7$ have no integer solutions, either $v(z) = 1$ or $v(z) = 21$ and $v(w) = 1$, and so either z or w is a unit. Since $v(q) = 21$ also, the same argument shows that q is irreducible.
Thus we have an element, 21, of $\mathbb{Z}[\sqrt{5}i]$ that can be factored into irreducible factors in two completely different ways. ◇

9.2.8 DEFINITION An integral domain D is said to be a **unique factorization domain (UFD)** if the following two conditions are satisfied:

(1) Every nonzero element a of D that is not a unit can be written as the product of irreducibles p_i of D:

$$a = p_1 p_2 \dots p_r$$

(2) If

$$a = p_1 p_2 \dots p_r = q_1 q_2 \dots q_s$$

where the p_i and q_j are all irreducible, then $r = s$ and the q_j can be renumbered so that p_i and q_i are associates in D for all i, $1 \le i \le r$. ○

9.2.9 EXAMPLE We know several examples already.

(1) $\mathbb{Z}$ is a UFD by Theorem 0.3.26.

(2) $F[x]$, where F is a field, is a UFD by Theorem 8.4.6.

(3) $\mathbb{Z}[\sqrt{5}i]$ is not a UFD by Example 9.2.7. ◇

9.2.10 PROPOSITION In a UFD a nonzero element is prime if and only if it is irreducible.

Proof By Proposition 9.2.4 we only need to show that if p is irreducible in D, where D is a UFD, then p is prime. So let $p \in D$ be irreducible in D and let $p|xy$ for some $x, y \in D$. Since D is assumed to be a UFD, $x = p_1 p_2 \dots p_r$ and $y = p_{r+1} p_{r+2} \dots p_s$, where the p_i are irreducible in D. Hence $xy = p_1 p_2 \dots p_r p_{r+1} p_{r+2} \dots p_s$. Since $p|xy$, there exists $c \in D$ such that $xy = pc$, and since D is a UFD, $c = q_1 q_2 \dots q_t$ for some irreducible q_j. Hence $p_1 p_2 \dots p_r p_{r+1} p_{r+2} \dots p_s = p q_1 q_2 \dots q_t$, and p must be an associate of some p_i, $1 \le i \le s$. If $i \le r$, then $p|x$, and if $r < i$, then $p|y$. Hence p is prime. □

9.2.11 COROLLARY Let D be a UFD and let $a_1, \dots, a_n$ be nonzero elements of D that are not units in D. Then there exist elements d and m in D such that

(1) d is a greatest common divisor of $a_1, \dots, a_n$ in D.

(2) m is a least common multiple of $a_1, \dots, a_n$ in D.

Proof See Exercises 21 and 22. □

We state now one of the main theorems of this section.

9.2.12 THEOREM Every PID is a UFD.

We break up the proof into four steps.

9.2.13 PROPOSITION Let D be a PID, and let $I_1 \subseteq I_2 \subseteq I_3 \subseteq \ldots$ be an ascending chain of ideals. Then there exists a positive integer n such that $I_m = I_n$ for all $m \geq n$.

Proof Let $I = \bigcup_i I_i$. First we will show that I is an ideal in D. Let a and b be in I. Then $a \in I_k$ and $b \in I_l$ for some k and l. Let m be the larger of k and l. Then a and b are both in I_m, and I_m is an ideal, hence $a - b \in I_m \subseteq I$. Thus by the subring text I is a subring of D. Also, for any $c \in D$, $ca = ac \in I_m \subseteq I$. Hence I is an ideal in D. But D is a PID, which implies that $I = \langle x \rangle$ for some $x \in D$. Since $x \in I = \bigcup_i I_i$, $x \in I_n$ for some n. Hence for all $m \geq n$ we have

$$\langle x \rangle \subseteq I_n \subseteq I_m \subseteq I = \langle x \rangle$$

Therefore, $I_m = I_n$ for all $m \geq n$. □

9.2.14 DEFINITION We say that the **ascending chain condition** (**ACC**) holds in an integral domain D if D contains no strictly increasing infinite chain of ideals. ○

Thus Proposition 9.2.13 says that the ACC holds in every PID.

9.2.15 LEMMA Let D be a PID and a a nonzero element of D that is not a unit. Then a has at least one irreducible divisor.

Proof Let a be a nonzero element in D that is not a unit. If a is irreducible we are done. Otherwise, $a = x_1 y_1$ for some x_1 and y_1 that are not units. Now consider the ideal $\langle x_1 \rangle$. Since $a \in \langle x_1 \rangle$, we have $\langle a \rangle \subseteq \langle x_1 \rangle$, and $\langle a \rangle \neq \langle x_1 \rangle$ because a and x_1 are not associates. If x_1 is irreducible we are done. Otherwise, $x_1 = x_2 y_2$ for some x_2 and y_2 that are not units, and we have $\langle x_1 \rangle \subseteq \langle x_2 \rangle$ and $\langle x_1 \rangle \neq \langle x_2 \rangle$. Continuing in this way, we obtain an ascending chain of ideals, while by Proposition 9.2.13 the ACC holds, so this chain must end at some ideal $\langle x_r \rangle$, where x_r is an irreducible divisor of a. □

9.2.16 PROPOSITION Let D be a PID and a a nonzero element of D that is not a unit. Then a is a product of irreducibles.

Proof Let a be a nonzero element of D that is not a unit. By Lemma 9.2.15, $a = q_1 y_1$, where q_1 is irreducible. If y_1 is a unit, then a is irreducible and we are done. Otherwise, by Lemma 9.2.15 again, $y_1 = q_2 y_2$, where q_2 is irreducible. If y_2 is a unit, then y_1 is irreducible and $a = q_1 y_1$ is a product of irreducibles and we are done. Otherwise, we apply Lemma 9.2.15 again. Continuing in this way, we obtain a chain of ideals:

$$\langle a \rangle \subseteq \langle y_1 \rangle \subseteq \langle y_2 \rangle \subseteq \ldots$$

Since by Proposition 9.2.13 the ACC holds, this chain must end with some r such that y_{r+1} is a unit, so that y_r is irreducible, and $a = q_1 q_2 \ldots q_r y_r$ is a product of irreducibles. □

9.2.17 PROPOSITION Let D be a PID and a a nonzero element of D that is not a unit. If

$$a = p_1p_2 \ldots p_r = q_1q_2 \ldots q_s$$

where the p_i and q_j are all irreducible, then $r = s$ and the q_j can be renumbered so that p_i and q_i are associates in D for all i, $1 \le i \le r$.

Proof p_1 is irreducible in D and therefore by Proposition 9.2.6 is prime in D. Since $p_1 | a = q_1q_2 \ldots q_s$, it follows that $p_1 | q_j$ for some j. (See Exercise 20.) Renumbering, we may assume $p_1 | q_1$, and hence p_1 and q_1 are associates, say $q_1 = u_1p_1$, where u_1 is a unit. Since the cancellation law holds in D, we have

$$p_2 \ldots p_r = u_1q_2 \ldots q_s$$

Repeating the process with each p_i we eventually obtain

$$1 = u_1u_2 \ldots u_rq_{r+1}q_{r+1} \ldots q_s$$

Since the q_j are irreducible and are not units, we must have $s = r$, and each p_i and q_i are associates. The proof of Theorem 9.2.12 is now complete. □

We end this section with a very important theorem: "If D is a UFD, then $D[x]$ is a UFD." We proved in the preceding chapter the special case where $D = F$, a field (Theorem 8.4.6). For the general case where D is any UFD, we begin by noting that since $f(x) \in D[x] \subseteq Q[x]$ can be seen as an element of $Q[x]$, where Q is the field of quotients of D, by Theorem 8.4.6 $f(x)$ has a unique factorization into irreducibles in $Q[x]$. The hard part of the proof is to show that this will give a unique factorization into irreducibles in $D[x]$. For this we need some definitions and lemmas, which we have already encountered in Chapter 8 for the special case $D = \mathbb{Z}$. Some of the proofs are left as exercises with references to the proofs of the special case in the preceding chapter.

9.2.18 DEFINITION Let D be a UFD and $f(x) \in D[x]$ a nonconstant polynomial. Then $f(x)$ is said to be **primitive** in $D[x]$ if 1 is a gcd of all the coefficients of $f(x)$. ○

9.2.19 EXAMPLE Let us describe all the irreducible elements in $\mathbb{Z}[x]$. First note that $f(x) = 2x^3 - 10 = 2(x^3 - 5)$ is not irreducible in $\mathbb{Z}[x]$ since 2 is not a unit in $\mathbb{Z}[x]$. The units in $\mathbb{Z}[x]$ are by Proposition 8.1.14 the constant polynomials 1 and -1. If $\deg g(x) = 0$ and $g(x)$ is irreducible in $\mathbb{Z}[x]$, then $g(x) = p$ where p is a prime in $\mathbb{Z}$. On the other hand, if $\deg g(x) > 0$ and the polynomial $g(x)$ is irreducible in $\mathbb{Z}[x]$, then $g(x)$ must be primitive, since if the gcd in $\mathbb{Z}$ of the coefficients of $g(x)$ is $c \neq 0$, then dividing through by c we obtain $g(x) = ch(x)$, where neither c nor $h(x)$ is a unit in $\mathbb{Z}[x]$. ◇

9.2.20 LEMMA Let D be a UFD and $f(x) \in D[x]$ a nonconstant polynomial. Then there exist an element $c \in D$, and a primitive polynomial $g(x) \in D[x]$, such that $f(x) = cg(x)$. Here c and $g(x)$ are unique up to multiplication by a unit; that is, if

$f(x) = dh(x)$, where $d \in D$ and $h(x) \in D[x]$ is primitive, then d and c are associates in D and $g(x)$ and $h(x)$ are associates in $D[x]$.

Proof Starting with $f(x) = cg(x)$, where c is a gcd of all the coefficients of $f(x)$ in D, the proof can be completed by showing that $g(x)$ is primitive and unique up to a unit. (See Exercise 24.) □

Note that the notion of a primitive polynomial is essential in the case where the coefficient domain D is not a field. For in the case of a field, all the nonzero coefficients are units.

9.2.21 LEMMA (Gauss's lemma) Let D be a UFD and let $f(x)$ and $g(x)$ be primitive polynomials in $D[x]$. Then $f(x)g(x)$ is also a primitive polynomial.

Proof Suppose $f(x)g(x)$ is not primitive. Then by Lemma 9.2.20, $f(x)g(x) = ck(x)$, where c is not a unit in D. So let p be a prime in D that divides c. The ideal $I = \langle p \rangle$ is a prime ideal in D, hence $D_p = D/\langle p \rangle$ is an integral domain, and $D_p[x]$ is also an integral domain (by Theorem 8.1.12). Let $\phi : D \to D_p$ be the natural homomorphism, and $\phi^* : D[x] \to D_p[x]$ the induced homomorphism (as in Section 8.1, Exercise 15). The proof can now be completed by following the proof of Gauss's lemma for $\mathbb{Z}$ (Lemma 8.4.15). (See Exercise 25.) □

We now have all the tools we need for the proof of the generalization of Theorem 8.4.16.

9.2.22 THEOREM Let D be a UFD, Q the field of quotients of D, and let $f(x) \in D[x]$. Then $f(x)$ factors into a product of polynomials of degrees r and s in $Q[x]$ if and only if $f(x)$ factors into a product of polynomials of degrees r and s in $D[x]$.

Proof The proof can be carried out by following the steps of the proof of Theorem 8.4.16, using Lemma 9.2.20 and the generalization of Gauss's lemma, Lemma 9.2.21. (See Exercise 26.) □

9.2.23 COROLLARY Let D be a UFD and Q the field of quotients of D, and $f(x) \in D[x]$ a nonconstant polynomial. If $f(x)$ is irreducible in $D[x]$, then $f(x)$ is irreducible in $Q[x]$.

Proof This follows directly from the theorem. (See Exercise 27.) □

9.2.24 THEOREM If D is a UFD, then $D[x]$ is a UFD.

Proof Let $f(x) \in D[x]$ be a nonzero polynomial that is not a unit in $D[x]$. If $\deg f(x) = 0$, then $f(x) = c$ a constant polynomial, where $c \in D$ is not a unit in D. In this case, since D is a UFD $f(x) = c$ factors, uniquely except for order and multiplication by units, into a product

$$f(x) = p_1 \ldots p_s \in D$$

where the p_i are irreducible elements of D, or equivalently irreducible polynomials of degree 0 in $D[x]$, and we are done.

Now suppose $\deg f(x) > 0$. Let Q be the field of quotients of D. Since $\deg f(x) > 0$, it follows that $f(x)$ is not a unit in $Q[x]$, and by Theorem 8.4.6 we have

$$f(x) = p_1(x) \ \dots \ p_s(x) \in Q[x]$$

where the $p_i(x)$ are irreducible in $Q[x]$. The coefficients of each $p_i(x)$ are of the form ed^{-1}, where $e, d \in D$. Let $d_i \in D$ be the least common multiple of all the denominators d of the coefficients ed^{-1} of the polynomial $p_i(x)$, let d be the product of all the d_i, and let $q_i(x) = d_ip_i(x)$. Then we have

$$df(x) = q_1(x) \ \dots \ q_s(x) \in D[x]$$

where $q_i(x) \in D[x]$. Since d_i is unit in Q, and $p_i(x)$ was irreducible in $Q[x]$, $q_i(x)$ is still irreducible in $Q[x]$. By Lemma 9.2.20, $f(x) = cg(x)$, where $c \in D$ and $g(x) \in D[x]$ is primitive. Likewise, each $q_i(x) = c_ih_i(x)$, where $c_i \in D$ and $h_i(x) \in D[x]$ is primitive. Thus we have

$$dcg(x) = (c_1 \ldots c_s)h_1(x) \ldots h_s(x) \in D[x]$$

Since c_i is a unit in Q, and $q_i(x)$ was irreducible in $Q[x]$, $h_i(x)$ is still irreducible in $Q[x]$. By the uniqueness assertion in Lemma 9.2.20, dc and $c_1 \ldots c_i$ are associates in D. Thus there is a unit $u \in D$ such that

$$df(x) = dcg(x) = (dcu)h_1(x) \ldots h_s(x)$$
$$f(x) = (cu)h_1(x) \ldots h_s(x)$$

where the $h_i(x)$ are primitive in $D[x]$ and irreducible in $Q[x]$ and hence in $D[x]$. Since D is a UFD, cu can be written as a product of primes $a_1, \ \dots \ , a_r$ in D. So we have the existence of a factorization of $f(x)$ as a product of irreducible polynomials in $D[x]$

$$f(x) = a_1 \ldots a_r h_1(x) \ldots h_s(x)$$

where the a_j are irreducible in D and the $h_i(x)$ are primitive and are irreducible in $D[x]$ and hence in $Q[x]$.

What remains to show is uniqueness. Suppose

$$f(x) = b_1 \ldots b_{r'} k_1(x) \ldots k_{s'}(x)$$

where the b_j are irreducible in D and the $k_i(x)$ are primitive and irreducible in $D[x]$ and hence in $Q[x]$ by Corollary 9.2.23. We want to show that $r' = r$, $s' = s$, and that reordering if necessary, a_j and b_j are associates in D and $h_i(x)$ and $k_i(x)$ are associates in $D[x]$.

Note first that in Q the a_j and b_j are units. It then follows from the fact that $Q[x]$ is a UFD that $s = s'$ and that reordering if necessary, $h_i(x)$ and $k_i(x)$ are associates in $Q[x]$, say $k_i(x) = v_iw_i^{-1}h_i(x)$ and hence $w_ik_i(x) = v_ih_i(x)$ for some $v_i, w_i \in D$. Note now that if we let v and w be the products of the v_i and w_i, respectively, and $h(x)$ and $k(x)$ the products of the $h_i(x)$ and $k_i(x)$, respectively, then $wk(x) = vh(x)$, and since $h(x)$ and $k(x)$ are primitive by Gauss's lemma, it follows by Lemma 9.2.20 that v and w are associates in D, say $w = uv$, where u is a unit in D.

We thus have

$$a_1 \ldots a_r uvk(x) = a_1 \ldots a_r wk(x) = a_1 \ldots a_r vh(x) = vf(x) = b_1 \ldots b_{r'} vk(x)$$

$$ua_1 \ldots a_r = b_1 \ldots b_{r'}$$

where u is a unit in D. Since D is a UFD, it follows that $r' = r$ and that, reordering if necessary , a_j and b_j are associates.□

9.2.25 EXAMPLE $\mathbb{Z}$ is a UFD, therefore, $\mathbb{Z}[x]$ is a UFD. We have shown (in Example 8.6.11) that $\mathbb{Z}[x]$ is not a PID, so we now have an example of a UFD that is not a PID. ◇

9.2.26 EXAMPLE $F[x]$, where F is a field, is a UFD. Therefore, $F[x, y] = F[x][y]$ is a UFD. We have shown (in Example 8.6.10) that $F[x, y]$ is not a PID, so this is another example of a UFD that is not a PID. ◇

Exercises 9.2

1. Determine all the units in the indicated domains:

(a) $\mathbb{Z}[\sqrt{2}i]$ (b) $\mathbb{Z}_7[x]$ (c) $\mathbb{Z}[i][x]$ (d) $\mathbb{C}[x]$

In Exercises 2 through 6 determine whether the indicated pairs of elements are associates in the indicated domains.

2. 1 and $2 + \sqrt{3}$ in $\mathbb{Z}[\sqrt{3}]$

3. $5x - 10$ and $x - 2$ in $\mathbb{Z}[x]$

4. $3 + 2\sqrt{2}$ and $1 - \sqrt{2}$ in $\mathbb{Z}[\sqrt{2}]$

5. $5x^3 + x + 3$ and $x^3 + 3x + 2$ in $\mathbb{Z}_7[x]$

6. $1 + \sqrt{5}i$ and $1 - \sqrt{5}i$ in $\mathbb{Z}[\sqrt{5}i]$

In Exercises 7 through 10 determine whether the indicated elements are prime in the indicated domains. If not, determine whether they are irreducible in the indicated domain.

7. $6x - 21$ in $\mathbb{Z}[x]$, in $\mathbb{Q}[x]$, in $\mathbb{Z}_7[x]$

8. 5 in $\mathbb{Z}[\sqrt{5}i]$ in $\mathbb{Z}$ in $\mathbb{Q}$

9. 11 in $\mathbb{Z}[\sqrt{3}]$ in $\mathbb{Z}$ in $\mathbb{R}$

10. 19 in $\mathbb{Z}[\sqrt{3}i]$ in $\mathbb{Z}[\sqrt{2}i]$ in $\mathbb{C}$

In Exercises 11 through 16 determine whether or not the indicated integral domains are UFDs.

11. $\mathbb{Z}_3[x]$ **12.** $\mathbb{Z}[x,y]$ **13.** $\mathbb{Z}[i]$

14. $\mathbb{Z}[\sqrt{2}]$ **15.** $\mathbb{Z}[\sqrt{2}i]$ **16.** $\mathbb{Z}[\sqrt{3}i]$

17. Show that in an integral domain D, a nonzero element $p \in D$ is prime if and only if $\langle p \rangle$ is a prime ideal in D.

18. Let D be a PID. Show that a nonzero element $p \in D$ is irreducible in D if and only if p is prime in D.

19. Does Proposition 9.2.5 hold if the condition "Let D be a PID" is replaced by "Let D be a UFD"? If so, give a proof; if not, give a counterexample.

20. Let D be a PID, p an irreducible element in D, and $b_1, \ldots, b_r$ elements in D such that p divides the product $b_1 \ldots b_r$. Show that p divides b_i for some $1 \le i \le r$.

21. Prove Corollary 9.2.11, part (1).

22. Prove Corollary 9.2.11, part (2).

23. Let a and b be elements in a UFD. Show that if c is a gcd of a and b and d is an lcm of a and b, then cd and ab are associates.

24. Prove Lemma 9.2.20.

25. Complete the proof of the generalization of Gauss's lemma, Lemma 9.2.21.

26. Complete the proof of Theorem 9.2.22.

27. Prove Corollary 9.2.23.

9.3 Gaussian Integers

The first Euclidean domain we encountered after the ring of integers $\mathbb{Z}$ and the polynomial ring $F[x]$ over a field F was $\mathbb{Z}[i]$, the ring of Gaussian integers. A further study of this historically important ring reveals a very interesting connection with number theory. More concretely, the characterization of the prime elements of $\mathbb{Z}[i]$ gives us a proof of a highly nontrivial theorem of Fermat.

As was shown in Proposition 9.1.4, $\mathbb{Z}[i]$ is a Euclidean domain with the function

$$v(z) = z \cdot \overline{z} = a^2 + b^2 \qquad \text{for } z = a + bi \in \mathbb{Z}[i]$$

Therefore, $\mathbb{Z}[i]$ is a PID and hence a UFD (by Theorems 9.1.6 and 9.2.12). Consequently, prime elements and irreducible elements in $\mathbb{Z}[i]$ coincide by Proposition 9.2.10. By Theorem 9.1.15, $z \in \mathbb{Z}[i]$ is a unit if and only if $v(x) = v(1) = 1$. Hence $1, -1, i$, and $-i$ are the only units in $\mathbb{Z}[i]$.

9.3.1 EXAMPLE Consider 5 as an element of $\mathbb{Z}[i]$. We can easily see that 5 is not a prime in $\mathbb{Z}[i]$, since $5 = (1 + 2i)(1 - 2i)$. But consider $1 + 2i$. If $1 + 2i = (a + bi)(c + di)$, then $5 = v(1 + 2i) = (a^2 + b^2)(c^2 + d^2)$, which implies that either

$a^2 + b^2 = 1$ or $c^2 + d^2 = 1$, and hence either $a + bi$ or $c + di$ is a unit in $\mathbb{Z}[i]$. In other words, $1 + 2i$ is a prime in $\mathbb{Z}[i]$, and similarly so is $1 - 2i$. ◇

9.3.2 PROPOSITION Let $z = a + bi \in \mathbb{Z}[i]$ be such that $v(z) = a^2 + b^2$ is prime in $\mathbb{Z}$. Then $z = a + bi$ is prime in $\mathbb{Z}[i]$.

Proof Let $z = a + bi$ be an element of $\mathbb{Z}[i]$ such that $v(z) = a^2 + b^2 = p$ is prime in $\mathbb{Z}$. If

(1) $z = (a + bi)(c + di)$ in $\mathbb{Z}[i]$

then

(2) $p = v(z) = (a^2 + b^2)(c^2 + d^2)$ in $\mathbb{Z}$

Thus since p is prime in $\mathbb{Z}$, one of the factors in (2) must be 1, which implies that one of the factors in (1) must be a unit. This shows that z is irreducible and hence prime in $\mathbb{Z}[i]$. □

9.3.3 COROLLARY If p is a prime in $\mathbb{Z}$ that can be written as a sum of two squares, $p = a^2 + b^2$, then $a + bi$ is prime in $\mathbb{Z}[i]$.

Proof Immediate from the preceding theorem. □

What we have just shown about primes in $\mathbb{Z}[i]$ can be used to prove Fermat's theorem in number theory on the characterization of primes in $\mathbb{Z}$ that can be written as a sum of two squares. Given any odd prime $p \in \mathbb{Z}$, then either $p \equiv 1 \bmod 4$ or $p \equiv 3 \bmod 4$. Fermat's theorem can be stated as follows.

9.3.4 THEOREM (Fermat's theorem on sums of squares) An odd prime p in $\mathbb{Z}$ can be written as the sum of two squares , $p = a^2 + b^2$, if and only if $p \equiv 1 \pmod 4$.

Proof Let p be an odd prime in $\mathbb{Z}$.

(⇒) Suppose $p = a^2 + b^2$. Now for any integers a and b in $\mathbb{Z}$, a^2 and b^2 are each congruent either to 0 or to 1 mod 4. Since p is odd, a^2 and b^2 cannot both be congruent to 0 or both congruent to 1. Hence one is congruent to 0 and the other to 1, and p is congruent to 1 mod 4.

(⇐) Assume $p \equiv 1 \bmod 4$. Consider the multiplicative group $\mathbb{Z}_p^*$, which is a cyclic group of order $p - 1$. Since 4 divides $p - 1$, there exists an element $x \in \mathbb{Z}_p^*$ with order $|x| = 4$ and $|x^2| = 2$. Hence $x^2 = -1$ in $\mathbb{Z}_p^*$ and $x^2 + 1 = 0$ in $\mathbb{Z}_p^*$. In other words, $p | x^2 + 1$. Now consider $x^2 + 1 = (x + i)(x - i)$ in $\mathbb{Z}[i]$. If p were prime in $\mathbb{Z}[i]$, it would divide one of the two factors, say $x + i$. In that case we would have $x + i = p(a + bi)$ in $\mathbb{Z}[i]$. But this would imply $pb = 1$ in $\mathbb{Z}$, which is impossible. Hence p is not a prime in $\mathbb{Z}[i]$, and $p = (a + bi)(c + di)$ in $\mathbb{Z}[i]$, where neither factor is a unit. Therefore, $p^2 = v(p) = (a^2 + b^2)(c^2 + d^2)$, where neither factor is 1, and hence each must be p. Hence $p = a^2 + b^2$ and p is the sum of two squares as desired. □

9.3.5 COROLLARY If p is a prime in $\mathbb{Z}$ with $p \equiv 3 \bmod 4$, then p is a prime in $\mathbb{Z}[i]$.

Proof Let p be a prime in $\mathbb{Z}$ with $p \equiv 3 \bmod 4$. If

(1) $\qquad p = (a + bi)(c + di)$ in $\mathbb{Z}[i]$

then

(2) $\qquad p^2 = v(p) = (a^2 + b^2)(c^2 + d^2)$ in $\mathbb{Z}$

Since by the preceding theorem, p is not a sum of two squares, neither factor in (2) can be equal to p. Hence one factor must be equal to 1, and hence one factor in (1) must be a unit. This shows that p is irreducible and hence prime in $\mathbb{Z}[i]$. □

We can now complete the characterization of the primes in $\mathbb{Z}[i]$.

9.3.6 LEMMA Let $a + bi$ be a prime in $\mathbb{Z}[i]$. Then $a - bi$ is a prime in $\mathbb{Z}[i]$.

Proof If $a - bi = xy$ were a nontrivial factorization of $a - bi$ in $\mathbb{Z}[i]$, then

$$a + bi = \overline{xy} = \overline{x} \cdot \overline{y}$$

would be a nontrivial factorization of $a + bi$. □

9.3.7 LEMMA Let $a + bi$ be a prime in $\mathbb{Z}[i]$ with $a \neq 0$ and $b \neq 0$. Then $a^2 + b^2$ is prime in $\mathbb{Z}$.

Proof First note that if the prime $z = a + bi$ divides an integer m, say, $zw = m$, then

$$m = \overline{m} = \overline{zw} = \overline{z} \cdot \overline{w} \in \mathbb{Z}[i]$$

and so $\overline{z} = a - bi$, which by the preceding lemma is also prime, also divides m. Now since $a \neq 0$ and $b \neq 0$, z and $\overline{z}$ can be associates only if $a = \pm b$, in which case $a + bi = a(1 \pm i)$ can be prime only if $a = \pm 1$, in which case $a^2 + b^2 = 2$, which is a prime in $\mathbb{Z}$. If the primes $a + bi$ and $a - bi$ are not associates, then since each divides m, their product $a^2 + b^2$ must also divide m. This means there cannot be a nontrivial factorization $a^2 + b^2 = mn$ in $\mathbb{Z}$ or, in other words, that $a^2 + b^2$ is a prime in $\mathbb{Z}$. □

9.3.8 PROPOSITION All the primes in the Gaussian integers $\mathbb{Z}[i]$ are exactly the following elements and their associates:

(1) $1 + i$ and $1 - i$

(2) p a prime in $\mathbb{Z}$ with $p \equiv 3 \bmod 4$

(3) $a + bi$ and $a - bi$, where $(a + bi)(a - bi) = a^2 + b^2 = p$, where p is a prime in $\mathbb{Z}$ with $p \equiv 1 \bmod 4$.

Proof (1) $1 \pm i$ is prime in $\mathbb{Z}[i]$ by Proposition 9.3.2, since $v(1 \pm i) = 2$.

(2) If p is a prime in $\mathbb{Z}$ with $p \equiv 3 \bmod 4$, then p is prime in $\mathbb{Z}[i]$ by Corollary 9.3.5.

(3) If p is a prime in $\mathbb{Z}$ with $p \equiv 1 \bmod 4$ and $p = a^2 + b^2$ in accordance with Theorem 9.3.4, then $a \pm bi$ is prime in $\mathbb{Z}[i]$ by Corollary 9.3.3.

This shows that all the numbers in (1) through (3) are prime in $\mathbb{Z}[i]$, as well as their associates.

Now conversely suppose $z = a + bi$ is prime in $\mathbb{Z}[i]$. If $b = 0$, then z is an integer and must be a prime in $\mathbb{Z}$ as in (2). If $a = 0$, then iz is also prime in $\mathbb{Z}[i]$, being an associate of z, and iz is an integer and therefore must be a prime in $\mathbb{Z}$ as in (2), and z will be an associate of such a prime. If $a \neq 0$ and $b \neq 0$, then by Lemma 9.3.7, $a^2 + b^2$ is prime in $\mathbb{Z}$, and $a + bi$ will be a prime as in (1) if $a^2 + b^2 = 2$, and a prime as in (3) if $a^2 + b^2$ is odd. Thus we have found all the primes in $\mathbb{Z}[i]$. □

9.3.9 EXAMPLE Now that we know which are the irreducible elements in $\mathbb{Z}[i]$, we can factor elements of $\mathbb{Z}[i]$ into irreducible factors. For instance, suppose we wish to factor $4 + 3i$ into irreducible factors. First note that if $4 + 3i = (a + bi)(c + di)$, then $25 = v(4 + 3i) = (a^2 + b^2)(c^2 + d^2)$. Hence $a^2 + b^2 = c^2 + d^2 = 5$ and either $a = \pm 2$ and $b = \pm 1$ or $a = \pm 1$ and $b = \pm 2$. Hence we obtain $4 + 3i = (2 - i)(1 + 2i)$. By Proposition 9.3.8, part (3), these are the irreducible factors in $\mathbb{Z}[i]$. ◇

Exercises 9.3

In Exercises 1 through 4 express the indicated primes in $\mathbb{Z}$ as the sum of two squares and write their factorizations into irreducibles in $\mathbb{Z}[i]$.

1. 17 **2.** 29 **3.** 37 **4.** 41

In Exercises 5 through 8 factor the indicated Gaussian integers into a product of irreducibles in $\mathbb{Z}[i]$.

5. 11 **6.** 13 **7.** $-1 + 5i$ **8.** $8 - i$

9. Prove that there are an infinite number of primes p in $\mathbb{Z}$ such that $p \equiv 3 \bmod 4$.

10. (*a*) Prove that there are an infinite number of primes $a + bi$ in $\mathbb{Z}[i]$ with $a \neq 0$ and $b \neq 0$.
(*b*) Prove that there are an infinite number of primes p in $\mathbb{Z}$ such that $p \equiv 1 \bmod 4$.

11. Let n be a positive integer with prime factorization

$$n = p_1{}^{a_1}p_2{}^{a_2}\ldots p_r{}^{a_r}q_1{}^{b_1}q_2{}^{b_2}\ldots q_s{}^{b_s}$$

where $p_i \equiv 1 \bmod 4$ for $1 \leq i \leq r$ and $q_j \equiv 3 \bmod 4$ for $1 \leq j \leq s$. Show that n can be written as a sum of two squares if and only if b_j is even for all $1 \leq j \leq s$.

12. Let I be a nontrivial ideal in $\mathbb{Z}[i]$. Show that $\mathbb{Z}[i]/I$ is a finite ring.

13. Let z be irreducible in $\mathbb{Z}[i]$. Show that $\mathbb{Z}[i]/\langle z\rangle$ is a field.

In Exercises 14 through 17 find the order and characteristic of the indicated fields.

14. $\mathbb{Z}[i]/\langle 3\rangle$ **15.** $\mathbb{Z}[i]/\langle 7\rangle$ **16.** $\mathbb{Z}[i]/\langle 1+i\rangle$ **17.** $\mathbb{Z}[i]/\langle 2+i\rangle$

18. Let q be a prime in $\mathbb{Z}$ such that $q \equiv 3 \bmod 4$. Show that $\mathbb{Z}[i]/\langle q\rangle$ is a field of order q^2.

19. Let p be a prime in $\mathbb{Z}$ such that $p \equiv 1 \bmod 4$. Show that

(a) $\mathbb{Z}[i]/\langle p\rangle$ is not an integral domain.

(b) $\mathbb{Z}[i]/\langle p\rangle$ has order p^2.

(c) $\mathbb{Z}[i]/\langle p\rangle \cong \mathbb{Z}_p \times \mathbb{Z}_p$

20. (Chinese remainder theorem) Let R be a commutative ring and K and L two proper ideals in R such that $K + L = R$. Show that

$$R/(K \cap L) \cong R/K \times R/L$$

21. In $\mathbb{Z}[i]$ let I be the ideal generated by $3 + 11i$ and $7 + 4i$.

(a) Find $d \in \mathbb{Z}[i]$ such that $I = \langle d\rangle$.

(b) Write d as the product of irreducibles in $\mathbb{Z}[i]$.

(c) Use the preceding exercise to show that

$$\mathbb{Z}[i]/I \cong F_1 \times F_2$$

for some fields F_1 and F_2.

(d) Describe these fields F_1 and F_2.

Chapter 10

Field Theory

After a brief introduction of *vector spaces* over a field F, we introduce the notion of a *field extension* E of F, which is a larger field containing F. The field extension E can be viewed both as a field and as a vector space over F. We prove Kronecker's theorem, stating that every nonconstant polynomial over a field F has a zero in some field extension E of F. We then study the smallest extension of F that contains all the zeros of the given polynomial, called the *splitting field* of the polynomial. In the last section we use the existence and uniqueness of the splitting field of a given polynomial to obtain a classification theorem for finite fields.

10.1 Vector Spaces

Linear algebra is a fundamental subject in mathematics. In this section we only give a brief introduction to this subject in order to introduce some basic notions of linear algebra that are essential for the study of fields in the later sections. The crucial notions are those of a *vector space* over a field, of a *basis* of a vector space, and of the *dimension* of a vector space.

10.1.1 EXAMPLE Consider the field $\mathbb{C}$ of complex numbers. In Section 0.4 we explained how a complex number $z = a + bi \in \mathbb{C}$ can be represented by a point $(a, b) \in \mathbb{R}^2$.

Elements of $\mathbb{C}$	Points in $\mathbb{R}^2$
$a + bi$	(a, b)
1	$(1, 0)$
i	$(0, 1)$
$(a + bi) + (c + di) = (a + c) + (b + d)i$	$(a, b) + (c, d) = (a + c, b + d)$
$c(a + bi) = ca + cbi$ for $c \in \mathbb{R}$	$c(a, b) = (ca, cb)$ for $c \in \mathbb{R}$
$a + bi = a \cdot 1 + b \cdot i$	$(a, b) = a(1, 0) + b(0, 1)$

We have singled out in this example properties of elements of $\mathbb{C}$ shared with the corresponding points in the plane $\mathbb{R}^2$. ◇

10.1.2 DEFINITION Let F be a field. A set V equipped with two operations, written as addition and multiplication, is said to be a **vector space** over F if

(1) V is an Abelian group under addition.

For all $a, b \in F$ and all $u, v \in V$,

(2) The product $av \in V$ is defined.

(3) $a(v + w) = av + aw$

(4) $a(bv) = (ab)v$

(5) $1v = v$

An element of $v \in V$ is called a **vector**. The identity element of V under addition is called the **zero vector** and written **0**. An element of $a \in F$ is called a **scalar**, and the operation of forming av is called **scalar multiplication**. ○

10.1.3 EXAMPLE $\mathbb{C}$ is a vector space over $\mathbb{R}$. ◇

10.1.4 EXAMPLE For any field F and any positive integer n, let

$$F^n = \{(a_1, a_2, \dots, a_n) \mid a_i \in F \text{ for } 1 \le i \le n\}$$

Then F^n is a vector space over F with the operations of componentwise addition:

$$(a_1, a_2, \dots, a_n) + (b_1, b_2, \dots, b_n) = ((a_1 + b_1, a_2 + b_2, \dots, a_n + b_n)$$

and scalar multiplication:

$$c(a_1, a_2, \dots, a_n) = (ca_1, ca_2, \dots, ca_n)$$

for $c \in F$. ◇

10.1.5 EXAMPLE For any field F the ring of polynomials $F[x]$ is a vector space over F, since

(1) $F[x]$ is a ring, and therefore an additive Abelian group.

(2) For any constant $c \in F$ and any polynomial $f(x) \in F[x]$, $cf(x) \in F[x]$, and thus conditions (3) through (5) of Definition 10.1.2 hold. ◇

10.1.6 EXAMPLE For any field F, $M(2, F)$, the ring of all 2×2 matrices with entries from F is a vector space over F with the usual addition operation on matrices and the scalar multiplication:

$$c\begin{bmatrix} a_{11} & a_{12} \\ a_{21} & a_{22} \end{bmatrix} = \begin{bmatrix} ca_{11} & ca_{12} \\ ca_{21} & ca_{22} \end{bmatrix}$$

for all $c \in F$. ◇

Several of these examples of vector spaces can be generalized as follows.

10.1.7 PROPOSITION Given $F \subseteq R$, where F is a field and R is a ring containing F as a subring, R is a vector space over F.

Proof See Exercise 8. □

For any vector space V over a field F, there are two additive identity elements involved, the zero vector $\mathbf{0} \in V$ and the zero scalar $0 \in F$. Also, any vector $v \in V$ has an additive inverse $-v \in V$, while any scalar $c \in F$ has an additive inverse $-c \in F$. The next proposition states the relationship between the two kinds of additive identities and the two kinds of additive inverses and implies, in particular, that $-v = (-1)v$.

10.1.8 PROPOSITION Let V be a vector space over a field F. Then for any scalar $c \in F$ and any vector $v \in V$

(1) $cv = \mathbf{0}$ if and only if $c = 0 \in F$ or $v = \mathbf{0} \in V$.

(2) $(-c)v = -(cv) = c(-v)$

Proof (1) ($\Leftarrow$) If $c = 0 \in F$, then $0v = (0 + 0)v = 0v + 0v$ by Definition 10.1.2, part (3). Hence $0v = \mathbf{0}$. Similarly, if $v = \mathbf{0} \in V$, then $c\mathbf{0} = c(\mathbf{0} + \mathbf{0}) = c\mathbf{0} + c\mathbf{0}$, and $c\mathbf{0} = \mathbf{0}$.
($\Rightarrow$) If $cv = \mathbf{0}$ and $c \neq 0 \in F$, then c is a unit in F. We have $c^{-1}\mathbf{0} = \mathbf{0}$ by what we have already proved. But we also have $c^{-1}\mathbf{0} = c^{-1}(cv) = (c^{-1}c)v = 1v = v$, using Definition 10.1.2, parts (4) and (5). Hence $v = c^{-1}\mathbf{0} = \mathbf{0}$.

(2) We have $(-c)v + cv = (-c + c)v = 0v = \mathbf{0}$ using part (1). Therefore, $(-c)v = -(cv)$. Similarly, $c(-v) + cv = c(-v + v) = c\mathbf{0} = \mathbf{0}$, and $c(-v) = -(cv)$. □

With every algebraic structure we have introduced (groups, rings, fields) we have defined a notion of substructure (subgroup, subring, subfield). In a similar manner we define a *subspace* of a vector space and state a *subspace test* whose proof is left as an exercise.

10.1.9 DEFINITION A nonempty subset U of a vector space V over a field F is a **subspace** of V if U is a vector space over F under the same addition and scalar multiplication operations as V. ○

10.1.10 THEOREM (Subspace test) A nonempty subset U of a vector space V over a field F is a subspace of V if and only if for all $c \in F$ and $u, w \in U$ we have

(1) $u - w \in U$

(2) $cu \in U$

Proof See Exercise 7. □

10.1.11 EXAMPLE As we saw in Example 10.1.5, for any field F the ring $F[x]$ is a vector space over F. Let

$$F^n[x] = \{f(x) \in F[x] \mid \deg f(x) < n \text{ or } f(x) = 0\}$$

Then $F^n[x]$ is a subspace of $F[x]$ because for any $c \in F$ and any $f(x)$ and $g(x)$ in $F^n[x]$ we have

(1) $\deg (f(x) - g(x)) < n$, and hence $f(x) - g(x) \in F^n[x]$, or $f(x) - g(x) = 0 \in F^n[x]$

(2) $\deg (cf(x)) = \deg f(x) < n$, and hence $cf(x) \in F^n[x]$, or $cf(x) = 0 \in F^n[x]$

and the subspace test applies. ◇

10.1.12 PROPOSITION In $F[x]$, where F is a field, any ideal I in $F[x]$ is a subspace of $F[x]$.

Proof See Exercise 9. □

10.1.13 EXAMPLE In F^n as defined in Example 10.1.4, and $s \leq n$, let

$$U^s = \{(a_1, a_2, \ldots, a_s, 0, 0, \ldots, 0) \mid a_i \in F \text{ for } 1 \leq i \leq s\}$$

Then U^s is a subspace of F^n. ◇

10.1.14 EXAMPLE Consider a system of linear equations:

$$\begin{aligned} &c_{11}x_1 + c_{12}x_2 + \ldots + c_{1m}x_m = 0 \\ &c_{21}x_1 + c_{22}x_2 + \ldots + c_{2m}x_m = 0 \\ &\vdots \\ &c_{n1}x_1 + c_{n2}x_2 + \ldots + c_{nm}x_m = 0 \end{aligned}$$

with $c_{ij} \in F$, where F is a field. Let U be the subset of the vector space

$$F^m = \{(a_1, a_2, \ldots, a_m) \mid a_i \in F \text{ for } 1 \leq i \leq m\}$$

consisting of all solutions of this system. Then U is a subspace of F^m. For if $(a_1, a_2, \ldots, a_m)$ and $(b_1, b_2, \ldots, b_m)$ are in U, then

(1) $$\begin{aligned} &c_{i1}(a_1 - b_1) + \ldots + c_{im}(a_m - b_m) = \\ &(c_{i1}a_1 + \ldots + c_{im}a_m) - (c_{i1}b_1 + \ldots + c_{im}b_m) = \\ &0 - 0 = 0 \end{aligned}$$

for all i, $1 \leq i \leq n$, hence $(a_1 - b_1, \ldots, a_m - b_m) \in U$.

(2) $$c_{i1}(da_1) + \ldots + c_{im}(da_m) = d(c_{i1}a_1 + \ldots + c_{im}a_m) = d0 = 0$$

for any $d \in F$, and hence $d(a_1, \ldots, a_m) = (da_1, \ldots, da_m) \in U$. ◇

10.1.15 EXAMPLE Let V be a vector space over a field F and let v_1, v_2 be two vectors in V. Let

$$U = \{c_1v_1 + c_2v_2 \mid c_i \in F\}$$

Then U is a subspace of V. For if $u = a_1v_1 + a_2v_2 \in U$ and $w = b_1v_1 + b_2v_2 \in U$, then

(1) $u - w = (a_1 - b_1)v_1 + (a_2 - b_2)v_2 \in U$

(2) $cu = (ca_1)v_1 + (ca_2)v_2 \in U$

for any $c \in F$. ◇

This last example introduces us to the following important notions.

10.1.16 DEFINITION Let V be a vector space over a field F and let $v_1, v_2, \ldots, v_n$ be vectors in V. Then

(1) $c_1v_1 + c_2v_2 + \ldots + c_nv_n$, where $c_i \in F$ for all i is called a **linear combination** of the vectors v_i.

(2) The **span** of the vectors v_i is the set

$$\text{span}\,\{v_1, v_2, \ldots, v_n\} = \{c_1v_1 + c_2v_2 + \ldots + c_nv_n \mid c_i \in F \text{ for all } i\}$$

of all linear combinations of the vectors v_i. ○

An immediate consequence of the preceding definition is the following proposition, whose two parts together say that span $\{v_1, \dots, v_n\}$ is the smallest subspace of V containing all the v_i.

10.1.17 PROPOSITION Let V be a vector space over a field F and $v_1, \dots, v_n$ vectors in V. Then

(1) span $\{v_1, \dots, v_n\}$ is a subspace of V.

(2) If U is a subspace of V and $v_1, \dots, v_n \in U$, then span $\{v_1, \dots, v_n\} \subseteq U$.

Proof (1) The case $n = 2$ is proved in Example 10.1.15, and the proof in the general case is the same.

(2) By Definition 10.1.16, span $\{v_1, \dots, v_n\}$ consists of all linear combinations of the v_i, and by the subspace test (Theorem 10.1.10), a subspace U contains all linear combinations of its elements.□

10.1.18 EXAMPLE In the vector space $F[x]$, the vectors $1, x, x^2, \dots, x^{n-1}$ span the subspace $F^n[x] = \{f(x) \in F[x] \mid \deg f(x) < n \text{ or } f(x) = 0\}$ ◇

10.1.19 EXAMPLE In $\mathbb{R}^3$ the vectors $v_1 = (-3, 0, 1)$ and $v_2 = (2, 1, 0)$ span the subspace

$$\text{span}\,\{v_1, v_2\} = \{(2s - 3t, s, t) \mid s, t \in \mathbb{R}\}$$

which is the plane in $\mathbb{R}^3$ going through the origin and defined by the equation $x - 2y + 3z = 0$. ◇

10.1.20 EXAMPLE In a vector space V a nonzero vector w is in the span of two given vectors $v_1, v_2 \in V$ if there exist scalars $c_1, c_2 \in F$ such that $w = c_1v_1 + c_2v_2$ or, equivalently, $w + (-c_1)v_1 + (-c_2)v_2 = 0$. Conversely, if there exist scalars d_0, d_1, d_2 in the field F, with $d_0 \neq 0$, such that $d_0w + d_1v_1 + d_2v_2 = 0$, then the vector w can be written as $w = (-d_1d_0^{-1})v_1 + (-d_2d_0^{-1})v_2$, and $w \in \text{span}\{v_1, v_2\}$. ◇

10.1.21 DEFINITION Let $v_1, \dots, v_n$ be vectors in a vector space V over a field F.

(1) The set $\{v_1, v_2, \dots, v_n\}$ is called **linearly independent** over F if

$$c_1v_1 + c_2v_2 + \dots + c_nv_n = 0$$

for $c_i \in F$, implies that $c_i = 0$ for all $1 \leq i \leq n$.

(2) The set $\{v_1, v_2, \dots, v_n\}$ is called **linearly dependent** over F if

$$c_1v_1 + c_2v_2 + \dots + c_nv_n = 0$$

for some $c_i \in F$, $1 \leq i \leq n$, with not all $c_i = 0$. ○

Note that the set $\{v_1, v_2, \dots, v_n\}$ is linearly dependent over F if and only if at least one of the vectors v_i is in the span of the remain $n - 1$ vectors in the set.

10.1.22 EXAMPLE In $F[x]$ the set of vectors $\{1, x, x^2, \dots, x^n\}$ for any $n \geq 1$ is linearly independent over F since

$$c_0 \cdot 1 + c_1 x + c_2 x^2 + \ldots + c_n x^n = 0 \in F[x]$$

if and only if $c_i = 0$ for all i. ◇

10.1.23 EXAMPLE In $\mathbb{C}$ the set $\{1, i\}$ is linearly independent over $\mathbb{R}$ since

$$a \cdot 1 + b \cdot i = 0 \in \mathbb{C}$$

if and only if $a = b = 0$. ◇

10.1.24 DEFINITION In a vector space V, a set of vectors $\{v_1, \ldots, v_n\}$ is called a **basis** for V over F if

(1) span $\{v_1, \ldots, v_n\} = V$

(2) $\{v_1, \ldots, v_n\}$ is linearly independent over F. ○

10.1.25 EXAMPLES

(1) The set $\{1, i\}$ is a basis for $\mathbb{C}$ over $\mathbb{R}$.

(2) The set $\{1, \sqrt{2}\}$ is a basis for $\mathbb{Q}(\sqrt{2})$ over $\mathbb{Q}$. ◇

10.1.26 EXAMPLE The set $\{1, x, x^2, \ldots, x^{n-1}\}$ is a basis for

$$F^n[x] = \{f(x) \in F[x] \mid \deg f(x) < n \text{ or } f(x) = 0\}$$

over F. ◇

10.1.27 EXAMPLE For any field F, in the vector space

$$F^n = \{(a_1, a_2, \ldots, a_n) \mid a_i \in F \text{ for } 1 \leq i \leq n\}$$

a basis is formed by the vectors

$$\begin{aligned} e_1 &= (1, 0, \ldots, 0) \\ e_2 &= (0, 1, \ldots, 0) \\ &\vdots \\ e_n &= (0, 0, \ldots, 1) \end{aligned}$$

This basis $\{e_1, e_2, \ldots, e_n\}$ is called the **standard basis** for F^n over F. ◇

10.1.28 EXAMPLE For any field F, in the vector space $M(2, F)$ of 2×2 matrices with entries from F, the matrices

$$E_1 = \begin{bmatrix} 1 & 0 \\ 0 & 0 \end{bmatrix}, E_2 = \begin{bmatrix} 0 & 1 \\ 0 & 0 \end{bmatrix}, E_3 = \begin{bmatrix} 0 & 0 \\ 1 & 0 \end{bmatrix}, E_4 = \begin{bmatrix} 0 & 0 \\ 0 & 1 \end{bmatrix}$$

form a basis for $M(2, F)$ over F. ◇

A very important concept that in a certain sense describes a vector space is its *dimension*. We define the dimension of a vector space in terms of the size of a basis of the vector space. But first we must show that all bases for a given vector space have the same size, which is our next theorem.

10.1.29 THEOREM Let V be a vector space over a field F, and let $\{v_1, \ldots, v_n\}$ and $\{u_1, \ldots, u_m\}$ be two bases V over F. Then $n = m$.

Proof Suppose $n \neq m$, let's say $n < m$. Consider the set $\{u_1, v_1, \ldots, v_n\}$. Since $u_1 \in$ span $\{v_1, v_2, \ldots, v_n\}$, there exist scalars $a_i \in F$, such that

$$u_1 = a_1 v_1 + a_2 v_2 + \ldots + a_n v_n$$

and the a_i, $1 \le i \le n$ are not all zero. Let's say (renumbering the v_i if necessary) that $a_1 \ne 0$. Then

$$v_1 = a_1^{-1}u_1 + (-a_1^{-1}a_2)v_2 + \dots + (-a_1^{-1}a_n)v_n$$

So $v_1 \in \text{span}\{u_1, v_2, \dots, v_n\}$, and since trivially $v_i \in \text{span}\{u_1, v_2, \dots, v_n\}$ for $2 \le i < n$, $\text{span}\{v_1, v_2, \dots, v_n\} \subseteq \text{span}\{u_1, v_2, \dots, v_n\}$ by Proposition 10.1.17 and $V = \text{span}\{u_1, v_2, \dots, v_n\}$. Now consider $u_2 \in V = \text{span}\{u_1, v_2, \dots, v_n\}$. We have

$$u_2 = b_1u_1 + b_2v_2 + b_3v_3 + \dots + b_nv_n$$

for some $b_i \in F$, and since u_1 and u_2 are linearly independent, the b_i for $2 \le i \le n$ are not all zero. Let's say $b_2 \ne 0$. Then

$$v_2 = (-b_2^{-1}b_1)u_1 + b_2^{-1}u_2 + (-b_2^{-1}b_3)v_3 + \dots + (-b_2^{-1}b_n)v_n$$

and $V = \text{span}\{u_1, u_2, v_3, \dots, v_n\}$. Continuing in this way, we obtain

$$V = \text{span}\{u_1, u_2, \dots, u_n\}$$

We made the assumption that $n < m$. Therefore, $u_{n+1} \in V$ is a linear combination of the vectors $\{u_1, \dots, u_n\}$, and the set $\{u_1, \dots, u_m\}$ is linearly dependent and so not a basis. This is a contradiction, and the proof is complete. □

10.1.30 DEFINITION Let V be a vector space over a field F. If there exists a finite set that forms a basis for V over F, then the number n of vectors in such a basis $\{v_1, \dots, v_n\}$ (which is the same for all such bases by the preceding theorem) is called the **dimension** of V over F, written $\dim_F V$. If there exists no finite basis for V over F, then V is said to be **infinite dimensional** over F. ◇

10.1.31 EXAMPLES

(1) $\dim_{\mathbb{R}} \mathbb{C} = 2$

(2) $\dim_{\mathbb{Q}} \mathbb{Q}(\sqrt{2}) = 2$

(3) $\dim_{\mathbb{R}} M(2, \mathbb{R}) = 4$

(4) $\dim_{\mathbb{R}} \mathbb{R}^n = n$

(5) $\dim_F F^n[x] = n$

(6) $F[x]$ is infinite dimensional over F. ◇

We conclude this section with a useful fact about linearly independent sets in a finite-dimensional vector space.

10.1.32 THEOREM Let V be a vector space over a field F with $\dim_F V = n$, and let $\{u_1, \dots, u_r\}$ be a linearly independent set of vectors in V. Then

(1) $r \le n$

(2) If $r < n$, then there exist vectors $u_{r+1}, \dots, u_n$ in V such that $\{u_1, \dots, u_n\}$ forms a basis for V over F.

Proof If $\text{span}\{u_1, \dots, u_r\} = V$, then $\{u_1, \dots, u_r\}$ is a basis, and $r = n$ by Theorem 10.1.29. Otherwise, let $\{v_1, \dots, v_n\}$ be a basis for V over F. Since

$$V = \text{span}\{v_1, \dots, v_n\} \ne \text{span}\{u_1, \dots, u_r\}$$

there must be some $v_i \notin \text{span}\{u_1, \dots, u_r\}$. Let's say (renumbering the v_i if necessary) that it is v_1. Then $\{u_1, \dots, u_r, v_1\}$ is a linearly independent set of vectors.

If span $\{u_1, \ldots, u_r, v_1\} = V$, then $\{u_1, \ldots, u_r, v_1\}$ is a basis, and $r = n - 1 < n$. Otherwise, using the same argument as before, there must be some vector $v_i \notin$ span $\{u_1, \ldots, u_r, v_1\}$, let's say that it is v_2. Then $\{u_1, \ldots, u_r, v_1, v_2\}$ is a linearly independent set of vectors. Repeating this process, it must terminate with some $k < n$ such that

$$\{u_1, \ldots, u_r, v_1, v_2, \ldots, v_k\}$$

is a basis, and $r = n - k < n$. □

10.1.33 COROLLARY Let V be a vector space over a field F with $\dim_F V = n$, and U a subspace of V. Then

(1) Any basis for U can be extended to a basis for V.

(2) $\dim_F U \leq \dim_F V$.

(3) $\dim_F U = \dim_F V$ if and only if $U = V$.

Proof See Exercise 23. □

Exercises 10.1

In Exercises 1 through 6 determine whether the indicated set of vectors is a basis for the indicated vector space V over the indicated field F.

1. $\{(1,1,1), (0,0,1), (1,1,0)\}$ $V = \mathbb{R}^3$ $F = \mathbb{R}$

2. $\{(1,0,1), (0,2,2), (3,3,0)\}$ $V = \mathbb{R}^3$ $F = \mathbb{R}$

3. $\{2 + 3i, -5\}$ $V = \mathbb{C}$ $F = \mathbb{R}$

4. $\{1 + i, 2 + i, 3i\}$ $V = \mathbb{C}$ $F = \mathbb{R}$

5. $\{2x + 3, -x^2 + x, x^2 + 5\}$ $V = \mathbb{Q}^3[x]$ $F = \mathbb{Q}$

6. $\{1 + 3\sqrt{2}, 2 + 5\sqrt{2}\}$ $V = \mathbb{Q}(\sqrt{2})$ $F = \mathbb{Q}$

7. Prove the subspace test, Theorem 10.1.10.

8. Let F be a field and R a ring such that $F \subseteq R$ is a subring of R. Show that R is a vector space over F.

9. Prove Proposition 10.1.12.

In Exercises 10 through 17 determine whether the indicated subset U is a subspace of the indicated vector space V over the indicated field F.

10. $U = \mathbb{Q}$ $V = \mathbb{Q}(\sqrt{3})$ $F = \mathbb{Q}$

11. $U = \mathbb{R}$ $V = \mathbb{C}$ $F = \mathbb{Q}$

12. $U = \{f(x) \in \mathbb{Q}[x] \mid f(1) = 0\}$ $V = \mathbb{Q}[x]$ $F = \mathbb{Q}$

13. $U = \{f(x) \in \mathbb{Q}[x] \mid f(1) = 2\}$ $V = \mathbb{Q}[x]$ $F = \mathbb{Q}$

14. $U = \{(x, 0, z) \mid x, z \in \mathbb{R}\}$ $V = \mathbb{R}^3$ $F = \mathbb{R}$

15. $U = \{(x, 1, z) \mid x, z \in \mathbb{R}\}$ $V = \mathbb{R}^3$ $F = \mathbb{R}$

16. $U = \{(a, b, 2b - 3a) \mid a, b \in \mathbb{R}\}$ $V = \mathbb{R}^3$ $F = \mathbb{R}$

17. $U = \{ \begin{bmatrix} a & 2a \\ 0 & 3a \end{bmatrix} \mid a \in \mathbb{R}\}$ $V = M(2, \mathbb{R})$ $F = \mathbb{R}$

18. Describe all the subspaces of $\mathbb{R}^3$.

In 19 and 20 find a basis of the indicated vector space V over the indicated field F.

19. $V = \mathbb{R}[x]/\langle x^2 + 1\rangle, F = \mathbb{R}$ **20.** $V = \mathbb{Q}[x]/\langle x^2 - 5\rangle, F = \mathbb{Q}$

21. Let c be any element of a field F. Show that

$$\{1, (x - c), (x - c)^2, \dots , (x - c)^{n-1}\}$$

is a basis for $F^n[x] = \{f(x) \in F[x] \mid \deg f(x) < n \text{ or } f(x) = 0\}$ over F.

22. Let V be a finite-dimensional vector space over a field F. Show that a subset $\{v_1, \dots , v_n\}$ of V is a basis for V over F if and only if for each $w \in V$ there exists a unique set of elements $c_i \in F$ such that $w = c_1v_1 + \dots + c_nv_n$.

23. Prove Corollary 10.1.33.

24. Let V be a vector space over a field F with $\dim_F V = n$, and U and W subspaces of V with $\dim_F U = m$ and $\dim_F W = k$. Show that if $m + k > n$, then $U \cap W \neq \{0\}$.

25. Determine whether the indicated subsets of $\mathbb{Q}[x]$ are linearly independent over $\mathbb{Q}$:

(a) $S = \{x^2 - 1, x^2 - 4\}$ (b) $T = \{x^2 - 1, x^2 - 4, x^2 - 9\}$

(c) $U = \{x^2 - 1, x^3 - 4, x^4 - 9\}$

26. Let V and W be vector spaces over the same field F. A function $T: V \to W$ is said to be a **linear transformation** from V to W if for all $c, d \in F$ and all $u, v \in V$ we have

$$T(cu + vd) = cT(u) + dT(v)$$

Let T be a linear transformation.

(a) Show that Im(T), the image of T, is a subspace of W.

(b) Show that Kern $T = \{v \in V \mid T(v) = \mathbf{0}\}$ is a subspace of V.

27. Let V and W be vector spaces over the same field F, and let $T: V \to W$ be a linear transformation. Show that

(a) For any subset $\{v_1, \dots , v_n\}$ of V with span $\{v_1, \dots , v_n\} = V$, we have span $\{T(v_1), \dots , T(v_n)\} = W$ if and only if T is onto.

(b) T is one to one if and only if for every subset $\{v_1, \dots , v_n\}$ of V that is linearly independent in V over F, $\{T(v_1), \dots , T(v_n)\}$ is linearly independent in W over F.

28. Let V and W be vector spaces over the same field F. An **isomorphism** from V to W is a linear transformation $T: V \to W$ that is one to one and onto. Suppose $\dim_F V = n$, and let $T: V \to W$ be a linear transformation. Show that

(a) If there is a basis $\{v_1, \ldots, v_n\}$ for V over F such that $\{T(v_1), \ldots, T(v_n)\}$ is a basis for W over F, then T is an isomorphism.

(b) If T is an isomorphism, then for any basis $\{v_1, \ldots, v_n\}$ for V over F, $\{T(v_1), \ldots, T(v_n)\}$ is a basis for W over F.

29. Two finite-dimensional vector spaces V and W over the same field F are **isomorphic** if there exists an isomorphism T from V to W. Show that two finite-dimensional vector spaces over a the same field F are isomorphic if and only if they have the same dimension.

30. Let V and W be vector spaces over the same field F. Suppose $\dim_F V = n$ and let $T: V \to W$ be a linear transformation. Show that

$$\dim_F \text{Kern } T + \dim_F \text{Im}(T) = n$$

10.2 Algebraic Extensions

The most basic question in the study of polynomials is finding their zeros. In this section we first show that any polynomial with coefficients in any field F always has a zero, though not necessarily in F. Then we study the structure of the smallest field that contains F and a zero of the given polynomial. This is where the notions of vector space, basis, and dimension come into play. In Section 8.7 we began the study of examples of quotient rings of the polynomial ring $F[x]$. In this section we resume the study of such quotient rings and complete the picture by showing that these quotient rings provide us with the fields that contain the zeros of a given polynomial.

10.2.1 EXAMPLES

(1) The polynomial $x^2 - 2 \in \mathbb{Q}[x]$ has no zero in $\mathbb{Q}$ but has the zero $\sqrt{2}$ in $\mathbb{R}$. There are, however, smaller subfields of $\mathbb{R}$ that contain $\mathbb{Q}$ and as well as this zero $\sqrt{2}$, and the smallest of these is $\mathbb{Q}(\sqrt{2}) = \{a + b\sqrt{2} \mid a, b \in \mathbb{Q}\}$.

(2) The polynomial $x^2 + 1$ has no zero in $\mathbb{R}$ but has the zero i in $\mathbb{C}$. In this case there is no smaller subfield of $\mathbb{C}$ containing $\mathbb{R}$ and i, since we already have $\mathbb{C} = \{a + bi \mid a, b \in \mathbb{R}\}$. ◇

Our first proposition guarantees, for any field F and any nonzero polynomial $f(x) \in F[x]$, that *if* we can find a field E of which F is a subfield containing an element α that is a zero of $f(x)$, then there will be a *smallest* subfield of E containing F and α.

Later we turn to the question of finding such a field E.

10.2.2 PROPOSITION Let E be a field, $F \subseteq E$ a subfield of E, and $\alpha \in E$ an element of E. In E let

$$F[\alpha] = \{f(\alpha) \mid f(x) \in F[x]\}$$
$$F(\alpha) = \{f(\alpha)/g(\alpha) \mid f(x), g(x) \in F[x], g(\alpha) \neq 0\}$$

Then

(1) $F[\alpha]$ is a subring of E containing F and α.
(2) $F[\alpha]$ is the smallest such subring of E.
(3) $F(\alpha)$ is a subfield of E containing F and α.
(4) $F(\alpha)$ is the smallest such subfield of E.

Proof (1) $F[\alpha]$ contains each element $a \in F$, since $a = f(\alpha)$, where $f(x)$ is the constant polynomial $f(x) = a \in F[x]$. Note that $F[\alpha]$ contains α, since $\alpha = f(\alpha)$ where $f(x)$ is the polynomial $f(x) = x \in F[x]$. Consider now any two elements $f_1(\alpha), f_2(\alpha) \in F[\alpha]$, where $f_1(x), f_2(x) \in F[x]$. Then $f_1(\alpha) - f_2(\alpha) = g(\alpha) \in F[\alpha]$ and $f_1(\alpha)f_2(\alpha) = h(\alpha) \in F[\alpha]$, where $g(x) = f_1(x) - f_2(x) \in F[x]$ and $h(x) = f_1(x)f_2(x) \in F[x]$. Thus $F[\alpha]$ is a subring of E by the subring test.

(2) What we are to show is that if R is any subring of E with $F \subseteq R$ and $\alpha \in R$, then $F[\alpha] \subseteq R$. But any such subring R will contain any power α^i of α and any linear combination $a_0 + a_1\alpha + \ldots + a_n\alpha^n$ of such powers, where $a_i \in F$. And every element $f(\alpha) \in F[\alpha]$, where

$$f(x) = a_0 + a_1x + \ldots + a_nx^n \in F[x]$$

is such a linear combination.

(3) and (4) are immediate from (1) and (2), since $F[\alpha] \subseteq E$ and E is a field, $F[\alpha]$ is an integral domain, and $F(\alpha)$ is simply the field of quotients of $F[\alpha]$. □

Note that if $F[\alpha]$ is already a field, then $F(\alpha) = F[\alpha]$. This happens in both cases in Example 10.2.1, where

$$\mathbb{Q}(\sqrt{2}) = \mathbb{Q}[\sqrt{2}] = \{a + b\sqrt{2} \mid a, b \in \mathbb{Q}\}$$
$$\mathbb{R}(i) = \mathbb{R}[i] = \{a + bi \mid a, b \in \mathbb{R}\} = \mathbb{C}$$

10.2.3 DEFINITION Let F and E be fields. Then if $F \subseteq E$ and F is a subfield of E, we also say that E is an **extension** of F. The extension $F(\alpha)$ in the preceding proposition is said to be obtained by **adjoining** α to F, and the notation $F(\alpha)$ is read "F adjoining α." ○

We now turn to the question of the existence of extension fields containing zeros of polynomials.

10.2.4 EXAMPLE Consider $f(x) = x^3 + x + 1 \in \mathbb{Q}[x]$. By Theorems 8.4.7 and 8.4.11, $f(x)$ is irreducible over $\mathbb{Q}$. We could look for a zero of $f(x)$ in $\mathbb{R}$ using the methods of Section 8.5, since $f(x)$ is a cubic polynomial. But the methods of Section 8.7 provide a different way to obtain an extension field of $\mathbb{Q}$ containing a zero of $f(x)$, one that works more generally than just for cubic polynomials.

Let us review those methods. Since $f(x)$ is irreducible over $\mathbb{Q}$, $I = \langle f(x) \rangle$ is a maximal ideal by Theorem 8.6.6, and $E = \mathbb{Q}[x]/I$ is a field. The elements of E can be written as cosets of I. Recall that I is the zero element in E and that $1 + I$ is unity in E. Now consider the element

$$\alpha = x + I$$

Since I is an ideal, $\alpha^2 = (x + I)(x + I) = x^2 + I$ and $\alpha^3 = x^3 + I$. Hence in E

$$\alpha^3 + \alpha + 1 = (x^3 + I) + (x + I) + (1 + I) = (x^3 + x + 1) + I$$

But since $I = \langle x^3 + x + 1 \rangle$, it follows that $\alpha^3 + \alpha + 1 = 0$ in E. In other words, $f(\alpha) = 0$, and the field E contains a zero of the polynomial $f(x) = x^3 + x + 1$, namely α. Note also that the set of elements $\{c + I \mid c \in \mathbb{Q}\}$ forms a subfield of E isomorphic to $\mathbb{Q}$, so that we can regard E as an extension field of $\mathbb{Q}$. ◇

10.2.5 THEOREM (Kronecker's theorem) Let F be a field and $p(x)$ a nonconstant polynomial in $F[x]$. Then there exists a field E and an element α of E such that E is an extension field of F and α is a zero of $p(x)$.

Proof First consider the special case where $p(x)$ is irreducible over F. Then by Theorem 8.6.6, $I = \langle p(x) \rangle$ is a maximal ideal in $F[x]$ by Theorem 8.6.6, and $E = F[x]/I$ is a field.

We first claim that $F' = \{c + I \mid c \in F\}$ is a subfield of E isomorphic to F. For since I is an ideal in $F[x]$, for any $c + I$ and $d + I$ in F', we have

$$(c + I) - (d + I) = (c - d) + I$$

and if $d \neq 0$ in F, then

$$(d + I)^{-1} = d^{-1} + I$$
$$(c + I)(d + I)^{-1} = cd^{-1} + I$$

Hence by the subfield test (Theorem 6.3.17), F' is a subfield of E. The map $\phi: F \to F'$ defined by $\phi(c) = c + I$ for all $c \in F$ is an isomorphism. (See Exercise 6.) If we identify $c \in F$ and $\phi(c) = c + I$, we may regard E as an extension field of F.

We now show that E contains a zero of $p(x)$. For since I is an ideal in $F[x]$, we have

$$(x + I)^i = (x^i + I)$$
$$c(x^i + I) = (c + I)(x^i + I) = (cx^i + I)$$

and more generally for any polynomial

$$f(x) = c_0 + c_1x + \dots + c_mx^m \in F[x]$$

we have

$$f(x + I) = c_0 + c_1(x + I) + \dots + c_m(x^m + I)$$
$$= (c_0 + c_1x + \dots + c_mx^m) + I = f(x) + I$$

Thus if we write α for $x + I$, then, in particular, we have

$$p(\alpha) = p(x + I) = p(x) + I = I = 0 \text{ in } E$$

Thus α is a zero of $p(x)$. Moreover, every element of E is of form $f(x) + I = f(\alpha)$ for some $f(x) \in F[x]$, so in the notation of Proposition 10.2.2 we have $E = F[\alpha] = F(\alpha)$. We have used the quotient ring construction to adjoin a zero α of a given irreducible polynomial $p(x)$ to F.

This completes the proof in the special case where $p(x)$ is irreducible. In the general case where $p(x)$ need not be irreducible, simply apply this special case to any irreducible factor of $p(x)$. □

Some examples of the construction in Kronecker's theorem have already been given in Section 8.7, where we considered, for instance, the cases $F = \mathbb{Q}$, $f(x) = x^2 + x + 1$ in Example 8.7.1 and the case $F = \mathbb{Z}_2$, $f(x) = x^2 + x + 1$ in Example 8.7.2. To understand the construction better, we consider a few more examples here.

10.2.6 EXAMPLE Consider $p(x) = x^2 + 1 \in \mathbb{R}[x]$. Let $\phi: \mathbb{R}[x] \to \mathbb{C}$ be the evaluation homomorphism $\phi(f(x)) = f(i)$ for all $f(x) \in \mathbb{R}[x]$. For any complex number $a + bi \in \mathbb{C}$ we have $a + bi = \phi(a + bx)$ and $a + bx \in \mathbb{R}[x]$. Hence ϕ is onto. Kern ϕ = $\{f(x) \in \mathbb{R}[x] \mid f(i) = 0\}$. Therefore, $p(x) = x^2 + 1 \in$ Kern ϕ and

$$I = \langle p(x)\rangle \subseteq \text{Kern } \phi \subseteq \mathbb{R}[x]$$

Since $I = \langle p(x)\rangle$ is a maximal ideal in $\mathbb{R}[x]$, either Kern $\phi = \mathbb{R}[x]$ or Kern $\phi = I = \langle p(x)\rangle$. Since $\phi(1) = 1$, ϕ is not the 0-homomorphism, and Kern $\phi \neq \mathbb{R}[x]$. Hence Kern $\phi = I = \langle p(x)\rangle$. Therefore, by the first isomorphism theorem for rings (Theorem 7.2.17) we have

$$\mathbb{R}[x]/\langle x^2 + 1\rangle \cong \mathbb{C} = \mathbb{R}(i)$$

Under this isomorphism we have the following:

$$\begin{array}{rcl} I & \leftrightarrow & 0 \\ x + I & \leftrightarrow & i \\ c + I & \leftrightarrow & c \in \mathbb{R} \\ (c + dx) + I & \leftrightarrow & c + di \in \mathbb{C} \end{array}$$

which completely describes the correspondence. ◇

10.2.7 EXAMPLE Consider the complex number $-{}^1/_2 + {}^{\sqrt{3}}/_2\, i$, which we have encountered several times. It is one of the third roots of unity or, equivalently, a zero of $x^3 - 1$. (See Example 8.5.1.) In $\mathbb{Q}[x]$ we have the factorization $x^3 - 1 = (x - 1)(x^2 + x + 1)$, and ω is a zero of the irreducible factor $(x^2 + x + 1)$. Let $\phi: \mathbb{Q}[x] \to \mathbb{C}$ be the evaluation homorphism $\phi(f(x)) = f(\omega)$. Then the image

$$\text{Im}(\phi) = \{f(\omega) \mid f(x) \in \mathbb{Q}[x]\} = \mathbb{Q}[\omega]$$

as defined in Proposition 10.2.2. Let us determine the kernel of ϕ. Since ω is a zero of $x^2 + x + 1$, we have $x^2 + x + 1 \in$ Kern ϕ and therefore $\langle x^2 + x + 1\rangle \subseteq$ Kern ϕ. Since $x^2 + x + 1$ is irreducible over $\mathbb{Q}$, the ideal $\langle x^2 + x + 1\rangle$ is a maximal ideal in $\mathbb{Q}[x]$. Therefore, either Kern $\phi = \mathbb{Q}[x]$ or Kern $\phi = \langle x^2 + x + 1\rangle$. Now since $\phi(1) = 1$, ϕ is not the 0-homomorphism, Kern $\phi \neq \mathbb{Q}[x]$. Therefore, Kern $\phi = \langle x^2 + x + 1\rangle$.

Let us now consider the image of ϕ. By the first isomorphism theorem for rings we have

$$\text{Im}(\phi) \cong \mathbb{Q}[x]/\langle x^2 + x + 1\rangle$$

Since the ideal $\langle x^2 + x + 1\rangle$ is maximal, the quotient $\mathbb{Q}[x]/\langle x^2 + x + 1\rangle$ is a field, and so $\mathrm{Im}(\phi) = \mathbb{Q}[\omega]$ is a field, so $\mathrm{Im}(\phi) = \mathbb{Q}[\omega]= \mathbb{Q}(\omega)$ and we have

$$\mathbb{Q}(\omega) \cong \mathbb{Q}[x]/\langle x^2 + x + 1\rangle$$

We can go on to give a more explicit description of the elements of $\mathbb{Q}(\omega)$. Every such element is a linear combination

$$c_0 + c_1\omega + \ldots + c_m\omega^m$$

over $\mathbb{Q}$ of powers of ω, where

$$c_0 + c_1x + \ldots + c_mx^m \in \mathbb{Q}[x]$$

But since ω is a zero of $x^2 + x + 1$ and a cube root of unity, we have

$$\omega^2 = -1 - \omega$$
$$\omega^3 = 1$$
$$\omega^4 = \omega$$
$$\vdots$$

so every power of ω is a linear combination over $\mathbb{Q}$ of 1 and ω. Hence every element of $\mathbb{Q}(\omega)$ is such a linear combination, and

$$\mathbb{Q}(\omega) = \{a + b\omega \mid a, b \in \mathbb{Q}\}$$

Note that we have

$$(a + b\omega)(c + d\omega) = ac + (bc + ad)\omega + bd\omega^2 =$$
$$ac + (bc + ad)\omega + bd(-1 - \omega) = (ac - bd) + (bc + ad - bd)\omega$$

as the rule for products in $\mathbb{Q}(\omega)$. ◇

10.2.8 DEFINITION Let F be a field and E an extension field, $F \subseteq E$. Then an element $\alpha \in E$ is said to be **algebraic over** F if there exists a nonzero polynomial $0 \neq f(x) \in F[x]$ such that $f(\alpha) = 0$. Otherwise, α is said to be **transcendental over** F. ○

10.2.9 THEOREM Let $F \subseteq E$ be fields and let $\alpha \in E$ be algebraic over F. Then there exists a unique monic polynomial $p(x) \in F[x]$ such that

(1) $p(\alpha) = 0$

(2) $p(x)$ is irreducible over F.

(3) If $f(x) \in F[x]$ is such that $f(\alpha) = 0$, then $p(x)$ divides $f(x)$.

Proof Consider $I = \{f(x) \in F[x] \mid f(\alpha) = 0\}$. Then I is a proper ideal in $F[x]$. (See Exercise 7.) Hence by Theorem 8.6.5, $I = \langle p(x)\rangle \in F[x]$, where $p(x)$ is of minimal degree in I. We can also choose $p(x)$ to be monic by multiplying by a suitable unit.

We next show that $p(x)$ is irreducible. Suppose $p(x) = g(x)h(x)$. Then $g(\alpha)h(\alpha) = p(\alpha) = 0$. Hence either $g(\alpha) = 0$ or $h(\alpha) = 0$, and either $g(x) \in I$ or $h(x) \in I$. Since $p(x)$ is of minimal degree in I, whichever factor $g(x)$ or $h(x)$ belongs to I must be a unit multiple of $p(x)$, and the other factor must be a unit, as required to show $p(x)$ irreducible. This proves property (2) of the theorem, and properties (1) and (3) are immediate from the definition of I.

Finally, to prove uniqueness, suppose $q(x)$ is another monic polynomial with the properties (1) through (3). Then by property (3) we have $p(x) \mid q(x)$ and $q(x) \mid p(x)$. Hence $q(x) = up(x)$ for some unit u. But since $p(x)$ and $q(x)$ are both monic, we must have $u = 1$ and $q(x) = p(x)$. □

The existence of this irreducible polynomial $p(x)$ for every algebraic element α helps us to understand the field extension $F(\alpha)$.

10.2.10 DEFINITION Let $F \subseteq E$ be fields and let $\alpha \in E$ be algebraic over F. Then the unique monic irreducible polynomial $p(x) \in F[x]$ such that $p(\alpha) = 0$ is called the **minimal polynomial** of α over F. The **degree of** α **over** F is defined as the degree of the minimal polynomial $p(x)$ and is written $\deg_F(\alpha)$. ○

Note that any $a \in F$ counts as algebraic over F of degree 1, the minimal polynomial in this case being the linear polynomial $p(x) = x - a$. Example 10.2.6 shows that i is algebraic over $\mathbb{R}$ and that $\deg_{\mathbb{R}}(i) = 2$. Example 10.2.7 shows that ω is algebraic over $\mathbb{Q}$ and that $\deg_{\mathbb{Q}}(\omega) = 2$. These two examples illustrate the following basic theorem.

10.2.11 THEOREM Let $F \subseteq E$ be fields, let $\alpha \in E$ be algebraic over F with $\deg_F(\alpha) = n$, and let $p(x)$ be the minimal polynomial of α over F. Then

(1) $F(\alpha) \cong F[x] / \langle p(x) \rangle$

(2) $\{1, \alpha, \alpha^2, \dots, \alpha^{n-1}\}$ is a basis for the vector space $F(\alpha)$ over F.

(3) $\dim_F F(\alpha) = \deg_F(\alpha) = \deg p(x)$

Proof (1) Consider the evaluation homomorphism $\phi\colon F[x] \to E$ defined by $\phi(f(x)) = f(\alpha)$. $\operatorname{Im}(\phi) = \{f(\alpha) \mid f(x) \in F[x]\} = F[\alpha]$ as defined in Proposition 10.2.2. $\operatorname{Kern} \phi = \{f(x) \in F[x] \mid f(\alpha) = 0\}$. By Theorem 10.2.9, part (1), $p(\alpha) = 0$ and so $f(\alpha) = 0$ for any $f(x) \in F[x]$ such that $p(x)$ divides $f(x)$; and by Theorem 10.2.9, part (3), if $f(x) \in F[x]$ and $f(\alpha) = 0$, then $p(x)$ divides $f(x)$. Thus

$$\operatorname{Kern} \phi = \{f(x) \mid p(x) \text{ divides } f(x)\} = \langle p(x) \rangle$$

By Theorem 10.2.9, part (2), $p(x)$ is irreducible and hence $\langle p(x) \rangle$ is a maximal ideal and $F[x]/\langle p(x) \rangle$ is a field. By the first isomorphism theorem

$$\operatorname{Im}(\phi) \cong F[x] / \langle p(x) \rangle$$

So $\operatorname{Im}(\phi) = F[\alpha]$ is a field, and $\operatorname{Im}(\phi) = F[\alpha] = F(\alpha)$, and (1) is proved.

(2) Consider $S = \operatorname{span}_F \{1, \alpha, \alpha^2, \dots, \alpha^{n-1}\}$. We saw in the proof of (1) that

$$F(\alpha) = F[\alpha] = \{f(\alpha) \mid f(x) \in F[x]\}$$

So $F(\alpha)$ consists of all elements of the form

$$a_m\alpha^m + \dots + a_2\alpha^2 + a_1\alpha + a_0$$

or, in other words, all linear combinations over F of powers of α. Since every element of S is such a linear combination (with $m < n$), we have $S \subseteq F(\alpha)$. For the opposite inclusion, let the minimal polynomial $p(x)$ be

$$p(x) = x^n + b_{n-1}x^{n-1} + \dots + b_1x + b_0$$

Since $p(\alpha) = 0$, we have

$$\alpha^n = (-b_0) + (-b_1\alpha) + \dots + (-b_{n-1}\alpha^{n-1})$$

This shows that α^n is a linear combination of the α^i for $i < n$, and so belongs to S. It follows that

$$\alpha^{n+1} = \alpha \cdot \alpha^n = (-b_0)\alpha + (-b_1)\alpha^2 + \dots + (-b_{n-1})\alpha^n$$

and this shows that α^{n+1} is a linear combination of elements of S and so also belongs to S. Continuing in this way, every power of α belongs to S, hence so does every linear combination of powers of α, and $F(\alpha) \subseteq S$. Thus

$$\text{span}_F \{1, \alpha, \alpha^2, \dots, \alpha^{n-1}\} = F(\alpha)$$

On the other hand, if there were elements $d_i \in F$ such that

$$d_0 + d_1\alpha + d_2\alpha^2 + \dots + d_{n-1}\alpha^{n-1} = 0$$

then

$$g(x) = d_0 + d_1x + d_2x^2 + \dots + d_{n-1}x^{n-1} \in F[x]$$

would be a polynomial with $\deg g(x) \leq n - 1 < n = \deg p(x)$ and $g(\alpha) = 0$. By Theorem 10.2.9, part (3), this is impossible unless all $d_i = 0$. This shows that the set $\{1, \alpha, \alpha^2, \dots, \alpha^{n-1}\}$ is linearly independent, and we have proved (2).

(3) follows immediately from (2). □

10.2.12 EXAMPLES

(1) $\mathbb{Q}(\sqrt{5}) = \mathbb{Q}(5^{1/2})$ is an extension field of $\mathbb{Q}$ and also a vector space over $\mathbb{Q}$ with basis $\{1, 5^{1/2}\}$. Any element of $\mathbb{Q}(5^{1/2})$ is of form $c_0 + c_1 5^{1/2}$ for some $c_i \in \mathbb{Q}$.

(2) $\mathbb{Q}(2^{1/3})$ is a field extension of $\mathbb{Q}$ and also a vector space over $\mathbb{Q}$ with basis $\{1, 2^{1/3}, 2^{2/3}\}$. Therefore, any element of $\mathbb{Q}(2^{1/3})$ is of the form $c_0 + c_1 2^{1/3} + c_2 2^{2/3}$ for some $c_i \in \mathbb{Q}$. ◇

10.2.13 EXAMPLE We can easily see that $3^{1/5}$ is algebraic over $\mathbb{Q}$ since $3^{1/5}$ is a zero of $x^5 - 3 \in \mathbb{Q}[x]$. By Eisenstein's criterion, $x^5 - 3$ is irreducible over $\mathbb{Q}$. Hence $3^{1/5}$ is algebraic of degree five. Now consider some element of $\mathbb{Q}(3^{1/5})$, say

$$\alpha = 3^{1/5} - 4 \cdot 3^{2/5} + 7 \cdot 3^{3/5}$$

Is α algebraic over $\mathbb{Q}$? We could try to show that it is by calculating various powers of α and trying various linear combinations of these powers to see if we could get zero. Our next theorem, however, immediately implies that α is algebraic, without our having to do any such calculations. ◇

Before we can present the next theorem, we need some definitions.

10.2.14 DEFINITION Let E be an extension field of a field F. Then

(1) E is said to be an **algebraic extension** of F if every element $\alpha \in E$ is algebraic over F.

(2) E is said to be a **finite extension** of F if E is a finite-dimensional vector space over F. In this case, we denote the dimension n of E over F by $[E : F] = n$, and we call n the **degree** of E over F. ○

10.2.15 EXAMPLES

(1) $[\mathbb{Q}(2^{1/3}): \mathbb{Q}] = 3$
(2) $[\mathbb{C} : \mathbb{R}] = 2$
(3) $[F : F] = 1$ for any field F
(4) $[F(\alpha) : F] = \deg_F(\alpha)$ by Theorem 10.2.11 ◇

The next theorem shows that the two types of field extensions we just introduced are closely related.

10.2.16 THEOREM Let E be a finite extension of a field F. Then

(1) E is an algebraic extension of F.
(2) $\deg_F(\alpha) \leq [E : F]$ for every $\alpha \in E$

Proof Let E be a finite extension of F with $[E : F] = n$. Let $\alpha \in E$ and consider the set $\{1, \alpha, \alpha^2, \ldots, \alpha^{n-1}, \alpha^n\}$. Since these are $n + 1$ elements while E is a vector space of dimension n over F, these elements must be linearly dependent by Theorem 10.1.32. This means there are $c_i \in F$, not all zero, such that

$$c_0 + c_1\alpha + c_2\alpha^2 + \ldots + c_{n-1}\alpha^{n-1} + c_n\alpha^n = 0$$

But then

$$g(x) = c_0 + c_1x + c_2x^2 + \ldots + c_{n-1}x^{n-1} + c_nx^n$$

is a nonzero polynomial $g(x) \in F[x]$ with $g(\alpha) = 0$. Hence α is algebraic over F and $\deg_F(\alpha) \leq \deg g(x) = n$. Since this is so for every element $\alpha \in E$, E is an algebraic extension of F. □

10.2.17 EXAMPLE By the preceding theorem, since

$$[\mathbb{Q}(3^{1/5}) : \mathbb{Q}] = \deg_{\mathbb{Q}} 3^{1/5} = 5$$

the element

$$\alpha = 3^{1/5} - 4{\cdot}3^{2/5} + 7{\cdot}3^{3/5} \in \mathbb{Q}(3^{1/5})$$

mentioned in Example 10.2.13 is indeed algebraic over $\mathbb{Q}$ of degree ≤ 5. ◇

10.2.18 EXAMPLE The polynomial $x^2 - 5$ is irreducible over $\mathbb{Q}$, the square root $5^{1/2}$ is algebraic of degree 2 over $\mathbb{Q}$, $\mathbb{Q}(5^{1/2})$ is an algebraic extension of $\mathbb{Q}$ of degree 2, and $\{1, 5^{1/2}\}$ forms a basis for $\mathbb{Q}(5^{1/2})$ over $\mathbb{Q}$. The polynomial $x^3 - 2$ is irreducible over $\mathbb{Q}$, the cube root $2^{1/3}$ is algebraic of degree 3 over $\mathbb{Q}$, $\mathbb{Q}(2^{1/3})$ is an algebraic extension of $\mathbb{Q}$ of degree 3, and $\{1, 2^{1/3}, 2^{2/3}\}$ forms a basis for $\mathbb{Q}(2^{1/3})$ over $\mathbb{Q}$.
Now consider $\mathbb{Q}(5^{1/2}, 2^{1/3}) = (\mathbb{Q}(5^{1/2}))(2^{1/3})$, the extension obtained by adjoining $2^{1/3}$ to $\mathbb{Q}(5^{1/2})$. Since every element of $\mathbb{Q}(5^{1/2})$ is of degree ≤ 2 over $\mathbb{Q}$, while $2^{1/3}$ is of degree 3 over $\mathbb{Q}$, $2^{1/3} \notin \mathbb{Q}(5^{1/2})$. It follows that the polynomial $x^3 - 2$ is still irreducible over $\mathbb{Q}(5^{1/2})$, $\mathbb{Q}(5^{1/2}, 2^{1/3})$ is an algebraic extension of $\mathbb{Q}(5^{1/2})$ of degree 3, and $\{1, 2^{1/3}, 2^{2/3}\}$ forms a basis for $\mathbb{Q}(5^{1/2}, 2^{1/3})$ over $\mathbb{Q}(5^{1/2})$.
An element of $\mathbb{Q}(5^{1/2}, 2^{1/3})$ has the form

$$\alpha = c_0 + c_12^{1/3} + c_22^{2/3}$$

where $c_i \in \mathbb{Q}(5^{1/2})$ and therefore have the form

$$c_0 = d_{00} + d_{01}5^{1/2} \qquad c_1 = d_{10} + d_{11}5^{1/2} \qquad c_2 = d_{20} + d_{21}5^{1/2}$$

Thus

$$\alpha = (d_{00} + d_{01}5^{1/2}) + (d_{10} + d_{11}5^{1/2})2^{1/3} + (d_{20} + d_{21}5^{1/2})2^{2/3} = d_{00} + d_{01}5^{1/2} + d_{10}2^{1/3} + d_{11}5^{1/2}2^{1/3} + d_{20}2^{2/3} + d_{21}5^{1/2}2^{2/3}$$

Hence

$$S = \{1, 5^{1/2}, 2^{1/3}, 2^{1/3}5^{1/2}, 2^{2/3}, 2^{2/3}5^{1/2}\}$$

spans $\mathbb{Q}(5^{1/2}, 2^{1/3})$ over $\mathbb{Q}$. The set S is actually a basis, because if

$$b_{00} + b_{01}5^{1/2} + b_{10}2^{1/3} + b_{11}5^{1/2}2^{1/3} + b_{20}2^{2/3} + b_{21}5^{1/2}2^{2/3} = 0$$

then since $\{1, 2^{1/3}, 2^{2/3}\}$ is a basis for $\mathbb{Q}(5^{1/2})(2^{1/3})$ over $\mathbb{Q}(5^{1/2})$, we must have

$$b_{00} + b_{01}5^{1/2} = 0 \qquad b_{10} + b_{11}5^{1/2} = 0 \qquad b_{20} + b_{21}5^{1/2} = 0$$

and since $\{1, 5^{1/2}\}$ is a basis for $\mathbb{Q}(5^{1/2})$ over $\mathbb{Q}$, we must have all $b_{ij} = 0$. Thus

$$[\mathbb{Q}(5^{1/2}, 2^{1/3}) : \mathbb{Q}] = 6 = 3 \cdot 2 = [\mathbb{Q}(5^{1/2}, 2^{1/3}) : \mathbb{Q}(2^{1/3})]\ [\mathbb{Q}(5^{1/2}) : \mathbb{Q}]$$

and we have obtained a basis for $\mathbb{Q}(5^{1/2}, 2^{1/3})$ over $\mathbb{Q}$ by "multiplying out" a basis for $\mathbb{Q}(5^{1/2})$ over $\mathbb{Q}$ and a basis for $\mathbb{Q}(5^{1/2}, 2^{1/3})$ over $\mathbb{Q}(5^{1/2})$. ◇

10.2.19 EXAMPLE The polynomial $x^2 - 2$ is irreducible over $\mathbb{Q}$, the square root $2^{1/2}$ is algebraic of degree 2 over $\mathbb{Q}$, $\mathbb{Q}(2^{1/2})$ is an algebraic extension of $\mathbb{Q}$ of degree 2, and $\{1, 2^{1/2}\}$ forms a basis for $\mathbb{Q}(2^{1/2})$ over $\mathbb{Q}$. The polynomial $x^4 - 2$ is irreducible over $\mathbb{Q}$, the fourth root $2^{1/4}$ is algebraic of degree 4 over $\mathbb{Q}$, $\mathbb{Q}(2^{1/4})$ is an algebraic extension of $\mathbb{Q}$ of degree 4, and $\{1, 2^{1/4}, 2^{2/4}, 2^{3/4}\}$ forms a basis for $\mathbb{Q}(2^{1/4})$ over $\mathbb{Q}$. But over $\mathbb{Q}(2^{1/2})$ the polynomial $x^4 - 2$ factors as $(x^2 + 2^{1/2})(x^2 - 2^{1/2})$. The fourth root $2^{1/4}$ is a zero of the second factor $x^2 - 2^{1/2}$. Since every element of $\mathbb{Q}(2^{1/2})$ is of degree ≤ 2 over $\mathbb{Q}$, while $2^{1/4}$ is of degree 4 over $\mathbb{Q}$, $2^{1/4} \notin \mathbb{Q}(2^{1/2})$. It follows that the polynomial $x^2 - 2^{1/2}$ is irreducible over $\mathbb{Q}(2^{1/2})$, and $2^{1/4}$ is algebraic of degree 2 over $\mathbb{Q}(2^{1/2})$. Since $2^{1/2} = (2^{1/4})^2 \in \mathbb{Q}(2^{1/4})$, it follows that $\mathbb{Q}(2^{1/2}, 2^{1/4})$, the extension obtained by adjoining $2^{1/4}$ to $\mathbb{Q}(2^{1/2})$, is just $\mathbb{Q}(2^{1/4})$. Note, however, that we still have the relationship

$$[\mathbb{Q}(2^{1/2}, 2^{1/4}) : \mathbb{Q}] = 4 = 2 \cdot 2 = [\mathbb{Q}(2^{1/2}, 2^{1/4}) : \mathbb{Q}(2^{1/2})]\ [\mathbb{Q}(2^{1/2}) : \mathbb{Q}]$$

Moreover, since

$$\{1, 2^{1/4}, 2^{2/4}, 2^{3/4}\} = \{1, 2^{1/4}, 2^{1/2}, 2^{1/4} \cdot 2^{1/2}\}$$

we still obtain a basis for $\mathbb{Q}(2^{1/2}, 2^{1/4})$ over $\mathbb{Q}$ when we "multiply out" a basis for $\mathbb{Q}(2^{1/2})$ over $\mathbb{Q}$ and a basis for $\mathbb{Q}(2^{1/2}, 2^{1/4})$ over $\mathbb{Q}(2^{1/2})$ ◇

10.2.20 THEOREM Let E be a finite extension field of a field F and K a finite extension field of E. Then K is a finite extension field of F and

$$[K : F] = [K : E]\ [E : F]$$

Proof We have $F \subseteq E \subseteq K$. Let $[K : E] = m$ and let $\{v_1, \ldots, v_m\}$ be a basis for K over E. Let $[E : F] = n$ and let $\{w_1, \ldots, w_n\}$ be a basis for E over F.

We prove the theorem by proving that the set of $m{\cdot}n$ elements

$$\{v_iw_j \mid 1 \le i \le m,\ 1 \le j \le n\}$$

forms a basis for K over F.

First we show that the v_iw_j span K over F. Let α be any element of K. Since the v_i span K over E, we have

$$\alpha = \beta_1v_1 + \dots + \beta_mv_m$$

for some $\beta_i \in E$. Since the w_j span E over F, we have

$$\begin{aligned}\beta_1 &= c_{11}w_1 + \dots + c_{1n}w_n\\ &\ \ \vdots\\ \beta_m &= c_{m1}w_1 + \dots + c_{mn}w_n\end{aligned}$$

for some $c_{ij} \in F$. From this it follows that

$$\alpha = (c_{11}w_1 + \dots + c_{1n}w_n)v_1 + \dots + (c_{m1}w_1 + \dots + c_{mn}w_n)v_m = \Sigma_{ij}c_{ij}(v_iw_j)$$

where $1 \le i \le m$ and $1 \le j \le n$. Thus the v_iw_j span K over F.

Next we show that the v_iw_j are linearly independent over F. Suppose we had

$$\Sigma_{ij}d_{ij}(v_iw_j) = 0$$

where $d_{ij} \in F$ for $1 \le i \le m$ and $1 \le j \le n$. Then

$$(d_{11}w_1 + \dots + d_{1n}w_n)v_1 + \dots + (d_{m1}w_1 + \dots + d_{mn}w_n)v_m = 0$$

Since the v_i are linearly independent over E, we must have

$$\begin{aligned}d_{11}w_1 + \dots + d_{1n}w_n &= 0\\ &\ \ \vdots\\ d_{m1}w_1 + \dots + d_{mn}w_n &= 0\end{aligned}$$

Since the w_j are linearly independent over F, we must have $d_{ij} = 0$ for all i and j. □

10.2.21 COROLLARY Let

$$F_1 \subseteq F_2 \subseteq \dots \subseteq F_i \subseteq \dots \subseteq F_r$$

be fields with $[F_{i+1} : F_i] = n_{i+1}$, $1 \le i \le r - 1$. Then

$$[F_r : F] = n_2n_3 \dots n_r$$

Proof Immediate from Theorem 10.2.20. □

10.2.22 COROLLARY Let E be a finite extension of a field F and let $\alpha \in E$. Then $\deg_F(\alpha)$ divides $[E : F]$.

Proof We have $F \subseteq F(\alpha) \subseteq E$. By Corollary 10.2.21,

$$[E : F] = [E : F(\alpha)]\ [F(\alpha) : F] = [E : F(\alpha)] \cdot \deg_F(\alpha)$$

so $\deg_F(\alpha) \mid [E : F]$ as asserted. □

10.2.23 DEFINITION Let $F \subseteq E$ be fields and let $\alpha_1, \alpha_2, \dots, \alpha_r$ be elements of E. Let $F_0 = F$, and $F_{i+1} = F_i(\alpha_{i+1})$ for $0 \le i < r$. Then we call F_r the field obtained by **successively adjoining** the elements $\alpha_1, \alpha_2, \dots, \alpha_r$ of E to F, and we write $F(\alpha_1, \alpha_2, \dots, \alpha_r)$ for F_r. ○

We have introduced this notation in the special case $r = 2$ in Examples 10.2.18 and 10.2.19.

10.2.24 COROLLARY Let $F \subseteq E$ be fields and α and β be elements of E that are algebraic over F, with $\deg_F(\alpha) = n$ and $\deg_F(\beta) = m$. Then $[F(\alpha, \beta) : F] \leq nm$.

Proof By Theorem 10.2.20 we have

$$[F(\alpha, \beta) : F] = [F(\alpha, \beta) : F(\alpha)]\, [F(\alpha) : F]$$

By Theorem 10.2.11

$$[F(\alpha) : F] = \deg_F(\alpha)$$

$$[F(\alpha, \beta) : F(\alpha)] = \deg_{F(\alpha)}(\beta)$$

It remains to show that $\deg_{F(\alpha)}(\beta) \leq \deg_F(\beta) = m$. We know that β is a zero of a polynomial $p(x) \in F[x]$ of degree m that is irreducible over F. If $p(x)$ is still irreducible over $F(\alpha)$, then $\deg_{F(\alpha)}(\beta) = m$. Otherwise, $p(x)$ factors over $F(\alpha)$ into a product of polynomials of degree $< m$, and β is a zero of one of these factors, and therefore $\deg_{F(\alpha)}(\beta) < m$. This completes the proof. □

10.2.25 COROLLARY Let $F \subseteq E$ be fields and α and β be elements of E that are algebraic over F. Then $\alpha \pm \beta$, $\alpha\beta$, and (if $\beta \neq 0$) α/β are all algebraic over F.

Proof By Corollary 10.2.24, $F(\alpha, \beta)$ is a finite extension of F, and thus by Theorem 10.2.16 $F(\alpha, \beta)$ is an algebraic extension of F. But $\alpha \pm \beta$, $\alpha\beta$, and (if $\beta \neq 0$) α/β are all elements of $F(\alpha, \beta)$, hence they are algebraic over F. □

10.2.26 COROLLARY Let $F \subseteq E$ be fields and let S be the set of elements of E that are algebraic over F. Then S is a subfield of E and an extension field of F.

Proof Immediate from Corollary 10.2.25. □

10.2.27 DEFINITION Let $\bar{\mathbb{Q}} = \{\alpha \in \mathbb{C} \mid \alpha \text{ is algebraic over } \mathbb{Q}\}$. Then $\bar{\mathbb{Q}}$ is called the field of **algebraic numbers**. ○

10.2.28 EXAMPLE The field $\bar{\mathbb{Q}}$ of algebraic numbers is an example of an infinite algebraic extension of $\mathbb{Q}$. By definition, $\bar{\mathbb{Q}}$ is an algebraic extension of $\mathbb{Q}$. Now consider for any positive integer n the elements $2^{1/n}$. They are algebraic over $\mathbb{Q}$, hence elements of $\bar{\mathbb{Q}}$, and since $[\mathbb{Q}(2^{1/n}) : \mathbb{Q}] = n$, it follows $[\bar{\mathbb{Q}} : \mathbb{Q}] \geq n$ for all n. ◇

10.2.29 THEOREM Let K be an algebraic extension of a field E, and E an algebraic extension of a field F. Then K is an algebraic extension of F.

Proof Let α be any element of K. Then α is algebraic over E. Therefore,

(1) $$c_n\alpha^n + \dots + c_1\alpha + c_0 = 0$$

for some $c_i \in E$, $0 \le i \le n$. Since E is algebraic over F, all the $c_i \in E$ are algebraic over F. Hence by Corollary 10.2.24, $L = F(c_0, c_1, \dots, c_n)$ is a finite extension of F. By (1) α is algebraic over L, so $L(\alpha)$ is a finite extension of L. By Theorem 10.2.20, $L(\alpha)$ is a finite extension of F, and hence by Theorem 10.2.16 an algebraic extension of F, and α is algebraic over F. □

There do exist elements $\alpha \in \mathbb{R}$ that are transcendental over $\mathbb{Q}$. (A proof is outlined in the last three exercises for this section.) It is much easier to prove that there exist transcendental real numbers than to prove that particular important real numbers encountered in calculus and other branches of mathematics, such as e and π, are transcendental. They are, but the proof requires methods of calculus and does not belong to algebra.

Exercises 10.2

In Exercises 1 through 5 show that the indicated $\alpha \in \mathbb{C}$ is algebraic over $\mathbb{Q}$.

1. $1 - \sqrt{5}$ **2.** $\sqrt{2} + \sqrt{3}$ **3.** $1 + i$

4. $\sqrt{1 + \sqrt{3}}$ **5.** $2^{1/3} + 1$

6. Show that the map ϕ defined in the proof of Theorem 10.2.5 is an isomorphism.

7. Let $F \subseteq E$ be fields and let $\alpha \in E$ be algebraic over F. Show that
$$I = \{f(x) \in F[x] \mid f(\alpha) = 0\}$$
is a proper ideal in $F[x]$.

In Exercises 8 through 10 show that the indicated $\alpha \in \mathbb{C}$ is algebraic over $\mathbb{Q}$, and determine $\deg_{\mathbb{Q}}(\alpha)$.

8. $\sqrt{3} - i$ **9.** $\sqrt{3} - \sqrt{3}\,i$ **10.** $\sqrt{2} + \sqrt{2}\,i$

11. Find $\deg_F(\sqrt{2} + \sqrt{3})$ for the indicated field F:

(a) $F = \mathbb{Q}$ (b) $F = \mathbb{Q}(\sqrt{6})$ (c) $F = \mathbb{Q}(\sqrt{5})$ (d) $F = \mathbb{Q}(\sqrt{2}, \sqrt{3})$

12. Find the minimal polynomial of $(\sqrt{2} + \sqrt{2}i)/2$ over the indicated field F:

(a) $F = \mathbb{Q}$ (b) $\mathbb{R}$ (c) $\mathbb{Q}(i)$

In Exercises 13 through 17 find a basis for the indicated extension field of $\mathbb{Q}$ over $\mathbb{Q}$.

13. $\mathbb{Q}(\sqrt{2}, \sqrt{3})$ **14.** $\mathbb{Q}(\sqrt{2}, i)$ **15.** $\mathbb{Q}(\sqrt{2}i)$

16. $\mathbb{Q}(2^{1/3}, 7^{1/2})$ **17.** $\mathbb{Q}(\sqrt{2} + i)$

18. Let E be a field, $F \subseteq E$ a subfield of E, $\alpha \in E$ an element of E. Show that $F(\alpha)$ is a finite-dimensional vector space over F if and only if α is algebraic over F.

19. Let E be a field, $F \subseteq E$ a subfield of E, $\alpha \in E$ an element of E. Show that α is algebraic over F if and only if $F[\alpha] = \{f(\alpha) \mid f(x) \in F[x]\}$ is a field.

20. Let E be a field, $F \subseteq E$ a subfield of E, $\alpha \in E$ an element of E. Show that α is transcendental over F if and only if $F[\alpha] = \{f(\alpha) \mid f(x) \in F[x]\}$ is isomorphic to $F[x]$.

21. Let E be a field, $F \subseteq E$ a subfield of E, $\alpha \in E$ an element of E. If α is algebraic over F of degree 15 and $\beta \in F(\alpha)$, what are the possible values of $[F(\beta) : F]$?

22. Let $f(x)$ and $g(x)$ be irreducible polynomials over a field F with $\deg f(x) = 15$ and $\deg g(x) = 14$. Let α be a zero of $f(x)$ in some extension field of F. Show that $g(x)$ is still irreducible over $F(\alpha)$.

23. Let E be a field, $F \subseteq E$ a subfield of E, $\alpha, \beta \in E$ elements of E. If α and β are algebraic over F with $\deg_F \alpha = n$ and $\deg_F \beta = m$, where $\gcd(n, m) = 1$, show that $[F(\alpha, \beta) : F] = nm$.

24. If $E = F(\alpha_1, \dots, \alpha_r)$, where each α_i is algebraic over F with $\deg_F \alpha_i = n_i$, show that $[E : F] \le n_1 \cdot \dots \cdot n_r$.

25. Let E be a field, and $F \subseteq E$ a subfield of E. Show that E is a finite extension of F if and only if there exist elements $\alpha_i \in E$ with each α_i algebraic over F and $E = F(\alpha_1, \dots, \alpha_r)$.

26. Let E be a field, $F \subseteq E$ a subfield of E, $\alpha \in E$ an element of E, and $f(x) \in F[x]$ a nonzero polynomial. Show that if $f(\alpha)$ is algebraic over F, then α is algebraic over F.

27. Let p and q be two distinct prime integers.
(a) Show that $\mathbb{Q}(\sqrt{p} + \sqrt{q}) = \mathbb{Q}(\sqrt{p}, \sqrt{q})$.
(b) Find the minimal polynomial of $\sqrt{p} + \sqrt{q}$ over $\mathbb{Q}$.

28. Let E be a field, $F \subseteq E$ a subfield of E, $\alpha, \beta \in E$ two elements of E. Show that if $\alpha + \beta$ and $\alpha\beta$ are both algebraic over F, then α and β are both algebraic over F.

Existence of Transcendental Numbers

Exercises 29 through 31 outline the proof of the existence of real numbers that are transcendental over $\mathbb{Q}$.

29. Write $\mathbb{N}$ for the set $\{n \in \mathbb{Z} \mid n > 0\}$ of positive integers. A set X is **countable** if there exists a map ϕ: $\mathbb{N} \to X$ from $\mathbb{N}$ to X that is onto.

(a) Show that the set $\mathbb{N}^2 = \{(i,j) \mid i, j \in \mathbb{N}\}$ of pairs of positive integers is countable by defining a map ϕ_2 from the set $\mathbb{N}$ of positive integers n onto the set $\mathbb{N}^2$ of pairs of positive integers (i, j). (*Hint:* Every positive integer n can be written uniquely as the product $2^{i-1}(2j - 1)$ of a power of 2 and an odd number.)

(b) Show that the set $\mathbb{N}^3$ of triples of positive integers is countable by defining a map ϕ_3 from the set $\mathbb{N}$ of positive integers onto the set $\mathbb{N}^3$ of triples of positive integers (i, j, k). (*Hint:* First use ϕ_2 to define a map from the set $\mathbb{N}^2$ of pairs of positive integers (n, k) onto the set of triples of positive integers (i, j, k).)

(c) Show for every positive integer r that the set $\mathbb{N}^r$ of r-tuples of positive integers is countable by showing how to define a map ϕ_r from $\mathbb{N}$ onto $\mathbb{N}^r$.

(d) Let $\mathbb{N}^{<\infty}$ be the set of all finite sequences of positive integers or, in other words, $\mathbb{N}^{<\infty} = \cup_r \mathbb{N}^r$. Show that $\mathbb{N}^{<\infty}$ is countable by showing how to define a map ϕ from $\mathbb{N}$ onto $\mathbb{N}^{<\infty}$. (*Hint:* Show that the map $\psi(r, n) = \phi_r(n)$ from $\mathbb{N}^2$ to $\mathbb{N}^{<\infty}$ is onto.)

30. Using the results of the preceding problem, show that the following sets are countable:

(a) $\mathbb{Q}$, the field of rational numbers (*Hint:* Every rational number can be represented as a quotient $(i - j)/k$, where i, j, k are positive integers.)

(b) $\mathbb{Q}[x]$, the ring of polynomials over the field of rational numbers (*Hint:* First show that the set $\mathbb{Q}^{<\infty}$ of finite sequences of rational numbers is countable.)

(c) $\overline{\mathbb{Q}}$, the field of algebraic numbers

31. Let ϕ: $\mathbb{N} \to I$, where I is the unit interval $\{a \in \mathbb{R} \mid 0 < a < 1\}$. For each $n \in \mathbb{N}$, write $\phi(n)$ in decimal notation:

$$\phi(n) = .c_{n1}c_{n2}c_{n3}\dots$$

where each $0 \leq c_{ni} \leq 9$ for all n and i. The **diagonal number** $\delta(\phi)$ of ϕ is the real number $b \in I$ with the decimal representation

$$\delta(\phi) = .d_{n1}d_{n2}d_{n3}\dots$$

where $d_{ni} = 1$ if $c_{ni} \neq 1$, and $d_{ni} = 2$ if $c_{ni} = 1$.

(a) Show that $\delta(\phi) \neq \phi(n)$ for any n.

(b) Show that I is uncountable.

Using these facts and the results of the preceding problem

(c) Show that the set of transcendental real numbers is nonempty.

(d) Show that the set of transcendental real numbers is uncountable.

10.3 Splitting Fields

As was shown in the preceding section, every polynomial $f(x) \in F[x]$ has a zero α in some extension field of F (Kronecker's theorem). We have studied the structure of $F(\alpha)$, the smallest extension field of F that contains α. In this section we construct the *splitting field* of $f(x)$ over F, which is the smallest extension field of F that contains *all* the zeros of $f(x)$. Kronecker's theorem is used to show that such an extension field exists for any polynomial $f(x)$.

10.3.1 EXAMPLE Adjoining to $\mathbb{Q}$ one zero $\sqrt{2} = 2^{1/2} \in \mathbb{C}$ of the irreducible polynomial $x^2 - 2$ in $\mathbb{Q}[x]$ also gives us the other zero $-2^{1/2}$, and over $\mathbb{Q}(2^{1/2})$ we have a complete factorization into a product of linear factors:

$$x^2 - 2 = (x - 2^{1/2})(x + 2^{1/2})$$

Similarly, adjoining to $\mathbb{Q}$ one zero $\sqrt{3}\, i = 3^{1/2}i \in \mathbb{C}$ of the irreducible polynomial $x^2 + 3$ in $\mathbb{Q}[x]$ also gives us the other zero $-3^{1/2}i$, and over $\mathbb{Q}(3^{1/2}i)$ we have a complete factorization into a product of linear factors:

$$x^2 + 3 = (x - 3^{1/2}i)(x + 3^{1/2}i)$$

Similarly, adjoining to $\mathbb{Q}$ one zero $\omega \in \mathbb{C}$ of the irreducible polynomial $x^2 + x + 1$ also gives us the other zero ω^2, and over $\mathbb{Q}(\omega)$ we have a complete factorization into a product of linear factors:

$$x^2 + x + 1 = (x - \omega)(x - \omega^2)$$

Actually, since $\omega = {}^1/_2(-1 + 3^{1/2}i)$ and $3^{1/2}i = 2\omega + 1$, the fields are the same in these last two cases, $\mathbb{Q}(3^{1/2}i) = \mathbb{Q}(\omega)$. ◇

10.3.2 EXAMPLE By contrast, adjoining to $\mathbb{Q}$ one zero $\sqrt[3]{2} = 2^{1/3} \in \mathbb{C}$ of the irreducible polynomial $x^3 - 2$ in $\mathbb{Q}[x]$ does *not* give us the other two zeros of this polynomial in $\mathbb{C}$, which are $\omega 2^{1/3}$ and $\omega^2 2^{1/3}$. Over $\mathbb{Q}(2^{1/3})$ we have the factorization

$$x^3 - 2 = (x - 2^{1/3})(x^2 + 2^{1/3}x + 2^{2/3})$$

but the second factor, whose zeros in $\mathbb{C}$ are $\omega 2^{1/3}$ and $\omega^2 2^{1/3}$, is irreducible. Further adjoining to $\mathbb{Q}(2^{1/3})$ one of these zeros does give the other, and over $\mathbb{Q}(2^{1/3}, \omega 2^{1/3})$ we have a complete factorization into a product of linear factors:

$$x^3 - 2 = (x - 2^{1/3})(x - \omega 2^{1/3})(x - \omega^2 2^{1/3})$$

Adjoining $\omega 2^{1/3}$ to $\mathbb{Q}(2^{1/3})$ clearly produces the same result as adjoining ω, which we have mentioned in the preceding example produces the same result as adjoining $3^{1/2}i$. That is, we have

$$\mathbb{Q}(2^{1/3}, \omega 2^{1/3}) = \mathbb{Q}(2^{1/3}, \omega) = \mathbb{Q}(2^{1/3}, 3^{1/2}i)$$

Note that ω and $3^{1/2}i$ have degree 2 over $\mathbb{Q}$, while by Corollary 10.2.22, the degree over $\mathbb{Q}$ of any element of $\mathbb{Q}(2^{1/3})$ must divide 3, so ω and $3^{1/2}i$ do *not* belong to $\mathbb{Q}(2^{1/3})$. ◇

The next few definitions and propositions introduce some terminology for describing the kind of situation encountered in the preceding example and state some basic facts about the notions introduced.

10.3.3 DEFINITION Let F be a field, $f(x) \in F[x]$ a nonconstant polynomial, and E an extension field of F. Then we say $f(x)$ **splits** over E if in $E[x]$ we have a factorization of $f(x)$ into the product of a unit times monic linear factors:

$$f(x) = u(x - \alpha_1)\ldots(x - \alpha_n)$$

Note that in such a factorization, since u is the leading coefficient in $f(x)$, we have $u \in F$. ○

10.3.4 PROPOSITION Let F be a field, $f(x) \in F[x]$ a nonconstant polynomial, and E an extension field of F.

(1) If in $E[x]$ we have a complete factorization of $f(x)$ into a product of linear factors:

$$f(x) = (a_1x + b_1)\ldots(a_nx + b_n)$$

then $f(x)$ splits over E.

(2) If E contains n distinct zeros $\alpha_1, \ldots, \alpha_n$ of $f(x)$, where n is the degree of $f(x)$, then $f(x)$ splits over E.

Proof (1) Given a factorization as in (1), we obtain a splitting as in Definition 10.3.3 by setting $u = a_1\ldots a_n$ and $\alpha_i = -a_i^{-1}b_i$.

(2) If α_1 is a zero of $f(x)$, then

$$f(x) = (x - \alpha_1)q_1(x) \in E[x]$$

where deg $q_1(x) = n - 1$. If α_2 is a zero of $f(x)$ distinct from α_1, then $x - \alpha_2$ divides $q_1(x)$ and

$$f(x) = (x - \alpha_1)(x - \alpha_2)q_2(x) \in E[x]$$

where deg $q_2(x) = n - 2$. Continuing in this way, since all the $x - \alpha_i$ are distinct, we have

$$f(x) = (x - \alpha_1)(x - \alpha_2)\ldots(x - \alpha_n)q_n(x) \in E[x]$$

where deg $q_n(x) = 0$, so that $q_n(x) = u \neq 0$ is a nonzero constant, and hence a unit. □

10.3.5 DEFINITION Let F be a field, $f(x) \in F[x]$ a nonconstant polynomial, and E an extension field of F over which $f(x)$ splits. A subfield K of E containing F is called the **splitting field** in E of $f(x)$ over F if

(1) $f(x)$ splits over K.

(2) K is the smallest subfield of E containing F over which $f(x)$ splits. ○

10.3.6 PROPOSITION Let F be a field, $f(x) \in F[x]$ a nonconstant polynomial, and E an extension field of F over which $f(x)$ splits as

$$f(x) = u(x - \alpha_1)\ldots(x - \alpha_n)$$

Then $K = F(\alpha_1, \ldots, \alpha_n)$ is the splitting field in E of $f(x)$ over F.

Proof (1) Clearly $f(x)$ splits over K.

(2) To show that K is the smallest subfield of E containing F over which $f(x)$ splits, we have to show that if L is any other subfield of E containing F over which $f(x)$ splits, then $K \subseteq L$. So let L be such a subfield. Over L we have a splitting:

$$f(x) = v(x - \beta_1)\ldots(x - \beta_n) \in L[x] \subseteq E[x]$$

But by unique factorization for $E[x]$, the factors $(x - \beta_j)$ must be the same as the factors $(x - \alpha_i)$ except for order. So each α_i is equal to one of the $\beta_j \in L$. Hence $\alpha_i \in L$ for all i, and since also L is assumed to contain F, we get

$$K = F(\alpha_1, \ldots, \alpha_n) \subseteq L$$

as required. □

Note that if adjoining some of the α_i give us the others, say

$$\alpha_{m+1}, \ldots, \alpha_n \in F(\alpha_1, \ldots, \alpha_m)$$

then

$$F(\alpha_1, \ldots, \alpha_m) = F(\alpha_1, \ldots, \alpha_n)$$

and is the splitting field. This was the situation with the field

$$\mathbb{Q}(2^{1/3}, \omega 2^{1/3}) = \mathbb{Q}(2^{1/3}, \omega) = \mathbb{Q}(2^{1/3}, 3^{1/2}i)$$

in Example 10.3.2, which by the preceding proposition is the splitting field in $\mathbb{C}$ of $x^3 - 2$ over $\mathbb{Q}$.

10.3.7 EXAMPLE Consider $f(x) = x^4 - 7x^2 + 1 \in \mathbb{Q}[x]$. We begin by checking for reducibility. If we look for a factorization

$$x^4 - 7x^2 + 1 = (x^2 + ax + b)(x^2 + cx + d)$$

in $\mathbb{Z}[x]$ we obtain the conditions

$$bd = 1$$
$$ad + bc = 0$$
$$b + d + ac = -7$$
$$a + c = 0$$

and $b = d = 1$, $a = -c = 3$ is a solution. Hence

$$x^4 - 7x^2 + 1 = (x^2 + 3x + 1)(x^2 - 3x + 1)$$

We can now use the quadratic formula to find that the zeros of $f(x)$ in $\mathbb{C}$ are the four real numbers

$$\frac{\pm 3 \pm \sqrt{5}}{2}$$

Obviously, if r is any of these four zeros, then $\sqrt{5} = 5^{1/2} \in \mathbb{Q}(r)$, and all four of these zeros belong to $\mathbb{Q}(5^{1/2})$, which is therefore the splitting field in $\mathbb{C}$ of $f(x)$ over $\mathbb{Q}$. ◇

10.3.8 EXAMPLE Consider $f(x) = x^4 + 2x^2 + 1 \in \mathbb{Q}[x]$. We have the factorization

$$x^4 + 2x^2 + 1 = (x^2 + 1)^2$$

The zeros of $f(x)$ in $\mathbb{C}$ are just $\pm i$. The splitting field in $\mathbb{C}$ of $f(x)$ over $\mathbb{Q}$ is just $\mathbb{Q}(i)$, over which we have the splitting

$$(x - i)^2(x + i)^2$$

This illustrates the case where the number of distinct zeros is less than the degree of the polynomial. ◇

Proposition 10.3.6 tells us that a splitting field K for a polynomial $f(x) \in F[x]$ exists provided there exists an extension field E of F over which $f(x)$ splits. In the examples considered so far, F has been the rational numbers $\mathbb{Q}$ and $f(x)$ has been at worst a quartic polynomial, and the polynomial has split over $\mathbb{C}$. The next theorem tells us that for any F and any nonconstant $f(x) \in F[x]$ there will exist some extension field E over which the polynomial splits, and therefore a splitting field exists.

10.3.9 THEOREM Let F be a field and $f(x) \in F[x]$ a nonconstant polynomial. Then there exists an extension field E of F over which $f(x)$ splits.

Proof We prove the theorem by induction on $n = \deg f(x)$. If $n = 1$, then $f(x)$ is already linear, and we may take $E = F$. Otherwise, assume as induction hypothesis that the theorem holds for $n - 1$ over any field, and consider $f(x)$ of degree n. By Kronecker's theorem (Theorem 10.2.5) there is an extension field K of F in which $f(x)$ has a zero β. Over K we have a factorization: $f(x) = (x - \beta)g(x)$, where $g(x)$ has degree $n - 1$. By the induction hypothesis there is an extension field E of K over which there is a splitting:

$$g(x) = u(x - \alpha_1)\dots(x - \alpha_{n-1})$$

But then over E we have

$$f(x) = u(x - \alpha_1)\dots(x - \alpha_{n-1})(x - \beta)$$

which is a splitting. □

Proposition 10.3.6 and Theorem 10.3.9 imply the existence for any nonconstant polynomial $f(x) \in F[x]$ of a finite extension field K of F that is a splitting field of $f(x)$. What can we say about the degree $[K : F]$ of K over F? Of course, if $f(x)$ already splits in F, we can take $K = F$, so $[K : F]$ can be as small as 1. How large can it be? An inductive argument like that used to prove Theorem 10.3.9 can be used to prove the following.

10.3.10 PROPOSITION Let F be a field and $f(x) \in F[x]$ a polynomial of degree $n > 1$. Then there exists a finite extension E of F of degree $[E : F] \le n!$ over which $f(x)$ splits.

Proof See Exercise 13. □

Even where none of the zeros of $f(x)$ belongs to F, the degree $[K : F]$ of a splitting field K may be much smaller than $n!$ as in the preceding proposition. For instance, in Examples 10.3.7 and 10.3.8 we have

$$[K : F] = 2 < 4 = \deg f(x)$$

In these examples, $f(x)$ was reducible. If $f(x)$ is an irreducible polynomial of degree n, then even adjoining just *one* zero of a polynomial $f(x)$ produces an extension of degree n, so we have the lower bound $n \le [K : F]$ to go with the upper bound

$[K : F] \leq n!$ implied by the preceding proposition. The next two examples illustrate the possibilities.

10.3.11 EXAMPLE Let us look again at Example 10.3.2, where we considered the splitting field $\mathbb{Q}(2^{1/3}, \omega)$ of $x^3 - 2$ over $\mathbb{Q}$. Figure 1 shows the lattice of subfields of this splitting field.

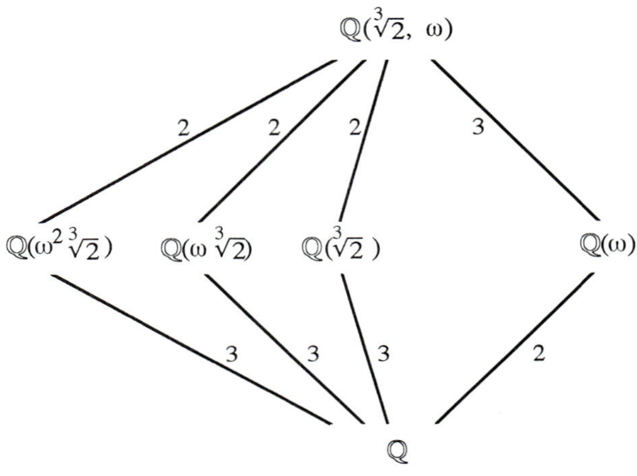

FIGURE 1

The numbers 2 or 3 beside the lines connecting a field and an extension field thereof indicate the degree of the extension. Thus we can obtain the splitting field by adjoining to $\mathbb{Q}$ any one of the zeros of $x^3 - 2$ and then adjoining a zero of $x^2 + x + 1$, or by doing the reverse. As for the degree, we have

$$[\mathbb{Q}(2^{1/3}, \omega) : \mathbb{Q}] = [\mathbb{Q}(2^{1/3}, \omega) : \mathbb{Q}(2^{1/3})]\,[\mathbb{Q}(2^{1/3}) : \mathbb{Q}] = 2{\cdot}3 = 6 = 3!$$

$$[\mathbb{Q}(2^{1/3}, \omega) : \mathbb{Q}] = [\mathbb{Q}(2^{1/3}, \omega) : \mathbb{Q}(\omega)]\,[\mathbb{Q}(\omega) : \mathbb{Q}] = 3{\cdot}2 = 6 = 3!$$

However we get there, the degree of the splitting field is the maximum possible for a polynomial of degree 3. ◇

10.3.12 EXAMPLE The zeros of $x^n - 1 \in \mathbb{Q}[x]$ are what we have called the nth roots of unity. They are the complex numbers

$$\zeta_n = \cos(\frac{2\pi}{n}) + i \sin(\frac{2\pi}{n})$$

$$\zeta_n{}^k = \cos(k\frac{2\pi}{n}) + i \sin(k\frac{2\pi}{n})$$

where $0 \leq k \leq n - 1$. They form a cyclic group

$$\{1, \zeta_n, \zeta_n{}^2, \ldots , \zeta_n{}^{n-1}\}$$

of order n under multiplication. Hence all n of these distinct zeros belong to $\mathbb{Q}(\zeta_n)$, which is therefore a splitting field of $x^n - 1 \in \mathbb{Q}[x]$ over $\mathbb{Q}$.

Note that if $n = p$, a prime, then

$$x^p - 1 = (x - 1)(x^{p-1} + \ldots + x^2 + x + 1)$$

and

$$\Phi_p(x) = x^{p-1} + \ldots + x^2 + x + 1$$

is irreducible by Corollary 8.4.22. Hence the splitting field $\mathbb{Q}(\zeta_p)$ of $x^p - 1$ has degree $[\mathbb{Q}(\zeta_p) : \mathbb{Q}] = p - 1$.

To illustrate the case where n is not a prime, consider $n = 6$. Then

$$x^6 - 1 = (x^3 - 1)(x^3 + 1) = (x - 1)(x^2 + x + 1)(x + 1)(x^2 - x + 1)$$

and the 6th roots of unity are 1 and

$$\zeta_6 = \cos(\tfrac{2\pi}{6}) + i\sin(\tfrac{2\pi}{6}) = \frac{1 + \sqrt{3}i}{2} = -\omega^2$$

$$\zeta_6^2 = \frac{-1 + \sqrt{3}i}{2} = \omega$$

$$\zeta_6^3 = -1$$

$$\zeta_6^4 = \frac{-1 - \sqrt{3}i}{2} = \omega^2$$

$$\zeta_6^5 = \frac{1 - \sqrt{3}i}{2} = -\omega$$

Note that 1 and ζ_6^2 and ζ_6^4 are the zeros of $x^3 - 1$, ζ_6^2 and ζ_6^4 being the zeros of $x^2 + x + 1$, while -1 and ζ_6, and ζ_6^5 are the zeros of $x^3 + 1$, ζ_6 and ζ_6^5 being the zeros of $x^2 - x + 1$.

$\mathbb{Q}(\zeta_6) = \mathbb{Q}(\sqrt{3}\,i)$ contains all these six zeros and is the splitting field of $x^6 - 1$ over $\mathbb{Q}$, so in this case the splitting field has degree $[\mathbb{Q}(\zeta_6) : \mathbb{Q}] = 2$. ◇

10.3.13 DEFINITION A generator of the cyclic group of nth roots of unity is called a **primitive *n*th root of unity**. There are $\phi(n)$ primitive nth roots of unity, where ϕ is the Euler function. (See Exercise 15.) ○

10.3.14 DEFINITION For any $n > 0$ the ***n*th cyclotomic polynomial** $\Phi_n(x)$ is the monic polynomial whose zeros are the primitive nth roots of unity:

$$\Phi_n(x) = (x - \zeta_1)\ldots(x - \zeta_{\phi(n)})$$

where $\zeta_1, \ldots, \zeta_{\phi(n)}$ are the $\phi(n)$ primitive nth roots of unity. ○

10.3.15 EXAMPLE In the notation of Example 10.3.12, the primitive 6th roots of unity are ζ_6 and ζ_6^5. Hence $\Phi_6(x) = x^2 - x + 1$. Note that $\Phi_6(x) \in \mathbb{Z}[x]$ and

$$\deg \Phi_6(x) = [\mathbb{Q}(\zeta_6) : \mathbb{Q}] = 2 = \phi(6)$$

$$x^6 - 1 = x^6 - 1 = (x^3 - 1)(x^3 + 1) = (x - 1)(x + 1)(x^2 + x + 1)(x^2 - x + 1)$$

$$= \Phi_1(x)\Phi_2(x)\Phi_3(x)\Phi_6(x) = \prod_{d|6} \Phi_d(x)$$

Our next proposition states that this last relationship holds generally. ◇

10.3.16 PROPOSITION For any $n > 0$ we have

$$x^n - 1 = \prod_{d|n} \Phi_d(x)$$

Proof See Exercise 22. □

10.3.17 EXAMPLE Let us consider what Definitions 10.3.13 and 10.3.14 amount to in the special case $n = p$, a prime.
Since every element but the identity in a cyclic group of prime order is a generator, all the pth roots of unity except the trivial root 1 are primitive pth roots of unity.
In Definition 8.4.21 we defined the cyclotomic polynomial $\Phi_p(x)$ in the special case of p a prime. Definition 10.3.14 applies to any n but agrees with our earlier Definition 8.4.21 for primes. For by the preceding proposition

$$x^p - 1 = \Phi_1(x)\Phi_p(x) = (x - 1)\Phi_p(x)$$

but

$$x^p - 1 = (x - 1)(x^{p-1} + x^{p-2} + \ldots + x^2 + x + 1)$$

and so

$$\Phi_p(x) = x^{p-1} + x^{p-2} + \ldots + x^2 + x + 1$$

which was our earlier definition in the special case of a prime. This polynomial was shown to be irreducible over $\mathbb{Q}$ in Corollary 8.4.22. In the case of p a prime, all pth roots of unity except the trivial root 1 itself are zeros of the irreducible polynomial $\Phi_p(x)$ of degree $p - 1$ over $\mathbb{Q}$, and hence all nontrivial pth roots have degree $p - 1$ over $\mathbb{Q}$ ◇

10.3.18 EXAMPLE Let p be a prime and consider the splitting field in $\mathbb{C}$ of $f(x) = x^p - 3$ over $\mathbb{Q}$. If α is a zero of $f(x)$, then $\alpha^p = 3$ and hence $(\alpha\zeta)^p = 3$ where ζ is any pth root of unity. Thus the splitting field in this case is $\mathbb{Q}(3^{1/p}, \zeta)$ where ζ is any nontrivial pth root of unity. Note that $3^{1/p}$ still has degree p over $\mathbb{Q}(\zeta)$, for if it had degree $m < p$ we would have

$$[\mathbb{Q}(3^{1/p}, \zeta) : \mathbb{Q}] = [\mathbb{Q}(3^{1/p}, \zeta) : \mathbb{Q}(\zeta)]\,[\mathbb{Q}(\zeta) : \mathbb{Q}] = m(p - 1)$$

which is impossible since $p = \deg_{\mathbb{Q}}(3^{1/p})$ divides $[\mathbb{Q}(3^{1/p}, \zeta) : \mathbb{Q}]$ by Corollary 10.2.22, but the prime p cannot divide a product $m(p - 1)$ of positive integers $< p$. Thus in this case we have

$$[\mathbb{Q}(3^{1/p}, \zeta) : \mathbb{Q}] = [\mathbb{Q}(3^{1/p}, \zeta) : \mathbb{Q}(\zeta)]\,[\mathbb{Q}(\zeta) : \mathbb{Q}] = p(p - 1)$$

and the degree of the splitting field is $p(p - 1)$, which is $< p!$ for primes $p > 3$. ◇

10.3.19 EXAMPLE We find the splitting field of $x^5 - 1$ over $\mathbb{Q}$ by two different methods.
First Method: Let

$$\zeta = \cos(\tfrac{2\pi}{5}) + i\sin(\tfrac{2\pi}{5}) = a + bi$$

Then $(a + bi)^5 = 1$. Taking real and imaginary parts, we obtain

$$a^5 - 10b^2a^3 + 5b^4a = 1$$

Since

$$a^2 + b^2 = \cos^2(\tfrac{2\pi}{5}) + \sin^2(\tfrac{2\pi}{5}) = 1$$

we have

$$a^5 - 10(1 - a^2)a^3 + 5(1 - a^2)^2 a = 1$$

Hence $a = \cos(\frac{2\pi}{5})$ is a zero of

$$f(x) = 16x^5 - 20x^3 + 5x - 1$$

But we have the factorization

$$f(x) = (x - 1)(4x^2 + 2x - 1)^2$$

Using the quadratic formula, we obtain

$$\cos(\tfrac{2\pi}{5}) = \frac{-1 + \sqrt{5}}{2}$$

$$\sin(\tfrac{2\pi}{5}) = \frac{\sqrt{10 + 2\sqrt{5}}}{4}$$

Therefore,

$$\zeta = \frac{-1 + \sqrt{5}}{2} + i\frac{\sqrt{10 + 2\sqrt{5}}}{4}$$

and the splitting field of $x^5 - 1$ is $K = \mathbb{Q}(\sqrt{5}, i\sqrt{10 + 2\sqrt{5}})$. Note that $[K : \mathbb{Q}] = 4$, in agreement with Example 10.3.17.

Second Method: We have

$$x^5 - 1 = (x - 1)\Phi_5(x) = (x - 1)(x^4 + x^3 + x^2 + x + 1)$$

and we want to solve the quartic $\Phi_5(x)$. Using the notation we developed in Section 8.5 for quartics, the auxiliary cubic (as defined in Definition 8.5.7) in this case is

$$g(k) = 8k^3 - 4k^2 - 6k + 2$$

and $k = 1$ is a zero of $g(k)$. Hence

$$h = \frac{1}{2} \qquad u = \frac{\sqrt{5}}{2} \qquad v = 0$$

and

$$\Phi_5(x) = (x^2 + \tfrac{1}{2}x + 1 + \tfrac{\sqrt{5}}{2}x)(x^2 + \tfrac{1}{2}x + 1 - \tfrac{\sqrt{5}}{2}x) =$$

$$(x^2 + \frac{1 + \sqrt{5}}{2}x + 1)(x^2 + \frac{1 - \sqrt{5}}{2}x + 1)$$

and finally, using the quadratic formula, we obtain the four zeros of $\Phi_5(x)$:

$$\zeta = \frac{-1 + \sqrt{5}}{2} + i\frac{\sqrt{10 + 2\sqrt{5}}}{4}$$

and $\zeta^2, \zeta^3, \zeta^4$. ◇

Let F be a field, $f(x) \in F[x]$ a nonconstant polynomial. We have shown that there exists an extension field of F over which $f(x)$ splits, and that within any such extension field there is a splitting field for $f(x)$ over F. If we have two such extension fields of F, then we have a splitting field of $f(x)$ over F in each. We prove next that

these two splitting fields are isomorphic: The splitting field of $f(x)$ over F is unique up to isomorphism.
The proof is broken up into two steps.
Given two fields F and F' and an isomorphism ϕ: $F \to F'$, the induced isomorphism ϕ^*: $F[x] \to F'[x]$ sends any element

$$f(x) = a_n x^n + \dots + a_1 x + a_0 \in F[x]$$

to

$$\phi^*(f(x)) = \phi(a_n)x^n + \dots + \phi(a_1)x + \phi(a_0) \in F'[x]$$

(See Section 8.1, Exercises 15 and 17.)

10.3.20 THEOREM Let F be a field, let $f(x) \in F[x]$ be irreducible, and let α be a zero of $f(x)$ in some extension field of F. If F' is another field and ϕ: $F \to F'$ is an isomorphism, and if β is a zero of $\phi^*(f(x)) \in F'[x]$ in some extension field of F', then there exists an isomorphism $\phi^\dagger : F(\alpha) \to F'(\beta)$ such that

(1) $\phi^\dagger$ agrees with ϕ on F, that is, $\phi^\dagger(c) = \phi(c)$ for all $c \in F$

(2) $\phi^\dagger(\alpha) = \beta$

Proof The proof refers to Figure 2.

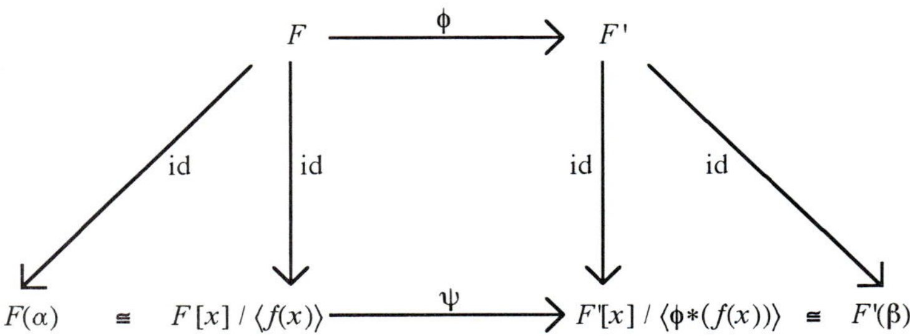

FIGURE 2

It is convenient, for any given $g(x) \in F[x]$, to write $g'(x)$ as an abbreviation for $\phi^*(g(x)) \in F'[x]$. Note that since ϕ^* is an isomorphism, $g(x)$ and $g'(x)$ have the same degree. We first note that $f'(x) = \phi^*(f(x))$ is irreducible over F'. For if we had a factorization $f'(x) = g'(x)h'(x)$ in $F'[x]$, where both factors have degree less than that of $f'(x)$, then since ϕ^* is an isomorphism we have

$$\phi^*(f(x)) = f'(x) = g'(x)h'(x) = \phi^*(g(x)\ \phi^*((h(x)) = \phi^*(g(x)h(x))$$

but then we would have a factorization $f(x) = g(x)h(x)$ in $F[x]$ where both factors have degree less than that of $f(x)$, which is impossible since $f(x)$ is irreducible over F.
Therefore, $f'(x)$ is irreducible over F', and $F[x]/\langle f(x)\rangle$ and $F'[x]/\langle f'(x)\rangle$ are both fields. We define a map

$$\psi: F[x]/\langle f(x)\rangle \to F'[x]/\langle f'(x)\rangle$$

by

$$\psi(g(x) + \langle f(x)\rangle) = g'(x) + \langle f'(x)\rangle$$

Then ψ is an isomorphism. (See Exercise 23.)

By Theorem 10.2.11, part (1), $F(\alpha)$ and $F[x]/\langle f(x)\rangle$ are isomorphic, as are $F'(\beta)$ and $F'[x]/\langle f'(x)\rangle$. Take the isomorphism in the direction from $F(\alpha)$ to $F[x]/\langle f(x)\rangle$ and the isomorphism in the direction from $F'[x]/\langle f'(x)\rangle$ to $F'(\beta)$, and define $\phi^\dagger$ to be their composition, as at the bottom of the Figure 2. For any element $h(\alpha)$ of $F(\alpha)$, $h(\alpha)$ is carried first to $h(x) + F[x]/\langle f(x)\rangle$ by the isomorphism from $F(\alpha)$ to $F[x]/\langle f(x)\rangle$, then to $h'(x) + \langle f'(x)\rangle$ by ψ, then to $h'(\beta)$ by the isomorphism from $F'[x]/\langle f'(x)\rangle$ to $F'(\beta)$ or, in a diagram

$$\phi^\dagger: \quad h(\alpha) \quad \to \quad h(x) + F[x]/\langle f(x)\rangle \quad \to \quad h'(x) + \langle f'(x)\rangle \quad \to \quad h'(\beta)$$

Applying this to a constant polynomial $h(x) = c \in F$, where $h'(x) = \phi^*(h(x)) = \phi(c)$, we have

$$\phi^\dagger: \quad c \quad \to \quad c + F[x]/\langle f(x)\rangle \quad \to \quad c + \langle f'(x)\rangle \quad \to \quad \phi(c)$$

Applying it to the polynomial $h(x) = x$, where $h'(x) = \phi^*(h(x)) = x$, we have

$$\phi^\dagger: \quad \alpha \quad \to \quad x + F[x]/\langle f(x)\rangle \quad \to \quad x + \langle f'(x)\rangle \quad \to \quad \beta$$

These are the conditions (1) and (2) required for $\phi^\dagger$ in the statement of the theorem. □

10.3.21 THEOREM Let F be a field, $f(x) \in F[x]$ a nonconstant polynomial, and K a splitting field for $f(x)$ over F. If F' is another field and $\phi: F \to F'$ is an isomorphism, and if K' is a splitting field for $\phi^*(f(x))$ over F', then there exists an isomorphism $\phi^\ddagger: K \to K'$ such that $\phi^\ddagger$ agrees with ϕ on F.

Proof We use induction on the degree n of $f(x)$. If $n = 1$, then $K = F$, $K' = F'$ and we may take $\phi^\ddagger = \phi$. So assume as induction hypothesis that the theorem holds for polynomials of degree $n - 1$. Let $p(x)$ be any irreducible factor of $f(x)$. By the preceding theorem, if $\alpha \in K$ is any zero of $p(x)$ and $\beta \in K'$ is any zero of $\phi^*(p(x))$, then there is an isomorphism $\phi^\dagger: F(\alpha) \to F'(\beta)$ that agrees with ϕ on F and has $\phi^\dagger(\alpha) = \beta$. This isomorphism gives rise to an induced isomorphism $\chi: F(\alpha)[x] \to F'(\beta)[x]$ that agrees with ϕ^* on $F[x]$.

Now consider the factorization

$$f(x) = (x - \alpha)g(x) \in F(\alpha)[x]$$

where $g(x)$ has degree $n - 1$. This corresponds to a factorization

$$\chi(f(x)) = \phi^*(f(x)) = (x - \beta)\chi(g(x)) \in F'(\beta)[x]$$

Now K is a splitting field of $g(x)$ over $F(\alpha)$ and K' is a splitting field of $\chi(g(x))$ over $F'(\beta)$. So the induction hypothesis implies there is an isomorphism $\phi^\ddagger: K \to K'$ that agrees with $\phi^\dagger$ on $F(\alpha)$ and therefore agrees with ϕ on F. □

The uniqueness of the splitting field up to isomorphism follows immediately.

10.3.22 COROLLARY Let F be a field, $f(x) \in F[x]$ a nonconstant polynomial, and K and K' two splitting fields for $f(x)$ over F. Then K and K' are isomorphic.

Proof Apply the preceding theorem with $F' = F$ and ϕ = the identity. □

Using Kronecker's theorem, we were able to show that every polynomial has a splitting field. In all our examples of polynomials from $\mathbb{Q}[x]$ and $\mathbb{R}[x]$, the splitting fields were all subfields of $\mathbb{C}$. We have encountered no algebraic extensions of $\mathbb{C}$ (other than $\mathbb{C}$ itself). In fact, there are none. This fact, known as the *fundamental theorem of algebra*, is proved in Chapter 12. We include the statement and a few immediate consequences of it in this section because of its close relation to the theorems and examples in this section.

10.3.23 DEFINITION A field F is **algebraically closed** if every nonconstant polynomial in $F[x]$ has a zero in F. ○

The following theorem allows us to use four equivalent definitions of an algebraically closed field.

10.3.24 THEOREM Let F be any field. Then the following statements are equivalent:

(1) F is algebraically closed.
(2) $f(x) \in F[x]$ is irreducible if and only if $\deg f(x) = 1$.
(3) Every nonconstant polynomial in $F[x]$ splits over F.
(4) If E is an algebraic extension of F, then $E = F$.

Proof (1) $\Rightarrow$ (2) See Exercise 24.

(2) $\Rightarrow$ (3) See Exercise 25.

(3) $\Rightarrow$ (4) Let E be an algebraic extension of F and let $\alpha \in E$. Then α is algebraic over F, and so $f(\alpha) = 0$ for some nonzero polynomial $f(x) \in F[x]$. Thus $f(x)$ is a nonconstant polynomial, and so by (3) we have a splitting over F:

$$f(x) = u(x - \alpha_1)\dots(x - \alpha_n)$$

where each $\alpha_i \in F$. Since $f(\alpha) = 0$, $x - \alpha$ must divide $f(x)$ and so must be equal to one of the factors $(x - \alpha_i)$. But this means $\alpha = \alpha_i \in F$. This shows $E \subseteq F$ and therefore $E = F$.

(4) $\Rightarrow$ (1) Let $f(x)$ be a nonconstant polynomial in $F[x]$ and let α be a zero of $f(x)$ in some extension of F. Then $F(\alpha)$ is an algebraic extension of F, and so by (4) we have $F(\alpha) = F$ and $\alpha \in F$. □

10.3.25 THEOREM (Fundamental theorem of algebra) The field $\mathbb{C}$ of complex numbers is algebraically closed.

Proof The proof is given in Chapter 12. □

10.3.26 COROLLARY For any nonconstant polynomial $f(x) \in \mathbb{Q}[x]$ there exists a splitting field $K \subseteq \mathbb{C}$ for $f(x)$ over $\mathbb{Q}$.

Proof This is immediate from Theorem 10.3.25, (using definition (3) of algebraically closed from Theorem 10.3.24) together with Proposition 10.3.6. □

10.3.27 COROLLARY If E is an extension field of $\mathbb{C}$ and $[E : \mathbb{C}] > 1$, then E is not an algebraic extension of $\mathbb{C}$ and $[E : \mathbb{C}]$ is infinite.

Proof This is immediate from Theorem 10.3.25 (using definition (4) of algebraically closed from Theorem 10.3.24) together with Theorem 10.2.16. □

10.3.28 COROLLARY Let $f(x) \in \mathbb{R}[x]$ be a nonconstant polynomial. If $f(x)$ is irreducible over $\mathbb{R}$, then $\deg f(x) = 1$ or 2.

Proof See Exercise 26. □

Exercises 10.3

In Exercises 1 through 11 find the splitting field K in $\mathbb{C}$ of the indicated polynomial $f(x)$ over $\mathbb{Q}$, and determine $[K : \mathbb{Q}]$. (*Hint for* 8 *through* 11: Use the methods of Section 8.5.)

1. $f(x) = x^4 - 1$ **2.** $f(x) = x^3 + 1$ **3.** $f(x) = x^4 - 4$

4. $f(x) = x^4 + 1$ **5.** $f(x) = x^3 - 5$ **6.** $f(x) = x^4 - 2x^2 + 1$

7. $f(x) = x^5 + x^4 + x^3 - x^2 - x - 1$ **8.** $f(x) = x^3 + x + 1$

9. $f(x) = x^3 - 3x + 2$ **10.** $f(x) = x^4 - 4x^3 + 6x^2 - 4x + 1$

11. $f(x) = x^4 - 2x^3 - x + 2$

12. Let α be a zero of $x^3 + x^2 + 1$ over $\mathbb{Z}_2$. Show that $f(x)$ splits over $\mathbb{Z}_2(\alpha)$.

13. Prove Proposition 10.3.11.

14. Let $E = \mathbb{Q}(\sqrt{3}, \sqrt[3]{2}, i)$. Construct a subfield lattice for E as in Example 10.3.12.

15. Show that the number of primitive nth roots of unity is $\phi(n)$.

16. Calculate the cyclotomic polynomials $\Phi_n(x)$ for all $1 \le n \le 8$.

17. Let $n = p_1^{a_1} p_2^{a_2} \ldots p_k^{a_k} \in \mathbb{Z}$, where the p_i are distinct primes. Show that

$$\phi(n) = \prod_{1 \le i \le k} p_i^{a_i - 1}(p_i - 1)$$

18. Find all the 8th roots of unity and identify which are primitive 8th roots of unity.

19. Find the splitting field in $\mathbb{C}$ of $f(x) = x^8 - 2$ over $\mathbb{Q}$.

20. Let n and m be relatively prime integers, ζ_n a primitive nth root of unity, and ζ_m a primitive mth root of unity. Show that $\zeta_n\zeta_m$ is a primitive (mn)th root of unity.

21. Let ζ_n be a primitive nth root of unity and d a divisor of n. Show that ${\zeta_n}^d$ is a primitive (n/d)th root of unity.

22. Prove Proposition 10.3.16.

23. Show that the map ψ defined in the proof of Theorem 10.3.20 is an isomorphism.

24. Prove (1) $\Rightarrow$ (2) in Theorem 10.3.24. **25.** Prove (2) $\Rightarrow$ (3) in Theorem 10.3.24.

26. Prove Corollary 10.3.28.

27. Show that the field $\bar{\mathbb{Q}} = \{\alpha \in \mathbb{C} \mid \alpha \text{ is algebraic over } \mathbb{Q}\}$ of algebraic numbers is algebraically closed.

10.4 Finite Fields

We end our chapter on field theory with a beautiful theorem on finite fields. We show that any finite field F is of order p^n for some prime p and positive integer n, and that such a field is a splitting field of the polynomial $x^{p^n} - x$ over $\mathbb{Z}_p$. We use the theorems on the existence and uniqueness of splitting fields proved in the preceding section to show that there exists a field, unique up to isomorphism, of order p^n for every prime p and positive integer n.

Before applying this work on splitting fields from the preceding section, let us review what we know about finite fields from earlier chapters and summarize what follows by the basic results of the first two sections of this chapter.

10.4.1 PROPOSITION Let F be a finite field. Then

(1) The order of F is a prime power $|F| = p^n$, where $p = \text{char } F$.

(2) F is an algebraic extension $\mathbb{Z}_p(\alpha)$, where α is a zero of an irreducible monic polynomial $q(x)$ of degree n over $\mathbb{Z}_p$.

Proof (1) Suppose F is a finite field. Since F is finite, the characteristic of F is a prime p, and $\mathbb{Z}_p$ is a subfield of F, by Theorem 7.3.12. Let $n = [F : \mathbb{Z}_p]$. Then since a vector space of dimension n over the field of $\mathbb{Z}_p$ of p elements has exactly p^n elements (being isomorphic as a vector space to ${\mathbb{Z}_p}^n$ by Exercise 29, Section 10.1), we have $|F| = p^n$.

(2) The multiplicative group F^* of nonzero elements of F is cyclic by Theorem 8.3.10, and if α is a generator, then every nonzero element of F, being equal to some power of α, belongs to $\mathbb{Z}_p(\alpha)$. Thus $F = \mathbb{Z}_p(\alpha)$, the smallest field containing $\mathbb{Z}_p$ and α. By Theorem 10.2.9 there is a unique monic polynomial $q(x) \in \mathbb{Z}_p[x]$, irreducible over $\mathbb{Z}_p$ and of degree n, such that α is a zero of $q(x)$, namely the minimal polynomial of α over $\mathbb{Z}_p$. □

10.4.2 EXAMPLE Let F be a field having between 65 and 124 elements. By the preceding proposition, since the only prime power in this range is $81 = 3^4$, F has characteristic 3, F has 81 elements, and $F = \mathbb{Z}_3(\alpha)$, where α is a zero of an irreducible monic polynomial $q(x) \in \mathbb{Z}_3[x]$ of degree 4. Note now that since the nonzero elements of F form a group of order 80, $a^{80} = 1$ for all nonzero $a \in F$ by Lagrange's theorem. Thus each such a is a zero of $x^{80} - 1$ and of $x^{81} - x$, and 0 is also a zero of the latter. Thus every element of F is a zero of $f(x) = x^{81} - x$, and F contains $81 = \deg f(x)$ distinct zeros of $f(x)$. It follows that F is a splitting field of $f(x)$ over $\mathbb{Z}_3$. If now F' is another field of 81 elements, then by the same argument F' is also a splitting field of $f(x)$ over $\mathbb{Z}_3$. By Corollary 10.3.23, we have $F \cong F'$. ◇

This example shows how we prove that any two finite fields of the same order are isomorphic, which is the uniqueness half of the main theorem of this section. For the existence half, we show that the splitting field of $x^{p^n} - x$ over $\mathbb{Z}_p$ is a field of order p^n. A key step is to show that the zeros of this polynomial are all distinct, and for this step the key notion that we need to introduce is that of the *derivative* of a polynomial.

10.4.3 DEFINITION Let F be a field and $f(x) \in F[x]$ a polynomial

$$f(x) = a_n x^n + \ldots + a_i x^i + \ldots + a_1 x + a_0$$

Then the **derivative** of $f(x)$ is the polynomial

$$f'(x) = n a_n x^{n-1} + (n - 1)a_{n-1} x^{n-2} + \ldots + a_1$$

in $F[x]$. Note that in the case of a constant polynomial $f(x) = a_0$ the derivative is the zero polynomial $f'(x) = 0$. ○

Note that while in calculus the derivative is defined for polynomial functions and other functions from $\mathbb{R}$ to $\mathbb{R}$, the algebraic definition of derivative applies to polynomials over any field and does not involve the notion of limit. Of course, we cannot rely on the proofs from calculus of basic rules for calculating derivatives but must give new proofs based on our new definition.

10.4.4 PROPOSITION Let F be a field, $c \in F$ an element of F, and $f(x), g(x) \in F[x]$ polynomials over F. Then

(1) $[cf(x)]' = cf'(x)$

(2) $[f(x) + g(x)]' = f'(x) + g'(x)$

(3) $[f(x)g(x)]' = f(x)g'(x) + f'(x)g(x)$

(4) $[(x - c)^n]' = n(x - c)^{n-1}$

Proof (1) and (2) are immediate from the preceding definition.

(3) For any polynomial $h(x) = a_n x^n + \ldots + a_i x^i + \ldots + a_1 x + a_0$ we define

$$h_0(u, v) = (h(u) - h(v)) / (u - v) =$$

$$(a_n(u^n - v^n) + \ldots + a_2(u^2 - v^2) + a_1(u - v)) / (u - v) =$$

$$a_n[u^{n-1} + u^{n-2}v + \ldots + uv^{n-2} + v^{n-1}] + \ldots + a_2(u + v) + a_1$$

Then $h'(x) = h_0(x, x)$.

We now apply this to $f(x)$, $g(x)$, and $h(x) = f(x)g(x)$. We have

$$h_0(u, v) = (f(u)g(u) - f(v)g(v)) / (u - v) =$$
$$(f(u)g(u) - f(u)g(v) + f(u)g(v) - f(v)g(v)) / (u - v) =$$
$$f(u)\big[(g(u) - g(v)) / (u - v)\big] + g(u)\big[(f(u) - f(v)) / (u - v)\big] =$$
$$f(u)g_0(u, v) + f_0(u, v)g(u)$$

And hence

$$h'(x) = h_0(x, x) = f(x)g_0(x, x) + f_0(x, x)g(x) = f(x)g'(x) + f'(x)g(x)$$

(4) We proceed by induction on n. For $n = 1$, if $f(x) = (x - c)^1 = x - c$, then $f'(x) = 1 = 1 \cdot (x - c)^0$. Now assume (4) holds for $n - 1$. If $f(x) = (x - c)^n = (x - c)(x - c)^{n-1}$, then by (3) we have

$$f'(x) = (x - c)(n - 1)(x - c)^{n-2} + 1 \cdot (x - c)^{n-1} = n(x - c)^{n-1}$$

as required. □

10.4.5 EXAMPLE Consider the polynomial

$$f(x) = (x - 2)^3(x - 1)^2(x + 1) \in \mathbb{Q}[x]$$

In the terminology of Definition 8.3.5, 2 is zero of $f(x)$ of multiplicity 3, 1 is a zero of multiplicity 2, and -1 is a zero of multiplicity 1. Let us calculate the derivative of $f(x)$ using parts (3) and (4) of the preceding proposition.

$$f'(x) = (x - 2)^3[(x - 1)^2 + 2(x - 1)(x + 1)] + 3(x - 2)^2[(x - 1)^2(x + 1)] =$$
$$(x - 2)^2(x - 1)[(x - 2)(x - 1) + 2(x - 2)(x + 1) + 3(x - 1)(x + 1)]$$

Hence $f'(2) = f'(1) = 0$, while $f'(-1) = -108 \neq 0$. Therefore, the zeros of $f(x)$ of multiplicity greater than 1 are exactly those zeros of $f(x)$ that are also zeros of $f'(x)$. ◇

10.4.6 THEOREM Let F be a field, $f(x) \in F[x]$ a polynomial, and α a zero of $f(x)$ in some extension field of F. Then α is a zero of $f(x)$ of multiplicity $s > 1$ if and only if α is a zero of $f'(x)$.

Proof ($\Rightarrow$) If α is an element of some extension field E of F and α is a zero of $f(x)$ of multiplicity $s > 1$, then according to Definition 8.3.5 this means that in $E[x]$ we have

$$f(x) = (x - \alpha)^s g(x)$$

where $g(\alpha) \neq 0$. Then

$$f'(x) = (x - \alpha)^s g'(x) + s(x - \alpha)^{s-1}g(x)$$

and $f'(\alpha) = 0$.

($\Leftarrow$) If α is a zero of $f(x)$ of multiplicity $s = 1$, then in $E[x]$ we have the factorization

$$f(x) = (x - \alpha)g(x)$$

where $g(\alpha) \neq 0$. Then

$$f'(x) = (x - \alpha)g'(x) + g(x)$$

and $f'(\alpha) = g(\alpha) \neq 0$. □

10.4.7 EXAMPLE Consider $f(x) = x^8 - 1 \in \mathbb{Z}_3[x]$. Then $f'(x) = 8x^7$ and $f'(\alpha) = 0$ if and only if $\alpha = 0$. But $f(0) = -1 \neq 0$. Hence $f(x)$ and $f'(x)$ have no zeros in common.

Hence by the preceding proposition, all zeros of $f(x)$ have multiplicity 1, and $f(x)$ has 8 distinct zeros in its splitting field over $\mathbb{Z}_3$, and $g(x) = x^9 - x$, which has the additional zero 0, therefore has 9 distinct zeros in its splitting field over $\mathbb{Z}_3$. ◇

This example illustrates how we prove the existence, for each prime p and positive integer n, of an extension of $\mathbb{Z}_p$ having at least p^n elements. To prove the existence of a field with *exactly* p^n, we make use of an important map. Recall from Section 7.1 (Exercise 19) that if F is a field of characteristic p, then the map $\phi: F \to F$ defined by $f(\alpha) = \alpha^p$ for all $\alpha \in F$ is called the *Frobenius map*.

10.4.8 PROPOSITION Let F be a finite field of charactersitic p. Then the Frobenius map $\phi: F \to F$ is an automorphism of F.

Proof ϕ is a homomorphism by Exercise 19 of Section 7.1. To show that ϕ is an isomorphism, it only remains to show that it is one to one (and therefore onto). For this see Exercise 21 at the end of this section. □

We now have all the tools needed to prove the main theorem on finite fields promised in our earlier discussion.

10.4.9 THEOREM Let p be a prime and n a positive integer. Then F is a finite field of order p^n if and only if F is a splitting field of $f(x) = x^{p^n} - x$ over $\mathbb{Z}_p$.

Proof ($\Rightarrow$) If F is a finite field with $|F| = p^n$, then $F^* = F - \{0\}$ is a multiplicative group of order $p^n - 1$, and by Lagrange's theorem, $\alpha^{p^n-1} = 1$ for all $\alpha \in F^*$. Hence $\alpha^{p^n} - \alpha = 0$ for all $\alpha \in F$, and the p^n elements of F are all distinct zeros of $f(x) = x^{p^n} - x$. Thus F is the splitting field of $f(x)$ over $\mathbb{Z}_p$.
($\Leftarrow$) Let F be a splitting field of $f(x) = x^{p^n} - x$ over $\mathbb{Z}_p$, and let

$$K = \{\alpha \in F \mid f(\alpha) = 0\}$$

the subset of F consisting of the zeros of $f(x)$. Note that $f'(x) = -1 \neq 0$; hence all the zeros of $f(x)$ have multiplicity 1, and $f(x)$ has p^n distinct zeros. That is, $|K| = p^n$. If we can show that K is a field, then we will have $F = K$, since then $K \subseteq F$ will both be splitting fields of $f(x)$. Consider the Frobenius automorphism $\phi: F \to F$. Then since $\phi^n(\alpha) = \alpha^{p^n}$ and $\alpha \in K$ if and only if $f(\alpha) = \alpha^{p^n} - \alpha = 0$, we have

$$K = \{\alpha \in F \mid \phi^n(\alpha) = \alpha\}$$

Then K is a subfield of F. (See Exercise 22.) Therefore, K is a field and, as remarked previously, $F = K$ and so $|F| = |K| = p^n$ as required. □

10.4.10 COROLLARY Given any prime p and any positive integer n,

(1) There exists a finite field F of order p^n.

(2) Any two fields of order p^n are isomorphic.

Proof Immediate from the preceding theorem and the Theorem 10.3.10 and Corollary 10.3.23 on the existence and uniqueness up to isomorphism of splitting fields. □

10.4.11 COROLLARY Given any positive integer k,

(1) There exists a finite field F of order k if and only if k is a prime power p^n.

(2) Any two fields of order k are isomorphic.

Proof Immediate from Corollary 10.4.10 and Proposition 10.4.1, part (1). □

10.4.12 COROLLARY Given any prime p and any positive integer n, there exists a monic polynomial $q(x) \in \mathbb{Z}_p[x]$ of degree n that is irreducible over $\mathbb{Z}_p$.

Proof Immediate from Corollary 10.4.10 and Proposition 10.4.1, part (2). □

10.4.13 DEFINITION Given any prime p and any positive integer n, the unique (up to isomorphism) field of order p^n is denoted $GF(p^n)$ and is called the **Galois field** of order p^n. ○

10.4.14 DEFINITION Given a finite field F, a generator of the cyclic group F^* is called a **primitive element** of F. ○

10.4.15 EXAMPLE

(1) In $F = GF(7) = \mathbb{Z}_7 = U(7) = \{1, 2, 3, 4, 5, 6\}$, the element 3 is a primitive element, since $3^2 = 2$, $3^3 = 6$, $3^4 = 4$, $3^5 = 5$, and $3^6 = 1$. A similar calculation shows that 5 is also a primitive element.

(2) We have examined $F = GF(4) = \mathbb{Z}_2[\alpha] \cong \mathbb{Z}_2[x]/\langle x^2 + x + 1\rangle$ in Example 8.7.2 and written out its multiplication table. The element α, satisfying $\alpha^2 + \alpha + 1 = 0$, is a primitive element, since $\alpha^2 = \alpha + 1$ and $\alpha^3 = 1$. ◇

10.4.16 EXAMPLE Let us consider $F = GF(2^{10})$ and work out its subfield lattice. $\mathbb{Z}_2 \subseteq F$ and $[F : \mathbb{Z}_2] = 10$. By Theorem 10.2.20, for any subfield $\mathbb{Z}_2 \subseteq E \subseteq F$ we have $[E : \mathbb{Z}_2] = 1, 2, 5$, or 10. In the first and last cases $E = \mathbb{Z}_2$ and $E = F$, respectively. In the other two cases, $E \cong GF(2^2)$ or $E \cong GF(2^5)$. And since E^* is a subgroup of F^*, and a cyclic group has only one subgroup of each possible order, there can be only one subfield of $GF(2^{10})$ of a given order. Hence the subfield lattice of $GF(2^{10})$ is as shown in Figure 3. ◇

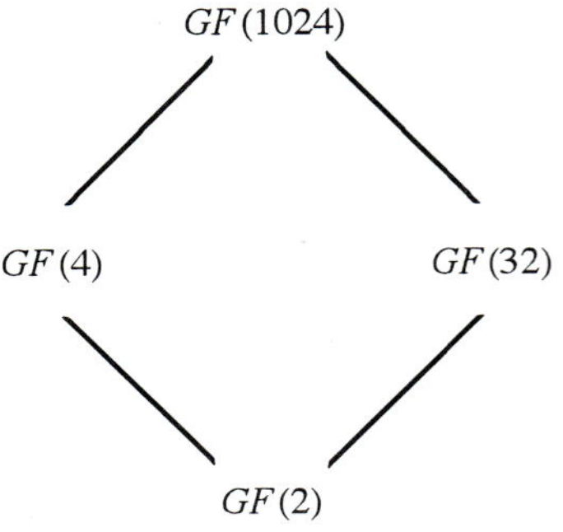

FIGURE 3

10.4.17 THEOREM Let p be a prime and n a positive integer.

(1) If E is a subfield of $GF(p^n)$, then $E \cong GF(p^r)$ for some r dividing n.

(2) If r divides n, then there exists a unique subfield E of $GF(p^n)$ having order p^r, given by

$$E = \{\beta \in GF(p^n) \mid \beta^{p^r} = \beta\}$$

Proof (1) If E is a subfield of $GF(p^n)$, then char $E = p$ and $\mathbb{Z}_p$ is a subfield of E. Thus by Theorem 10.2.20

$$n = [GF(p^n) : \mathbb{Z}_p] = [GF(p^n) : E]\,[E : \mathbb{Z}_p]$$

Hence $r = [E : \mathbb{Z}_p]$ divides n, $|E| = p^r$, and hence $E \cong GF(p^r)$.

(2) If r divides n, then $p^r - 1$ divides $p^n - 1$. (See Exercise 23.) The multiplicative group F^* of nonzero elements of F, being a cyclic group of order $p^n - 1$, has a unique subgroup of order d for each divisor d of $p^n - 1$, consisting of all elements whose orders divide d. In particular, $GF(p^n)^*$ has a unique subgroup of order $p^r - 1$, namely

$$E^* = \{\beta \in GF(p^n)^* \mid \beta^{p^r-1} = 1\}$$

Then $E = E^* \cup \{0\}$ consists of p^r distinct zeros of $g(x) = x^{p^r} - x$ and, as in the proof of Theorem 10.4.9, forms a subfield of F. □

Exercises 10.4

In Exercises 1 through 4 construct a field F of the indicated order N if possible.

1. $N = 9$ **2.** $N = 10$ **3.** $N = 15$ **4.** $N = 16$

In Exercises 5 through 8 find a primitive element for the indicated field F.

5. $F = \mathbb{Z}_5$ **6.** $F = \mathbb{Z}_{17}$ **7.** $F = GF(8)$ **8.** $F = GF(9)$

In Exercises 9 through 12 find the number of primitive elements in the indicated field F.

9. $F = GF(9)$ **10.** $F = GF(19)$ **11.** $F = GF(27)$ **12.** $F = GF(32)$

In Exercises 13 through 18 construct the subfield lattice of the indicated field F.

13. $F = GF(8)$ **14.** $F = GF(16)$ **15.** $F = GF(2^6)$

16. $F = GF(2^{12})$ **17.** $F = GF(3^{18})$ **18.** $F = GF(5^{30})$

19. **(Chain rule)** Let $f(x)$ and $g(x)$ be polynomials in $F[x]$. Using Definition 10.4.3, show that

$$[f(g(x))]' = f'(g(x))\cdot g'(x)$$

20. Let F be a field with char $F = p$. Show that for any $a, b \in F$ and any positive integer i, we have

$$(a + b)^{p^i} = a^{p^i} + b^{p^i}$$

21. Let F be a finite field with char $F = p$. Show that the Frobenius homomorphism $\phi: F \to F$ defined by $\phi(\alpha) = \alpha^p$ for all $\alpha \in F$ is an isomorphism.

22. Let F be a finite field with char $F = p$ and $\phi: F \to F$ the Frobenius automorphism as in the preceding problem. Show that

(a) $\phi^i(\alpha) = \alpha^{p^i}$ for all $\alpha \in F$.

(b) $K = \{\alpha \in F \mid \phi^n(\alpha) = \alpha\}$ for a fixed n is a subfield of F.

23. (a) Given positive integers r and s show that

$$x^{rs} - 1 = (x^r - 1)(x^{rs-r} + x^{rs-2r} + x^{rs-3r} + \ldots + x^r + 1)$$

(b) Use (1) to show that if r divides n, then x^{p^r-1} divides x^{p^n-1}.

24. Calculate:

(a) $[GF(16) : GF(4)]$ (b) $[GF(64) : GF(8)]$ (c) $[GF(p^n) : GF(p^r)]$

25. Let α be a zero of $x^3 + x + 1$ in some extension field of $\mathbb{Z}_2$ and let β be a zero of $x^3 + x^2 + 1$ in some field extension of $\mathbb{Z}_2$. Show that $\mathbb{Z}_2(\alpha) \cong \mathbb{Z}_2(\beta)$.

26. List all monic irreducible polynomials of degrees 1, 2, and 4 over $\mathbb{Z}_2$. Show that their product is $x^{16} - x$.

27. Show that $x^{p^n} - x$ is the product of all monic irreducible polynomials over $\mathbb{Z}_p$ whose degrees divide n.

28. Let $f(x) \in F[x]$ be a nonconstant polynomial. Show that $f(x)$ and $f'(x)$ are relatively prime in $F[x]$ if any only if every zero of $f(x)$ in any extension field of F has multiplicity $s = 1$.

29. Show that if char $F = 0$, and $f(x)$ is irreducible over F, then all zeros of $f(x)$ in any extension on F have multiplicity $s = 1$.

30. Show that if F is an algebraically closed field, then F is infinite.

Chapter 11

Geometric Constructions

One of the most beautiful aspects of mathematics is how seemingly unrelated subjects turn out to be connected. A striking example is the way the theory of field extensions, developed in the preceding chapter, can be applied to geometric constructions. This beautiful example of interplay between algebra and geometry, in which modern algebraic ideas are used to solve ancient geometric problems, is the subject of this chapter. In the first section we analyze geometric *straightedge and compass* constructions from an algebraic point of view and describe the field of *constructible* real numbers. In the second section we apply this analysis to some classical geometric construction problems. In the third and fourth sections we allow ourselves the use of a stronger tool, the *marked ruler*, and show how this enables us to carry out some constuctions impossible with straightedge and compass alone and to solve cubics and quartics geometrically. In the next chapter, after developing Galois theory, we see how it can be used to settle completely the kinds of questions studied in this chapter.

11.1 Constructible Real Numbers

In Euclidean geometry we learn to perform a number of constructions of figures in the plane with a straightedge and compass (for instance, constructing the midpoint of a given line segment or the bisector of a given angle).

In these constructions the **straightedge** is used only to draw the line between two given points A and B, that is, to draw the segment between them and extend it as far as necessary to intersect some other line or circle. The straightedge has no markings for measuring distances.

The **compass** can be used in two ways: first, to draw a circle with center at one given point C and passing through another given point D; second, to draw a circle with a center at a given point C, having radius equal to the distance between two other given points P and Q. That is, we assume that we can set the two ends of the compass on points P and Q, thus opening it to a length equal to PQ, and then lift the compass, keeping it open to this length, and place one end on point C to draw a circle with this length as radius and center at C. An instrument with which this can be done is called a **noncollapsing** compass. The ancient geometers used a **collapsing** compass, which closes as soon as it is lifted and so can only be used in the first of our two ways. It can be shown that any construction that can be performed with a straightedge and noncollapsing compass can also be performed with a straightedge and collapsing compass, though generally more steps are required in the construction. (See Exercise 22.)

The use of these instruments will become clearer as we review a few basic constructions in Euclidean geometry.

11.1.1 EXAMPLE Given two points A and B, we can construct the midpoint C of the line segment AB as follows. First use the straightedge to draw the line segment AB itself. Then use the compass to draw two circles with radius AB, one with center at A, the other with center at B. (Only two short arcs of each circle are indicated in Figure 1.) Let the two circles intersect in points P and Q. Now use the straightedge again to draw the line segment PQ. Let PQ and AB intersect at C. Then C is the required midpoint. For since $AP = AQ = BP = BQ$, the two triangles APQ and BPQ are congruent, hence $\angle APC = \angle BPC$. Hence the two triangles APC and BPC are congruent, and $AC = CB$. ◇

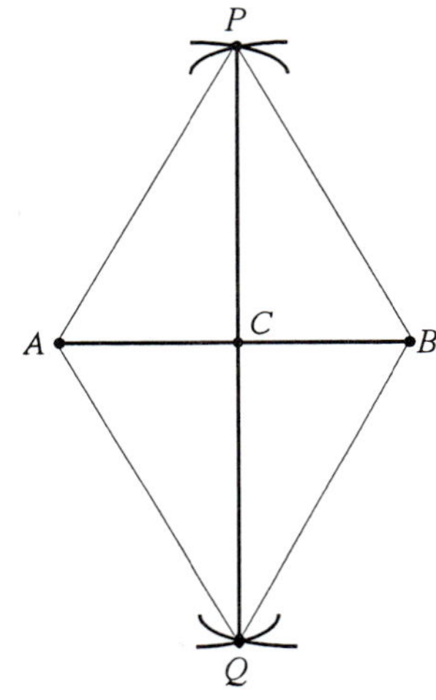

FIGURE 1

11.1.2 EXAMPLE The construction of Example 11.1.1 actually accomplishes more than was demanded. For the line PQ is actually the perpendicular bisector of AB. ◇

11.1.3 EXAMPLE Indeed, the construction of Example 11.1.1 accomplishes even more. For the triangle PAB is equilateral, and hence $\angle PAB$ is a 60° angle. Thus we have shown how to construct a 60° angle at a given point on a given line segment. ◇

11.1.4 EXAMPLE A variation on the same construction can be used to draw at a point A on a line AB a perpendicular line $AP \perp AB$.

To do this we simply draw a circle with center A passing through B and let its other point of intersection with line AB be Q, so that A is the midpoint of BQ, and then construct a perpendicular bisector AP to QB by the construction of the preceding examples. (Only the result of the construction has been shown in Figure 2.)

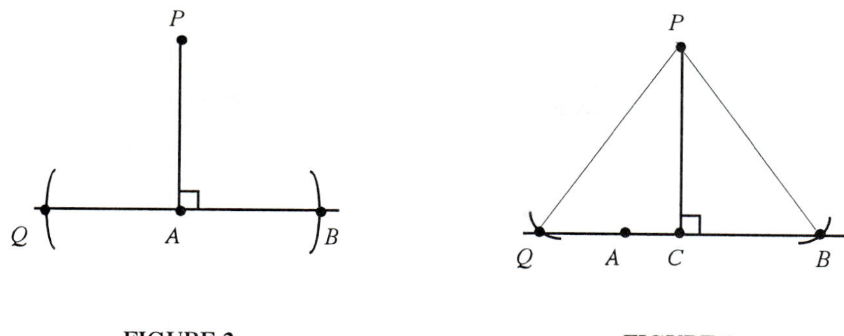

FIGURE 2 FIGURE 3

A further variation can be used to drop a perpendicular from a given point P to a given line segment AB.
Draw a circle with center at P passing through B, and let its other point of intersection with AB be Q. Then construct the midpont C by the method of Example 11.1.1. (Only the result of the construction has been shown in Figure 3.)
Since $PQ = PB$ and $QC = CB$, the triangles PCQ and PCB are congruent, hence the $\angle PCQ = \angle PCB$ and hence both are right angles. Thus PC is the required perpendicular to AB through P. ◇

11.1.5 EXAMPLE Combining the two constructions in the preceding example, we can construct through any point P a line PQ parallel to a given line AB.

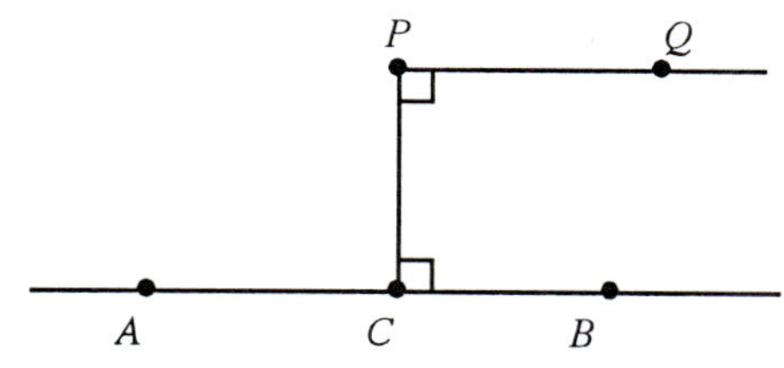

FIGURE 4

The construction is shown in Figure 4. We first use the construction illustrated in Figure 3 of the preceding example to drop a perpendicular PC to AB. Then we use the construction illustrated in Figure 2 of the same example to construct a line PQ perpendicular to PC. This line PQ is the required parallel to the given line AB passing through the given point P. ◇

11.1.6 EXAMPLE To bisect a given angle $\angle AOB$, draw a circle with center O and radius OA and let it intersect OB at P. Then draw circles one with center at A, the other with center at P, both with radius $OA = OP$. Let these circles intersect at Q. Then since $OA = OP$ and $AQ = PQ$, the triangles OAQ and OPQ are congruent, and hence $\angle AOQ = \angle POQ = {}^1/_2\ \angle AOB$. (See Figure 5.) ◇

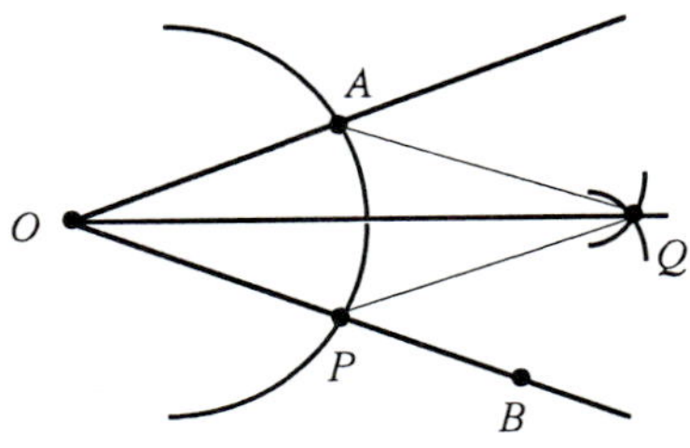

FIGURE 5

11.1.7 EXAMPLE We have not yet used the noncollapsing feature of our compass, which allows us to transfer a line segment from one location to another — that is, given AB and CD, to construct on CD a segment CQ equal to AB. We use it in the next contstruction, transferring an angle.

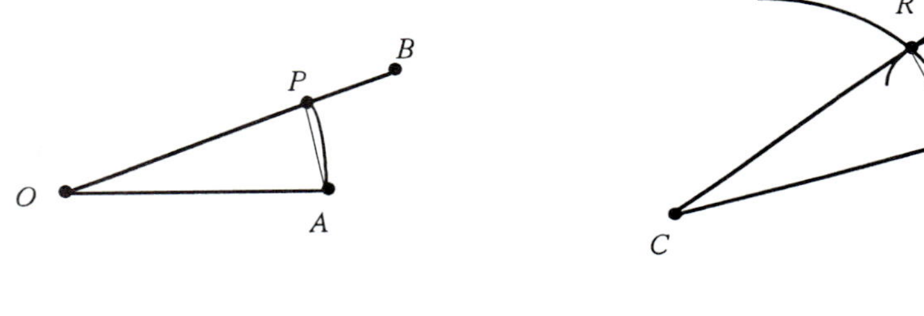

FIGURE 6

FIGURE 7

Let $\angle AOB$ and points C and D be given. We wish to find a point R such that $\angle DCR = \angle AOB$. To do so, draw the circle with center O and radius OA and let the circle intersect OB at P. Now draw the circle with center at C and radius OA, and let it intersect CD at Q. Then draw the circle with center at Q and radius AP, and let it intersect the first circle at R. Then $OA = OP = CQ = CR$ and $AP = QR$ and therefore the triangles AOP and QCR are congruent, and $\angle AOB = \angle AOP = \angle QCR = \angle DCR$ as required. (See Figures 6 and 7.) ◇

The basic steps out of which all such constructions as we have been reviewing are composed are listed in the next definition.

11.1.8 DEFINITION The basic operations in the plane used in **straightedge and compass constructions** are as follows:

(1) to draw a line through two given points

(2) to draw a circle with center at a given point and radius equal to the distance between two other given points

(3) to mark the point of intersection of two straight lines

(4) to mark the points of intersection of a straight line and a circle
(5) to mark the points of intersection of two circles

Any straightedge and compass construction starts from given points, lines, and circles and involves a finite sequence of steps of these kinds to obtain some other points, lines, or circles. ○

To relate geometric constructions with the algebra of real numbers, we arbitrarily choose two points O and X in the plane and take the length of the segment OX as our unit of measure, so that OX has length 1.

11.1.9 DEFINITION A real number α is said to be **constructible** by straightedge and compass if a segment of length $|\alpha|$ can be obtained starting from our unit segment by a straightedge and compass construction. ○

The next propositions establish that the set of constructible real numbers is closed under the field operations of $\mathbb{R}$, and thus we start building a bridge between the field theory of the preceding chapter and geometric constructions.

Proof We first show that if γ and δ are constructible real numbers with $0 < \delta < \gamma$, then so are $\gamma + \delta$ and $\gamma - \delta$.

FIGURE 8

Since γ and δ are constructible, segments PC and $P'D$ of lengths γ and δ are constructible. Draw a circle with center C and radius equal to $P'D$, and let Q and R be its points of intersection with PC, Q on the side nearer P, and R on the other side. Then PR has length $\gamma + \delta$ and PQ has length $\gamma - \delta$. (See Figure 8.)

To prove the proposition as stated, given α and β, let γ be the larger and δ the smaller of $|\alpha|$ and $|\beta|$. Then $|\alpha + \beta|$ is equal to one, and $|\alpha - \beta|$ to the other of $\gamma + \delta$ and $\gamma - \delta$ and hence are constructible by what we have just shown. Hence $\alpha + \beta$ and $\alpha - \beta$ are constructible. □

11.1.11 COROLLARY Let C be the set of constructible real numbers. Then $\mathbb{Z} \subseteq C$.

Proof Immediate from the preceding proposition, since $1 \in C$. □

11.1.12 PROPOSITION If α and β are constructible real numbers, then so is $\alpha \cdot \beta$ and so is α/β provided $\beta \neq 0$.

Proof We first show that if γ, δ, ζ are positive constructible real numbers, then so is $(\gamma\cdot\delta)/\zeta$. Let PC be a segment of length γ. Construct a perpendicular to PC. Draw circles with center P of radius δ and radius ζ and let them intersect the perpendicular at D and Z, respectively, as in Figure 9. Draw CD and draw the line through Z parallel to CD. Let this parallel intersect PC at H. The length η of PH is a constructible real number. The triangles DPC and ZPH are similar, so $\eta/\gamma = PH/PC = PZ/PD = \zeta/\delta$, so $\eta = (\gamma\cdot\zeta)/\delta$ and is constructible as asserted.

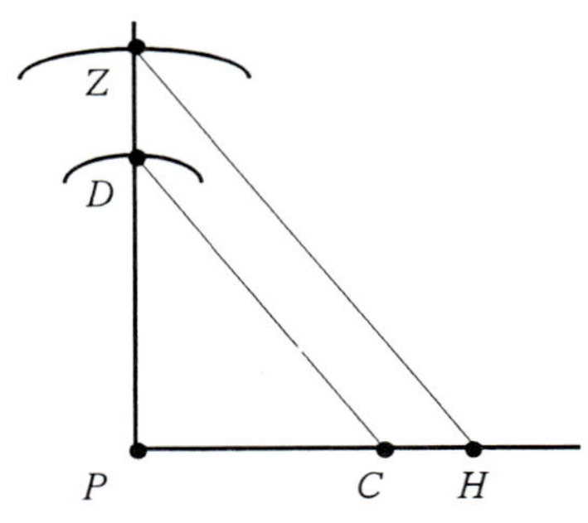

FIGURE 9

To prove the proposition as stated, let α and β be constructible, and apply what we have just proved with $\gamma = |\alpha|$, $\delta = 1$, $\zeta = |\beta|$ in order to show that $|\alpha\cdot\beta|$ and hence that $\alpha\cdot\beta$ is constructible, and with $\gamma = |\alpha|$, $\delta = |\beta|$, $\zeta = 1$ in order to show that $|\alpha/\beta|$ and hence that α/β is constructible. □

11.1.13 COROLLARY Let C be the set of constructible real numbers. Then C is an extension field of $\mathbb{Q}$ and a subfield of $\mathbb{R}$.

Proof By definition, C is a subset of $\mathbb{R}$. By Propositions 11.1.10 and 11.1.12, C is a field and thus a subfield of $\mathbb{R}$. By Corollary 11.1.11, $\mathbb{Z}$ is a subring of C, and hence its field of quotients $\mathbb{Q}$ is a subfield of $\mathbb{C}$ by Corollary 7.3.9. □

Now that we know the constructible real numbers form an extension field of $\mathbb{Q}$, we would like to know what kind of field extension is involved.

11.1.14 PROPOSITION Let α be a constructible real number with $\alpha > 0$. Then $\sqrt{\alpha}$ is constructible.

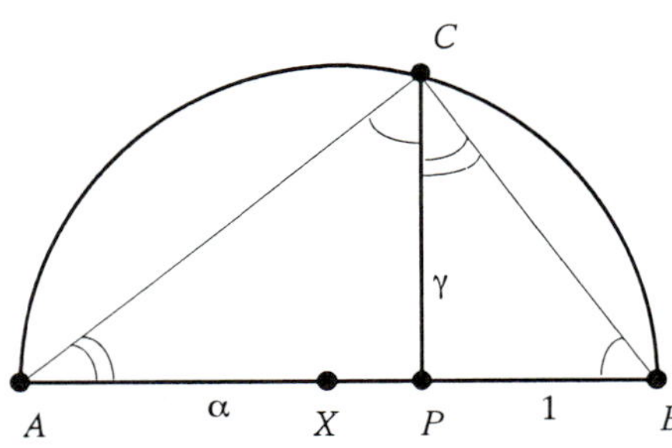

FIGURE 10

Proof On the same line construct a line segment PA of length α and on the opposite side of P a line segment PB of length 1. (See Figure 10.) Draw the circle with diameter AB. (This can be done by constructing the midpoint X of AB and drawing the circle with center X and radius $AX = XB$.) Draw the line perpendicular to AB at P, and let their point of intersection be C. Now let γ be the length of PC, so γ is constructible.

The triangles APC and CPB are similar, so $PC/PB = PA/PC$ or $\gamma/1 = \alpha/\gamma$. Thus $\gamma^2 = \alpha$ or $\gamma = \sqrt{\alpha}$ and $\sqrt{\alpha}$ is constructible as asserted. □

11.1.15 PROPOSITION Let F be a subfield of $\mathbb{R}$ and let

$$f(x) = x^2 + ax + b \in F[x]$$

be a quadratic polynomial with coefficients in F. If $f(x)$ has a zero in $\mathbb{R}$, then both zeros α_1 and α_2 of $f(x)$ belong to $\mathbb{R}$, and either $\alpha_1, \alpha_2 \in F$ or

$$\alpha_1, \alpha_2 \in F(\sqrt{\gamma}) \subseteq \mathbb{R}$$

for some $\gamma \in F$ with $\gamma > 0$.

Proof By the quadratic formula, the zeros of $f(x)$ in $\mathbb{C}$ are

$$\frac{-a \pm \sqrt{a^2 - 4b}}{2}$$

If either is real, then $\gamma = a^2 - 4b > 0$, and so both are real and belong to $F(\sqrt{\gamma})$. □

11.1.16 DEFINITION Let F be a subfield of $\mathbb{R}$. Then a **square root tower** over F is a sequence of fields $K_0, K_1, \dots, K_n$ such that

(1) $F = K_0 \subseteq K_1 \subseteq \dots \subseteq K_n \subseteq \mathbb{R}$

(2) $K_{i+1} = K(\sqrt{\rho})$ for some $\rho \in K_i$ with $\sqrt{\rho} \notin K_i$, for all $1 \leq i < n$

The field K_n is called the **top** of the tower. ○

We now have all the tools to give a complete algebraic description of the field C of constructible real numbers. We first introduce a coordinate system in the plane, taking the end points of our chosen unit section as the origin $O = (0, 0)$ and the unit point $X = (1, 0)$ on the x-axis.

11.1.17 LEMMA In any straightedge and compass construction starting from our unit segment, there is a square root tower over $\mathbb{Q}$ such that all

(1) the coordinates α, β of any point marked

$$(\alpha, \beta)$$

(2) the coefficients γ, δ of the equation of any line drawn

$$y = \gamma x + \delta$$

(3) the coefficients κ, λ, μ of the equation of any circle drawn

$$x^2 + \kappa x + y^2 + \lambda y + \mu = 0$$

in the course of the construction all belong to the top of the tower.

Proof We proceed by induction on the number of operations of kinds (1) through (5) described in Definition 11.1.8 performed in the course of the construction. At the beginning when we have performed no operations we have only the points (0, 0) and (0, 1), which belong to the top of the trival tower $\mathbb{Q} = K_0 \subseteq \mathbb{R}$. Suppose we have performed m operations and that all the coordinates and coefficients belong to the top K_n of some tower $K_0, K_1, \dots, K_n$, and consider performing one more operation.

There are five cases to consider depending on which of the five kinds of operation (1) through (5) is involved.

For (1), suppose we want to draw the line through two points (α_1, β_1) and (α_2, β_2) we have already marked. Since we have already marked these points, the coefficients $\alpha_1, \beta_1, \alpha_2, \beta_2$ belong to K_n. The equation of the line is

$$(y - \beta_1)/(\beta_2 - \beta_1) = (x - \alpha_1)/(\alpha_2 - \alpha_1)$$

and multiplying through to put this in the form

$$y = \gamma x + \delta$$

the coefficients γ, δ are obtained from $\alpha_1, \beta_1, \alpha_2, \beta_2$ by rational operations (that is, addition, subtraction, multiplication, and division), and so belong to K_n. Therefore, we may keep the same tower, and all coordinates and coefficients after this one additional operation of drawing the line still belong to the top of this tower.

For (2), suppose we want to draw the circle with center at one point (α_0, β_0) we have already marked and having radius equal to the segment between two other points (α_1, β_1) and (α_2, β_2) we have already marked. Then again the α_i, β_i belong to K_n. The equation of the circle is

$$(x - \alpha_0)^2 + (y - \beta_0)^2 = (\alpha_2 - \alpha_1)^2 + (\beta_2 - \beta_1)^2$$

and multiplying through to put this in the form

$$x^2 + \kappa x + y^2 + \lambda y + \mu = 0$$

again the coefficients κ, λ, μ are obtained from the α_i, β_i by rational operations, and we may keep the same tower.

For (3), suppose we want to mark the point of intersection of lines

$$y = \gamma_1 x + \delta_1 \text{ and } y = \gamma_2 x + \delta_2$$

that we have already marked. Then $\gamma_1, \delta_1, \gamma_2, \delta_2 \in K_n$. By the usual methods for solving simultaneous linear equations, the coordinates (α, β) of this point are obtained from $\gamma_1, \delta_1, \gamma_2, \delta_2 \in K_n$ by rational operations:

$$\alpha = (\delta_2 - \delta_1)/(\gamma_1 - \gamma_2) \qquad \beta = (\gamma_1\delta_2 - \gamma_2\delta_1)/(\gamma_1 - \gamma_2)$$

So again we may keep the same tower.

For (4), suppose we want to mark the point of intersection of a line and a circle

$$y = \gamma x + \delta \qquad x^2 + \kappa x + y^2 + \lambda y + \mu = 0$$

we have already drawn. Then $\gamma, \delta, \kappa, \lambda, \mu \in K_n$. If the coordinates of the points of intersection belong to K_n, we are done. Suppose they do not. To find them we substitute $\gamma x + \delta$ for y in the equation for the circle to get

$$x^2 + \kappa x + (\gamma x + \delta)^2 + \lambda(\gamma x + \delta) + \mu = 0$$

This is a quadratic equation with coefficients from F, so by Proposition 11.1.15 its solutions x_1, x_2 belong to the extension of K_n obtained by adjoining the square root of some positive $\rho \in K_n$. The points of intersection are

$$(x_1, \gamma x_1 + \delta) \text{ and } (x_2, \gamma x_2 + \delta)$$

and all their coordinates belong to $K_n(\sqrt{\rho})$, which we may take for the new top of tower K_{n+1}.

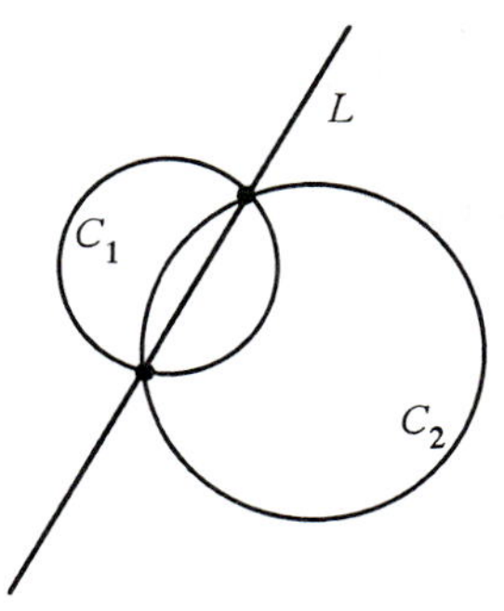

FIGURE 11

For (5), we use the fact that the intersection of two circles

$$C_1: x^2 + \kappa_1 x + y^2 + \lambda_1 y + \mu_1 = 0$$
$$C_2: x^2 + \kappa_2 x + y^2 + \lambda_2 y + \mu_2 = 0$$

is the same as the intersection of either of the circle with the line

$$L: y = \gamma x + \delta$$

where

$$\gamma = (\kappa_1 - \kappa_2)/(\lambda_2 - \lambda_1)$$
$$\delta = (\mu_1 - \mu_2)/(\lambda_2 - \lambda_1)$$

This reduces case (5) to case (4). □

11.1.18 THEOREM Let α be a real number. Then α is constructible if and only if α belongs to the top of some square root tower over $\mathbb{Q}$.

Proof ($\Leftarrow$) Let $\alpha \in K_n$, where

$$\mathbb{Q} = K_0 \subseteq K_1 \subseteq \ldots \subseteq K_n \subseteq \mathbb{R}$$

be a square root tower over $\mathbb{Q}$, and let $C \subseteq \mathbb{R}$ be the field of constructible real numbers. Then $\mathbb{Q}=K_0 \subseteq C$, and if $K_i \subseteq C$, then

$$K_{i+1} = K_i(\sqrt{\rho})$$

for some $\rho \in K_i \subseteq C$, and since then $\sqrt{\rho} \in C$ by Proposition 11.1.14, we have $K_{i+1} \subseteq C$. Thus by induction $K_i \subseteq C$ for all i, and since $\alpha \in K_n$, α is constructible.
($\Rightarrow$) Let α be constructible. Then starting from our unit segment OX the points on the x-axis OX at distance $|\alpha|$ from the origin O can be constructed with straightedge and compass, and therefore their coordinates belong to the top of some square root tower by the preceding lemma. But these coordinates are $(\pm\alpha, 0)$, so α belongs to the top of some square root tower. □

11.1.19 COROLLARY Let α be a real number. If α is constructible, then $[\mathbb{Q}(\alpha) : \mathbb{Q}]$ is a power of 2.

Proof If α is constructible, then by the preceding theorem, $\alpha \in K_n$ for some square root tower:

$$\mathbb{Q} = K_0 \subseteq K_1 \subseteq \ldots \subseteq K_n \subseteq \mathbb{R}$$

In such a tower, $K_{i+1} = K_i(\sqrt{\rho})$ for some $\rho \in K_i$, and therefore $[K_{i+1} : K_i] = 2$. By Theorem 10.2.20 we have

$$[K_n : \mathbb{Q}(\alpha)]\,[\mathbb{Q}(\alpha) : \mathbb{Q}] = [K_n : \mathbb{Q}] = [K_n : K_{n-1}] \ldots [K_2 : K_1]\,[K_1 : \mathbb{Q}] = 2^n$$

Hence $[\mathbb{Q}(\alpha) : \mathbb{Q}] = 2^m$ for some $m \leq n$. □

11.1.20 COROLLARY Transcendental real numbers are not constructible.

Proof If α is constructible, then $\mathbb{Q}(\alpha)$ is of finite degree over $\mathbb{Q}$ and hence is algebraic over $\mathbb{Q}$ by Theorem 10.2.16 and is not transcendental. □

11.1.21 EXAMPLE Consider the polynomial

$$f(x) = x^5 - 6x^3 + 4x - 10 \in \mathbb{Q}[x]$$

$f(x)$ is irreducible over $\mathbb{Q}$ by Eistenstein's criterion. Hence if α is a zero of this polynomial, then $[\mathbb{Q}(\alpha) : \mathbb{Q}] = 5$, and α is not constructible. ◇

11.1.22 EXAMPLE Consider the polynomial $f(x) = x^4 - 4x^2 + 2 \in \mathbb{Q}[x]$
The zeros of the quadratic obtained by setting $x^2 = y$

$$y^2 - 4y + 2$$

can be found by the quadratic formula, and we have

$$x^2 = y = 2 \pm \sqrt{2} \qquad\qquad x = \pm\sqrt{2 \pm \sqrt{2}}$$

Thus the zeros of $f(x)$ are constructible real numbers. We have, for instance, the square root tower

$$\mathbb{Q} \subseteq \mathbb{Q}(\sqrt{2}) \subseteq \mathbb{Q}(\sqrt{2+\sqrt{2}})$$

and $\sqrt{2+\sqrt{2}}$ belongs to the top of this tower. ◇

11.1.23 EXAMPLE The converse of Corollary 11.1.19 is not true. Let

$$p(x) = x^4 - 6x + 3$$

Then $p(x)$ is irreducible over $\mathbb{Q}$ by Eisenstein's criterion.
But using some calculus, we can see that $p(x)$ has two real zeros, call them r and s. We show that at least one of them is not constructible.

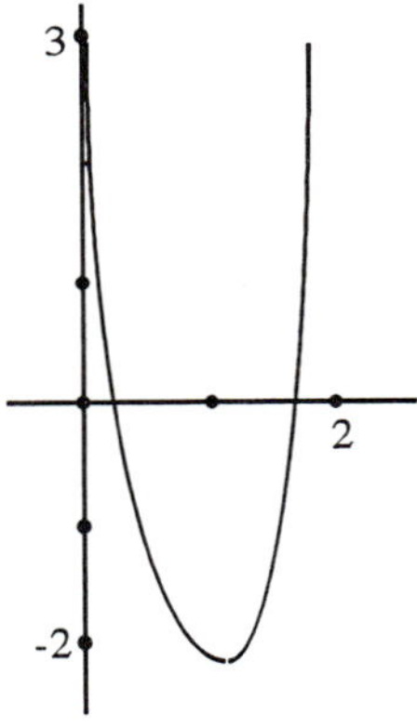

FIGURE 12

Over $\mathbb{R}$, $p(x)$ factors as

$$x^4 - 6x + 3 = (x - r)(x - s)(x^2 + cx + d) =$$
$$(x^2 + ax + b)(x^2 + cx + d)$$

where $a, b, c, d \in \mathbb{R}$ and

$$a = -(r+s) \qquad b = rs$$

Comparing coefficients, we see that

$$a + c = 0 \qquad b + d + ac = 0 \qquad ad + bc = -6 \qquad bd = 3$$

Hence

$$a = -c \qquad b + d = a^2 \qquad a(b - d) = -6$$

We now construct a polynomial of degree 3, irreducible over $\mathbb{Q}$, having $b + d$ as its real zero:

$$a^2(d - b)^2 = 36$$
$$(b + d)(d - b)^2 = 36$$
$$(b + d)((d + b)^2 - 4bd) = 36$$
$$(b + d)((b + d)^2 - 12) = 36$$

Hence $b + d$ is a zero of $g(y) = y^3 - 12y - 36$. Now $g(y)$ is irreducible over $\mathbb{Q}$ by Theorem 8.4.11. Hence $[\mathbb{Q}(b + d) : \mathbb{Q}] = 3$ and is not a power of 2, so $b + d$ is not constructible. On the other hand, the two real zeros r and s of $p(x)$ satisfy the relation

$$(r + s)^2 = a^2 = b + d$$

Hence if r and s were both constructible, then $b + d$ would be constructible by Propositions 11.1.10 and 11.1.12, which we have just shown it not to be. Hence $x^4 - 6x + 3$ has a real zero α that is not constructible, even though $[\mathbb{Q}(\alpha) : \mathbb{Q}] = 4$. ◇

Exercises 11.1

In Exercises 1 through 8 determine whether the indicated real number is constructible.

1. $\sqrt[4]{3}$ **2.** $\sqrt{3 + \sqrt[4]{5}}$ **3.** $\sqrt[6]{3 + \sqrt{5}}$ **4.** $3 - \dfrac{2}{\sqrt[8]{5}}$

5. $\dfrac{3}{1 + \sqrt[8]{5}}$ **6.** $\dfrac{3}{1 + \sqrt[5]{2}}$ **7.** $\sqrt[4]{2 - \sqrt[4]{3}}$ **8.** $(\sqrt{2} - 3\sqrt{3})^5$

In Exercises 9 through 13 given a line segment of length 1, construct with straightedge and compass a line segment of the indicated length.

9. 2/3 **10.** $\dfrac{\sqrt{3}}{\sqrt{5}}$ **11.** $\sqrt{1 + \sqrt{3}}$ **12.** $\sqrt{1 - \sqrt{2}}$ **13.** $\sqrt[4]{3}$

14. Find the points of intersection of the circles

$$(x - a_1)^2 + (y - b_1)^2 = r_1{}^2 \qquad (x - a_2)^2 + (y - b_2)^2 = r_2{}^2$$

15. Show geometrically (that is, by a straightedge and compass construction) that for any angle ϕ, the real number $\cos\phi$ is constructible if and only if the real number $\sin\phi$ is constructible.

16. Show algebraically (that is, using an algebraic expression) that for any angle ϕ, the real number $\cos\phi$ is constructible if and only if the real number $\sin\phi$ is constructible.

17. Let z be a complex number

$$z = a + bi = r(\cos\phi + i\sin\phi)$$

where a, b, r, ϕ are real numbers. Show that a and b are both constructible if and only if r and $\cos\phi$ and $\sin\phi$ are all constructible.

18. Find positive integers a and b such that

$$\sqrt{13 + 4\sqrt{3}} = a + b\sqrt{3}$$

19. Show that $[\mathbb{Q}(\alpha) : \mathbb{Q}] = 2$, where

$$\alpha = \sqrt{4 + 2\sqrt{3}}$$

20. Show that all the zeros of $f(x) = x^4 - x^2 + 1$ are constructible.

21. Show that $f(x) = x^4 + 2x^2 - 2$ has a real zero that is not constructible. (*Hint:* Imitate Example 11.1.23.)

22. Given three points A, B, C, construct a circle with center C and radius AB with a collapsing compass. (*Hint:* Start with two circles of radius CA. From the two points of intersection R and S, draw circles with radius RB and RS, respectively.)

11.2 Classical Problems

In the previous section we used the theory of finite field extensions to show that a constructible real number is algebraic over $\mathbb{Q}$ with degree a power of 2. In this section we show how our characterization of constructible real numbers can be used to answer some ancient and famous construction problems.

11.2.1 DEFINITION The problem of **squaring the circle** is, given a circle in the plane, to construct a square of the same area. ○

11.2.2 THEOREM The problem of squaring the circle cannot be solved using only a straightedge and compass.

Proof If a given circle has radius r, then its area is πr^2 and the side a of a square having the same area would have to satisfy $a^2 = \pi r^2$ so that $\pi = a^2/r^2$. This means that, taking the radius of the circle as our unit, we would have to construct a segment whose length is π. But such a segment is impossible by Corollary 11.1.20, since π is transcendental over $\mathbb{Q}$. □

11.2.3 DEFINITION The problem of **duplicating the cube** is, given the edge of a cube, to construct the edge of a cube of twice the volume.

11.2.4 THEOREM The problem of duplicating the cube cannot be solved using only straightedge and compass.

Proof If the given cube has volume $V = a^3$, where a is the length of the edge, then the cube to be constructed would have volume $W = 2a^3$, with an edge length

$$b = \sqrt[3]{2}\, a = 2^{1/3}a$$

This means that taking the edge of the original cube as our unit, we would have to construct a length of $2^{1/3}$, which is impossible using only a straightedge and compass by Corollary 11.1.19, since $2^{1/3}$ is a zero of $x^3 - 2$, a polynomial that is irreducible over $\mathbb{Q}$ by Eisenstein's criterion, and therefore $[\mathbb{Q}(2^{1/3}):\mathbb{Q}] = 3$, not a power of 2. □

Before taking up the third classical problem, we need the following result relating the problem of constructing an angle to the problem of constructing a segment.

11.2.5 PROPOSITION An angle of magnitude ϕ can be constructed using only a straightedge and compass if and only if $\sin\phi$ and $\cos\phi$ are constructible real numbers.

Proof ($\Rightarrow$) See Figure 13. Suppose that we can construct an angle $\angle ABC$ of magnitude ϕ. Draw the circle with center B and radius 1 and let it intersect AB at A'. Draw a line BB' perpendicular to BC and drop perpendiculars $A'C'$ and $A'D$ from A' to BC and BB', respectively. (See Example 11.1.4.) Then segments $A'C'$ and BC' have lengths $\sin\phi$ and $\cos\phi$, respectively. Hence these are constructible real numbers.

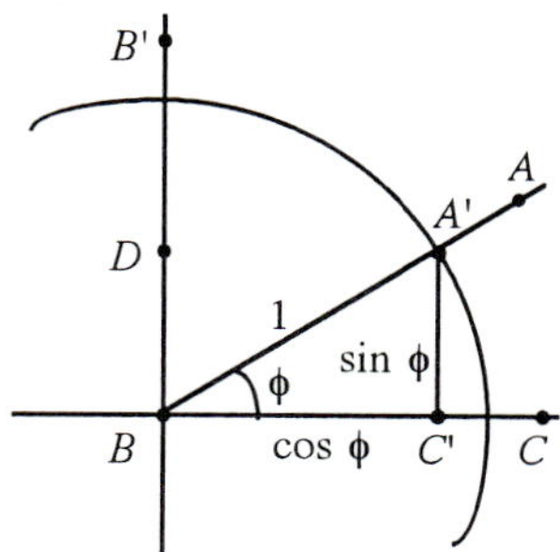

FIGURE 13

($\Leftarrow$) If $\sin\phi$ and $\cos\phi$ are constructible real numbers, then constructing a segment BC' of length $\cos\phi$ and a perpendicular segment $C'A'$ of length $\sin\phi$, $\angle A'BC'$ has magnitude ϕ. □

11.2.6 DEFINITION The general problem of the **trisection of an angle** is, given an angle of magnitude ϕ, to construct an angle of magnitude $\phi/3$. ○

11.2.7 THEOREM The general problem of the trisection of the angle cannot be solved using only a straightedge and compass. In particular, an angle of magnitude 20° cannot be constructed using only a straightedge and compass.

Proof We have seen in Example 11.1.3 how to construct with a straightedge and compass an angle of magnitude 60°. If the general problem of the trisection of the angle could be solved using only as straightedge and compass, then we would also be able to construct an angle of magnitude 20°, in which case by the preceding proposition sin 20° and cos 20° would be constructible real numbers. But we show that cos 20° is *not* a constructible real number. To do this we use the following consequence of the sum angle formulas from trigonometry:

$$\cos 3\theta = 4\cos^3\theta - 3\cos\theta$$

(See Exercise 1.) Since cos 60° = $^1/_2$, the number $\alpha = \cos 20°$ satisfies $4\alpha^3 - 3\alpha = {}^1/_2$ and hence is a zero of the polynomial $p(x) = 8x^3 - 6x - 1$. This polynomial is irreducible over $\mathbb{Q}$ by Theorem 8.4.11, since none of the the numbers $\pm^1/_8, \pm^1/_4, \pm^1/_2, \pm 1$ is a zero of $p(x)$. Hence $[\mathbb{Q}(\alpha) : \mathbb{Q}] = 3$, not a power of 2, and $\alpha = \cos 20°$ is not constructible by Corollary 11.1.19, hence an angle of magnitude 20° is not constructible by Proposition 11.2.5. □

Construction of Regular Polygons

The problem of constructing regular polygons is another classical problem and one closely related to the problem of constructing angles of a specified magnitude. The specific relationship is stated in the next proposition.

11.2.8 PROPOSITION A regular n-gon for $n \geq 3$ is constructible using only straightedge and compass if and only if $\cos(360°/n)$ is a constructible real number.

Proof ($\Rightarrow$) If a regular n-gon is constructible, then so of course are its interior and exterior angles.
But the magnitude of the interior angle is $180° - (360°/n)$ and of the exterior angle is $(360°/n)$. (See Exercise 2.)
If these angles are constructible, then so is

$$\cos(180° - (360°/n) = -\cos(360°/n)$$

by Proposition 11.2.5.

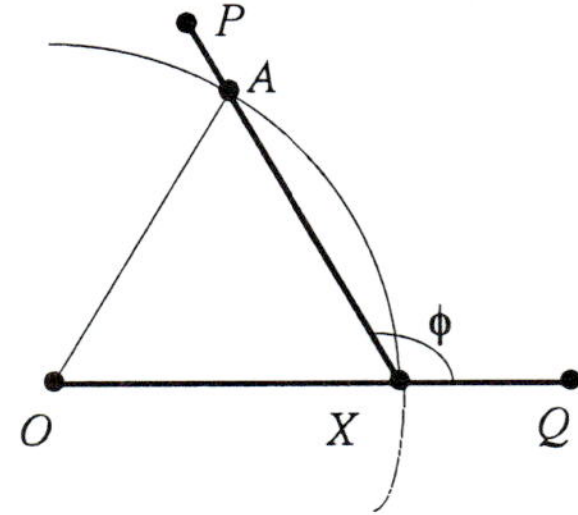

FIGURE 14

($\Leftarrow$) Starting with the unit segment OX, draw a circle with center O and radius 1. If an angle of magnitude $\phi = 360°/n$ is constructible, then such an angle $\angle PXQ$ can be construct having X as vertex and the extension of OX as side. (See Example 11.1.7.) Let XP intersect the unit circle at A. Then XA is one side of the required n-gon. Repeat the process with OA in place of OX to construct the next side, and continue until the entire n-gon is constructed. (Figure 14 illustrates the case of the regular hexagon.) □

11.2.9 EXAMPLES

(1) Since $\cos 60° = 1/2$ is a constructible real number, 60° is a constructible angle, and in Example 11.1.3 we constructed such an angle and a regular 3-gon, which is to say, an equilateral triangle.

(2) Also, $\sin 90° = 1$ is, of course, a constructible real number, and 90° is a constructible angle, and in Example 11.1.4 we constructed such an angle and provided all that is needed to construct a regular 4-gon, which is to say a square. ◇

11.2.10 THEOREM The real number $\cos {}^{2\pi}/_5$ is constructible, and hence the regular pentagon is constructible with straightedge and compass.

Proof The complex number

$$z = \cos {}^{2\pi}/_5 + i \sin {}^{2\pi}/_5$$

is a fifth root of unity, hence a zero of

$$x^5 - 1 = (x - 1)(x^4 + x^3 + x^2 + x + 1)$$

and since $z \neq 1$, z is a zero of the second factor. Since

$$z^4 = z^{-1} = \cos {}^{2\pi}/_5 - i \sin {}^{2\pi}/_5$$

we obtain

$$z + z^4 = 2 \cos {}^{2\pi}/_5$$

Let $\alpha = 2 \cos {}^{2\pi}/_5 > 0$. Then

$$\alpha^2 = (z + z^4)^2 = z^2 + 2 + z^3$$
$$\alpha^2 + \alpha = z^2 + 2 + z^3 + z + z^4 = 1$$

Thus α is a zero of $y^2 + y - 1$, which we can solve by the quadratic formula, obtaining one positive and one negative solution:

$$y = {}^1/_2(-1 \pm \sqrt{5})$$

Since $\alpha > 0$, we conclude that

$$\cos {}^{2\pi}/_5 = {}^1/_4(-1 + \sqrt{5})$$

which is a constructible real number. That the regular 5-gon is constructible then follows by Proposition 11.2.8. □

More generally, it was known already to Euclid that regular n-gons with

$$n = 2^k \text{ or } 2^k{\cdot}3 \text{ or } 2^k{\cdot}5 \text{ or } 2^k{\cdot}3{\cdot}5$$

are constructible using only straightedge and compass.

These were the only constructible regular polygons known for over two thousand years, until it was proved by C. F. Gauss in 1796 that the regular 17-gon is constructible. A partial answer to the question exactly which regular n-gons are constructible is given in the last two exercises at the end of this section. A complete answer is provided by Galois theory, to be discussed in Chapter 12.

Exercises 11.2

1. Use the necessary trigonometric identities to show that for any angle θ we have

$$\cos 3\theta = 4\cos^3 \theta - 3 \cos \theta$$

2. Show that for any $n \geq 3$, the interior angle θ of a regular n-gon has magnitude

$$\theta = 180° - (360°/n)$$

3. Show algebraically that an angle of magnitude ϕ is constructible if and only if an angle of magnitude $\phi/2$ is constructible.

In Exercises 4 through 10 determine in which cases an angle of the indicated magnitude is constructible.

4. 40° **5.** 30° **6.** 45° **7.** 72° **8.** 42° **9.** 10° **10.** 18°

In Exercises 11 through 15 determine for which of the indicated n a regular n-gon is constructible with straightedge and compass.

11. $n = 8$ **12.** $n = 9$ **13.** $n = 10$ **14.** $n = 20$ **15.** $n = 30$

16. With a straightedge and compass construct a regular 5-gon with sides of length 1.

17. Show that in a regular 5-gon with sides of length 1, any diagonal has length

$$\alpha = {}^1/_2(1 + \sqrt{5})$$

18. Show that a regular heptagon is not constructible using only a straightedge and compass. (*Hint:* Show that $2 \cos {}^{2\pi}/_7$ is a zero of $x^3 + x^2 - 2x - 1$.)

19. Prime numbers of the form $2^{2^n} + 1$ are called **Fermat primes**. Show that if a regular p-gon is constructible, where p is a prime, then p must be a Fermat prime. (*Hint:* Use the pth root of unity that is a zero of the cyclotomic polynomial Φ_p.)

20. (a) Show that if a regular n-gon is constructible and $q > 2$ divides n, then a regular q-gon is also constructible.

(b) Find the minimal polynomial over $\mathbb{Q}$ of

$$\cos {}^{2\pi}/_{p^2} + i \sin {}^{2\pi}/_{p^2}$$

for any prime p.

(c) Show that if a regular n-gon is constructible, then

$$n = 2^k p_1 \ldots p_s$$

where $p_1, \ldots, p_s$ are distinct Fermat primes.

11.3 Constructions with Marked Ruler and Compass

In all the geometric constructions considered so far, we have allowed ourselves to use only the straightedge and the compass, which enable us to draw lines and circles. It is not really surprising that the degrees of the real numbers constructible in this way turn out to be powers of 2. Now we see that if we allow ourselves to use just one more tool, we will be able to duplicate the cube and trisect angles, and more generally to solve cubic and quartic equations. In much of this section we follow the elegant exposition in R. Hartshorne, *Geometry: Euclid and Beyond* (New York: Springer-Verlag, 2000).

11.3.1 DEFINITION A construction with a **marked ruler and compass** allows, in addition to the operations (1) through (5) allowed in straightedge and compass constructions (as in Definition 11.1.8), the following operation:

(6) Given points A and B, lines L and M, and a point P, to construct points C on L and D on M such that P lies on the line through C and D, and the length of segment CD equals that of segment AB.

A real number α is **constructible using a marked ruler and compass** if a segment of length $|\alpha|$ is so constructible. ○

The idea is that we lay the straightedge along AB, mark on it the points corresponding to A and B, move it to touch P, and, while keeping it in contact with P, slide it until the first marked point touches L, and continue to slide the first marked point along L until the second marked point touches M.

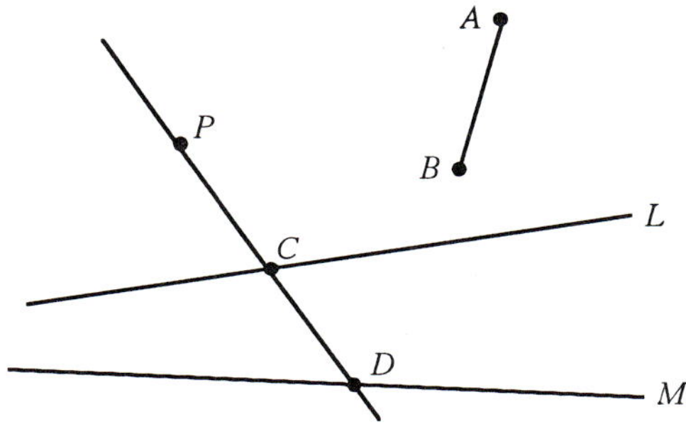

FIGURE 15

In Proposition 11.1.4 we saw that with the ordinary straightedge and compass we can construct square roots. We soon see that with the addition of the marked ruler we can construct cube roots as well. But first we need a couple of theorems from Euclidean geometry.

11.3.2 LEMMA Let F be a point outside a circle, and let the line FBA cut the circle at A and B. Let the line FGH cut the circle at G and H. Then $AF \cdot BF = HF \cdot GF$.

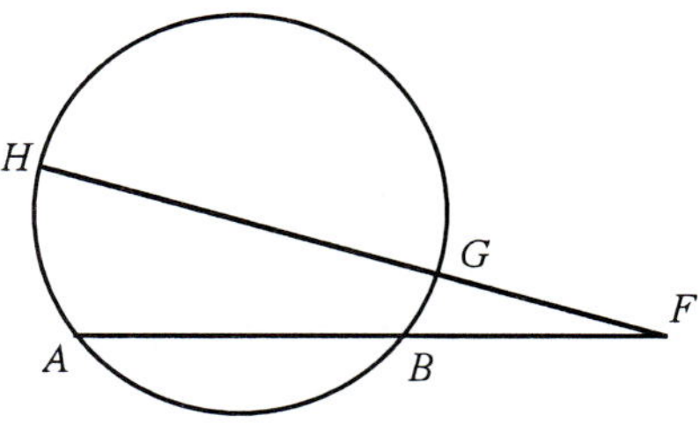

FIGURE 16

Proof See Euclid, *Elements of Geometry* (New York: Dover, 1956), Book III, Proposition 32. □

11.3.3 LEMMA (Menelaus' theorem) Let ACF be any triangle and let a line L cut the sides or their extensions at points D, B, E. Then

$$AD/AB \cdot BF/EF \cdot EC/DC = 1$$

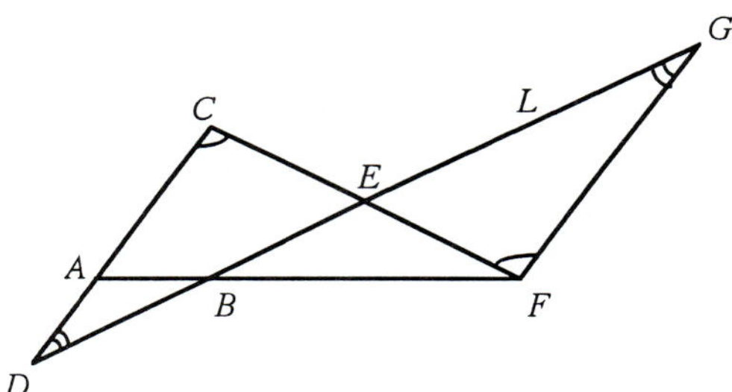

FIGURE 17

Draw a line through F parallel to AC and let G be its intersection with line L, as in Figure 17. We obtain two pairs of similar triangles, EFG and ECD, and hence $FG/EF = DC/EC$ and BFG is similar to BAD, and hence $FG/BF = AD/AB$. Therefore, $FG = (DC/EC) \cdot EF = (AD/AB) \cdot BF$ and the desired relation follows. □

We now have everything we need to prove that the marked ruler and compass enable us to extract cube roots.

11.3.4 PROPOSITION Let α be a real number that is constructible with a marked ruler and compass. Then the real number

$$\sqrt[3]{\alpha} = \alpha^{1/3}$$

is constructible with the marked ruler and compass.

Proof It is enough to consider the case where α is positive. Let $\beta > \alpha$ be a natural number of form 2^{3k-1}. All natural numbers are constructible, and it is enough to construct a segment of length $2^{2k}\alpha^{1/3}$ since then by repeated bisection we can construct a segment of length $\alpha^{1/3}$.

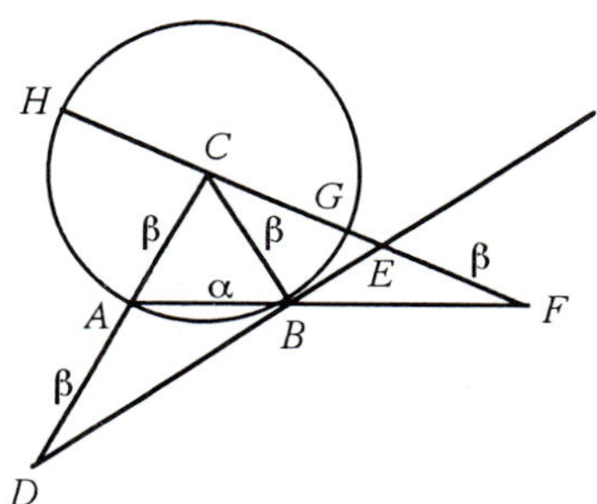

FIGURE 18

We construct first an isosceles triangle ABC with $AC = BC = \beta$, as in Figure 18. Then we extend the side AC to D so that $AD = \beta$ and also extend the side AB. We now use the marked ruler to draw the line CEF through C intersecting DB at E and AB at F, where $EF = \beta$. Let $y = BF$. We claim that $y = 2^{2k}\alpha^{1/3}$.

To prove this claim, we first apply Lemma 11.3.3 to the triangle ACF. This gives the relationship

$$AD/_{AB} \cdot BF/_{EF} \cdot EC/_{DC} = 1$$

If we let $x = EC$, this can be written as $(\beta/\alpha)(y/\beta)(x/2\beta) = 1$ or

(1) $$xy = 2\alpha\beta$$

Now we draw the circle with center C and radius β. Note that since $x = EC$ and $CG = \beta$, and $EF = \beta$, we have $GF = EC = x$. We then apply Lemma 11.3.2 to this circle. This gives us the relationship

(2) $$y(y + \alpha) = x(x + 2\beta)$$

Using (1), we can eliminate x from (2) to obtain

$$y(y + \alpha) = (2\alpha\beta/_y)^2 + (4\alpha\beta^2/_y) =$$
$$(4\alpha\beta^2/_y)(\alpha/_y + 1) = (4\alpha\beta^2/_y)((\alpha + y)/_y)$$

It follows that

$$y^2(y + \alpha) = (4\alpha\beta^2/_y)(y + \alpha)$$

and

$$y^3 = 4\alpha\beta^2 = 4\alpha \cdot 2^{6k-2} = 2^{6k}\alpha$$

recalling that β was chosen equal to 2^{3k-1}. So $y = 2^{2k}\alpha^{1/3}$ and $\alpha^{1/3}$ is constructible with the marked ruler and compass. □

11.3.5 COROLLARY The problem of the duplication of the cube can be solved using the marked ruler and compass.

Proof We have seen that the problem of the duplication of the cube amounts to the problem of constructing $2^{1/3}$, which can be done with the marked ruler and compass by the preceding proposition. □

This result has probably convinced the reader that the marked ruler is an important additional tool. The next result is therefore not too surprising.

11.3.6 PROPOSITION The general problem of the trisection of the angle can be solved using the marked ruler and compass.

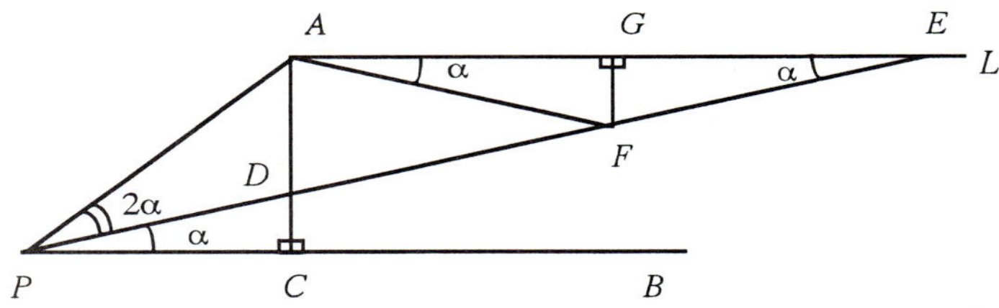

FIGURE 19

Proof Let an angle $\angle APB$ be given. From A drop the perpendicular to the side PB, and let C be the point of intersection, as in Figure 19. Draw a line L through A parallel to PB. At this point we use the marked ruler to draw the line PDE through P intersecting AC at D and L at E, such that $DE = 2PA$. Let G be the midpoint of AE, draw the perpendicular to AE at G, and let the point of intersection with DE be F. Then F is the midpoint of DE, so $DF = FE = PA$. By the side-angle-side criterion, triangles AGF and EGF are congruent, and hence $AF = FE = PA$, and $\angle GAF = \angle GEF$, while also $\angle GEF = \angle BPD$ since PB and AE are parallel. Call the magnitude of this angle α. Then the exterior angle

$$\angle AFP = \angle GAF + \angle GEF = 2\alpha$$

Triangle APF is an isosceles triangle with $AF = PA$ and hence $\angle APF = \angle AFP = 2\alpha$, and PF is the trisector of the original angle $\angle APB$. □

Being able to extract square and cube roots and to trisect angles turns out to be equivalent to being able to solve cubics and quartics, as we will see in the next section, where we also give a complete analysis of the real numbers constructible with marked ruler and compass, analogous to Theorem 11.1.18.

Exercises 11.3

1. Show that the regular heptagon is constructible using the marked ruler and compass, in the following steps:

(a) Let $\zeta = \cos(^{2\pi}/_7) + i\ \sin(^{2\pi}/_7)$ and $\alpha = \zeta + \zeta^{-1} = 2\cos(^{2\pi}/_7)$. Show that α is a zero of $f(x) = x^3 + x^2 - 2x - 1$.

(b) By an appropriate substitution, reduce $f(x)$ to the form $g(y) = y^3 - (3\sqrt{7}/7)$.

(c) Show that for any angle θ, $2\cos{}^{\theta}/_3$ is a zero of $y^3 - 3y - 2\cos\theta$.

(d) Show that if $2\cos\theta = (\sqrt{7}/7)$, then $\alpha = (2\sqrt{7}/3)\cdot\cos{}^{\theta}/_3 - {}^1/_3$.

(e) Show that $\cos{}^{2\pi}/_7$ is constructible using the marked ruler and compass, and hence the regular heptagon is so constructible.

2. In the real Cartesian plane, find the coordinates of the points of intersection of the parabola $y = x^2$ with the circle with center $(^a/_2, {}^1/_2)$ passing through $(0, 0)$, where $a \in \mathbb{R}$.

3. In the real Cartesian plane, suppose we are given the fixed parabola $y = x^2$ and, in addition to the other operations in straightedge and compass constructions, we are allowed to mark the intersection of lines and circles that have been drawn with this parabola. Show that then the problem of the duplication of the cube can be solved, and more generally the cube root of any real number is constructible using the parabola.

4. In the Cartesian plane, let $O = (0, 0)$, $X = (1, 0)$, $OA = 1$ and suppose that $\angle XOA = \theta$. Draw through A the line perpendicular to $y = 2$ and call the point of intersection C. (See Figure 20.) Draw the circle with center C and radius OC, and let D be its intersection with the parabola $y = x^2$. Express in terms of the angle θ

(a) The coordinates of A and C

(b) The coordinates of D

(*Hint:* Use Exercise 1 in Section 11.2.)

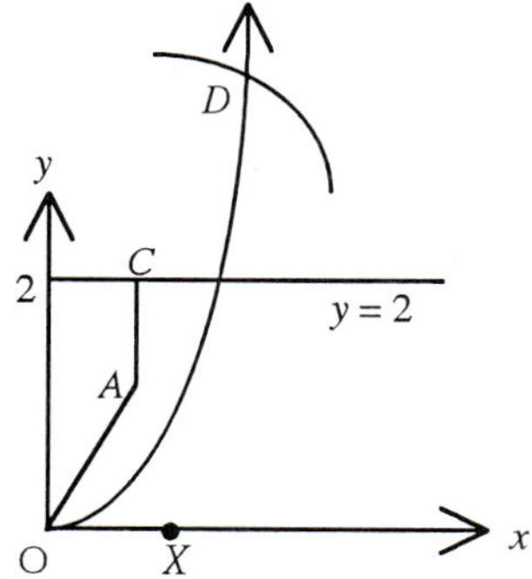

FIGURE 20

5. In the real Cartesian plane, suppose we are given the fixed parabola $y = x^2$ and, in addition to the other operations in straightedge and compass constructions, we are allowed to mark the intersection of lines and circles that have been drawn with this parabola. Show that in this case the general problem of the trisection of the angle can be solved.

6. Show how to construct $\sqrt[3]{2}$ as the intersection of a parabola with the hyperbola $xy = 1$.

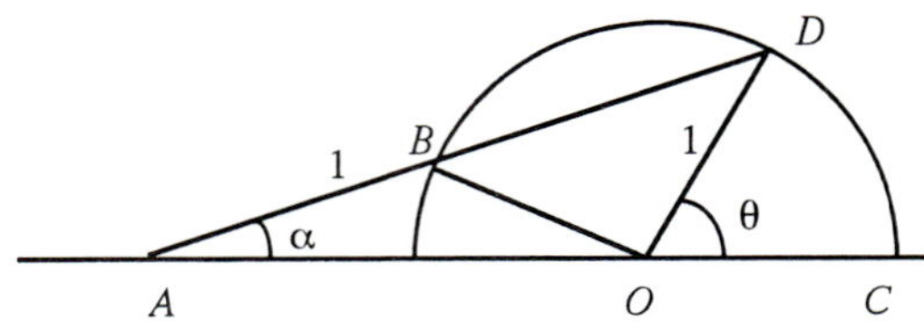

FIGURE 21

7. (Archimedes' trisection) Draw a circle of radius 1 with a central angle $\angle COD = \theta$, as in Figure 21. Extend the line OC and then with a distance of 1 marked on the ruler, keeping it in contact with D, slide it until it reaches a position where, if B and A are its intersection with the circle and the extension of OC, respectively, then $AB = 1$. Let $\angle OAB = \alpha$. Show that $\alpha = \theta/3$.

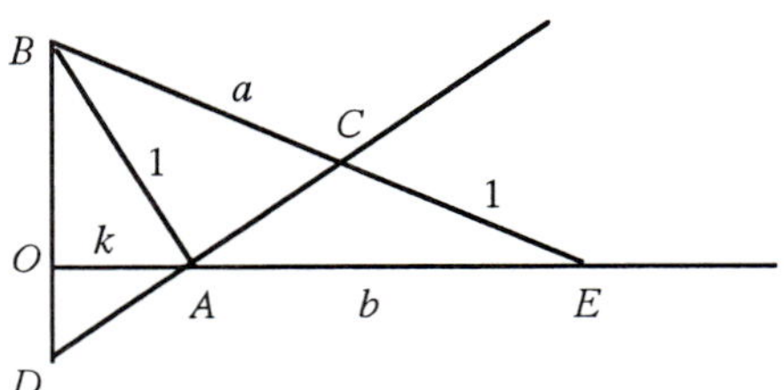

FIGURE 22

8. (J. H. Conway's cube root construction) In the Cartesian plane for some $0 < k < 1$ let

$$O = (0, 0), A = (k, 0), D = (0, -{}^1/_3\sqrt{1 - k^2}\,)$$

Let B be the point on the y-axis such that $AB = 1$, as in Figure 22. Use the marked ruler to draw a line through B such that if C and E are its intersections with AD and the x-axis, respectively, then $CE = 1$.

Let $BC = a$ and $AE = b$ and let (x, y) be the coordinates of C. Show that

(a) $y = \dfrac{\sqrt{1 - k^2}}{1 + a}$

(b) $x = a \cdot \dfrac{b + k}{1 + a}$

(c) $\dfrac{y}{x - k} = \dfrac{\sqrt{1 - k^2}}{3k}$

(d) $(1 - k^2) + (b + k)^2 = (1 + a)^2$

(e) Solve for a and b.

11.4 Cubics and Quartics Revisited

In Theorem 11.1.18 we saw that the real numbers constructible using only the straightedge and compass are just those obtainable from the rationals by repeatedly adjoining solutions to quadratic equations or, equivalently, repeatedly adjoining square roots. In this section we show that the real numbers constructible using the marked ruler and compass are just those obtainable from the rationals by repeatedly adjoining solutions of quadratic, cubic, and quartic equations. We rely on, and also refine, our analysis of cubics in Section 8.5.

We first study the real zeros of three special cubics. We saw in the proof of Theorem 11.2.7 that, given a subfield F of $\mathbb{R}$, adjoining $\cos {}^{\theta}/_{3}$ for some $\cos\theta \in F$ amounts to adjoining a zero of a certain cubic with coefficients in F. Conversely, a certain class of cubic equations over F can be solved by adjoining $\cos {}^{\theta}/_{3}$ for some $\cos\theta \in F$, according to the following proposition.

11.4.1 PROPOSITION Let F be a subfield of $\mathbb{R}$, and let

$$f(x) = x^3 - 3x - b \text{ with } |b| < 2$$

be a cubic in $F[x]$. Then a zero of $f(x)$ can be obtained by trisecting an angle, that is, by adjoining $\cos {}^{\theta}/_{3}$ for some $\cos\theta \in F$.

Proof Let θ be such that $\cos\theta = {}^{b}/_{2}$, as is possible since $|b| < 2$. Then $\alpha = 2\cos {}^{\theta}/_{3}$ is a zero of $f(x)$. This is immediate from the trigonometric identity:

$$\cos(3\phi) = 4\cos^3\phi - 3\cos\phi$$

letting $\theta = 3\phi$. □

In Propostion 8.5.6 we gave the general formula for a real zero of a cubic

$$f(x) = x^3 + px + q$$

in $\mathbb{R}[x]$. In general, that formula involves square roots of numbers that may be negative, and so involves complex numbers, even for the real root of the cubic. But for a certain class of cubic equations, only square roots of positive real numbers and real cube roots of real numbers are required, according to the next two propositions.

11.4.2 PROPOSITION Let F be a subfield of $\mathbb{R}$,

$$f(x) = x^3 - 3x - b \text{ with } |b| \geq 2$$

be a cubic in $F[x]$. Then a real zero of $f(x)$ can be obtained by adjoining a square root

$$\sqrt{d} = d^{1/2}$$

of a positive element d of F and then adjoining a real cube root of an element of $F(d^{1/2})$.

Proof For any number μ, the number $\mu + {}^1/_\mu$ is a zero of

$$x^3 - 3x - (\mu^3 + {}^1/_{\mu^3})$$

Setting

$$\mu^3 + \mu^{-3} = b$$

gives a quadratic equation in μ^3 whose solution is

$$\mu^3 = \frac{b \pm \sqrt{b^2 - 4}}{2}$$

It follows that if we let μ be a real solution of this last equation, then

$$\mu + {}^1/_\mu = \sqrt[3]{\frac{b + \sqrt{b^2 - 4}}{2}} - \sqrt[3]{\frac{b - \sqrt{b^2 - 4}}{2}}$$

is one of the real zeros of the original cubic. Notice that since $|b| \geq 2$, the square root we took was that of a positive real number, and then the cube root we took was the real cube root of a real number. □

11.4.3 PROPOSITION Let F be a subfield of $\mathbb{R}$, and let

$$f(x) = x^3 + 3x - b$$

be a cubic in $F[x]$. Then a real zero of $f(x)$ can be obtained by adjoining a square root

$$\sqrt{d} = d^{1/2}$$

of a positive element d of F and then adjoining a real cube root of an element of $F(d^{1/2})$.

Proof For any number μ, the number $\mu - {}^1/_\mu$ is a zero of

$$x^3 + 3x - (\mu^3 - {}^1/_{\mu^3})$$

Setting

$$\mu^3 - \mu^{-3} = b$$

gives a quadratic equation in μ^3 whose solution is

$$\mu^3 = \frac{b \pm \sqrt{b^2 + 4}}{2}$$

It follows that if we let μ be a real solution of this last equation then

$$\mu - {}^1/_\mu = \sqrt[3]{\frac{b + \sqrt{b^2 + 4}}{2}} - \sqrt[3]{\frac{b - \sqrt{b^2 + 4}}{2}}$$

is one of the real zeros of the original cubic. □

The next propositions tell us that the three special cases we have studied are not so special after all, since the general cubic and quartic can be reduced to them.

11.4.4 Proposition Let F be a subfield of $\mathbb{R}$, and

$$f(x) = x^3 + a_2x^2 + a_1x + a_0$$

a cubic in $F[x]$. Then a real zero of $f(x)$ can be obtained by taking square roots of positive real numbers and either trisecting an angle or taking a real cube root of a real number. In fact, all real zeros of $f(x)$ can be so obtained.

Proof In each of the three special cases we have considered, we can find one real zero of $f(x)$ by taking square roots of positive real numbers and either trisecting an angle or taking a real cube root of a real number. We show that any cubic reduces to one of these three cases. First we eliminate the x^2 by making the substitution

$$z = x + (a_2/3)$$

which reduces the equation $f(x) = 0$ to the form

$$g(z) = z^3 + b_1z + b_0 = 0$$

where the b_i are rational combinations of the a_i and therefore belong to F. If $b_1 = 0$ we are done, since we need only take

$$z = -\sqrt[3]{b_0}$$

Otherwise, we reduce the coefficient of the z term to ±3.
If $b_1 > 0$, we make the substitution

$$z = {}^1/_3\sqrt{3b_1}\, y$$

which reduces the equation to the form

$$h(y) = y^3 + 3y - b = 0$$

where b belongs to $F(\sqrt{3b_1}\,)$. Note that the square root involved is that of a positive real number. In this case, by Proposition 11.4.3 a solution to $h(y) = 0$ can be found by taking a square root of a positive real number and a real cube root of a real number.
If $b_1 < 0$, we make the substitution

$$z = {}^1/_3\sqrt{-3b_1}\, y$$

which reduces the equation to the form

$$h(y) = y^3 - 3y - b = 0$$

where b belongs to $F(\sqrt{-3b_1}\,)$. Again the square root involved is that of a positive real number. If now $|b| \geq 2$, then by Proposition 11.4.2, a solution to $h(y) = 0$ can be found by taking a square root of a positive real number and a real cube root of a real number. If instead $|b| < 2$, then by Proposition 11.4.1, a solution can be found by trisecting an angle.
Once we have one real zero r of $f(x)$, we have the factorization $f(x) = (x - r)f_1(x)$, where $f_1(x)$ is quadratic and the remaining zeros of $f(x)$ in $\mathbb{C}$ are just the zeros of $f_1(x)$. By Theorem 8.3.12, if one of them is real they both are and they can be obtained by taking a square root of a positive real number. □

11.4.5 COROLLARY Let F be a subfield of $\mathbb{R}$ and

$$f(x) = x^4 + a_3x^3 + a_2x^2 + a_1x + a_0$$

a quartic in $F[x]$. If $\alpha \in \mathbb{R}$ is a zero of $f(x)$, then α can be obtained by taking square roots of positive real numbers and either trisecting an angle or taking a real cube root of a real number. In fact, all real zeros of $f(x)$ can be so obtained.

Proof Let $x = y - a^3/4$ to obtain

$$f(x) = h(y) = y^4 + py^2 + qy + r \in F[x]$$

If $f(x)$ has a real zero, then so does $h(y)$. Let $\alpha \in \mathbb{R}$ be a real zero of $h(y)$. By Theorem 8.3.12, $h(y)$ has another real zero β, and α and β are zeros of a quadratic polynomial $y^2 + ay + b$. Hence

$$h(y) = (y^2 + ay + b)(y^2 - ay + c)$$

where $a, b, c \in \mathbb{R}$. By comparing the coefficients of $h(y)$, we obtain

$$p = b + c - a^2 \qquad q = a(c - b) \qquad r = bc$$

from which we can derive

(1) $$b = {}^1/_2\,(p + a^2 - q/a) \text{ and}$$

$$a^6 + 2pa^4 + (p^2 - 4r)a^2 = q^2 = 0$$

Therefore, a^2 is a positive real zero of a cubic polynomial with coefficients in F. Furthermore, (1) implies that $b \in F(a)$. Since $\alpha \in \mathbb{R}$ is a zero of $y^2 + ay + b$, it is obtained by adjoining to $F(a)$ the square root of a positive real number in $F(a)$. The corollary now follows from Proposition 11.4.4. □

We are now ready for a characterization of the real numbers constructible with a marked ruler and compass analogous to the characterization in Theorem 11.1.18 of the real numbers constructible with a straightedge and compass.

11.4.6 DEFINITION Let F be a subfield of $\mathbb{R}$. A **square root/cube root/trisection tower** over F is a sequence of fields $K_0, K_1, \ldots, K_n$ with

(1) $F = K_0 \subseteq K_1 \subseteq \ldots \subseteq K_n \subseteq \mathbb{R}$

(2) For all i with $1 \le i \le n$, $K_i = K_{i-1}(\sigma_i)$, where either

$\sigma_i = \sqrt{\rho}$ for some $\rho \in K_{i-1}$ with $\rho > 0$ and $\sqrt{\rho} \notin K_{i-1}$, or

$\sigma_i = \sqrt[3]{\rho}$ for some $\rho \in K_{i-1}$ with $\sqrt[3]{\rho} \notin K_{i-1}$, or

$\sigma_i = \cos{}^{\phi}/_3$ for some ϕ with $\cos\phi \in K_{i-1}$ and $\cos{}^{\phi}/_3 \notin K_{i-1}$

A **quadratic/cubic/quartic tower** is a similar sequence of fields where for each i with $1 \le i \le n$, $K_i = K_{i-1}(\sigma_i)$, where σ_i is a real zero of a quadratic, cubic, or quartic polynomial with coefficients from K_{i-1}. In either case, field K_n is called the **top** of the tower. ○

11.4.7 LEMMA Consider any marked ruler and compass construction starting from our unit segment, there is a square root/cube root/trisection tower over $\mathbb{Q}$. Then the top of the tower contains

(1) the coordinates α, β of any point (α, β)
(2) the coefficients γ, δ of the equation of any line $y = \gamma x + \delta$
(3) the coefficients κ, λ, μ of the equation of any circle

$$x^2 + \kappa x + y^2 + \lambda y + \mu = 0$$

that is drawn in the course of the construction.

Proof Just as in the proof of Lemma 11.1.17, we proceed by induction on the number of steps in the construction. Suppose we have performed m operations and that all the coordinates and coefficients belong to the top K_n of some tower $K_0, K_1, \ldots, K_n$, and consider performing one more operation. By Definition 11.3.1, there are six cases to consider depending on which of the six kinds of operation (1) through (6) is involved, but the five cases (1) through (5) have already been handled in the proof of Lemma 11.1.17.

For case (6), suppose that, as in Definition 11.3.1, we have points A and B, lines L and M, and a point P, and we want to mark the line N such that N passes through P, and if C and D are the points of intersection of N with L and M respectively, then the distance between C and D equals that between A and B. And suppose the coordinates of A, B, P, and the coefficients in the equations of L and M all belong to the top K_n of our tower. We want to show that the tower can be extended by taking further square roots of positive real numbers and either taking a further real cube root of a real numbers or trisecting a further angle, so that the coefficients in the equation of N will belong to the top of the extended tower.

We begin with three preliminary simplifications of the problem. First, what matters about A and B is not their coordinates (α_1, α_2) and (β_1, β_2) but just the distance ρ between them, for which we have

$$\rho^2 = (\beta_1 - \alpha_1)^2 + (\beta_2 - \alpha_2)^2 \in K_n$$

Second, if the coordinates of P are (π_1, π_2) and we make the substitution

$$x = x' + \pi_1 \qquad y = y' + \pi_2$$

then the coordinates of P become $(0, 0)$ in the new x', y'-coordinate system, and the coefficients of the equation of any line in the old coordinate system will be rational combinations of the coefficients of the same line in the new coordinate system and the numbers $\pi_1, \pi_2 \in K_n$, so if the coefficients in the old coordinate system belong to the top of a tower extending K_n, so do the coefficients in the new coordinate system. So we may assume that P is the origin $(0, 0)$ from the beginning. Since the line N passes through the origin P, its equation will have the form

$$y = \mu x$$

for some real number μ, so there is only one coefficient μ to be determined.

Third, by a similar substitution, we may assume that one of the lines, say L, is parallel to the y-axis and therefore has equation of the form $y = \eta$. Let the equation of the other line M be

$$y = \gamma x + \delta$$

So we have $\gamma, \delta, \eta, \rho^2 \in K_n$, and we want to show that if $y = \mu x$ is the line through the origin such that the distance between the points of intersection of this line with the lines $y = \eta$ and $y = \gamma x + \delta$ is ρ, then the coefficient μ can be obtained by starting

with K_n and taking further square roots of positive real numbers, and either taking a further real cube root of a real number or trisecting a further angle. By Corollary 11.4.5, to show this it will be enough to show that μ is a zero of some quartic whose coefficients are rational combinations of γ, δ, η, ρ^2. This we now do.
We begin by noting that the coordinates of the point C of intersection between the lines $y = \mu x$ and $y = \eta$ are

$$C = (\eta/\mu, \mu)$$

The coordinates of the point D of intersection between the lines $y = \mu x$ and $y = \gamma x + \delta$ are

$$D = (\delta/(\mu-\gamma), \mu\delta/(\mu-\gamma))$$

The condition that the distance between C and D is ρ therefore amounts to the condition

$$(\eta/\mu - \delta/(\mu-\gamma))^2 + (\mu - \mu\delta/(\mu-\gamma))^2 = \rho^2$$

Multiplying out, this gives us a quartic equation in μ with coefficients that are rational combinations of γ, δ, η, ρ^2. □

11.4.8 THEOREM Let α be a real number. Then the following are equivalent:

(1) α is constructible from our unit segment using the marked ruler and compass.

(2) α belongs to the top of some square root/cube root/trisection tower over $\mathbb{Q}$.

(3) α belongs to the top of some quadratic/cubic/quartic tower over $\mathbb{Q}$.

Proof (1) $\Rightarrow$ (2) Let α be constructible using the marked ruler and compass. Then the point $(\alpha, 0)$ can be constructed, and its coordinate α belongs to the top of some square root/cube root/trisection tower over $\mathbb{Q}$ by the preceding lemma.

(2) $\Rightarrow$ (1) Let

$$\mathbb{Q} = K_0 \subseteq K_1 \subseteq \ldots \subseteq K_n \subseteq \mathbb{R}$$

be a square root/cube root/trisection tower over $\mathbb{Q}$, and let $C' \subseteq \mathbb{R}$ be the field of real numbers constructible from our unit segment using the marked ruler and compass. Then $K_0 = \mathbb{Q} \subseteq C'$ and if $K_i \subseteq C'$, then $K_{i+1} = K_i(\sigma)$ where either

$\sigma = \sqrt{\rho}$ for some $\rho \in K_i \subseteq C'$ with $\rho > 0$, or

$\sigma = \sqrt[3]{\rho}$ for some $\rho \in K_i \subseteq C'$, or

$\sigma = \cos\phi/3$ for some ϕ with $\cos\phi \in K_i \subseteq C'$

In the first case $\sigma \in C'$ by the proof of Proposition 11.1.14. In the second case $\sigma \in C'$ by Proposition 11.3.4. In the third case, $\sigma \in C'$ by Proposition 11.3.6. So in every case $K_{i+1} \subseteq C'$, and by induction $K_i \subseteq C'$ for all i, so every element α of the top K_n of the tower is constructible using the marked ruler and compass.

(2) $\Rightarrow$ (3) This implication holds because adjoining a square root or a cube root is adjoining a zero of a quadratic $x^2 - \rho$ or cubic $x^3 - \rho$, and trisecting an angle is also adjoining a zero of a cubic by the proof of Theorem 11.2.7.

(3) $\Rightarrow$ (2) This implication holds because a real zero of a quadratic, cubic, or quartic can always be obtained by taking square roots, cube roots, and trisecting angles, according to Proposition 11.1.15 for the quadratic case, Proposition 11.4.4 for the cubic case, and Corollary 11.4.5 for the quartic case. □

The previous result was analogous to Theorem 11.1.18. The next is analogous to Corollary 11.1.19.

11.4.9 COROLLARY Let α be a real number. If α is constructible with the marked ruler and compass, then $[\mathbb{Q}(\alpha):\mathbb{Q}]$ is the product of a power of 2 and a power of 3.

Proof If α is constructible using the marked ruler and compass, then $\alpha \in K_n$ for some square root/cube root/trisection tower

$$\mathbb{Q} = K_0 \subseteq K_1 \subseteq \ldots \subseteq K_n \subseteq \mathbb{R}$$

In such a tower $[K_{i+1} : K_i] = 2$ if the extension involves adjoining a square root of a positive real number, and $[K_{i+1} : K_i] = 3$ if the extension involves adjoining a real cube root of a real number, or trisecting an angle. Supposing the first case arises p times and the second and third cases together q times, by Theorem 10.2.20 we have

$$[K_n{:}\mathbb{Q}(\alpha)]\cdot[\mathbb{Q}(\alpha){:}\mathbb{Q}] = [K_n{:}\mathbb{Q}] = [K_n{:}K_{n-1}]\cdot\ldots\cdot[K_1{:}\mathbb{Q}] = 2^p3^q$$

Hence $[\mathbb{Q}(\alpha){:}\mathbb{Q}] = 2^r\,3^s$ for some $r \leq p$ and $s \leq q$. □

11.4.10 COROLLARY The problem of squaring the circle cannot be solved using the marked ruler and compass.

Proof The proof is the same as for Theorem 11.2.2, using the fact that π is transcendental. □

Just as the condition $[\mathbb{Q}(\alpha) : \mathbb{Q}] = 2^r$ was necessary but not sufficient for α to be constructible with straightedge and compass, so also the condition $[\mathbb{Q}(\alpha) : \mathbb{Q}] = 2^r3^s$ is necessary but not sufficient for α to be constructible with the marked ruler and compass. A complete analysis with necessary and sufficient conditions will be supplied in Chapter 12, using Galois theory.

Exercises 11.4

In Exercises 1 through 5 determine which of the indicated real numbers are constructible using marked ruler and compass.

1. $\sqrt[6]{5}$ **2.** $\sqrt[4]{5}$ **3.** $\sqrt[5]{2}$

4. $\dfrac{1}{\sqrt{1+\sqrt[3]{2}}}$ **5.** $\sqrt[3]{1+3\sqrt{5}}$

In Exercises 6 through 9 determine which of the indicated cubics need square roots and angle trisections to be solved, and which need square roots and cube roots.

6. $x^3 + x - 1$ **7.** $x^3 - 2x + 1$ **8.** $x^3 - 3x + 2$ **9.** $x^3 - x + 1$

10. Show that the angle $\theta = 24°$ is constructible using the marked ruler and compass or, equivalently, that the regular 15-gon is so constructible.

11. Show that the regular 11-gon is not constructible using the marked ruler and compass.

12. Let $f(x) = x^3 + px + q$. Show that

(a) If $(^q/_2)^2 + (^p/_3)^3 \geq 0$, then a real zero of $f(x)$ can be found by taking real square roots and cube roots.

(b) If $(^q/_2)^2 + (^p/_3)^3 \geq 0$, then a real zero of $f(x)$ can be found by taking real square roots and trisecting an angle.

13. Show that the cubics that can be solved with square roots and angle trisections are those with two real zeros, and that the cubics for which one needs square roots and cube roots are those with one real and two complex zeros.

14. Call a complex number $a + bi$ constructible using the marked ruler and compass if its real and imaginary parts a and b are both constructible using the marked ruler and compass. Show that if the complex number $a + bi$ is constructible using the marked ruler and compass, then so are its cube roots.

15. Show that finding the points of intersection of a line or circle with the parabola $y = x^2$ amounts to solving a quadratic or a quartic equation whose coefficients are rational combinations of the coefficients of the equation of the line or circle in question.

16. Using the preceding exercise and the results of Exercises 3 and 5 of the preceding section, show that if we are given the fixed parabola $y = x^2$ and we are allowed, in addition to the other operations in straightedge and compass constructions, to mark the intersection of lines and circles that have been drawn with this parabola, then exactly the same real numbers are constructible as are constructible using the marked ruler and compass.

17. Show that solving $y^4 + py^2 + qy + r$ is equivalent to finding the intersection of a suitable parabola with the fixed parabola $y = x^2$.

Chapter 12

Galois Theory

The material we have studied so far, starting with Chapter 1, has fallen into two parts. In Part A we studied groups, and in Part B we concentrated on rings and fields. In this chapter we establish a connection between the theory of groups and the theory of fields, and we use our knowledge of groups to obtain information about field extensions. For a given polynomial, we have defined its splitting field, which contains all the zeros of the polynomial. In this chapter we construct a group of permutations of the zeros, and the structure of this group gives us information about the structure of the splitting field. We apply this correspondence between field extensions and groups to cubic and quartic polynomials, and to geometric construction problems of the kind we considered in the preceding chapter. We close this chapter with the proof of the *unsolvability by radicals* of the quintic: This means that we cannot hope to find a general formula, involving roots, for the zeros of a polynomial of degree 5, comparable to those we have found for polynomials of degrees ≤ 4.

12.1 Galois Groups

As mentioned in the preceding paragraph, we are going to be studying some groups of permutations of the zeros of polynomials. In this section we define the groups in question and establish some of their basic properties. We are especially concerned with irreducible polynomials and, more specifically, with irreducible polynomials all of whose roots are distinct — in other words, irreducible polynomials with no roots of multiplicity > 1.

12.1.1 EXAMPLE Consider the polynomial

$$f(x) = x^4 + x + 1 \in \mathbb{Z}_2[x]$$

Since its derivative is $f'(x) = 1$, by Theorem 10.4.6 it follows that $f(x)$ has no zeros of multiplicity $s > 1$. ◇

12.1.2 EXAMPLE Consider the polynomial

$$\Phi_5(x) = x^4 + x^3 + x^2 + x + 1 \in \mathbb{Q}[x]$$

which is irreducible over $\mathbb{Q}$. (See Corollary 8.4.22.) Its derivative is

$$g(x) = 4x^3 + 3x^2 + 2x + 1 \in \mathbb{Q}[x]$$

$\Phi_5(x)$ and $g(x)$ cannot have a zero in common by Theorem 10.2.9. Therefore, by Theorem 10.4.6 the zeros of $\Phi_5(x)$ are all of multiplicity $s = 1$. ◇

12.1.3 DEFINITION Let $f(x) \in F[x]$ be a polynomial over a field F. Then $f(x)$ is said to be a **separable polynomial** over F if all its zeros in its splitting field have multiplicity 1. In an extension E of F an algebraic element $\alpha \in E$ is said to be **separable over** F if its minimal polynomial is separable over F. An algebraic extension E of F is said to be a **separable extension** of F if every element of the E is separable over F. A field F is said to be **perfect** if every algebraic extension E of F is a separable extensions of F. ○

12.1.4 PROPOSITION Let $f(x) \in F[x]$ be an irreducible polynomial over a field F, and let $f'(x) \in F[x]$ be its derivative. Then $f(x)$ is separable over F if and only if its derivative is not the zero polynomial, that is, $f'(x) \neq 0 \in F[x]$.

Proof Let E be the splitting field of $f(x)$ over F. First note that by Theorem 10.4.6, $f(x)$ is separable, that is, every zero of $f(x)$ in E has multiplicity $s = 1$, if and only if no zero of $f(x)$ is also a zero of $f'(x)$.

($\Rightarrow$) If the derivative is the zero polynomial, $f'(x) = 0 \in F[x]$, then *every* zero of $f(x)$ is a zero of $f'(x)$, and $f(x)$ is not separable over F.

($\Leftarrow$) If $f'(x) \neq 0 \in F[x]$ and if $\alpha \in E$ were a zero of both $f(x)$ and $f'(x)$, then $x - \alpha$ would divide both $f(x)$ and $f'(x)$. But by the definition of the derivative (Definition 10.4.3) $\deg f'(x) < \deg f(x)$, and so $f(x)$ does not divide $f'(x)$; and since $f(x)$ is irreducible, it follows that $f(x)$ and $f'(x)$ are relatively prime. Hence by Theorem 8.2.8 there exist $u(x), v(x) \in F[x]$ such that

$$u(x)f(x) + v(x)f'(x) = 1 \in F[x]$$

and so $x - \alpha$ would have to divide the constant polynomial $1 \in F[x]$, which is impossible. So no zero $\alpha \in E$ of $f(x)$ can be a zero of $f'(x)$, and $f(x)$ is separable over F. □

12.1.5 EXAMPLE Consider the polynomial

$$f(x) = x^9 + 2x^3 + 1 \in \mathbb{Z}_3[x]$$

We have $f'(x) = 0 \in \mathbb{Z}_3[x]$. Hence by Theorem 10.4.6, $f(x)$ has zeros of multiplicity $s > 1$ in its splitting field. Note that if we let

$$g(y) = y^3 + 2y + 1$$

then $f(x) = g(x^3)$. Furthermore,

$$[g(x)]^3 = (x^3 + 2x + 1)^3 = x^9 + 2x^3 + 1 = f(x)$$

so, in particular, $f(x)$ is not irreducible in this example. ◇

We come now to the basic fact about separable polynomials that we will be using throughout.

12.1.6 THEOREM Let $f(x) \in F[x]$ be an irreducible polynomial over a field F. Then

(1) If char $F = 0$, then $f(x)$ is separable over F.

(2) If char $F = p$, then $f(x)$ is separable over F if and only if $f(x) \neq g(x^p)$ for any polynomial $g(x) \in F[x]$.

Proof (1) Suppose char $F = 0$ and $f(x) \in F[x]$ is irreducible over F of degree n. If $n = 1$, then $f(x)$ has exactly one zero, and it is of multiplicity 1. If $n > 1$, then since char $F = 0$, $\deg f'(x) = n - 1 > 0$. Hence $f'(x) \neq 0$ and $f(x)$ is separable over F by Proposition 12.1.4.

(2) Suppose char $F = p$ and $f(x) \in F[x]$ is irreducible over F.

($\Rightarrow$) Assume $f(x) = g(x^p)$ for some polynomial $g(x) \in F[x]$. Then

$$f(x) = a_n x^{pn} + a_{n-1} x^{p(n-1)} + \dots + a_1 x^p + a_0 \in F[x]$$

and $f'(x) = 0$. Hence $f(x)$ is not separable over F by Proposition 12.1.4.

($\Leftarrow$) Assume $f(x) \neq g(x^p)$ for any polynomial $g(x) \in F[x]$. Then

$$f(x) = a_m x^m + \dots + a_i x^i + \dots + a_1 x + a_0 \in F[x]$$

where there is at least one power of x, say x^i, where i is not a multiple of p and the coefficient a_i is not zero. But then in the derivative $f'(x)$ there is at least one power of x, namely x^{i-1}, where the coefficient ia_i is not zero. So $f'(x) \neq 0$, and $f(x)$ is separable over F by Proposition 12.1.4. □

12.1.7 COROLLARY Every field of characteristic 0 is perfect.

Proof Immediate from Definition 12.1.3 and Theorem 12.1.6. □

12.1.8 COROLLARY Every finite field is perfect.

Proof Let F be a finite field with char $F = p$, E an algebraic extension of F, and α an element of E. Let $f(x) \in F[x]$ be the minimal polynomial of α over F. If $f(x)$ is not separable over F, then by Theorem 12.1.6 there exists $g(x) \in F[x]$ such that

$$f(x) = g(x^p) = a_n x^{pn} + a_{n-1} x^{p(n-1)} + \dots + a_1 x^p + a_0 \in F[x]$$

Since F is finite, by Proposition 10.4.8 for each coefficient we have $a_j = (b_j)^p$ for some $b_j \in F$. Hence by Exercise 20 in Section 10.4, we have

$$f(x) = (b_n x^n)^p + \dots + (b_1 x)^p + (b_0)^p =$$
$$= (b_n x^n + \dots + b_1 x + b_0)^p \in F[x]$$

and $f(x)$ is not irreducible over F. We assumed, though, that $f(x)$ was the minimal polynomial of α over F, and hence irreducible over F. Thus the assumption that $f(x)$ is not separable over F leads to a contradiction, which completes the proof. □

12.1.9 EXAMPLE Let us construct an example of a field F that is not perfect. From the preceding corollaries we know that F will have to be of characteristic $p > 0$ and will have to be infinite. The simplest such field is $F = \mathbb{Z}_p(x)$, the field of rational functions over the integers mod p (the field of quotients of $\mathbb{Z}_p[x]$). Now consider $F(\alpha)$, where α is a zero of $f(y) = y^p - x \in F[y]$. Over $F(\alpha)$ we have

$$f(y) = y^p - \alpha^p = (y - \alpha)^p \in F(\alpha)[y]$$

Since $\alpha = x^{1/p} \notin F$, it follows that $F(\alpha) \neq F$, $f(x)$ is irreducible over F, and F has an algebraic extension that is not separable. Therefore, F is not a perfect field. ◇

For any polynomial $f(x) \in \mathbb{Q}[x]$ of degree n that is irreducible over $\mathbb{Q}$, Corollary 12.1.7 implies that $f(x)$ has n distinct zeros in its splitting field. Similarly, for any polynomial $f(x) \in \mathbb{Z}_p[x]$ that is irreducible over $\mathbb{Z}_p$, Corollary 12.1.8 implies that the number of distinct zeros of $f(x)$ in its splitting field is equal to the degree of $f(x)$.

12.1.10 EXAMPLE Let $f(x) = x^2 - 2 \in \mathbb{Q}[x]$. Then $f(x)$ has two distinct zeros $\pm\sqrt{2}$ and the splitting field of $f(x)$ is $\mathbb{Q}(\sqrt{2})$. Let us find all possible isomorphisms

$$\phi\colon \mathbb{Q}(\sqrt{2}) \to \mathbb{Q}(\sqrt{2})$$

Observe that the following must hold for any such isomorphism.
First, $\phi(1) = 1$ by Proposition 7.1.9, part (3), and therefore $\phi(a) = a$ for all $a \in \mathbb{Q}$. (See Exercise 9.)
Second, any element $u \in \mathbb{Q}(\sqrt{2})$ is of the form $u = a + b\sqrt{2}$ for some $a, b \in \mathbb{Q}$. Hence

$$\phi(u) = \phi(a) + \phi(b)\phi(\sqrt{2}) = a + b\phi(\sqrt{2})$$

and therefore ϕ is completely determined by $\phi(\sqrt{2})$.
Third, $\phi(\sqrt{2})\phi(\sqrt{2}) = \phi(2) = 2$. Hence

$$\phi(\sqrt{2}) = \pm\sqrt{2}$$

This shows that there are exactly two such isomorphisms:
(1) $\phi(\sqrt{2}) = \sqrt{2}$, in which case $\phi = \mathrm{id}$
(2) $\phi(\sqrt{2}) = -\sqrt{2}$, in which case

$$\phi(a + b\sqrt{2}) = a - b\sqrt{2} \text{ for all } a + b\sqrt{2} \in \mathbb{Q}(\sqrt{2})$$

If we let ϕ be the latter automorphism, then $G = \{\mathrm{id}, \phi\}$ is a group under composition of maps, of order 2. ◇

12.1.11 EXAMPLE Let $\phi\colon \mathbb{C} \to \mathbb{C}$ be an isomorphism such that $\phi(a) = a$ for all $a \in \mathbb{R}$. Then for any $z = a + bi \in \mathbb{C}$ we have

$$\phi(a + bi) = \phi(a) + \phi(b)\phi(i) = a + b\phi(i)$$

and since

$$\phi(i)\phi(i) = \phi(i^2) = \phi(-1) = -1$$

by Proposition 7.1.8, part (2), ϕ is completely determined by $\phi(i)$. Therefore, there are exactly two such isomorphisms:
(1) $\phi(i) = i$, in which case $\phi = \mathrm{id}$
(2) $\phi(i) = -i$ in which case $\phi(z) = \overline{z}$
and again we obtain a group of order 2. ◇

12.1.12 DEFINITION

(1) Let E be a field. An isomorphism $\phi\colon E \to E$ is called an **automorphism** of E.

(2) Let F be a field and E an extension field of F. Then an automorphism $\phi: E \to E$ of E is said to **fix** F if $\phi(a) = a$ for all $a \in F$. An automorphism of E that fixes F is also called an **F-automorphism** of E. ○

12.1.13 THEOREM Let F be a field and E an extension field of F. Then

(1) The set Aut(E) of all automorphisms of E is a group under composition of maps.

(2) The set of all F-automorphisms of E is a subgroup of Aut(E).

Proof (1) Let ϕ and τ be elements of Aut(E). By Theorem 0.1.15, composition of maps is an associative operation, and $\phi\circ\tau$ is one to one and onto, since ϕ and τ are. Furthermore, $\phi\circ\tau$ is a ring homomorphism since for any $a, b \in E$ we have

$$\phi \circ \tau(a + b) = \phi(\tau(a + b)) = \phi(\tau(a) + \tau(b)) = \phi(\tau(a)) + \phi(\tau(b)) = \phi \circ \tau(a) + \phi \circ \tau(b)$$
$$\phi \circ \tau(ab) = \phi(\tau(ab)) = \phi(\tau(a) \cdot \tau(b)) = \phi(\tau(a)) \cdot \phi(\tau(b)) = \phi \circ \tau(a) \cdot \phi \circ \tau(b)$$

Therefore $\phi \circ \tau \in$ Aut(E). The identity map is an isomorphism of E, and id $\in$ Aut(E) acts as an identity element under composition of maps. Finally, by Theorem 0.1.15 the inverse map ϕ^{-1} is one to one and onto, and since

$$\phi(\phi^{-1}(a + b)) = a + b$$

we have

$$\phi(\phi^{-1}(a) + \phi^{-1}(b)) = a + b$$

and using the fact that ϕ is one to one we then have

$$\phi^{-1}(a + b) = \phi^{-1}(a) + \phi^{-1}(b)$$

Similarly, $\phi^{-1}(ab) = \phi^{-1}(a)\phi^{-1}(b)$, hence $\phi^{-1} \in$ Aut(E), and ϕ^{-1} is an inverse element to ϕ under composition of maps. Thus we have shown that Aut(E) is a group under composition of maps.

(2) Let ϕ and τ be elements of Aut(E) that are F-automorphisms or, in other words, $\phi(a) = a$ and $\tau(a) = a$ for all $a \in F$. Since τ is one to one and onto, $\tau^{-1}(a) = a$ for all $a \in F$. Hence

$$\phi \circ \tau^{-1}(a) = \phi(\tau^{-1}(a)) = \phi(a) = a$$

for all $a \in F$, and $\phi \circ \tau^{-1}$ is also an F-automorphism. Hence the set of F-automormphisms is a subgroup of Aut(E) by the subgroup test (Theorem 1.2.10), and the proof is complete. □

12.1.14 DEFINITION Let F be a field and E an extension field of F. Then the group of F-automorphisms of E is called the **Galois group** of E over F and is denoted by Gal(E/F). ○

12.1.15 EXAMPLES

(1) $\text{Gal}(\mathbb{Q}(\sqrt{2})/\mathbb{Q}) \cong \mathbb{Z}_2$ (2) $\text{Gal}(\mathbb{C}/\mathbb{R}) \cong \mathbb{Z}_2$ ◇

12.1.16 PROPOSITION Let F be a field, E an extension field of F, and let $\phi \in$ Gal(E/F). Then

$$\phi[f(\alpha)] = f(\phi(\alpha))$$

for all $\alpha \in E$ and $f(x) \in F[x]$.

Proof Let $f(x) = a_nx^n + \dots + a_1x + a_0 \in F[x]$. Then

$$\phi[f(\alpha)] = \phi(a_n\alpha^n + \dots + a_1\alpha + a_0) =$$
$$\phi(a_n\alpha^n) + \dots + \phi(a_1\alpha) + \phi(a_0) = \phi(a_n)\phi(\alpha)^n + \dots + \phi(a_1)\phi(\alpha) + \phi(a_0) =$$
$$= a_n\phi(\alpha)^n + \dots + a_1\phi(\alpha) + a_0 = f(\phi(\alpha))$$

as asserted. □

12.1.17 COROLLARY Let F be a field, E a field extension of F, $\phi \in \mathrm{Gal}(E/F)$, $f(x) \in F[x]$, and $\alpha \in E$. Then α is a zero of $f(x)$ if and only if $\phi(\alpha)$ is a zero of $f(x)$.

Proof Immediate from the preceding proposition, since ϕ is one to one. □

12.1.18 LEMMA Let F be a field, E a finite field extension of F, and $\{v_1, \dots, v_n\}$ a basis for E as a vector space over F. Then for any $\phi \in \mathrm{Gal}(E/F)$, ϕ is completely determined by $\phi(v_i)$ for $1 \le i \le n$.

Proof Since $\{v_1, \dots, v_n\}$ is a basis for E as a vector space over F, every element $u \in E$ can be uniquely expressed as a linear combination

$$u = c_1v_1 + \dots + c_nv_n$$

where $c_i \in F$ for $1 \le i \le n$. Since $\phi \in \mathrm{Gal}(E/F)$, then

$$\phi(u) = c_1\phi(v_1) + \dots + c_n\phi(v_n)$$

Hence ϕ is a linear transformation (see Exercise 26 in Section 10.1) and is uniquely determined by $\phi(v_i)$ for $1 \le i \le n$. □

In Examples 12.1.10 and 12.1.11 we saw that any $\phi \in \mathrm{Gal}\,(\mathbb{Q}(\sqrt{2}\,)/\mathbb{Q})$ is completely determined by $\phi(\sqrt{2})$ and any element of $\mathrm{Gal}(\mathbb{C}/\mathbb{R})$ is completely determined by $\phi(i)$. The next proposition states that this happens for any algebraic extension $E = F(\alpha)$.

12.1.19 PROPOSITION Let F be a field and $E = F(\alpha)$ a finite extension of F. Then for any $\phi \in \mathrm{Gal}(E/F)$, ϕ is completely determined by $\phi(\alpha)$.

Proof $E = F(\alpha)$ is a vector space over F of dimension $\dim_F E = [E : F] = n$, where n is the degree of α over F and $\{1, \alpha, \alpha^2, \dots, \alpha^{n-1}\}$ is a basis for E over F by Theorem 10.2.11. Hence by the preceding lemma, for any $\phi \in \mathrm{Gal}(E/F)$, ϕ is completely determined by $\phi(\alpha^i)$ for $0 \le i \le n - 1$. But since $\phi(\alpha^i) = \phi(\alpha)^i$, all $\phi(\alpha^i)$ for $0 \le i \le n - 1$ are completely determined by $\phi(\alpha)$: If ϕ and τ are elements of $\mathrm{Gal}(E/F)$ such that $\phi(\alpha) = \tau(\alpha)$, then for any

$$u = c_0 + c_1\alpha + \dots + c_{n-1}\alpha^{n-1} \in F(\alpha)$$

we have

$$\phi(u) = c_0 + c_1\phi(\alpha) + \dots + c_{n-1}\phi(\alpha)^{n-1} =$$
$$c_0 + c_1\tau(\alpha) + \dots + c_{n-1}\tau(\alpha)^{n-1} = \tau(u)$$

and $\phi = \tau$. □

12.1.20 COROLLARY Let F be a field and $E = F(\alpha_1, \dots, \alpha_s)$ a finite extension of F. Then for any $\phi \in \mathrm{Gal}(E/F)$, ϕ is completely determined by $\phi(\alpha_i)$ for $1 \le i \le s$.

Proof E is a vector space over F with a basis consisting of the elements

$$\alpha_1^{n_1}\alpha_2^{n_2}\dots\alpha_s^{n_s},\ 0 \le n_i < \deg \alpha_i,\ 1 \le i \le s$$

by Theorem 10.2.20. Hence by Lemma 12.1.18, ϕ is completely determined by the values

$$\phi(\alpha_1^{n_1}\alpha_2^{n_2}\dots\alpha_s^{n_s}) = \phi(\alpha_1)^{n_1}\phi(\alpha_2)^{n_2}\dots\phi(\alpha_s)^{n_s}$$

and hence ϕ is completely determined by $\phi(\alpha_i)$ for $1 \le i \le s$. □

12.1.21 EXAMPLE Let us calculate the Galois group of the splitting field in $\mathbb{C}$ of $p(x) = x^3 - 2$ over $\mathbb{Q}$. This splitting field is $E = \mathbb{Q}(2^{1/3}, \omega)$ where

$$\omega = -\frac{1}{2} + \frac{\sqrt{3}}{2}i$$

is a primitive cube root of unity, with minimal polynomial $x^2 + x + 1$. Note that

(1) The three zeros of $x^3 - 2$ are $2^{1/3}$, $2^{1/3}\omega$, $2^{1/3}\omega^2$.

(2) The two zeros of $x^2 + x + 1$ are ω, ω^2.

(3) A basis for E as a vector space over $\mathbb{Q}$ can be obtained as follows:

$$\{1, 2^{1/3}, 2^{2/3}, \omega, 2^{1/3}\omega, 2^{1/3}\omega^2\}$$

(4) By the preceding corollary any $\phi \in \mathrm{Gal}(E/\mathbb{Q})$ is completely determined by $\phi(2^{1/3})$ and $\phi(\omega)$.

(5) By Corollary 12.1.17

$$\phi(2^{1/3}) = 2^{1/3} \text{ or } 2^{1/3}\omega \text{ or } 2^{1/3}\omega^2$$
$$\phi(\omega) = \omega \text{ or } \omega^2$$

and therefore $|\mathrm{Gal}(E/\mathbb{Q})| \le 3 \cdot 2 = 6 = [E:\mathbb{Q}]$.

Consider now the elements ϕ and τ of $\mathrm{Gal}(E/\mathbb{Q})$ determined by

$$\phi(2^{1/3}) = 2^{1/3}\omega \quad \phi(\omega) = \omega$$
$$\tau(2^{1/3}) = 2^{1/3} \quad \tau(\omega) = \omega^2$$

Then

$$\phi\circ\tau(2^{1/3}) = \phi(2^{1/3}) = 2^{1/3}\omega$$
$$\tau\circ\phi(2^{1/3}) = \tau(2^{1/3}\omega) = 2^{1/3}\omega^2$$

Therefore, $\phi\circ\tau \ne \tau\circ\phi$ and $\mathrm{Gal}(E/\mathbb{Q})$ is a non-Abelian group of order ≤ 6, and hence $\mathrm{Gal}(E/\mathbb{Q}) \cong S_3$. ◇

12.1.22 EXAMPLE Let us calculate the Galois group of the splitting field in $\mathbb{C}$ of $f(x) = (x^2 - 2)(x^2 - 3)$. This splitting field is

$$E = \mathbb{Q}(\sqrt{2}, \sqrt{3})$$

and by Corollary 12.1.17, for any $\phi \in \mathrm{Gal}(E/\mathbb{Q})$ we have

$$\phi(\sqrt{2}) = \pm\sqrt{2} \qquad \phi(\sqrt{3}) = \pm\sqrt{3}$$

Therefore, $|\mathrm{Gal}(E/\mathbb{Q})| \le 2 \cdot 2 = 4 = [E:\mathbb{Q}]$. Now consider

$$H = \mathrm{Gal}(E/\mathbb{Q}[\sqrt{2}]) \qquad K = \mathrm{Gal}(E/\mathbb{Q}[\sqrt{3}])$$

Any $\psi \in H$ is completely determined by $\psi(\sqrt{3})$, and there are two possibilities:

$$\psi = \text{id} \quad \text{or} \quad \psi(\sqrt{3}) = -\sqrt{3}$$

So H is a subgroup of $\text{Gal}(E/\mathbb{Q})$ of order 2, and similarly so is K. So $\text{Gal}(E/\mathbb{Q})$ is a group of order ≤ 4 with two distinct subgroups of order 2, and hence $\text{Gal}(E/\mathbb{Q}) \cong \mathbb{Z}_2 \otimes \mathbb{Z}_2$. ◇

Note that in both the preceding examples we first found that $\text{Gal}(E/\mathbb{Q}) \leq [E:\mathbb{Q}]$ and then showed that equality held. The next theorem tells us that we always have equality in such cases. Its proof makes use of theorems from Section 10.3 that we originally used as part of the proof of the uniqueness of splitting fields.

12.1.23 LEMMA Le F be a field, $f(x) \in F[x]$ a polynomial over F, E a splitting field of $f(x)$ over F, $p(x) \in F[x]$ an irreducible polynomial dividing $f(x)$, and α_1 and α_2 two zeros of $p(x)$ in E. Then there exists an F-automorphism $\psi \in \text{Gal}(E/F)$ such that $\psi(\alpha_1) = \alpha_2$.

Proof Theorem 10.3.20 says that if ϕ: $F_1 \to F_2$ is an isomorphism between fields F_1 and F_2, and if $p_1(x) \in F_1[x]$ is irreducible over F_1 and $p_2(x) \in F_2[x]$ is the corresponding polynomial irreducible over F_2, and if α_1 is a zero of $p_1(x)$ in an extension field E_1 of F_1 and α_2 a zero of $p_2(x)$ in an extension field E_2 of F_2, then there exists an isomorphism τ: $F_1(\alpha_1) \to F_2(\alpha_2)$ that agrees with ϕ on F_1 and has $\tau(\alpha_1) = \alpha_2$. We apply this theorem with $F_1 = F_2 = F$, $\phi = \text{id}$, $p_1(x) = p_2(x) = p(x)$ to conclude that there exists an isomorphism τ: $F(\alpha_1) \to F(\alpha_2)$ that fixes F and has $\tau(\alpha_1) = \alpha_2$.
Theorem 10.3.21 says that if τ: $F_1 \to F_2$ is an isomorphism between fields F_1 and F_2, and if $f_1(x) \in F_1[x]$ is any polynomial over F_1 and $f_2(x) \in F_2[x]$ is the corresponding polynomial over F_2, and if E_1 is a splitting field of $f_1(x)$ over F_1 and E_2 is a splitting field of $f_2(x)$ over F_2, then there is an isomorphism ψ: $E_1 \to E_2$ that agrees with τ on F_1. We apply this theorem with $F_1 = F(\alpha_1)$, $F_2 = F(\alpha_2)$, $f_1(x) = f_2(x) = f(x)$, and $E_1 = E_2 = E$ to conclude that there exists an automorphism ψ: $E \to E$ that agrees with τ on $F_1(\alpha_1)$, and therefore fixes F and has $\psi(\alpha_1) = \alpha_2$. □

12.1.24 THEOREM Let F be a field, $f(x) \in F[x]$ a separable polynomial over F, and E a splitting field of $f(x)$ over F. Then

$$|\text{Gal}(E/F)| = [E:F]$$

Proof We use induction on the degree $[E:F]$ of the extension. If $[E:F] = 1$, then $E = F$ and $\text{Gal}(E/F)$ is the trivial group. So assume the theorem holds when the degree of the extension is $< n$ and consider the case where $[E:F] = n$. Since $[E:F] > 1$, $f(x)$ does not split already over F and so has a divisor $p(x)$ that is irreducible over F and of degree $k > 1$.
Since $f(x)$ is separable and E is the splitting field of $f(x)$ over F, $f(x)$ splits over E and all its zeros are of multiplicity 1. Since $p(x)$ divides $f(x)$, $p(x)$ also splits in E and all

its zeros also have multiplicity 1. Thus there are k distinct zeros $\alpha_1, \ldots, \alpha_k$ of $p(x)$ in E. Now we apply the preceding lemma. It tells us that for each i, $1 \leq i \leq k$ there exists at least one F-automorphism $\psi \in \text{Gal}(E/F)$ with $\psi(\alpha_1) = \alpha_i$. Pick one such F-automorphism and call it ψ_i.

We have $[F(\alpha_1):F] = k$, and $[E:F] = [E:F(\alpha_1)][F(\alpha_1):F]$. Therefore, $[E:F(\alpha_1)] = m$, where $m = n/k$. Since E is the splitting field of the separable polynomial $f(x)$ over $F(\alpha_1)$, and $[E:F(\alpha_1)] < n$, we can apply our induction hypothesis, and it tells us that $|\text{Gal}(E/F(\alpha_1)| = [E:F(\alpha_1)] = m$. Let the elements of $\text{Gal}(E/F(\alpha_1)$ be $\theta_1, \ldots, \theta_m$.

Claim 1 The $km = n$ compositions $\psi_i \circ \theta_r$ for $1 \leq i \leq k$ and $1 \leq r \leq m$ are all distinct, and therefore $|\text{Gal}(E/F)| \geq km = n$.

Proof of Claim 1 Suppose $\psi_i \circ \theta_r = \psi_j \circ \theta_s$. Then

$$\alpha_j = \psi_j(\alpha_1) = \psi_j(\theta_s(\alpha_1)) = \psi_j \circ \theta_s(\alpha_1) =$$
$$\psi_i \circ \theta_r(\alpha_1) = \psi_i(\theta_r(\alpha_1)) = \psi_i(\alpha_1) = \alpha_i$$

So $i = j$ and we have $\psi_i \circ \theta_r = \psi_j \circ \theta_s = \psi_i \circ \theta_s$ and hence $\theta_r = \theta_s$ and $r = s$.

Claim 2 Every F-automorphism $\phi: E \to E$ is equal to $\psi_i \circ \theta_r$ for some $1 \leq i \leq k$ and $1 \leq r \leq m$, and therefore $|\text{Gal}(E/F)| \leq km = n$.

Proof of Claim 2 Let $\phi: E \to E$ be an F-automorphism, and let $\phi(\alpha_1) = \alpha_i$, where $1 \leq i \leq k$. Setting $\theta = \psi_i^{-1} \circ \phi$, we have

$$\theta(\alpha_1) = \psi_i^{-1} \circ \phi(\alpha_1) = \psi_i^{-1}(\phi(\alpha_1)) = \psi_i^{-1}(\alpha_i) = \alpha_1$$

Thus θ fixes F and θ fixes α_1, and therefore θ fixes $F(\alpha_1)$, which is to say $\theta \in \text{Gal}(E/F(\alpha_1))$ and so $\psi_i^{-1} \circ \phi = \theta = \theta_r$ for some $1 \leq r \leq m$. Thus $\phi = \psi_i \circ \theta_r$ as required to complete the proof of the claim and of the theorem. □

12.1.25 EXAMPLE Let us calculate $\text{Gal}(\mathbb{Q}(\zeta)/\mathbb{Q})$, where ζ is a primitive pth root of unity for a prime p. The minimal polynomial of ζ is the cyclotomic polynomial $\Phi_p(x) \in \mathbb{Q}[x]$, and ζ^i, $1 \leq i \leq p - 1$ are all the $p - 1$ distinct zeros.

$$|\text{Gal}(\mathbb{Q}(\zeta)/\mathbb{Q})| = [\mathbb{Q}(\zeta):\mathbb{Q}] = p - 1$$

Furthermore, if $\psi \in \text{Gal}(\mathbb{Q}(\zeta)/\mathbb{Q})$ is defined by $\psi(\zeta) = \zeta^k$, where $1 \leq k \leq p - 1$, then $\psi^i(\zeta) = \zeta^{k^i}$, $|\psi| = p - 1$, and $\text{Gal}(\mathbb{Q}(\zeta)/\mathbb{Q})$ is a cyclic group of order $p - 1$. ◇

12.1.26 PROPOSITION Let F be a field, $f(x) \in F[x]$ a separable polynomial of degree n, and E a splitting field of $f(x)$ over F. Then

(1) $\text{Gal}(E/F)$ is isomorphic to a subgroup of S_n.

(2) $|\text{Gal}(E/F)|$ divides $n!$

Proof (1) Since $f(x)$ is separable and E is its splitting field over F, it will have n distinct zeros $\alpha_1, \ldots, \alpha_n$ in E. To each $\phi \in \text{Gal}(E/F)$ we associate a map $\chi(\phi)$ from $\{1, \ldots, n\}$ to $\{1, \ldots, n\}$ as follows. Given $1 \leq i \leq n$, by Corollary 12.1.17 $\phi(\alpha_i)$ must be one of the α_j. We let $\chi(\phi)(i) = j$. Since ϕ is one to one, this gives a permutation $\chi(\phi) \in S_n$. It follows from Corollary 12.1.20 that $\chi: \text{Gal}(E/F) \to S_n$ is one to one, and it is also a group homomorphism. (See Exercise 10.) Hence by the first isomorphism theorem (Theorem 7.2.17), Gal(E/F) is isomorphic to a subgroup of S_n.

(2) Follows immediately by Lagrange's theorem. □

In Theorem 12.1.24 we assumed both that $f(x)$ was separable and that E was its splitting field. Let us see what can happen if either of these assumptions is dropped.

12.1.27 EXAMPLE Let us calculate the Galois group of $\mathbb{Q}(2^{1/3})$ over $\mathbb{Q}$. This is a separable extension since char $\mathbb{Q} = 0$, but $\mathbb{Q}(2^{1/3})$ is not the splitting field of the minimal polynomial $x^3 - 2$ of $2^{1/3}$ since it does not contain the complex zeros of this polynomial. Any $\phi \in \text{Gal}(\mathbb{Q}(2^{1/3})/\mathbb{Q})$ is completely determined by $\phi(2^{1/3})$, which must be a zero of $x^3 - 2$, and hence the only possibility is $\phi = \text{id}$. Thus $\text{Gal}(\mathbb{Q}(2^{1/3})/\mathbb{Q})$ is trivial and not a group of order $[\mathbb{Q}(2^{1/3}):\mathbb{Q}] = 3$. ◇

12.1.28 EXAMPLE Consider the polynomial $f(y) = y^p - x \in F[y]$, where $F = \mathbb{Z}_p(x)$ as in Example 12.1.9. Then the splitting field of $f(y)$ over F is $F(\alpha)$, and in $F(\alpha)[y]$ we have $f(y) = (y - \alpha)^p$. This means that $f(y)$ is an irreducible polynomial of degree p in $F[y]$ that has only one zero in its splitting field. Hence $f(y)$ is not separable. As in the preceding example, because there is only one zero, Gal($F(\alpha)/F$) is trivial and not a group of order $[F(\alpha):F] = p$. ◇

We conclude this section with another important and useful property of separable extensions.

12.1.29 DEFINITION Let F be a field, E a finite extension of F. If there exists an element $\gamma \in E$ such that $E = F(\gamma)$, then E is called a **simple** extension of F, and any such γ is called a **primitive element** of E over F. ○

12.1.30 THEOREM (Primitive element theorem) Let F be a field and E a finite separable extension of F. Then E is a simple extension of F.

Proof Suppose first that F is finite. Then E is also finite, and so E^* is a cyclic group by Theorem 8.3.10. If $E^* = \langle\alpha\rangle$, then $E = F(\alpha)$.

Now suppose that F is infinite and E is a finite separable extension of F. So $E = F(u_1, \dots, u_s)$, where each u_i is separable over F. If $s = 1$ we are done. It is enough to prove the theorem in the case $s = 2$, since then we have

$$F(u_1, u_2) = F(v) \text{ for some } v$$
$$F(u_1, u_2, u_3) = F(u_1, u_2)(u_3) = F(v)(u_3) = F(v, u_3) = F(v') \text{ for some } v'$$

and so on.

So let $E = F(\alpha, \beta)$. Let $p(x)$ and $q(x)$ be the minimal polynomials over F of α and β, respectively. Since E is a separable extension, $p(x)$ and $q(x)$ are separable polynomials. Let K be a splitting field of $p(x)q(x)$ over E. Then $p(x)$ has $n = \deg p(x)$ distinct zeros $\alpha = \alpha_1, \dots, \alpha_n$ in K and $q(x)$ has $m = \deg q(x)$ distinct zeros $\beta = \beta_1, \dots, \beta_m$ in K. Now consider all the quotients

$$(\alpha_i - \alpha_1)/(\beta_1 - \beta_j) \in K \text{ for } 1 < i \leq n,\ 1 < j \leq m$$

Since there are only finitely many such quotients, while F is infinite, there must exist a nonzero element $u \in F$ such that

$$u \neq (\alpha_i - \alpha_1)/(\beta_1 - \beta_j) \text{ for } 1 < i \leq n,\ 1 < j \leq m$$

Let $\gamma = \alpha_1 + u\beta_1 \in E$. We show $E = F(\gamma)$. Since we have $F \subseteq F(\gamma) \subseteq E$ and $E = F(\alpha_1, \beta_1)$, it is enough to show $\alpha_1, \beta_1 \in F(\gamma)$. In fact, it is enough to show $\beta_1 \in F(\gamma)$, since $\alpha_1 \in F(\gamma)$ then follows because $\alpha_1 = \gamma - u\beta_1$, where $u \in F$.

So let $h(x)$ be the minimal polynomial of β_1 over $F(\gamma)$. We want to show deg $h(x) = 1$. First note that since $q(\beta) = 0$ and $h(x)$ is the minimal polynomial of β_1 over $F(\gamma)$, $h(x)$ divides $q(x)$, and so over K the polynomial $h(x)$ splits as a product of some of the factors $(x - \beta_j)$ of $q(x)$. To show that in fact $h(x) = x - \beta_1$, it is therefore enough to show that $h(\beta_j) \neq 0$ for $j \neq 1$.

Toward showing this, consider another polynomial

$$k(x) = p(\gamma - ux) \in F(\gamma)[x]$$

We have

$$k(\beta_1) = p(\gamma - u\beta_1) = p(\alpha_1) = 0$$

And so since $h(x)$ is the minimal polynomial of β_1 over $F(\gamma)$, $h(x)$ divides $k(x)$. Thus to show that $h(\beta_j) \neq 0$ for $j \neq 1$, it is enough to show that $k(\beta_j) \neq 0$ for $j \neq 1$. But if we had $k(\beta_j) = p(\gamma - u\beta_j) = 0$, we would have $\gamma - u\beta_j = \alpha_i$ for some $i \neq 1$, and since $\gamma = \alpha_1 + u\beta_1$, this would give $u = (\alpha_i - \alpha_1)/(\beta_1 - \beta_j)$, contrary to our choice of u. □

12.1.31 COROLLARY Let F be a field of characteristic 0. Then any finite extension of F is a simple extension of F.

Proof If E is a finite extension of F, then E is an algebraic extension of F by Theorem 10.2.16, and E is a separable extension of F by Corollary 12.1.7. Hence E is a simple extension of F by the preceding theorem. □

Let us see how the construction of Theorem 12.1.30 works in a specific example.

12.1.32 EXAMPLE Let us find a primitive element γ over $\mathbb{Q}$ for the field $\mathbb{Q}(2^{1/3}, \omega)$ considered in Example 12.1.21. In this case, $p(x) = x^3 - 2$ and the α_i are $2^{1/3}$, $2^{1/3}\omega$, and $2^{1/3}\omega^2$, while $q(x)$ is $x^2 + x + 1$ and the β_j are ω and ω^2. We want a nonzero $u \in \mathbb{Q}$ such that

$$u \neq (2^{1/3}\omega - 2^{1/3})/(\omega - \omega^2) = -2^{1/3}/\omega = -2^{1/3}\omega^2 \text{ and}$$
$$u \neq (2^{1/3}\omega^2 - 2^{1/3})/(\omega - \omega^2) = -2^{1/3}(\omega + 1)/\omega = -2^{1/3}(1 + \omega^2)$$

Clearly $u = 1$ will do. This gives for the primitive element $\gamma = 2^{1/3} + \omega$. To check that this works, note that $(\gamma - \omega)^3 = 2$, and therefore

$$2 = \gamma^3 - 3\gamma^2\omega + 3\gamma\omega^2 - 1$$
$$3 = \gamma^3 - 3\gamma^2\omega + 3\gamma\omega^2 = \gamma^3 - 3\gamma^2\omega + 3\gamma(-\omega - 1)$$

Since $\gamma \neq 0$ and $\gamma \neq -1$, it follows that

$$\omega = (\gamma^3 - 3 - 3\gamma)/(3\gamma^2 + 3\gamma) \in \mathbb{Q}(\gamma) \ \diamond$$

Exercises 12.1

In Exercises 1 through 5 calculate the Galois group $\mathrm{Gal}(E/\mathbb{Q})$ for the indicated fields E.

1. $E = \mathbb{Q}(\sqrt{3}, \sqrt{5})$

2. $E = \mathbb{Q}(\omega)$, where ω is the primitive cube root of unity

3. $E = \mathbb{Q}(\sqrt{3}i)$

4. $E = \mathbb{Q}(\sqrt{3}, i)$

5. $E = \mathbb{Q}(\zeta)$, where ζ is a primitive 4th root of unity

In Exercises 6 through 8 calculate the $\mathrm{Gal}(E/\mathbb{Q})$, where E is the splitting field in $\mathbb{C}$ of the indicated polynomial $f(x) \in \mathbb{Q}[x]$.

6. $f(x) = x^3 - 5$

7. $f(x) = x^4 - 2x$

8. $f(x) = x^4 - 4x^2 + 4$

9. Show that the identity map is the only automorphism of $\mathbb{Q}$.

10. Show that the map χ: $\mathrm{Gal}(E/F) \to S_n$ defined in the proof of Proposition 12.1.26 is a one-to-one group homomorphism.

In Exercises 11 through 13 use the method of the proof of Theorem 12.1.30 to find a primitive element over $\mathbb{Q}$ for the indicated fields E.

11. $E = \mathbb{Q}(\sqrt{3}, i)$

12. $E = \mathbb{Q}(\sqrt{2}, \sqrt{3}, \sqrt{5})$

13. $E = \mathbb{Q}(\sqrt{p}, \sqrt{q})$, where p and q are distinct primes.

14. Let ζ be a primitive 6th root of unity over $\mathbb{Q}$. Find the Galois group $\mathrm{Gal}(\mathbb{Q}(\zeta)/\mathbb{Q})$.

15. Show that if E is a separable extension of F of degree $[E:F] = 2$, then $\mathrm{Gal}(E/F) \cong \mathbb{Z}_2$.

16. Let ϕ: $\mathbb{C} \to \mathbb{C}$ be defined by $\phi(a + bi) = a - bi$ (complex conjugation). Show that ϕ is an automorphism of $\mathbb{C}$.

17. Show that $\mathbb{Q}(\sqrt{3})$ and $\mathbb{Q}(\sqrt{5})$ are not isomorphic fields.

18. Show that an algebraic extension of a perfect field is a perfect field.

19. Let F be a field of characteristic p. Show that F is a perfect field if and only if for every element $a \in F$ there is an element $b \in F$ such that $a = b^p$.

20. Let $f(x) \in F[x]$ be an irreducible polynomial of degree n over a field F, and let E be its splitting field. Show that

(a) n divides $[E:F]$.

(b) If $f(x)$ is separable, then n divides $|\mathrm{Gal}(E/F)|$

21. Let F be a field, E an extension field of F, $\phi: E \to E'$ a field isomorphism that fixes F, and $F' = \phi(F) \subseteq E'$. Define χ: $\mathrm{Gal}(E/F) \to \mathrm{Gal}(E'/F')$ by $\chi(\tau) = \phi\tau\phi^{-1}$. Show that χ is a group isomorphism.

22. Let F be a field and $f(x) \in F[x]$, and E and E' two splitting fields of $f(x)$ over F. Show that $\mathrm{Gal}(E/F) \cong \mathrm{Gal}(E'/F)$. (*Hint:* Use Corollary 10.3.22 and the preceding exercise.)

12.2 The Fundamental Theorem of Galois Theory

We are now ready to undertake the proof of the most beautiful theorem of the subject, which gives us a correspondence between field extensions and Galois groups. We establish a one to one correspondence between the intermediate fields between a given field F and the splitting field E of a given polynomial over F and the subgroups of the Galois group $\mathrm{Gal}(E/F)$.The lattice of these intermediate fields is the inverted lattice of the subgroups of its Galois group. The three remaining sections of this chapter are all devoted to working out various applications of the correspondence established in this section.

12.2.1 EXAMPLE Let us consider the Galois group of the splitting field in $\mathbb{C}$ of the polynomial $f(x) = x^3 - 2 \in \mathbb{Q}[x]$. This splitting field is $E = \mathbb{Q}(2^{1/3}, \omega)$, where ω is the familiar primitive cube root of unity. We looked at this field extension in Example 12.1.21 and found that $\mathrm{Gal}(E/\mathbb{Q}) \cong S_3$. We looked at this field extension earlier in Example 10.3.11 and displayed the lattice of all intermediate fields K, where $\mathbb{Q} \subseteq K \subseteq E$. Now let us look at the Galois groups associated with these intermediate fields.

(1) Let us begin with $K_1 = \mathbb{Q}(\omega)$ and $\mathrm{Gal}(E/K_1)$. The elements of $\mathrm{Gal}(E/K_1)$ are the automorphisms of E that leave every element of K_1 fixed. Since $\mathbb{Q} \subseteq K_1$, they also leave every element of $\mathbb{Q}$ fixed and so are elements of $\mathrm{Gal}(E/\mathbb{Q})$. Since they are both groups under the same operation, composition of maps, $\mathrm{Gal}(E/K_1)$ is a subgroup of $\mathrm{Gal}(E/\mathbb{Q})$. Now E is the splitting field of $x^3 - 2$ over K_1 and therefore by Theorem 12.1.24

$$|\mathrm{Gal}(E/K_1)| = [E:K_1] = 3$$

So $\mathrm{Gal}(E/K_1)$ is isomorphic to the normal subgroup A_3 of S_3.

K_1 is the splitting field of $x^2 + x + 1$ over $\mathbb{Q}$, and therefore by Theorem 12.1.24
$$|\mathrm{Gal}(K_1/\mathbb{Q})| = [K_1 : \mathbb{Q}] = 2$$
So $\mathrm{Gal}(K_1/\mathbb{Q})$ is isomorphic to $\mathbb{Z}_2$.
Now let us look at the relationship among the groups $\mathrm{Gal}(E/\mathbb{Q})$, $\mathrm{Gal}(E/K_1)$, and $\mathrm{Gal}(K_1/\mathbb{Q})$ more closely. If $\phi \in \mathrm{Gal}(E/\mathbb{Q})$, then ϕ leaves every element of $\mathbb{Q}$ fixed and either $\phi(\omega) = \omega$ or $\phi(\omega) = \omega^2$ and in either case, $\phi(\alpha) \in K_1$ for every $\alpha \in K_1$, so the restriction $\phi|_{K_1}$ is an element of $\mathrm{Gal}(E/K_1)$. Now consider the map
$$\psi\colon \mathrm{Gal}(E/\mathbb{Q}) \to \mathrm{Gal}(K_1/\mathbb{Q})$$
defined by $\psi(\phi) = \phi|_{K_1}$. Then ψ is a group homomorphism, and by Theorem 10.3.21, for every automorphism of K_1 there is an automorphism of E that agrees with it on K_1, so ψ is onto. Kern ψ consists of the elements $\phi \in \mathrm{Gal}(E/\mathbb{Q})$ whose restriction to K_1 is the identity or, in other words, that leave every element of K_1 fixed, so Kern $\psi = \mathrm{Gal}(E/K_1)$. It follows that $\mathrm{Gal}(E/K_1)$ is a normal subgroup of $\mathrm{Gal}(E/\mathbb{Q})$ and
$$\mathrm{Gal}(E/\mathbb{Q})/_{\mathrm{Gal}(E/K_1)} \cong \mathrm{Gal}(K_1/\mathbb{Q})$$
Of course, we have already found what these groups are, and indeed $S_3/A_3 \cong \mathbb{Z}_2$.

(2) Now let us turn to $K_2 = \mathbb{Q}(2^{1/3})$. As in the case of K_1, $\mathrm{Gal}(E/K_2)$ is a subgroup of $\mathrm{Gal}(E/\mathbb{Q})$. Since E is the splitting field of $x^2 + x + 1$ over K_2, we have
$$|\mathrm{Gal}(E/K_2)| = [E : K_2] = 2$$
So $\mathrm{Gal}(E/K_2) \cong \mathbb{Z}_2$.
However, unlike K_1, K_2 is not a splitting field, and, as we saw in Example 12.1.27, $\mathrm{Gal}(K_2/\mathbb{Q})$ is trivial, and there is no isomorphism relating the groups $\mathrm{Gal}(E/\mathbb{Q})$, $\mathrm{Gal}(E/K_2)$, and $\mathrm{Gal}(K_2/\mathbb{Q})$ such as we found in the case of K_1.

(3) For the other intermediate fields $\mathbb{Q}(2^{1/3}\omega)$ and $\mathbb{Q}(2^{1/3}\omega^2)$, the Galois groups are just as in the case of $K_2 = \mathbb{Q}(2^{1/3})$.

(4) Each automorphism in $\mathrm{Gal}(E/\mathbb{Q})$ is completely determined by its values for $2^{1/3}$ and ω. Since we are very familiar with the group S_3, let us use the same names we have been using for elements of S_3 for the corresponding elements of $\mathrm{Gal}(E/\mathbb{Q})$, writing ρ_0 for the identity, to begin with. Then the elements of $\mathrm{Gal}(E/\mathbb{Q})$ are

$\rho_0(2^{1/3}) = 2^{1/3}$	$\rho_0(\omega) = \omega$	$\mu_1(2^{1/3}) = 2^{1/3}$	$\mu_1(\omega) = \omega^2$
$\rho(2^{1/3}) = 2^{1/3}\omega$	$\rho(\omega) = \omega$	$\mu_2(2^{1/3}) = 2^{1/3}\omega^2$	$\mu_2(\omega) = \omega^2$
$\rho^2(2^{1/3}) = 2^{1/3}\omega^2$	$\rho^2(\omega) = \omega$	$\mu_3(2^{1/3}) = 2^{1/3}\omega$	$\mu_3(\omega) = \omega^2$

In this notation we see that $\mathrm{Gal}(E/K_1)$, which consists of the automorphisms that keep ω fixed, is equal to $\langle\rho\rangle$, while $\mathrm{Gal}(E/K_2)$, which consists of those that keep $2^{1/3}$ fixed, is equal to $\langle\mu_1\rangle$. It is easily computed that
$$\mu_2(2^{1/3}\omega) = \mu_2(2^{1/3})\mu_2(\omega) = 2^{1/3}\omega^2\omega^2 = 2^{1/3}\omega^4 = 2^{1/3}\omega$$
It follows that $\mathrm{Gal}(E/\mathbb{Q}(2^{1/3}\omega)) = \langle\mu_2\rangle$, and similarly $\mathrm{Gal}(E/\mathbb{Q}(2^{1/3}\omega^2)) = \langle\mu_3\rangle$.

Thus the lattice of subfields of E displayed in Example 10.3.11 corresponds to the lattice of subgroups of S_3 as follows.

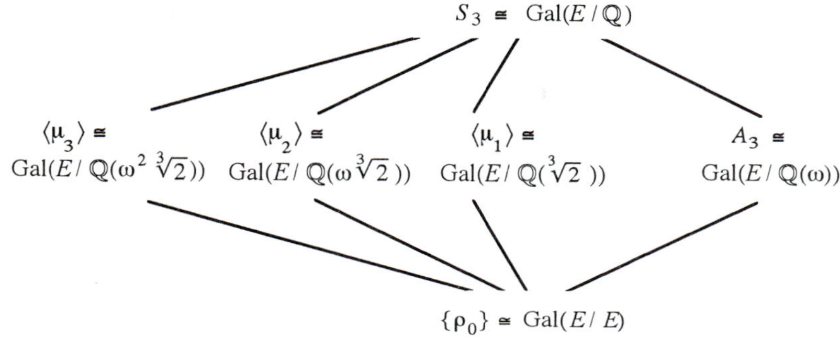

FIGURE 1

Note that in considering the groups $\mathrm{Gal}(E/K)$, the case of the smallest field $K = \mathbb{Q}$ is at the top and the case of the largest field $K = E$ is at the bottom, which is the *inverse* of the way the subfields were displayed in Example 10.3.11. Larger fields K correspond to smaller groups $\mathrm{Gal}(E/K)$. ◇

Having in mind the example just described, we next prove some features it illustrates.

12.2.2 PROPOSITION Let $F \subseteq K \subseteq E$ be a tower of fields. Then $\mathrm{Gal}(E/K)$ is a subgroup of $\mathrm{Gal}(E/F)$.

Proof An element of $\mathrm{Gal}(E/K)$ is an automorphism of E that keeps every element of K fixed. Since $F \subseteq K$, such an automorphism keeps every element of F fixed and so is an element of $\mathrm{Gal}(E/F)$. Since $\mathrm{Gal}(E/K) \subseteq \mathrm{Gal}(E/F)$ are both groups under the same operation, composition of maps, $\mathrm{Gal}(E/K)$ is a subgroup of $\mathrm{Gal}(E/F)$. □

12.2.3 PROPOSITION Let $F \subseteq K \subseteq E$ be a tower of fields, where K is a splitting field of some polynomial $f(x) \in F[x]$. Then for any $\phi \in \mathrm{Gal}(E/F)$, if $\phi|_K$ is the restriction of ϕ to K, then $\phi|_K \in \mathrm{Gal}(K/F)$.

Proof Let $\phi \in \mathrm{Gal}(E/F)$. We need to show that for any $\alpha \in K$, $\phi(\alpha) \in K$. By Proposition 10.3.6, $K = F(\alpha_1, \ldots, \alpha_n)$, where the α_i are the zeros of $f(x)$. Since $\phi \in \mathrm{Gal}(E/F)$, $\phi(\beta) = \beta$ for all $\beta \in F$ and by Corollary 12.1.17, for each $1 \le i \le n$, $\phi(\alpha_i) = \alpha_j$ for some $1 \le j \le n$. It follows that for any $\alpha \in F(\alpha_1, \ldots, \alpha_n) = K$, $\phi(\alpha) \in F(\alpha_1, \ldots, \alpha_n) = K$. □

We now show that the relationship illustrated in part (1) of Example 12.2.1 holds more generally.

12.2.4 THEOREM Let $F \subseteq K \subseteq E$ be a tower of fields, where E is a splitting field of some polynomial $f(x) \in F[x]$ and K is a splitting field of some polynomial $g(x) \in F[x]$. Then

(1) $\text{Gal}(E/K) \lhd \text{Gal}(E/F)$

(2) $\text{Gal}(E/F)\,/\,\text{Gal}(E/K) \cong \text{Gal}(K/F)$

Proof (1) Let $\psi: \text{Gal}(E/F) \to \text{Gal}(K/F)$ be the map defined by $\psi(\phi) = \phi|_K$. Then ψ is a group homomorphism, since for any $\rho, \tau \in \text{Gal}(E/F)$ and any $\alpha \in K$ we have

$$\psi(\rho \circ \tau)(\alpha) = (\rho \circ \tau)|_K(\alpha) = (\rho \circ \tau)\alpha = \rho(\tau(\alpha)) =$$
$$\rho|_K(\tau|_K(\alpha)) = \psi(\rho)(\psi(\tau)(\alpha)) = (\psi(\rho) \circ \psi(\tau))(\alpha)$$

We have $\phi \in \text{Kern}\,\psi$ if and only if $\phi|_K$ is the identity on K, that is, $\phi(\alpha) = \alpha$ for all $\alpha \in K$. Hence $\text{Kern}\,\psi = \text{Gal}(E/K)$, and (1) follows.

(2) Since E is the splitting field of $f(x)$ over K, by Theorem 10.3.21, for every automorphism τ of K, there is an automorphism ϕ of E with $\phi|_K = \tau$. If $\tau \in \text{Gal}(K/F)$, for every element $\beta \in F$ we have $\phi(\beta) = \phi|_K(\beta) = \tau(\beta) = \beta$, so $\phi \in \text{Gal}(E/F)$. This shows $\text{Im}\,\psi = \text{Gal}(K/F)$. By the first isomorphism theorem (Theorem 7.2.17), (2) follows. □

The correspondence between subfields and subgroups illustrated in part (4) of Example 12.2.1 also holds more generally, but before we can define the general correspondence, we need the following proposition.

12.2.5 PROPOSITION Let F be a field, let E be an extension field of F, and let $G = \text{Gal}(E/F)$. For any subgroup $H \leq G$ let

$$E^H = \{\alpha \in E \mid \phi(\alpha) = \alpha \text{ for all } \phi \in H\}$$

Then

(1) E^H is a subfield of E containing F: $F \subseteq E^H \subseteq E$.

(2) $E^{H_1} \subseteq E^{H_2}$ if $H_2 \leq H_1 \leq G$

Proof (1) Let $\alpha, \beta \in E^H$, and let $\phi \in H$. By definition of E^H, $\phi(\alpha) = \alpha$ and $\phi(\beta) = \beta$. But also, since ϕ is an automorphism, we have

$$\phi(\alpha - \beta) = \phi(\alpha) - \phi(\beta) = \alpha - \beta$$
$$\phi(\alpha\beta^{-1}) = \phi(\alpha)\phi(\beta)^{-1} = \alpha\beta^{-1}$$

So $\alpha - \beta \in E^H$ and $\alpha\beta^{-1} \in E^H$ and by the subfield test (Theorem 6.3.17) E^H is a subfield of E. Also, if $\alpha \in F$, then for any $\phi \in H$, since $H \leq \text{Gal}(E/F)$, we have $\phi(\alpha) = \alpha$. Hence $\alpha \in E^H$.

(2) If $\phi \in H_2 \leq H_1$, then for any $\alpha \in E^{H_1}$, since $\phi \in H_1$, we have $\phi(\alpha) = \alpha$. Hence $\alpha \in E^{H_2}$. □

12.2.6 DEFINITION Let F be a field, E an extension field of F, $G = \text{Gal}(E/F)$, and H a subgroup of G. Then

$$E^H = \{\alpha \in E \mid \phi(\alpha) = \alpha \text{ for all } \phi \in H\}$$

is called the **fixed field** of H. ○

12.2.7 EXAMPLE In Example 12.2.1, where $F = \mathbb{Q}$, $E = \mathbb{Q}(2^{1/3}, \omega)$, $G \cong S_3$ we obtained

$$\begin{aligned} \mathbb{Q}(\omega) &= E^{A_3} \\ \mathbb{Q}(2^{1/3}) &= E^{\langle\mu_1\rangle} \\ \mathbb{Q}(2^{1/3}\omega) &= E^{\langle\mu_2\rangle} \\ \mathbb{Q}(2^{1/3}\omega^2) &= E^{\langle\mu_3\rangle} \end{aligned}$$

as the fixed fields. ◇

12.2.8 EXAMPLE Let F be a field, E an extension field of F, and $G = \text{Gal}(E/F)$. Then $F \subseteq E^G$, but equality need not hold.

(1) For $K_2 = \mathbb{Q}(2^{1/3})$ in Example 12.2.1, we found in Example 12.1.27 that $G = \text{Gal}(K_2 / \mathbb{Q}) = \{\text{id}\}$. Thus we have $K_2{}^G = K_2 \neq \mathbb{Q}$. This contrasts with $K_1 = \mathbb{Q}(\omega)$ in Example 12.2.1, where $G = \text{Gal}(K_1 / \mathbb{Q}) \cong A_3$ and $K_1{}^G = \mathbb{Q}$. The main difference is that K_1 is a splitting field while K_2 is not.

(2) Also, for the extension $F(\alpha)$ of $F = \mathbb{Z}_p[x]$, where α is a zero of $y^p - x$, we found in Example 12.1.28 that $G = \text{Gal}(F(\alpha)/F) = \{\text{id}\}$. Here again we have the fixed field $F(\alpha)^G = F(\alpha) \neq F$. In this case the extension is inseparable. ◇

Having these examples in mind, it will not be surprising that in the fundamental theorem of Galois theory we restrict ourselves to certain kinds of field extensions.

12.2.9 THEOREM Let E be a finite extension of a field F and $G = \text{Gal}(E/F)$. Then the following are equivalent:

(1) $E^G = F$

(2) Every irreducible polynomial $p(x) \in F[x]$ that has a zero in E is separable and has all its zeros in E.

(3) E is the splitting field of a separable polynomial $f(x) \in F[x]$.

Proof (1) ⇒ (2) Assume (1) and let $p(x) \in F[x]$ be an irreducible polynomial and $\alpha \in E$ a zero of $p(x)$. Let the number of distinct elements in the set $\{\sigma(\alpha) \mid \sigma \in G\}$ be n and list them as $\alpha = \alpha_1, \alpha_2, \ldots, \alpha_n$. The product of all $(x - \alpha_i)$ is a polynomial $h(x)$ of degree n:

$$(x - \alpha_1)(x - \alpha_2)\ldots(x - \alpha_n) =$$
$$x^n + a_{n-1}x^{n-1} + \ldots + a_1 x + a_0$$

Applying any $\sigma \in G$, we have

$$(x - \sigma(\alpha_1))(x - \sigma(\alpha_2))\dots(x - \sigma(\alpha_n)) =$$
$$x^n + \sigma(a_{n-1})x^{n-1} + \dots + \sigma(a_1)x + \sigma(a_0)$$

But since σ simply permutes the α_i, the product of the $(x - \alpha_i)$ and the product of the $(x - \sigma(\alpha_i))$ are the same, from which it follows that $\sigma(a_i) = a_i$ and the coefficients a_i of $h(x)$ all belong to E^G. Since $E^G = F$, $h(x) \in F[x]$, and since $p(x) \in F[x]$ is irreducible and $p(\alpha) = h(\alpha) = 0$, it follows that $p(x)$ divides $h(x)$ in $F[x]$. It follows that $p(x)$ factors over E into a product of some of the factors $(x - \alpha_i)$ of $h(x)$, which are all distinct, which is to say that $p(x)$ is separable and has all its zeros in E.

(2) $\Rightarrow$ (3) Assume (2). If $E = F$, there is nothing to prove. So assume there exists an $\alpha_1 \in E$ such that $\alpha_1 \notin F$. Let $p_1(x) \in F[x]$ be the minimal polynomial of α_1 over F. By (2), $p_1(x)$ is separable and splits over E. Let E_1 be the splitting field of $p_1(x)$ over F in E. We have $F \subseteq E_1 \subseteq E$. If $E_1 = E$ we are done. Otherwise, there exists an $\alpha_2 \in E$ such that $\alpha_2 \notin E_1$. Let $p_2(x) \in F[x]$ be the minimal polynomial of α_2 over F. Again by (2), $p_2(x)$ is separable and splits over E. Let E_2 be the splitting field of $p_1(x)p_2(x)$ over F in E. If $E_2 = E$ we are done. Otherwise, repeat the process. Since the extension is finite, the process cannot go on forever, and eventually we obtain a separable polynomial $p_1(x)p_2(x)\dots p_r(x)$ whose splitting field over F is E.

(3) $\Rightarrow$ (1) Assume (3), so that E is the splitting field of the separable polynomial $f(x) \in F[x]$ over F. Then E is the splitting field of $f(x)$ over E^G. By Theorem 12.1.24 we have

$$[E:E^G] = |\mathrm{Gal}(E/E^G)|$$
$$[E:F] = |\mathrm{Gal}(E/F)|$$

Now since $F \subseteq E^G \subseteq E$, every automorphism of E that leaves each element of E^G fixed leaves each element of F fixed, and $\mathrm{Gal}(E/E^G) \leq \mathrm{Gal}(E/F)$. But conversely, by definition of the fixed field, every automorphism of E that leaves each element of F fixed leaves each element of E^G fixed, and hence $\mathrm{Gal}(E/F) \leq \mathrm{Gal}(E/E^G)$. So we get $\mathrm{Gal}(E/E^G) = \mathrm{Gal}(E/F) = G$ and $[E:E^G] = [E:F]$. It follows that $[E^G:F] = 1$ and $E^G = F$. □

12.2.10 DEFINITION A finite extension field E of a field F is called a **Galois extension** (or **normal extension**) if it satisfies any and hence all of the three equivalent conditions in the preceding theorem. ○

12.2.11 COROLLARY If E is the splitting field of a separable polynomial $f(x)$, where $f(x) \in F[x]$, then E is a finite separable extension of F.

Proof Let E be the splitting field of a separable polynomial $f(x) \in F[x]$. Let $n = \deg f(x)$. Then by Theorem 10.3.10, $[E:F] \leq n!$ and E is a finite extension of F. By assumption, E satisfies condition (3) of Theorem 12.2.9, and so by that theorem E also satisfies condition (2). But this condition implies that if $\alpha \in E$ and $p(x)$ is the minimal polynomial of α over F, then $p(x)$ is separable. Hence every element $\alpha \in E$ is separable over F and E is a separable extension of F. □

12.2.12 COROLLARY Let $F \subseteq K \subseteq E$ be a tower of fields. If E is a Galois extension of F, then E is a Galois extension of K.

Proof Immediate from Theorem 12.2.9, condition (3). □

12.2.13 COROLLARY Let $F \subseteq K \subseteq E$ be a tower of fields. If E is a Galois extension of F, then $K = E^{\mathrm{Gal}(E/K)}$.

Proof Immediate from the preceding Corollary and Theorem 12.2.9, condition (1). □

12.2.14 EXAMPLES

(1) $\mathbb{Q}(2^{1/3}, \omega)$ is a Galois extension of $\mathbb{Q}$.

(2) $\mathbb{C} = \mathbb{R}(i)$ is a Galois extension of $\mathbb{R}$.

(3) $\mathbb{Q}(2^{1/3})$ is not a Galois extension of $\mathbb{Q}$. ◇

12.2.15 EXAMPLE Let E be a finite field of order $|E| = p^n$. We have seen in Section 10.4 that E is a finite extension of $\mathbb{Z}_p$ with $[E:\mathbb{Z}_p] = n$ and that E is the splitting field of $f(x) = x^{p^n} - x$ over $\mathbb{Z}_p$. Since $f'(x) = -1 \neq 0$, E is a separable extension of $\mathbb{Z}_p$. Hence E is a Galois extension of $\mathbb{Z}_p$, and $|\mathrm{Gal}(E/\mathbb{Z}_p)| = n$. ◇

The fundamental theorem of Galois theory gives a correspondence between subfields and subgroups similar to that we found in Example 12.2.1 in the case of any Galois extension.

12.2.16 THEOREM (Fundamental theorem of Galois theory) Let E be a Galois extension of a field F. For any intermediate field K, $F \subseteq K \subseteq E$, let $\chi(K) = \mathrm{Gal}(E/K)$. Then

(1) χ is a one-to-one map from the set of all intermediate fields K to the set of all subgroups of $\mathrm{Gal}(E/F)$.

(2) $K = E^{\mathrm{Gal}(E/K)}$

(3) $\chi(E^H) = H$ for all $H \leq \mathrm{Gal}(E/F)$

(4) $[E:K] = |\mathrm{Gal}(E/K)|$

(5) $[K:F] =$ the index of $\mathrm{Gal}(E/K)$ in $\mathrm{Gal}(E/F)$

(6) K is a Galois (or normal) extension of F if and only if $\mathrm{Gal}(E/K) \lhd \mathrm{Gal}(E/F)$, in which case

$$\mathrm{Gal}(K/F) \cong \mathrm{Gal}(E/F) / \mathrm{Gal}(E/K)$$

(7) For any two intermediate fields K_1, K_2 we have $K_1 \subseteq K_2$ if and only if $\chi(K_2) \leq \chi(K_1)$, and thus the lattice of subgroups $H \leq \mathrm{Gal}(E/F)$ is the lattice of intermediate fields $F \subseteq K \subseteq E$, inverted.

Proof Several parts of the theorem follow easily from things we have already done.

(2) This is Corollary 12.2.13.

(4) This is immediate from Theorem 12.1.24 and Corollary 12.2.12.

(5) Follows from Theorem 12.1.24, which tells us

$$|\mathrm{Gal}(E/F)| = [E{:}F]$$
$$|\mathrm{Gal}(E/K)| = [E{:}K]$$

and Theorem 10.2.20, which tells us

$$[E{:}F] = [K{:}F][E{:}K]$$

(6) ($\Rightarrow$) This is Theorem 12.2.4.

(7) ($\Rightarrow$) This is Proposition 12.2.2.

(7) ($\Leftarrow$) This is Proposition 12.2.5, part (2).

(1) (one to one) Suppose $\chi(K_1) = \chi(K_2)$, that is, $\mathrm{Gal}(E/K_1) = \mathrm{Gal}(E/K_2)$. Then by (2)

$$K_1 = E^{\mathrm{Gal}(E/K_1)} = E^{\mathrm{Gal}(E/K_2)} = K_2$$

(1) (onto) This will be immediate once we prove (3).

It remains to prove (3) and the ($\Leftarrow$) direction of (6).

(3) Let H be a subgroup of $\mathrm{Gal}(E/F)$. By definition of χ, $\chi(E^H) = \mathrm{Gal}(E/E^H)$. By definition of E^H, we have $\rho(\alpha) = \alpha$ for any $\rho \in H$ and any $\alpha \in E^H$, and so $H \leq \mathrm{Gal}(E/E^H)$. It is therefore enough to show $|H| \geq |\mathrm{Gal}(E/E^H)| = [E{:}E^H]$. Now E is a finite separable extension of E^H, and therefore by the primitive element theorem (Theorem 12.1.30), $E = E^H(\alpha)$ for some $\alpha \in E$. Let n be the degree of α over E^H. Then $[E{:}E^H] = [E^H(\alpha){:}E^H] = n$, so it is enough to show that $|H| \geq n$. Let $|H| = s$ and let the elements of H be $\rho_1, \ldots, \rho_s$. Consider the polynomial $f(x)$ of degree s given by

$$f(x) = (x - \rho_1(\alpha))(x - \rho_2(\alpha))\ldots(x - \rho_s(\alpha))$$

Since one of the ρ_i is the identity, one of the factors is $(x - \alpha)$, and $f(\alpha) = 0$. Since the ρ_i are distinct elements of H for any ρ_i, we have

$$(x - \rho_i\rho_1(\alpha))(x - \rho_i\rho_2(\alpha))\ldots(x - \rho_i\rho_s(\alpha)) = f(x)$$

since the factors are the same except for order. As in the proof of the direction (1) $\Rightarrow$ (2) of Theorem 12.2.9, this implies that the coefficients of $f(x)$ belong to E^H. That is, $f(x)$ is an element of $E^H[x]$ of degree s having $f(\alpha) = 0$. Since n is the degree of α over E^H, it follows that $s \geq n$ as required to complete the proof.

(6) ($\Leftarrow$) Suppose that $\mathrm{Gal}(E/K) \lhd \mathrm{Gal}(E/F)$. We want to show K is a Galois extension of F. We first note that if $\tau \in \mathrm{Gal}(E/F)$, then for any $\rho \in \mathrm{Gal}(E/K)$ we have $\tau^{-1} \circ \rho \circ \tau \in \mathrm{Gal}(E/K)$ and so for any $\alpha \in K$ we have $\tau^{-1} \circ \rho \circ \tau(\alpha) = \alpha$ or $\rho(\tau(\alpha)) = \tau(\alpha)$, and therefore $\tau(\alpha) \in E^{\mathrm{Gal}(E/K)} = K$. Now K is a finite separable extension of F since E is and $K \subseteq E$. So by the primitive element theorem, $K = F(\alpha)$ for some $\alpha \in K$. Let $p(x)$ be the minimal polynomial of α over F. By Theorem 12.2.9, condition (2), all the zeros of $p(x)$ belong to E. If β is any zero of $p(x)$, then by Lemma 12.1.23 there is an element $\tau \in \mathrm{Gal}(E/F)$ with $\beta = \tau(\alpha)$. But we have just shown that $\tau(\alpha) \in K$ for all $\alpha \in K$ and $\tau \in \mathrm{Gal}(E/F)$, so $\beta \in K$. So all the zeros of

$p(x)$ belong to K, and K is the splitting field of $p(x)$ over F and hence a Galois extension of F. □

The fundamental theorem of Galois theory, which we have just proved, builds a bridge between the two parts of algebra we have been studying, the theory of groups and the theory of fields. Our knowledge of group theory can now be put to work to help us describe the structure of field extensions, as is illustrated in the next examples.

12.2.17 EXAMPLE In Example 11.1.22 we considered the polynomial $f(x) = x^4 - 4x^2 + 2 \in \mathbb{Q}[x]$, which is irreducible by Eisenstein's criterion, and found that its four zeros in $\mathbb{C}$ are

$$\alpha = \sqrt{2+\sqrt{2}} \qquad \beta = \sqrt{2-\sqrt{2}} \qquad -\alpha = -\sqrt{2+\sqrt{2}} \qquad -\beta = -\sqrt{2-\sqrt{2}}$$

Let us calculate the Galois group $\mathrm{Gal}(\mathbb{Q}(\alpha)/\mathbb{Q})$. We have

$$\sqrt{2} = \alpha^2 - 2 \in \mathbb{Q}(\alpha) \qquad \alpha\beta = \sqrt{2} \in \mathbb{Q}(\alpha)$$

and therefore all four zeros belong to $\mathbb{Q}(\alpha)$, which is therefore the splitting field of $x^4 - 4x^2 + 2$. Hence $|\mathrm{Gal}(\mathbb{Q}(\alpha)/\mathbb{Q})| = [\mathbb{Q}(\alpha):\mathbb{Q}] = 4$. There must be an automorphism $\phi \in \mathrm{Gal}(\mathbb{Q}(\alpha)/\mathbb{Q})$ with $\phi(\alpha) = \beta$. For such an automorphism we have

$$\phi(\sqrt{2}) = \phi(\alpha^2 - 2) = \beta^2 - 2 = -\sqrt{2}$$

$$\beta\phi^2(\alpha) = \beta\phi(\beta) = \phi(\alpha)\phi(\beta) = \phi(\alpha\beta) = \phi(\sqrt{2}) = -\sqrt{2} \neq \sqrt{2} = \beta\alpha$$

and therefore $\phi^2(\alpha) \neq \alpha$, ϕ is not of order 2, and $\mathrm{Gal}(\mathbb{Q}(\alpha)/\mathbb{Q})$ is the cyclic group of order 4. We then have

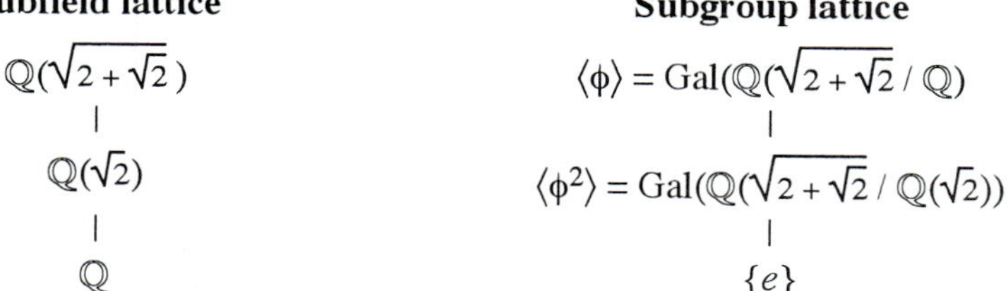

for the subfield and subgroup lattices. ◇

12.2.18 EXAMPLE Consider $E = \mathbb{Q}(\sqrt{2} + \sqrt{3}) = \mathbb{Q}(\sqrt{2}, \sqrt{3})$. E is the splitting field of $(x^2 - 2)(x^2 - 3)$. Let us calculate the Galois group $\mathrm{Gal}(E/\mathbb{Q})$. The subfield lattice is as follows:

$$\mathbb{Q}(\sqrt{2} + \sqrt{3}) = \mathbb{Q}(\sqrt{2}, \sqrt{3}).$$

/ \

$\mathbb{Q}(\sqrt{2})$ $\mathbb{Q}(\sqrt{3})$

\ /

$\mathbb{Q}$

Since $\mathbb{Q}(\sqrt{2})$ and $\mathbb{Q}(\sqrt{3})$ are the splitting fields of $(x^2 - 2)$ and $(x^2 - 3)$, respectively, any automorphism $\mathrm{Gal}(E/\mathbb{Q})$ is completely determined by the values $\phi(\sqrt{2}) = \pm\sqrt{2}$

and $\phi(\sqrt{3}) = \pm\sqrt{3}$. Hence the Galois group has order 4 with no element of order 4. Therefore, it is the Klein 4-group. ◇

12.2.19 EXAMPLE Now we work out a slightly more complicated example. Consider $f(x) = x^4 - 2x^2 - 2 \in \mathbb{Q}[x]$, which is irreducible by Eisenstein's criterion. Let us first work out its splitting field E. The four zeros of $f(x)$ can be found by setting $y = x^2$, using the quadratic formula to find the zeros of $g(y) = y^2 - 2y - 2$, and then taking square roots. The zeros are

$$\alpha = \sqrt{1+\sqrt{3}} \qquad \beta = \sqrt{1-\sqrt{3}}$$

and $-\alpha$ and $-\beta$. Thus the splitting field in $\mathbb{C}$ of $f(x)$ over $\mathbb{Q}$ is $E = \mathbb{Q}(\alpha, \beta)$. Note that

(1) $\alpha \in \mathbb{R}$, $\beta \notin \mathbb{R}$, hence $\beta \notin \mathbb{Q}(\alpha)$

(2) $\alpha^2 - 1 = 1 - \beta^2 = \sqrt{3}$, hence $\mathbb{Q}(\sqrt{3}) \subsetneq \mathbb{Q}(\alpha) \cap \mathbb{Q}(\beta) \subsetneq E$

(3) $[E:\mathbb{Q}] = [E:\mathbb{Q}(\alpha)][\mathbb{Q}(\alpha):\mathbb{Q}(\sqrt{3})][\mathbb{Q}(\sqrt{3}):\mathbb{Q}] = 2 \cdot 2 \cdot 2 = 8$

(4) $|\text{Gal}(E/\mathbb{Q})| = [E:\mathbb{Q}] = 8$

And finally note that since $\mathbb{Q}(\alpha)$ contains only 2 of the 4 zeros of the irreducible polynomial $f(x)$, $\mathbb{Q}(\alpha)$ is not a Galois extension of $\mathbb{Q}$, and hence the corresponding subgroup H of $\text{Gal}(E/\mathbb{Q})$ is not a normal subgroup, so $\text{Gal}(E/\mathbb{Q})$ is non-Abelian. Thus $\text{Gal}(E/\mathbb{Q})$ is a non-Abelian group of order 8.

Now let $\phi \in \text{Gal}(E/\mathbb{Q})$. Then $\phi(\sqrt{3}) = \pm\sqrt{3}$. If $\phi(\sqrt{3}) = \sqrt{3}$ then $(\phi(\alpha))^2 - 1 = \phi(\alpha^2 - 1) = \phi(\sqrt{3}) = \sqrt{3} = \alpha^2 - 1$, so $\phi(\alpha) = \pm\alpha$, and similarly $\phi(\beta) = \pm\beta$. If $\phi(\sqrt{3}) = -\sqrt{3}$, then $\phi(\alpha) = \pm\beta$ and $\phi(\beta) = \pm\alpha$.

TABLE 1 Correspondence between $\text{Gal}(E/\mathbb{Q})$ and D_4.

	$\text{Gal}(E/\mathbb{Q})$	$\cong$	D_4
(1)	id		ρ_0
(2)	$\alpha \to \alpha$ $\beta \to -\beta$		τ
(3)	$\alpha \to -\alpha$ $\beta \to \beta$		$\rho^2\tau$
(4)	$\alpha \to -\alpha$ $\beta \to -\beta$		ρ^2
(5)	$\alpha \to \beta$ $\beta \to \alpha$		$\rho^3\tau = \tau\rho$
(6)	$\alpha \to -\beta$ $\beta \to \alpha$		ρ
(7)	$\alpha \to \beta$ $\beta \to -\alpha$		ρ^3
(8)	$\alpha \to -\beta$ $\beta \to -\alpha$		$\rho\tau = \tau\rho^3$

Table 1 lists the elements of Gal($E/\mathbb{Q}$) on the left and the corresponding elements of D_4 on the right. It is not hard to check that the correspondence between elements of Gal($E/\mathbb{Q}$) and elements of D_4 indicated in the table is an isomorphism. We call the elements of $G = \text{Gal}(E/\mathbb{Q})$ by the names of the corresponding elements

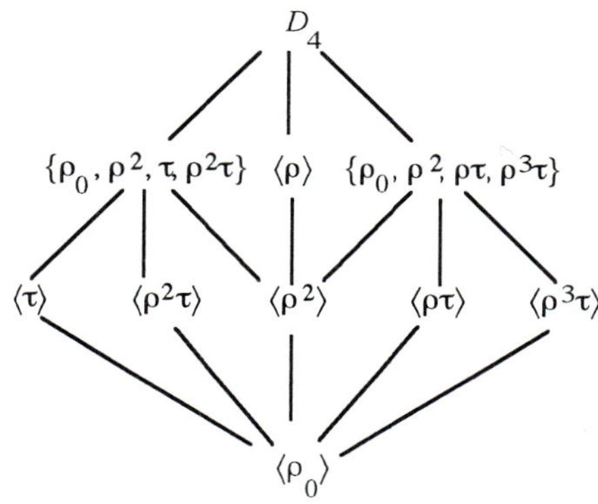

FIGURE 2

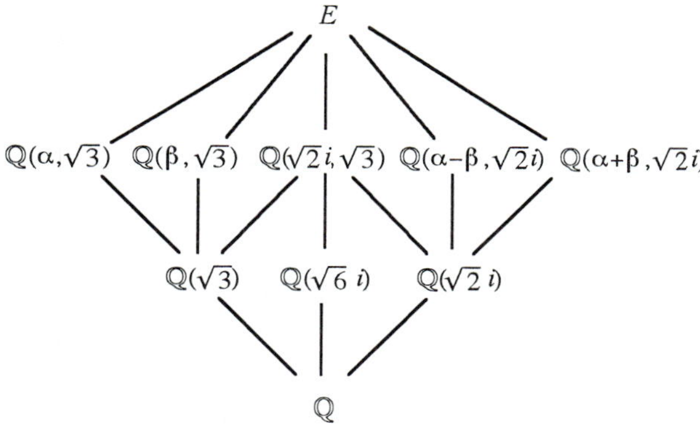

FIGURE 3

Now let us consider the correspondence between subgroups and intermediate fields. Let $H = \langle\rho\rangle \leq D_4$, and let us determine E^H. We have

$$\rho(\sqrt{2}i) = \rho(\alpha\beta) = (-\beta)\alpha = -\alpha\beta = -\sqrt{2}i$$
$$\rho(2\sqrt{3}) = \rho(\alpha^2 - \beta^2) = (-\beta)^2 - \alpha^2 = -2\sqrt{3}$$
$$\rho(-\sqrt{6}i) = \rho((\alpha^2 - \beta^2)/\alpha\beta) = (-2\sqrt{3})/(-\sqrt{2}i) = -\sqrt{6}i$$

Hence $\sqrt{6}i \in E^H$ and indeed $\mathbb{Q}(\sqrt{6}i) = E^H$, since $[E^H : \mathbb{Q}] = \text{index}_G H = 2$. So in this case E^H is the splitting field of $x^2 + 6$.

Now let $H = \langle\rho^2\rangle$ instead, and let us determine E^H. We have

$$\rho^2(\sqrt{2}i) = \rho^2(\alpha\beta) = (-\alpha)(-\beta) = \alpha\beta = \sqrt{2}i$$
$$\rho^2(2\sqrt{3}) = \rho^2(\alpha^2 - \beta^2) = (-\alpha)^2 - (-\beta)^2 = 2\sqrt{3}$$

Hence $\sqrt{2}i,\ 2\sqrt{3} \in E^H$ and indeed we have $\mathbb{Q}(\sqrt{2}i, 2\sqrt{3}) = E^H$, since we have $[E^H : \mathbb{Q}] = \text{index}_G H = 4$. So in this case E^H is the splitting field of $(x^2 + 2)(x^2 - 3)$. The resulting correspondence between subfields and subgroups is indicated in Figures 2 and 3. Note that the four nontrivial normal subgroups of D_4 correspond to the four intermediate fields that are splitting fields of polynomials over $\mathbb{Q}$:

Normal subgroup	**Polynomial**
$\langle\rho\rangle$	$x^2 + 6$
$\langle\rho^2\rangle$	$(x^2 + 2)(x^2 - 3)$
$\{\rho_0, \rho^2, \tau, \rho^2\tau\}$	$x^2 - 3$
$\{\rho_0, \rho^2, \rho\tau, \rho^3\tau\}$	$x^2 + 2$

The splitting fields of these polynomials are the intermediate fields that are Galois extensions of $\mathbb{Q}$.◇

The fundamental theorem of Galois theory has an important implication about the structure of finite fields, as shown in our next example.

12.2.20 EXAMPLE Let E be a finite field with $|E| = p^n$. Then by Example 12.2.15, E is a Galois extension of $\mathbb{Z}_p$ and $|\text{Gal}(E/\mathbb{Z}_p)| = n$. Furthermore, by the primitive element theorem (Theorem 12.1.30), E is a simple extension of $\mathbb{Z}_p$. Hence $E = \mathbb{Z}_p(\gamma)$, where γ is a zero of $f(x) = x^{p^n} - x$. Now consider the Frobenius automorphism $\sigma: E \to E$ defined by $\sigma(\alpha) = \alpha^p$. Then $E^{\langle\sigma\rangle} = \mathbb{Z}_p$ since $\alpha^p = \alpha$ for all $\alpha \in \mathbb{Z}_p$ by Fermat's theorem, while $x^p - x$ can have at most p zeros. Hence by the fundamental theorem of Galois theory,

$$\langle\sigma\rangle = \chi(E^{\langle\sigma\rangle}) = \chi(\mathbb{Z}_p) = \text{Gal}(E/\mathbb{Z}_p)$$

and $\text{Gal}(E/\mathbb{Z}_p)$ is a cyclic group of order n generated by the Frobenius automorphism. ◇

We end this important section of the book with the proof of the fundamental theorem of algebra, which was promised in Chapter 10, Theorem 10.3.25. The proof uses both the powerful tools of Galois theory and the Sylow theorems from group theory. This is another instance where we put our knowledge of group structures and field extensions to work in combination with each other.

Any proof of the fundamental theorem of algebra must use some facts from calculus. Our proof uses just one such fact, which we now state without proof. A proof can be found in most calculus textbooks.

12.2.21 THEOREM (Intermediate value theorem) Let $f(x)$ be a polynomial with real coefficients, $f(x) \in \mathbb{R}[x]$, and suppose there exist $a, b \in \mathbb{R}$ such that $f(a) < 0$ and $f(b) > 0$. Then there exists some $c \in \mathbb{R}$ such that $f(c) = 0$. □

12.2.22 EXAMPLE Consider any $a \in \mathbb{R}$ with $a > 0$, and let $f(x) = x^2 - a$. Since

$$f(0) = -a < 0 \text{ and } f(a + 1) = a^2 + a + 1 > 0$$

the intermediate value theorem guarantees the existence of some $c \in \mathbb{R}$ such that $f(c) = 0$, that is, $c^2 = a$. In other words, the fact that every positive real number has a square root is a special case of the theorem.

In guaranteeing the existence of a square root for any positive real number, the theorem also guarantees the existence of a square root for any complex number. For given $z = a + bi$, let

$$c = \sqrt{\frac{a + \sqrt{a^2 + b^2}}{2}} \qquad d = \sqrt{\frac{-a + \sqrt{a^2 + b^2}}{2}}$$

Note that since $\sqrt{a^2 + b^2} \geq |a|$, the quantities under both radical signs are positive, and c and d are positive real numbers. If we now let $w = c + di$ if $b > 0$, or $w = c - di$ if $b < 0$, then direct computation shows $w^2 = z$.

Since the quadratic formula expresses the zeros of any quadratic polynomial in terms of square roots, by guaranteeing the existence of the square roots, the intermediate value theorem guarantees that every polynomial $f(x) \in \mathbb{C}[x]$ of degree 2 has a zero in $\mathbb{C}$. ◇

12.2.23 COROLLARY Every polynomial $f(x) \in \mathbb{R}[x]$ of odd degree has a zero in $\mathbb{R}$.

Proof Let

$$f(x) = x^n + a_{n-1}x^{n-1} + \ldots + a_1x + a_0 \in \mathbb{R}[x]$$

where n is odd, and let

$$u = 1 + |a_{n-1}| + \ldots + |a_1| + |a_0|$$

and consider

$$f(u) = u^n + a_{n-1}u^{n-1} + \ldots + a_1u + a_0$$

We have

$$|a_{n-1}u^{n-1} + \ldots + a_1u + a_0| \leq |a_{n-1}|u^{n-1} + \ldots + |a_1|u + |a_0|$$

and since

$$|a_i| \leq |a_{n-1}| + \ldots + |a_1| + |a_0| = u - 1$$

we obtain

$$|a_{n-1}u^{n-1} + \ldots + a_1u + a_0| \leq (u - 1)(u^{n-1} + \ldots + u + 1) = u^n - 1 < u^n$$

Therefore,

$$f(u) \geq u^n - |a_{n-1}u^{n-1} + \ldots + a_1u + a_0| > 0$$

Since n is odd, $(-u)^n = -u^n$ and

$$f(-u) \leq -u^n + |a_{n-1}u^{n-1} + \ldots + a_1u + a_0| < 0$$

Hence Theorem 12.2.21 guarantees the existence of some $c \in \mathbb{R}$ with $f(c) = 0$. □

We now have all the information needed for the proof.

Fundamental theorem of algebra (Theorem 10.3.25) $\mathbb{C}$ is algebraically closed.

Proof By Theorem 10.3.24 it suffices to show that any nonconstant polynomial $f(x) \in \mathbb{C}[x]$ has a zero in $\mathbb{C}$. Let

$$f(x) = a_n x^n + a_{n-1}x^{n-1} + \ldots + a_1 x + a_0$$

and let

$$\overline{f(x)} = \bar{a}_n x^n + \bar{a}_{n-1} x^{n-1} + \ldots + \bar{a}_1 x + \bar{a}_0 \in \mathbb{C}[x]$$

where $\bar{a}_i$ is the complex conjugate of a_i. Let $g(x) = f(x)\overline{f(x)}$. Then $g(x) \in \mathbb{R}[x]$, and $f(x)$ has a zero in $\mathbb{C}$ if and only if $g(x)$ does. (See Exercise 19.) Therefore, it suffices to show that any nonconstant polynomial in $\mathbb{R}[x]$ has a zero in $\mathbb{C}$.

Let $g(x)$ be an irreducible polynomial in $\mathbb{R}[x]$. Let

$$h(x) = (x^2 + 1)g(x) \in \mathbb{R}[x]$$

and let E be an extension field of $\mathbb{C}$ that is a splitting field of $h(x)$. E is a Galois extension of $\mathbb{R}$ by Theorem 12.1.6. Consider now the Galois group $\mathrm{Gal}(E/\mathbb{R})$. Let $|\mathrm{Gal}(E/\mathbb{R}| = 2^k m$, where m is odd.

By the first Sylow theorem (Theorem 4.6.5), $\mathrm{Gal}(E/\mathbb{R})$ has a 2-Sylow subgroup P of order 2^k. By Theorem 12.2.16, part (5),

$$[E^P : \mathbb{R}] = \text{index } P = m$$

Hence E^P is an extension of $\mathbb{R}$ of degree m, which implies that there exists an irreducible polynomial of degree m over $\mathbb{R}$. By Corollary 12.2.23, the only odd m for which this is possible is $m = 1$. Thus $|\mathrm{Gal}(E/\mathbb{R})| = 2^k$.

Since $\mathbb{C} = \mathbb{R}(i) \subseteq E$, $\mathrm{Gal}(E/\mathbb{C})$ is a subgroup of $\mathrm{Gal}(E/\mathbb{R})$ and hence $|\mathrm{Gal}(E/\mathbb{C})| = 2^n$ for some $n \leq k$. By Theorem 4.6.13, every p-group of order p^n has a subgroup of order p^i for every $1 \leq i \leq n$. Thus if $n > 0$, $\mathrm{Gal}(E/\mathbb{C})$ has a subgroup H of order 2^{n-1}. Then

$$[E^H : \mathbb{C}] = \text{index } H = 2$$

Hence E^H is an extension of $\mathbb{C}$ of degree 2, which implies that there exists an irreducible polynomial of degree 2 over $\mathbb{C}$. But as remarked in Example 12.2.22, this is impossible. So $n = 0$, $E = \mathbb{C}$, and $\mathbb{C}$ contains the zeros of the polynomial $g(x) \in \mathbb{R}[x]$. □

Exercises 12.2

In Exercises 1 through 4 determine whether the indicated field extension is a Galois extension.

1. $\mathbb{Q}(\sqrt{3}\, i)$ over $\mathbb{Q}$

2. $\mathbb{Q}(\sqrt[3]{3}\,)$ over $\mathbb{Q}$

3. $\mathbb{Q}(\cos\frac{2\pi}{7} + i\sin\frac{2\pi}{7})$ over $\mathbb{Q}$

4. $\mathbb{Q}(\sqrt[3]{2}, i\sin\frac{2\pi}{3})$ over $\mathbb{Q}$

In Exercises 5 through 8 let $E = \mathbb{Q}(\sqrt{2}, \sqrt{3}, i)$, and consider the intermediate fields $\mathbb{Q} \subseteq K \subseteq E$.

5. Construct the subfield lattice for E over $\mathbb{Q}$.

6. For each K, calculate $|\mathrm{Gal}(E/K)|$.

7. For each K, calculate $|\chi(K)|$.

8. Describe $\mathrm{Gal}(E/\mathbb{Q})$.

9. Calculate $\mathrm{Gal}(\mathbb{Q}(\sqrt{3+\sqrt{3}})/\mathbb{Q})$.

10. Calculate the Galois group of the splitting field of $x^3 + 1$ over $\mathbb{Q}$.

In Exercises 11 through 14 let $f(x) = x^4 - 2x^2 - 1 \in \mathbb{Q}[x]$.

11. Find the splitting field E of $f(x)$ over $\mathbb{Q}$.

12. Find the Galois group $\mathrm{Gal}(E/\mathbb{Q})$.

13. Construct the subfield lattice for E and the corresponding subgroup lattice for $\mathrm{Gal}(E/\mathbb{Q})$.

14. Identify the normal subgroups $H \triangleleft \mathrm{Gal}(E/\mathbb{Q})$ and show that in each case

(a) E^H is a Galois extension of $\mathbb{Q}$.

(b) $\mathrm{Gal}(E^H/\mathbb{Q}) \cong \mathrm{Gal}(E/\mathbb{Q}) / \mathrm{Gal}(E/E^H)$

In Exercises 15 through 17 give an example of a Galois extension of $\mathbb{Q}$ such that the Galois group is as indicated.

15. $\mathrm{Gal}(E/\mathbb{Q}) \cong \mathbb{Z}_3$

16. $\mathrm{Gal}(E/\mathbb{Q}) \cong \mathbb{Z}_4$

17. $\mathrm{Gal}(E/\mathbb{Q}) \cong \mathbb{Z}_6$

18. Give an example of a field F with two nonisomorphic extensions E_1 and E_2 such that $\mathrm{Gal}(E_1/F) \cong \mathrm{Gal}(E_2/F)$.

19. Let $f(x) \in \mathbb{C}[x]$. Show that $\overline{f(x)}$ has a complex zero if and only if $f(x)$ does.

20. Let E be a Galois extension of a field F such that $\mathrm{Gal}(E/F)$ is Abelian. Show that for any intermediate field $F \subseteq K \subseteq E$, K is a Galois extension of F.

21. Let E be a Galois extension of a field F and K an intermediate field $F \subseteq K \subseteq E$. Show that if $\mathrm{Gal}(E/F)$ is cyclic, then K is a Galois extension of F and $\mathrm{Gal}(K/F)$ and $\mathrm{Gal}(E/K)$ are both cyclic.

22. Let E be a Galois extension of a field F with $\mathrm{Gal}(E/F)$ cyclic. Show that for each divisor d of $[E:F]$ there exists exactly one intermediate field $F \subseteq K \subseteq E$ such that $[K:F] = d$.

Norm and Trace

Let E be a Galois extension of a field F with Galois group $G = \text{Gal}(E/F)$. For any element $\alpha \in E$ we define the **norm** and **trace** of α to be

$$N(\alpha) = \prod_{\sigma \in G} \sigma(\alpha) \qquad \text{Tr}(\alpha) = \sum_{\sigma \in G} \sigma(\alpha)$$

For Exercises 23 through 26 consider $E = \mathbb{Q}(2^{1/3}, \sqrt{3}i)$ over $\mathbb{Q}$ and calculate the norm and trace of the indicated elements $\alpha \in E$.

23. $\alpha = 2^{1/3}$ **24.** $\alpha = \sqrt{3}i$ **25.** $\alpha = 2^{1/3} + \sqrt{3}i$ **26.** $\alpha = -\frac{1}{2} + \frac{\sqrt{3}}{2}i$

27. Show that for any Galois extension E over F and any $\alpha \in E$, $\text{N}(\alpha) \in F$ and $\text{Tr}(\alpha) \in F$.

12.3 Galois Groups of Polynomials

We understand by the *Galois group* of a given polynomial over a given field the Galois group of its splitting field. In this section we are concerned with calculating the Galois groups of polynomials, concentrating on cubics and quartics. We introduce the notion of the *discriminant* of a polynomial and, in the case of a quartic polynomial, the notion of its *resolvant cubic*. This is a cubic having the same discriminant, whose Galois group is a quotient of the Galois group of the quartic, so that knowing the Galois group for the resolvant cubic helps to determine that for the original quartic. To understand better both the discriminant and the resolvant, we introduce the *elementary symmetric functions* of the zeros of a polynomial. As a consequence of the *fundamental theorem of symmetric functions*, we show that the *general polynomial* of degree n has Galois group S_n.

12.3.1 DEFINITION We saw in Section 10.3 that the splitting field E of a given polynomial $f(x) \in F[x]$ over a given field F is unique up to isomorphism, and in Exercise 22 in Section 12.1 that the Galois group $\text{Gal}(E/F)$ is also unique up to isomorphism, and we call it the **Galois group** of the polynomial. ○

We saw in Proposition 12.1.26 that the Galois group of a polynomial can always be viewed as a subgroup of S_n.

12.3.2 EXAMPLE In Example 12.2.17, where we considered the polynomial $f(x) = x^4 - 4x^2 + 2$, we found its four zeros α, β, $-\alpha$, $-\beta$ and its splitting field E. We saw that $\text{Gal}(E/\mathbb{Q}) \cong \mathbb{Z}_4$.

Now $\mathbb{Z}_4$ can be viewed as a subgroup of S_4, and if we number the zeros $\alpha_1 = \alpha$, $\alpha_2 = \beta$, $\alpha_3 = -\alpha$, $\alpha_4 = -\beta$, then we can see the correspondence between elements of $\text{Gal}(E/\mathbb{Q})$ and elements of S_4 directly.

	Galois group Gal($E/\mathbb{Q}$)	**Subgroup of** S_4
ρ_0	identity	ρ_0 = identity
ϕ	$\alpha_1 \to \alpha_2 \to \alpha_3 \to \alpha_4 \to \alpha_1$	(1 2 3 4)
ϕ^2	$\alpha_1 \leftrightarrow \alpha_3 \quad \alpha_2 \leftrightarrow \alpha_4$	(1 3)(2 4)
ϕ^3	$\alpha_1 \to \alpha_4 \to \alpha_3 \to \alpha_2 \to \alpha_1$	(1 4 3 2)

We found the one intermediate field between $\mathbb{Q}$ and E to be $\mathbb{Q}(\alpha_1\alpha_2)$. ◇

12.3.3 EXAMPLE In Example 12.2.19 we considered in $\mathbb{Q}[x]$ the polynomial $f(x) = x^4 - 2x^2 - 2$. Again we found its four zeros α, β, $-\alpha$, $-\beta$ and its splitting field E. We saw this time that Gal($E/\mathbb{Q}$) $\cong D_4$. In Example 1.4.24 we saw how D_4 can be viewed as a subgroup of S_4. Again if we number the zeros $\alpha_1 = \alpha$, $\alpha_2 = \beta$, $\alpha_3 = -\alpha$, $\alpha_4 = -\beta$, then we can see the correspondence between elements of Gal($E/\mathbb{Q}$) and elements of S_4 directly. In this case we found the various intermediate fields between $\mathbb{Q}$ and E by considering expressions like $\alpha_1\alpha_2$ and $\alpha_1^2 - \alpha_2^2$ and seeing which permutations in S_4 kept which of these fixed. ◇

We now undertake a more general study of such polynomial expressions in the zeros of a polynomial. First we need some definitions.

12.3.4 DEFINITION Let F be a field and E an extension field of F. The elements $\alpha_1, \alpha_2, \ldots, \alpha_n \in E$ are said to be **indeterminates** over F if

$$f(\alpha_1, \alpha_2, \ldots, \alpha_n) \neq 0$$

for every nonzero polynomial $f(x_1, x_2, \ldots, x_n) \in F[x_1, x_2, \ldots, x_n]$. ○

12.3.5 DEFINITION Let F be a field and $E = F(\alpha_1, \alpha_2, \ldots, \alpha_n)$, where the α_i are indeterminates over F. Then the polynomial

$$f(x) = (x - \alpha_1)(x - \alpha_2)\ldots(x - \alpha_n)$$

is called the **general polynomial** of degree n over F. ○

12.3.6 DEFINITION Let F be a field, $E = F(\alpha_1, \alpha_2, \ldots, \alpha_n)$, where the α_i are indeterminates over F, and $f(x)$ the general polynomial. If

$$f(x) = x^n - s_1x^{n-1} + s_2x^{n-2} - \ldots + (-1)^n s_n$$

then

$$\begin{aligned}
s_1 &= \alpha_1 + \alpha_2 + \ldots + \alpha_n \\
s_2 &= \alpha_1\alpha_2 + \alpha_1\alpha_3 + \ldots + \alpha_i\alpha_j + \ldots + \alpha_{n-1}\alpha_n \\
s_2 &= \alpha_1\alpha_2\alpha_3 + \alpha_1\alpha_2\alpha_4 + \ldots + \alpha_i\alpha_j\alpha_k + \ldots + \alpha_{n-2}\alpha_{n-1}\alpha_n \\
&\vdots \\
s_n &= \alpha_1\alpha_2\ldots\alpha_n
\end{aligned}$$

and the functions s_i, $1 \leq i \leq n$ are called the **elementary symmetric functions** of the α_i, $1 \leq i \leq n$. ○

For example, if $E = F(x_1, x_2, \dots, x_n)$, the field of rational functions in n variables over F, then the elements $x_1, x_2, \dots, x_n$ are indeterminates over F. The next proposition says that up to isomorphism this is the only example.

12.3.7 PROPOSITION Let F be a field, E an extension field of F, and $\alpha_1, \alpha_2, \dots, \alpha_n \in E$ indeterminates over F. Then

$$F(x_1, x_2, \dots, x_n) \cong F(\alpha_1, \alpha_2, \dots, \alpha_n)$$

Proof The elements of $F(x_1, x_2, \dots, x_n)$ are of the form

$$g_0(x_1, x_2, \dots, x_n) / g_1(x_1, x_2, \dots, x_n)$$

where g_1 is not the zero polynomial. Define a map ψ from $F(x_1, x_2, \dots, x_n)$ to $F(\alpha_1, \alpha_2, \dots, \alpha_n)$ by sending such an element to

$$g_0(\alpha_1, \alpha_2, \dots, \alpha_n) / g_1(\alpha_1, \alpha_2, \dots, \alpha_n)$$

This map is well defined since $g_1(\alpha_1, \alpha_2, \dots, \alpha_n) \neq 0$ if g_1 is not the zero polynomial. Moreover, only the zero element of $F(x_1, x_2, \dots, x_n)$ is carried to the zero element of $F(\alpha_1, \alpha_2, \dots, \alpha_n)$, since $g_0(\alpha_1, \alpha_2, \dots, \alpha_n) \neq 0$ if g_0 is not the zero polynomial. Since ψ carries sums, products, differences, and quotients of elements of $F(x_1, x_2, \dots, x_n)$ to the sums, products, differences, and quotients of the corresponding elements of $F(\alpha_1, \alpha_2, \dots, \alpha_n)$, ψ is an isomorphism. □

For example, in $\mathbb{Q}(\pi)$, the element π is an indeterminate over $\mathbb{Q}$ since π is transcendental over $\mathbb{Q}$. Therefore, $\mathbb{Q}(\pi) \cong \mathbb{Q}(x)$, the field of rational functions in one variable over $\mathbb{Q}$.

The permutation group S_n acts on $F(\alpha_1, \alpha_2, \dots, \alpha_n)$ by permuting subscripts. That is, $\sigma \in S_n$ applied to an element

$$g_0(\alpha_1, \alpha_2, \dots, \alpha_n) / g_1(\alpha_1, \alpha_2, \dots, \alpha_n)$$

gives the element

$$g_0(\alpha_{\sigma(1)}, \alpha_{\sigma(2)}, \dots, \alpha_n) / g_1(\alpha_{\sigma(1)}, \alpha_{\sigma(2)}, \dots, \alpha_{\sigma(n)})$$

The same proof as was given in the preceding proposition shows that this map is an automorphism. Thus, acting this way, the permutation σ can be viewed as an element of $\mathrm{Gal}(F(\alpha_1, \alpha_2, \dots, \alpha_n) / F)$.

12.3.8 DEFINITION Let F be a field and $E = F(\alpha_1, \alpha_2, \dots, \alpha_n)$, where the α_i are indeterminates over F. An element of E is called a **symmetric function** of the α_i, $1 \leq i \leq n$ if it is fixed by every permutation σ in S_n. ○

By inspection of the expressions for the elementary symmetric functions s_i in Definition 12.3.6, each of them is fixed under permutation of subscripts and is therefore a symmetric function.

12.3.9 EXAMPLES

(1) In the field $F(x_1, x_2, x_3)$ of rational functions, consider the function

$$f(x_1, x_2, x_3) = x_1^2 + x_2^2 + x_3^2$$

Then $f(x_1, x_2, x_3)$ remains fixed under any permutation $\rho \in S_3$ of subscripts and thus is a symmetric function. Note that

$$x_1^2 + x_2^2 + x_3^2 = (x_1 + x_2 + x_3)^2 - 2(x_1x_2 + x_1x_3 + x_2x_3)$$

Thus $f(x_1, x_2, x_3) = s_1^2 - 2s_2 \in F(s_1, s_2, s_3)$.

(2) Now consider the function

$$f(x_1, x_2, x_3) = x_1^2x_2^2 + x_1^2x_3^2 + x_2^2x_3^2$$

Again it is a symmetric function. We have

$$x_1^2x_2^2 + x_1^2x_3^2 + x_2^2x_3^2 = (x_1x_2 + x_1x_3 + x_2x_3)^2 - 2(x_1 + x_2 + x_3)(x_1x_2x_3)$$

Thus $f(x_1, x_2, x_3) = s_2^2 - 2s_1s_3 \in F(s_1, s_2, s_3)$. ◇

The importance of the elementary symmetric functions comes from the fact that *all* other symmetric functions can be obtained from them. This is our next theorem. But first we need a lemma that is also of interest because it shows that the general polynomial (Definition 12.3.5) gives us an "abstract" example of a polynomial of degree n whose Galois group is S_n.

12.3.10 LEMMA Let F be a field and $E = F(\alpha_1, \alpha_2, \dots, \alpha_n)$, where the α_i are indeterminates over F. Then E is a Galois extension of $F(s_1, s_2, \dots, s_n)$, where the s_i are the elementary symmetric functions of the α_i, and

$$\text{Gal}(E / F(s_1, s_2, \dots, s_n)) \cong S_n$$

Proof E is the splitting field over $F(s_1, s_2, \dots, s_n)$ of the general polynomial of degree n, which is separable since the α_i are all distinct, and therefore E is a Galois extension of $F(s_1, s_2, \dots, s_n)$. All elementary symmetric functions are symmetric functions, that is, are fixed by each automorphism $\rho \in S_n$ of E. Therefore,

$$S_n \leq \text{Gal}(E / F(s_1, s_2, \dots, s_n))$$

On the other hand, since the general polynomial has degree n, by Theorem 12.1.26

$$|\text{Gal}(E / F(s_1, s_2, \dots, s_n))| \leq n! = |S_n|$$

Therefore, $\text{Gal}(E / F(s_1, s_2, \dots, s_n)) \cong S_n$. □

12.3.11 THEOREM (Fundamental theorem of symmetric functions) Let F be a field and $E = F(\alpha_1, \alpha_2, \dots, \alpha_n)$, where the α_i are indeterminates over F. Then any symmetric function of the α_i can be written as a rational function of the elementary symmetric functions s_i.

Proof The symmetric functions are by definition the elements of the fixed field K of the subgroup S_n of $\text{Gal}(E / F)$. Since all elementary symmetric functions are symmetric functions, $F(s_1, s_2, \dots, s_n) \subseteq K$. Let $G = \text{Gal}(E / F(s_1, s_2, \dots, s_n))$. By the preceding lemma, the subgroup S_n of $\text{Gal}(E / F)$ is precisely G. By Theorem 12.2.9, condition (1), it follows that $K = F(s_1, s_2, \dots, s_n)$. To prove the theorem, we must show that if $h(\alpha_1, \alpha_2, \dots, \alpha_n)$ is any element of K, then there is some rational function $g(x_1, x_2, \dots, x_n)$ in $F(x_1, x_2, \dots, x_n)$ such that

$$h(\alpha_1, \alpha_2, \dots, \alpha_n) = g(s_1, s_2, \dots, s_n)$$

But this is now immediate, since the elements of $F(s_1, s_2, \dots, s_n)$ are precisely the rational functions $g(s_1, s_2, \dots, s_n)$ of the s_i. □

12.3.12 COROLLARY Let F be a field and $K = F(\beta_1, \beta_2, \dots, \beta_n)$, where the β_i are indeterminates over F. Then there is a separable polynomial of degree n in $K[x]$ whose Galois group over K is S_n.

Proof Consider the polynomial

$$f(x) = x^n - \beta_1 x^{n-1} + \beta_2 x^{n-2} - \dots + (-1)^n \beta_n$$

Let E be its splitting field over K, where we have the factorization

$$f(x) = (x - \alpha_1)(x - \alpha_2)\dots(x - \alpha_n)$$

Then

$$\beta_i = s_i(\alpha_1, \alpha_2, \dots, \alpha_n)$$

the elementary symmetric functions of the α_i. So, having Lemma 12.3.10, it is enough to show that the α_j are indeterminates over F. So suppose there were a nonzero polynomial $h(x_1, x_2, \dots, x_n)$ in $F[x_1, x_2, \dots, x_n]$ with $h(\alpha_1, \alpha_2, \dots, \alpha_n) = 0$. Consider the polynomial

$$H(x_1, x_2, \dots, x_n) = \prod_{\sigma \in S_n} h(x_{\sigma(1)}, x_{\sigma(2)}, \dots, x_{\sigma(n)})$$

This is a nonzero symmetric polynomial with $H(\alpha_1, \alpha_2, \dots, \alpha_n) = 0$. By the preceding theorem, there is a rational function

$$g(x_1, x_2, \dots, x_n) = g_0(x_1, x_2, \dots, x_n) \,/\, g_1(x_1, x_2, \dots, x_n)$$

in $F(x_1, x_2, \dots, x_n)$ such that

$$H(x_1, x_2, \dots, x_n) = g(s_1, s_2, \dots, s_n)$$

and hence

$$0 = H(\alpha_1, \alpha_2, \dots, \alpha_n) = g(\beta_1, \beta_2, \dots, \beta_n)$$

But this implies $g_0(\beta_1, \beta_2, \dots, \beta_n) = 0$, contrary to the assumption that the β_i are indeterminates. □

We now introduce a symmetric function that plays a very important role in the study of zeros of a polynomial.

12.3.13 EXAMPLE Let F be a field and let $f(x) \in F[x]$ be a polynomial of degree 3 with zeros $\alpha_1, \alpha_2, \alpha_3$ in its splitting field E. Consider the product

$$\Delta = (\alpha_1 - \alpha_2)(\alpha_1 - \alpha_3)(\alpha_2 - \alpha_3)$$

Let $\rho \in S_3$ act on E by permutation of subscripts, and consider $\rho(\Delta)$. We have

$$\rho(\Delta) = (\alpha_{\rho(1)} - \alpha_{\rho(2)})(\alpha_{\rho(1)} - \alpha_{\rho(3)})(\alpha_{\rho(2)} - \alpha_{\rho(3)})$$

and hence

$\rho(\Delta) = \Delta$ if ρ is an even permutation
$\rho(\Delta) = -\Delta$ if ρ is an odd permutation

or

$$\rho(\Delta) = (-1)^{\varepsilon(\rho)}\Delta$$

where $\varepsilon(\rho) = 0$ if ρ is even and 1 if ρ is odd. [A product like Δ with n numbers α_i was considered already in Example 1.4.29, where it was noted that $\rho \in A_n$ if and only if $\rho(\Delta) = \Delta$.] If $\Delta \in F$, then every permutation ρ in Gal(E/F) must keep Δ fixed and therefore must be an even permutation. Conversely, if there is any odd permuation ρ in Gal(E/F), it does not keep Δ fixed and therefore $\Delta \notin F$. Thus we have

$$\text{Gal}(E/F) \subseteq A_3 \text{ if and only if } \Delta \in F$$

But $D = \Delta^2$ is fixed by every ρ even or odd, and hence $D \in F$ in any case. ◇

12.3.14 DEFINITION Let F be a field of characteristic zero and $f(x) \in F[x]$. Let E be a splitting field of $f(x)$ over F and let $f(x)$ factor in $E[x]$ as

$$f(x) = u(x - \alpha_1)(x - \alpha_2)\ldots(x - \alpha_n)$$

Then

$$D = \prod_{i<j} (\alpha_i - \alpha_j)^2$$

is called the **discriminant** of $f(x)$. ○

Note that since this product is symmetric in the α_i, $D \in F$. Also, $D = 0$ if and only if $f(x)$ has a root of multiplicity > 1, that is, if and only if $f(x)$ is inseparable.

12.3.15 PROPOSITION Let F be a field of characteristic zero and $f(x) \in F[x]$ a separable polynomial of degree n with discriminant D, and E the splitting field of $f(x)$.

Then if $\Delta = \sqrt{D}$ we have

(1) If $H = \text{Gal}(E/F) \cap A_n$, then $E^H = F(\Delta)$.

(2) $\Delta \in F$ if and only if $\text{Gal}(E/F) \leq A_n$.

Proof (1) We have

$$\Delta = \sqrt{D} = \prod_{i<j} (\alpha_i - \alpha_j) \neq 0$$

since $f(x)$ is separable. Hence by Theorem 1.4.31, for any $\rho \in \text{Gal}(E/F)$ we have $\rho(\Delta) = \Delta$ if and only if ρ is an even permutation, which is to say if and only if $\rho \in H$. Hence $F(\Delta) \subseteq E^H$. By Exercise 26 in Section 1.4, the index of H in Gal(E/F) is 1 or 2.

Consider first the case where the index of H in Gal(E/F) is 1. In this case $\text{Gal}(E/F) \leq A_n$ and $H = \text{Gal}(E/F)$, and we have

$$F \subseteq F(\Delta) \subseteq E^H = F$$

We can conclude that $E^H = F(\Delta)$ and also that $F(\Delta) = F$ and hence $\Delta \in F$.

Consider now the case where the index of H in Gal(E/F) is 2. In this case we do not have $\text{Gal}(E/F) \leq A_n$. Therefore, there is some odd permutation $\rho \in \text{Gal}(E/F)$. Then

$\rho(\Delta) = -\Delta \neq \Delta$ since $\Delta \neq 0$. Hence $\Delta \notin F$, and since $\Delta^2 = D \in F$ we have $[F(\Delta):F] = 2$. Now we know

$$[E^H:F(\Delta)][F(\Delta):F] = [E^H:F] = \text{index of } H \text{ in } \mathrm{Gal}(E/F) = 2$$

So we can conclude $[E^H:F(\Delta)] = 1$ and $E^H = F(\Delta)$.
(2) By Theorem 12.2.16, $F(\Delta) = E^H = F$ if $H = \mathrm{Gal}(E/F)$, which is equivalent to the statement that $\mathrm{Gal}(E/F)$ is a subgroup of A_n. □

12.3.16 EXAMPLE Let $f(x) = x^2 + ax + b \in \mathbb{Q}[x]$. Then if α_1, α_2 are the zeros of $f(x)$ in its splitting field E over $\mathbb{Q}$, we have

$$x^2 + ax + b = f(x) = (x - \alpha_1)(x - \alpha_2) = x^2 - (\alpha_1 + \alpha_2)x + \alpha_1\alpha_2$$

and so

$$\alpha_1 + \alpha_2 = -a \qquad \alpha_1\alpha_2 = b$$

The discriminant is

$$D = (\alpha_1 - \alpha_2)^2 = \alpha_1{}^2 - 2\alpha_1\alpha_2 + \alpha_2{}^2 = (\alpha_1 + \alpha_2)^2 - 4\alpha_1\alpha_2$$

Hence

$$D = a^2 - 4b$$

(1) If $D = 0$, then $\alpha_1 = \alpha_2 = -a/2 \in \mathbb{Q}$, and $E = \mathbb{Q}$ and $\mathrm{Gal}(E/\mathbb{Q}) = \{\mathrm{id}\}$.
Otherwise, consider $\Delta = \sqrt{D}$ and apply the preceding proposition.
(2) If $\Delta \in \mathbb{Q}$, then $\mathrm{Gal}(E/\mathbb{Q}) = \{\mathrm{id}\}$ and again $E = \mathbb{Q}$.
(3) If $\Delta \notin \mathbb{Q}$, then $\mathrm{Gal}(E/\mathbb{Q}) \cong \mathbb{Z}_2$ and $E = \mathbb{Q}(\Delta)$. ◇

Let us now consider a cubic

$$f(x) = x^3 + ax^2 + bx + c \in \mathbb{Q}[x]$$

and find an expression for its discriminant comparable to the expression we have just found in the case of the quadratic. We first eliminate the x^2 term by setting $y = x + (a/3)$. Then

$$f(x) = f(y - (a/3)) = g(y)$$

where $g(y)$ has the form

$$g(y) = y^3 + py + q \in \mathbb{Q}[x]$$

where

$$p = {}^1/_3(3b - a^2)$$
$$q = {}^1/_{27}(2a^3 - 9ab + 27c)$$

12.3.17 THEOREM Let

$$f(x) = x^3 + ax^2 + bx + c \in \mathbb{Q}[x]$$

be a cubic and let

$$g(y) = y^3 + py + q \in \mathbb{Q}[x]$$

be the associated cubic without squared term. Then
(1) $f(x)$ and $g(y)$ have the same discriminant D.
(2) $D = -4p^3 - 27q^2$
(3) $D = a^2b^2 - 4b^3 - 4a^3c - 27c^2 + 18abc$

Proof (1) If the zeros of $f(x)$ are $\alpha_1, \alpha_2, \alpha_3$, then the zeros of $g(x)$ are $\beta_1, \beta_2, \beta_3$, where $\beta_i = \alpha_i + (a/3)$. But then $\beta_i - \beta_j = \alpha_i - \alpha_j$ and

$$D = \prod_{i<j} (\alpha_i - \alpha_j)^2 = \prod_{i<j} (\beta_i - \beta_j)^2$$

(2) Consider now

$$g(y) = y^3 + py + q = (y - \beta_1)(y - \beta_2)(y - \beta_3)$$

If we calculate the derivative of $g(y)$, from the first expression for $g(y)$ we obtain

$$g'(y) = 3y^2 + p$$

while from the second expression for $g(y)$ we obtain

$$g'(y) = (y - \beta_1)(y - \beta_2) + (y - \beta_1)(y - \beta_3) + (y - \beta_2)(y - \beta_3)$$

Thus

$$3y^2 + p = (y - \beta_1)(y - \beta_2) + (y - \beta_1)(y - \beta_3) + (y - \beta_2)(y - \beta_3)$$

and, in particular,

$$3\beta_1^2 + p = (\beta_1 - \beta_2)(\beta_1 - \beta_3)$$
$$3\beta_2^2 + p = (\beta_2 - \beta_1)(\beta_2 - \beta_3)$$
$$3\beta_3^2 + p = (\beta_3 - \beta_1)(\beta_3 - \beta_2)$$

It follows that

$$D = -(3\beta_1^2 + p)(3\beta_2^2 + p)(3\beta_3^2 + p) =$$
$$-[27\beta_1^2\beta_2^2\beta_3^2 + 9p(\beta_1^2\beta_2^2 + \beta_1^2\beta_3^2 + \beta_2^2\beta_3^2) + 3p^2(\beta_1^2 + \beta_2^2 + \beta_3^2) + p^3]$$

Now the coefficients of $g(y)$ are the elementary symmetric functions of its zeros, so

$$s_1 = \beta_1 + \beta_2 + \beta_3 = 0$$
$$s_2 = \beta_1\beta_2 + \beta_1\beta_3 + \beta_2\beta_3 = p$$
$$s_3 = \beta_1\beta_2\beta_3 = -q$$

But from Example 12.3.9 we know

$$\beta_1^2 + \beta_2^2 + \beta_3^2 = s_1^2 - 2s_2 = -2p$$
$$\beta_1^2\beta_2^2 + \beta_1^2\beta_3^2 + \beta_2^2\beta_3^2 = s_2^2 - 2s_1s_3 = p^2$$

Therefore,

$$D = -[27q^2 + 9p^3 - 6p^3 + p^3] = -4p^3 - 27q^2$$

(3) Follows on substituting the expressions for p and q in terms of a and b and c into the expression for D in terms of p and q we have just obtained. □

12.3.18 COROLLARY Let $f(x) = x^3 + px + q \in \mathbb{Q}[x]$ be irreducible, and let E be the splitting field of $f(x)$ over $\mathbb{Q}$. Then

(1) $\mathrm{Gal}(E/\mathbb{Q}) \cong A_3$ or S_3

(2) If $p = 0$, then $\mathrm{Gal}(E/\mathbb{Q}) \cong S_3$.

Proof (1) If $\alpha \in E$ is a zero of $f(x)$, then $\mathbb{Q} \subseteq \mathbb{Q}(\alpha) \subseteq E$ and $[\mathbb{Q}(\alpha):\mathbb{Q}] = 3$, so 3 divides $|\mathrm{Gal}(E/\mathbb{Q})|$. On the other hand, $\mathrm{Gal}(E/\mathbb{Q}) \leq S_3$, so the only possibilities are $\mathrm{Gal}(E/\mathbb{Q}) \cong S_3$ or $\mathrm{Gal}(E/\mathbb{Q}) \cong A_3$.

(2) If $p = 0$, then the discriminant $D = -27q^2 < 0$, and so if $\Delta = \sqrt{D}$, then $\Delta \notin \mathbb{Q}$. By Proposition 12.3.15, we do not have $\text{Gal}(E/\mathbb{Q}) \leq A_3$, and (1) then implies $\text{Gal}(E/\mathbb{Q}) \cong S_3$. □

12.3.19 EXAMPLE Let $f(x) = x^3 - 3x + 1 \in \mathbb{Q}[x]$. By Theorem 8.4.11, $f(x)$ is irreducible over $\mathbb{Q}$. We have $D = -4(-3)^3 - 27 = 81$, so if $\Delta = \sqrt{D}$, then $\Delta = \pm 9 \in \mathbb{Q}$, and the Galois group of $f(x)$ is A_3. Therefore, if α is a zero of $f(x)$, then since $[\mathbb{Q}(\alpha):\mathbb{Q}] = 3$, $\mathbb{Q}(\alpha)$ must be the splitting field of $f(x)$.
We use the formula of Proposition 8.5.6 (with the same notation) to calculate the three zeros. We have

$$u^3 = -\tfrac{1}{2} + \frac{\sqrt{3}}{2} i = \omega$$

Hence u is a 9th root of unity,

$$u = \cos {}^{2\pi}/_{9} + i \sin {}^{2\pi}/_{9}$$

and $v = u^{-1} = u^8$. Therefore, the three zeros are

$$\alpha_1 = u + u^{-1}$$
$$\alpha_2 = u\omega^2 + u^{-1}\omega = u^7 + u^2 = u^2 + u^{-2}$$
$$\alpha_3 = u\omega + u^{-1}\omega^2 = u^4 + u^5 = u^4 + u^{-4}$$

First note that α_1, α_2, α_3 are all real numbers, since

$$u^{-1} = \cos {}^{2\pi}/_{9} - i \sin {}^{2\pi}/_{9}$$

is the complex conjugate of u, and similarly u^{-4} is the conjugate of u^4 and u^{-2} is the conjugate of u^2. Furthermore,

$$\alpha_1{}^2 = u^2 + 2 + u^{-2} = \alpha_2 + 2$$
$$\alpha_2{}^2 = u^4 + 2 + u^{-4} = \alpha_3 + 2$$

hence $\alpha_2 \in \mathbb{Q}(\alpha_1)$ and $\alpha_3 \in \mathbb{Q}(\alpha_2) = \mathbb{Q}(\alpha_1)$, and $\mathbb{Q}(\alpha_1)$ is indeed the splitting field of $f(x)$ over $\mathbb{Q}$. ◇

12.3.20 THEOREM Let $f(x) \in \mathbb{Q}[x]$ be an irreducible cubic polynomial with discriminant D and splitting field $E \subseteq \mathbb{C}$. Then $D \neq 0$ and

(1) If $f(x)$ has three real zeros, then $D > 0$ and
$\text{Gal}(E/\mathbb{Q}) \cong A_3$ if $\Delta = \sqrt{D} \in \mathbb{Q}$
$\text{Gal}(E/\mathbb{Q}) \cong S_3$ if $\Delta = \sqrt{D} \notin \mathbb{Q}$

(2) If $f(x)$ has only one real zero, then $D < 0$ and $\text{Gal}(E/\mathbb{Q}) \cong S_3$.

Proof Since char $\mathbb{Q} = 0$ and $f(x)$ is irreducible over $\mathbb{Q}$, $f(x)$ is separable over $\mathbb{Q}$, and hence $D \neq 0$.

(1) If $f(x)$ has three real zeros, then Δ being a product of the differences of those zeros, $\Delta \in \mathbb{R}$, and by Proposition 12.3.15, $\text{Gal}(E/\mathbb{Q}) \cong A_3$ or S_3 according as $\Delta \in \mathbb{Q}$ or $\Delta \notin \mathbb{Q}$.

(2) If $f(x)$ has only one real zero α_1, then the other zeros are complex conjugates:

$$\alpha_2 = a + bi \qquad \alpha_3 = a - bi$$

In this case

$$\Delta = (\alpha_1 - \alpha_2)(\alpha_1 - \alpha_3)(\alpha_2 - \alpha_3) =$$
$$(\alpha_1 - a - bi)(\alpha_1 - a + bi)(2bi) = |\alpha_1 - a - bi|(2b\, i)$$

Hence

$$D = \Delta^2 = -4b^2\,|\alpha_1 - a - bi|^2 < 0$$

Furthermore, in this case $\Delta \notin \mathbb{Q}$, so by Proposition 12.3.15, $\mathrm{Gal}(E/\mathbb{Q}) \cong S_3$. □

12.3.21 EXAMPLES

(1) $x^3 - 3x + 1$. In Example 12.3.19 we illustrated the case in which $D > 0$ and $\Delta \in \mathbb{Q}$, so we had three real zeros and $\mathrm{Gal}(E/\mathbb{Q}) \cong A_3$.

(2) $x^3 - 4x + 2$ has discriminant $D = 88 > 0$ and $\Delta \notin \mathbb{Q}$, hence in this case there are three real zeros and $\mathrm{Gal}(E/\mathbb{Q}) \cong S_3$.

(3) $x^3 + x + 1$ has discriminant $D = -31 < 0$, hence in this case there is one real zero and $\mathrm{Gal}(E/\mathbb{Q}) \cong S_3$. ◇

Let us now consider a quartic

$$f(x) = x^4 + ax^3 + bx^2 + cx + d \in \mathbb{Q}[x]$$

We first eliminate the x^3 term by setting $y = x + (a/4)$. Then

$$f(x) = f(y - (a/4)) = g(y)$$

where $g(y)$ has the form

$$g(y) = y^4 + py^2 + qy + r \in \mathbb{Q}[x]$$

where

$$p = {}^1/_8(8b - 3a^2)$$
$$q = {}^1/_8(a^3 - 4ab + 8c)$$
$$r = {}^1/_{256}(-3a^4 + 16a^2b - 64ac + 256d)$$

Exactly as in Theorem 12.3.17, part (1), the zeros of $f(x)$ and $g(y)$ differ by the constant $a/4 \in \mathbb{Q}$, and the two polynomials have the same splitting field and the same Galois group as well as the same discriminant. So we can limit our attention to quartics without an x^3 term. We first dispose of the reducible case.

12.3.22 THEOREM Let $f(x) \in \mathbb{Q}[x]$ be a reducible quartic with splitting field $E \subseteq \mathbb{C}$. Then

(1) $\mathrm{Gal}(E/\mathbb{Q}) \cong S_3$ or A_3 if $f(x)$ has an irreducible factor of degree 3.

(2) $\mathrm{Gal}(E/\mathbb{Q}) \cong V_4$ or $\mathbb{Z}_2$ if $f(x)$ has two irreducible factors of degree 2, where V_4 is the Klein 4-group.

Proof (1) This is immediate from Theorem 12.3.20, since if $f(x) = (x - \alpha)g(x)$, where $\alpha \in \mathbb{Q}$ and $g(x)$ is an irreducible cubic, then $f(x)$ has the same splitting field and Galois group as $g(x)$.

(2) If $f(x) = g_1(x)g_2(x)$, where $g_1(x)$ and $g_2(x)$ are irreducible quadratics, let their discriminants be $D_1 = \Delta_1{}^2$ and $D_2 = \Delta_2{}^2$. Then by Example 12.3.16, the splitting

field of $g_1(x)$ over $\mathbb{Q}$ is $\mathbb{Q}(\Delta_1)$ and $\mathrm{Gal}(\mathbb{Q}(\Delta_1)/\mathbb{Q}) \cong \mathbb{Z}_2$. If $\Delta_2 \in \mathbb{Q}(\Delta_1)$, then the splitting field of $f(x)$ is $E = \mathbb{Q}(\Delta_1)$ and $\mathrm{Gal}(E/\mathbb{Q}) \cong \mathbb{Z}_2$. If $\Delta_2 \notin \mathbb{Q}(\Delta_1)$, then the splitting field of $f(x)$ is $E = \mathbb{Q}(\Delta_1, \Delta_2)$ and, as in Example 12.1.22, the Galois group is isomorphic to $\mathbb{Z}_2 \otimes \mathbb{Z}_2 \cong V_4$. □

Now let us consider an irreducible quartic

$$f(x) = x^4 + px^2 + qx + r \in \mathbb{Q}[x]$$

Let its zeros in its splitting field $E \subseteq \mathbb{C}$ be $\alpha_1, \alpha_2, \alpha_3, \alpha_4$. Consider the following elements of E:

$$\theta_1 = (\alpha_1 + \alpha_2)(\alpha_3 + \alpha_4)$$
$$\theta_2 = (\alpha_1 + \alpha_3)(\alpha_2 + \alpha_4)$$
$$\theta_3 = (\alpha_1 + \alpha_4)(\alpha_2 + \alpha_3)$$

Any element of S_4 permutes the θ_i, and therefore

$$\theta_1 + \theta_2 + \theta_3$$
$$\theta_1\theta_2 + \theta_1\theta_3 + \theta_2\theta_3$$
$$\theta_1\theta_2\theta_3$$

which are the elementary symmetric functions of the θ_i, are fixed by all elements of S_4 and hence belong to $\mathbb{Q}$. Thus the polynomial

$$g(x) = (x - \theta_1)(x - \theta_2)(x - \theta_3)$$

has rational coefficients.

12.3.23 DEFINITION Given an irreducible quartic

$$f(x) = x^4 + px^2 + qx + r \in \mathbb{Q}[x]$$

the polynomial

$$g(x) = (x - \theta_1)(x - \theta_2)(x - \theta_3) \in \mathbb{Q}[x]$$

is called its **resolvant cubic**. ○

12.3.24 THEOREM Let

$$f(x) = x^4 + px^2 + qx + r \in \mathbb{Q}[x]$$

be an irreducible quartic. Then its resolvant cubic is

$$g(x) = x^3 - 2px^2 + (p^2 - 4r)x + q^2 \in \mathbb{Q}[x]$$

Proof We go back to our discussion of quartics in Section 8.5, where we showed how obtaining the zeros of a quartic could be reduced to first obtaining the zeros of an auxiliary cubic and then the zeros of quadratics. (The a, b, c, d of Section 8.5 are now $0, p, q, r$.) Specifically, we showed that if k is any zero of

$$h(x) = 8x^3 - 4px^2 - 8rx + (4pr - q^2)$$

then there are u and v satisfying the conditions

$$u^2 = 2k - p$$
$$2uv = -q$$
$$k^2 - v^2 = r$$

from which we get a factorization of $f(x)$ into a product of quadratics:

$$f(x) = [x^2 + ux + (k + v)][x^2 - ux + (k - v)]$$

An entirely equivalent procedure would be to let t be any zero of

$$g(x) = -h((-x + p)/2) = x^3 - 2px^2 + (p^2 - 4r)x + q^2 \in \mathbb{Q}[x]$$

and let k be determined by the condition

$$k = (-t + p)/2$$

and then proceed as before to determine u and v and the preceding factorization. As there are three possible zeros of $h(x)$ to consider, we get three possible factorizations of $f(x)$ into a product of quadratics. In terms of the zeros $\alpha_1, \alpha_2, \alpha_3, \alpha_4$ of $f(x)$, the three possible factorizations are (apart from order of factors)

$$[(x - \alpha_1)(x - \alpha_2)][(x - \alpha_3)(x - \alpha_4)]$$
$$[(x - \alpha_1)(x - \alpha_3)][(x - \alpha_2)(x - \alpha_4)]$$
$$[(x - \alpha_1)(x - \alpha_4)][(x - \alpha_2)(x - \alpha_3)]$$

With the first factorization we have

$$(x - \alpha_1)(x - \alpha_2) = x^2 + ux + (k + v)$$
$$(x - \alpha_3)(x - \alpha_4) = x^2 - ux + (k - v)$$

which give

$$-(\alpha_1 + \alpha_2) = u$$
$$-(\alpha_3 + \alpha_4) = -u$$

and therefore in terms of the θ_i appearing in the definition of the resolvant cubic

$$\theta_1 = (\alpha_1 + \alpha_2)(\alpha_3 + \alpha_4) = -u^2 = -2k + p = t$$

Thus θ_1 is one of the zeros of $g(x)$. Considering the other factorizations similarly shows that θ_2 and θ_3 are the other zeros of $g(x)$. Hence

$$x^3 - 2px^2 + (p^2 - 4r)x + q^2 = (x - \theta_1)(x - \theta_2)(x - \theta_3)$$

and is the resolvant cubic, as claimed. □

12.3.25 PROPOSITION Let

$$f(x) = x^4 + px^2 + qx + r \in \mathbb{Q}[x]$$

be an irreducible quartic. Then $f(x)$ and its resolvant cubic $g(x)$ have the same discriminant.

Proof Let $\alpha_1, \alpha_2, \alpha_3, \alpha_4$ be the zeros of $f(x)$ and $\theta_1, \theta_2, \theta_3$ the zeros of $g(x)$. Then

$$\theta_1 - \theta_2 = (\alpha_1 + \alpha_2)(\alpha_3 + \alpha_4) - (\alpha_1 + \alpha_3)(\alpha_2 + \alpha_4) =$$
$$\alpha_1\alpha_3 + \alpha_2\alpha_4 - \alpha_1\alpha_2 - \alpha_3\alpha_4 = -(\alpha_1 - \alpha_4)(\alpha_2 - \alpha_3)$$

Similarly,

$$\theta_1 - \theta_3 = -(\alpha_1 - \alpha_3)(\alpha_2 - \alpha_4)$$
$$\theta_2 - \theta_3 = -(\alpha_1 - \alpha_2)(\alpha_3 - \alpha_4)$$

Thus the discriminant of $g(x)$ is

$$(\theta_1 - \theta_2)^2(\theta_1 - \theta_3)^2(\theta_2 - \theta_3)^2 =$$
$$(\alpha_1 - \alpha_4)^2(\alpha_2 - \alpha_3)^2(\alpha_1 - \alpha_3)^2(\alpha_2 - \alpha_4)^2(\alpha_1 - \alpha_2)^2(\alpha_3 - \alpha_4)^2$$

which is the discriminant of $f(x)$. □

We determined the possible Galois groups of a reducible quartic over $\mathbb{Q}$ in Theorem 12.3.22. We are now ready to study the possible Galois groups of an irreducible quartic $f(x)$ over $\mathbb{Q}$. The resolvant cubic plays an important role.

12.3.26 THEOREM Let $f(x) \in \mathbb{Q}[x]$ be a quartic polynomial irreducible over $\mathbb{Q}$. Let E be the splitting field of $f(x)$ over $\mathbb{Q}$, and let $G = \mathrm{Gal}(E/\mathbb{Q})$. Let $g(x)$ be the resolvant cubic of $f(x)$, let K be the splitting field of $g(x)$ over $\mathbb{Q}$, and let $H = \mathrm{Gal}(K/\mathbb{Q})$. Let D be the discriminant of $f(x)$ and $g(x)$, and let $\Delta = \sqrt{D}$. Then

(1) If $g(x)$ is irreducible and $\Delta \notin \mathbb{Q}$, then $G \cong S_4$.

(2) If $g(x)$ is irreducible and $\Delta \in \mathbb{Q}$, then $G \cong A_4$.

(3) If $g(x)$ has exactly one zero in $\mathbb{Q}$ and $f(x)$ is irreducible over $\mathbb{Q}(\Delta)$, then $G \cong D_4$.

(4) If $g(x)$ has exactly one zero in $\mathbb{Q}$ and $f(x)$ is reducible over $\mathbb{Q}(\Delta)$, then $G \cong \mathbb{Z}_4$.

(5) If $g(x)$ splits over $\mathbb{Q}$, then $G \cong V_4$.

Proof Note that since $f(x)$ is irreducible over F, any zero of $f(x)$ has degree 4 over F, and hence $|G| = [E:F]$ is divisible by 4. Note also that since the zeros θ_i of $g(x)$ are obtained by addition and multiplication from the zeros α_j of $f(x)$, $K \subseteq E$, $\mathrm{Gal}(E/K) \triangleleft G$ and $H \cong G/\mathrm{Gal}(E/K)$, so $|H|$ divides $|G|$.

(1) Suppose $g(x)$ is irreducible over F. If $\Delta \notin \mathbb{Q}$, then Proposition 12.3.15 applied to $f(x)$ implies that G is not contained in A_4, and applied to $g(x)$ implies that $H \cong S_3$ and $|H| = 6$. Thus G is a subgroup of S_4 of order divisible by 6 and by 4 and hence by 12, but not contained in A_4. But the only such group is S_4 itself.

(2) Suppose $g(x)$ is irreducible over F. If $\Delta \in \mathbb{Q}$, then Proposition 12.3.15 applied to $f(x)$ implies that $G \leq A_4$, and applied to $g(x)$ implies that $H \cong A_3$ and $|H| = 3$. Thus G is a subgroup of S_4 of order divisible by 3 and by 4 and hence by 12, and contained in A_4. But the only such group is A_4 itself.

(3) and (4) If $g(x)$ has exactly one zero in $\mathbb{Q}$, then, renumbering if necessary, we may assume

$$\theta_1 = (\alpha_1 + \alpha_2)(\alpha_3 + \alpha_4) \in \mathbb{Q}$$

Over $\mathbb{Q}$ we have the factorization

$$g(x) = (x - \theta_1)h(x) = (x - \theta_1)(x^2 - ax + b) =$$

where

$$\theta_2 + \theta_3 = a \in \mathbb{Q} \qquad \theta_2\theta_3 = b \in \mathbb{Q}$$

Since

$$\Delta = (\theta_1 - \theta_2)(\theta_1 - \theta_3)(\theta_2 - \theta_3) = h(\theta_1)(\theta_2 - \theta_3)$$

and $h(\theta_1) \in \mathbb{Q}$, we have $\theta_2 - \theta_3 \in \mathbb{Q}(\Delta)$, and since $\theta_2 + \theta_3 \in \mathbb{Q}$, we have $\theta_2, \theta_3 \in \mathbb{Q}(\Delta)$, and $\mathbb{Q}(\Delta)$ is the splitting field of $g(x)$. Since $g(x)$ does not split over $\mathbb{Q}$, we have $\Delta \notin \mathbb{Q}$, $[\mathbb{Q}(\Delta):\mathbb{Q}] = 2$ and $\mathrm{Gal}(\mathbb{Q}(\Delta)/\mathbb{Q}) \cong \mathbb{Z}_2$. Now the group of permutations of the zeros α_j that leave θ_1 fixed correspond to the following subgroup of S_4:

$$T = \{\text{id}, (1\,3\,2\,4), (1\,2)(3\,4), (1\,4\,2\,3), (1\,2), (1\,4)(2\,3), (3\,4), (1\,3)(2\,4)\}$$
$$= \{\rho_0, \rho, \rho^2, \rho^3, \tau, \rho\tau, \rho^2\tau, \rho^3\tau\} \cong D_4$$

Its subgroups of order 4 are

$$U = \{\text{id}, (1\,3\,2\,4), (1\,2)(3\,4), (1\,4\,2\,3)\} = \{\rho_0, \rho, \rho^2, \rho^3\} \cong \mathbb{Z}_4$$
$$V = \{\text{id}, (1\,2)(3\,4), (1\,2), (3\,4)\} = \{\rho_0, \rho^2, \tau, \rho^2\tau\} \cong V_4$$
$$W = \{\text{id}, (1\,2)(3\,4), (1\,4)(2\,3), (1\,3)(2\,4)\} = \{\rho_0, \rho^2, \rho\tau, \rho^3\tau\} \cong V_4$$

Since $\Delta \notin \mathbb{Q}$, G is not a subgroup of A_4, so W is ruled out. Also, since $f(x)$ is irreducible over $\mathbb{Q}$, for any of its zeros α_i there must be a $\psi \in G$ with $\psi(\alpha_1) = \alpha_i$, so V is ruled out. $\text{Gal}(E/\mathbb{Q}(\Delta))$ consists of the elements of G that fix

$$\Delta = (\alpha_1 - \alpha_4)(\alpha_2 - \alpha_3)(\alpha_1 - \alpha_3)(\alpha_2 - \alpha_4)(\alpha_1 - \alpha_2)(\alpha_3 - \alpha_4)$$

and these are precisely the even permutations in G. So the two surviving candidates for $\text{Gal}(E/\mathbb{Q}(\Delta))$ are

$$T \cap A_4 = \{\text{id}, (1\,2)(3\,4), (1\,4)(2\,3), (1\,3)(2\,4)\}$$
$$U \cap A_4 = \{\text{id}, (1\,2)(3\,4)\}$$

Now if $f(x)$ is still irreducible over $\mathbb{Q}(\Delta)$, it must still be that for each of its zeros α_j there is a $\psi \in \text{Gal}(E/\mathbb{Q}(\Delta))$ with $\psi(\alpha_1) = \alpha_j$. But if $f(x)$ becomes reducible over $\mathbb{Q}(\Delta)$, this cannot be the case, since if $f(x) = h(x)k(x)$, then for any $\psi \in \text{Gal}(E/\mathbb{Q}(\Delta))$, $\psi(\alpha_1)$ can only be another zero of whichever factor $h(x)$ or $k(x)$ has α_1 as a zero. Thus in the irreducible case we must have $G = T \cong D_4$, while in the reducible case we must have $G = U \cong \mathbb{Z}_4$.

(5) If $g(x)$ splits completely over $\mathbb{Q}$, then all the $\theta_i \in \mathbb{Q}$ are fixed by every element of G. Thus G must be contained not only in the group T, but also the corresponding groups for θ_2 and θ_3, which are

$$T' = \{\text{id}, (1\,2\,3\,4), (1\,3)(2\,4), (1\,4\,3\,2), (1\,3), (1\,4)(2\,3), (2\,4), (1\,2)(3\,4)\}$$
$$T'' = \{\text{id}, (1\,2\,4\,3), (1\,4)(2\,3), (1\,3\,4\,2), (1\,4), (1\,3)(2\,4), (2\,3), (1\,2)(3\,4)\}$$

But the intersection of these three groups is the group W, so $G \leq W$ and since $|G| \geq 4$, in fact $G = W \cong V_4$. □

12.3.27 EXAMPLE We have worked out the Galois group G over $\mathbb{Q}$ for

$$f(x) = x^4 - 2x^2 - 2$$

in Example 12.2.19, but let us have another look. Applying Theorem 12.3.24, the resolvant cubic works out to be

$$g(x) = x^3 + 4x^2 + 12x = x(x^2 + 4x + 12)$$

For the zeros of $g(x)$ we have

$$\theta_1 = 0 \qquad \theta_2 = -2 + \sqrt{2}i \qquad \theta_3 = -2 - \sqrt{2}i$$

and we are in case (3) or (4) of the preceding theorem. For the square root of the discriminant we have

$$\Delta = (\theta_1 - \theta_2)(\theta_1 - \theta_3)(\theta_2 - \theta_3) = 24\sqrt{2}i$$

so

$$\mathbb{Q}(\Delta) = \mathbb{Q}(\theta_1, \theta_2, \theta_3) = \mathbb{Q}(\sqrt{2}i)$$

Whether we have $G \cong D_4$ or $G \cong \mathbb{Z}_4$ depends, according to the preceding theorem, on whether $f(x)$ is irreducible over $\mathbb{Q}(\Delta)$. Since 3 does not divide the order of G in either case, $f(x)$ cannot factor into an irreducible cubic and a linear factor, so if it factors at all, it is into a product of two quadratic factors, as in the proof of Theorem 12.3.24. But for the factorization corresponding to the zero $\theta_1 = 0$ of $g(x)$ we have

$$k = (-\theta_1 + p)/2 = -1 \qquad \text{and} \qquad v^2 = k^2 - r = 3$$

while for the factorization corresponding to any θ_i we have $u^2 = -\theta_i$. Since none of $3, -\theta_2, -\theta_3$ has a square root in $\mathbb{Q}(\Delta)$, $f(x)$ is irreducible over $\mathbb{Q}(\Delta)$, and $G \cong D_4$, in agreement with what we found in Example 12.2.19 by explicitly working out the zeros of $f(x)$ and the automorphisms of its splitting field. ◇

12.3.28 EXAMPLE Let us find the Galois group G over $\mathbb{Q}$ for

$$f(x) = x^4 + 4x^2 - 2$$

which is irreducible by Eisenstein's criterion. The resolvant cubic works out to be

$$g(x) = x^3 + 8x + 16$$

This reduces to $x^3 + 2x + 1$ mod 3, which has no zeros in $\mathbb{Z}_3$ and is therefore irreducible over $\mathbb{Z}_3$. It follows that $g(x)$ is irreducible over $\mathbb{Q}$ and we are in case (1) or (2) of the preceding theorem. Which it is depends, according to that theorem, on the discriminant, for which we have

$$D = -4(8)^3 - 27(16)^2 < 0$$

So the square root Δ of the discriminant is not in $\mathbb{Q}$, and $G \cong S_4$. ◇

We close this section by restating what Theorem 12.3.26 tells us about the Galois group of a quartic, adding what was seen in the proof of that theorem about the Galois group of the resolvant cubic. So let $f(x) \in \mathbb{Q}[x]$ be a quartic polynomial irreducible over $\mathbb{Q}$, let E be the splitting field of $f(x)$ over $\mathbb{Q}$, and let $G = \mathrm{Gal}(E/\mathbb{Q})$. Let $g(x)$ be the resolvant cubic of $f(x)$, let K be the splitting field of $g(x)$ over $\mathbb{Q}$, and let $H = \mathrm{Gal}(K/\mathbb{Q})$. Let D be the discriminant of $f(x)$ and $g(x)$, and let $\Delta = \sqrt{D}$. Then the possible Galois groups G of $f(x)$ are as in Table 2.

TABLE 2 The Galois group G of a quartic

$g(x)$		H	G
irreducible over $\mathbb{Q}$	$\Delta \notin \mathbb{Q}$	S_3	S_4
irreducibe over $\mathbb{Q}$	$\Delta \in \mathbb{Q}$	A_3	A_4
exactly one zero in $\mathbb{Q}$	$f(x)$ irreducible over $\mathbb{Q}(\Delta)$	$\mathbb{Z}_2$	D_4
exactly one zero in $\mathbb{Q}$	$f(x)$ reducible over $\mathbb{Q}(\Delta)$	$\mathbb{Z}_2$	$\mathbb{Z}_4$
splits over $\mathbb{Q}$		trivial	V_4

Exercises 12.3

In Exercises 1 through 3 express the indicated symmetric functions of n indeterminates in terms of the elementary symmetric functions in n indeterminates.

1. $n = 2$ $(x_1 - x_2)^2$

2. $n = 3$ $(x_1 - x_2)^2(x_1 - x_3)^2(x_2 - x_3)^2$

3. $n = 3$ $(x_1 + x_2)(x_1 + x_3)(x_2 + x_3)$

In Exercises 4 through 7 calculate the discriminant of the indicated cubic polynomial in $\mathbb{Q}[x]$.

4. $x^3 + 2x - 1$

5. $x^3 - 3x^2 + 2$

6. $x^3 + 2x^2 - 3x + 1$

7. $x^3 - 5x + 1$

In Exercises 8 through 11 calculate the discriminant of the indicated quartic polynomial in $\mathbb{Q}[x]$.

8. $x^4 + 2x + 1$

9. $x^4 + 2x^2 + 2$

10. $x^4 + 2$

11. $x^4 - 3x^2 + 3x + 3$

In Exercises 12 through 16 determine the Galois group over $\mathbb{Q}$ of the indicated cubic polynomial.

12. $x^3 - 3x + 1$

13. $x^3 - 5x + 4$

14. $x^3 + x + 1$

15. $x^3 - x^2 + 2x + 1$

16. $x^3 + x^2 + 2x + 1$

In Exercises 17 through 25 for the indicated quartic polynomial find

(a) the resolvant cubic
(b) the discriminant D
(c) the Galois group over $\mathbb{Q}$ of the resolvant cubic
(d) the Galois group over $\mathbb{Q}$ of the quartic

17. $x^4 + 1$

18. $x^4 + 2$

19. $x^4 - 2$

20. $x^4 + x + {}^3/_4$

21. $x^4 + 2x - {}^1/_4$

22. $x^4 + x^2 + 1$

23. $x^4 + x^3 + x^2 + x + 1$

24. $x^4 - x + 3$

25. $16x^4 - 8x^2 + 9$

26. (Newton's formula) Let $f(x)$ be a monic polynomial of degree n with zeros $\alpha_1, \alpha_2, \ldots, \alpha_n$, and let s_i for $1 \leq i \leq n$ be the elementary symmetric functions of the α_j, and let $s_i = 0$ for $i > n$. Let $p_k = \alpha_1{}^k + \ldots + \alpha_n{}^k$ for $k \geq 0$. Show that

$$\begin{aligned} p_0 &= n \\ p_1 &= s_1 \\ &\vdots \\ p_k &= p_{k-1}s_1 - p_{k-2}s_2 + \ldots + (-1)^k p_1 s_{k-1} + (-1)^{k+1} k s_k \text{ for all } k > 1 \end{aligned}$$

In Exercises 27 through 29 calculate p_k as defined in the preceding problem for all $k \geq 0$ for the following polynomials.

27. $x^4 + x^3 + x^2 + x + 1$

28. $x^4 - 2x^3 + 3x^2 + x - 1$

29. $x^5 + x^2 - 2x + 3$

For Exercises 30 through 31 we need the notion of a determinant for a square matrix of any size. Let $A = \{a_{ij}\}$ be an $n \times n$ matrix with entries a_{ij} in a field F. For any $\sigma \in S_n$ let $\varepsilon(\sigma) = 0$ if σ is an even permutation, and $\varepsilon(\sigma) = 1$ if σ is an odd permutation. The **determinant** of A is defined to be

$$\det A = \Sigma_{\sigma \in S_n} (-1)^{\varepsilon(\sigma)} \cdot a_{1\sigma(1)} \cdot a_{2\sigma(2)} \cdot \ldots \cdot a_{n\sigma(n)}$$

30. Let $f(x)$ be a monic polynomial of degree n with zeros $\alpha_1, \alpha_2, \ldots, \alpha_n$. Let D be the discriminant

$$D = \Pi_{i<j} (\alpha_i - \alpha_j)^2$$

The **Vandermonde matrix** is the matrix $V = \{a_{ij}\}$, where $a_{ij} = \alpha_j{}^{i-1}$, so that

$$\det V = \begin{vmatrix} 1 & \alpha_1 & \alpha_1{}^2 & \cdots & \alpha_1{}^{n-1} \\ 1 & \alpha_2 & \alpha_2{}^2 & \cdots & \alpha_2{}^{n-1} \\ \vdots & \vdots & \vdots & \ddots & \vdots \\ 1 & \alpha_n & \alpha_n{}^2 & \cdots & \alpha_n{}^{n-1} \end{vmatrix}$$

Show that $D = (\det V)^2$.

31. For an $n \times n$ matrix $A = \{a_{ij}\}$ the **transpose** is $A^T = \{b_{ij}\}$, where $b_{ij} = a_{ji}$ for all $1 \leq i, j \leq n$. Let the notation be as in the preceding exercise. Show that

$$D = \det(V^T V) = \begin{vmatrix} p_0 & p_1 & \cdots & p_{n-1} \\ p_1 & p_2 & \cdots & p_n \\ \vdots & \vdots & \ddots & \vdots \\ p_{n-1} & p_n & \cdots & p_{2n-2} \end{vmatrix}$$

where the p_i are as defined in Newton's formula.

12.4 Geometric Constructions Revisited

In this section we apply the fundamental theorem of Galois theory to the geometric construction problems we investigated in the preceding chapter. We again study constructions with a straightedge and compass, as well as constructions with a marked ruler and compass. The towers of fields we introduced in the preceding chapter correspond to towers of subgroups of a Galois group, and knowing the Galois group of the minimal polynomial of an algebraic real number will prove to be important for determining whether that number can be constructed. In particular, we apply Galois theory to the problem of which regular polygons are constructible with straightedge and compass or with marked ruler and compass.

12.4.1 EXAMPLE In Example 11.1.23 we showed that $f(x) = x^4 - 6x + 3$, an irreducible quartic over $\mathbb{Q}$, has a real zero α that is not constructible by straightedge and compass, even though $[\mathbb{Q}(\alpha):\mathbb{Q}] = 4$. Let us now apply the classification theorem we proved at the end of the preceding section to find the Galois group $\mathrm{Gal}(E/\mathbb{Q})$, where E is the splitting field of $f(x)$ over $\mathbb{Q}$. The resolvant cubic works out, using Theorem 12.3.24, to be $g(x) = x^3 - 12x + 36$. This reduces mod 7 to $x^3 + 2x + 1$, which has no zeros in $\mathbb{Z}_7$, and therefore $g(x)$ is irreducible over $\mathbb{Z}_7$ and over $\mathbb{Q}$. The discriminant is $D = 4(12)^3 - 27(36)^2 < 0$, and so by Theorem 12.3.26, $\mathrm{Gal}(E/\mathbb{Q}) \cong S_4$, and we have the tower of subfields

$$\mathbb{Q} \subseteq \mathbb{Q}(\Delta) \subseteq E$$

where $\Delta = \sqrt{D}$ and this corresponds to the tower of subgroups

$$\{\rho_0\} \subseteq A_4 \subseteq S_4$$

of the Galois group. Note that in this tower the intermediate group A_4 is a normal subgroup of the Galois group S_4; hence in the tower of fields the intermediate field is a normal extension of $\mathbb{Q}$. Note also that the order of the intermediate group is 12 and its index in the Galois group is 2. ◇

12.4.2 EXAMPLE In Example 11.2.10 we showed that the regular pentagon is constructible by showing that the primitive 5th root of unity ζ is constructible. Here ζ is a zero of the cyclotomic polynomial

$$\Phi_5(x) = x^4 + x^3 + x^2 + x + 1$$

whose splitting field we worked out in 10.3.19 to be

$$K = \mathbb{Q}(\sqrt{5}, \sqrt{10 + 2\sqrt{5}}\, i)$$

The elements of this field are constructible since K is the top of a tower of subfields

$$\mathbb{Q} \subseteq \mathbb{Q}(\sqrt{5}) \subseteq K$$

in which each extension field has degree 2 over the field below it. In Example 12.1.25 we showed the Galois group $G = \mathrm{Gal}(K/\mathbb{Q})$ to be isomorphic to $\mathbb{Z}_4$. The tower of subfields of K corresponds to a tower of subgroups of G

$$\{\rho_0\} \subseteq \mathbb{Z}_2 \subseteq \mathbb{Z}_4$$

where the intermediate group has order 2 and its index in the Galois group is 2. ◇

12.4.3 EXAMPLE $\sqrt[4]{2}$ is constructible by straightedge and compass since $2^{1/4} = (2^{1/2})^{1/2}$ and we have the square root tower

$$\mathbb{Q} \subseteq \mathbb{Q}(\sqrt{2}) \subseteq \mathbb{Q}(\sqrt{\sqrt{2}}) = \mathbb{Q}(\sqrt[4]{2})$$

The intermediate field $\mathbb{Q}(2^{1/2})$ contains both zeros $\pm 2^{1/2}$ of $x^2 - 2$ and is the splitting field of that polynomial over $\mathbb{Q}$ and a Galois extension of $\mathbb{Q}$. But $\mathbb{Q}(2^{1/4})$ contains only the square roots of one of the zeros $2^{1/2}$ of $x^2 - 2$ and so contains only two zeros $\pm 2^{1/4}$ of the minimal polynomial $x^4 - 2$ of $2^{1/4}$ and therefore is not a Galois extension of $\mathbb{Q}$. To get a Galois extension, we need to add the other two zeros of $x^4 - 2$, namely $\pm 2^{1/4}i$. But this amounts to adding the square roots of the other zero $-2^{1/2}$ of $x^2 - 2$. Thus the original square root tower can be extended to a longer square root tower

$$\mathbb{Q} \subseteq \mathbb{Q}(\sqrt{2}) \subseteq \mathbb{Q}(\sqrt[4]{2}) \subseteq \mathbb{Q}(\sqrt[4]{2}, \sqrt[4]{2}\, i)$$

so that the top of the extended tower is a Galois extension of $\mathbb{Q}$. Note that since the order of the splitting field over $\mathbb{Q}$ in this case is $2 \cdot 2 \cdot 2 = 8$, the Galois group must be D_4, the only group of order 8 allowed by Theorem 12.3.26. ◇

12.4.4 THEOREM A real number α is constructible by straightedge and compass if and only if α is algebraic over $\mathbb{Q}$ and the order of the Galois group of its minimal polynomial over $\mathbb{Q}$ is 2^r for some $r \geq 0$.

Proof ($\Rightarrow$) Let $\alpha \in K_r$, where K_r is the top of a tower of extensions of degree 2 over $\mathbb{Q}$

$$\mathbb{Q} = K_0 \subseteq K_1 \subseteq K_2 \subseteq \ldots \subseteq K_{r-1} \subseteq K_r \subseteq \mathbb{C}$$

with $[K_{i+1} : K_i] = 2$ for $0 \leq i < r$. Recall that α is a constructible real number if and only if there is such a tower with $\alpha \in K_r$ and $K_r \subseteq \mathbb{R}$, but we do not need the information that $K_r \subseteq \mathbb{R}$ for this direction of the proof. Note that $[K_r : \mathbb{Q}] = 2^r$. If K_r is a Galois extension of $\mathbb{Q}$, then since $\alpha \in K_r$, the whole splitting field E of the minimal polynomial of α over $\mathbb{Q}$ will be contained in K_r. Since E is a Galois extension of $\mathbb{Q}$, by the fundamental theorem of Galois theory, $\mathrm{Gal}(K_r/E)$ is a normal subgroup of $\mathrm{Gal}(K_r/\mathbb{Q})$ and $\mathrm{Gal}(E/\mathbb{Q}) \cong \mathrm{Gal}(K_r/\mathbb{Q}) / \mathrm{Gal}(K_r/E)$ and hence $|\mathrm{Gal}(E/\mathbb{Q})|$ divides $|\mathrm{Gal}(K_r/\mathbb{Q})|$ and is a power of 2. So it suffices to prove that if the top K_r of the given tower of extensions of degree 2 is *not* a Galois extension of $\mathbb{Q}$, then the given tower can be extended to a higher tower of degree 2 whose top L *is* a Galois extension of $\mathbb{Q}$.

To prove this we proceed by induction on the height r of the given tower. If $r = 1$, then K_1 is itself a Galois extension of $\mathbb{Q}$. For the induction step we imitate the idea of the preceding example. Suppose the claim holds for $r - 1$. Then there is a field F such that F is the top of a tower of extensions of degree 2 over $\mathbb{Q}$, $K_{r-1} \subseteq F$, and F is a Galois extension of $\mathbb{Q}$. Since $[K_r : K_{r-1}] = 2$, K_r is the splitting field over K_{r-1} of some polynomial $x^2 + bx + c$ in $K_{r-1}[x]$, and hence $K_r = K_{r-1}(\Delta)$, where $\Delta^2 = D = b^2 - 4c$. It suffices to show that there is a field L such that

(1) L is the top of a tower of extensions of degree 2 over F.
(2) $\Delta \in L$
(3) L is a Galois extension of $\mathbb{Q}$.

For (1) implies that F is the top of a tower of extensions of degree 2 over $\mathbb{Q}$, and (2) implies that $K_r = K_{r-1}(\Delta) \subseteq F(\Delta) \subseteq L$. So let $\mathrm{Gal}(F/\mathbb{Q}) = \{\sigma_1, \ldots, \sigma_k\}$, where $\sigma_1 = \mathrm{id}$. Let $D_i = \sigma_i(D)$, and consider the polynomial

$$g(x) = \prod_{1 \le i \le k}(x^2 - D_i) \in F[x]$$

and let L be the splitting field of $g(x)$ over F.

(1) L is the top of a tower of extensions of degree 2 over F, since the zeros of $g(x)$ are $\pm\Delta_i$, where $\Delta_i^2 = D_i \in F$ and we have the tower

$$F \subseteq F(\Delta_1) \subseteq F(\Delta_1, \Delta_2) \subseteq \ldots \subseteq F(\Delta_1, \Delta_2, \ldots, \Delta_k) = L$$

(2) $\Delta \in L$, since $\sigma_1 = \mathrm{id}$ and therefore $D_1 = D$ and $\pm\Delta_1 = \pm\Delta$.

(3) L is a Galois extension of $\mathbb{Q}$, as can be shown as follows. Each $\sigma_i \in \mathrm{Gal}(F/\mathbb{Q})$ merely permutes the factors of $g(x)$ and therefore leaves the coefficients of $g(x)$ fixed. Hence these coefficients belong to the fixed field $\mathbb{Q}$ or, in other words $g(x) \in \mathbb{Q}[x]$. Since F is a Galois extension of $\mathbb{Q}$, it is the splitting field of some $h(x) \in \mathbb{Q}[x]$ over $\mathbb{Q}$. But then L, the splitting field of $g(x)$ over F, is the splitting field of $h(x)g(x)$ over $\mathbb{Q}$, and hence a Galois extension of $\mathbb{Q}$ to complete the proof.

($\Leftarrow$) Let α be algebraic with minimal polynomial $f(x)$ over $\mathbb{Q}$, let E be the splitting field of $f(x)$ over $\mathbb{Q}$, let $G = \mathrm{Gal}(E/\mathbb{Q})$, and assume that $|G| = 2^r$ for some $r \ge 0$. We must show that α is contained in the top of a tower of extensions of degree 2 over $\mathbb{Q}$ contained in $\mathbb{R}$. We show that $\mathbb{Q}(\alpha) \subseteq \mathbb{R}$ is itself the top of a tower of extension of degree 2 over $\mathbb{Q}$. Let $H = \mathrm{Gal}(E/\mathbb{Q}(\alpha))$. We claim there is a tower of subgroups

$$H = N_s \le \ldots \le N_2 \le N_1 \le N_0 = G$$

with N_{i+1} having index 2 in N_i for all $0 \le i < s$. For if $H = G$ we have a trivial tower with $s = 0$ and are done. Otherwise, apply Exercise 17, Section 4.6 to H and $N_0 = G$ to obtain an $N_1 \le N_0$ of index 2 with $H \le N_1$. If $H = N_1$, we have a tower with $s = 1$ and are done. Otherwise, apply the same result to H and N_1 to obtain an N_2. Continuing in this way, since the groups get smaller, we must eventually arrive at an s with $H = N_s$. The chain of groups corresponds, according to the fundamental theorem of Galois theory, with a chain of subfields

$$\mathbb{Q} = F_0 \subseteq F_1 \subseteq \ldots \subseteq F_s = \mathbb{Q}(\alpha)$$

with $[F_{i+1}:F_i] = 2$ for all $0 \le i < s$. Note that all these fields are contained in $\mathbb{R}$ since α is real. Thus α is constructible. □

12.4.5 EXAMPLE The real number

$$\alpha = \sqrt{1 + \sqrt{3}}$$

is constructible with the square root tower $\mathbb{Q} \subseteq \mathbb{Q}(\sqrt{3}) \subseteq \mathbb{Q}(\alpha)$. In Example 12.2.19 we found that the minimal polynomial of α is $f(x) = x^4 - 2x^2 - 2 \in \mathbb{Q}[x]$ and that the splitting field of $f(x)$ over $\mathbb{Q}$ is $E = \mathbb{Q}(\alpha, \beta)$, where

$$\beta = \sqrt{1 - \sqrt{3}}$$

and that $\mathrm{Gal}(E/\mathbb{Q}) \cong D_4$. Thus $|\mathrm{Gal}(E/\mathbb{Q})| = 8 = 2^3$ in agreement with the preceding theorem. ◇

12.4.6 COROLLARY Let α be a real number algebraic over $\mathbb{Q}$ with minimal polynomial $f(x)$ over $\mathbb{Q}$ of degree 4. Then α is not constructible by straightedge and compass if and only if the resolvant cubic $g(x)$ of $f(x)$ is irreducible over $\mathbb{Q}$.

Proof Immediate from Theorem 12.4.4 and Theorem 12.3.26. □

In Section 11.3 we defined the notion of a real number being constructible with marked ruler and compass, and in Theorem 11.4.8 we showed that this notion is equivalent to α belonging to the top of a tower of fields

$$\mathbb{Q} = K_0 \subseteq K_1 \subseteq K_r \subseteq \mathbb{R}$$

where each $[K_{i+1} : K_i] = 2$ or 3. We can now say what this implies about the Galois group of the minimal polynomial of α over $\mathbb{Q}$.

12.4.7 THEOREM Let α be a real number that is constructible with marked ruler and compass. Then α is algebraic and the order of the Galois group of its minimal polynomial over $\mathbb{Q}$ is of the form 2^r3^s for some $r \geq 0$ and $s \geq 0$.

Proof. Use Theorem 11.4.8 and argue as in the ($\Rightarrow$) direction of the proof of Theorem 12.4.4. (See Exercise 19.) □

In Section 11.2 we discussed the constructibility of regular polygons by straightedge and compass and showed that the regular n-gon is constructible if and only if $\cos(2\pi/n)$ is a constructible real number. We soon apply Theorem 12.4.4 to settle for which n the regular n-gon is constructible, but first we need some information on the cyclotomic polynomials $\Phi_n(x)$. Recall from Definition 10.3.14 that

$$\Phi_n(x) = (x - \zeta_1)\dots(x - \zeta_{\phi(n)})$$

where the ζ_i are the primitive nth roots of unity.

12.4.8 THEOREM $\Phi_n(x) \in \mathbb{Z}[x]$

Proof First we show $\Phi_n(x) \in \mathbb{Q}[x]$. Let ζ be a primitive nth root of unity. Since all the other nth roots of unity are powers of ζ, $\mathbb{Q}(\zeta)$ contains all the zeros of $\Phi_n(x)$ and is its splitting field over $\mathbb{Q}$. Any $\rho \in \mathrm{Gal}(\mathbb{Q}(\zeta)/\mathbb{Q})$ merely permutes the factors $(x - \zeta_i)$ of $\Phi_n(x)$ and thus keeps the coefficients of $\Phi_n(x)$ fixed. Therefore, the coefficients belong to the fixed field of $\mathrm{Gal}(\mathbb{Q}(\zeta)/\mathbb{Q})$, which by the fundamental theorem of Galois theory is $\mathbb{Q}$ itself.

Now we use induction on n to show that in fact $\Phi_n(x) \in \mathbb{Z}[x]$. For $n = 1$ we have $\Phi_1(x) = x - 1 \in \mathbb{Z}[x]$. Suppose $\Phi_k(x) \in \mathbb{Z}[x]$ for all $k < n$. Let $l(x) \in \mathbb{Z}[x]$ be the product of all $\Phi_d(x)$ for $d < n$ that divide n. Since each $\Phi_d(x)$ is monic, $l(x)$ is also monic. By Proposition 10.3.16, $l(x)\Phi_n(x) = (x^n - 1) \in \mathbb{Z}[x]$, and $(x^n - 1)$ is monic, of course. By Theorem 8.4.16 it follows that $\Phi_n(x) \in \mathbb{Z}[x]$.□

The main property of $\Phi_n(x)$ that we need is its irreducibility over $\mathbb{Q}$, which we prove next.

12.4.9 THEOREM $\Phi_n(x)$ is irreducible over $\mathbb{Q}$.

Proof Let ζ be a primitive nth root of unity and $f(x)$ its minimal polynomial over $\mathbb{Q}$. Since $\Phi_n(\zeta) = 0$, $f(x)$ divides $\Phi_n(x)$ in $\mathbb{Q}[x]$. By Theorem 8.4.16 there exist monic $g(x)$ and $h(x)$ in $\mathbb{Z}[x]$ such that $g(x)$ is irreducible over $\mathbb{Q}$, $\deg g(x) = \deg f(x)$ and $\Phi_n(x) = g(x)h(x)$. It is enough to show that every zero of $\Phi_n(x)$ is a zero of $g(x)$, for then we must have $\Phi_n(x) = g(x)$, which is irreducible over $\mathbb{Q}$. Now the nth roots of unity are the powers ζ^k of ζ for $1 \le k \le n$, and the zeros of $\Phi_n(x)$ are the primitive nth roots of unity, which are the powers ζ^k with $\gcd(k, n) = 1$.
We wish to show $g(\zeta^k) = 0$ for all such k. It is enough to show that $g(\zeta^p) = 0$ for any prime p not dividing n, since if this fact can be established, applying it with ζ^p in place of ζ and p' a prime not dividing n (possibly $= p$ and possibly $\neq p$), it will follow that $g(\zeta^{pp'}) = 0$; and repeating the process $g(\zeta^{pp'p''})$, and so on. Since any k with $\gcd(k, n) = 1$ can be written as a product of (not necessarily distinct) primes $pp'p''\ldots$ not dividing n, it follows that $g(\zeta^k) = 0$.
To show $g(\zeta^p) = 0$, where p is a prime not dividing n, we assume $g(\zeta^p) \neq 0$ and derive a contradiction. First, since $0 = \Phi_n(\zeta^p) = g(\zeta^p)h(\zeta^p)$, assuming $g(\zeta^p) \neq 0$ we must have $h(\zeta^p) = 0$, so ζ is a zero of $h(x^p)$. Since $g(x)$ is irreducible over $\mathbb{Q}$ and $g(\zeta) = 0$, $g(x)$ must divide $h(x^p)$ in $\mathbb{Q}[x]$. Moreover, since $g(x)$ and $h(x^p)$ are monic polynomials in $\mathbb{Z}[x]$, we have $h(x^p) = g(x)q(x)$ for some monic $q(x) \in \mathbb{Z}[x]$. Now we reduce mod p and obtain

$$\overline{g(x)}\ \overline{q(x)} = \overline{h(x^p)} = (\ \overline{h(x)}\ )^p$$

in $\mathbb{Z}_p[x]$. Therefore, $\overline{g(x)}$ and $\overline{h(x)}$ have a common irreducible factor in $\mathbb{Z}_p[x]$. Now, letting $l(x)$ be as in the proof of the preceding theorem, in $\mathbb{Z}_p[x]$ we have

$$x^n - 1 = \overline{x^n - 1} = \overline{\Phi_n(x)}\ \overline{l(x)} = \overline{g(x)}\ \overline{h(x)}\ \overline{l(x)}$$

and since $\overline{g(x)}$ and $\overline{h(x)}$ have a common factor in $\mathbb{Z}_p[x]$, $x^n - 1$ has a zero of multiplicity > 1 in its splitting field over $\mathbb{Z}_p$ or, in other words, is inseparable over $\mathbb{Z}_p$. But this is impossible by Theorem 12.1.6, since its degree n is not divisible by p. This contradiction completes the proof. □

12.4.10 COROLLARY Let ζ be a primitive nth root of unity. Then $\mathrm{Gal}(\mathbb{Q}(\zeta)/\mathbb{Q}) \cong U(n)$, and $|\mathrm{Gal}(\mathbb{Q}(\zeta)/\mathbb{Q})| = \phi(n)$, where ϕ is the Euler ϕ-function.

Proof The elements of the cyclic group $\langle\zeta\rangle = \{1, \zeta,, \ldots , \zeta^{n-1}\}$ of all nth roots of unity are the zeros of $x^n - 1$, and the generators of $\langle\zeta\rangle$ are the primitive nth roots of unity or, in other words, the zeros of $\Phi_n(x)$. By the preceding theorem, $\Phi_n(x)$ is irreducible over $\mathbb{Q}$ and is therefore the minimal polynomial of ζ over $\mathbb{Q}$, and since all its zeros belong to $\langle\zeta\rangle$, the splitting field of $\Phi_n(x)$ is $\mathbb{Q}(\zeta)$. Any field automorphism $\rho \in \mathrm{Gal}(\mathbb{Q}(\zeta)/\mathbb{Q})$ must take zeros of $x^n - 1$ to zeros of $x^n - 1$, that is, elements of $\langle\zeta\rangle$ to elements of $\langle\zeta\rangle$, and its restriction to $\langle\zeta\rangle$ is a group automorphism $\psi(\rho) \in \mathrm{Aut}(\langle\zeta\rangle)$. The map ψ taking $\rho \in \mathrm{Gal}(\mathbb{Q}(\zeta)/\mathbb{Q})$ to its restriction $\psi(\rho) \in \mathrm{Aut}(\langle\zeta\rangle)$ is a homomorphism, since the group operation is the same, composition of maps, in both $\mathrm{Gal}(\mathbb{Q}(\zeta)/\mathbb{Q})$ and $\mathrm{Aut}(\langle\zeta\rangle)$. $\psi(\rho)$ is the identity only if $\rho(\zeta) = (\psi(\rho))(\zeta) = \zeta$, in which case ρ is the identity, since ρ is completely determined by $\rho(\zeta)$, according to Theorem 12.1.19. It follows that ψ is one to one. Every $\sigma \in \mathrm{Aut}(\langle\zeta\rangle)$ is likewise completely determined by $\sigma(\zeta)$, which must be another generator of $\langle\zeta\rangle$, that is, another zero of $\Phi_n(x)$. But then there is a $\rho \in \mathrm{Gal}(\mathbb{Q}(\zeta)/\mathbb{Q})$ with $\rho(\zeta) = \sigma(\zeta)$ and therefore $\sigma = \psi(\rho)$, according to Theorem 12.1.23. It follows that ψ is onto. Thus ψ is an isomorphism and $\mathrm{Gal}(\mathbb{Q}(\zeta)/\mathbb{Q}) \cong \mathrm{Aut}(\langle\zeta\rangle) \cong U(n)$, and $|\mathrm{Gal}(\mathbb{Q}(\zeta)/\mathbb{Q})| = |U(n)| = \phi(n)$. □

Now we have everything we need to settle the question of the constructibility of regular polygons.

12.4.11 THEOREM The regular n-gon is constructible with straightedge and compass if and only if $n = 2^r p_1 \ldots p_t$ for some $r \geq 0$ and some number $t \geq 0$ of distinct Fermat primes $p_1, \ldots , p_s$ (that is, distinct primes each of the form $2^{2^k} + 1$ for some $k \geq 0$).

Proof Consider any integer $n > 2$. A regular n-gon is constructible if and only if $\alpha = \cos(^{2\pi}/_n)$ is a constructible real number. Now

$$\zeta = \cos(^{2\pi}/_n) + i \sin(^{2\pi}/_n)$$

is a primitive nth root of unity and we have

$$\zeta^{-1} = \zeta^{n-1} = \cos(^{2\pi}/_n) - i \sin(^{2\pi}/_n)$$
$$2\alpha = \zeta + \zeta^{-1} \in \mathbb{Q}(\zeta)$$
$$\mathbb{Q} \subseteq \mathbb{Q}(\alpha) \subseteq \mathbb{Q}(\zeta)$$

$\mathbb{Q}(\zeta)$ is the splitting field of $x^n - 1$ over $\mathbb{Q}$, hence a Galois extension of $\mathbb{Q}$, and $[\mathbb{Q}(\zeta) : \mathbb{Q}(\alpha)] = 2$ since ζ and ζ^{-1} are the two zeros of $x^2 - 2\alpha x + 1$, so

$$|\mathrm{Gal}(\mathbb{Q}(\zeta)/\mathbb{Q})| = [\mathbb{Q}(\zeta) : \mathbb{Q}] = 2[\mathbb{Q}(\alpha) : \mathbb{Q}]$$

It follows by Theorem 12.4.4 that α is constructible if and only if $|\mathrm{Gal}(\mathbb{Q}(\zeta)/\mathbb{Q})|$ is a power of 2, and by the preceding corollary $|\mathrm{Gal}(\mathbb{Q}(\zeta)/\mathbb{Q})| = \phi(n)$, so α is constructible if and only if $\phi(n)$ is a power of 2.

Now let

$$n = 2^r p_1^{a_1} p_2^{a_2} \ldots p_s^{a_s}$$

Then

$$\phi(n) = 2^{r-2} p_1^{a_1-1}(p_1 - 1) p_2^{a_2-1}(p_2 - 1) \ldots p_s^{a_s-1}(p_s - 1)$$

(See Exercise 17 in Section 10.3.) It follows that $\phi(n)$ is a power of 2 if and only if for all i, $a_i = 1$ and $p_i - 1$ is a power of 2, that is, $p_i = 2^{b_i} + 1$ for some b_i. Now a number of the form $2^b + 1$ can be a prime only if b is a power of 2. (See Exercise 14.) So we must have $p_i = 2^{2^{k_i}} + 1$ for some k_i. □

The regular n-gons that are constructible with marked ruler and compass can now be completely characterized, as the next theorem shows.

12.4.12 THEOREM The regular n-gon is constructible with marked ruler and compass if and only if $n = 2^r 3^s p_1 \ldots p_t$ for some $r \geq 0$ and $s \geq 0$ and some number $t \geq 0$ of distinct primes $p_i > 3$ each of the form $2^k 3^l + 1$ for some $k > 0$ and $l \geq 0$.

Proof Let α and ζ be as in the proof of the preceding theorem, so that the regular n-gon is constructible with marked ruler and compass if and only if α is so constructible, and

$$\mathbb{Q} \subseteq \mathbb{Q}(\alpha) \subseteq \mathbb{Q}(\zeta)$$

$$\phi(n) = |\mathrm{Gal}(\mathbb{Q}(\zeta)/\mathbb{Q})| = [\mathbb{Q}(\zeta):\mathbb{Q}] = 2[\mathbb{Q}(\alpha):\mathbb{Q}]$$

($\Rightarrow$) Assume α is constructible with marked ruler and compass. We then know from the preceding chapter that $[\mathbb{Q}(\alpha):\mathbb{Q}]$ must be a product of a power of 2 times a power of 3, hence $\phi(n)$ must be a product of a power of 2 times a power of 3. That n is of the form specified in the statement of the theorem then follows as in the proof of the preceding theorem.

($\Leftarrow$) Assume that n is of the specified form. It follows that $\phi(n)$ is a product of a power of 2 times a power of 3. By Corollary 12.4.10, $\mathrm{Gal}(\mathbb{Q}(\zeta)/\mathbb{Q})$ is a finite Abelian group, hence every subgroup is normal and every intermediate field is a Galois extension of $\mathbb{Q}$. In particular, $\mathbb{Q}(\alpha)$ is a Galois extension of $\mathbb{Q}$, and if $G = \mathrm{Gal}(\mathbb{Q}(\alpha)/\mathbb{Q})$, then

$$G \cong \mathrm{Gal}(\mathbb{Q}(\zeta)/\mathbb{Q}) / \mathrm{Gal}(\mathbb{Q}(\zeta)/\mathbb{Q}(\alpha))$$

is an Abelian group of order $\phi(n)/2$, a product of a power of 2 and a power of 3. We claim that there is a chain of subgroups

$$\{\mathrm{id}\} = H_m \leq \ldots \leq H_1 \leq H_0 = G$$

where each H_{i+1} has index 2 or 3 in H_i. To prove this, note that if $G = H_0$ is trivial we are done, since we have a trivial such tower with $m = 0$. Otherwise, $|H_0|$ is divisible by 2 or 3. By Theorem 3.4.7, for any divisor d of $|H_0|$ there exists a subgroup of H_0 of order d and hence of index $|H_0|/d$. Apply this theorem to $d = |H_0|/2$ or $|H_0|/3$ to obtain H_1 of index 2 or 3 in H_0. If H_1 is trivial we are done. Otherwise, repeat the process to obtain the required chain. By the fundamental theorem of Galois theory, the chain of subgroups corresponds to a chain of subfields:

$$\mathbb{Q} = K_0 \subseteq K_1 \subseteq \ldots \subseteq K_m = \mathbb{Q}(\alpha)$$

where each $[K_{i+1}:K_i] = 2$ or 3. So α belongs to the top of a tower of extensions of degrees 2 or 3 and is constructible by marked ruler and compass. □

Exercises 12.4

In Exercises 1 through 7 determine for the indicated n whether or not the regular n-gon is constructible by straightedge and compass.

1. $n = 22$ **2.** $n = 24$ **3.** $n = 26$ **4.** $n = 34$

5. $n = 48$ **7.** $n = 56$ **6.** $n = 52$

In Exercises 8 through 13 determine for the indicated n whether or not the regular n-gon is constructible by marked ruler and compass.

8. $n = 12$ **9.** $n = 14$ **10.** $n = 22$

11. $n = 52$ **12.** $n = 56$ **13.** $n = 92$

14. Show that if $p = 2^k + 1$ is a prime, then k is a power of 2.

In Exercises 15 through 17 determine whether the real zeros of the indicated polynomial are constructible by straightedge and compass.

15. $x^4 - x^2 - 1$ **16.** $x^4 - 4x + 2$ **17.** $x^4 + 4x - 2$

18. Show that if α is a real number algebraic over $\mathbb{Q}$ of degree 4, then α is not constructible by straightedge and compass if and only if the Galois group of the splitting field of its minimal polynomial over $\mathbb{Q}$ is isomorphic to A_4 or S_4.

19. Prove Theorem 12.4.7.

20. Show that if $n = rs$, where r and s are relatively prime, then

$$\mathrm{Gal}(\mathbb{Q}(\zeta_n)/\mathbb{Q}) \cong \mathrm{Gal}(\mathbb{Q}(\zeta_r)/\mathbb{Q}) \times \mathrm{Gal}(\mathbb{Q}(\zeta_s)/\mathbb{Q})$$

where ζ_i is a primitive ith root of unity. (Use Exercise 18 in Section 3.1.)

21. Let $n = p_1^{a_1}p_2^{a_2}\ldots p_s^{a_s}$ be the decomposition of the positive integer n into a product of powers of distinct primes. Show that

$$\mathrm{Gal}(\mathbb{Q}(\zeta_n)/\mathbb{Q}) \cong \mathrm{Gal}(\mathbb{Q}(\zeta_{p_1^{a_1}})/\mathbb{Q}) \times \ldots \times \mathrm{Gal}(\mathbb{Q}(\zeta_{p_s^{a_s}})/\mathbb{Q})$$

22. Let G be a finite Abelian group. By Theorem 3.4.1,

$$G \cong \mathbb{Z}_{n_1} \times \ldots \times \mathbb{Z}_{n_s}$$

for some $n_1, \ldots, n_s$. Let p_i be primes such that $p_i \equiv 1 \bmod n_i$ for $1 \le i \le s$. Show that there exists a subfield K of $\mathbb{Q}(\zeta_{p_1p_2\cdots p_s})$ with $\mathrm{Gal}(K/\mathbb{Q}) \cong G$.

12.5 Radical Extensions

In this section we prove one of the most important consequences of the fundamental theorem of Galois theory, the insolubility of the quintic by radicals. Solving a polynomial equation by radicals means expressing the zeros of the polynomial using the basic field operations and extracting nth roots. We know how to express the zeros of quadratics, cubics, and quartics in terms of square roots and cube roots. Such operations allow us to express the splitting field of the polynomial as a square root/cube root tower, as we saw in Section 11.4. In Section 12.4 we saw how such a tower gives us a Galois group of a certain order. In this section we show that a polynomial is solvable by radicals if and only if its Galois group is a solvable group. We studied solvable groups in Section 5.3, where we showed that S_n for $n \geq 5$ is not a solvable group. Therefore, polynomials of degree $n \geq 5$ with Galois group isomorphic to S_n are not solvable by radicals.

We begin by reviewing what we know about quadratics, cubics, and quartics.

12.5.1 EXAMPLE For an irreducible quadratic $f(x) = x^2 + bx + c$ in $\mathbb{Q}[x]$, we know that the splitting field of $f(x)$ over $\mathbb{Q}$ is $\mathbb{Q}(\Delta)$, where Δ^2 is the discriminant $\Delta^2 = D = b^2 - 4c \in \mathbb{Q}$. The Galois group of $f(x)$ is isomorphic to $\mathbb{Z}_2$. A reducible quadratic splits and its Galois group is trivial. ◇

12.5.2 EXAMPLE For an irreducible cubic $f(x) = x^3 + px + q$ in $\mathbb{Q}[x]$, let α_1 be such that

$$\alpha_1{}^2 = (q/2)^2 + (p/3)^3 \in \mathbb{Q}$$

and let α_2 be such that

$$\alpha_2{}^3 = (-(q/2) + \alpha_1) \in \mathbb{Q}(\alpha_1)$$

and finally let ω be a primitive cube root of unity. Then we obtain a tower of fields

$$K_0 = \mathbb{Q} \subseteq K_1 = \mathbb{Q}(\alpha_1) \subseteq K_2 = \mathbb{Q}(\alpha_1, \alpha_2) \subseteq K_3 = \mathbb{Q}(\alpha_1, \alpha_2, \omega)$$

whose top contains the splitting field of $f(x)$ by Proposition 8.5.6, and each extension field K_{i+1} is obtained by adjoining an element whose square or cube belongs to the field K_i below. The Galois group of $f(x)$ over $\mathbb{Q}$ is isomorphic to A_3 or S_3. A reducible cubic either splits, in which case its Galois group is trivial, or is a product of a linear and an irreducible quadratic factor, in which case its Galois group is $\mathbb{Z}_2$. ◇

12.5.3 EXAMPLE For an irreducible quartic $f(x) = x^4 + px^2 + qx + r$ in $\mathbb{Q}[x]$, we first take a tower

$$\mathbb{Q} = K_0 \subseteq K_1 \subseteq K_2 \subseteq K_3$$

as in the preceding example to get a zero k of the auxiliary cubic. We next take u and v such that

$$u^2 = 2k - p \qquad v^2 = k^2 - r$$

to obtain a factorization

$$[x^2 + ux + (k + v)][x^2 - ux + (k - v)]$$

We then take the discriminants D_1 and D_2 of the two quadratic factors and take Δ_1 and Δ_2, where $\Delta_1{}^2 = D_1$ and $\Delta_2 = D_2$, to get a tower

$$\mathbb{Q} = K_0 \subseteq \ldots \subseteq K_3 \subseteq K_4 = K_3(u) \subseteq K_5 = K_4(v) \subseteq K_6 = K_5(\Delta_1) \subseteq K_7 = K_6(\Delta_2)$$

where K_7 contains the splitting field of $f(x)$ and each extension field K_{i+1} is obtained by adjoining an element whose square or cube belongs to the field K_i below. The Galois group of $f(x)$ over $\mathbb{Q}$ is isomorphic to one of $\mathbb{Z}_4$, V_4, D_4, A_4, S_4 by Theorem 12.3.26. A reducible quartic either splits, in which case its Galois group is trivial, or else its Galois group is one of $\mathbb{Z}_2$, V_4, A_3, S_3 by Theorem 12.3.22. ◇

Before we proceed further, we need to recall from Section 5.3 the definition of a solvable group (Definition 5.3.1). A group G is solvable if it has a chain of subgroups

$$G = G_0 \geq G_1 \geq G_2 \geq \ldots \geq G_r = \{\mathrm{id}\}$$

such that for all $0 \leq i < r$ we have

(1) $G_{i+1} \triangleleft G_i$

(2) G_i / G_{i+1} is Abelian

12.5.4 EXAMPLE The Galois groups of any quadratic, cubic, or quartic over $\mathbb{Q}$ are solvable. The possible Galois groups have been listed in the preceding examples. All except three are Abelian, therefore solvable, and furthermore

$S_3 \geq A_3 \geq \{\mathrm{id}\}$

$D_4 \geq \mathbb{Z}_4 \geq \{\mathrm{id}\}$

$S_4 \geq A_4 \geq \mathbb{Z}_4 \geq \{\mathrm{id}\}$

to show S_3, D_4, A_4, S_4 solvable.◇

We soon define solving a polynomial by radicals over a field to mean obtaining the zeros of the polynomial by successively adjoining nth roots of elements already adjoined to the field, as was done for the cubic and quartic in Examples 12.5.2 and 12.5.3. We go on to prove Galois's theorem, which says that a polynomial is solvable by radicals if and only if its Galois group is solvable. Hence if we can construct an example of a polynomial of degree 5 with Galois group S_5, this is an example of a quintic that, unlike a quadratic, cubic, or quartic, is not solvable by radicals, since we know from Theorem 5.3.9 that S_5 is not a solvable group.

12.5.5 DEFINITION Let F be a field. An extension field K of F is a **simple radical extension** of F if $K = F(\rho)$ for some ρ such that $\rho^n \in F$ for some $n > 0$. A **radical tower** over F is a tower of fields

$$F = K_0 \subseteq K_1 \subseteq K_2 \subseteq \ldots \subseteq K_r$$

where K_i is a simple radical extension of K_{i-1} for all $1 \leq i \leq r$. Here r is called the **height** and K_r the **top** of the tower. If $K_i = K_{i-1}(\rho_i)$, where $\rho_i{}^{n_i} \in K_{i-1}$, then $K_i = F(\rho_1, \rho_2, \ldots, \rho_i)$ and we can write the preceding tower as

$$F \subseteq F(\rho_1) \subseteq F(\rho_1, \rho_2) \subseteq \ldots \subseteq F(\rho_1, \rho_2, \ldots, \rho_r)$$

An extension field K of F is an **extension by radicals** of F if it is the top of some radical tower over F. Finally, a polynomial $f(x) \in F[x]$ is **solvable by radicals** over F if the splitting field E of $f(x)$ is contained in an extension by radicals of F. ○

12.5.6 EXAMPLE Examples 12.5.1 through 12.5.3 show that any quadratic, cubic, or quartic polynomial in $\mathbb{Q}[x]$ is solvable by radicals over $\mathbb{Q}$. ◇

12.5.7 EXAMPLE $f(x) = x^5 - 2 \in \mathbb{Q}[x]$ is solvable by radicals over $\mathbb{Q}$. Let ζ be the primitive fifth root of unity

$$\zeta = \frac{-1+\sqrt{5}}{2} + i\frac{\sqrt{10+2\sqrt{5}}}{4}$$

as in Example 10.3.20. Then the splitting field E of $f(x)$ is the top of the radical tower

$$\mathbb{Q} \subseteq \mathbb{Q}(\sqrt{5}) \subseteq \mathbb{Q}(\sqrt{5}, \sqrt{10+2\sqrt{5}}\, i) \subseteq \mathbb{Q}(\sqrt{5}, \sqrt{10+2\sqrt{5}}\, i, \sqrt[5]{2}\ ) = E$$

Actually, under Definition 12.5.5, $\mathbb{Q}(\zeta)$ counts as a simple radical extension of $\mathbb{Q}$ since $\zeta^5 = 1 \in \mathbb{Q}$. Therefore,

$$\mathbb{Q} \subseteq \mathbb{Q}(\zeta) \subseteq \mathbb{Q}(\zeta, 2^{1/5}) = E$$

is also a radical tower. ◇

In much of our work to follow, we want to assume that roots of unity are in the ground field at the bottom of the tower. The following lemma and proposition are useful when we need this assumption.

12.5.8 LEMMA (Auxiliary irrationalities lemma) Let F be a field and F' any extension field of F. Let $f(x) \in F[x]$, and let E' be a splitting field of $f(x)$ over F' and $E \subseteq E'$ the splitting field of $f(x)$ over F. Then the restriction map sending $\sigma \in G' =$ Gal(E'/F') to $\sigma|_E \in G =$ Gal(E/F) is a one-to-one homomorphism, and G' is isomorphic to a subgroup of G.

Proof The restriction of an automorphism of a field E' to a subfield E is, in general, an isomorphism between E and some other subfield K. In the present case, let $\alpha_1, \ldots \alpha_n$ be the zeros of $f(x)$ in E so that

$$E = F(\alpha_1, \ldots \alpha_n) \quad \text{and} \quad E' = F'(\alpha_1, \ldots \alpha_n)$$

Then $\sigma \in G'$ permutes the α_i while keeping each element of F' and therefore each element of $F \subseteq F'$ fixed. So the image of E under σ is E itself, and in this case the restriction is an automorphism of E and an element of G. The restriction map is a homomorphism because the group operation is the same, composition of maps, in both cases. If $\sigma \in G'$ is not the identity, then we must have $\sigma(\alpha_i) \neq \alpha_i$ for some i. Since $\sigma|_E(\alpha_i) = \sigma(\alpha_i)$, it follows that $\sigma|_E$ is not the identity either. It then follows that the restriction map is one to one, and is an isomorphism between G' and its image, which is a subgroup of G. □

12.5.9 PROPOSITION Let F be a field of characteristic 0 and let ζ be a primitive nth root of unity. Then Gal$(F(\zeta)/F)$ is Abelian, hence solvable.

Proof Since F is of characteristic 0, by Theorem 7.3.12 F is an extension field of $\mathbb{Q}$. $\text{Gal}(\mathbb{Q}(\zeta)/\mathbb{Q})$ is Abelian by Corollary 12.4.10. By the preceding lemma, $\text{Gal}(F(\zeta)/F)$ is therefore isomorphic to a subgroup of an Abelian group and so is Abelian and hence solvable. □

12.5.10 DEFINITION Let F be a field and E and extension field of F. Then E is called a **cyclic** extension of F if E is a Galois extension of F and $\text{Gal}(E/F)$ is a cyclic group. ○

12.5.11 PROPOSITION Let F be a field of characteristic 0 containing a primitive nth root of unity ζ, and let $K = F(\rho)$ where $\rho^n \in F$. Then K is a cyclic extension of F.

Proof Let $\rho^n = a \in F$. The zeros of $x^n - a$ are precisely the $\zeta^i\rho$ for $0 \leq i < n$, so since $\zeta \in F$, $F(\rho)$ is the splitting field of $x^n - a$ over F. Any $\sigma \in G = \text{Gal}(F(\rho)/F)$ is completely determined by the value $\sigma(\rho)$, which must equal $\zeta^i\rho$ for some i. Moreover, since $\zeta^n = 1$, if $i \equiv j \bmod n$, then $\zeta^i\rho = \zeta^j\rho$. Thus for each $\sigma \in G$ there is a unique $i \in \mathbb{Z}_n$ with $\sigma(\rho) = \zeta^i\rho$, and if we let $\chi(\sigma)$ be this i, then $\chi\colon G \to \mathbb{Z}_n$ is one to one. Moreover, χ is a homomorphism, since if $\chi(\sigma) = i$ and $\chi(\tau) = j$, then since $\zeta \in F$ and is fixed by every element of G, we have

$$\sigma\circ\tau(\rho) = \sigma(\zeta^j\rho) = \sigma(\zeta)^j\sigma(\rho) = \zeta^j\sigma(\rho) = \zeta^j\zeta^i\rho = \zeta^{i+j}\rho$$

and $\chi(\sigma\circ\tau) = \chi(\sigma) + \chi(\tau) \in \mathbb{Z}_n$. Thus χ is an homomorphism. Moreover, $\chi(\sigma) = 0$ if and only if $\sigma(\rho) = \rho$ and $\sigma = \text{id}$, so in fact χ is an isomorphism between G and its image, which is a subgroup of the additive group of $\mathbb{Z}_n$, which is cyclic, and therefore G is cyclic. □

The converse of the preceding proposition also holds, but for the proof we need a new notion.

12.5.12 DEFINITION Let E be a field and let $\sigma_1, \sigma_2, \ldots, \sigma_n$ be maps from E to E. We say the σ_i are **linearly dependent** over E if there are $a_i \in E$, not all 0, such that

$$a_1\sigma_1(\alpha) + a_2\sigma_2(\alpha) + \ldots + a_n\sigma_n(\alpha) = 0$$

for all $\alpha \in E$. Otherwise we say the σ_i are **linearly independent** over E. ○

12.5.13 LEMMA Let F be a field and E an extension field of F, and let $\sigma_1, \ldots, \sigma_n$ be distinct elements of $\text{Gal}(E/F)$. Then the σ_i are linearly independent over E.

Proof Suppose the σ_i are linearly dependent, and

$$b_1\sigma_1(\alpha) + b_2\sigma_2(\alpha) + \ldots + b_n\sigma_n(\alpha) = 0$$

for all $\alpha \in E$, where the $b_i \in E$ are not all zero. Renumbering if necessary, we may suppose the nonzero b_i come first, so for some $s \leq n$ we have

$$b_1\sigma_1(\alpha) + b_2\sigma_2(\alpha) + \ldots + b_s\sigma_s(\alpha) = 0$$

for all $\alpha \in E$, and with all b_i nonzero for $1 \leq i \leq s$. Consider now the smallest number $r \leq s$ such that there exist $a_i \in E$, for $1 \leq i \leq r$, all nonzero, such that

(1) $$a_1\sigma_1(\alpha) + a_2\sigma_2(\alpha) + \ldots + a_r\sigma_r(\alpha) = 0$$

for all $\alpha \in E$. Since the σ_i are distinct automorphisms, there exists some $\beta \in E$ such that $\sigma_1(\beta) \neq \sigma_r(\beta)$. Furthermore,

$$a_1\sigma_1(\alpha\beta) + a_2\sigma_2(\alpha\beta) + \ldots + a_r\sigma_r(\alpha\beta) = 0$$

or, equivalently,

(2) $$a_1\sigma_1(\alpha)\sigma_1(\beta) + a_2\sigma_2(\alpha)\sigma_2(\beta) + \ldots + a_r\sigma_r(\alpha)\sigma_r(\beta) = 0$$

Now multiplying (1) by $\sigma_r(\beta)$ and subtracting (2) from it, we obtain

(3) $$[\sigma_r(\beta) - \sigma_1(\beta)]a_1\sigma_1(\alpha) + \ldots + [\sigma_r(\beta) - \sigma_{r-1}(\beta)]a_{r-1}\sigma_{r-1}(\alpha) = 0$$

where $[\sigma_r(\beta) - \sigma_1(\beta)]a_1 \neq 0$ since all $a_i \neq 0$ and $\sigma_r(\beta) \neq \sigma_1(\beta)$. But then (3) is a relation of the same form as (1) but with $r - 1 < r$ terms, contrary to our choice of r, and the contradiction completes the proof. □

Now for the promised converse to Proposition 12.5.11.

12.5.14 PROPOSITION Let F be a field of characteristic 0 containing a primitive nth root of unity ζ, and let E be a cyclic extension of F of degree n. Then E is a simple radical extension $E = F(\rho)$, where $\rho^n \in F$.

Proof $G = \mathrm{Gal}(E/F)$ is a cyclic group of order n. Let $\sigma \in G$ be any generator. Then for any $\alpha \in E$ let

$$g(\alpha) = \alpha + \zeta\sigma(\alpha) + \zeta^2\sigma^2(\alpha) + \ldots + \zeta^{n-1}\sigma^{n-1}(\alpha) \in E$$

Then since $\zeta \in F$, we have

$$\sigma(g(\alpha)) = \sigma(\alpha) + \zeta\sigma^2(\alpha) + \zeta^2\sigma^3(\alpha) + \ldots + \zeta^{n-1}\sigma^n(\alpha) \in E$$

and then since $\zeta^{n-1} = \zeta^{-1}$ and $\sigma^n = \mathrm{id}$, we have

$$\sigma(g(\alpha)) = \zeta^{-1}[\alpha + \zeta\sigma(\alpha) + \zeta^2\sigma^2(\alpha) + \ldots + \zeta^{n-1}\sigma^{n-1}(\alpha)] = \zeta^{-1}g(\alpha)$$

It follows that $\sigma(g(\alpha)^2) = \zeta^{-2}g(\alpha)^2$ and

$$\sigma(g(\alpha)^n) = \zeta^{-n}g(\alpha)^n = g(\alpha)^n$$

which implies that $g(\alpha)^n \in F$ for all $\alpha \in E$. Since σ is a generator, the elements $\mathrm{id}, \sigma, \sigma^2, \ldots, \sigma^{n-1}$ are all distinct and therefore by the preceding lemma are linearly independent over F. Hence there exists an $\alpha \in E$ such that $g(\alpha) \neq 0$, and

$$\sigma^i(g(\alpha)) = \zeta^{-i}g(\alpha) \neq g(\alpha)$$

for all $1 \leq i < n$. This implies that $g(\alpha)$ does not belong to the fixed field of any nontrivial subgroup of the Galois group $G = \langle\sigma\rangle$, which is to say does not belong to any intermediate field. Hence setting $\rho = g(\alpha)$, we have $E = F(\rho)$, where $\rho^n \in F$, and E is a simple radical extension of F. □

Our definition of solvable by radicals was that the splitting field of $f(x)$ over F is contained in a radical extension K of F. We show next that if the splitting field of $f(x)$ over F is contained in a radical extension of F, then in fact it is contained in a radical

Galois extension L of F, and indeed one whose Galois group over F is solvable. The next lemma implies this in the special case where K is a simple radical extension of F, and the following theorem and corollary extend this to the general case. The proofs of the lemma and the theorem resemble one direction of the proof of Theorem 12.4.4.

12.5.15 LEMMA Let $F \subseteq E \subseteq K$ be fields of characteristic 0 such that E is a Galois extension of F and K is a simple radical extension of E. Then there exists a field L such that

(1) $K \subseteq L$
(2) L is a Galois extension of F.
(3) L is a radical extension of E.
(4) $\mathrm{Gal}(L/E)$ is solvable.

Proof Let $K = E(\rho)$, where $\rho^n = \beta \in E$, and let $\mathrm{Gal}(E/F) = \{\sigma_1, \dots, \sigma_r\}$, where σ_1 is the identity. Let

$$f(x) = \Phi_n(x)\prod_{1 \le i \le r}(x^n - \sigma_i(\beta)) \in E[x]$$

and let L be the splitting field of $f(x)$ over $K = E(\rho)$.

(1) This is immediate.

(2) Since each $\sigma \in \mathrm{Gal}(E/F)$ merely permutes the factors of $f(x)$, the coefficients of $f(x)$ are fixed by $\mathrm{Gal}(E/F)$ and so $f(x) \in F[x]$. Since $\rho^n = \beta = \sigma_1(\beta)$, ρ is a zero of $f(x)$, and L is the splitting field of $f(x)$ over E. Since E is a Galois extension of F, it is the splitting field of some polynomial $h(x) \in F[x]$ over F. Hence L is the splitting field of $h(x)f(x) \in F[x]$ over F and is a Galois extension of F.

(3) Let ζ be any zero of $\Phi_n(x)$ in L, that is, a primitive nth root of unity in L. Then $E(\zeta)$ counts as a simple radical extension of E since $\zeta^n = 1 \in E$. By Theorem 10.3.21, each σ_i, $1 \le i \le r$, extends to an automorphism $\sigma_i^* \in \mathrm{Gal}(L/F)$, where we may take σ_1^* to be the identity. Each $\sigma_i^*(\rho)$ satisfies

$$(\sigma_i^*(\rho))^n = \sigma_i^*(\rho^n) = \sigma_i^*(\beta) = \sigma_i(\beta) \in E$$

Therefore, $\sigma_i^*(\rho)$ is a zero of $f(x)$, and all the zeros of $f(x)$ are powers of ζ or are of the form $\zeta^j\sigma_i^*(\rho)$ for some $0 \le j < n$ and $1 \le i \le r$. Thus L is the top of the radical tower

$$E \subseteq E(\zeta) \subseteq E(\zeta, \sigma_1^*(\rho)) \subseteq E(\zeta, \sigma_1^*(\rho), \sigma_2^*(\rho)) \subseteq \dots$$
$$\dots \subseteq E(\zeta, \sigma_1^*(\rho), \dots, \sigma_r^*(\rho)) = L$$

(4) Write E_1 for $E(\zeta)$, E_2 for $E(\zeta, \sigma_1^*(\rho)) = E_1(\sigma_1^*(\rho))$, ... , and E_{r+1} for $E_r(\sigma_r^*(\rho)) = L$. Write G_i for $\mathrm{Gal}(L/E_i)$. Corresponding to the preceding tower of subfields, we have the tower of subgroups

$$\{\mathrm{id}\} \le G_n \le \dots \le G_2 \le G_1 \le \mathrm{Gal}(L/E)$$

Since $E_1 = E(\zeta)$ is the splitting field of $\Phi_n(x)$ over E, E_1 is a Galois extension of E and $G_1 = \mathrm{Gal}(L/E_1) \triangleleft \mathrm{Gal}(L/E)$ and $G/G_1 \cong \mathrm{Gal}(E_1/E)$, which is Abelian by Proposition 12.5.9. For every other case, $E_{i+1} = E_i(\sigma_i^*(\rho))$ is a cyclic extension of E_i, meaning that G_{i+1}/G_i is cyclic, hence Abelian, by Proposition 12.5.11. Hence G is a solvable group. □

12.5.16 THEOREM Let $F \subseteq E \subseteq K$ be fields of characteristic 0 such that E is a Galois extension of F and K is a radical extension of E. Then there exists a field L such that

(1) $K \subseteq L$
(2) L is a Galois extension of F.
(3) L is a radical extension of E.
(4) $\mathrm{Gal}(L/E)$ is solvable.

Proof Since K is a radical extension of E, there is a radical tower

$$E = E_0 \subseteq E_1 \subseteq \dots \subseteq E_{r-1} \subseteq E_r = K$$

where each E_{i+1} is a simple radical extension of E_i for $0 \le i < r$. The proof is by induction on r. The case $r = 1$ is the preceding lemma, so suppose the theorem holds for towers of height $< r$ and consider a tower of height r as before. By our induction hypothesis, there is a field L_{r-1} such that $E_{r-1} \subseteq L_{r-1}$, L_{r-1} is a Galois extension of F, L_{r-1} is a radical extension of E, and $\mathrm{Gal}(L_{r-1}/E)$ is solvable. Let $E_r = E_{r-1}(\rho)$, where $\rho^n \in E_{r-1}$ for some $n > 0$, and apply the preceding lemma to F, L_{r-1} (in the role of E) and $L_{r-1}(\rho)$ (in the role of K) to obtain a field L such that $L_{r-1}(\rho) \subseteq L$, L is a Galois extension of F, L is a radical extension of L_{r-1}, and $\mathrm{Gal}(L/L_{r-1})$ is solvable. We claim that (1) through (4) all hold.

(1) We have

$$K = E_r = E_{r-1}(\rho) \subseteq L_{r-1}(\rho) \subseteq L$$

(2) It is given that L is a Galois extension of F.

(3) Since there is a radical tower from E up to L_{r-1}, and $L_{r-1}(\rho)$ is a simple radical extension of L_{r-1}, and there is a radical tower from $L_{r-1}(\rho)$ up to L, combining, there is a radical tower from E up to L.

(4) We have

$$\mathrm{Gal}(L/E)\ /\ \mathrm{Gal}(L/L_{r-1}) \cong \mathrm{Gal}(L_{r-1}/E)$$

and since $\mathrm{Gal}(L/L_{r-1})$ and $\mathrm{Gal}(L_{r-1}/E)$ are solvable, it follows that $\mathrm{Gal}(L/E)$ is solvable by Theorem 5.3.11. □

12.5.17 COROLLARY Let $F \subseteq K$ be fields of characteristic 0, where K is a radical extension of F. Then there is a field L such that $K \subseteq L$ and L is a Galois radical extension of F and $\mathrm{Gal}(L/F)$ is solvable.

Proof Put $E = F$ in the preceding lemma. □

We have now done most of the necessary work to prove Galois's celebrated theorem.

12.5.18 THEOREM (Galois's theorem) Let F be a field of characteristic 0 and $f(x) \in F[x]$. Then $f(x)$ is solvable by radicals over F if and only if the Galois group of $f(x)$ over F is a solvable group.

Proof ($\Rightarrow$) Assume $f(x) \in F[x]$ is solvable by radicals, and let E be its splitting field. Then $F \subseteq E \subseteq K$, where K is a radical extension of F. By the preceding corollary, there is a field L such that $K \subseteq L$, L is a Galois radical extension of F, and $\mathrm{Gal}(L/F)$ is solvable. Since E is a splitting field, $\mathrm{Gal}(L/E) \triangleleft \mathrm{Gal}(L/F)$ and

$$\mathrm{Gal}(E/F) \cong \mathrm{Gal}(L/E) / \mathrm{Gal}(L/F)$$

It follows that $\mathrm{Gal}(E/F)$ is solvable by Theorem 5.3.11.
($\Leftarrow$) Assume $G = \mathrm{Gal}(E/F)$ is a solvable group, where E is the splitting field of $f(x) \in F[x]$. Let $n = |\mathrm{Gal}(E/F)|$ and let E' be the splitting field of $\Phi_n(x)$ over E, and F' the splitting field of $\Phi_n(x)$ over F, $F' \subseteq E'$. Then $\mathrm{Gal}(F'/F)$ is Abelian, hence solvable, by Proposition 12.5.9, and $\mathrm{Gal}(E'/F')$ is isomorphic to a subgroup of G by Lemma 12.5.8, and hence $|\mathrm{Gal}(E'/F')|$ divides n and $\mathrm{Gal}(E'/F')$ is solvable by Theorem 5.3.10. Since

$$\mathrm{Gal}(F'/F) \cong \mathrm{Gal}(E'/F) \,/\, \mathrm{Gal}(E'/F')$$

it follows that $\mathrm{Gal}(E'/F)$ is solvable by Theorem 5.3.11. Since $F' = F(\zeta)$, where ζ is a primitive nth root of unity, F' counts as a radical extension of F since $\zeta^n = 1 \in F$. So it is enough to show that E' is a radical extension of F', using the facts that $\mathrm{Gal}(E'/F')$ is solvable and that F' contains a primitive nth root of unity ζ and hence contains a primitive dth root of unity $\zeta^{n/d}$ for any d that divides n, and hence for any d that divides $|\mathrm{Gal}(E'/F')|$.
Now solvability implies by Exercise 19 of Section 5.3 that we have a chain of subgroups

$$\{\mathrm{id}\} \leq G_r \leq G_{r-1} \leq \ldots \leq G_2 \leq G_1 \leq G_0 = \mathrm{Gal}(E'/F')$$

where each $G_{i+1} \triangleleft G_i$ and G_{i+1}/G_i is a cyclic group of order $|G_{i+1}/G_i|$, a prime dividing $|\mathrm{Gal}(E'/F')|$. This corresponds to a chain of subfields

$$F' = F_0 \subseteq F_1 \subseteq F_2 \subseteq \ldots \subseteq F_{r-1} \subseteq F_r = E'$$

where each F_{i+1} is a cyclic extension of F_i of order some prime p for which there is a primitive pth root of unity in $F' \subseteq F_i$. By Proposition 12.5.14, each F_{i+1} is a simple radical extension of F_i, and we have our required radical tower. □

12.5.19 COROLLARY (Abel's theorem) The general polynomial of degree $n \geq 5$ over $\mathbb{Q}$ is not solvable by radicals over $\mathbb{Q}$.

Proof By Lemma 12.3.10, the Galois group of the general polynomial of degree n is S_n, and by Theorem 5.3.9 this is not a solvable group for $n \geq 5$. The corollary is then immediate from the preceding theorem. □

12.5.20 PROPOSITION Let $f(x) \in \mathbb{Q}[x]$ be an irreducible quintic having exactly three real zeros. Then $f(x)$ is not solvable by radicals over $\mathbb{Q}$.

Proof Let $E \subseteq \mathbb{C}$ be the splitting field of $f(x)$ over $\mathbb{Q}$, and let $\alpha_1, \alpha_2, \alpha_3$ be the real zeros of $f(x)$. Then the other zeros α_4, α_5 are complex conjugates of each other. Now complex conjugation is an automorphism of $\mathbb{C}$ fixing $\mathbb{R}$ and $\mathbb{Q}$, and so its restriction σ to E is an element of $G = \text{Gal}(E/\mathbb{Q})$. When G is viewed as a subgroup $G \leq S_5$, σ will be precisely the 2-cycle (4 5). Since $f(x)$ is irreducible, $[\mathbb{Q}(\alpha_i):\mathbb{Q}] = 5$, and so $|G| = [E:\mathbb{Q}]$ is divisible by 5. Hence by Cauchy's theorem (Theorem 4.6.10), G has an element of order 5, and the elements of S_5 of order 5 are precisely the 5-cycles. Hence G contains both a 2-cycle and a 5-cycle, and by Exercise 36 in Section 1.4 it follows that $G = S_5$, which is not a solvable group; hence $f(x)$ is not solvable by radicals over $\mathbb{Q}$. □

12.5.21 EXAMPLE To get a specific example of an irreducible quintic that is not solvable by radicals over $\mathbb{Q}$, we use some calculus. Let $f(x) = x^5 - 10x - 5 \in \mathbb{Q}[x]$, which is irreducible by Eisenstein's criterion. The derivative $f'(x) = 5x^4 - 10$ has exactly two real zeros $\pm 2^{1/4}$. Hence $f(x)$ has exactly two critical points. One of them, $(2^{1/4}, f(2^{1/4}))$, where $f(2^{1/4}) = -5 - 8 \cdot 2^{1/4}$, is below the x-axis, and the other, $(-2^{1/4}, f(-2^{1/4}))$, where $f(-2^{1/4}) = -5 + 8 \cdot 2^{1/4}$, is above the x-axis. The proof of Corollary 12.2.23 shows that for $x \geq 16$, $f(x) > 0$ and $f(-x) < 0$, so the graph of $f(x)$ crosses the x-axis exactly three times, once between -16 and $-2^{1/4}$, once between $-2^{1/4}$ and $2^{1/4}$, and once between $2^{1/4}$ and 16, so the preceding proposition applies and tells us that $f(x)$ is not solvable by radicals. ◇

Exercises 12.5

In Exercises 1 through 8 express the splitting field of the indicated polynomial $f(x) \in \mathbb{Q}[x]$ as a radical extension of $\mathbb{Q}$.

1. $x^2 + 2x + 2$

2. $x^3 - 5$

3. $x^3 - 3x + 1$

4. $x^4 + 1$

5. $x^5 - x^4 + x + 1$

6. $x^4 + 2x^2 + 1$

7. $x^3 + x^2 + x + 1$

8. $x^4 + x^3 + 2x^2 + x + 1$

In Exercises 9 through 11 find a radical extension of $\mathbb{Q}$ that contains the indicated α.

9. $\alpha = \sqrt{\dfrac{1 - \sqrt{-3}}{2}}$

10. $\alpha = i(\sqrt{2} - 3)(\sqrt[3]{5} + 1)$

11. $\alpha = \dfrac{1 + \sqrt{-5}}{\sqrt{-2}}$

In Exercises 12 through 17 determine whether or not the indicated quintic polynomial in $\mathbb{Q}[x]$ is solvable by radicals over $\mathbb{Q}$.

12. $x^5 - 2$

13. $x^5 + 1$

14. $x^5 - x^4 - 4x + 4$

15. $x^5 - 4x + 2$

16. $x^5 - 6x + 3$

17. $x^5 + x^4 + x^3 - 2x^2 - 2x + 5$

18. Let $f(x) \in \mathbb{Q}[x]$ be an irreducible quintic over $\mathbb{Q}$ that is not solvable by radicals over $\mathbb{Q}$. Show that if E is the splitting field of $f(x)$ over $\mathbb{Q}$, then there exists a Galois extension K of $\mathbb{Q}$ with $K \subseteq E$ such that

(a) $[K:\mathbb{Q}] = 2$

(b) For any field F with $\mathbb{Q} \subseteq K \subseteq F \subseteq E$, F is not a Galois extension of $\mathbb{Q}$.

19. Let $f(x) \in \mathbb{Q}[x]$ be irreducible over $\mathbb{Q}$, and E its splitting field over $\mathbb{Q}$. Show that if $[E:\mathbb{Q}] = 9$, then $f(x)$ is solvable by radicals over $\mathbb{Q}$.

20. Let α be a positive real number that is constructible by straightedge and compass, and $f(x)$ its minimal polynomial over $\mathbb{Q}$. Show that $f(x)$ is solvable by radicals over $\mathbb{Q}$.

21. Let $f(x) \in \mathbb{Q}[x]$ be irreducible over $\mathbb{Q}$, and E its splitting field over $\mathbb{Q}$. Show that if $[E:\mathbb{Q}] = p^k$, for some prime p and some $k \geq 0$, then $f(x)$ is solvable by radicals over $\mathbb{Q}$.

Chapter 13

Historical Notes

These notes indicate briefly the history of the development of algebra of the kind presented in this book. We begin with the oldest surviving mathematical documents and end with the first textbook in which algebra is presented in its modern, abstract form. The two most important milestones on the way are the work of Cardano and the work of Galois, each of which opened a new period in the history of algebra.

From Ahmes the Scribe to Omar Khayyam

Egyptian and Babylonian mathematics. Two papyrus scrolls representing Egyptian mathematics have survived, both based on still older material that is lost. The scrolls date from between 2000 and 1500 B.C. (All dates this far back are doubtful.) The longer of the two is a copy by a scribe named Ahmes of a lost original by an author whose name is unknown. Thousands of clay tablets representing Mesopotamian mathematics have also survived, the oldest being even older than the Egyptian scrolls, but the most important being Babylonian tablets, of which many date from about 600 B.C. to 300 B.C.

The Egyptian papyri and one class of Babylonian tablets consist of problems with solutions. Another class of Babylonian tablets consists of tables of squares and cubes, of approximate square and cube roots, and the like. Closely associated with the mathematical tablets are astronomical ones, but these brief notes will have to leave astronomy and the trigonometry associated with it entirely out of account. In the problem documents the intent seems clearly to be to enable the reader to solve similar problems by going through the same steps with different numbers, but no terminology or notation had been evolved that would permit the explicit statement of general rules. There is no indication of how the methods used were arrived at: The oldest documents must have been preceded by a long prehistory of trial and error and insight. Nor is there anything like a proof that any of the methods works: Some only give approximate answers.

Many problems involve only performing arithmetic operations on known quantities (for instance, calculating areas and volumes of figures from known dimensions). Note, however, that even division can be quite difficult if one does not have a system of numerals like the one we use today, based on position and with a zero symbol.

Egyptian numerals were nothing like this. The Babylonians used different systems at different times, and in the end got close to positional system with zero, though on a base of 60 rather than of 10; but they did not quite get all the way.

Most interesting in the context of this book are the problems that we would express in our modern notation as one or more equations to be solved for one or more "unknowns" x or y. Since negative numbers had not been thought of, much less complex numbers, only positive solutions were sought. In some problems, such as where what is to be found is the number of people in some group, only integer solutions would make sense, while in other problems, such as where what is to be found is the length of some object, there is no such restriction. From our modern point of view, the restricted problems belong to number theory, and the others to algebra: to linear algebra if the unknowns appear only to the first power, and otherwise to general algebra, the subject of this book.

In number theory, the Babylonians knew how to sum arithmetic and geometric series and the like. They were familiar with the so-called Pythagorean theorem in geometry, and perhaps the highest development in their number theory is represented by tables of triples of positive integers (a, b, c), such as $(3, 4, 5)$ or $(5, 12, 13)$, that satisfy the condition

$$a^2 + b^2 = c^2$$

and could be the lengths of the sides and the hypotenuse of a right triangle. The highest development in Babylonian algebra is represented by problems that are solved by a method equivalent to applying the quadratic formula. Millennia were to pass before a similar formula for the cubic was found.

Pythagoras. The first mathematicians whose names are known are the Greek philosophers Thales of Miletus (active around 600 B.C.) and Pythagoras of Samos (active around 500 B.C.), the latter being the more important. A great many stories are told about him, but all are in the nature of dubious and unverifiable legends. He is said to have traveled extensively in the ancient Near East, and this is possible; but he might also have acquired what there was to be learned from Egyptian and Babylonian mathematics without leaving home, since Samos, an island off the coast of Asia Minor, was near the edge of the Greek world, and the Greeks there had extensive commerical contacts with older civilizations.

Pythagoras migrated to the area of Greek settlement in southern Italy, and a century and more later there was a Pythagorean Society in existence in that area, claiming to have been founded by him and teaching mathematical results claimed to have been discovered by him. Modern scholars suspect that most of the mathematics is really due to later members of the society, especially Archytas of Tarentum.

Pythagorean mathematics included many results in number theory and one great discovery in geometry. Pythagorean interest in number theory was connected with a research program attempting to understand all phenomena in terms of ratios of positive integers. To mention one example of what is supposed to have been a Pythagorean discovery, the musical interval of an octave corresponds to the ratio 2 to 1 in the sense that halving the length of a vibrating string raises its tone an octave, while other musical intervals correspond to other simple ratios of positive integers.

The great discovery in geometry was that this program cannot be carried out in general, because the ratio of the hypotenuse to the side in an isosceles right triangle is not equal to any ratio of positive integers. This fact follows by number-theoretic reasoning from the fact that the area of the square on the hypotenuse is twice the area of the square on a side, a special case of the so-called Pythagorean theorem. This consequence could only be discovered by something like proof. The Babylonians already must have tried many particular fractions without finding one whose square is exactly equal to 2, but this does not in itself show that there can be *no* such fraction: Only a proof can show that.

Euclid. The Pythagorean discovery seems to have convinced Greek mathematicians both of the centrality of geometry, which deals in more general kinds of ratios than does number theory, and above all of the importance of proof. However, the early steps in the process of the development of Greek mathematical ideas cannot be traced, since little remains from the first 300 years of Greek mathematical writing beyond occasional passages in later writers that quote fragments from or offer testimonies about works now lost. In the earliest extended Greek mathematical work that survives as a whole, the emphasis on proof and the emphasis on geometry are already firmly established.

The work in question is the famous *Elements of Geometry*, written by Euclid of Alexandria around 300 B.C. By that time Persia had conquered Egypt and Babylon, had attempted and failed to conquer Greece, and had then been conquered by the Greeks under Alexander of Macedon. The latter founded new cities in former Persian territory, the most important being Alexandria on the coast of Egypt, which became the main center of scholarly studies, as Greek became the main language of scholarly writing for the whole eastern Mediterranean world. Euclid lived and worked in the city, and that is about all that is known of his life.

His *Elements* became the greatest mathematical best-seller of all time, being translated into dozens of languages and used as a textbook for 2000 years. It is organized into a series of definitions and theorems with their proofs, a style very unlike that of any earlier surviving mathematical work and quite like that of present-day texts, this book included, except that Euclid does not include exercises for the student. It is in Euclid's work that mathematics first appears in the form of an organized science.

As its title indicates, the work is mainly devoted to geometry. It includes, for instance, the construction of the regular pentagon. There is also a good deal of number theory, including the proofs that every positive integer can be written uniquely as a product of primes, that there are infinitely many primes, and so on. Euclid's theorem and Euclid's algorithm are also to be found in the work. Scholars suspect that most of the number theory, and even perhaps some of the geometry, derives from Pythagorean sources. Euclid himself did not include a section of historical notes in his textbook to indicate his sources.

By contrast with the brilliant geometric and number-theoretic portions, the algebraic portion of the book is disappointing, for reasons worth stopping to explain. While we would say that the ratio of the hypotenuse to the side of an isosceles right triangle is an irrational real number, the Greeks did not think of ratios of lengths as numbers, as items that can be added and multiplied. They did have a notion of multiplying two lengths to produce an area, or a length times an area to produce a volume. But since higher-dimensional geometry had not been thought of, one could not raise a length to a higher power than its square or its cube, and of course one could not meaningfully add a length plus an area or an area plus a volume. Despite these limitations, procedures equivalent to solving quadratic equations by the quadratic formula do appear, in geometrical disguise, in Euclid, but not much more in the way of algebra.

Diophantus. Euclid's work was intended as an introduction to mathematics and did not include everything the author knew, let alone the whole of Greek mathematics. Euclid does not discuss geometric constructions involving conic sections, that is, ellipses and parabolas and hyperbolas, nor does he discuss constructions with marked ruler and compass, though the Greeks knew that using such constructions one could duplicate the cube and trisect the angle. But since our focus is on algebra, we pass over later Greek mathematics in the same tradition of geometrical proof as Euclid, including the work of Archimedes of Syracuse, the greatest of the Greek mathematicians, and of Apollonius of Perga (both active in the 200s B.C.).

More directly significant for algebra, there were mathematical works written in Greek that seem much more in the numerical, computational tradition of the ancient Near East. The most important of these was the *Arithmetica* of Diophantus of Alexandria, of which only about half survives. In form, the work is more similar to Babylonian mathematics than to Euclid, in that it consists of series of problems with solutions, with no statements of general methods or proofs of general theorems, and the work is entirely numerical and not geometrical and freely admits powers higher than squares and cubes. But many of the problems are more sophisticated than anything in the surviving Babylonian material.

The problem that was to have the greatest historical interest was to write a given square number such as 16, for example, as a sum of two squares. In our modern notation, this would be written as an equation in two unknowns:

$$x^2 + y^2 = 16$$

There are, of course, infinitely many solutions, but Diophantus is interested only in *rational* solutions and works out the answer $x = 16/5$ and $y = 12/5$. A modern number theorist would prefer to clear fractions and write the equation as

$$x^2 + y^2 = 16z^2$$

where only *integral* solutions are sought. The solution would then be written $x = 16$, $y = 12$, $z = 5$. In modern number theory, an equation where one is interested only in integral solutions is called a *Diophantine* equation, after Diophantus.

Diophantus goes a considerable way in the direction of introducing special algebraic symbolism. Unfortunately, later mathematicians did not preserve, let alone improve, this notation, but went back to writing everything out in words, and a symbolism had to be reinvented a millennium later. On the whole, Diophantus seems to have been less influential in his own day than he became when his work was rediscovered in modern times.

Diophantus wrote sometime between 150 B.C. and 350 A.D., by which time the whole Mediterranean world was under Roman rule. Mathematical works from this period were, however, still written in Greek, not Latin, regardless of the nationality of the author, which in the case of Diophantus, for example, is unknown. Work in Greek mathematics persisted through the Roman period and well after the adoption of Christianity (all dates from this point on are therefore A.D.). But in later years there was less and less in the way of original work, and more and more in the way of commentaries on earlier writers.

Indian algebra. By the middle 500s the western provinces of the Roman empire had come under the rule of Germanic tribes, and western Europe would make no contribution to mathematics for most of the next millennium. By the middle 600s the Arabs, after their conversion to Islam, had conquered the empire's southern provinces, leaving only the northeast quarter, which we today refer to as the Byzantine empire. The main development of Arabic mathematics, however, only began after 800 and the foundation of the University of Baghdad by the caliph Harun al-Rashid and his successor. Before considering that development, something must be said about mathematics further east.

Chinese mathematics lies largely outside our story, which is only concerned with the line of development leading to modern algebra. Chinese mathematical works contain much number-theoretic and algebraic material, incuding the Chinese remainder theorem, the so-called Pascal's triangle (centuries before its rediscovery by the French mathematician Pascal), and more. But unfortunately, though the Chinese were not as isolated as, say, the Maya (whose entire civilization rose and fell before there was contact between the old and new worlds), no more than bits and pieces of Chinese work ever became known outside east Asia. We see shortly that Arab mathematicians acquired and translated Indian and Greek mathematical texts, and later European mathematicians acquired and translated Greek and Arabic works. Chinese works were not similarly available; and so, for example, while China and India both had decimal positional systems of numerals like the one we use today, it was from India rather than China that the rest of the world got this system.

The roots of Indian mathematics go back to an early date, and Babylonian mathematical and astronomical procedures became known there. But the high period of Indian mathematics began with Aryabhata (476-550) and Brahmagupta (598-670). (It should be understood that until we reach modern times, dates of birth and death are somewhat doubtful.) This was just the time when Greek mathematics was finally winding down and Arabic mathematics had hardly begun to gear up. The decimal positional system of numerals appears about this time. The high period continued, contemporaneously with Arabic mathematics, down through Bhaskara (1114-1185).

At least in its later phases, Indian mathematics freely admitted irrational numbers as numbers, manipulating them according to the same rules that hold for positive integers. Bhaskara, for instance, gives rules for manipulating square roots that in our modern notation would be written

$$\sqrt{a} + \sqrt{b} = \sqrt{(a+b) + 2\sqrt{ab}}$$

Likewise, Indian mathematicians freely manipulated negative numbers. Unfortunately, like the use of special algebraic notation by Diophantus, this feature of Indian mathematics was not fully adopted in Arabic mathematics and had to be reinvented in Europe.

In number theory, perhaps the most impressive achievement was the study of the so-called Pell's equation. This is the Diophantine equation

$$x^2 - Ny^2 = 1$$

which we consider here only in the case $N = 2$. A solution (a, b) in this case gives a good approximation to an important irrational number, since we have

$$|\sqrt{2} - a/b| < 1/b^2$$

and thus we can get a close approximation without using too large a denominator. From our modern point of view, such a solution also gives a unit in an important ring, since we have

$$(a + b\sqrt{2})(a - b\sqrt{2}) = 1 \in \mathbb{Z}[\sqrt{2}]$$

The deepest fact about these units is that they can all be obtained from the unit α corresponding to the solution (1,1):

$$(a + b\sqrt{2}) = (1 + \sqrt{2})^n$$

for some positive integer n. The fact was discovered, needless to say not in modern notation, by Indian mathematicians and appears in different versions in different writers, though none give proofs.

Arabic algebra. After the Arab conquests, Arabic replaced Greek as the language of scholarship over a large area including the whole of the Near East and beyond. Translations were made of writers like Brahmagupta and Euclid from the Sanskrit or Greek into Arabic. In fact, the translators were so thorough that some Greek works survive *only* in their Arabic translations, just as some Arabic works survive *only* in Latin translations made later in Europe. Corresponding to its two main sources, Arabic mathematics had two sides, one concerned with numerical computation, the other with geometrical proof.

The first important writer in Arabic mathematics, Muhammad al-Khwarizmi (780-850), represents the numerical, computational side. Al-Khwarizmi's family was apparently from central Asia, though he lived and worked in Baghdad. His importance was less as an original mathematician, since there is little in his work that cannot be found in earlier Indian sources, than as a popularizer, in which role he was immensely successful.

He wrote a book on Indian numerals that introduced the decimal positional system to the Arabic world. Still more influential was his book on calculating solutions to equations by performing similar operations to both sides. He referred to the operations in question, such as adding equal quantities, by the words *al-jabr* and *al-muqabala*, of whose exact meaning different scholars give differing accounts. An indication of the popularity of his work is that the very name of our subject, "algebra," comes from the word *al-jabr* in the title to the book just mentioned, while our word "algorithm" comes from the name of the author.

The other side of Arabic mathematics was something like a resumption of the interrupted Greek geometrical tradition. Passing over most of the writers in this tradition, as we passed over the Greek successors of Euclid, the next writer it is important to note here was the famous Persian poet and mathematician Omar al-Hayyam, or Khayyam (1048-1131).

In his work we find the culmination of a tradition of looking for geometric solutions to cubic equations by means of conic sections. It should be noted that, if only positive coefficients are allowed, we have to consider not just one form of cubic, but three forms:

$$x^3 + px = q \qquad x^3 = px + q \qquad x^3 + q = px$$

and indeed if the trick for eliminating the square term is not used, there are many more cases. He enumerates all the cases and gives a geometric solution for each.

We also find in his work a trick that gets round the difficulties that result if one thinks in the Greek way of the product of two lengths being an area. He introduces a unit length w, more or less as we do in our discussion of geometric constructions in the body of this book, and then defines the product, *relative to this unit* of two lengths x and y to be a *length*, not an area, namely the length z such that z is to y as x is to w. Euclid knew how to construct such a length z but did not have the idea of using the construction to define a kind of multiplication. Logically, the procedure is equivalent to treating *ratios* of lengths as items that can be multiplied, which is to say as *numbers*, and in later Arabic mathematics and early European mathematics real numbers are understood as precisely such ratios, until a new foundation for the real number system was developed in the nineteenth century.

We have now about reached the period when the Mongol conquests were beginning to affect all Asia, after which leadership in mathematics passed elsewhere. Western Europe had by this time sufficiently recovered that scholarship was being resumed, and the first universities were being founded. Arabic mathematics began to become known in Europe partly through Spain, and especially from commerce between various Italian city-states and the east. Eventually Greek works, which became available either in Arabic versions or in Greek originals from Byzantium, were to exert the most profound influence; but at the beginning of European mathematics, it was translations and popularizations of al-Khwarizmi's *Indian Numerals* and *Algebra* that had the most immediate impact.

From Gerolamo Cardano to C. F. Gauss

Cardano. A popular account of Hindu-Arabic numerals was written in 1202 by the Italian merchant Leonardo of Pisa, called Fibonacci (1180-1240), and already one begins to find here some original material that goes beyond the writer's Arabic sources. Over the next three centuries there was more original work: Negative numbers, interpreted as debts in business arithmetic, were readmitted into mathematics; several tricks for solving various special kinds of cubic and higher-degree equations were developed; and there was also an advance beyond "rhetorical algebra," in which everything is written out in full words, to "syncopated algebra," in which various abbreviations are introduced. Nonetheless, notations remained clumsy compared with those we use today, which allow the symbols to do some of the thinking for us. Where we would write x for the unknown, the Italians would either write out the word "cosa," meaning "thing," or write an abbreviation "co." For an expression like

$$\sqrt{40 - \sqrt{320}}$$

for example, they might write something like

$$RV\,40\ \tilde{m}\ R\,320$$

This particular example comes from a textbook by Fra Luca Pacioli (1445-1514), whose notation is already considerably improved over that of earlier writers.

Pacioli's work was written in 1487, just before Columbus, and published in 1494, just after Columbus. From this point on "published" means "printed," since the printing press had been introduced, which meant that mathematical works could be produced in larger numbers and could circulate more widely. Pacioli's book ends with the pessimistic remark that thus far it had proved impossible to give any general rule for solving cubic equations. This would soon change.

Scipione del Ferro (1465-1526), a professor at the University of Bologna, who actually knew Pacioli, solved the first of the three types of cubic equation listed previously. That is, he found a general method for expressing a zero of a polynomial

$$x^3 + px - q$$

where p and q are positive, in terms of square and cube roots. But at the time of his death he still had not published this discovery.

In fact, Italian mathematicians of the period sometimes kept their results secret, because there were prizes to be won in problem-solving contests. An ex-student of del Ferro challenged another mathematician, Niccolò Fontana, called Tartaglia (1500-1557), to just such a contest in 1535, hoping to win by using his secret knowledge of del Ferro's formula. When Tartaglia saw the list of problems proposed by his opponent, he realized that del Ferro must have possessed a general method of solution. Sometimes the mere knowledge that someone else has solved a mathematical problem makes it psychologically easier to solve it oneself, and in one all-night session Tartaglia found the general formula for a solution, solved the particular word problems proposed, and so won the contest.

Gerolamo Cardano (1501-1576) a well-known physician and mathematician, heard of Tartaglia's victory and in 1539 invited Tartaglia to Milan, where Cardano lived, and coaxed and cajoled him into revealing his method. Before revealing the secret, however, Tartaglia insisted that Cardano swear a solemn oath never to publish it. As soon as Tartaglia was out the door, Cardano set to work and soon found the solution to the other two cases, and so had a completely general solution to the cubic. To cap this, Ludovico Ferrari (1522-1565), who had been taken on by Cardano originally as a servant but had become his student, then found the method for reducing the solution of the quartic to solving an auxiliary cubic.

The oath seemed to prevent Cardano and Ferrari from publishing their work, since it would be impossible to do so without revealing Tartaglia's method, but in 1543 they traveled to Bologna and were invited by del Ferro's family to inspect his papers. They confirmed what they had been led to suspect, that del Ferro had found the solution to cubic equations of the first kind even earlier than Tartaglia. Cardano considered that he could now go ahead and publish, since he would be revealing del Ferro's method, not Tartaglia's; and in 1545 he published his *Ars Magna*, containing the solution to the cubic and quartic and a brief but accurate statement as to what was due to del Ferro, to Tartaglia, to himself, and to Ferrari. This hardly satisfied Tartaglia, but the quarrels that ensued, though a colorful chapter in Renaissance history, need not concern us here.

The achievement of these mathematicians is more impressive when one realizes that they still did not have available anything like our convenient notation for expressing and manipulating equations. Cardano is also noteworthy for introducing, somewhat uneasily, "imaginary" or complex numbers, allowing square roots to be taken of negative numbers in the quadratic formula, for instance. Though Cardano does not emphasize the fact, this is crucial to his method for solving cubics, since in one case his formula expresses real zeros of the polynomial in terms of cube roots of complex numbers. Rafael Bombelli (1526-1573), in his algebra text of 1572, gives a fuller account of this role of complex numbers, with explicit rules for computing with them.

Descartes and Fermat. The trigonometric formula for the real zeros in the case of the cubic in question was given in a book published in 1593 by François Viète (1540-1603, French). Already the new mathematics, which was eventually to spread to the whole world, was spreading from Italy to the rest of Europe, and from this point we give nationalities along with dates. Viète also showed in the same work how the marked ruler and compass could be used to give geometric solutions to cubics and quartics, and therewith solutions to the problems of duplicating the cube, trisecting the angle, and constructing a regular heptagon.

Viète is credited with some improvements in algebraic notation, but the notation we still use today, with a few very minor changes, first appeared in the *Geometry* of the philosopher René Descartes (1596-1650, French) published in 1637. This work introduces coordinate systems, which allow the methods of algebra to be applied systematically to geometric problems, and thus unifies the geometric tradition of the Greeks with the algebraic tradition that stretches back through so many civilizations to the Babylonians. We still call rectangular coordinates Cartesian, though Descartes himself did not insist that the coordinate axes be at right angles. The work includes, among others, the result that equations of degree 2 in two variables x and y correspond to conic sections.

Coordinates were independently introduced at almost the same time by the lawyer Pierre de Fermat (1601-1665, French), who in 1643 wrote up his work in the form of a letter to a foreign mathematician. Fermat is also known for his work in founding modern probability theory, in an exchange of letters with Blaise Pascal (1623-1662, French). Fermat was most important, however, for his revival of number theory, inspired by his reading of a Latin translation of the *Arithmetica* of Diophantus.

We have occasion in this book to mention just one of his number-theoretic results, which we call Fermat's theorem. This, too, was circulated only by a letter to another mathematician. It is sometimes called Fermat's little theorem to distinguish it from Fermat's great theorem. The latter is an assertion written by Fermat in the margin of his copy of the *Arithmetica*, on the page containing the problem we cited previously, about representing a square as a sum of squares. He says this cannot be done for cubes or higher powers or, in other words, that the Diophantine equation

$$x^n + y^n = z^n$$

has no solutions with $x, y \neq 0$ for any $n > 2$.

Fermat wrote that he had a proof that the margin was too narrow to contain, but it is extremely unlikely that whatever he had in mind amounted to a valid proof. The theorem was only proved 350 years later by Andrew Wiles (b. 1953), making use of the work from the intervening period of dozens of mathematicians from several continents. Much work in number theory and algebra during the centuries between Fermat and Wiles was inspired, however, by attempts to prove the theorem, so that its statement without proof contributed materially to the progress of the subject.

The greatest achievement of seventeenth century mathematics was the development of the differential and integral calculus. The pursuit of this subject, and of applications to physics, became the main concern of mathematicians through the eighteenth century. As a result, some of the greatest names in the history of mathematics hardly appear in the history of algebra, and the next figures we have to consider in this short account come about a century after Fermat.

The letter in which Fermat stated his little theorem contained only the statement of the theorem that if p does not divide a, then

$$a^{p-1} - 1$$

is divisible by p. The proof was only given a century later by Leonhard Euler (1707-1783, Swiss), who introduced the ϕ-function to prove the generalization that if a and n are relatively prime, then

$$a^{\phi(n)} - 1$$

is divisible by n. Number theory and algebra were, however, only sidelines in the work of Euler, who was the most prolific mathematician of all time.

Lagrange. Joseph Louis Lagrange (1746-1813, French), a leading mathematician of the generation after Euler, was notable for his work in number theory and the theory of equations. His number-theoretic results included a complete solution to Pell's equation, supplying the proofs that all the solutions are generated by one fundamental solution. In European mathematics attention had been drawn to this type of equation by some work of John Pell (1611-1685, English). Lagrange also, extending work of Euler, supplied a complete solution to the problem of representing numbers as sums of squares. Any prime of the form $4n + 1$ can be represented as a sum of two squares, and trivially so can the prime 2; and any prime of the form $4n + 3$ can be represented as a sum of no more than four squares; if two positive integers can each be represented as a sum of four squares, then so can their product;

and hence any positive integer can be represented as a sum of four squares. This was another result that had been asserted without proof by Fermat.

Our concern here, however, is more with Lagrange's work on the theory of equations. It was during the century between Fermat and Lagrange that scientific journals began to be published and academies of science were established in various European countries, and Lagrange's work on the theory of equations was communicated to the Berlin Academy in 1771. Another mathematician, Alexandre-Théophile Vandermonde (1735-1796, French), presented similar work to the Paris Academy in 1770, but they didn't get around to publishing it until 1774.

Lagrange's work concerns the so-called general equation, in which the zeros of the polynomial are indeterminates over the rationals. The coefficients of the general equation are the elementary symmetric functions of the zeros, as had been observed by Viète. Also, any symmetric function of the zeros, which is to say any quantity left unchanged when the zeros are permuted, is obtainable by rational operations of addition, subtraction, multiplication, and division from the coefficients of the equation, as had been proved some few years before Lagrange by Edward Waring (1734-1798, English).

Lagrange considers intermediate quantities that are left fixed by some but not all of the permutations of the zeros, and he shows that if two such quantities α and β are left fixed by exactly the same permutations, then either can be obtained from the other by rational operations. In our modern terminology and notation, what Lagrange is proving in the case of

$$E = \mathbb{Q}(x_1, \dots, x_n) \qquad F = \mathbb{Q}(s_1, \dots, s_n)$$

is that

$$\text{if } \mathrm{Gal}(E/F(\alpha)) = \mathrm{Gal}(E/F(\beta)), \text{ then } F(\alpha) = F(\beta)$$

But of course, modern terminology and notation is precisely what Lagrange did not have. What Lagrange then does is to give a complete analysis, in terms of such intermediate quantities, of what goes on in finding the zeros of the general cubic or quartic.

In the course of his investigations Lagrange proves, by an argument involving cosets, the result that in our terminology and notation would be stated thus: the order of any element of a group of permutations divides the order of the group. But in Lagrange's day the notion of "group" had not yet been defined. From our modern point of view, Lagrange's result is merely a special case of the the fact that the order of *any* element of *any* finite group divides the order of the group, and indeed that the order of any *subgroup* of any finite group divides the order of the group. This last result has come to be called Lagrange's theorem, since Lagrange's coset argument is what is essential to its proof. But Lagrange himself would not have understood its statement.

Abel. Similar remarks apply to Cauchy's theorem, which is named after Augustin Louis Cauchy (1789-1857, French) on the strength of a paper from 1815 proving a special case. Indeed, similar remarks apply to *all* attributions of general, abstract results on groups to any writer before the latter part of the nineteenth century. Let that be understood as we proceed to describe some of the work of Lagrange's successors.

A former student of Lagrange's, Paolo Ruffini (1765-1822, Italian), claimed several times to be able to prove, using ideas about permutations of roots and quantities fixed by various of them, such as had been investigated by Lagrange, that the general quintic equation cannot be solved by radicals. In fact, his proof only shows that it cannot be solved by radicals *that are rational functions of the zeros*. It remained to show that if the equation can be solved by radicals at all, then it can be solved by radicals that are rational functions of the zeros, and this was a rather large gap in the proof.

Niels Henrik Abel (1802-1829, Norwegian) filled in the gap and reproved Ruffini's result, of which he was unaware, to produce a proof of the unsolvability of the general quintic in 1824. In 1829 he proved the further result that we would state by saying that if all the zeros of the minimal polynomial of α are in $F(\alpha)$, and if $\mathrm{Gal}(F(\alpha)/F)$ is an Abelian group, then the minimal polynomial of α is solvable by radicals over F. Needless to say, Abel was unacquainted with the word "Abelian": groups whose operation is commutative were named after him much later.

Gauss. Carl Friedrich Gauss (1777-1855, German), the leading mathematician of the first half of the nineteenth century, worked in many areas of pure and applied mathematics, including algebra and especially number theory. In algebra, he was the first to give a fully rigorous proof of the fundamental theorem of algebra. He offered one proof in his doctoral dissertation and eventually gave four more. Since he did not have available the tools of Galois theory, all the proofs are either considerably more complicated than that used in this book or draw on deeper facts from calculus.

Most of his other work on algebra (for instance, his work on the factorization of polynomials), along with the related work of his student Ferdinand Eisenstein (1823-1852, German), was incidental to his work on number theory. Gauss revolutionized that subject with the publication of his *Disquisitiones Arithmeticæ* of 1801. In this work Gauss systematically explores congruences or arithmetic mod n, and among many other results establishes that the nonzero integers mod p, where p is a prime, are all powers of some one of them, called a *primitive root* mod p. This fact was crucial to his treatment of cyclotomic polynomials.

Given the geometric representation of complex numbers and operations on them, which Gauss was one of the first to stress, finding the zeros of the nth cyclotomic polynomial is equivalent to constructing a regular n-gon. In particular, the n-gon can be constructed using only straightedge and compass if and only if the zeros of the nth cyclotomic polynomial can be obtained using only rational operations and square roots. As early as 1796 Gauss had found a straightedge and compass construction of the 17-gon, the first new constuction of a regular polygon since ancient times. In the *Disquisitiones Arithmeticæ*, Gauss's first published account of this work, he actually gives a general account of finding zeros of the cyclotomic polynomial by extracting roots, and observes that for n of the form

$$n = 2^r pp'p'' \ldots$$

where $p, p', p'', \ldots$ are distinct primes such that $p - 1$ is divisible by 2, then the zeros can be found taking only square roots, and the regular n-gon can be constructed by straightedge and compass.

He also states, as a warning to would-be polygon constructors, that if n is *not* of this form, then the n-gon is *not* constructible. He does not, however, give a proof of this last assertion. Gauss's statement implies the impossibility of constructing by straightedge and compass the regular 9-gon, and therefore of trisecting an angle of $2\pi/3$, and he was also aware of the impossibility of duplicating the cube. But the first published proofs were by Pierre Laurent Wantzel (1814-1848, French).

From Evariste Galois to Emmy Noether

Galois. Galois theory, which permits a unified treatment of the results of Abel, Gauss, and others on the solvability and insolubility of equations and the possibility and impossibility of geometric constructions and has many other applications besides, is named after Evariste Galois (1811-1831, French). The dates just given are not a mistake: Galois was killed in a duel at the age of 20. He had three times written up his work for the Academy of Sciences in Paris. Twice his submissions were lost, and the last time they were misunderstood. He wrote up his work again the night before he died. The first publication of his major contributions came only in 1846.

His main results include the primitive element theorem and an independent proof of Abel's theorem on the unsolvability of the general quintic, based on the more general theorem that we would state by saying that a polynomial equation $f(x) = 0$ is solvable by radicals if and only if its Galois group is a solvable group. His own terminology was naturally not completely modern. He did not have the concept of an abstract group, but used the term "group" to mean what we would call a group *of permutations*. He recognized the importance of the concept of normal subgroup (of a group of permutations) but had no special term for it. He did have the concept of an abstract "field," though he used a more cumbersome expression.

In addition to his main work, he also considered finite fields. Gauss had already introduced the fields of integers mod p, for p a prime, and Galois considered extending these fields by adding symbolic solutions to equations, that is, by adjoining zeros of polynomials, and determined the structure of all finite fields.

The definitive nineteenth century treatment of Galois theory was the 600+ page *Treatise on Substitution Groups and Algebraic Equations* by Camille Jordan (1838-1922, French), first published in 1870. Here "substitutions" is another word for permutations. The book contains an extended discussion of finite groups of permutations and, in particular, includes the theorem that every such group has a composition series:

$$G = G_0 \geq H_1 \geq \ldots \geq H_r = \{\text{id}\}$$

where each group is a normal subgroup of the group before it, and that in any two such series the indices of each group in the one before are the same, apart from order. The result that in fact the quotient groups are the same, apart from order, was first published by Otto Hölder (1859-1937, German) in 1889.

From permutation groups to abstract groups. During the nineteenth century mathematics developed rapidly in many directions. Geometry was enriched by the discovery of non-Euclidean and higher-dimensional geometries, as well as by a shift in point of view, away from the construction of particular figures, and toward the study of general transformations of a space that leave some of its properties unchanged, as, for instance, translations, rotations and reflections of the plane leave distances between points unchanged, though they move the points themselves. Linear algebra also grew, with a vast extension of the theory of matrices and their determinants, and this theory is related to geometry, since matrices can represent not only the coefficients in a system of linear equations, but also the change of coordinates that results from a geometric transformation.

We have occasion in this book to mention only a few examples of geometric transformations and matrices, by way of illustrating the concepts of abstract group theory. Historically, the realization that group theory could be applied to geometry was one inspiration for generalizing the concept of group, which was expanded to take in not just groups of permutations but also groups of geometric transformations. Jordan, though the title of his book seems to refer only to groups of permutations, studied groups of geometric transformations, and specifically groups of symmetries.

But with this expansion we have still not quite arrived at the abstract concept of a group as a set of elements subject to an operation obeying certain laws, since groups of permutations and of geometric transformations are both groups whose elements are maps and whose operation is composition, for which the associative law is automatic. The abstract notion of *Abelian* group was arrived at before the abstract notion of group, and appears in work of Leopold Kronecker (1823-1891, German). In a paper of 1853 he completely analyzes the structure of finite Abelian groups. His interest is still in Galois theory, and he uses his analysis to prove a deep theorem to the effect that if the Galois group of an irreducible polynomial over the rationals is Abelian, then a zero of the polynomial can be obtained by adjoining to the rationals a primitive nth root of unity for some n.

Kronecker, by the way, had disctinctive philosophical views that led him to reject anything in mathematics that did not have an immediate computational content, and, in particular, to reject the concept of a real number. For instance, when his colleague Carl Lindemann (1852-1939, German) showed squaring the circle to be impossible by proving that π is transcendental, Kronecker described the proof as beautiful but useless, since according to him π did not exist. With such views, it was impossible for Kronecker to be satisfied with Gauss's proof of the fundamental theorem of algebra, and he sought to show that zeros of a polynomial can be obtained simply by adding symbols to be computed with according to specified laws. His construction was equivalent to what we would describe as taking the quotient of a polynomial ring by a principal ideal, though neither "ring" nor "ideal" had as yet entered the mathematical vocabulary.

As often happens when the time is ripe for the formulation of a new mathematical notion, when the abstract notion of group was finally introduced in the early 1880s, 50 years after Galois, it was introduced simultaneously and independently by several mathematicians. Note that *the historical order of development of the subject was the reverse of the logical order in which topics are presented in textbooks*, including this one. The applications of group theory to the theory of equations came first, and the development of the pure theory of groups came later, after a half-century-long process of generalization and abstraction isolated the group concept.

The theory of abstract groups. Examples of abstract groups had actually appeared in mathematics well before that time, but without the name. For instance, the quaternion group was in effect introduced by William Rowan Hamilton (1805-1865, Irish) in 1843. If one thinks of complex numbers as represented by points in the plane and remembers that such points can be represented by pairs of real numbers, it is a short step to thinking of complex numbers as Hamilton did, namely, simply as pairs of real numbers added and multiplied according to certain rules. Hamilton then asked himself if it was possible to define addition and multiplication operations for *triples* of real numbers, and after several years of experimenting convinced himself that a reasonable notion of multiplication could not be defined for triples, but could be for *quadruples* of real numbers, which could then be thought of as a kind of "hypercomplex" number

$$a + bi + cj + dk \qquad a, b, c, d \in \mathbb{R}$$

that would obey all the usual laws *except the commutative law for multiplication*. The rule of multiplication for these hypercomplex numbers, called *quaternions*, is completely determined by the rule of multiplication for the group consisting of ± 1, $\pm i$, $\pm j$, $\pm k$.

Once the abstract group concept was introduced, it was found that there were quite a few examples already introduced that constituted groups in the new, generalized sense. It was also found that almost all theorems that had been proved about groups of permutations actually applied to abstract finite groups. This follows from an off-hand remark of Arthur Cayley (1821-1895, English), who mentioned that if one writes out the multiplication table of a finite group of geometric transformations each row will be a permutation of the original list of elements of the group. This remark applies to abstract groups as well as groups of geometric transformations and shows that every such group is isomorphic to a group of permutations. Hence the important theorems of Peter Ludvig Mejdell Sylow (1832-1918, Norwegian), for instance, on the subgroups of a group whose order is a power of a prime p, though originally proved for permutation groups, hold for all finite groups.

Cayley, at the same time he remarked that a group *could* be represented as a group of permutations, added that this was not always the *best* way to think of the group. As it happens, a finite group can also be represented as a group of *matrices*, and this representation is often a more useful one. An entire theory of such group representations was introduced, mainly by Georg Frobenius (1848-1917, German) and William Burnside (1852-1927, English). The latter used the theory to prove that any group whose order is a product of powers of two primes is solvable. (This implies that if the order of the Galois group of the minimal polynomial of a real number α over the rationals is the product of a power of 2 and a power of 3, then α can be constructed with marked ruler and compass.)

As the twentieth century began, the emphasis in group theory had shifted to trying to understand the structure of finite groups in general. According to the Jordan-Hölder theorem, the finite simple groups are the building blocks out of which all other finite groups are composed. The problem of classifying all finite simple groups took most of the twentieth century to solve.

The theory of abstract rings: Kummer and Dedekind. Number theory continued in the nineteenth century to inspire the introduction of new algebraic methods and structures. We have occasion in this book only to give a hint of the importance for number theory of rings of the form

$$\mathbb{Z}[\sqrt{N}] = \{a + b\sqrt{N} \mid a, b \in \mathbb{Z}\}$$

though we have only looked closely at the case of the Gaussian integers $N = -1$ and the case $N = 2$. We do note that not all cases are as well behaved as these two cases and, in particular, that unique factorization into irreducible elements sometimes fails, as in the case $N = -5$, where we have

$$21 = 3 \cdot 7 = (1 + 2\sqrt{5}i)(1 - 2\sqrt{5}i)$$

Ernst Kummer (1810-1893, German), who was working with such rings in an attempt to prove Fermat's great theorem, had the idea of trying to restore unique factorization by adding "ideal numbers" to the ring. For instance, there would be ideal numbers I, J, K, L such that

$$IJ = 3 \qquad KL = 7 \qquad IK = 1 + 2\sqrt{5}i \qquad JL = 1 - 2\sqrt{5}i$$

Then the two preceding factorizations would really be the same factorization, apart from order: $(IJ)(KL) = (IK)(JL)$. This idea can be made to work and to give a proof of Fermat's great theorem for many, though not all, values of n.

Kummer's procedure was analyzed by Richard Dedekind (1831-1916, German), who considered the *ideals* not to be extra elements added to the ring but certain special subsets of the ring, just as they are defined in this book. Multiplication of ideals is simply intersection. Given elements of the ring *generate* an ideal in a ring R,

$$\langle a, b, c, \ldots \rangle = \{xa + yb + zc + \ldots \mid x, y, z \in R\}$$

In the Kummer example, we would have

$$I = \langle 3, 1 + 2\sqrt{5}i\rangle \qquad J = \langle 3, 1 - 2\sqrt{5}i\rangle \qquad I \cap J = \langle 3\rangle$$

and so on. Dedekind was able to prove a general unique factorization theorem for ideals defined in this way.

The theory of abstract rings: Hilbert and Noether. Dedekind did not, however, have a fully abstract notion of ring, and indeed did not even use the word "ring." This first appeared in the writings of David Hilbert (1862-1945, German), the leading mathematician of the first half of the twentieth century, though Hilbert does not yet use the word in its fully general, abstract sense. Of Hilbert's many important theorems there is space here only to mention the earliest, which dates from 1888. We today call a ring a PID if every ideal in it is generated by a single element. The integers form a PID, and so does the ring of polynomials in one variable over any field. However, the ring of polynomials in one variable over the integers and the ring polynomials in two variables over a field are not PIDs, each having ideals generated by two elements that cannot be generated by any one element.

$$\langle 2, x\rangle \subseteq \mathbb{Z}[x] \qquad \langle x, y\rangle \subseteq F[x, y] = F[x][y]$$

Hilbert showed, however, that if every ideal in a ring R is generated by finitely many elements, then every ideal in the ring of polynomials over R is likewise generated by finitely many elements.

Hilbert's services to mathematics were not limited to proving important theorems but included also finding faculty positions for promising younger mathematicians at the University of Göttingen, where he, like Gauss before him, taught. Later, in the 1930s, many of these mathematicians were forced out of Germany and came to the United States. For the history of algebra, the most important of them was Emmy Noether (1882-1935). She developed an abstract theory of rings, ideals in rings, and *modules* over rings, the last being a notion that includes ideals in rings and vector spaces over fields. Among other results, she proved a far-reaching generalization of Dedekind's theorem about unique factorization of ideals.

While the content of her research is beyond the scope of this book, she more than anyone promoted the abstract and general *style* of doing algebra that is reflected on almost every page of this or any modern textbook. This influence was exerted mainly through her teaching. (In this connection it may be mentioned that as a woman she was for some years not allowed to hold a paid faculty position: The subterfuge had to be adopted of announcing a course by Hilbert, which actually turned out to be taught by his "assistant" Noether.) Her lectures inspired others to undertake abstract and generalized reformulations of other branches of algebra, a notable case being the treatment of Galois theory by Emil Artin (1898-1962, Austrian), which emphasizes the correspondence between groups of automorphisms and fields.

The first textbook presenting algebra in the modern, abstract, and general way was *Modern Algebra*, first published (in German) in 1930, by B. L. van der Waerden (1903-1996, Dutch). The title page describes the work as "based on the lectures of Emmy Noether and Emil Artin." This was a graduate-level text, but undergraduate-level expositions followed in the 1940s. Van der Waerden was also, as it happens, one of comparatively few mathematicians to take a serious, scholarly interest in the history of his subject, and his historical works, listed in the following readings, have been our main source for these notes.

Further reading

For further study of the history of mathematics, here are some excellent sources.

Kline, Morris *Mathematical Thought from Ancient to Modern Times* Oxford: Oxford University Press, 1972.

Neugebauer, O. *The Exact Sciences in Antiquity* New York: Dover, 1969.

van der Waerden, B. L. *Geometry and Algebra in Ancient Times* Berlin: Springer-Verlag, 1983.

——— *A History of Algebra: From al-Khwarizmi to Emmy Noether* Berlin: Springer-Verlag, 1985.

Part C

Selected Topics

Chapter 14

Symmetries

In this introductory chapter we only touch on a beautiful subject in the applications of group theory to geometry. We calculate the groups of rotations of the five *Platonic solids* and then show that these are exactly the possible finite subgroups of the *group of rotations* of $\mathbb{R}^3$. We give first the absolutely necessary background on linear transformations and define the *orthogonal group* of matrices and the *special orthogonal group*. We use extensively the theory of group action we developed in Chapter 4.

Linear Transformations

Given a vector space V over a field F, a linear transformation $T: V \to V$ as defined in Exercise 26 of Section 10.1 is a map such that for any two vectors $u, v \in V$ and any two scalars $c, d \in F$ we have

$$T(cu + dv) = cT(u) + dT(v)$$

14.1 EXAMPLE Let e_1, e_2, e_3 be the standard basis for $\mathbb{R}^3$ and let $T: \mathbb{R}^3 \to \mathbb{R}^3$ be a linear transformation given by

$$T(e_1) = 2e_1 - e_2 = v_1 \qquad T(e_2) = e_1 + 2e_3 = v_2 \qquad T(e_3) = e_1 + e_2 + e_3 = v_3$$

Let A be the 3×3 matrix obtained by taking v_i as its ith column, $1 \leq i \leq 3$. Thus

$$A = \begin{bmatrix} 2 & 1 & 1 \\ -1 & 0 & 1 \\ 0 & 2 & 1 \end{bmatrix}$$

Given any vector $v \in \mathbb{R}^3$,

$$v = c_1e_1 + c_2e_2 + c_3e_3 = \begin{bmatrix} c_1 \\ c_2 \\ c_3 \end{bmatrix}$$

for some $c_i \in \mathbb{R}$, we have

$$Av = \begin{bmatrix} 2c_1 + c_2 + c_3 \\ -c_1 + c_3 \\ 2c_2 + c_3 \end{bmatrix}$$

Hence we have

$$Av = c_1(2e_1 - e_2) + c_2(e_1 + 2e_3) + c_3(e_1 + e_2 + e_3)$$
$$= c_1T(e_1) + c_2T(e_2) + c_3T(e_3) = T(c_1e_1 + c_2e_2 + c_3e_3) = T(v)$$

So the linear transformation can be represented as multiplying column vectors by a matrix. ◇

Given a linear transformation $T\colon V \rightarrow V$, where V is a vector space over a field F with $\dim_F V = n$ and a basis $\{e_j\}$ for V, $1 \leq j \leq n$, we define the associated $n \times n$ matrix A as follows: Let $T(e_j) = v_j$ and let A be the matrix whose jth column is v_j. Then for any vector

$$v = c_1e_1 + \ldots + c_ne_n \in V$$
$$T(v) = c_1T(e_1) + \ldots + c_nT(e_n) = [v_1]c_1 + \ldots + [v_n]c_n = Av$$

For a fixed basis of V, if A is the associated matrix of T, then we write the linear transformation as T_A.
Note that

$$T_B \circ T_A = T_{BA}$$

Hence A^{-1} exists if and only if T_A is an isomorphism, since $T_{A^{-1}} \circ T_A = T_{A^{-1}A} = \text{id}$. Furthermore, by Exercise 28, Section 10.1, given a basis $\{v_j\}$, T is an isomorphism if and only if $\{w_j = T(v_j)\}$ is also a basis of V. Hence given two bases $\{v_j\}$ and $\{w_j\}$ in V, there exists an invertible matrix B such that if

$$\begin{bmatrix} c_1 \\ \vdots \\ c_n \end{bmatrix}$$

are the coordinates of a vector in terms of the basis $\{v_j\}$, then

$$B\begin{bmatrix} c_1 \\ \vdots \\ c_n \end{bmatrix}$$

are the coordinates of the same vector in terms of the basis $\{w_j\}$. Hence given a linear transformation $T\colon V \rightarrow V$, if $T = T_A$ in terms of the basis $\{v_j\}$, then $T = T_{BAB^{-1}} = T_BT_AT_B{}^{-1}$ in terms of the basis $\{w_j\}$.
One of the most basic notions in linear algebra is the determinant of an $n \times n$ matrix. We defined the determinant for 2×2 matrices in Definition 0.5.12, and the general definition for $n \times n$ matrices in the Exercises after Section 12.3. For convenience, we repeat the general definition here.

14.2 DEFINITION Let $A = \{a_{ij}\}$ be an $n \times n$ matrix with entries from a ring R. Then the **determinant** of A is

$$\det A = \sum_{\sigma \in S_n} (-1)^{\varepsilon(\sigma)} \cdot a_{1\sigma(1)} \cdot a_{2\sigma(2)} \cdot \ldots \cdot a_{n\sigma(n)}$$

where $\varepsilon(\sigma) = 0$ if σ is an even permuation, and $\varepsilon(\sigma) = 1$ if σ is an odd permutation. ○

14.3 EXAMPLE If $A = \{a_{ij}\}$ is a 3×3 matrix, then we have the formula

$$\det A = a_{11}a_{22}a_{33} + a_{12}a_{23}a_{31} + a_{13}a_{21}a_{32} - a_{11}a_{23}a_{32} - a_{12}a_{21}a_{33} - a_{13}a_{22}a_{31} =$$
$$a_{11}(a_{22}a_{33} - a_{23}a_{32}) - a_{12}(a_{21}a_{33} - a_{23}a_{31}) + a_{13}(a_{21}a_{32} - a_{22}a_{31})$$

for the determinant of A. ◇

One of the properties of determinants we use freely in this chapter is that the determinant is multiplicative or, in other words, that for any two $n \times n$ matrices A and B

$$\det(AB) = \det A \det B$$

(See Exercises 1 through 4.)
Some of the linear transformations we consider have fixed vectors, as do the rotations around a fixed axis.

14.4 DEFINITION Let $T = T_A: V \to V$ be a linear transformation, where V is a finite-dimensional vector space over a field F, and let $0 \neq v \in V$ be such that $T_A(v) = \lambda v$ for some $\lambda \in F$. Then v is called an **eigenvector** of T (and of A), and λ the corresponding **eigenvalue** of T (and of A). ○

Observe that if for some $0 \neq v \in V$ we have $T(v) = \lambda v$, then $(T - \lambda I_n)v = 0$. In other words, v is in the kernel of the linear transformation $T - \lambda I_n$.

14.5 THEOREM Let F be a field, V a vector space of dimension $\dim_F V = n$ over F, and $T_A: V \to V$ a linear transformation. Then the following conditions are equivalent:

(1) T_A is one to one.
(2) T_A is onto.
(3) $\det A \neq 0$
(4) 0 is not an eigenvalue of T_A.

Proof (1) ⇔ (2) by Section 10.1, Exercise 30.

(2) ⇔ (3) T_A is isomorphism if and only if there exists a matrix B such that $BA = I_n$, which by Exercise 5 is equivalent to $\det A \neq 0$.

(1) ⇔ (4) Let $0 \neq v \in V$. Then $v \in \text{Kern } T_A$ if and only if $T_A(v) = 0 = 0 \cdot v$ and hence v is an eigenvector of T_A with corresponding eigenvalue 0. □

14.6 COROLLARY The eigenvalues of an $n \times n$ matrix are exactly the scalars $\lambda \in F$ such that $\det(A - \lambda I_n) = 0$.

Proof See Exercise 6. □

14.7 DEFINITION Given an $n \times n$ matrix $A = \{a_{ij}\}$, the **transpose** of A is the matrix $A^T = \{b_{ij}\}$, where $b_{ij} = a_{ji}$.○

Note that

(1) $(AB)^T = B^T A^T$
(2) $(A^T)^T = A$
(3) $\det(A^T) = \det A$

(See Exercise 7.)

We concentrate for the rest of this chapter on the vector spaces $\mathbb{R}^2$ and $\mathbb{R}^3$. Let us recall some basic terminology.

14.8 DEFINITION Let

$$u = \begin{bmatrix} x_1 \\ x_2 \\ x_3 \end{bmatrix} \quad \text{and} \quad v = \begin{bmatrix} y_1 \\ y_2 \\ y_3 \end{bmatrix}$$

be two vectors in $\mathbb{R}^3$.

(1) The **dot product** of v and u is

$$v \cdot u = v^T u = x_1 y_1 + x_2 y_2 + x_3 y_3$$

(2) The **length** $|v|$ is given by

$$|v|^2 = v \cdot v = x_1 x_1 + x_2 x_2 + x_3 x_3$$

(3) The **distance** between v and u is the length $|v - u|$ of their difference $v - u$.

(Similar definitions apply to $\mathbb{R}^2$.) ○

An important property of the dot product of two vectors in $\mathbb{R}^2$ or $\mathbb{R}^3$ follows from the law of cosines:

$$v \cdot u = |v||u| \cos \theta$$

where θ is the angle between the vectors v and u. Hence

$$v \text{ and } u \text{ are orthogonal if and only if } v \cdot u = 0$$

14.9 DEFINITION A matrix A is said to be **orthogonal** if $A^T = A^{-1}$. ○

Observe that if $A = \{a_{ij}\}$, then $A^T A = \{c_{ij}\}$, where

$$c_{ij} = a_{i1} a_{j1} + \dots + a_{in} a_{jn} = (i\text{th row of } A) \cdot (j\text{th row of } A)$$

Since $A^T A = I_n$ if and only if $c_{ii} = 1$ for $1 \le i \le n$ and $c_{ij} = 0$ for $i \ne j$, this means that $A^T = A^{-1}$ if and only if each row of A is a vector of length 1 and distinct rows of A are orthogonal vectors.

14.10 PROPOSITION Let

$$O(n, \mathbb{R}) = \{A \in GL(n, \mathbb{R}) \mid A^T = A^{-1}\}$$

be the set of all orthogonal matrices in the general linear group $GL(n, \mathbb{R})$. Then $O(n, \mathbb{R})$ is a subgroup of $GL(n, \mathbb{R})$.

Proof Given $A, B \in O(n, \mathbb{R})$, we have $A^T = A^{-1}$, $B^T = B^{-1}$ and therefore

$$(AB^{-1})^T = (AB^T)^T = (B^T)^T A^T = BA^{-1} = (AB^{-1})^{-1}$$

Hence $AB^{-1} \in O(n, \mathbb{R})$, and by the subgroup test (Theorem 1.2.10), $O(n, \mathbb{R})$ is a subgroup of $GL(n, \mathbb{R})$. □

For any matrix $A \in O(n, \mathbb{R})$, $A^TA = I_n$, hence

$$(\det A)^2 = \det A \det A^T = \det(A^TA) = \det I_n = 1$$

so $\det A = \pm 1$. Furthermore, if $\det A = \det B = 1$, then $\det(AB^{-1}) = 1$. Hence by the subgroup test, the subset

$$SO(n, \mathbb{R}) = \{A \in O(n, \mathbb{R}) \mid \det A = 1\}$$

of $O(n, \mathbb{R})$ forms a subgroup of $O(n, \mathbb{R})$.

14.11 DEFINITION

(1) $O(n, \mathbb{R})$ is called the **orthogonal group** of $n \times n$ invertible matrices.
(2) $SO(n, \mathbb{R})$ is called the **special orthogonal group**.○

To simplify notation, we write just $O(n)$ and $SO(n)$ from now on and leave it to be understood that we are working over $\mathbb{R}$.

14.12 EXAMPLE Let $A \in O(2)$,

$$A = \begin{bmatrix} a & b \\ c & d \end{bmatrix}$$

Case 1. $\det A = 1$, so $A \in SO(2)$. Then

$$A^T = \begin{bmatrix} a & c \\ b & d \end{bmatrix} = A^{-1} = \begin{bmatrix} d & -b \\ -c & a \end{bmatrix}$$

So $a = d$, $b = -c$, and $1 = \det A = ad - bc = a^2 + b^2$. Hence (a, b) is a point on the unit circle in $\mathbb{R}^2$. Thus $(a, b) = (\cos\theta, \sin\theta)$ for some angle $0 \leq \theta < 2\pi$ and

$$A = A(\theta) = \begin{bmatrix} \cos\theta & -\sin\theta \\ \sin\theta & \cos\theta \end{bmatrix}$$

In Example 4.1.1 we showed that any such matrix $A(\theta)$ is a rotation of the plane by an angle of θ. Furthermore, we showed that given any two points in $\mathbb{R}^2$ lying on a circle with center at the origin, there exists a matrix $A(\theta)$ that rotates the first point to the second point.

Case 2. $\det A = -1$. Then

$$A^T = A^{-1} = \begin{bmatrix} -d & b \\ c & -a \end{bmatrix}$$

So $a = -d$, $b = c$, and $-1 = ad - bc$, so $a^2 + b^2 = 1$, and

$$A = A'(\theta) = \begin{bmatrix} \cos\theta & \sin\theta \\ \sin\theta & -\cos\theta \end{bmatrix}$$

Given a point

$$P = \begin{bmatrix} x \\ y \end{bmatrix} = \begin{bmatrix} r\cos\phi \\ r\sin\phi \end{bmatrix} \in \mathbb{R}^2$$

then

$$A'(\theta)\, P = \begin{bmatrix} r\cos(\phi - \theta) \\ -r\sin(\phi - \theta) \end{bmatrix}$$

which gives us a **reflection** in a line at angle $\theta/2$ with the positive x-axis. ◇

14.13 EXAMPLE Let

$$B = B(\psi) = \begin{bmatrix} \cos\psi & -\sin\psi & 0 \\ \sin\psi & \cos\psi & 0 \\ 0 & 0 & 1 \end{bmatrix}$$

which we write

$$B(\psi) = \begin{bmatrix} \begin{matrix} & A(\psi) & \end{matrix} & \begin{matrix} 0 \\ 0 \end{matrix} \\ \begin{matrix} 0 & 0 \end{matrix} & 1 \end{bmatrix}$$

Then $B^T B = I_3$, $\det B = 1$, and $B \in SO(3)$. Furthermore, given a point

$$P(\rho, \phi, \theta) = \begin{bmatrix} \rho\sin\phi\cos\theta \\ \rho\sin\phi\sin\theta \\ \rho\cos\phi \end{bmatrix} \in \mathbb{R}^3$$

we have

$$B(\psi)\, P(\rho, \phi, \theta) = P(\rho, \phi, \theta + \psi)$$

Hence $B(\psi)$ leaves the z-coordinate fixed and rotates the points in $\mathbb{R}^3$ around the fixed z-axis. Note that if

$$C = \begin{bmatrix} 1 & \begin{matrix} 0 & 0 \end{matrix} \\ \begin{matrix} 0 \\ 0 \end{matrix} & A(\psi) \end{bmatrix} \in SO(3)$$

then C rotates the points in $\mathbb{R}^3$ around the x-axis instead. ◇

14.14 PROPOSITION For any matrix $A \in SO(3)$, $\lambda = 1$ is an eigenvalue.

Proof By Corollary I.6, we only need to show that $\det(A - I_3) = 0$. Since $A \in SO(3)$, $\det A = 1 = \det A^T$. Furthermore,

$$A^T(A - I_3) = I_3 - A^T = (I_3 - A)^T$$

Hence since $1 = \det A^T$,

$$\det(A - I_3) = \det(A^T(A - I_3)) = \det((I_3 - A)^T) = \det(I_3 - A) = -\det(A - I_3) \in \mathbb{R}$$

It follows that $\det(A - I_3) = 0$. □

14.15 THEOREM Let T_A: $\mathbb{R}^3 \to \mathbb{R}^3$ be a linear transformation. Then T_A preserves the dot product, that is, $T_A(v) \cdot T_A(w) = v \cdot w$ for all $v, w \in \mathbb{R}^3$, if and only if $A \in O(3)$.

Proof ($\Rightarrow$) Suppose T_A preserves the dot product. Since

$$Ae_i = i\text{th column of } A = i\text{th row of } A^T = (Ae_i)^T$$

we have

$$(Ae_i)^T(Ae_j) = (Ae_i) \cdot (Ae_j) = T_A(e_i) \cdot T_A(e_j) = e_i \cdot e_j = e_i^T e_j = \begin{cases} 1 \text{ if } i = j \\ 0 \text{ if } i \neq j \end{cases}$$

Hence $A^TA = I_3$ and $A \in O(3)$.

($\Leftarrow$) Suppose $A \in O(3)$. Then for any $v, w \in \mathbb{R}^3$

$$T_A(v) \cdot T_A(w) = (Av) \cdot (Aw) = (Av)^T(Aw) = v^T(A^TA)w = v^Tw = v \cdot w$$

and T_A preserves the dot product. □

Note that if a linear transformation preserves the dot product, then it preserves lengths, distances, and orthogonality.

Isometries

14.16 DEFINITION A **rigid motion** or **isometry** of $\mathbb{R}^n$ is a map

S: $\mathbb{R}^n \to \mathbb{R}^n$

that preserves distances, that is,

$$|S(p) - S(q)| = |p - q|$$

for any pair of points $p, q \in \mathbb{R}^n$. ○

14.17 THEOREM Let S: $\mathbb{R}^3 \to \mathbb{R}^3$ be a map. Then the following conditions are equivalent:

(1) S is an isometry that fixes the origin.
(2) S preserves the dot product.
(3) $S = T_A$ for some $A \in O(3)$

Proof (1) $\Rightarrow$ (2) Suppose S fixes the origin, so $S(0) = 0$, and S is an isometry, so $|S(v) - S(w)| = |v - w|$ for all $v, w \in \mathbb{R}^3$. This implies

$$(S(v) - S(w)) \cdot (S(v) - S(w)) = (v - w) \cdot (v - w)$$

or, expanding,

$$S(v)\cdot S(v) - 2S(v)\cdot S(w) + S(w)\cdot S(w) = v\cdot v - 2v\cdot w + w\cdot w$$

Taking first $w = 0$ and then $v = 0$, and using $S(0) = 0$, we get

$$S(v)\cdot S(v) = v\cdot v \quad \text{and} \quad S(w)\cdot S(w) = w\cdot w$$

Substituting back in the previous equation, we get

$$S(v)\cdot S(w) = v\cdot w$$

(2) $\Rightarrow$ (3) Assume that S preserves the dot product. Let A be the matrix with $S(e_i)$ as its ith column. Since

$$S(e_i)\cdot S(e_j) = 1 \text{ if } i = j, \qquad S(e_i)\cdot S(e_j) = 0 \text{ if } i \neq j$$

we have

$$A^TA = I_3$$

Hence $A \in O(3)$. It follows that A preserves dot products, and therefore the composite map $R = T_{A^{-1}} \circ S$ also preserves dot products. We have

$$T_{A^{-1}} \circ S(e_i) = A^{-1}S(e_i) = A^{-1}(i\text{th column of } A) = e_i$$

for all $1 \leq i \leq n$, For any $v = \mathbb{R}^3$ we have

$$v = (v\cdot e_1)e_1 + (v\cdot e_2)e_2 + (v\cdot e_3)e_3$$

Applying this to $R(v)$ and using the fact that

$$R(v)\cdot e_i = R(v)\cdot R(e_i) = R(v\cdot e_i)$$

we obtain

$$R(v) = (R(v)\cdot e_1)e_1 + (R(v)\cdot e_2)e_2 + (R(v)\cdot e_3)e_3 = (v\cdot e_1)e_1 + (v\cdot e_2)e_2 + (v\cdot e_3)e_3 = v$$

Hence $T_{A^{-1}} \circ S = R$ is the identity, and $S = T_A$.

(3) $\Rightarrow$ (1) Suppose $S = T_A$, where $A \in O(3)$. By Theorem 14.15, T_A preserves the dot product and hence lengths and distances, so S is an isometry. Furthermore, $T_A(0) = 0$ for *any* A, so S fixes the origin. □

14.18 COROLLARY An isometry that fixes the origin is a linear transformation.

Proof Immediate from (1) $\Rightarrow$ (3) in the preceding theorem. □

In Example 14.12 we showed how the matrices in $SO(2)$ correspond to rotations of $\mathbb{R}^2$. We show next that matrices in $SO(3)$ correspond to rotations of $\mathbb{R}^3$. But first we must define what is meant by a rotation in $\mathbb{R}^3$.

14.19 DEFINITION A **rotation** (fixing the origin) of $\mathbb{R}^3$ is a map $\rho: \mathbb{R}^3 \to \mathbb{R}^3$ such that

(1) ρ is an isometry that fixes the origin.
(2) ρ fixes some nonzero vector $v \in \mathbb{R}^3$.
(3) ρ restricted to the plane P orthogonal to the fixed vector v is a rotation of P. ○

14.20 THEOREM A map $\rho: \mathbb{R}^3 \to \mathbb{R}^3$ is a rotation if and only if $\rho = T_A$ for some $A \in SO(3)$.

Proof ($\Leftarrow$) Assume $A \in SO(3)$ and $\rho = T_A$ under the standard basis e_i of $\mathbb{R}^3$.

(1) By Theorem 14.17, ρ is an isometry that fixes the origin.

(2) By Proposition 14.14, A has an eigenvector v with eigenvalue 1, hence

$$\rho(v) = Av = v$$

(3) If $A = B(\psi) \in SO(3)$, as in Example 14.13, then A fixes the vector e_3 and rotates the plane P determined by $\{e_1, e_2\}$, which is orthogonal to e_3. We show that with a change of basis of $\mathbb{R}^3$, ρ is given by a matrix of this form.

Let v be a nonzero vector fixed by ρ and let $v_3 = v/|v|$, so v_3 is a unit vector fixed by ρ. Let v_1, v_2 be two orthogonal vectors in the plane P, where P is orthogonal to v_3. Then $\{v_1, v_2, v_3\}$ is a basis for $\mathbb{R}^3$. (See Exercise 10.) If

$$v_i = c_{i1}e_1 + c_{i2}e_2 + c_{i3}e_3$$

let

$$D^{\mathrm{T}} = \begin{bmatrix} c_{11} & c_{21} & c_{31} \\ c_{12} & c_{22} & c_{32} \\ c_{13} & c_{23} & c_{33} \end{bmatrix}$$

Since the v_i are orthogonal unit vectors, $DD^T = I_3 = D^TD$ and $\rho = T_A$ in terms of the standard basis and $\rho = T_{DADT}$ in terms of the basis $\{v_i\}$. Furthermore,

$$(DAD^T)^T(DAD^T) = DA^T(D^TD)AD^T = DAA^TD^T = I_3$$

and

$$\det(DAD^T) = \det D \det A \det D^T = 1$$

Hence $DAD^T \in SO(3)$. In terms of the basis $\{v_i\}$

$$T_{DADT}\begin{bmatrix} 0 \\ 0 \\ 1 \end{bmatrix} = DAD^T\begin{bmatrix} 0 \\ 0 \\ 1 \end{bmatrix} = DAv_3 = Dv_3 = \begin{bmatrix} 0 \\ 0 \\ 1 \end{bmatrix}$$

since $v_i \cdot v_3 = 0$ for $i \neq 3$ and $v_i \cdot v_3 = 1$ for $i = 3$. Therefore, the third column and third row of DAD^T are

$$\begin{bmatrix} 0 \\ 0 \\ 1 \end{bmatrix} \text{ and } \begin{bmatrix} 0 & 0 & 1 \end{bmatrix}$$

respectively. Hence

$$DAD^T = \begin{bmatrix} & C & & 0 \\ & & & 0 \\ 0 & & 0 & 1 \end{bmatrix}$$

where C is a 2×2 matrix with $\det C = 1$. And since

$$DA^TD^T = \begin{bmatrix} & C^T & 0 \\ & & 0 \\ 0 & 0 & 1 \end{bmatrix}$$

we have $C^TC = I_2$ and $C \in SO(2)$ determines a rotation of the plane P orthogonal to v_3, by some angle $-\pi \le \theta \le \pi$. Choosing instead the basis $\{v_1, v_2, -v_3\}$ changes the sign of θ and of $\det D$. If we insist on the choice that makes $\det D = 1$, an angle $-\pi < \theta \le \pi$ is uniquely determined.
($\Rightarrow$) Assume ρ is a rotation of $\mathbb{R}^3$. Then ρ is an isometry that fixes the origin and hence with respect to the standard basis, $\rho = T_A$ for some $A \in O(3)$ by Theorem 14.17. As in the ($\Leftarrow$) direction of the proof, let $v \in \mathbb{R}^3$ be a nonzero vector such that $Av = v$. Let $v_3 = v/|v|$, so v_3 is a unit vector with $Av_3 = v_3$. Let v_1, v_2 be two orthogonal vectors in the plane P orthogonal to v_3. Then with respect to the basis $\{v_i\}$, $\rho = T_B$, where

$$B = \begin{bmatrix} & G & 0 \\ & & 0 \\ 0 & 0 & 1 \end{bmatrix}$$

and G determines a rotation of the plane P and therefore has $\det G = 1$. It follows that $\det B = 1$. Letting D be as in the the ($\Leftarrow$) direction of the proof, $B = DAD^T$. Since $\det D = 1$, it follows that $\det A = 1$ and $A \in SO(3)$. □

Let us consider possible isometries of $\mathbb{R}^2$ or $\mathbb{R}^3$. These include

(1) **Translations**: These are each of the form $t_b(v) = v + b$ for some $b \in \mathbb{R}^n$.

(2) **Rotations**: These are each of the form T_A for some $A \in SO(n)$. We write ρ_θ for a rotation by an angle of θ (about some axis through the origin).

(3) **Reflections**: These are each given by T_A for some $A \in O(n)$ with $\det A = -1$, that is, $A \notin SO(n)$. We write r for the reflections given by the matrices

$$\begin{bmatrix} 1 & 0 \\ 0 & -1 \end{bmatrix} \in O(2)$$

$$\text{and} \quad \begin{bmatrix} 1 & 0 & 0 \\ 0 & 1 & 0 \\ 0 & 0 & -1 \end{bmatrix} \in O(3)$$

Since $SO(n)$ is a subgroup of $O(n)$ of index 2 (see Exercise 13), every reflection is of the form $\rho_\theta \circ r$.

14.21 THEOREM An isometry S of $\mathbb{R}^2$ or $\mathbb{R}^3$ can be uniquely expressed as

$$S = t_b \circ \rho_\theta \circ r^i \quad \text{where } i = 0 \text{ or } 1$$

Proof To prove existence, let $b = S(0)$. Then $t_{-b} \circ S$ fixes the origin and so $t_{-b} \circ S = T_A$ for some $A \in O(n)$. If $\det A = 1$, then T_A is a rotation ρ_θ and $S = t_b \circ \rho_\theta$. If $\det A = -1$, then T_A is a reflection $\rho_\theta \circ r$ and $S = t_b \circ \rho_\theta \circ r$.
To prove uniqueness, suppose

$$t_b \circ \rho_\theta \circ r^i = S = t_c \circ \rho_\phi \circ r^j$$

So

$$\rho_\theta = t_{-b} \circ t_c \circ \rho_\phi \circ r^j \circ r^{-i} = t_{-b+c} \circ \rho_\phi \circ r^{j-i}$$

Then t_{-b+c} fixes 0 and therefore is the identity t_0 and $b = c$. Thus

$$\rho_\theta = \rho_\phi \circ r^{j-i}$$

The left-hand side is represented by an orthogonal matrix with determinant 1, so the right-hand side must also be, so r^{j-i} is the identity. Hence $i = j$ and $\rho_\theta = \rho_\phi$. □

14.22 COROLLARY An isometry S of $\mathbb{R}^2$ or $\mathbb{R}^3$ can be uniquely expressed as

$$S(v) = Av + b$$

where $A \in O(n)$ and $b \in \mathbb{R}^3$.

Proof Immediate from the preceding theorem. □

14.23 THEOREM The set of all isometries of $\mathbb{R}^3$ forms a group, M_3, under composition of maps.

Proof See Exercise 14. □

The following useful rules are also left as exercises.

I.24 PROPOSITION For any translation t_a, rotation ρ_θ, and reflection r of $\mathbb{R}^2$ or $\mathbb{R}^3$, the following hold:

(1) $\rho_\theta \, t_a = t_b \, \rho_\theta$ where $b = \rho_\theta(a)$

(2) $r \, t_a = t_b \, r$ where $b = r(a)$

(3) $r \, \rho_\theta = \rho_{-\theta} \, r$

Proof See Exercise 15. □

Symmetry Groups

Given a subset H of $\mathbb{R}^n$, consider all the isometries $S \in M_n$ such that S restricted to H maps H to H. If we have two such isometries R and S, then $R \circ S^{-1}$ is also an isometry of $\mathbb{R}^n$ that maps H to H. Thus such isometries form a subgroup of M_n.

14.25 DEFINITION

(1) Given a subset H of $\mathbb{R}^n$, an isometry of $\mathbb{R}^n$ that maps H to H is called a **symmetry** of H.

(2) Given a subset H of $\mathbb{R}^n$, the group of all isometries of $\mathbb{R}^n$ that map H to H is called the **group of symmetries** of H in $\mathbb{R}^n$. ○

From our previous discussions we know that symmetries, since they are isometries of $\mathbb{R}^n$, are determined by matrices in $O(n)$ together with a vector that gives rise to a translation component. For finite groups of symmetries, the translation component must be the identity by the following theorem, and therefore the group of symmetries is isomorphic to a subgroup of $O(n)$.

14.26 THEOREM (Fixed point theorem) Let G be a finite subgroup of the group M_n of isometries of $\mathbb{R}^n$, where $n = 2$ or 3. Then there exists a point $p \in \mathbb{R}^n$ such that

$$g(p) = p \text{ for all } g \in G$$

Proof If G is the trivial subgroup $G = \{\text{id}\}$, then every point in M_n is fixed. So let us assume G is a finite nontrivial subgroup of M_n, so there is a point $x \in M_n$ such that $g(x) \neq x$ for some $g \in G$. Consider the orbit of x under the action of G:

$$O_x = \{g_1(x), g_2(x), \dots, g_s(x)\}, \text{ for all } g_i \in G$$

where the $g_i(x)$ are distinct points. Note that for any $g \in G$ and any point $g_i(x)$ in O_x, $gg_i(x) = g(g_i(x)) = g_j(x) \in O_x$ for some $1 \leq j \leq s$, and $gg_i(x) \neq gg_k(x)$ if $i \neq k$. Hence for any $g \in G$,

$$\{gg_1(x), gg_2(x), \dots, gg_s(x)\} = O_x$$

The **center of gravity** of the points $g_i(x)$ is defined as follows:

$$p = {}^1/_s\,[g_1(x) + g_2(x) + \dots + g_s(x)]$$

By Theorem 14.21, any $g \in G$ can be written as $g = t_b\rho_\theta r^i$, $i = 0$ or 1. Therefore, $t_{-b}g$ is a linear transformation and

$$t_{-b}g(p) = {}^1/_s\,[t_{-b}gg_1(x) + t_{-b}gg_2(x) + \dots + t_{-b}gg_s(x)]$$

Since the $g(g_i(x))$ are just the $g_i(x)$, possibly in a different order, we obtain

$$t_{-b}g(p) = {}^1/_s\,[t_{-b}g_1(x) + t_{-b}g_2(x) + \dots + t_{-b}g_s(x)] =$$
$${}^1/_s\,[(g_1(x) - b) + (g_2(x) - b) + \dots + (g_s(x) - b)] = p - b$$

Hence $g(p) = t_b(p - b) = p$ for all $g \in G$, and p is a fixed point of G. □

Let G be a finite nontrivial subgroup of M_2. By the preceding theorem, there exists a point in $\mathbb{R}^2$ that is fixed by every element of G. Therefore, if we change coordinates, we may place the fixed point at the origin and consider G as a subgroup of $O(2)$.

14.27 EXAMPLE In $\mathbb{R}^2$ let ρ_θ be a rotation around the origin by an angle $\theta = {}^{2\pi}/_n$. Then $G = \langle \rho_\theta \rangle \cong \mathbb{Z}_n$ is a finite subgroup of $O(2)$. ◇

14.28 EXAMPLE If $H \subseteq \mathbb{R}^2$ is a regular n-gon centered at the origin, then the group of symmetries of H is the dihedral group D_n. Hence D_n is isomorphic to a subgroup of $O(2)$. ◇

We show next that these last two examples are the *only* examples of finite subgroups of $O(2)$.

14.29 THEOREM Every nontrivial finite subgroup of $O(2)$ is isomorphic to one of the groups

$$\mathbb{Z}_n \text{ or } D_n$$

Proof Let G be a nontrivial finite subgroup of $O(2)$.
Case 1. Suppose G is a subgroup of $SO(2)$.
Hence every element of G is a rotation of $\mathbb{R}^2$ given by a matrix $A(\theta) \in SO(2)$, representing a rotation by some angle θ about the origin, $0 \le \theta < 2\pi$. Choose $A(\theta) \in G$ with the minimum value of θ among elements of G. Now consider any other element $A(\psi) \in G$, and let $\psi = q\theta + \phi$ where $0 \le \phi < \theta$. Then

$$A(\psi) = A(q\theta + \phi) = A(q\theta)A(\phi)$$

(See Example 4.1.1.) Therefore,

$$A(\psi) = A(\theta)^q A(\phi) \text{ and}$$
$$A(\phi) = A(\psi)A(\theta)^{-q} \in G$$

By our choice of θ, since $0 \le \phi < \theta$, we conclude that $\phi = 0$ and $A(\psi) = A(\theta)^q$. In other words, $A(\theta)$ generates G, and since G is finite, $|A(\theta)| = n$, $G \cong \mathbb{Z}_n$.
Case 2. Suppose G is not a subgroup of $SO(2)$.
Since $SO(2)$ is a subgroup of $O(2)$ of index 2 (see Exercise 13), by Theorem 5.1.6, $H = G \cap SO(2)$ is a normal subgroup of G of index 2 in G. Hence $H \le SO(2) \le O(2)$ and $|H| = |G|/2 = n$ for some $n \ge 1$. Therefore, by case (1), $H = \langle \rho \rangle$ is cyclic of order n. Let $\tau \in G$ be such that $\tau \notin H$. Then by Proposition I.24, part (3), $\tau^2 = \text{id}$. Then

$$G = \{\rho_0 = \text{id}, \rho, \rho^2, \dots, \rho^{n-1}, \tau, \rho\tau, \rho^2\tau, \dots, \rho^{n-1}\tau\}$$

where $\rho^n = \tau^2 = \text{id}$ and $\tau\rho = \rho^{-1}\tau$ by Proposition I.24, part (3), so $G \cong D_n$. □

The finite subgroups of $O(2)$ correspond to the groups of symmetries of the regular n-gon and their subgroups. To study the finite subgroups of $O(3)$ we need to look first at the three-dimensional analogues of the regular n-gons, the regular polyhedra in $\mathbb{R}^3$.

Platonic Solids

A **regular polyhedron** or **Platonic solid** is a polyhedron in which all the faces are regular n-gons for some n, and all vertices are incident with the same number k of faces. Since the interior angle of a regular n-gon has measure $\pi - {}^{2\pi}/_{n}$ and the sum of the k angles at a given vertex must be less than 2π, we obtain

$$k[\pi - 2\pi/n] < 2\pi$$

Therefore, $3 \leq k \leq 5$ and the five possible regular polyhedra are

(1)	$k = 3$ and	$n = 3$	a tetrahedron
		$n = 4$	a cube
		$n = 5$	a dodecahedron
(2)	$k = 4$ and	$n = 3$	an octahedron
(3)	$k = 5$ and	$n = 3$	an icosahedron

These figures may be pictured as in Figure 1.

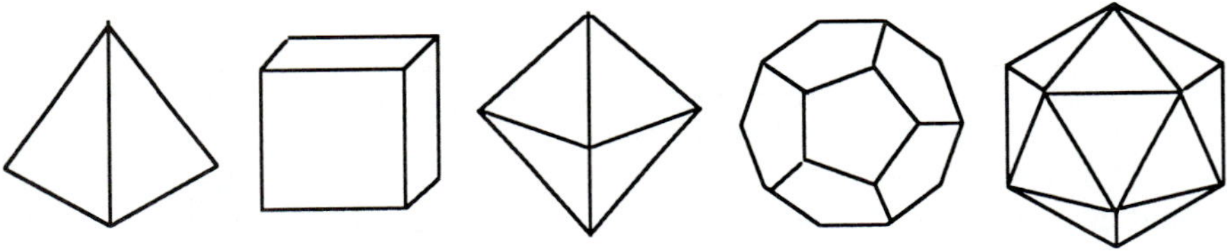

FIGURE 1

Their properties may be tabulated as in Table 1.

TABLE 1 The Platonic Solids

Polyhedron	V	E	F	n	k
Tetrahedron	4	6	4	3	3
Cube	8	12	6	4	3
Octahedron	6	12	8	3	4
Dodecahedron	20	30	12	5	3
Icosahedron	12	30	20	3	5

Here V is the number of vertices, E the number of edges, and F the number of faces. Given a regular polyhedron, let $p_1, \ldots, p_F$ be the centers of each of its F faces, and join by an edge any pair of the p_i whose corresponding faces meet at an edge. The polyhedron so constructed is the **dual** of the original polyhedron. The number of vertices of the dual is equal to the number of faces of the original polyhedron and conversely. The dual of a regular polyhedron is also a regular polyhedron, and any symmetry of the original polyhedron is also a symmetry of the dual and conversely. The tetrahedron is dual to itself, the square and the octahedron are dual to each other, and the dodecahedron is dual to the icosahedron. The groups of rotations of the regular solids are finite subgroups of $SO(3)$. We now identify them.

14.30 THEOREM The group of rotations of the regular tetrahedron is isomorphic to A_4.

Proof See Example 4.3.6. □

14.31 THEOREM The group of rotations of the cube and of its dual, the regular octahedron, is isomorphic to S_4.

Proof Let G be the group of rotations of the cube. By Example 4.2.13, $|G| = 24$. Let d_1, d_2, d_3, d_4 be the four diagonals of the cube. Any rotation of the cube permutes the d_i. This gives us a homomorphism $\phi: G \to S_4$, and by the first isomorphism theorem (Theorem 5.1.1) $G/\text{Kern } \phi \cong \phi(G) \leq S_4$. Now observe that

(1) $\phi(G)$ contains a 2-cycle (1 2), which corresponds to the rotation around the axis going through the diagonal d_3, which exchanges d_1 and d_2 but sends d_3 to d_3 and d_4 to d_4.

(2) $\phi(G)$ contains a 4-cycle (1 2 3 4), which corresponds to a rotation of the cube around an axis orthogonal to a pair of parallel faces.

Therefore, by Section 1.4, Exercise 36, $\phi(G) = S_4$, and since $|G| = |S_4|$, we obtain $G \cong S_4$. □

14.32 THEOREM The group of rotations of the regular dodecahedron and of its dual, the regular icosahedron, is isomorphic to A_5.

Proof Let G be the group of rotations of the regular dodecadehron. Let x be a vertex of the dodecahedron. Since x can be carried to any other vertex by an element of G, the orbit of x under the action of G has size $|O_x| = 20$, the number of vertices. The elements of G that fix x are the identity and the rotations around the axis through x and the opposite vertices that carry faces with which x is incident to other such faces. Since there are $k = 3$ such faces, these rotations are through angles of $2\pi/3$ and $4\pi/3$. Thus the stabilizer of x has order $|G_x| = 3$. Therefore, by Theorem 4.2.8, $|G| = 60$. We prove the theorem by constructing a homomorphism $\phi: G \to S_5$ and then showing that $A_5 \leq \phi(G)$.

The dodecahedron has pentagonal faces. Each vertex is incident to $k = 3$ faces. From any vertex we can draw three diagonals, one in each of the incident pentagons, so that the three diagonals are orthogonal to each other. In this way we construct a cube inscribed in the dodecahedron so that the 8 vertices of the cube are among the 20 vertices of the dodecahedron, and the 12 edges of the cube are diagonals on the 12 faces of the dodecahedron. The dodecahedron contains 5 distinct such inscribed cubes, since a pentagonal face has 5 diagonals. G acts on the 5 cubes. Therefore, we have a homomorphism $\phi: G \to S_5$.

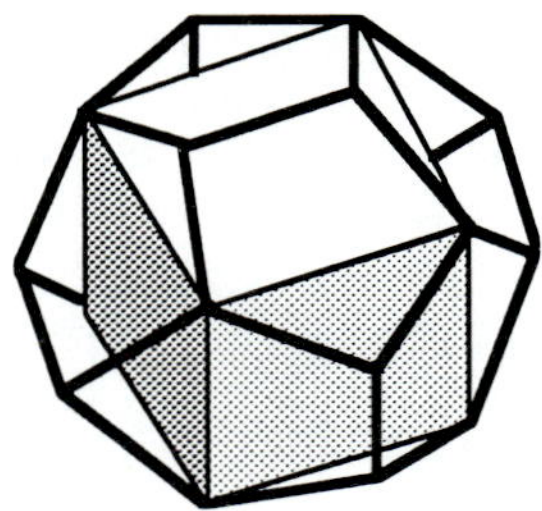

FIGURE 2

Observe that

(1) S_5 contains 20 3-cycles, and these generate A_5. (See Section 1.4, Exercise 33.)

(2) Given any vertex x of the dodecahedron, there are three vertices joined to x by edges. The line joining any pair of these is a diagonal of one of the three faces to which x is incident. These three lines are edges of three distinct inscribed cubes. The rotations through angles of $2\pi/3$ and $4\pi/3$ about the axis through x and the vertex opposite x are elements of G of order 3 that permute these three cubes, and if the cubes are numbered 1, 2, 3, correspond to the 3-cycles $(1\ 2\ 3)$ and $(1\ 3\ 2)$. Each pair of opposite vertices therefore determine 2 3-cycles in S_5, and since there are 20 vertices, or 10 pairs of opposite vertices, we conclude that $\phi(G)$ contains 20 3-cycles. Hence by our first observation, $A_5 \le \phi(G) \le S_5$. Since $|G| = 60 = |A_5|$, we have $\phi(G) = A_5$ and $G \cong A_5$. □

Subgroups of the Special Orthogonal Group

We showed in Theorem 14.29 that the group of symmetries of the regular n-gon in the plane and its subgroups are the only finite subgroups of $O(2)$. In the case of $\mathbb{R}^3$ we show that the groups of rotations of the Platonic solids, which we have just calculated, together with $\mathbb{Z}_n$ and D_n give all the finite subgroups of $SO(3)$.

14.33 THEOREM Let G be a nontrivial finite subgroup of $SO(3)$. Then G is isomorphic to one of the following groups:

$$\mathbb{Z}_n \qquad D_n \qquad A_4 \qquad S_4 \qquad A_5$$

Proof Let G be a nontrivial finite subgroup of $SO(3)$. By Theorem 14.20, every nontrivial element of G gives a rotation of $\mathbb{R}^3$ about an axis that passes through the origin. For any such rotation let p and q be the points in $\mathbb{R}^3$ where the unit sphere intersects the axis of rotation. We call these points the **poles** of the rotation. Let S be the set of the poles of all nontrivial elements of G.

Claim G acts on S.

Proof of claim Let $g \in G$ and $p \in S$. Since g is an isometry, $g(p)$ is still a point on the unit sphere. To show $g(p) \in S$ it therefore suffices to show that $g(p)$ is fixed by some nonidentity element of G. But p is a pole for some nonidentity element $h \in G$. That is, $h(p) = p$ and hence

$$(ghg^{-1})(g(p)) = gh(p) = g(p)$$

and the claim is proved.

Now let N be the number of distinct orbits in this action, and let $p_1, p_2, \dots, p_N$ be one element from each orbit. In the notation of Chapter 4

$O_p = \{g(p) \mid g \in G\} \subseteq S$ is the orbit of p

$G_p = \{g \in G \mid g(p) = p\} \leq G$ is the stabilizer of p

$S_g = \{x \in S \mid g(x) = x\}$

Then Burnside's theorem (Theorem 4.3.3) gives us

$$N = {}^1/_{|G|} \sum_{g\in G} |S_g|$$

Since every nontrivial element in G fixes exactly two poles, while the identity fixes every element, Burnside's theorem gives us

$$N = {}^1/_{|G|}\ [2(|G| - 1) + |S|] = {}^1/_{|G|}\ [2(|G| - 1) + \sum_{1\leq i\leq N} |O_{p_i}|]$$

Furthermore, by Theorem 4.2.8,

(1)
$$|O_{p_i}| = {}^{|G|}/_{|G_{p_i}|}$$

So we obtain

$$2(1 - {}^1/_{|G|}) = N - \sum_{1\leq i\leq N} {}^1/_{|G_{p_i}|}$$

Finally, since $|G_{p_i}| \geq 2$ and

$$1 \leq 2(1 - {}^1/_{|G|}) < 2$$

we derive that

$$2 \leq N < 4$$

Therefore, $N = 2$ or 3.

If $N = 2$, then

$$2(1 - {}^1/_{|G|}) = 2 - {}^1/_{|G|} \sum_{1\leq i\leq 2} |O_{p_i}|$$

which implies that

$$2 = |O_{p_1}| + |O_{p_2}|$$

So we have two orbits with one element in each. In other words, there are exactly two poles, and every rotation in G is a rotation about the axis through these two poles. Therefore, G is isomorphic to a subgroup of $SO(2)$. Hence by the proof of Theorem 14.29 (Case 1), G is isomorphic to $\mathbb{Z}_n$ for some n.

If $N = 3$, then

$$2(1 - {}^1/_{|G|}) = 3 - ({}^1/_{|G_{p_1}|} + {}^1/_{|G_{p_2}|} + {}^1/_{|G_{p_3}|})$$

which can be rewritten as

(2) $$1 + {}^{2}/|G| = {}^{1}/|G_{p_1}| + {}^{1}/|G_{p_2}| + {}^{1}/|G_{p_3}|$$

Hence the right hand side is less than 2 and greater than 1. Therefore, we have the following four possibilities for the $|G_{p_i}|$ as shown in Table 2.

TABLE 2 Stabilizers of G

Case	$\|G_{p_1}\|$	$\|G_{p_2}\|$	$\|G_{p_3}\|$	$\|G\|$	$\|O_{p_1}\|$	$\|O_{p_2}\|$	$\|O_{p_3}\|$
(1)	2	2	$n \geq 2$	$2n$	n	n	2
(2)	2	3	3	12	6	4	4
(3)	2	3	4	24	12	8	6
(4)	2	3	5	60	30	20	12

The order $|G|$ in the table has been determined by equation (2), which implies that $|G|$ equals

$$\frac{2|G_{p_1}||G_{p_2}||G_{p_3}|}{|G_{p_2}||G_{p_3}| + |G_{p_1}||G_{p_3}| + |G_{p_1}||G_{p_2}| - |G_{p_1}||G_{p_2}||G_{p_3}|}$$

The values of $|O_{p_i}|$ have been determined using equation (1). Note that since all of the elements of $|G_{p_i}|$ are rotations about the axis through the origin and p_i, and fix p_i, arguing as in the case $N = 2$, the G_{p_i} are cyclic.

Case 1. Since $|O_{p_3}| = 2$, the third orbit consists of two elements p and q. If $g \in G$ exchanges p and q, g must be a rotation of angle π, and p and q must lie on the same line through the origin, and equidistant from the origin, since g preserves distances and fixes the origin. In other words, we must have $q = -p$, and a nontrivial element g of G is either a rotation about the axis determined by p and $-p$ or else is a rotation of angle π that exchanges p and $-p$. In either case, g is completely determined by its restriction to the plane P through the origin that is orthogonal to the axis through p and $-p$. The restriction of g to P is a rotation of the plane if p and $-p$ are fixed and a reflection of the plane if p and $-p$ are exchanged. Then, arguing as in Theorem 14.29, we conclude that G is isomorphic to D_n.

Case 2. Since $|O_{p_3}| = 4$, we can label its elements as x_1, x_2, x_3, x_4. Since $G_{x_i} \cong G_{p_3}$ for $1 \leq i \leq 4$, all the stabilizers G_{x_i} are cyclic groups of order 3. Let $g \in G_{x_1}$ be nontrivial. Then $g(x_1) = x_1$, and g rotates to each other the x_i, $2 \leq i \leq 4$, which therefore form the vertices of an equilateral triangle. Since g preserves distances,

$$|x_1 - x_2| = |x_1 - g(x_2)| = |x_1 - g^2(x_2)|$$

Hence x_2, x_3, x_4 are all equidistant from x_1. If we repeat the argument with $g \in G_{x_2}$ we find that all four points x_1, x_2, x_3, x_4 are equidistant from each other and are the vertices of a regular tetrahedron, which is sent to itself by every rotation in G. Hence by Theorem I.30, G has a subgroup isomorphic to A_4. But since $|A_4| = 12 = |G|$, $G \cong A_4$.

Case 3. Since $|O_{p_3}| = 6$ while G_{p_3} is a cyclic group of order 4, a generator of G_{p_3} and hence every element of G_{p_3} must leave another point q fixed besides $p = p_3$. Arguing as in Case 1, we conclude that $q = -p$. Similarly, the other four elements must occur in opposite pairs. If we label p, $-p$ as x_1, x_6, and the other four elements of O_{p_3} as x_2, x_3, and $x_4 = -x_2$, $x_5 = -x_3$, arguing as in Case 2, we conclude first that the x_i, $2 \le i \le 5$ are all equidistant from x_1, and lie at the vertices of a square in the plane through the origin orthogonal to the axis through p and $-p$. Replacing G_p by G_{x_2}, similarly $\pm p$, $\pm x_3$ are equidistant from x_2 and form the vertices of a square. It follows that these poles are the vertices of a regular octahedron, and since $|G| = 24$, by Theorem 14.31 we obtain $G \cong S_4$.

Case 4. Since $|O_{p_3}| = 12$ while G_{p_3} is a cyclic group of order 5, any element of G_{p_3} must leave another point fixed besides $p = p_3$, which must be $-p$, and must permute the remaining 10 elements of O_{p_3} in two sets of 5, which we may label $\{x_2, x_3, x_4, x_5, x_6\}$ and $\{x_7, x_8, x_9, x_{10}, x_{11}\}$. (See Figure 3.) Arguing as in previous cases, the x_i for $2 \le i \le 6$ are equidistant from p and from $-p$ and so lie in the same plane orthogonal to the axis between p and $-p$, where they form the vertices of a regular pentagon (similarly for the x_i for $7 \le i \le 11$). (See Figure 4.)

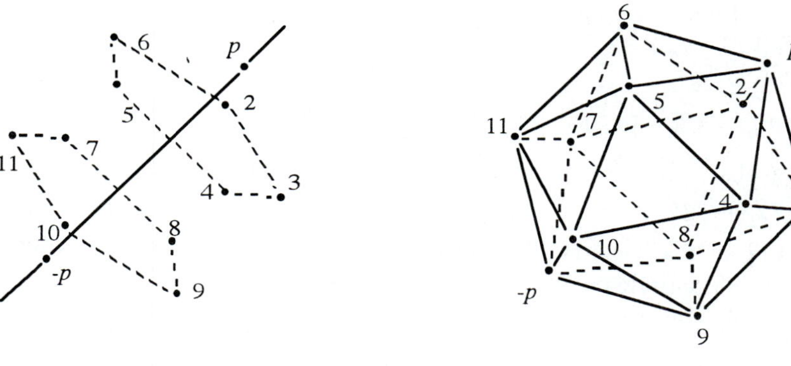

FIGURE 3 FIGURE 4

Then taking G_{x_i} for $i > 1$ in place of G_{p_3} and arguing similarly, we conclude that the twelve poles in O_{p_3} are the vertices of a regular icosahedron, and since $|G| = 60$, by Theorem 14.32 we obtain $G \cong A_5$. □

Further Reading

For further study of the subject of symmetries and subgroups of $O(3)$, here are some excellent sources.

Armstrong, M. A. *Groups and Symmetry* New York: Springer-Verlag, 1988.

Coxeter, H. S. M. *Introduction to Geometry* New York: Wiley, 1969.

Lockwood, E. H. and R. H. Macmillan *Geometric Symmetry* Cambridge: Cambridge University Press, 1978.

Martin, G. E. *Transformation Geometry* New York: Springer-Verlag, 1983.

Weyl, H. *Symmetry* Princeton: Princeton University Press, 1952.

Exercises 14

Linear transformations

All matrices considered have entries from $\mathbb{R}$. Let e_{ij} denote the $n \times n$ matrix with 1 at the (i, j)th entry and 0 elsewhere, and let I_n be the $n \times n$ identity matrix. We define three types of $n \times n$ **elementary matrices**:

(a) $E^1 = I_n + c\, e_{ij},\ 0 \neq c \in \mathbb{R},\ i \neq j$

(b) $E^2 = I_n + e_{ij} + e_{ji} - e_{ii} - e_{jj},\ i \neq j$

(c) $E^3 = I_n + (c - 1)e_{ii},\ 0 \neq c \in \mathbb{R}$

1. Show that elementary matrices are invertible and that their inverses are also elementary matrices.

2. Let A be an $n \times n$ matrix. Show that the following two conditions are equivalent:

(a) A is a finite product of elementary matrices.

(b) A is invertible.

3. For any $n \times n$ elementary matrix E and any $n \times n$ matrix A, show that

$$\det(EA) = \det E \det A$$

4. For any two $n \times n$ matrices A and B, show that

$$\det(AB) = \det A \det B$$

5. Show that a linear transformation $T_A: \mathbb{R}^n \to \mathbb{R}^n$ is an isomorphism if and only if $\det A \neq 0$.

6. Prove Corollary 14.6.

7. For any $n \times n$ matrices A and B, show that

(a) $\det(A^T) = \det A$ (b) $(AB)^T = B^T A^T$

8. Let $T: \mathbb{R}^3 \rightarrow \mathbb{R}^3$ be defined by

$$T\begin{bmatrix} x_1 \\ x_2 \\ x_3 \end{bmatrix} = \begin{bmatrix} x_1 - x_2 \\ x_2 + x_3 \\ 2x_3 + x_1 \end{bmatrix}$$

Show that $T = T_A$ by finding the 3×3 matrix A representing T with respect to the standard basis. Is T an isomorphism?

9. Find all linear transformations $T: \mathbb{R}^2 \rightarrow \mathbb{R}^2$ that map the line $y = -x$ to the line $y = x$.

10. Let $\{v_1, v_2, \ldots, v_s\}$ be a set of nonzero vectors in $\mathbb{R}^n$ such that

$$v_i \cdot v_j = 0 \quad \text{if } i \neq j$$

Show that the v_i are linearly independent over $\mathbb{R}$.

11. Find all linear transformations $T_A: \mathbb{R}^3 \rightarrow \mathbb{R}^3$ such that

$$T_A\begin{bmatrix} 1 \\ 1 \\ 1 \end{bmatrix} = \begin{bmatrix} 0 \\ 1 \\ 1 \end{bmatrix} \quad \text{and} \quad T_A\begin{bmatrix} 0 \\ 0 \\ 1 \end{bmatrix} = \begin{bmatrix} 0 \\ 0 \\ 1 \end{bmatrix}$$

Isometries

12. Let G be a subgroup of $O(n)$. Show that either every element $A \in G$ has $\det A = 1$ or exactly half do.

13. Show that $SO(n)$ is a subgroup of $O(n)$ of index 2, hence normal.

14. Prove Theorem 14.23.

15. Prove Proposition 14.24.

16. Find all isometries of $\mathbb{R}^2$ that map the line $y = x$ to the line $y = 1 - 2x$.

17. Find the locus of points $P = (x, y, z) \in \mathbb{R}^3$ such that $P = S((1, -1, 1))$, where S is any isometry of $\mathbb{R}^3$ that fixes the points $(1, 1, 1)$ and $(-1, -1, -1)$.

18. Describe all the isometries of $\mathbb{R}^3$ that map the point $(1, 1, 1)$ to $(1, 0, 0)$, and express them in the form $t_b \circ \rho_\theta \circ r^i$ as in Theorem 14.21, and in terms of $S(v) = Av + b$ as in Corollary 14.22.

Symmetries

19. Find a subgroup of $O(2)$ isomorphic to D_5.

20. Find the complete group of symmetries of the regular tetrahedron.

21. Find the group of rotations of the regular tetrahedron where one face is painted black and the other three white.

22. Find the group of rotations of a cube where three faces incident to the same vertex are painted black and the other three faces white.

23. Find the order of the complete group of symmetries of the cube.

24. Let (x_1, x_2, x_3) with $x_i = \pm 1$, $1 \le i \le 3$ be the vertices of a cube in $\mathbb{R}^3$.

(a) Write the equations of the four lines that go through the diagonals of the cube, and number them d_1, d_2, d_3, d_4.

(b) Find a matrix $A \in SO(3)$ that corresponds to the cycle (1 2 3 4).

(c) Find a matrix $B \in SO(3)$ that corresponds to the cycle (1 2).

Chapter 15

Gröbner Bases

As we saw in Chapter 8, the division algorithm in $F[x]$, the ring of polynomials in one variable over a field F, allows us to show that $F[x]$ is a PID. Furthermore, given a polynomial $f(x) \in F[x]$, $f(x)$ is in the ideal $I = \langle g(x) \rangle$ if and only if $g(x)$ divides $f(x)$. If we want to consider $F[x_1, \dots, x_n]$, a ring of polynomials in n variables, to be able to determine whether a polynomial $f \in F[x_1, \dots, x_n]$ is an element of an ideal I in $F[x_1, \dots, x_n]$, we need to

(1) Establish a division algorithm in $F[x_1, \dots, x_n]$, and

(2) Construct appropriate generators for the ideal I.

In this introductory chapter on *Gröbner bases* for ideals in $F[x_1, \dots, x_n]$, we show how a division algorithm in $F[x_1, \dots, x_n]$ can be defined; prove the *Hilbert basis theorem* for $F[x_1, \dots, x_n]$, which tells us that every ideal I in $F[x_1, \dots, x_n]$ is finitely generated; show that every such ideal I has a Gröbner basis; and show how applying the division algorithm with a Gröbner basis allows us to determine whether a given polynomial f is in I.

Lexicographic Order

For a polynomial in one variable $f(x) \in F[x]$, we use the degrees of its terms to order the terms, starting with the leading term, the term of highest degree. When we divide $f(x)$ by another polynomial $g(x) \in F[x]$, we start the long division by comparing the two leading terms. In order to do something comparable for polynomials in several variables, we need to agree first on some ordering of the variables. Given a monomial

$$x^{\alpha} = x_1^{\alpha_1} \dots x_n^{\alpha_n} \in F[x_1, \dots, x_n]$$

it determines an n-tuple of natural numbers $\alpha = (\alpha_1, \dots, \alpha_n) \in \mathbb{N}^n$, an ordering on $\mathbb{N}^n$ provides us with an ordering on the monomials in $F[x_1, \dots, x_n]$:

$$x^{\alpha} > x^{\beta} \text{ if and only if } \alpha > \beta \text{ in } \mathbb{N}^n$$

Throughout this chapter, F is any field.

15.1 DEFINITION A **monomial ordering** on $F[x_1, \dots, x_n]$ is a relation $>$ on the monomials x^{α}, $\alpha \in \mathbb{N}^n$ or, equivalently, on $\mathbb{N}^n$ such that

(1) For any $\alpha, \beta \in \mathbb{N}^n$, exactly one of the following holds:

$$\alpha > \beta \quad \text{or} \quad \alpha = \beta \quad \text{or} \quad \beta > \alpha$$

(2) If $\alpha > \beta$ and $\gamma \in \mathbb{N}^n$, then $\alpha + \gamma > \beta + \gamma$.

(3) Every nonempty subset U of $\mathbb{N}^n$ has a least element under $>$, that is, an element α such that $\beta > \alpha$ for all $\alpha \neq \beta \in U$. ○

A relation with properties (1) and (3) is a wellordering. Note that (1) and (3) imply (4):

(4) If $\alpha > \beta$ and $\beta > \gamma$, then $\alpha > \gamma$.

For if $\alpha > \beta$ and $\beta > \gamma$, by (1) α, β, γ must all be distinct, and by (3) $\{\alpha, \beta, \gamma\}$ must have a least element, and since by (1) again we do not have $\beta > \alpha$ or $\gamma > \beta$, the least element can only be γ, so $\alpha > \gamma$. (1)-(3) also imply (5):

(5) If $\alpha_i \geq \beta_i$ for $1 \leq i \leq n$, then $\alpha = (\alpha_1, \dots, \alpha_n) \geq (\beta_1, \dots, \beta_n) = \beta$.

For if $\alpha_i \geq \beta_i$ for $1 \leq i \leq n$, then $\gamma = \alpha - \beta \in \mathbb{N}^n$. If we do not have $\alpha \geq \beta$, then by (1) we have $\beta > \alpha = \beta + \gamma$. Then by (2) we have

$$\beta > \beta + \gamma > \beta + 2\gamma > \beta + 3\gamma > \dots$$

and the set $\{\beta + n\gamma \mid n \in \mathbb{N}\}$ has no least element, contrary to (3). In terms of monomials, (5) means that if x^β divides x^α, then $x^\alpha \geq x^\beta$.

15.2 DEFINITION The **lexicographic order** on $\mathbb{N}^n$ is defined by $\alpha >_{\text{lex}} \beta$ if the left-most nonzero entry in $\alpha - \beta$ is positive. ○

15.3 EXAMPLES

(1) $(3, 2, 2, 0) >_{\text{lex}} (3, 1, 2, 1)$ (2) $(3, 4, 1, 1, 0) >_{\text{lex}} (3, 4, 1, 0, 5)$ ◇

First we establish that the lexicographic order satisfies all the useful properties of a monomial ordering.

15.4 PROPOSITION The lexicographic order $>_{\text{lex}}$ on $\mathbb{N}^n$ is a monomial ordering.

Proof (1) If $\alpha \neq \beta$, they differ in some entry, so some entry of $\alpha - \beta$ is nonzero. Then $\alpha > \beta$ or $\beta < \alpha$ according as that entry is positive or negative.

(2) Since $(\alpha + \gamma) - (\beta + \gamma) = \alpha - \beta$, their left-most entries are the same; hence $\alpha + \gamma > \beta + \gamma$ if and only if $\alpha > \beta$.

(3) Suppose U is some nonempty subset of $\mathbb{N}^n$. We want to show that U has a least element under $>_{\text{lex}}$. Let

$$U_1 = \{\alpha_1 \in \mathbb{N} \mid \alpha_1 \text{ is the first entry of some element of } U\}$$

U_1 is a nonempty subset of $\mathbb{N}$, so by the well-ordering principle (0.3.1) U_1 has a least element $\beta_1 \in \mathbb{N}$. Now consider all elements of U having β_1 as their first entry, and let

$$U_2 = \{\alpha_2 \in \mathbb{N} \mid \alpha_2 \text{ is the second entry of some element of } U_1 \text{ with } \beta_1 \text{ as first entry}\}$$

U_2 has a least element β_2. Now consider all elements of U_1 having β_1 as their first entry and β_2 as their second entry, and repeat the process until we get

$$\beta = (\beta_1, \beta_2, \dots, \beta_n) \in U$$

For any element $\alpha = (\alpha_1, \alpha_2, \dots, \alpha_n) \in U$,

$$\alpha - \beta = (\alpha_1 - \beta_1, \dots, \alpha_n - \beta_n)$$

and by construction the first nonzero entry of $\alpha - \beta$ is positive. Hence $\alpha > \beta$ for all $\alpha \in U$, and β is the least element of U under $>_{\text{lex}}$. □

15.5 EXAMPLE Let

$$x^\alpha = x_1^{\alpha_1} x_2^{\alpha_2} x_3^{\alpha_3} \in F[x_1, x_2, x_3]$$

where $\alpha = (\alpha_1, \alpha_2, \alpha_3) \in \mathbb{N}^3$. Then if $f = \sum c_\alpha x^\alpha$, where only a finite number of the coefficients c_α are nonzero, then f is a polynomial in $F[x_1, x_2, x_3]$. Consider such a polynomial, say

$$f = x_2^5 x_3^2 - x_1^2 x_3^4 + x_2^3 x_3^5 + x_1 x_3$$

Then with respect to the lexicographic order with $x_1 > x_2 > x_3$, we reorder the terms of f so that the terms of higher order come first. The result is as follows:

$$f = -x_1^2 x_3^4 + x_1 x_3 + x_2^5 x_3^2 + x_2^3 x_3^5$$

If we write each of the four terms of f as $c_\alpha x^\alpha$, the exponents α are

$$(2, 0, 4) > (1, 0, 1) > (0, 5, 2) > (0, 3, 5)$$

in lexicographic order. ◇

15.6 DEFINITION Let $f = \sum c_\alpha x^\alpha$ be a polynomial in $F[x_1, \dots, x_n]$ and $>$ a monomial order. Then

(1) The **multidegree** of f is

$$\text{multideg } f = \max \{\alpha \in \mathbb{N}^n \mid c_\alpha \neq 0\}$$

(2) The **leading coefficient** of f is

$$\text{LC}(f) = c_{\text{multideg } f} \in F$$

(3) The **leading monomial** of f is

$$\text{LM}(f) = x^{\text{multideg } f}$$

(4) The **leading term** of f is

$$\text{LT}(f) = \text{LC}(f) \cdot \text{LM}(f)$$

Here max means "maximum with respect to >." ○

15.7 EXAMPLE For

$$f = 3x_1^2 x_2 - x_2^3 x_3^2 + 2x_1^3 x_2 + 7x_3^5 + x_1 x_2 x_3 \in \mathbb{R}[x_1, x_2, x_3]$$

if we use the lexicographic order with $x_1 > x_2 > x_3$, we reorder the terms of f as follows:

$$f = 2x_1^3 x_2 + 3x_1^2 x_2 + x_1 x_2 x_3 - x_2^3 x_3^2 + 7x_3^5$$

Then

$$\text{multideg } f = (3, 1, 0)$$
$$\text{LC}(f) = 2$$
$$\text{LM}(f) = x_1^3 x_2$$
$$\text{LT}(f) = 2x_1^3 x_2$$

with respect to the lexicographic order. ◇

There are many examples of monomial orders, but for simplicity from this point we consider only the lexicographic order, which we write simply as $>$. In $F[x, y, z]$ we always take $x > y > z$.

A Division Algorithm

Given a finite set of polynomials $\{g_1, \ldots, g_s\}$ in $F[x_1, \ldots, x_s]$, they generate the ideal $I = \langle g_1, \ldots, g_s \rangle$, which consists of all polynomials f of the form

$$f = a_1 g_1 + \ldots + a_s g_s$$

for some $a_i \in F[x_1, \ldots, x_s]$, $1 \le i \le s$. As we saw in Chapter 8, in the case of one variable, for any $f(x), g(x) \in F[x]$ the division algorithm for $F[x]$ enables us to write $f(x) = q(x)g(x) + r(x)$, where $q(x)$ and $r(x)$ are unique and $r(x) = 0$ or $\deg r(x) < \deg g(x)$, and $f(x)$ belongs to the ideal $I = \langle g(x) \rangle$ if and only if $r(x) = 0$. To answer the question, in the case of several variables whether $f \in F[x_1, \ldots, x_s]$ belongs to the ideal $I = \langle g_1, \ldots, g_s \rangle$, we need first to establish a division algorithm. Let us first try to illustrate some of the obstacles to doing so.

15.8 EXAMPLE Let us divide $f = x^2y + 1$ by $g_1 = x + 1$ and $g_2 = xy + 1$. If we divide first by g_1, we obtain

$$\begin{array}{r|l} & xy \;-\; y \\ \hline x+1 & x^2y + 1 \\ & \underline{x^2y + xy} \\ & \quad -xy + 1 \\ & \quad \underline{-xy - y} \\ & \qquad\quad y + 1 \end{array}$$

Hence

(1) $\quad f = (xy - y)g_1 + (y + 1)$

The $\text{LT}(g_2) = xy$, which does not divide any of the terms of $y + 1$, so we stop.

If instead we divide first by g_2 we obtain

$$\begin{array}{r|l} & x \\ \hline xy+1 & x^2y + 1 \\ & \underline{x^2y + x} \\ & \quad -x + 1 \end{array}$$

Hence

$$f = xg_2 + (-x + 1)$$

In this case the $\text{LT}(g_1) = x$ divides the leading term of $(-x + 1)$, so we can proceed with

$$-x + 1 = -g_1 + 2$$

Therefore,

(2) $\quad f = xg_2 - g_1 + 2$

Thus we have

$$f = a_1g_1 + a_2g_2 + r$$

in two different ways.
In (1), $a_1 = xy - y$, $a_2 = 0$, $r = y + 1$.
In (2), $a_1 = -1$, $a_2 = x$, $r = 2$. ◇

15.9 EXAMPLE Consider $f(x) = x^2y + y$ and $g_1 = xy + 1$, $g_2 = x^2 + 1$. If we divide by g_1 first, we obtain

(1) $\quad f = x\,g_1 + 0\,g_2 + (-x + y)$

If we divide by g_2 first, we obtain

(2) $\quad f = 0\,g_1 + y\,g_2 + 0$

So we get a nonzero remainder in one case and a remainder 0 in the other. ◇

15.10 EXAMPLE Let $f = x^2y + x^2 + xy^2$ and $g_1 = xy + 1$, $g_2 = x^2 - 1$. Let us start dividing f by g_1:

$$\begin{array}{r|l} & x \\ \hline xy + 1 & x^2y + x^2 + xy^2 \\ & \underline{x^2y + x} \\ & \quad\quad\; x^2 + xy^2 - x \end{array}$$

The leading term of the remainder is x^2, which is not divisible by $\mathrm{LT}(g_1) = xy$ but is divisible by $\mathrm{LT}(g_2) = x^2$.
So let us divide it now by g_2:

$$\begin{array}{r|l} & 1 \\ \hline x^2 - 1 & x^2 + xy^2 - x \\ & \underline{x^2 - 1} \\ & \quad\quad xy^2 - x + 1 \end{array}$$

The leading term of this remainder is not divisible by $\mathrm{LT}(g_2) = x^2$ but is divisible by $\mathrm{LT}(g_1) = xy$.
So let us divide it by g_1 again:

$$\begin{array}{r|l} & y \\ \hline xy + 1 & xy^2 - x + 1 \\ & \underline{xy^2 + y} \\ & \quad\quad -x - y + 1 \end{array}$$

The leading term $-x$ of this remainder is not divisible by either $\mathrm{LT}(g_1)$ or $\mathrm{LT}(g_2)$, and indeed no term is.
So we stop the process with

$$f = (x + y)g_1 + g_2 + (-x - y + 1)$$

as our final result. ◇

15.11 EXAMPLE Let $f = x^2y + xy^2 + y^2$ and $g_1 = xy + 1$, $g_2 = y + 1$. Then

$$\begin{array}{r|l}
 & x + y \\ \hline
xy + 1 & x^2y + xy^2 + y^2 \\
 & x^2y + x \\ \hline
 & \qquad xy^2 - x + y^2 \\
 & \qquad xy^2 + y \\ \hline
 & \qquad\qquad -x + y^2 - y
\end{array}$$

Hence

$$f = (x + y)g_1 + (-x + y^2 - y)$$

The leading term $-x$ of this remainder is not divisible by either $\mathrm{LT}(g_1) = xy$ or $\mathrm{LT}(g_2) = y$, but $\mathrm{LT}(g_2)$ divides other terms of the remainder. So we first remove $-x$ and store it as a remainder and then proceed.

$$\begin{array}{r|l}
 & y - 2 \\ \hline
y + 1 & y^2 - y \qquad \rightarrow \qquad -x \\
 & y^2 + y \\ \hline
 & \quad -2y \\
 & \quad -2y - 2 \\ \hline
 & \qquad\quad 2
\end{array}$$

Hence

$$f = (x + y)g_1 + (y - 2)g_2 + (-x + 2)$$

Note that none of the terms of the remainder we found are divisible by either $\mathrm{LT}(g_1)$ or $\mathrm{LT}(g_2)$. ◇

The last two examples illustrate the division algorithm we will be using, which we now describe.

Let $f \in F[x_1, \ldots, x_n]$ and let $g_1, g_2, \ldots, g_s$ be an ordered s-tuple of polynomials in $F[x_1, \ldots, x_n]$. We find a_i, $1 \leq i \leq s$ and r in $F[x_1, \ldots, x_n]$ such that

(1) $\qquad f = a_1g_1 + \ldots + a_sg_s + r$

where no monomial term of r is divisible by any of the $\mathrm{LT}(g_i)$, $1 \leq i \leq s$.

We work with a sequence of expressions for f, which at step t will be given as

$$f = p_t + a_{t,1}g_1 + \ldots + a_{t,s}g_s + r_t$$

Step 0. We start with

$$f = f + 0\,g_1 + \ldots + 0\,g_s + 0$$

That is, $p_0 = f$, $a_{0,i} = r = 0$, $1 \leq i \leq s$.

Step $t + 1$. Suppose we have

$$f = p_t + a_{t,1}g_1 + \ldots + a_{t,s}g_s + r_t$$

Consider $\mathrm{LT}(p_t)$ and check whether it is divisible by any of the $\mathrm{LT}(g_i)$, starting with $\mathrm{LT}(g_1)$, then proceeding to $\mathrm{LT}(g_2)$, and so on.

(1) If $\mathrm{LT}(p_t)$ is divisible by one of these leading terms, let $\mathrm{LT}(g_i)$ be the first one, and let $\mathrm{LT}(p_t) = a\ \mathrm{LT}(g_i)$. Then let

$p_{t+1} = p_t - ag_i$
$a_{t+1,i} = a_{t,i} + a$
$a_{t+1,j} = a_{t,j}$ for all $j \neq i$, and
$r_{t+1} = r_t$

So we have

$$f = (p_t - ag_i) + a_{t,1}g_1 + \dots + (a_{t,i} + a)g_i + \dots + a_{t,s}g_s + r_t =$$
$$= p_{t+1} + a_{t+1,1}g_1 + \dots + a_{t+1,i}g_i + \dots + a_{t+1,s}g_s + r_{t+1}$$

Note that if $\mathrm{LM}(p_t) = x^\alpha$ and $\mathrm{LM}(g_i) = x^\beta$, then $\alpha = \beta + \gamma$, where $\mathrm{LM}(a) = x^\gamma$. Every other term of g_i after the leading term has the form $c_{\beta'}x^{\beta'}$ with $\beta > \beta'$, and every term of a has the form $d_{\gamma'}x^{\gamma'}$ with $\gamma \geq \gamma'$. Hence every term of ag_i after the leading term has the form $c_{\beta'}d_{\gamma'}x^{\beta'+\gamma'}$, where $\alpha = \beta + \gamma > \beta' + \gamma \geq \beta' + \gamma'$. Hence $p_{t+1} = p_t - ag_i$ is a sum of terms $e_\delta x^\delta$ with $\alpha > \delta$, and $\mathrm{LT}(p_t) > \mathrm{LT}(p_{t+1})$. Note also that the remainder term r has not changed.

(2) If $\mathrm{LT}(p_t)$ is divisible by none of these leading terms $\mathrm{LT}(g_i)$, $1 \leq i \leq s$, let

$p_{t+1} = p_t - \mathrm{LT}(p_t)$
$a_{t+1,i} = a_{t,i}$ for all $1 \leq i \leq s$, and
$r_{t+1} = r_t + \mathrm{LT}(p_t)$

So we have

$$f = (p_t - \mathrm{LT}(p_t)) + a_{t,1}g_1 + \dots + a_{t,s}g_s + (r_t + \mathrm{LT}(p_t)) =$$
$$= p_{t+1} + a_{t+1,1}g_1 + \dots + a_{t+1,s}g_s + r_{t+1}$$

Note again that $\mathrm{LT}(p_t) > \mathrm{LT}(p_{t+1})$ and that the monomial term we have added to r_t is not divisible by any $\mathrm{LT}(g_i)$, $1 \leq i \leq s$.

Since we have

$$\mathrm{LT}(p_0) > \mathrm{LT}(p_1) > \mathrm{LT}(p_2) > \dots$$

in a finite number of steps N the process stops with

$p_N = 0$
$r_N =$ a sum of monomials not divisible by any $\mathrm{LT}(g_i)$, $1 \leq i \leq s$, and
$f = a_{N,1}g_1 + \dots + a_{N,s}g_s + r_N$

15.12 THEOREM (Division algorithm) Let $g_1, g_2, \dots, g_s$ be an ordered s-tuple of polynomials in $F[x_1, \dots, x_n]$. Then for any polynomial $f \in F[x_1, \dots, x_n]$ there exist $a_i, r \in F[x_1, \dots, x_n]$ such that

(1) $f = a_1g_1 + \dots + a_sg_s + r$

(2) $r = 0$ or r is a linear combination with coefficients from F of monomials not divisible by any of the leading terms $\mathrm{LT}(g_i)$, $1 \leq i \leq s$.

(3) If $a_ig_i \neq 0$, then multideg $f \geq$ multideg a_ig_i.

Here r is called the **remainder** in the expression (1) for f.

Proof (1) and (2) have been proved in the course of the description of the division algorithm. To prove (3), we claim that for all t, if $a_{t,i} \neq 0$, then multideg $f \geq$

multideg $a_{t,i}\, g_i$. This is proved by induction on t. For $t = 0$, $a_{0,i} = 0$ and there is nothing to prove. Suppose the claim holds for t. If $a_{t+1,i} = a_{t,i}$ the claim holds for $t + 1$ also. Otherwise, $a_{t+1,i} = a_{t,i} + a$, where $\mathrm{LT}(p_t) = a\,\mathrm{LT}(g_i)$, and

$$a_{t+1,i}g_i = a_{t,i}g_i + a\,g_i$$

Now note that since $f = p_0$ and $\mathrm{LT}(p_t) > \mathrm{LT}(p_{t+1})$ for all t, multideg $f \geq$ multideg p_t for all $t < N$. Furthermore,

$$\text{multideg } p_t = \text{multideg LT}(p_t) = \text{multideg } a + \text{multideg LT}(g_i) =$$
$$= \text{multideg } a + \text{multideg } g_i = \text{multideg } a\,g_i$$

(See Exercise 10.) Therefore, multideg $f \geq$ multideg $a\,g_i$, while by our induction hypothesis multideg $f \geq$ multideg $a_{t,i}g_i$. We use now the fact that if h and k are polynomials and $h + k \neq 0$, then

$$\text{multideg}(h + k) \leq \max\{\text{multideg } h, \text{multideg } k\}$$

(See Exercise 11.) From this we conclude that multideg $f \geq$ multideg$(a_{t,i}g_i + a\,g_i)$ = multideg $a_{t+1,i}g_i$. □

The major difference between the preceding theorem and the division algorithm for polynomials in one variable is that we were not able to assert that the remainder r is unique. To remedy this problem, instead of considering division by the specific polynomials $g_1, g_2, \ldots, g_s$, we consider division by the ideal $I = \langle g_1, g_2, \ldots, g_s\rangle$ they generate and look for different polynomials generating the same ideal I that could give us a division algorithm with a unique remainder.

Dickson's Lemma

We pay special attention now to ideals that are generated by a set of monomials.

15.13 DEFINITION An ideal $I \subseteq F[x_1, \ldots, x_n]$ is a **monomial ideal** if there exists a subset $S \subseteq \mathbb{N}^n$ such that

$$I = \langle x^\alpha \mid \alpha \in S\rangle$$

In other words, if $f \in I$, then $f = \Sigma c_\alpha x^\alpha$, where $c_\alpha \in F[x_1, \ldots, x_n]$, $\alpha \in S$, and at most finitely many of the $c_\alpha \neq 0$. ○

15.14 EXAMPLE Let $g_1 = x^3y$, $g_2 = xy^2$, $g_3 = x^5$, $g_4 = y^3$. These are all monomials, so $I = \langle g_1, g_2, g_3, g_4\rangle$ is a monomial ideal in $\mathbb{R}[x, y]$. The polynomial

$$f = 5x^4y^2 - 3x^2y^4 + 2x^6y^2 - x^4y^4$$

is in the ideal I since

$$f = (5xy)g_1 - (3xy^2)g_2 + (2xy^2)g_3 - (x^4y)g_4$$

and all the coefficients are in $F[x, y]$. ◇

15.15 PROPOSITION Let $I = \langle x^\alpha \mid \alpha \in S \rangle$ be a monomial ideal in $F[x_1, \ldots, x_n]$. Then

(1) A monomial $x^\beta \in I$ if and only if x^α divides x^β for some $\alpha \in S$.

(2) A polynomial $f \in I$ if and only if every monomial term of f is in I.

(3) If J is another monomial ideal, then $J = I$ if and only if they contain the same monomials.

Proof See Exercise 14. □

Any ideal in $F[x]$ is principal, that is, $I = \langle g(x) \rangle$ for some $g(x) \in F[x]$, while $F[x_1, \ldots, x_n]$ contains ideals that cannot be generated by one polynomial only. However, we can show that every ideal in $F[x_1, \ldots, x_n]$ has a *finite* set of generators. We prove this property first for monomial ideals.

15.16 THEOREM (Dickson's lemma) Let I be a monomial ideal in $F[x_1, \ldots, x_n]$ and $G \subseteq F[x_1, \ldots, x_n]$ a set of monomials that generate I. Then there exists a *finite* subset of G that generates I.

We break up the proof into three steps by first giving two lemmas.

15.17 LEMMA Let

$$a_1, a_2, a_3, \ldots$$

be an infinite sequence of natural numbers. Then it has an infinite subsequence

$$b_1 = a_{i_1}, b_2 = a_{i_2}, b_3 = a_{i_3}, \ldots$$

for some

$$i_1 < i_2 < i_3 < \ldots$$

that is nondecreasing, meaning that $b_m \leq b_n$ whenever $m \leq n$.

Proof In fact, $\{a_i\}$ either has an infinite subsequence that is constant, with $b_m = b_n$ for all m, n, or else it has an infinite subsequence that is strictly increasing, with $b_m < b_n$ whenever $m < n$.

(1) If only finitely many values appear in the sequence $\{a_i\}$, then at least one of these values must occur infinitely often, say $a \in N$. Then if $i_1 < i_2 < i_3 < \ldots$ are all the i with $a_i = a$, we get a constant subsequence $\{a_{i_m}\}$.

(2) If infinitely many values appear in the sequence $\{a_i\}$, then for any i, only finitely many have appeared among the a_j with $j \leq i$, and of the infinitely many others, only finitely many are $\leq a_i$. So there must be a $j > i$ with $a_i < a_j$. If we let $i_1 = 1$, $i_2 =$ the least $j > i_1$ with $a_{i_1} < a_j$, $i_3 =$ the least $j > i_2$ with $a_{i_2} < a_j$, and so on, we get an infinite strictly increasing subsequence $\{a_{i_m}\}$. □

15.15 LEMMA In an infinite sequence of monomials in $F[x_1, \dots, x_n]$

$$M_1, M_2, M_3, \dots$$

there exist $i < j$ such that M_i divides M_j.

Proof We use induction on n. For $n = 1$, let

$$x_1^{\alpha_1}, x_1^{\alpha_2}, x_1^{\alpha_3}, \dots$$

be an infinite sequence of monomials. By the preceding lemma there exist $i < j$ such that $\alpha_i \leq \alpha_j$. But then $x_1^{\alpha_i}$ divides $x_1^{\alpha_j}$.

Now assume the lemma holds for $n = k$. An infinite sequence of monomials in $F[x_1, \dots, x_k, x_{k+1}]$ can be expressed as

$$M_1 x_{k+1}^{\alpha_1}, M_2 x_{k+1}^{\alpha_2}, M_3 x_{k+1}^{\alpha_3}, \dots$$

where the $\{M_i\}$ form an infinite sequence of monomials in $F[x_1, \dots, x_k]$. By the preceding lemma, there is an infinite subsequence

$$M_{i_1} x_{k+1}^{\alpha_{i_1}}, M_{i_2} x_{k+1}^{\alpha_{i_2}}, M_{i_3} x_{k+1}^{\alpha_{i_3}}, \dots$$

with

$$\alpha_{i_1} \leq \alpha_{i_2} \leq \alpha_{i_3} \leq \dots.$$

Now consider the infinite sequence of monomials

$$M_{i_1}, M_{i_2}, M_{i_3}, \dots$$

in $F[x_1, \dots, x_k]$. By our induction hypothesis there exist $i_m < i_n$ such that

$$M_{i_m} \text{ divides } M_{i_n}$$

Since $i_m < i_n$, $\alpha_{i_m} \leq \alpha_{i_n}$, also

$$x_{k+1}^{\alpha_{i_m}} \text{ divides } x_{k+1}^{\alpha_{i_n}}$$

But then

$$M_{i_m} x_{k+1}^{\alpha_{i_m}} \text{ divides } M_{i_n} x_{k+1}^{\alpha_{i_n}}$$

and the lemma is proved. □

Proof of Theorem 15.16. Let G be a set of generators of the monomial ideal I, and suppose that no finite subset of G can generate I. Let g_1 be the least element of G in the lexicographic order. By our assumption $I \neq \langle g_1 \rangle$, so let g_2 be the least element of G not in $\langle g_1 \rangle$. Similarly, since $I \neq \langle g_1, g_2 \rangle$, let g_3 be the least element of G not in $\langle g_1, g_2 \rangle$, and so on.

Thus we have an infinite sequence of monomials

$$g_1, g_2, g_3, \dots$$

where for $j > 1$, g_j is not in the ideal $\langle g_1, \dots, g_{j-1} \rangle$, and so in particular g_j is not a multiple of g_i for any $1 \leq i < j$. But this contradicts the preceding lemma. This contradiction completes the proof of Dickson's lemma. □.

The Hilbert Basis Theorem

We have just shown that every monomial ideal in $F[x_1, \ldots, x_n]$ has a finite basis. Actually, *every* ideal in $F[x_1, \ldots, x_n]$ has a finite basis. This is our next theorem (the Hilbert basis theorem for polynomial rings over a field). But first we need to associate to every ideal I in $F[x_1, \ldots, x_n]$ a monomial ideal determined by I.

15.19 DEFINITION Let $I \neq \{0\}$ be an ideal in $F[x_1, \ldots, x_n]$. Define

$$\mathrm{LT}(I) = \{c_\alpha x^\alpha \mid \text{there exists } f \in I \text{ with } \mathrm{LT}(f) = c_\alpha x^\alpha\}$$

the set of all leading terms of elements of the ideal I. We call $\langle \mathrm{LT}(I) \rangle$ the monomial ideal **associated** with I. ○

15.20 PROPOSITION Let $I \neq \{0\}$ be an ideal in $F[x_1, \ldots, x_n]$. Then there exist finitely many $g_1, \ldots, g_s$ in I such that the $\mathrm{LT}(g_i)$, $1 \leq i \leq s$ generate the monomial ideal associated with I: $\langle \mathrm{LT}(I) \rangle = \langle \mathrm{LT}(g_1), \ldots, \mathrm{LT}(g_s) \rangle$.

Proof The proof follows directly from Theorem 15.16 and is left to the reader. (See Exercise 15.) □

We can now proceed with the proof of the Hilbert basis theorem for $F[x_1, \ldots, x_n]$.

15.21 THEOREM (Hilbert basis theorem) Let I be an ideal in $F[x_1, \ldots, x_n]$. Then there exist finitely many $g_1, \ldots, g_s$ in I such that the g_i, $1 \leq i \leq s$ generate the ideal: $I = \langle g_1, \ldots, g_s \rangle$.

Proof If $I = \{0\}$, then $I = \langle 0 \rangle$. If $I \neq \{0\}$, then by Proposition 15.20

$$\langle \mathrm{LT}(I) \rangle = \langle \mathrm{LT}(g_1), \ldots, \mathrm{LT}(g_s) \rangle$$

for some finitely many $g_1, \ldots, g_s$ in I. Let $f \in I$ and let us perform the division algorithm to divide f by the $g_1, \ldots, g_s$. We obtain $a_i, r \in F[x_1, \ldots, x_n]$ such that

$$f = a_1 g_1 + \ldots + a_s g_s + r$$

where $r = 0$ or r is a linear combination with coefficients from F of monomials not divisible by any of the leading terms $\mathrm{LT}(g_i)$. Since

$$r = f - a_1 g_1 - \ldots - a_s g_s \in I$$

we conclude that

$$\mathrm{LT}(r) \in \langle \mathrm{LT}(I) \rangle = \langle \mathrm{LT}(g_1), \ldots, \mathrm{LT}(g_s) \rangle$$

Therefore, by Proposition 15.15, part (1), $\mathrm{LT}(r)$ is divisible by some $\mathrm{LT}(g_i)$. It follows that $r = 0$ and

$$f = a_1 g_1 + \ldots + a_s g_s \in \langle g_1, \ldots, g_s \rangle$$

Thus every element f of I is in $\langle g_1, \ldots, g_s \rangle$, and since $\langle g_1, \ldots, g_s \rangle \subseteq I$ we conclude that $I = \langle g_1, \ldots, g_s \rangle$. □

The more general Hilbert basis theorem applies to any commutative ring R with the property that every ideal in R is finitely generated. Such a ring R is called a **Noetherian ring**. In its general form, the Hilbert basis theorem says that if R is Noetherian, then $R[x]$ is Noetherian. Theorem 15.21 is a special case, since starting with a field F by induction on n the general theorem implies that $F[x_1, \ldots, x_n]$ is Noetherian for all $n > 0$. For our present purposes, however, the proof just given for the special case of rings of polynomials in several variables is important because it demonstrates the existence of a finite basis of a special kind, which we now give a name.

15.22 DEFINITION A finite set of polynomials $\{g_1, \ldots, g_s\}$ is called a **Gröbner basis** for an ideal I in $F[x_1, \ldots, x_n]$ if $\langle \text{LT}(I) \rangle = \langle \text{LT}(g_1), \ldots, \text{LT}(g_s) \rangle$. ○

We next state an immediate consequence of Theorem 15.21 that is important enough to be called a theorem rather than a corollary.

15.23 THEOREM Every nonzero ideal $I \subseteq F[x_1, \ldots, x_n]$ has a Gröbner basis, and every Gröbner basis of I is a basis of I.

Proof See Exercise 16. □

Gröbner Bases and the Division Algorithm

Intuitively, $\{g_1, \ldots, g_s\}$ is a Gröbner basis for I if for every $f \in I$, $\text{LT}(f)$ is divisible by one or more of the $\text{LT}(g_i)$.

15. 24 EXAMPLE The two polynomials $g_1 = x + 1$ and $g_2 = xy + 1$ in Example II.8 do *not* form a Gröbner basis for $I = \langle g_1, g_2 \rangle$ since $h = yg_1 - g_2 = y - 1 \in I$ but $\text{LT}(g_1) = x$ and $\text{LT}(g_2) = xy$ do not divide $\text{LT}(h) = y$. We can also easily check that the polynomials g_1, g_2 in Example 15.9 do not form a Gröbner basis either. ◇

15.25 EXAMPLE Let $g_1 = x + y$ and $g_2 = y + z$ in $\mathbb{R}[x, y, z]$ and let $I = \langle g_1, g_2 \rangle$. Now consider a nonzero polynomial $f \in I$, $f = h_1 g_1 + h_2 g_2$ for some $h_i \in \mathbb{R}[x, y, z]$. Assume that $\text{LT}(f)$ is not divisible by either $\text{LT}(g_1) = x$ or $\text{LT}(g_2) = y$. Since $x > y > z$ in the lexicographic order, f must be a polynomial in one variable, namely z, $f = f(z)$. The polynomial equations

$$g_1 = x + y = 0 \quad \text{and} \quad g_2 = y + z = 0$$

determine two planes in $\mathbb{R}^3$ that intersect at the line $(t, -t, t)$, where $t \in \mathbb{R}$. Therefore, $f(z) = f = h_1 g_1 + h_2 g_2$ has an infinite number of zeros, which is impossible since f is a nonzero polynomial of one variable. This argument shows that $\text{LT}(f)$ must be divisible by one of the $\text{LT}(g_i)$. Therefore, $\{g_1, g_2\}$ is a Gröbner basis of $I = \langle g_1, g_2 \rangle$. ◇

A Gröbner basis $\{g_1, \dots, g_s\}$ for an ideal I has two important properties listed in our next theorem.

15.26 THEOREM Let $\{g_1, \dots, g_s\}$ be a Gröbner basis for an ideal I in $F[x_1, \dots, x_n]$. Then for any $f \in F[x_1, \dots, x_n]$

(1) The remainder r upon division by $g_1, \dots, g_s$ is unique.

(2) The remainder r upon division by $g_1, \dots, g_s$ is independent of the order in which the g_i are listed.

Proof By Theorem 15.12 there exist $a_i, r \in F[x_1, \dots, x_n]$ such that

$$f = a_1 g_1 + \dots + a_s g_s + r$$

where $r = 0$ or no term of r is divisible by $\mathrm{LT}(g_i)$ for any $1 \le i \le s$. Suppose two such remainders r_1 and r_2 exist. Then

$$f = h_1 + r_1 = h_2 + r_2$$

for some $h_1, h_2 \in I$. If $r_1 - r_2 \neq 0$, then $\mathrm{LT}(r_1 - r_2) \subseteq \mathrm{LT}(I)$. Since $\{g_1, \dots, g_s\}$ is a Gröbner basis, it follows that

$$\mathrm{LT}(r_1 - r_2) \in \langle \mathrm{LT}(g_1), \mathrm{LT}(g_2) \rangle$$

and by Proposition 15.15, $r_1 - r_2$ is divisible by some $\mathrm{LT}(g_i)$. But since r_1 and r_2 are remainders, none of their terms are so divisible. Hence we must have $r_1 - r_2 = 0$, and (1) is proved. Furthermore, $h_1 = h_2$, and (2) follows. □

The uniqueness of the remainder r on division of a polynomial f by a Gröbner basis $\{g_1, \dots, g_s\}$ of an ideal I gives us a criterion for when f is an element of I.

15.27 COROLLARY Let $\{g_1, \dots, g_s\}$ be a Gröbner basis for an ideal I in $F[x_1, \dots, x_n]$, and let $f \in F[x_1, \dots, x_n]$. Then $f \in I$ if and only if f has remainder 0 upon division by $\{g_1, \dots, g_s\}$.

Proof See Exercise 17.□

Given a polynomial $f \in F[x_1, \dots, x_n]$ and a set $\{f_i\} \subseteq F[x_1, \dots, x_n]$, our initial question was whether we can perform division of f by the set $\{f_i\}$, which allows us to determine whether f belongs to the ideal generated by the $\{f_i\}$. With the theory we have just developed, to answer the question we would do the following:

(1) Define the ideal I generated by the $\{f_i\}$.

(2) Find a Gröbner basis $\{g_1, \dots, g_s\}$ for I.

(3) Perform the division of f by the $\{g_1, \dots, g_s\}$.

We end this introductory chapter on Gröbner bases with a description of Buchberger's algorithm, which allows us to construct a Gröbner basis from a given basis of the ideal.

15.28 DEFINITION Let $f, g \in F[x_1, \dots, x_n]$ be nonzero polynomials. Let

$$\text{multideg } f = \alpha = (\alpha_1, \dots, \alpha_n)$$
$$\text{multideg } g = \beta = (\beta_1, \dots, \beta_n)$$

and

$$\gamma = (\gamma_1, \dots, \gamma_n), \text{ where } \gamma_i = \max(\alpha_i, \beta_i)$$

Then we define the S-**polynomial** of f and g as

$$S(f, g) = [x^\gamma / \text{LT}(f)]\cdot f - [x^\gamma / \text{LT}(g)]\cdot g$$

Buchberger's algorithm is the following sequence of operations.
Let $I = \langle f_1, \dots, f_r \rangle \subseteq F[x_1, \dots, x_n]$.

(1) Calculate $S(f_i, f_j)$ for all $i \neq j$. If the remainder upon division of $S(f_i, f_j)$ by $\{f_1, \dots, f_r\}$ is *not* zero, add this nonzero remainder to the set $\{f_i\}$ of generators of I.

(2) Calculate $S(f_i', f_j')$ for all $i \neq j$ in the augmented set of generators of I. Again add to the set of generators any nonzero remainders of the S-polynomials.

(3) Repeat the process until a set of generators $\{g_1, \dots, g_s\}$ is obtained such that the remainder on dividing any $S(g_i, g_j)$ by the $\{g_1, \dots, g_s\}$ is 0. ○

For the proof that this process terminates and produces a Gröbner basis, we refer the reader to the following list of readings.

Further Reading

The following are recommended for in-depth study of Gröbner bases.

Adams, W. W. and P. Loustaunau *An Introduction to Gröbner Bases* (Graduate Studies in Mathematics, vol. 3) Providence, RI: American Mathematical Society, 1994.

Becker, T. and V. Weispfenning *Gröbner Bases: A Computational Approach to Commutative Algebra* New York: Springer-Verlag, 1993.

Cox, David, John Little, and Donald O'Shea *Ideals, Varieties and Algorithms* New York: Springer-Verlag, 1992.

Exercises 15

In Exercises 1 through 4 write the indicated polynomials in $\mathbb{R}[x, y, z]$ in decreasing term order using the lexicographic order with $x > y > z$.

1. $3xy - 5yz + 7xz$

2. $3z - 2x + y^2 - z^2 + xy$

3. $5 + 3x^2z - 2xy^4z^3 + 3z - 5x + 2y$

4. $xyz^4 - xy^2z + x^2yz + x^3z - x^5$

In Exercises 5 through 8 determine the multideg f, LM(f), and LT(f) of f using the lexicographic order with $x > y > z$, where

5. f is as in Exercise 1.

6. f is as in Exercise 2.

7. f is as in Exercise 3.

8. f is as in Exercise 4.

9. For $\alpha, \beta \in \mathbb{N}^n$ let $|\alpha| = \Sigma_{1 \le i \le n} \alpha_i$ and let $\alpha >_{\text{grlex}} \beta$ if $|\alpha| > |\beta|$ or $|\alpha| = |\beta|$ and $\alpha >_{\text{lex}} \beta$. Show that $>_{\text{grlex}}$ is a monomial ordering.

10. Let $f, g \in F[x_1, \dots, x_n]$ be nonzero polynomials. Show that

$$\text{multideg}(fg) = \text{multideg}\, f + \text{multideg}\, g$$

11. Let $f, g \in F[x_1, \dots, x_n]$ be nonzero polynomials. Show that if $f + g \neq 0$, then

$$\text{multideg}(f + g) \le \max\,\{\text{multideg}\, f, \text{multideg}\, g\}$$

In Exercises 12 and 13 calculate the remainder on dividing f by the given sets of polynomials S, using the lexicographic order.

12. $f = x^2yz + xz^2 - yz$ $\qquad S = \{x^2 - y, y - z\}$

13. $f = x^3y^2 - xyz + yz^2$ $\qquad S = \{x^2 - yz, x + z^2, y - z\}$

14. Prove Proposition 15.15.

15. Prove Proposition 15.20.

16. Prove Theorem 15.23

17. Prove Corollary 15.27.

In Exercises 18 through 22 calculate the S-polynomials $S(f, g)$ of the indicated polynomials f, g using the lexicographic order with $x > y > z$.

18. $f = xy - z,\ g = x^2 + yz$

19. $f = xy^2 + z^4,\ g = x^2y - z^2$

20. $f = x^4z - y^2,\ g = xy^2 - z$

21. $f = xy^2z + 3xy^4,\ g = x^2y - z^2$

22. $f = x^3y^2z - x + y,\ g = x^2z^3 + z$

In Exercises 23 through 27 construct a Gröbner basis for the following ideals in $\mathbb{R}[x, y, z]$ with $x > y > z$.

23. $I = \langle x - y, x + y\rangle$

24. $I = \langle xy - z, x - yz\rangle$

25. $I = \langle x - y + z, x + y - 2z, 3x - y + 3z\rangle$

26. $I = \langle x^4 + x^3 + x^2 - x + 1, x^3 - x^2 - x - 1\rangle$

27. $I = \langle x^2y - xy^2, xy - x\rangle$

Chapter 16

Coding Theory

Coding theory was developed to deal with practical problems in transmitting information. The goal of coding theory is the detection and correction of errors that occur during the transmission of information. This is usually achieved by creating a *code word* that contains the intended message as well as some additional information that helps to detect errors and possibly correct them.

The purpose of this chapter is to familiarize the reader with some basic notions and terminology of algebraic coding theory. A *linear code* is defined as a subspace of a vector space over a finite field F. Different methods of decoding are introduced, such as *coset decoding* using a *standard array* or the *syndrome method*, and *parity-check* decoding. In the second part of the exercises, *cyclic codes* are defined, making use of the construction of quotient rings of polynomial rings from Section 8.7.

Linear Binary Codes

A message can be transmitted as a finite sequence of digits since we can assign a number to each letter of our alphabet. A computer uses the binary system, so each number is written as a sequence of 0s and 1s. For example, the number written 13 in our usual decimal system is written 1101 in the binary system. Therefore, we can use $\mathbb{Z}_2 = \{0, 1\}$ as our alphabet and express any message as a sequence of 0s and 1s. In this introduction to coding theory, we study the possible encoding of such message sequences so that errors occurring during transmission can be detected and corrected.

16.1 EXAMPLE Let us form all possible four-digit words with the alphabet $\mathbb{Z}_2 = \{0, 1\}$. There are 16 of them, as shown in Table 1.

TABLE 1 Four-digit Words in $\mathbb{Z}_2$

0000	0001	0010	0011	0100	0101	0110	0111
1000	1001	1010	1011	1100	1101	1110	1111

We denote one such word as $x_1x_2x_3x_4$, where $x_i \in \mathbb{Z}_2$, $1 \le i \le 4$. Now let us define a fifth digit for each word as follows:

$$x_5 = x_1 + x_2 + x_3 + x_4 \bmod 2$$

This means that $x_5 = 0$ if and only if an even number of the x_i are equal to 1. Furthermore, notice that

$$x_1 + x_2 + x_3 + x_4 + x_5 = 0 \bmod 2$$

Therefore, if $x_1x_2x_3x_4$ is the message we want to send and we encode it as $x_1x_2x_3x_4x_5$, then if exactly one error occurs during the transmission, the receiver will be able to realize that an error has occurred because the number of 1s in the received word is odd instead of even. For example, if the message is 1101, then the encoded word is 11011. If the word received is 10011, then the receiver knows that an error occurred but cannot identify the place that it occurred. ◇

16.2 EXAMPLE Here is another way of encoding the four-digit word 1101 of the preceding example. Suppose we transmit each word three times, so, for example, the message 1101 is encoded as 110111011101. If only one error occurs, then the receiver can not only realize that an error has occurred, but actually can correct it. ◇

Let us now establish some terminology.

16.3 DEFINITION Let A be any finite set, which we call an **alphabet**.

(1) An element $u \in A^n = A \times \ldots \times A$ (n copies) is called a **word** of **length** n from the alphabet A.

(2) A subset $C \subseteq A^n$ is called a **code** over the alphabet A.

(3) An element $u \in C \subseteq A^n$ is called a **code word** of the code C.

(4) If A is a field, then A^n is a vector space over A, and $C \subseteq A^n$ is called a **linear code** over A if C is a subspace of A^n. Furthermore, if $\dim_A C = k$, then C is called an (n, k) **linear code**. If $A = \mathbb{Z}_2$, then C is called a **linear binary code**. ○

Note that an (n, k) linear binary code C over $A = \mathbb{Z}_2$ is a subgroup of A^n of order $|C| = 2^k$, and any subgroup of A^n of order 2^k is an (n, k) linear binary code.

16.4 EXAMPLE In Example16.1 we described a (5, 4) linear binary code where

$$C = \{x_1x_2x_3x_4x_5 \mid x_i \in \mathbb{Z}_2 \text{ and } x_5 = x_1 + x_2 + x_3 + x_4 \bmod 2\} \subseteq A^5$$

Since $A = \mathbb{Z}_2$ is a field, A^5 is a vector space of dimension 5 over $\mathbb{Z}_2$. For any $a \in \mathbb{Z}_2$ and $u \in C$, either $au = 0 \in C$ or $au = u \in C$. Also, if $u \in C$ and $v \in C$, then the sum of the five components of u and the sum of the five components of v are both $0 \bmod 2$, hence the same holds for $u - v$. Thus by the subspace test (Theorem 10.1.10), C is a subspace of A^5. Since

$$x_5 = x_1 + x_2 + x_3 + x_4 \bmod 2$$

$\dim_A C = 4$. ◇

16.5 EXAMPLE In Example 16.2 again $A = \mathbb{Z}_2$ and

$$C = \{x_1x_2 \ldots x_{12} \mid x_i \in \mathbb{Z}_2 \text{ and } x_i = x_{i+4} = x_{i+8},\ 1 \le i \le 4\} \subseteq A^{12}$$

Using the subspace test, it can easily be shown that C is a $(12, 4)$ linear binary code. ◇

16.6 DEFINITION Let u and v be two words in A^n.

(1) The **Hamming distance** between u and v is the number of components in which u and v differ, and it is denoted by $d(u, v)$.

(2) The **Hamming weight** of u is the number of components of u different from 0, and it is denoted by $\mathrm{wt}(u)$.

(3) For $r \geq 0$ a real number, the set

$$S_r(u) = \{v \in A^n \mid \mathrm{d}(u, v) \leq r\}$$

is called the r-**sphere** about u.

(4) If C is a code in A^n, the number

$$d = \min \{\mathrm{d}(u, w) \mid u, w \in C, u \neq w\}$$

is called the **minimum distance** of C. ○

16.7 THEOREM Let u, v, w be words in A^n, where $A = \mathbb{Z}_2$. Then

(1) $\mathrm{wt}(u) = d(u, 0)$

(2) $d(u, v) = \mathrm{wt}(u - v)$

(3) $d(u, v) = \mathrm{d}(v, u)$

(4) $d(u, v) = 0$ if and only if $u = v$

(5) $d(u, w) \leq d(u, v) + d(v, w)$ (**triangle inequality**)

Proof (1) This is immediate from Definition 16.6.

(2) The word $u - v$ has a 1 as its ith component if and only if u and v differ in the ith component, hence $\mathrm{wt}(u - v) = d(u, v)$.

(3) This is immediate from Definition 16.6.

(4) $d(u, v) = 0$ if and only if u and v have the same ith component for all $1 \leq i \leq n$, hence if and only if $u = v$.

(5) Let $x = u - v$ and $y = v - w$. Then by (2) $d(u, w) = \mathrm{wt}(u - w) = \mathrm{wt}(x + y)$, while $d(u, v) = \mathrm{wt}(x)$ and $d(v, w) = \mathrm{wt}(y)$. The inequality (5) to be proved now becomes

$$\mathrm{wt}(x + y) \leq \mathrm{wt}(x) + \mathrm{wt}(y)$$

To prove this, let x_i and y_i be the ith components of x and y, respectively, and z_i the ith component of $x + y$. Then

$$\mathrm{wt}(x + y) = \Sigma_{1 \leq i \leq n} z_i \qquad \mathrm{wt}(x) = \Sigma_{1 \leq i \leq n} x_i \qquad \mathrm{wt}(y) = \Sigma_{1 \leq i \leq n} y_i$$

So it will suffice to prove that $z_i \leq x_i + y_i$ for all $1 \leq i \leq n$. But since

$$z_i = x_i + y_i \bmod 2$$

this is immediate. □

16.8 EXAMPLE [The Hamming (7, 4) linear binary code] In a single error transmission the distance between the received word and the code word is 1. Therefore, if we make the distance between any two code words in our code C to be at least 3, then the correct code word is the only code word with distance 1 from the received word. In Example 16.1 we listed all 16 elements of A^4, where $A = \mathbb{Z}_2$. We then added a fifth component that allowed us to detect an odd number of errors. We now add instead three more components to ensure that the distance between any two code words is at least 3. Consider the code $C \subseteq A^7$ defined as follows:

$$x_1x_2x_3x_4x_5x_6x_7 \in C \text{ if and only if}$$
$$x_5 = x_1 + x_2 + x_3 \bmod 2$$
$$x_6 = x_1 + x_3 + x_4 \bmod 2$$
$$x_7 = x_2 + x_3 + x_4 \bmod 2$$

It can be checked that C is a subspace of A^7 with $\dim_A C = 4$. We list in Table 2 all the message words in A^4 and their corresponding code words in $C \subseteq A^7$.

TABLE 2 Code words in A^7

Message word	Code word
0000	0000000
0001	0001011
0010	0010111
0011	0011100
0100	0100101
0101	0101110
0110	0110010
0111	0111001
1000	1000110
1001	1001101
1010	1010001
1011	1011010
1100	1100011
1101	1101000
1110	1110100
1111	1111111

If, for example, we want to transmit the message 1101, then the code word we transmit is $u = 1101000 \in C$, while for 1001 the code word is $v = 1001101$. The two message words differ in only one place, but the two code words differ in three places, $d(u, v) = 3$. In general, if two message words differ in one component, since that component also appears in at least two of the defining equations for x_5, x_6, x_7, the distance between the two code words is at least 3.

Now let us look at message words with distance 2, say 1101 and 0111. The corresponding code words are $u = 1101000$ and $v = 0111001$, and $d(u, v) = 3$. In general, if two message words differ in exactly two components, say the ith and the jth components, then at least one of the defining equations for x_5, x_6, x_7 either contains x_i but not x_j or else contains x_j but not x_i. Therefore, the distance between the two code words is at least 3. We have also shown that the minimum distance in this example is $d = 3$. ◇

16.9 DEFINITION A code word $x_1x_2 \ldots x_k \ldots x_n$ in an (n, k) linear code consists of the first k components that make up the **message word**, and the last $n - k$ components, which are called the **parity-check** portion. ○

In Example 16.8, $x_5x_6x_7$ was the parity-check portion of a code word $x_1x_2x_3x_4x_5x_6x_7$, and the defining equations for x_5, x_6, x_7 are called the **parity-check** equations.

16.10 EXAMPLE If we use the Hamming (7, 4) linear binary code as defined in Example 16.8 and we receive the word 010110, since this is a code word, we decode it as 0101. If, however, we receive the word $u = 1110010$, since u is not a code word, we know that an error has occurred. Furthermore, $v = 0110010$ is a code word with $d(u, v) = 1$, while the distance of u from any other code word is at least 3. Therefore, if only a single error occurred in the transmission, the message word must have been 0110, and u will be decoded as this message word. ◇

Error Correction and Coset Decoding

The Hamming (7,4) linear binary code is an example that illustrates the **nearest neighbor decoding** in a single-error transmission, which means that a received word is decoded as the code word closest to it. We defined the minimum distance d of a code C as the smallest distance between two distinct code words in C. As we will see, d gives us an upper bound on the number of errors that can be detected and corrected.

16.11 THEOREM Let C be a code in A^n with minimum distance d. If the nearest neighbor decoding is used, then for any positive integer t

(1) If $t + 1 \leq d$, then C can detect t or fewer errors.

(2) If $2t + 1 \leq d$, then C can correct t or fewer errors.

Proof (1) For each code word $u \in C$, the t-sphere $S_t(u)$ contains all possible received words if the word being sent was u and at most t errors occurred in transmission. If d is the minimum distance in C and $t < d$, then $S_t(u)$ contains no other code word besides u. Hence if errors occur, but no more than t of them, then the received word w will not be a code word. Thus the occurrence of t or fewer errors can be detected.

(2) If d is the minimum distance in C and $2t < d$, then a received word w cannot be in both the t-spheres $S_t(u)$ and $S_t(v)$ of two distinct code words $u, v \in C$, since otherwise by the triangle inequality we would have

$$d(u, v) \leq d(u, w) + d(w, v) \leq t + t < d$$

contrary to the minimality of d. Hence if w is received, the message sent must have been the unique code word u with $w \in S_t(u)$. Thus t or fewer errors can be corrected. □

For the rest of this chapter we study binary codes. In other words, we assume that our alphabet is $A = \mathbb{Z}_2$.

16.12 EXAMPLE The Hamming (7, 4) linear binary code was shown in Example 16.8 to have minimum distance $d = 3$. Therefore, for $t = 1$ the code can, as we have seen, correct a single error in transmission. But for $t = 2$, the code can detect two errors in transmission but cannot correct them. For example, suppose the code word $u = 1010001$ was transmitted and was received as $w = 1011101$. Since w is not a code word, we can detect that the transmission was not without error. But $d(u, w) = 2$ and we also have $d(v, w) = 2$, where v is the code word $1111111 \in C$, and we even have $d(z, w) = 1$, where z is the code word $1001101 \in C$. Therefore, z or v could be mistaken for the correct code word. ◇

The size of an error correcting code has a bound that depends on the number of errors it can correct.

1613 THEOREM (The Hamming bound) Let C be an (n, k) linear binary code with minimum distance $d \geq 2t + 1$, so that C can correct t errors by the nearest neighbor decoding. Then

$$2^k = |C| \leq \frac{2^n}{\binom{n}{0} + \binom{n}{1} + \ldots + \binom{n}{t}}$$

where

$$\binom{n}{r} = \frac{n!}{r!(n-r)!}$$

is the binomial coefficient.

Proof C is a vector space over $\mathbb{Z}_2$ of dimension k. Hence $|C| = 2^k$. If $u \in C$, then $w \in S_t(u)$ if and only if $d(u, w) \leq t$. Since u has n components, the binomial coefficient $n!/r!(n - r)!$ gives the number of words that differ from u in r components. Hence

$$|S_t(u)| = \binom{n}{0} + \binom{n}{1} + \ldots + \binom{n}{t}$$

Since $2t + 1 \leq d$, the t-spheres about distinct code words are pairwise disjoint, hence they contain a total of $|S_t(u)|\,|C|$ distinct words. The total number of words in the vector space A^n is 2^n, hence $|S_t(u)|\,|C| \leq 2^n$, and the desired inequality follows. □

Let C be an (n, k) linear binary code. The vector space

$$\mathbb{Z}_2{}^n = \mathbb{Z}_2 \times \ldots \times \mathbb{Z}_2 \text{ (n copies)}$$

over $\mathbb{Z}_2$ is an Abelian group under addition of order 2^n, and

$$C \cong \mathbb{Z}_2{}^k = \mathbb{Z}_2 \times \ldots \times \mathbb{Z}_2 \ (k \text{ copies})$$

is a subgroup under addition of order 2^k. Hence C has 2^{n-k} cosets in $\mathbb{Z}_2{}^n$. Another method of coding can be devised as follows, using the cosets of C: In each coset of C choose a word u_i of minimum weight. So we can write the cosets as

$$u_1 + C, u_2 + C, \ldots, u_N + C$$

where $N = 2^{n-k}$. The coset representatives u_i are called the **coset leaders**. If a word w is received, then w must belong to one of the cosets, say $w \in u_i + C$. Then we decode w as $w - u_i \in C$. This is called the method of **coset decoding**. This method can be carried out by constructing a **standard array** for C, which is a table where the rows are the cosets of C with the coset leaders in the first column.

16.14 EXAMPLE Let {10001, 11100, 01011} be a basis of a (5, 3) linear binary code C. Then the standard array for C may be displayed as in Table 3.

TABLE 3 Standard Array

00000	10001	11100	01011	01101	11010	10111	00110
10000	00001	01100	11011	11101	01010	00111	10110
00010	10011	11110	01001	01111	11000	10101	00100
01000	11001	10100	00011	00101	10010	11111	01110

The first row contains the 2^3 elements of C and the first column contains the coset leaders of each of the four cosets of C. The entry in any row i and any column j is the sum of the element of C at the top of column j and the coset leader at the left of row i.

Now to illustrate the method of coset decoding, suppose we receive the word 10010. It appears in the fourth row, sixth column. The word in the first column of the fourth row is 01000, and the word in the first row of the sixth column is 11010: Thus the received word 10010 belongs to the coset $01000 + C$ and is equal to $01000 + 11010$ and will be decoded as 11010.

Note that all the nonzero coset leaders in this example have weight 1. Therefore, for any code word $v \in C$ and any word w in the column with v in the first row, $d(v, w) \leq 1$. Therefore, using this standard array and the coset decoding, we are still using the nearest neighbor decoding. ◇

16.15 THEOREM Coset decoding is nearest neighbor decoding.

Proof Let C be an (n, k) linear binary code. If we receive the word w and $w \in u + C$, where u is the coset leader, then we decode w as $w - u = x \in C$. We want to show that $d(w, x) \leq d(w, y)$ for any $y \in C$. And indeed we have

$$d(w, x) = \text{wt}(w - x) = \text{wt}(u) \leq \text{wt}(w - y) = d(w, y)$$

Here we have used Theorem 16.7, part (2), as well as the fact that for any $y \in C$, $w - y \in w + C = u + C$. □

Standard Generator Matrices

An (n, k) linear binary code C consists of 2^k code words of length n. Listing all the elements of C as we did in Examples 16.8 and 16.14 can become very impractical. We can, however, put all the needed information into a $k \times n$ matrix.

16.16 EXAMPLE Consider the following 4×7 matrix.

$$G = \begin{bmatrix} 1 & 0 & 0 & 0 & 1 & 1 & 0 \\ 0 & 1 & 0 & 0 & 1 & 0 & 1 \\ 0 & 0 & 1 & 0 & 1 & 1 & 1 \\ 0 & 0 & 0 & 1 & 0 & 1 & 1 \end{bmatrix}$$

Then given any message word $x_1x_2x_3x_4$ written as a 1×4 matrix

$$[x_1\ x_2\ x_3\ x_4]\, G = [x_1\ x_2\ x_3\ x_4\ x_5\ x_6\ x_7]$$

where

$$x_5 = x_1 + x_2 + x_3 \qquad x_6 = x_1 + x_3 + x_4 \qquad x_7 = x_2 + x_3 + x_4$$

as in Example 16.8. In other words, we obtain the Hamming (7, 4) linear binary code we described in Example 16.8 . For example, given the message word 1010 we have

$$[1\ \ 0\ \ 1\ \ 0]\, G = [1\ \ 0\ \ 1\ \ 0\ \ 0\ \ 0\ \ 1]$$

and 1010001 is the corresponding code word in the table of Example 16.8. Thus

$$C = \{wG \mid w \in A^k\}$$

and G contains all the needed information to obtain all the code words. ◇

16.17 DEFINITION A $k \times n$ **standard generator matrix** is a matrix G of the form $G = [I_k\ \ B]$, where I_k is the $k \times k$ identity matrix and B is a $k \times (n - k)$ binary matrix. ○

Thus in Example 16.16, G is a standard generator matrix with

$$B = \begin{bmatrix} 1 & 1 & 0 \\ 1 & 0 & 1 \\ 1 & 1 & 1 \\ 0 & 1 & 1 \end{bmatrix}$$

as its last columns. ○

16.18 EXAMPLE In Example 16.14

$$G = \begin{bmatrix} 1 & 0 & 0 & 0 & 1 \\ 0 & 1 & 0 & 1 & 1 \\ 0 & 0 & 1 & 1 & 0 \end{bmatrix}$$

is the standard generator matrix. ◇

16.19 DEFINITION An (n, k) linear binary code C is called a **systematic** code if each message word in A^k constitutes the first k components of exactly one code word in C. ○

Thus the codes in Examples 16.8 and 16.16 are systematic codes.
The following theorem shows how the two notions, standard generator matrix and systematic code, are interrelated.

16.20 THEOREM

(1) Let G be a $k \times n$ standard generator matrix. Then the code $C = \{vG \mid v \in A^k\} \subseteq A^n$ is a systematic (n, k) linear binary code.

(2) Let $C \subseteq A^n$ be a systematic (n, k) linear binary code. Then there exists a $k \times n$ standard generator matrix G such that $C = \{vG \mid v \in A^k\}$.

Proof (1) Let G be a $k \times n$ standard generator matrix and let $C = \{vG \mid v \in A^k\}$. Recall that for each $r \geq 0$, A^r is a group under addition of order 2^r. Consider the map ϕ: $A^k \to A^n$ defined by $\phi(v) = Gv \in A^n$. This is a group homomorphism since for all $v, u \in A^k$,

$$\phi(v + v) = (v + u)G = vG + uG = \phi(v) + \phi(u)$$

By definition of C, Im $\phi = C$. Furthermore, $G = [I_k \;\; B]$, where I_k is the $k \times k$ identity matrix and B is a $k \times (n - k)$ matrix. But then

$$\phi(v) = [v \;\; vB] = [u \;\; uB] = \phi(u)$$

if and only if $v = u$ in A^k. Hence ϕ is one to one. Therefore, by Theorem 2.4.15, $A^k \cong C$, and C is a subgroup of A^n with $|C| = 2^k$, hence by definition C is an (n, k) linear binary code. Furthermore, C is systematic because ϕ is a well-defined map onto C.

(2) Let C be a systematic (n, k) linear binary code. We construct the required matrix G as follows. Let $\{e_i\}$ be the standard basis for A^k. (See Example 10.1.27.) Since C is systematic, for each $1 \leq i \leq k$ there exists a unique $c_i \in C$ such that $c_i = [e_i \;\; d_i]$, where the first k components of c_i are the vector e_i and d_i is some vector in A^{n-k}. Let B be the $k \times (n - k)$ matrix with the d_i as rows, and let $G = [I_k \;\; B]$. Then G is a $k \times n$ standard generator matrix. We want to show that $C = \{vG \mid v \in A^k\}$. Note that for each $1 \leq i \leq k$, e_iB is the ith row of B, hence $e_iB = d_i$. Therefore, for all $1 \leq i \leq k$

$$e_iG = e_i[I_k \;\; B] = [e_i \;\; e_iB] = [e_i \;\; d_i] = c_i \in C$$

and since the $\{e_i\}$ form a basis for A^k, we obtain

$$\{vG \mid v \in A^k\} \subseteq C$$

Now let $c \in C$, where $c = [u \;\; w]$ with $u \in A^k$, $w \in A^{n-k}$. Then

$$uG = u[I_k \;\; B] = [u \;\; uB] = [u \;\; w'] \in C$$

Since C is systematic, by definition $[u \;\; w'] = [u \;\; w] = c$. Hence $c = uG$, which proves the opposite inclusion

$$C \subseteq \{vG \mid v \in A^k\}$$

and completes the proof that equality holds. □

The Syndrome Method

The standard generator matrix gives us a convenient way to encode a message word. Now we need to find a way to decode the received message.

16.21 DEFINITION Let C be a systematic (n, k) linear binary code with standard generator matrix $G = [I_k \ \ B]$. Then an $n \times (n - k)$ matrix

$$H = \begin{bmatrix} B \\ I_{n-k} \end{bmatrix}$$

is called the **parity-check matrix** for C, and for every word $w \in A^n$, $wH \in A^{n-k}$ is called the **syndrome** of w. ○

16.22 EXAMPLE In Examples 16.14 and 16.8 the parity-check matrices are

$$H = \begin{bmatrix} 0 & 1 \\ 1 & 1 \\ 1 & 0 \\ 1 & 0 \\ 0 & 1 \end{bmatrix} \quad \text{and } H = \begin{bmatrix} 1 & 1 & 0 \\ 1 & 0 & 1 \\ 1 & 1 & 1 \\ 0 & 1 & 1 \\ 1 & 0 & 0 \\ 0 & 1 & 0 \\ 0 & 0 & 1 \end{bmatrix}$$

respectively. ◇

16.23 THEOREM Let C be a systematic (n, k) linear binary code with standard generator matrix G and parity-check matrix H. Then for any word $v \in A^n$

$$vH = 0 \in A^{n-k} \text{ if and only if } v \in C$$

Proof As in the proof of Theorem 16.20 we use the fact that a vector space over a field is an Abelian group under addition. Define a map $\phi: A^n \to A^{n-k}$ by letting $\phi(v) = vH$ for $v \in A^n$. Then for any v and u in A^n

$$\phi(v + u) = (v + u)H = vH + uH = \phi(u) + \phi(v)$$

Therefore, ϕ is a group homomorphism. Also, ϕ is onto. For if $w \in A^{n-k}$, then let $v = [0 \ \ w] \in A^n$ be the vector whose first k components are 0 and whose last $n - k$ components are the components of w. Then

$$\phi(v) = vH = [0 \quad w]\begin{bmatrix} B \\ I_{n-k} \end{bmatrix} = 0 \cdot B + wI_{n-k} = w$$

Hence by the first isomorphism theorem (Theorem 2.4.15)

$$A^n / \text{Kern } \phi \cong A^{n-k}$$

Hence $|\text{Kern } \phi| = |A|^k = |C|$. To finish the proof it suffices to show that $C \subseteq \text{kern } \phi$. So let $v \in C$. Since C is a systematic code, by Theorem 16.20 $v = wG$ for some message word $w \in A^k$. Hence $\phi(v) = vH = wGH = 0$ since

$$GH = [I_k \;\; B]\begin{bmatrix} B \\ \\ I_{n-k} \end{bmatrix} = B + B = 0$$

This completes the proof. □

16.24 COROLLARY Under the same hypotheses as in the preceding theorem, two words u and v lie in the same coset of C if and only if u and v have the same syndrome.

Proof Let $u, v \in A^n$. Then

$$u + C = v + C \qquad \text{if and only if} \qquad u - v \in C$$

By Theorem 16.23 we have

$$u - v \in C \qquad \text{if and only if} \qquad (u - v)H = 0$$

Hence

$$u - v \in C \qquad \text{if and only if} \qquad uH = vH$$

to complete the proof. □

16.25 EXAMPLE Consider the (5, 3) linear binary code in Example 16.14 with parity-check matrix H as shown in Example 16.22. Then the coset leaders v with their corresponding syndromes vH are as in Table 4.

TABLE 4 Syndromes

v	00000	10000	00010	01000
vH	00	01	10	11

Now suppose we receive the word $u = 01101$. We calculate its syndrome and obtain $uH = 10$, so w is in the coset led by $v = 00010$, so we decode w by $w - v = 10111$. ◇

The advantage of this **syndrome method** of coset decoding over the standard array method is that we only have to identify the coset leaders and calculate their syndromes. We do not have to construct the whole 4×8 table, as we did in Example 16.14.

16.26 EXAMPLE For the Hamming (7, 4) linear binary code C, the parity-check matrix H was given in Example 16.22, and the parity-check equations were introduced in Example 16.8. Let $x_1x_2x_3x_4x_5x_6x_7 \in A^7$. Then

$$[x_1\, x_2\, x_3\, x_4\, x_5\, x_6\, x_7]\, H =$$
$$= [x_1 + x_2 + x_3 + x_5 \quad x_1 + x_3 + x_4 + x_6 \quad x_2 + x_3 + x_4 + x_7]$$

Therefore, given a word $v \in A^7$, vH is a word in A^3 where the three last components are given by the three parity-check equations. As defined in Example 16.8, the code words are exactly those words whose components satisfy the three parity-check equations. Hence $v \in C$ if and only if its syndrome $vH = [0\ 0\ 0] \in A^3$, as Theorem 16.23 states. Seen as a subgroup of A^7, C has index $2^{7-4} = 8$, and the eight coset leaders can be taken to be the 0-vector and the seven vectors $u_i = [0\ 0 \ldots\ 1\ \ldots\ 0\ 0]$ with 1 in the ith place only. Hence the coset $u_i + C$ consists of all elements in C with their ith components changed. Every word in A^7 belongs to exactly one coset $u_i + C$. By Corollary 16.24, if $v \in u_i + C$, then $vH = u_iH$. Let us calculate u_4H:

$$[0\ 0\ 0\ 1\ 0\ 0\ 0]\, H = [0\ 1\ 1]$$

which is the fourth row of H. Similarly, u_iH is the ith row of H. Therefore, if we use the Hamming (4, 7) linear binary code C and v is the received word, then

(1) If $vH = 0$, $v \in C$ and we decode v as v.

(2) If $vH \neq 0$, then vH is the ith row of H and we code v by changing its ith component.

Note that in this example the nonzero coset leaders u_i form a basis of A^7 and therefore are linearly independent. That is why for any $v \in A^7$ we found that vH is either equal to 0 or to some row of H and not a linear combination of rows of H. ◇

16.27 DEFINITION Let C be a systematic (n,k) linear binary code with parity-check matrix H. A **parity-check matrix decoding** consists of the following steps. If a word $v \in A^n$ is received, compute its syndrome vH and then

(1) If $vH = 0$, decode v as v.

(2) If $vH \neq 0$ and vH is the ith row of H for exactly one i, decode v by changing its ith component.

(3) If $vH \neq 0$ and vH is not the ith row of H for exactly one i, ask for retransmission. ○

16.28 THEOREM Let C be a systematic (n, k) linear binary code with parity-check matrix H. Then parity-check decoding will correct any single-error transmission if and only if no row of H is zero and no two rows of H are identical.

Proof ($\Leftarrow$) Assume no row of H is zero and no two rows of H are identical. Consider the vectors e_i with 1 in the ith component and 0 elsewhere, $1 \leq i \leq n$. Then e_iH is the ith row of H and by our assumption $e_iH \neq 0$ and if $i \neq j$, then $e_iH \neq e_jH$. Suppose there has been at most one error in transmission and let v be the received word.

(1) If $vH = 0$, then no error has occurred and by Theorem 16.23, $v \in C$ and v is correctly decoded as v.

(2) If $vH \neq 0$, then a single error must have occurred, say in the ith component, so that $v = c + e_i$ for some $c \in C$. Hence

$$vH = (c + e_i)H = cH + e_iH = 0 + e_iH = e_iH$$

and e_iH is the ith row of H. Then v is correctly decoded as c, which is the result of changing the ith component of v.

($\Rightarrow$) Suppose that parity-check decoding correctly decodes all received words with at most one error.

(1) Suppose that for some i the ith row of H is 0. If the code word $c = 0\ldots0$ is received with one error as e_i, then since in this case $e_iH = 0$, the received word will be incorrectly decoded as e_i, contrary to our assumption that all words are decoded correctly when there is at most one error. Therefore, H cannot have a zero row.

(2) Suppose that for some $i \neq j$ the ith and jth rows of H are identical. Then if the code word $c \in C$ is received with one error in the ith component, it will be received as $c + e_i$. Hence

$$(c + e_i)H = cH + e_iH = 0 + e_iH = e_iH = e_jH$$

and our parity-check decoding will tell us not to decode, again contrary to our assumption that all words are decoded correctly when there is at most one error. Therefore, H cannot have two rows identical. □

Further reading

For the interested reader, here are some excellent references.

Lidl, Rudolf, and Pilz Günter *Applied Abstract Algebra*, 2nd ed. New York: Springer-Verlag, 1997.

MacWilliams, F. J. and N. J. A. Sloane *The Theory of Error-Correcting Codes*, I and II Amsterdam: North-Holland, 1977.

Peterson, W. W. "Error-Correcting Codes," *Scientific American*, vol. 206 (1962), 96-108.

Peterson, W. W. and E. J. Weldon, Jr. *Error-Correcting Codes*, 2nd ed. Cambridge: MIT Press, 1972.

Exercises 16

In Exercises 1 through 4 find the Hamming weight of the indicated word.

1. 1001011

2. 01011101011

3. 00110111011

4. 11100110111

In Exercises 5 through 8 find the Hamming distance $d(u, v)$ between the indicated words.

5. $u = 1101, v = 0110$

6. $u = 110001, v = 011110$

7. $u = 1110101011$, $v = 0101101101$

8. $u = 000111010101$, $v = 101010101010$

In Exercises 9 through 12 using nearest neighbor decoding in the Hamming (7, 4) linear binary code of Example 16.8, decode the indicated words.

9. 1101101

10. 1100101

11. 1110010

12. 0110001

13. For any words u, v, and w in A^n, show that

$$d(u, v) = d(u + w, v + w)$$

In Exercises 14 through 17 determine the (n, k) linear binary code with the indicated standard generator matrix.

14. $G = \begin{bmatrix} 1 & 0 & 1 & 0 \\ 0 & 1 & 1 & 1 \end{bmatrix}$

15. $G = \begin{bmatrix} 1 & 0 & 0 & 1 & 0 & 1 \\ 0 & 1 & 0 & 0 & 1 & 1 \\ 0 & 0 & 1 & 1 & 1 & 0 \end{bmatrix}$

16. $G = \begin{bmatrix} 1 & 0 & 0 & 0 & 1 & 1 & 1 \\ 0 & 1 & 0 & 0 & 0 & 1 & 1 \\ 0 & 0 & 1 & 0 & 1 & 0 & 1 \\ 0 & 0 & 0 & 1 & 1 & 1 & 0 \end{bmatrix}$

17. $G = \begin{bmatrix} 1 & 0 & 0 & 1 & 0 & 0 & 1 & 0 & 0 \\ 0 & 1 & 0 & 0 & 1 & 0 & 0 & 1 & 0 \\ 0 & 0 & 1 & 0 & 0 & 1 & 0 & 0 & 1 \end{bmatrix}$

In Exercises 18 through 20 find the minimum distance d of the code generated by the indicated G.

18. G is the standard generator matrix in Exercise 16.

19. G is the standard generator matrix in Exercise 17.

20. G is the standard generator matrix in Example III.18.

In Exercises 21 through 23 determine the error-detecting capability and the error-correcting capability of the indicated codes, if nearest neighbor decoding is used.

21. C consisting of

000000	001111	010111	110101
011000	111010	100010	101101

22. The code generated by G in Exercise 15.

23. The code generated by G in Exercise 16.

In Exercises 24 through 26 use Theorem 16.13 to determine the maximum number of errors that the indicated code C can correct.

24. C is a Hamming (7, 4) linear binary code.

25. C is a (5, 2) linear binary code. **26.** C is an (8, 3) linear binary code.

27. Show that if C is an $(n, 3)$ linear binary code that corrects two errors, then $n \geq 9$.

28. A **Hamming (n, k) linear binary code** is defined as follows. Let $n + 1 = 2^r$ and $k + r + 1 = 2^r$ for some $r \geq 0$. Let H be the parity-checking matrix consisting of all $n = 2^r - 1$ nonzero elements of A^r as its rows, with the identity matrix I_r as its last r rows, so

$$H = \begin{bmatrix} B \\ I_r \end{bmatrix}$$

where B is an $n \times (n - k) = n \times r$ matrix. Show that every (n, k) Hamming linear binary code can correct one error.

29. Construct the standard generator matrix G and the parity-checking matrix H of a Hamming (15, 11) linear binary code C.

In Exercises 30 and 31 let C be the linear binary code generated by the indicated set of vectors. Determine whether C is a systematic code.

30. {10011, 11001, 10001} in A^5

31. {100110, 110010, 100011, 010011} in A^6

32. Let C be the (6, 3) linear binary code generated by the basis {100110, 010101, 001011} and suppose we have received the words 010111, 111010, 101101, and 010011. Decode each word using the four different methods illustrated in the text:

(a) nearest neighbor decoding
(b) coset decoding using the standard array
(c) coset decoding using the syndrome method
(d) parity-check matrix decoding

33. Show that a (6, 3) linear binary code can correct a single-error transmission.

34. Show that no (6, 3) linear binary code can correct a two-error transmission.

In Exercises 35 and 36 a nonempty subset S of a code C in A^n is called a **subcode** of C is S is itself a code in A^n.

35. Let C be a linear binary code in A^n and let

$$S = \{v \in C \mid \text{wt}(v) \text{ is even}\}$$

Show that S is a subcode of C.

36. Let C be a linear binary code in A^n and for each $1 \leq i \leq n$ let

$$C_i = \{ v \in C \mid \text{the } i\text{th component of } v \text{ is } 0\}$$

Show that each C_i, $1 \leq i \leq n$ is a subcode of C.

37. Let C be a linear binary code. Show that either every code word in C has even weight or exactly half of the code words in C have even weight.

38. Let C be a linear binary code. Show that either all the code words in C end with a 0 or exactly half of the code words in C end with a 0.

Cyclic Codes

Let $A = \mathbb{Z}_2$ and let C be a code in A^n. Then C is said to be a **cyclic code** if it is closed under **cyclic shifts**, that is,

$$\text{if } c_0c_1c_2\ldots c_{n-1} \in C, \text{ then } c_{n-1}c_0c_1\ldots c_{n-2} \in C$$

Let $f(x) = 1 - x^n \in \mathbb{Z}_2[x]$, let $I = \langle f(x)\rangle$, the ideal generated by $f(x)$ in $\mathbb{Z}_2[x]$, and let $A_n = \mathbb{Z}_2[x]/I$, the quotient ring. Then by Section 8.7, if we let $\alpha = x + I \in A_n$, then

$$A_n = \{c_0 + c_1\alpha + \ldots + c_{n-1}\alpha^{n-1} \mid c_i \in \mathbb{Z}_2, f(\alpha) = 0\}$$

39. Show that A_n is a vector space over A that is isomorphic, as an additive group, to A^n, hence $\dim_A A_n = n$.

40. Let C be a code in A_n. Show that C is a cyclic code if and only if

$$\alpha C = \{\alpha g(\alpha) \mid g(\alpha) \in C\} \subseteq C$$

41. Let C be a code in A_n. Show that C is a cyclic code if and only if C is an ideal of the quotient ring A_n.

42. Let C be a cyclic code in A_n. By the preceding exercise, C is an ideal in A_n. Show that there exists a unique polynomial $g(x)$ of lowest degree such that $g(\alpha) \in C$ and $C = \langle g(\alpha)\rangle$, the principal ideal generated by $g(\alpha)$ in A_n.

43. Let C be a cyclic code in A_n, and let $C = \langle g(\alpha)\rangle$ as in the preceding problem. Show that

(a) $g(x)$ divides $f(x) = 1 - x^n$ in $\mathbb{Z}_2[x]$.

(b) If $D = \langle h(\alpha)\rangle$ is another cyclic code in A_n, then $C \subseteq D$ if and only if $h(x)$ divides $g(x)$ in $\mathbb{Z}_2[x]$.

44. Determine all the cyclic codes in A_4 and A_5.

Consider now the isomorphism of vector spaces

A_n	$\leftrightarrow$	A^n
$c_0 + c_1\alpha + \ldots + c_{n-1}\alpha^{n-1}$	$\leftrightarrow$	$c_0c_1\ldots c_{n-1}$
polynomial form		word form

Then if $g(\alpha) = c_0 + c_1\alpha + \ldots + c_{n-1}\alpha^{n-1}$ a matrix in polynomial form with the ith row $[g(\alpha)]$ will correspond to a matrix in word form with the ith row $[c_0\, c_1\, \ldots\, c_{n-1}]$.

45. Let $C = \langle g(\alpha) \rangle$ be a cyclic code in A_n where

$$g(\alpha) = c_0 + c_1\alpha + \ldots + c_{n-k}\alpha^{n-k}$$

Write the following $k \times n$ matrix in word form.

$$G = \begin{bmatrix} g(\alpha) \\ \alpha\, g(\alpha) \\ \alpha^2\, g(\alpha) \\ \vdots \\ \alpha^{k-1}\, g(\alpha) \end{bmatrix}$$

G is called the **generator matrix** of C.

46. Let $C = \langle g(\alpha) \rangle$ be a cyclic code in A_n such that $f(x) = 1 - x^n = g(x)h(x)$, where

$$h(\alpha) = d_0 + d_1\alpha + \ldots + d_k\alpha^k \in A_n$$

Define an $n \times (n - k)$ matrix H as follows:

$$H = \begin{bmatrix} 0 & 0 & \cdots & 0 & d_k \\ 0 & 0 & \cdots & d_k & d_{k-1} \\ \vdots & \vdots & \iddots & \iddots & \vdots \\ 0 & d_k & \iddots & \vdots & \vdots \\ d_k & d_{k-1} & & \vdots & \vdots \\ d_{k-1} & d_{k-2} & & \vdots & \vdots \\ \vdots & \vdots & & \vdots & \vdots \\ \vdots & \vdots & & d_2 & d_1 \\ \vdots & \vdots & & d_1 & d_0 \\ \vdots & \vdots & \iddots & d_0 & 0 \\ \vdots & \iddots & \iddots & \vdots & \vdots \\ d_1 & d_0 & \cdots & 0 & 0 \\ d_0 & 0 & \cdots & 0 & 0 \end{bmatrix}$$

Let G be the generator matrix of $C = \langle g(\alpha) \rangle$ defined in the preceding exercise. Show that $GH = 0$ and H is a parity-check matrix of C.

Chapter 17

Boolean Algebras

The use of algebra to simplify switching circuits was one of the first applications of *Boolean algebras*, which can now be seen as the appropriate mathematical tool to describe such diverse problems as computer designs, road systems, telephone circuits, and others. Boolean algebras are a special type of *lattices*, which in turn are a special type of *partially ordered sets*. In this chapter we introduce the basic notions pertaining to partially order sets, lattices, and Boolean algebras. After giving examples, we prove that every finite Boolean algebra is isomorphic to the power set of some finite set. We then illustrate how Boolean expressions can be used to describe *series-parallel circuits*.

Lattices

There are two basic notions that we want to define in this section, that of a *partially ordered set* and that of a *lattice*.

17.1 EXAMPLE Consider the set of all integers $\mathbb{Z}$. The usual order relation on $\mathbb{Z}$ has the following properties.

(1) For any $a \in \mathbb{Z}$, $a \leq a$.

(2) For any $a, b \in \mathbb{Z}$, if $a \leq b$ and $b \leq a$, then $a = b$.

(3) For any $a, b, c \in \mathbb{Z}$, if $a \leq b$ and $b \leq c$, then $a \leq c$. ◇

17.2 EXAMPLE Let G be any group and consider the set $S = \{H \mid H \leq G\}$ of all its subgroups. Then

(1) For any $H \in S$, $H \leq H$.

(2) For any $H, K \in S$, if $H \leq K$ and $K \leq H$, then $H = K$.

(3) For any $H, K, L \in S$, if $H \leq K$ and $K \leq L$, then $H \leq L$. ◇

17.3 DEFINITION A set S with a relation $\leq$ is called a **partially ordered set** if the following three axioms are satisfied:

(1) **(reflexivity)** For any $x \in S$, $x \leq x$.

(2) **(antisymmetry)** For any $x, y \in S$, if $x \leq y$ and $y \leq x$, then $x = y$.

(3) **(transitivity)** For any $x, y, z \in S$, if $x \leq y$ and $y \leq z$, then $x \leq z$. ○

In Examples 17.1 and 17.2 we have two examples of partially ordered sets.

17.4 EXAMPLE For any set S, let $P(S)$ be the **power set** of S, the set of all subsets of S, and for any $A, B \in P(S)$, let $A \leq B$ if and only if $A \subseteq B$, A is a subset of B. Then $P(S)$ equipped with $\leq$ is a partially ordered set. ◇

17.5 EXAMPLE Any finite partially ordered set can be given in the form of a diagram, as in Figure 1. Here the set S is the set of all positive divisors of 18, $S = \{1, 2, 3, 6, 9, 18\}$, and for a and b in S, $a \leq b$ if and only if a is a divisor of b. In the diagram, $a \leq b$ if there is a sequence of line segments connecting them, with a below and b above. Such diagrams are called **Hasse diagrams**. ◇

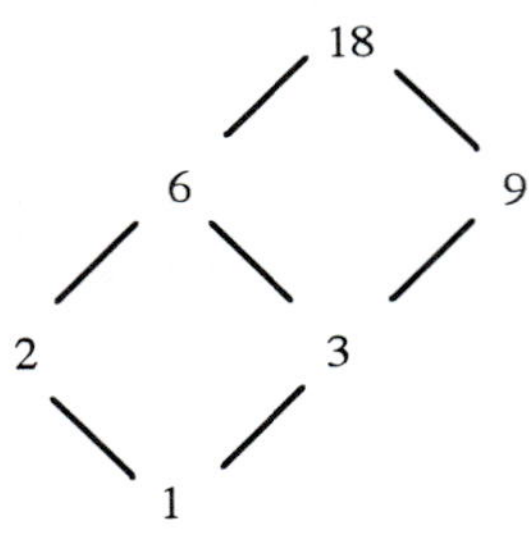

FIGURE 1

17.6 DEFINITION Let $\langle S, \leq \rangle$ be a partially ordered set and let X be a subset of S. Then $y \in S$ is an **upper bound** for X if $x \leq y$ for all $x \in X$, and $z \in S$ is a **lower bound** for X if $z \leq x$ for all $x \in X$. Furthermore, $u \in S$ is called a **least upper bound (lub)** for X if u is an upper bound for X and $u \leq y$ for every upper bound y for X. Similarly, $v \in S$ is called a **greatest lower bound (glb)** for X if v is a lower bound for X and $z \leq v$ for every lower bound z for X. ○

17.7 EXAMPLES

(1) In Example 17.5, if $X = \{3, 6, 9\}$, then the lub of X is 18 and the glb of X is 3.

(2) Let S be the set of all subgroups of D_4, where $H \leq K$ for $H, K \in S$ means that H is a subgroup of K. (See Example 5.1.9.) If X is $\{\langle \tau \rangle, \langle \rho^2 \rangle, \langle \rho \rangle\}$, then the lub of X is D_4 and the glb of X is $\langle \rho_0 \rangle$. ◇

17.8 PROPOSITION Let $\langle S, \leq \rangle$ be a partially ordered set and let X be a subset of S. Then

(1) If a lub of X exists, it is unique.
(2) If a glb of X exists, it is unique.

Proof (1) If each of u_1 and u_2 is a lub of X, then both are upper bounds of X, and since u_1 is a *least* upper bound, $u_1 \leq u_2$, while since u_2 is a *least* upper bound, $u_2 \leq u_1$. By antisymmetry, $u_1 = u_2$.

(2) If each of v_1 and v_2 is a glb of X, then both are lower bounds of X, and since v_1 is a *greatest* upper bound, $v_2 \leq v_1$, while since v_2 is a *greatest* upper bound, $v_1 \leq v_2$. By antisymmetry, $v_1 = v_2$. ◇

17.9 DEFINITION A **lattice** is a partially ordered set $\langle S, \leq \rangle$ such that for any $a, b \in S$

(1) The lub of $\{a, b\}$ exists. It is denoted $a \vee b$.

(2) The glb of $\{a, b\}$ exists. It is denoted $a \wedge b$. ○

The examples of partially ordered sets we have given so far are all examples of lattices.

If $\langle L, \leq \rangle$ is a lattice, observe that since the relation $\leq$ is reflexive, antisymmetric, and transitive, so is the relation $\geq$. Furthermore, if $\leq$ is replaced by $\geq$, then the lub in the first relation ($\leq$) becomes the glb in the second relation ($\geq$) and the glb in the first relation ($\leq$) becomes the lub in the second relation ($\leq$). Hence $\langle L, \geq \rangle$ is also a lattice. This observation underlies the following principle.

17.10 PRINCIPLE (Duality) A statement true for every lattice remains true if

(1) $\leq$ and $\geq$ are interchanged, and (2) $\vee$ and $\wedge$ are interchanged. ☆

17.11 DEFINITION Let $\langle L, \leq \rangle$ be a lattice. Then

(1) An element $1 \in L$ is called **unity** if $a \leq 1$ for all $a \in L$.

(2) An element $0 \in L$ is called **zero** if $0 \leq a$ for all $a \in L$.

Furthermore, if $\langle L, \leq \rangle$ is a lattice with unity 1 and zero 0, then

(3) For any $a \in L$, an element $a' \in L$ such that

$$a \vee a' = 1 \qquad \text{and} \qquad a \wedge a' = 0$$

is called a **complement** of a. ○

We usually write just L for a lattice $\langle L, \leq \rangle$ when it is understood what partial order relation $\leq$ the set L is equipped with.

17.12 THEOREM Let L be a lattice and $a, b, c \in L$. Then

(1) **(Commutativity)**

$$a \vee b = b \vee a \qquad \text{and} \qquad a \wedge b = b \wedge a$$

(2) **(Associativity)**

$$a \vee (b \vee c) = (a \vee b) \vee c \qquad \text{and} \qquad a \wedge (b \wedge c) = (a \wedge b) \wedge c$$

(3) **(Idempotence)**

$$a \vee a = a \qquad \text{and} \qquad a \wedge a = a$$

(4) **(Absorption)**

$$(a \vee b) \wedge a = a \qquad \text{and} \qquad (a \wedge b) \vee a = a$$

Proof See Exercises 8 and 9. □

The converse of this theorem can be used as an equivalent definition of a lattice.

17.13 THEOREM Let L be a set with two operations $\vee$ and $\wedge$ satisfying the axioms (1) through (4) stated in the preceding theorem. Let $\leq$ be the relation on L defined by

$$a \leq b \text{ if and only if } a \vee b = b \text{ or } a \wedge b = a$$

Then L is a lattice.

Proof First we note that the condition $a \vee b = b$ and the condition $a \wedge b = a$ are equivalent, since if $a \vee b = b$, then

$$a \wedge b = a \wedge (a \vee b) = (a \vee b) \wedge a = a$$

using commutativity (1) and absorption (4). And conversely, if $a \wedge b = a$, then

$$a \vee b = (a \wedge b) \vee b = (b \wedge a) \vee b = b$$

Next we show that the relation $\leq$ is a partial order. Let $a, b, c \in L$.
(Reflexivity) By idempotence (3) $a \vee a = a$, hence $a \leq a$.
(Antisymmetry). If $a \leq b$ and $b \leq a$, then $a \vee b = b$ and $b \vee a = a$. By commutativity (1),

$$b = a \vee b = b \vee a = a$$

Hence $a = b$.
(Transitivity) If $a \leq b$ and $b \leq c$, then $a \vee b = b$ and $b \vee c = c$. By associativity (2),

$$a \vee c = a \vee (b \vee c) = (a \vee b) \vee c = b \vee c = c$$

Hence $a \leq c$.
Now that we have shown that L is a partially ordered set, to show that L is actually a lattice, we show that $a \vee b$ is the lub of $\{a, b\}$ and $a \wedge b$ is the glb of $\{a, b\}$. The proofs for the lub and the glb are the same (interchanging $\leq$ and $\geq$ and interchanging $\vee$ and $\wedge$), so we give only the proof for lub.
First note that using (1) and (4),

$$a \wedge (a \vee b) = (a \vee b) \wedge a = a$$

$$b \wedge (a \vee b) = (a \vee b) \wedge b = (b \vee a) \wedge b = b$$

So $a \leq a \vee b$ and $b \leq a \vee b$ and $a \vee b$ is an upper bound for a and b.
To show that $a \vee b$ is the *least* upper bound, let $c \in L$ be such that

$$a \leq c \text{ and } b \leq c$$

Then

$$a \vee c = c \text{ and } b \vee c = c$$

Using (2),

$$(a \vee b) \vee c = a \vee (b \vee c) = a \vee c = c$$

So $a \vee b \leq c$ as required to complete the proof. □

17.14 EXAMPLE The lattice given by the Hasse diagram in Figure 2 can also be given by the operation tables, Tables 1 and 2.

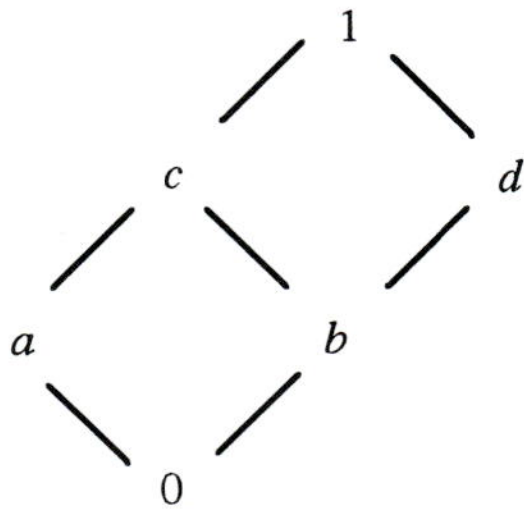

FIGURE 2

TABLE 1 Least upper bounds

$\vee$	0	a	b	c	d	1
0	0	a	b	c	d	1
a	a	a	c	c	1	1
b	b	c	b	c	d	1
c	c	c	c	c	1	1
d	d	1	d	1	d	1
1	1	1	1	1	1	1

TABLE 2 Greatest lower bounds

$\wedge$	0	a	b	c	d	1
0	0	0	0	0	0	0
a	0	a	0	a	0	a
b	0	0	b	b	b	b
c	0	a	b	c	b	c
d	0	0	b	b	d	d
1	0	a	b	c	d	1

Boolean Algebras

17.15 DEFINITION A **Boolean algebra** is a lattice B with zero and unity such that

(1) (**Distributivity**) For all $a, b, c \in B$

$$a \wedge (b \vee c) = (a \wedge b) \vee (a \wedge c) \qquad \text{and} \qquad a \vee (b \wedge c) = (a \vee b) \wedge (a \vee c)$$

(2) (**Complements**) For each $a \in B$ there exists an element $a' \in L$ such that

$$a \vee a' = 1 \qquad \text{and} \qquad a \wedge a' = 0$$

Recall that according to Definition 17.11, part (3) such an a' is called a **complement** of a. ○

17.16 EXAMPLE For any set S, the power set $P(S)$ is a Boolean algebra with the operations of union $\cup$ and intersection $\cap$. S is the unity element and the empty set $\varnothing$ is the zero element. ◇

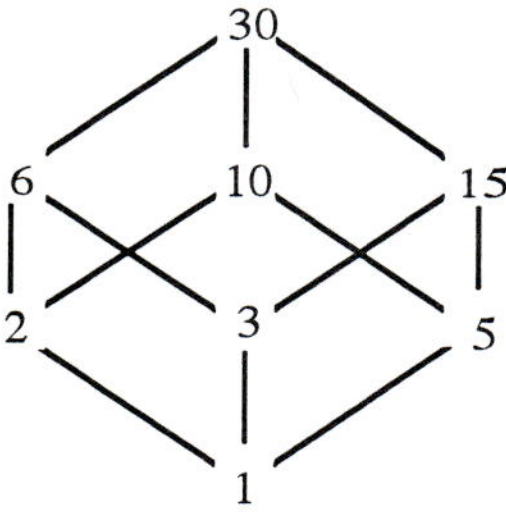

FIGURE 3

17.17 EXAMPLE Let

$$S = \{1, 2, 3, 5, 6, 10, 15, 30\}$$

be the set of all positive divisors of 30, and let $a \leq b$ if a divides b. This is a partially ordered set that can be represented by the Hasse diagram in Figure 3. For any x and y in S, $\gcd(x, y)$ and $\operatorname{lcm}(x, y)$ are in S. We have

$$x \vee y = \operatorname{lcm}(x, y)$$

$$x \wedge y = \gcd(x, y)$$

and S is a lattice.

Furthermore, the operations $\vee$ and $\wedge$ are distributive. For if $x, y, z \in S$, then

$$x = 2^{r_2}3^{r_3}5^{r_5} \qquad y = 2^{s_2}3^{s_3}5^{s_5} \qquad z = 2^{t_2}3^{t_3}5^{t_5}$$

where the exponents are 0 or 1. The exponent of a prime $p = 2, 3,$ or 5 in

$$\gcd(x, \operatorname{lcm}(y, z))$$

is

$$\min(r_p, \max(s_p, t_p)) = \max(\min(r_p, s_p), \min(r_p, t_p)$$

which is the exponent on p in

$$\text{lcm}(\gcd(x, y), \gcd(x, z))$$

Thus $x \wedge (y \vee z) = (x \wedge y) \vee (x \wedge x)$. The proof of the other form of the distributive law is similar.

To show that S is a Boolean algebra, we need to show that every element x has a complement. For any $x \in S$ let $x' = 30/x$. Then

$$x \vee x' = \text{lcm}(x, 30/x) = 30 \qquad x \wedge x' = \gcd(x, 30/x) = 1$$

Hence x' is the complement of x. 30 is the unity and 1 is the zero element of S. ◇

17.18 EXAMPLE Let $S = \{1, 2, 3, 6, 9, 18\}$ as in Example17.5, with $x \leq y$ if x divides y. 1 must be the zero and 18 must be unity. Consider now $6 \in S$ and suppose $\gcd(6, y) = 1$. By checking the six elements of S, we see that we must have $y = 1$. But then $\text{lcm}(6, y) = \text{lcm}(6, 1) = 6 \neq 18$. So no element y satisfies *both* the conditions that need to hold for y to be a complement of 6. Thus this lattice is *not* a Boolean algebra. The difference between this example and the preceding one is that $18 = 2 \cdot 3 \cdot 3$ has a repeated factor, while $30 = 2 \cdot 3 \cdot 5$ does not. ◇

17.19 PROPOSITION Let B be a Boolean algebra. Then for any element $a \in B$

(1) $\qquad a \wedge 1 = a \qquad$ and $\qquad a \vee 0 = a$

Proof Immediate from Definition 17.11. □

Thus 1 plays the role of an identity element under $\wedge$ and 0 plays the role of an identity element under $\vee$.

17.20 PROPOSITION In a Boolean algebra B

(1) The 1 and 0 elements are unique.

(2) The complement $a' \in B$ of any element $a \in B$ is unique.

Proof (1) Let e_1 and e_2 be identity elements under the same operation $*$ on a set B. Then

$$e_1 = e_1 * e_2 = e_2$$

Hence the 1 and 0 elements in a Boolean algebra B are unique by our remarks immediately preceding the statement of the theorem.

(2) Let $a \in B$ and let b and c both be complements of a. Then

$$b = b \vee 0 = b \vee (a \wedge c) = (b \vee a) \wedge (b \vee c)$$
$$= (a \vee b) \wedge (b \vee c) = 1 \wedge (b \vee c) = (b \vee c)$$

So $b = b \vee c$. Similarly, $c = c \vee b$, and hence $b = b \vee c = c \vee b = c$. □

We have seen that if $\langle B, \leq \rangle$ is a lattice, then so is $\langle B, \geq \rangle$, and that interchanging $\leq$ and $\geq$ amounts to interchanging $\wedge$ and $\vee$. If there is a zero element and a unity, interchanging $\leq$ and $\geq$ will also interchange 0 and 1. Since the two forms of the distributive law on the left and right sides of Definition 17.15, part (1), are the same except for interchanging $\wedge$ and $\vee$, if $\langle B, \leq \rangle$ satisfies one form of the law, $\langle B, \geq \rangle$ will satisfy the other, and if $\langle B, \leq \rangle$ satisfies both forms, so will $\langle B, \geq \rangle$. Similar remarks apply to Definition 17.15, part (2). Hence if $\langle B, \leq \rangle$ is a Boolean, so is $\langle B, \geq \rangle$. This observation underlies the following principle.

17.21 PRINCIPLE (Duality) A statement true for every Boolean algebra remains true if

(1) $\vee$ and $\wedge$ are interchanged, and (2) 0 and 1 are interchanged. ☆

17.22 PROPOSITION Let B be a Boolean algebra. Then for any $a, b \in B$

(1)	$a \wedge 0 = 0$	and	$a \vee 1 = 1$
(2)	$(a \wedge b)' = a' \vee b'$	and	$(a \vee b)' = a' \wedge b'$
(3)		$(a')' = a$	
(4)	$0' = 1$	and	$1' = 0$

Proof By the duality principle in (1), (2), and (4) we need only prove the version on the left, and the version on the right will follow.

(1) $a \wedge 0 = a \wedge (a \wedge a') = (a \wedge a) \wedge a' = a \wedge a' = 0$

(2) First note using (1) that

$$(a \wedge b) \wedge (a' \vee b') = [(a \wedge b) \wedge a'] \vee [(a \wedge b) \wedge b'] =$$
$$[(a \wedge a') \wedge b] \vee [a \wedge (b \wedge b')] = (0 \wedge b) \vee (a \wedge 0) = 0 \vee 0 = 0$$

Then note

$$(a \wedge b) \vee (a' \vee b') = [a \vee (a' \vee b')] \wedge [b \vee (a' \vee b')] =$$
$$[(a \vee a') \vee b'] \wedge [a' \vee (b \vee b')] = (1 \vee b') \wedge (a' \vee 1) = 1 \wedge 1 = 1$$

Hence $a' \vee b'$ is the complement of $a \wedge b$.

(3) Since $a' \wedge a = a \wedge a' = 0$ and $a' \vee a = a \vee a' = 1$, a is a complement of a', and since complements are unique, $(a')' = a$.

(4) This follows from Proposition 17.19. □

In Example 17.17 we showed that the set S of divisors of $30 = 2 \cdot 3 \cdot 5$ is a Boolean algebra. Using the fact that each element of S is a product of some of the prime divisors of S, let us reconstruct the diagram of S in a more abstract way.

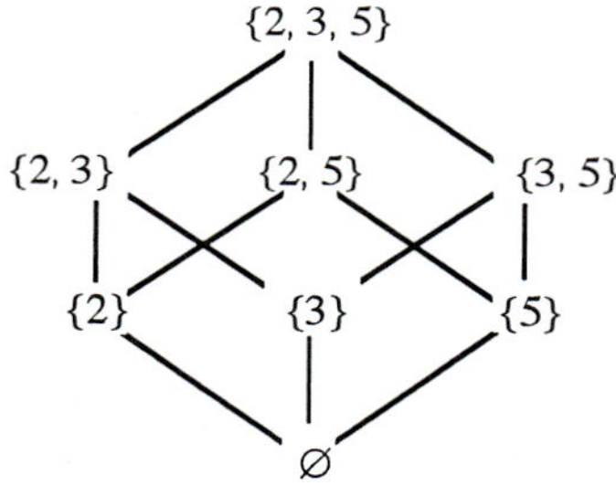

FIGURE 4

17.23 EXAMPLE Let B consist of all the subsets of the set

$$A = \{2, 3, 5\}$$

Thus the elements of $B = P(A)$ are

$$\varnothing, \{2\}, \{3\}, \{5\},$$
$$\{2, 3\}, \{2, 5\}, \{3, 5\}, \{2, 3, 5\}$$

Let $\vee$ and $\wedge$ be union and intersection of sets.

Then B can be represented by the Hasse diagram in Figure 4 . ◇

Observe that we obtained the "same" diagram as in Example 17.17.

17.24 DEFINITION Given two Boolean algebras A and B, a map $\phi: A \to B$ is said to be an **isomorphism** if

(1) ϕ is one to one and onto

(2) ϕ is **order preserving**, that is, for all a and b in A

$$a \le b \text{ in } A \qquad \text{if and only if} \qquad \phi(a) \le \phi(b) \text{ in } B$$

If such an isomorphism exists, then A and B are said to be **isomorphic**. ○

17.25 EXAMPLE If $B = P(A)$, where A is any set with three elements, then B would be isomorphic to the Boolean algebra in Example 17.23 (where A was the particular three-element set $\{2, 3, 5\}$). It is not hard to see that the Boolean algebra in Example 17.17 is another isomorphic Boolean algebra. In Example 17.17, the prime divisors $2, 3, 5$ play the same role as the singletons $\{2\}, \{3\}, \{5\}$ in Example 17.23: They are the elements closest to the zero element. In Example 17.17, the number 30 is square free, that is, the prime divisors of 30 appear all with exponent 1 in the factorization of 30, and of course there are no "repeated elements" in a given set in Example 17.23. ◇

17.26 DEFINITION Let B be a Boolean algebra. We write $a < b$ to mean that $a \le b$ and $a \ne b$. An element $a \in B$ is called an **atom** if $0 < a$ and there is no $x \in B$ such that $0 < x < a$. ○

In Example 17.17, the atoms are $2, 3, 5$; in Example 17.23, they are $\{2\}, \{3\}, \{5\}$.
We prove that the isomorphism between these two examples can be generalized: Any finite Boolean algebra is isomorphic to the Boolean algebra of subsets of some finite set. We do this by showing that the atoms of a given finite Boolean algebra play the same role as the singletons in the algebra of subsets of a finite set: They are the building blocks out of which all other elements are made up. For the remainder of this section we always assume that B is a finite Boolean algebra.
We break into several steps the proof that B is isomorphic to the Boolean algebra of subsets of some finite set. In each step when we make a statement about an abstract finite Boolean algebra B we try to point out the corresponding statement about the Boolean algebra $B(n)$ of divisors of a square-free integer n. The first step is to show that every finite Boolean algebra *has* atoms.

17.27 LEMMA Let b be any nonzero element of a finite Boolean algebra B. Then there exists an atom $a \in B$ such that $a \le b$.

Proof If b itself is an atom, we may take $a = b$ and we are done. Otherwise, there must exist an element $a_1 \in B$ with $0 < a_1 < b$. If a_1 is an atom, we are done. Otherwise, there must exist an element $a_2 \in B$ with $0 < a_2 < a_1$. Continuing in this way, we obtain distinct elements

$$0 < \ldots < a_2 < a_1 < b$$

Since B is finite, this chain must terminate with some element $a_n \le b$ that is an atom. □

In the case of the divisors $B(n)$ of a square-free integer n, this lemma says that if $1 \neq b \in B(n)$, then there exists a prime $p \in B(n)$ such that p divides b.

17.28 LEMMA If a_1 and a_2 are atoms in B and $a_1 \wedge a_2 \neq 0$, then $a_1 = a_2$.

Proof Since $a_1 \wedge a_2 \leq a_1$, and a_1 is an atom, we must have either $a_1 \wedge a_2 = 0$, which we are assuming is not the case, or else $a_1 \wedge a_2 = a_1$. Similarly, $a_1 \wedge a_2 = a_2$, so $a_1 = a_2$. □

In the case of $B(n)$ the lemma says that if p_1 and p_2 are prime divisors of n and are not relatively prime, then $p_1 = p_2$.

17.29 LEMMA Let $b, c \in B$. Then the following conditions are equivalent:

(1) $b \leq c$
(2) $b \wedge c' = 0$
(3) $b' \vee c = 1$

Proof (1) ⇒ (2) Assuming $b \leq c$, it follows that $c = b \vee c$ and hence

$$b \wedge c' = b \wedge (b \vee c)' = b \wedge (b' \wedge c') = (b \wedge b') \wedge c' = 0 \wedge c' = 0$$

(2) ⇒ (3) Assuming $b \wedge c' = 0$, it follows that

$$b' \vee c = b' \vee (c')' = (b \wedge c')' = 0' = 1$$

(3) ⇒ (1) Assuming $b' \vee c = 1$, it follows that

$$b = b \wedge 1 = b \wedge (b' \vee c) = (b \wedge b') \vee (b \wedge c) = 0 \vee (b \wedge c) = b \wedge c$$

and hence $b \leq c$. □

In the case of $B(n)$, this lemma says that if b and c are divisors of the square-free integer n, then the following are equivalent:

(1) b divides c.
(2) $\gcd(b, n/c) = 1$
(3) $\mathrm{lcm}(n/b, c) = n$

17.30 LEMMA Let $b, c \in B$ and suppose we do not have $b \leq c$. Then there exists an atom $a \in B$ such that $a \leq b$ but not $a \leq c$.

Proof Since we do not have $b \leq c$, by the preceding lemma $b \wedge c' \neq 0$, and so by Lemma 17.27 there exists an atom $a \leq b \wedge c'$. It follows that $a \leq b$. It also follows that $a \leq c'$, and hence we cannot have $a \leq c$, for then we would have $a \leq c \wedge c' = 0$ and $a = 0$, which is impossible since a is an atom. □

In the case of $B(n)$, this means that if b does not divide c, there is a prime p such that p divides b but does not divide c.
The next lemma says that, in the case of $B(n)$, if $p_1, p_2, \ldots, p_s$ are all the prime divisors of b, then $b = p_1 p_2 \ldots p_s$.

17.31 LEMMA Let $b \in B$ and let $a_1, \ldots, a_s$ be all the atoms $\leq b$ in B, then $b = a_1 \vee a_2 \vee \ldots \vee a_s$.

Proof Let $c = a_1 \vee a_2 \vee \ldots \vee a_s$. Then since each $a_i \leq b$, $c \leq b$. We want to prove $c = b$. If not, then we do not have $b \leq c$, and by the preceding lemma there is an atom a such that $a \leq b$ for which we do not have $a \leq c$. But if a is an atom and $a \leq b$, then a is one of the a_i, and therefore $a \leq c$ after all, a contradiction that completes the proof. □

The last lemma corresponds to Euclid's lemma: If p is a prime that divides the product of some primes, then p must be one of the factors.

17.32 LEMMA Let $b \in B$ and suppose that $a_1, \ldots, a_s$ are atoms in B such that $b = a_1 \vee a_2 \vee \ldots \vee a_s$. Then if a is an atom in B and $a \leq b$, then $a = a_i$ for some $1 \leq i \leq s$.

Proof We have $a \leq b$ and hence $a \wedge b = a$, which can be rewritten as

$$a = a \wedge (a_1 \vee \ldots \vee a_s) = (a \wedge a_1) \vee \ldots \vee (a \wedge a_s)$$

Hence for some $1 \leq i \leq s$, $a \wedge a_i \neq 0$ and so by Lemma 17.28, $a = a_i$. □

We have now all the necessary background to state and prove the main theorem of this section.

17.33 THEOREM (Representation theorem for finite Boolean algebras) Let B be a finite Boolean algebra. Then B is isomorphic to the power set $P(S)$ of some non-empty finite set S.

Proof Let S be the set of all atoms in B. By Lemma 17.27 atoms exist in B, so S is nonempty. S is finite since S is a subset of B. Consider the power set $P(S)$. We show that B is isomorphic to the Boolean algebra $P(S)$. Define a map $\phi: B \to S$ by sending $b \in B$ to the set of all atoms $a \in S$ with $a \leq b$. In particular, $\phi(0) = \varnothing$ since no atom is ≤ 0, and $\phi(1) = S$, since every atom is ≤ 1.

(1) ϕ is onto since for every subset $x = \{a_1, \ldots, a_s\}$ of S, if we let $b = a_1 \vee \ldots \vee a_s$, then we have $a_i \leq b$ for every $a_i \in x$, while by Lemma 17.32 we do not have $a \leq b$ for any atom $a \notin x$. Thus x is precisely the set of atoms $\leq b$, and $\phi(b) = x$.

(2) ϕ is order preserving. For clearly if $b \leq c$, then every atom $\leq b$ is an atom $\leq c$, so $\phi(b) \subseteq \phi(c)$. Inversely, if we do not have $b \leq c$, then by Lemma 17.30 there is an atom a with $a \leq b$ but not $a \leq c$ and therefore $a \notin \phi(c)$ but $a \in \phi(b)$. Hence we do not have $\phi(b) \subseteq \phi(c)$.

(3) It follows immediately from the fact that ϕ is order preserving that ϕ is one to one, since if $\phi(b) = \phi(c)$, then $b \leq c$ and $c \leq b$, and hence $b = c$ by antisymmetry. □

An immediate consequence of the representation theorem is a determination of the possible sizes of finite Boolean algebras.

17.34 COROLLARY If B is a finite Boolean algebra, then $|B| = 2^n$ for some positive integer n.

Proof By the preceding theorem, if B is a finite Boolean algebra, there exists a one-to-one, onto map $\phi: B \to P(S)$ for some finite set S. But then $|B| = |P(S)|$, and by Example 0.3.4, if $|S| = n$, then $|P(S)| = 2^n$. □

Circuits

We end this chapter with an illustration of how Boolean algebras can be used to simplify circuit diagrams. A **switch** is a device that can be in either of two positions. If it is **closed**, then current can pass through it, and if it is **open**, current cannot pass. A switching circuit consists of a certain configuration of switches. We illustrate how such circuits are composed with some examples.

17.35 EXAMPLE

(1) In Figure 5 is a diagram showing two switches, labeled a and b, that are connected in **series**. This means that a and b must *both* be closed in order for current to pass. Such a circuit is denoted by $a \wedge b$.

FIGURE 5

FIGURE 6

(2) In Figure 6 is a diagram showing two switches that are connected in **parallel**. This means that it is enough for either (or both) of a or b to be closed in order for current to pass. Such a circuit is denoted by $a \vee b$. ◇

A **series-parallel circuit** consists of two terminals (one on the left, one on the right) and a number of switches connected in the two ways indicated in the preceding example. Such a circuit is considered closed when current can pass from one terminal to the other, and open when it cannot.

In the diagram for a circuit there may be several switches all bearing the same label a. These are to be thought of as all controlled by a single master switch A, so that they are either all closed or all open at the same time. A switch that is open if a is closed and closed if a is open is said to be **opposite** to a and is denoted by a'. Again there may be many switches labeled a' in a single diagram.

Two circuits are **equivalent** if they permit the passage of current under the same circumstances or, in other words, are always either both closed or both open. Equivalence classes of circuits form a Boolean algebra under the operations $\wedge$ and $\vee$. The zero element corresponds to a switch that is always open, and unity to a switch that is always closed, as in Figures 7 and 8.

FIGURE 7 FIGURE 8

17.36 EXAMPLE

(1) The circuits in Figures 9 and 10, which are equivalent, illustrate the distributive law.

FIGURE 9 FIGURE 10

In either case, current can pass just in case the a is closed and either b or c is closed, which is to say, just in case either a and b are closed or a and c are closed.

(2) Complements are illustrated by the circuits in Figures 11 and 12.

FIGURE 11 FIGURE 12

In order for current to pass through the circuit on the left, one or the other of a or a' must be closed. But this will always be the case, since a' is closed whenever a is open. Hence the circuit on the left is equivalent to unity. In order for current to pass through the circuit on the right, both a and a' would have to be closed. But this will never be the case, since a' is open whenever a is closed. Hence the circuit on the right is equivalent to zero. ◇

What the abstract notions of order $\leq$, lub $\vee$, glb $\wedge$, and so on amount to in various concrete Boolean algebras is indicated in Table 3.

TABLE 3 Boolean Algebra Terminology

Boolean algebra	$P(S)$	$B(n)$	Series-parallel circuits
$\wedge$	$\cap$	gcd	series
$\vee$	$\cup$	lcm	parallel
a'	$S - a$	n/a	opposite
0	$\varnothing$	1	open
1	S	n	closed
=	=	=	equivalent
$a \leq b$	$a \subseteq b$	$a \mid b$	a closed $\Rightarrow$ b closed

The laws of Boolean algebras can be used to simplify circuits, as illustrated by the following examples.

17.37 EXAMPLE Consider the circuit in Figure 13.

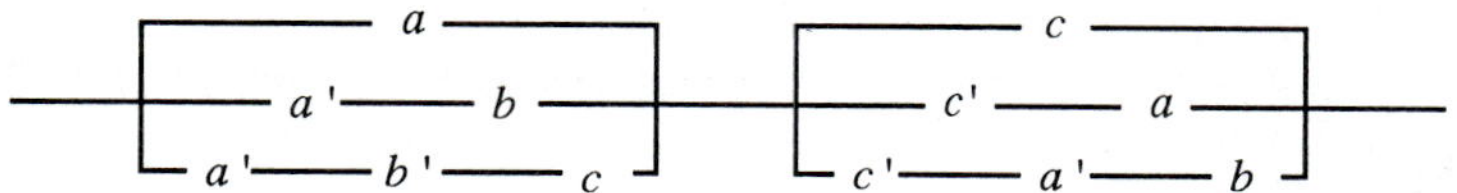

FIGURE 13

In order to simplify it — that is, to find an equivalent circuit with fewer switches — let us write it as in Figure 14.

FIGURE 14

and proceed to simplify each part C and D. Taking C first, we have

$$C = a \vee (a' \wedge b) \vee (a' \wedge b' \wedge c) = [(a \vee a') \wedge (a \vee b)] \vee (a' \wedge b' \wedge c) = (a \vee b) \vee (a' \wedge b' \wedge c) = (a \vee b) \vee [(a \vee b)' \wedge c] = [(a \vee b) \vee (a \vee b)'] \wedge [(a \vee b) \vee c] = a \vee b \vee c$$

Similarly, we obtain $D = a \vee b \vee c$ and hence $C \wedge D = a \vee b \vee c$, and the whole original circuit is equivalent to the one in Figure 15.

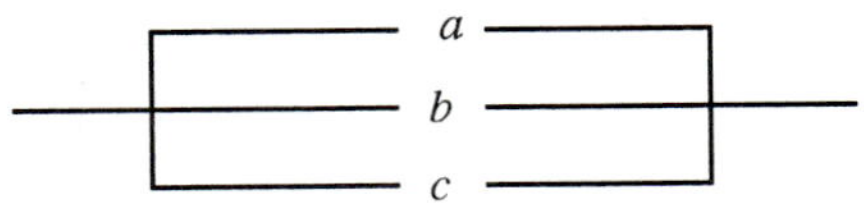

FIGURE 15

and 12 switches can be replaced by 3. ◇

17.38 EXAMPLE Suppose a room has three entrances, and we wish to place three master switches A, B, C one at each entrance, so that a person entering or leaving the room can turn on or off the lights by flipping the master switch beside the door. We wish, therefore, to design a circuit involving three kinds of switches a, b, c and their opposites, such that if the circuit is open, changing the setting of any one of a or b or c will close it, and vice versa. If we assume that current passes when all three kinds of switches are closed, then it should not pass if exactly two of the three are closed, but should pass again if exactly one of the three is closed, and should not pass again if none of the three are closed. Thus a circuit doing what we require can be represented by

$$(a \wedge b \wedge c) \vee (a \wedge b' \wedge c') \vee (a' \wedge b \wedge c') \vee (a' \wedge b' \wedge c)$$

This, however, involves 12 switches. We can reduce the number to 10 by using the equivalent form

$$\{a \wedge [(b \wedge c) \vee (b' \wedge c')]\} \vee \{a' \wedge [(b \wedge c') \vee (b' \wedge c)]\}$$

For a room with four entrances, a straightforward approach would yield a design with 32 switches, which by use of the laws of Boolean algebra can be cut in half, to 16. (See Exercise 29.) In more complicated problems, if switches are expensive, simplification can result in significant savings. ◇

Further Reading

For further reading, here are some excellent sources.

Gilbert, W. J. *Modern Algebra with Applications* New York: Wiley, 1976.

Halmos, Paul *Boolean Algebras* Princeton: Van Nostrand, 1967.

Lidl, Rudolf and Pilz Günter *Applied Abstract Algebra*, 2nd ed. New York: Springer-Verlag, 1997, Chapters 1 and 2.

Mendelson, E. *Boolean Algebras and Switching Circuits* New York: McGraw-Hill, 1970.

Whitesitt, J. E. *Boolean Algebras and Applications* Reading, MA: Addison-Wesley, 1961.

Exercises 17

In Exercises 1 through 4 determine whether the indicated sets and relations give examples of partially ordered sets.

1. The set of integers $\mathbb{Z}$ with $a \le b$ to mean a divides b.

2. The set of natural numbers $\mathbb{N}$ with $a \le b$ to mean a divides b.

3. The set of all subgroups of a group G with $H \le K$ to mean that H is a normal subgroup of K.

4. The set of all ideals in a ring R with $I \le J$ to mean that I is an ideal in J.

In Exercises 5 through 7 determine whether the indicated set with the indicated relation is a lattice.

5. The set of all positive divisors of 70 with $a \le b$ to mean a divides b.

6. The set of all positive divisors of 60 with $a \le b$ to mean a divides b.

7. The set $L \times M = \{(a, b) \mid a \in L,\ b \in M\}$, where L and M are lattices, with $(a, b) \le (c, d)$ to mean that $a \le c$ in L and $b \le d$ in M.

8. Prove Theorem 17.12, parts (1) and (2).

9. Prove Theorem 17.12, parts (3) and (4).

10. Determine the operation tables for $\wedge$ and $\vee$ for the lattice with the Hasse diagram in Figure 16.

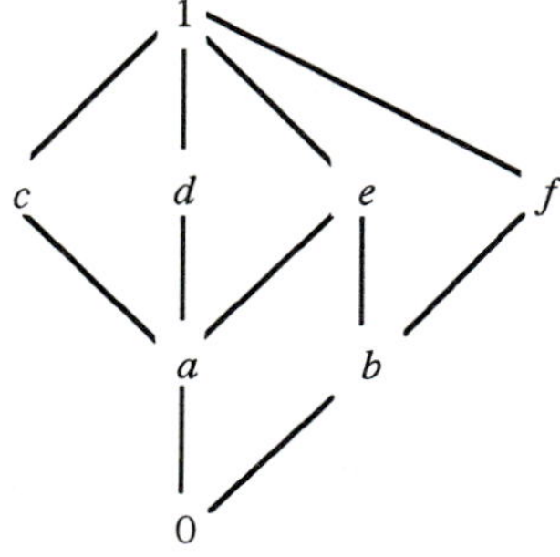

FIGURE 16

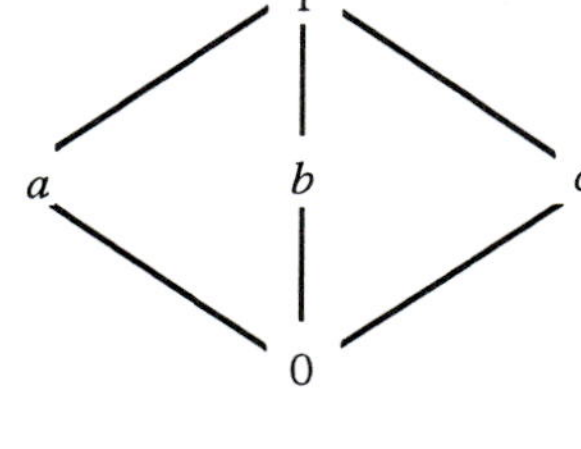

FIGURE 17

11. Show that the lattice with the Hasse diagram in Figure 17 is not distributive.

12. Show that a lattice L is distributive if and only if for all $a, b, c \in L$
(Cancellation) $a \wedge b = a \wedge c$ and $a \vee b = a \vee c$ implies $b = c$

13. Construct the Hasse diagram for the lattice of subgroups of S_3.

14. Construct the Hasse diagram for the lattice of subgroups of D_4.

15. Draw the Hasse diagrams of all nonisomorphic Boolean algebras of orders $|B| = 2, 4$, or 8.

16. Let $n \in \mathbb{N}$ and let $B(n)$ be the set of all positive divisors of n. Show that $B(n)$ with $\wedge = \gcd$ and $\vee = \text{lcm}$ is a Boolean algebra if and only if in the prime factorization of n no prime appears with an exponent ≥ 2.

17. Let B be a Boolean algebra and $a, b \in B$. Show that
(a) $a \leq b$ if and only if $b' \leq a'$
(b) $a \leq b'$ if and only if $a \wedge b = 0$
(c) $a \leq b$ if and only if $a' \vee b = 1$

18. Show that if $\phi: B \to C$ is a Boolean algebra isomorphism, then for all $a, b \in B$
(a) $\phi(a \vee b) = \phi(a) \vee \phi(b)$
(b) $\phi(a \wedge b) = \phi(a) \wedge \phi(b)$

19. Show that if $\phi: B \to C$ is a Boolean algebra isomorphism, then
(a) $\phi(1_B) = 1_C$
(b) $\phi(0_B) = 0_C$
(c) $\phi(a') = (\phi(a))'$ for all $a \in B$

For Exercises 20 and 21 recall from Section 6.2, Exercise 19, that a ring R is a Boolean ring if $a^2 = a$ for all $a \in R$, and that a Boolean ring is commutative and has $2a = 0$ for all $a \in R$.

20. Let R be a Boolean ring with unity 1 and for $a, b \in R$ define
$$a \vee b = a + b - ab$$
$$a \wedge b = ab$$
Show that R with $\vee$ and $\wedge$ is a Boolean algebra with unity 1, zero element 0, and $a' = 1 - a$.

21. Let B be a Boolean algebra and for $a, b \in B$ define
$$a + b = (a \wedge b') \vee (a' \wedge b)$$
$$ab = a \wedge b$$
(The operation $a + b$ thus defined is called the **symmetric difference** of a and b.) Show that B with these two operations is a Boolean ring with unity.

In Exercises 22 and 23 draw circuits that correspond to the indicated Boolean expressions.

22. $[a \wedge (b \vee c')] \vee b'$

23. $[(a \wedge b \wedge c') \vee (a' \wedge b' \wedge c) \vee (a \wedge b \wedge c)]$

24. Let C be a circuit with the Boolean expression $(a \vee b') \wedge (c \vee a')$.

(a) Draw the circuit C.

(b) If a switch is given the value 0 when it is open and 1 when it is closed, write a table with all possible values of the **switching function**

$$f(a, b, c) = (a \vee b') \wedge (c \vee a')$$

(Note that the domain of the function has 2^3 points.)

In Exercises 25 through 27 give Boolean expressions for the indicated circuit.

25.

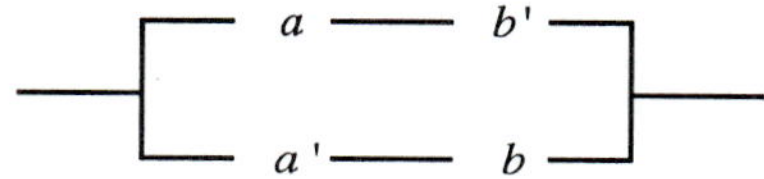

FIGURE 18

26.

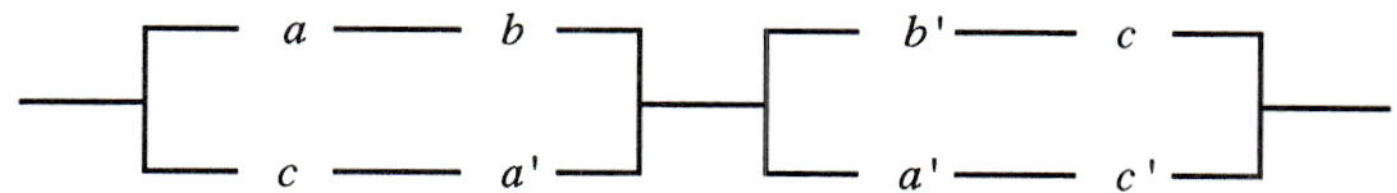

FIGURE 19

27.

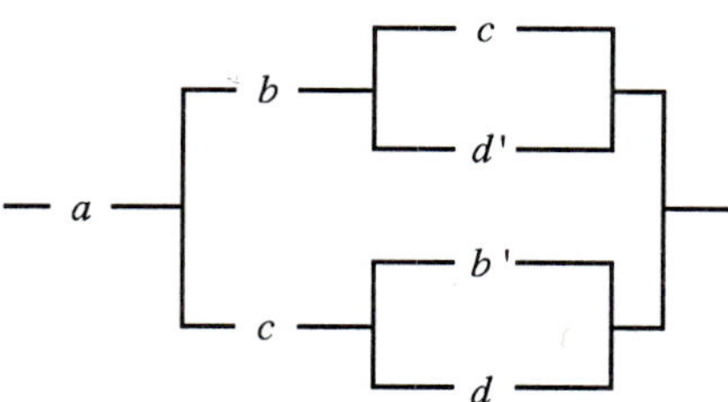

FIGURE 20

28. Simplify the circuit in Figure 21.

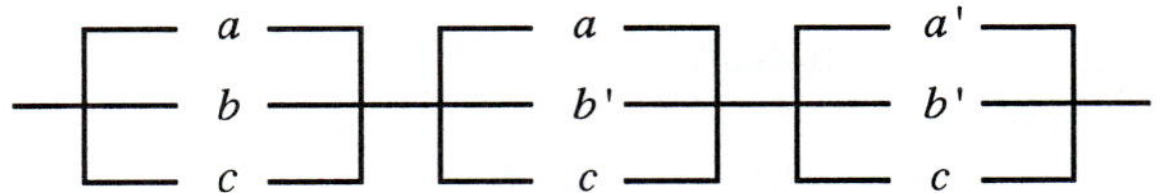

FIGURE 21

29. Show that, as asserted at the end of Example 17.38, 16 switches suffice for a circuit controlling the lighting of a room with four doors.

Answers and Hints to Selected Exercises

Chapter 0 Background

Section 0.1

3. Yes
6. No
7. No
13. Yes
16. *Hint:* Show that the map $\phi((n, m)) = (2n, 2m)$ is one to one and onto.

Section 0.2

3. Yes, [0] = all even integers and [1] = all odd integers.
7. Yes, all circles centered at (0, 0).
11. *Hint:* Use Theorem 0.2.4, part (4).

Section 0.3

6. Use induction on n and the fact that $(F_{n+1})^2 - F_nF_{n+2} = (F_{n+1})^2 - F_n^2 - F_nF_{n+1}$.
15. $c_1 = c_{10} = 11$, $c_2 = c_9 = 55$, $c_4 = 330$, $c_6 = 462$
23. $\gcd(9750, 59400) = 2\cdot3\cdot5^2$ and $\text{lcm}(9750, 59400) = 2^3\cdot3^3\cdot5^3\cdot11\cdot13$
32. *Hint:* Use Proposition 0.3.31, part (3), and Euclid's lemma.

Section 0.4

3. $1 + i$
5. -1
7. 1
15. $-{}^1/_2 + {}^1/_2 i$
17. $\sqrt{2}\,(\cos {}^{7\pi}/_4 + i \sin {}^{7\pi}/_4)$
23. $z_1 = 1$, $z_2 = \omega = -\frac{1}{2}+\frac{\sqrt{3}}{2}i$, $z_3 = -\frac{1}{2}-\frac{\sqrt{3}}{2}i$

Section 0.5

3. $\begin{bmatrix} 1+2i & 4 \\ i & 4+2i \end{bmatrix}$ **7.** $\begin{bmatrix} 4 & 2 \\ 4 & 2 \end{bmatrix}$ **9.** $\begin{bmatrix} i & 0 \\ 0 & i \end{bmatrix}$

11. $3 - 2i$ **15.** $A^{-1} = {}^1/_2 \begin{bmatrix} 1 & 1 \\ 1 & -1 \end{bmatrix}$

Part A Group Theory

Chapter 1 Groups

Section 1.1

3. Use Proposition 0.3.34 and Example 1.1.16.

8. Let $e = I$, $a = \begin{bmatrix} -1 & 0 \\ 0 & 1 \end{bmatrix}$ $b = \begin{bmatrix} 1 & 0 \\ 0 & -1 \end{bmatrix}$ $c = \begin{bmatrix} -1 & 0 \\ 0 & -1 \end{bmatrix}$, then

	e	a	b	c
e	e	a	b	c
a	a	e	c	b
b	b	c	e	a
c	c	b	a	e

and V is Abelian.

13. $\rho_0{}^2 = \mu_1{}^2 = \mu_2{}^2 = \mu_3{}^2 = \rho_0$ and $\rho^3 = (\rho')^3 = \rho_0$
25. Use Theorem 0.3.16, part (2).

Section 1.2

5. 2 **7.** 4 **11.** $2\mathbb{Z}$ and $3\mathbb{Z}$ **15.** $\langle\rho\rangle$ and $\langle\mu_1\rangle$

Section 1.3

1. (a) 5 (d) 30
3. (b) a^2, a^5, a^4 **7.** 3, 6, 9, 12
13. ($\Rightarrow$) If $a^k = e$, then $a^k = a^0$. Hence by Theorem 1.3.11, part (2), $n \mid k$.
($\Leftarrow$) If $n \mid k$, then $k = nl$ and $a^k = (a^n)^l = e$.
19. If G is not the trivial group, let $a \in G$, $a \neq e$, and consider $\langle a\rangle \leq G$.
21. 30

Section 1.4

7. $1 \to 6 \to 8 \to 2 \to 5 \to 1, \quad 3 \to 9 \to 3, \quad 4 \to 4, \quad 7 \to 7$
11. $\phi = (2\ 8\ 3)(4\ 9\ 10\ 6\ 5\ 7)$
15. Use Proposition 1.4.22 and the fact that $\langle \rho \rangle$ and $\langle \sigma \rangle$ intersect trivially.
19. 4
21. 6
23. 6
29. (*b*) $|H| = (n - 1)!$
33. *Hint:* $(i\ j)(j\ k) = (j\ k\ i)$ and $(i\ j)(k\ l) = (i\ j)(i\ k)(k\ i)(k\ l) = (i\ k\ j)(k\ l\ i)$
35. *Hint:* $(k\ i) = (k\ \ i-1)(i-1\ \ i)(k\ \ i-1)$

Chapter 2 Group Homomorphisms

Section 2.1

5. 2
7. 4
11. Use Theorem 0.2.4, part (4).
13. Use Lemma 2.1.7, part (2).
21. Use Theorem 2.1.16.
25. $9^{1573} \equiv 9^3 \equiv (-2)^3 \equiv 3 \bmod 11$
27. $(p - 1)(q - 1)$

Section 2.2

1. No
3. Yes, Kern $\phi = \{1, -1\}$
5. Yes, Kern $\phi = A_3$
11. 2
13. 0
15. ρ^2
17. $\phi_0(n) = n$ and $\phi_1(n) = -n$
27. $\phi(\rho) = (1\ 2\ 3\ 4)$ and $\phi(\tau) = (1\ 2)(3\ 4)$
33. Use Corollary 1.3.12.

Section 2.3

1. Yes
3. Yes
5. No
7. *Hint:* $(i\ k)(i\ j)(k\ l)(i\ k) = (i\ l)(k\ j)$
11. *Hint:* Use Exercise 48 in Section 2.2.
17. $xH = Hy$ implies $xHx^{-1} = Hyx^{-1}$, hence $e \in Hyx^{-1}$ and $Hyx^{-1} = H$.
19. Show that $\phi: H \to gHg^{-1}$ defined by $\phi(h) = ghg^{-1}$ is one to one and onto.
25. Q_8
27. Use Theorem 2.3.17 and Proposition 2.3.22.

Section 2.4

1. 2
5. 2
9. 3
11. 2
13. $(xH)(yH) = xyH = yxH = (yH)(xH)$
15. $\phi(\rho) = 0$ and $\phi(\mu_i) = 3$, $1 \le i \le 3$
19. Use Proposition 2.2.15, part (6).
27. (*b*) $xyx^{-1}y^{-1} \in N$ implies $Nxy = Nyx$.
(*d*) $ghg^{-1}h^{-1} \in N \subseteq H$ implies $ghg^{-1} \in H$.

Section 2.5

1. Use Proposition 0.3.31.

3. $\phi_i(a) = a^i$, where $i = 1, 3, 7$, or 9.

5. $\mathbb{Z}_2$

9. Show first that $Z(S_3)$ is trivial.

13. Use the fact that $|\phi(\rho)| = 4$ and $|\phi(\tau)| = 2$.

17. *Hint:* For any $g \in G$, the inner automorphism τ_g determines an automorphism of H.

Chapter 3 Direct Products and Abelian Groups

Section 3.1

5. Consider $\phi: \mathbb{Z} \times \mathbb{Z} \to \mathbb{Z}$ defined by $\phi((a, b)) = a - b$.

13. $(x, y) \in Z(G_1 \times G_2)$ if and only if $(ax, by) = (xa, yb)$ for all $a \in G_1, b \in G_2$.

17. (a) Show first that $Z(S_3)$ is trivial and that $Z(D_4) = \langle \rho^2 \rangle$.
(b) Use Exercises 28 and 29 of Section 2.4.

19. Factor 105 into primes and use Exercise 18.

Section 3.2

3. 6

7. H, $(1, 0) + H$, $(0, 1) + H$, and $(1, 1) + H$

11. 2

15. Use Proposition 2.2.23, part (4).

21. (a) Use Exercise 18 in Section 1.1 (b) Use Theorem 2.4.20.

Section 3.3

3. $H = \langle 2 \rangle$, $K = \langle 9 \rangle$

7. Use Corollary 3.3.11

11. Use Exercise 18 in Section 3.1.

15. Show that $(H_1 H_2 \ldots H_i) \triangleleft G$. Then use induction and Theorem 3.3.4.

19. *Hint:* Let $x \in G$ and consider $x - \phi(x)$.

23. *Hint:* For any $g \in G$ consider $g - \phi(g)$.

Section 3.4

5. $\mathbb{Z}_{16}$, $\mathbb{Z}_2 \times \mathbb{Z}_8$, $\mathbb{Z}_4 \times \mathbb{Z}_4$, $\mathbb{Z}_2 \times \mathbb{Z}_2 \times \mathbb{Z}_4$, $\mathbb{Z}_2 \times \mathbb{Z}_2 \times \mathbb{Z}_2 \times \mathbb{Z}_2$

9. $\mathbb{Z}_{72} \cong \mathbb{Z}_8 \times \mathbb{Z}_9$, $\mathbb{Z}_2 \times \mathbb{Z}_4 \times \mathbb{Z}_9$, $\mathbb{Z}_2 \times \mathbb{Z}_2 \times \mathbb{Z}_2 \times \mathbb{Z}_9$, $\mathbb{Z}_8 \times \mathbb{Z}_3 \times \mathbb{Z}_3$, $\mathbb{Z}_2 \times \mathbb{Z}_4 \times \mathbb{Z}_3 \times \mathbb{Z}_3$, and $\mathbb{Z}_2 \times \mathbb{Z}_2 \times \mathbb{Z}_2 \times \mathbb{Z}_3 \times \mathbb{Z}_3$

19. Use Theorem 3.4.17 and Corollary 3.3.11.

21. Use Theorem 1.3.26, part (2), and Exercise 26 in Section 1.3.

23. (b) Show that $g \to pg$ determines a group homomorphism from G onto pG with kernel $G^{(p)}$.

Chapter 4 Group Actions

Section 4.1

5. (c) $\chi((1, 0)) = (1\ 3)$, $\chi((0, 1)) = (2\ 4)$, $\chi((1, 1)) = (1\ 3)(2\ 4)$

7. (c) $\chi((1, 0, 0, 0)) = \chi((0, 1, 0, 0)) = (1\ 3)$, $\chi((0, 0, 1, 0)) = \chi((0, 0, 0, 1)) = (2\ 4)$

11. $X = \{\langle\rho_0\rangle, \langle\rho\rangle, \langle\mu_1\rangle, \langle\mu_2\rangle, \langle\mu_3\rangle, S_3\}$, $g\langle\rho_0\rangle g^{-1} = \langle\rho_0\rangle$, $g\langle\rho\rangle g^{-1} = \langle\rho\rangle$, $gS_3g^{-1} = S_3$, and $g\langle\mu_i\rangle g^{-1} = \langle\mu_j\rangle$, where for some $g \in S_3$, $g\langle\mu_i\rangle g^{-1} \neq \langle\mu_i\rangle$.

17. Let $H = \{e\}$ in Exercise 16.

Section 4.2

3. For $a = 1$ we have $G_1 = \{\sigma \in S_4 \mid \sigma(1) = 1\} \cong S_3$, $O_1 = X$, and $4 = |O_1| = [S_4 : S_3] = 4!/3!$.

7. For $a = 1$ we have $G_1 = \{\text{id} = \rho_0\}$, $O_1 = X$, and $4 = |O_1| = [G : G_1] = 4/1$.

9. (a) $T_1{}^2 = T_2$ and $T_1{}^3 = T_0$ (c) $-\frac{1}{2}+\frac{\sqrt{3}}{2}i$ and $-\frac{1}{2}-\frac{\sqrt{3}}{2}i$

11. (a) $xaH = aH$ if and only if $a^{-1}xa \in H$ if and only if $x \in aHa^{-1}$

(b) $|O_{aH}| = [G : G_{aH}] = [G : aHa^{-1}] = [G : H]$

Section 4.3

1. 2 **3.** 1 **5.** 210 **7.** 6

Section 4.4

3. $\{(0, \rho_0)\}$, $\{(1, \rho_0)\}$, $\{(0, \rho), (0, \rho^2)\}$, $\{(1, \rho), (1, \rho^2)\}$, $\{(0, \mu_1), (0, \mu_2), (0, \mu_3)\}$, and $\{(1, \mu_1), (1, \mu_2), (1, \mu_3)\}$. The class equation is $|\mathbb{Z}_2 \times S_3| = 2 + 2 + 2 + 3 + 3$.

9. Use the fact that $(gag^{-1})^k = ga^kg^{-1}$

13. If $x = gng^{-1} \in gN(s)g^{-1}$, then we have $gng^{-1}(gSg^{-1})gn^{-1}g^{-1} = gnSn^{-1}g^{-1} = gSg^{-1}$ and $x \in N(gSg^{-1})$. If $y \in N(gSg^{-1})$, then $y(gSg^{-1})y^{-1} = gSg^{-1}$ and $(g^{-1}yg)S(g^{-1}y^{-1}g) = S$ and $gyg^{-1} \in N(S)$.

15. Show first that if $b = gag^{-1}$, then $x \in C(s)$ if and only if $gxg^{-1} \in C(b)$.

17. Use Exercise 16. **23.** Use Exercises 18 and 19.

Section 4.5

5. *Hint:* $\rho\mu_i \neq \mu_i\rho$ for all $i = 1, 2, 3$.

7. *Hint:* ${}^{n!}/_{m(n-m)!} = (m-1)!$ times the binomial coefficient $\binom{m}{n}$.

9. *Hint:* Given four numbers x_1, x_2, x_3, x_4, we can construct three permutations conjugate to τ, namely $(x_1\ x_2)(x_3\ x_4)$, $(x_1\ x_3)(x_2\ x_4)$, and $(x_1\ x_4)(x_2\ x_3)$.

13. (a) 4! (b) $\langle\sigma\rangle$ (c) $\langle\sigma\rangle$ (d) 12

Section 4.6

1. (d) 27

3. See Example 4.6.3.

7. *Hint:* Consider the dihedral groups D_4 generated by
$\{\rho = (1\ 2\ 3\ 4), \tau = (1\ 2)(3\ 4)\}$, $\{\rho = (1\ 2\ 4\ 3), \tau = (1\ 2)(4\ 3)\}$,
$\{\rho = (1\ 3\ 2\ 4, \tau = (1\ 3)(2\ 4)\}$

13. If p divides $|G|$, then $|G| = p^n m$, where $n \geq 1$ and by Theorem 4.6.5 G has a subgroup H of order $|H| = p$. Hence $H = \langle a \rangle$ for some $a \in G$ and $|a| = p$.

15. If $p > q$, then $n_p = 1$.

17. Use Lemma 4.6.12.

Section 4.7

1. By Corollary 4.4.8, if $|G| = 9$, then $G \cong \mathbb{Z}_9$ or $\mathbb{Z}_3 \times \mathbb{Z}_3$.

7. $n_5 = 1$

11. *Hint:* $a_2 = a_1^{-1} = a_1^2$

17. *Hint:* D_{15} is generated by ρ and τ with $|\rho| = 15$, $|\tau| = 2$, $\rho\tau = \tau\rho^{-1}$.

27. $|G| = 8$. If G is non-Abelian, then show that there exists $a \in G$ with $|a| = 4$. Let $b \in G$ be not in $\langle a \rangle$. Construct the two cases.

Chapter 5 Composition Series

Section 5.1

1. (c) $\{0, 2, 4\}$

3. (c) and (d) *Hint:* Find a group you are familiar with that is isomorphic to D_6/M. Then use Proposition 5.1.3.

5. *Hint:* Consider $H_1 = \{(g_1, e_2, e_3) \mid g_1 \in G_1\}$.

7. Construct first a homomorphism from G onto $M/(M \cap N) \times N/(M \cap N)$.

Section 5.2

7. Composition factors: $\mathbb{Z}_2, \mathbb{Z}_3, \mathbb{Z}_7$

9. Composition factors: $\mathbb{Z}_2, \mathbb{Z}_2, \mathbb{Z}_3, \mathbb{Z}_4$

11. Use Proposition 5.1.3.

13. Use Exercise 9 in Section 5.1.

Section 5.3

3. Use the appropriate theorem from Section 4.6.

5. Do first Exercise 15 in Section 4.6.

9. See Example 5.2.7 and Proposition 5.1.3.

11. Use Theorem 5.3.10 and Theorem 5.3.8.

15. For some $\rho \in H$ construct $\sigma\rho\sigma^{-1} \in H$ such that $\rho \neq \sigma\rho\sigma^{-1}$ but $\rho(i) = \sigma\rho\sigma^{-1}(i)$ for some i.

17. Show first that for all $\rho \in S_n$, $\rho G_i \rho^{-1} = G_{\rho(i)}$.

19. Use Proposition 5.1.3. and Theorem 3.4.1.

Part B Rings and Fields

Chapter 6 Rings

Section 6.1

3. Yes **5.** No **7.** No
19. Show first that for all $a \in R$, $a + a = 0$.

Section 6.2

3. None
11. Do Exercise 11 in Section 6.1 first.
15. Use Proposition 6.2.11 and Example 6.2.9.
19. *Hint:* If $ax = 0$ and $aba = a$, then $a(x + b)a = a$.

Section 6.3

3. ± 1, $\pm i$
7. All nonzero elements
13. *Hint:* $(a, b)(c, d) = (ac, bd)$.
15. Use Theorem 1.2.10.
23. Do Exercise 11 in Section 6.1 first.
27. 0
31. Do Exercise 18 in Section 6.1 first.
33. Recall Corollary 2.1.14.

Chapter 7 Ring Homomorphisms

Section 7.1

7. $\phi_0 =$ 0-homomorphism, $\phi_1 =$ identity, $\phi_2(a + b\sqrt{2}) = a - b\sqrt{2}$
11. Use the fact that $\phi((1, 1)) = (a, b)$, where $(a, b) = (a^2, b^2)$.
13. (b) *Hint:* If $\phi(a)\phi(b) = 0 \in R'$, then $\phi(ab) = 0$, hence $ab \in \text{Kern } \phi$.
15. Use the four steps described in Section 2.2.
17. *Hint:* Use Proposition 7.1.8, part (4).
19. Do Exercise 31 in Section 6.3 first.

Section 7.2

3. Yes **5.** No **9.** No
11. Use first Theorem 6.1.13.
15. Use first Propositoion 7.1.9, parts (2) and (5).
19. Show first that $\mathbb{Z}_n / \langle m \rangle \cong \mathbb{Z}_d$, where $d = \gcd(n, m)$.

Section 7.3

7. $\mathbb{Q}(\sqrt{2})$ **9.** Does not exist **11.** $\mathbb{Q}$

21. (a) Use the fact that $\phi(a) = a$ for all $a \in D$ and $\phi(ab^{-1}) = \phi(a)\phi(b)^{-1}$.
(b) Use the definition of addition and multiplication in F.

Chapter 8 Rings of Polynomials

Section 8.1

5. sum = 2, product = 1 **11.** 1 **13.** $-i$

17. Use Exercise 15. **19.** (c) 1, 2, 3, 4 **21.** (c) $f(x) = 3x$

23. (b) With the notation as in Definition 8.1.7, let $g(x)f(x) = \Sigma m_i x^i$. Then

$$m_i = \Sigma_{0 \le k \le i} b_k a_{i-k} = b_0 a_i + \ldots + b_i a_0 = \Sigma_{0 \le k \le i} a_k b_{i-k} = d_i$$

Hence $g(x)f(x) = \Sigma m_i x^i. = \Sigma d_i x^i. = f(x)g(x)$.

27. If a_0 is a unit in R, let $g(x) = \Sigma_{0 \le i < \infty} b_i x^i$, where $b_0 = a_0^{-1}$ and

$$b_i = a_0^{-1}(-a_1 b_{i-1} - a_2 b_{i-2} - \ldots - a_i b_0)$$

for all $i > 0$. Then calculate $f(x)g(x)$.

31. Use the construction of the inverse in Exercise 27.

Section 8.2

5. $q(x) = x^2 + 8x + 3$, $r(x) = 3x + 4$ **9.** $1 = (x^3 + 2x + 1) + (2x^2 + 2x)(x + 2)$

11. Find $\gcd(x^2 + x + 1, x + 1)$ first. **13.** $m(x) = {}^1/_2(x^2 + x)$, $n(x) = {}^1/_2(x - x^2 - x^3)$

17. $f(x) = g^3(x) - 2g^2(x) + (2x + 1)g(x) + (1 - x)$

Section 8.3

3. $-1, \pm i$ **7.** $\mathbb{Z}_5{}^*$

9. Imitate the proof of Theorem 8.3.12 with $g(x) = (x - (a + b\sqrt{c}))(x - (a - b\sqrt{c}))$.

11. $\{1, i, -1, -i\}$ **15.** Use Exercise 14. **17.** 0

25. $f(x) = -{}^1/_2(x - 1)(x - 3)(x - 4) + {}^1/_6(x - 1)(x - 2)(x - 3)$

Section 8.4

5. $(x - 2)(x - 3)$ **9.** Calculate $f(a)$ for all $a \in \mathbb{Z}_5$. **13.** -1, 2, -3

17. $f(x) = (x^2 - 3x + 1)(x^2 + 2)$ and the zeros are $\frac{3 \pm \sqrt{5}}{2}, \pm\sqrt{2}i$.

23. Calculate $\overline{f(x)} \in \mathbb{Z}_5[x]$. **25.** Use Theorem 8.4.16.

31. First show that if $n = uv$, then $x^n - 1 = (x^u - 1)(x^{u(v-1)} + x^{u(v-2)} + \ldots)$.

Section 8.5

3. $\pm 2^{1/4}(\cos(\pi/8) + i\sin(\pi/8))$ **7.** $-1 \pm i$

11. *Hint:* $x^2 + bx + c = (x - r)(x - s) \in \mathbb{C}[x[$. Compare coefficients to express $(r - s)^2$ in terms of b and c.

15. $D = 81 > 0$ **19.** $D = -2349/27 < 0$

23. $u = \sqrt[3]{\frac{1+\sqrt{5}}{2}}$, $v = \sqrt[3]{\frac{1-\sqrt{5}}{2}}$, and the three zeros are $u + v$, $u\omega + v\omega^2$, $u\omega^2 + v\omega$.

27. $u = \cos(2\pi/9) + i\sin(2\pi/9)$, $v = \bar{u}$, $u + v = 2\cos(2\pi/9)$, $u\omega + v\omega^2 = 2\cos(8\pi/9)$, and $u\omega^2 + v\omega = 2\cos(14\pi/9)$.

33. $1, 2, \frac{-1+\sqrt{3}i}{2}, \frac{-1-\sqrt{3}i}{2}$

35. *Hint:* $u^3 + v^3 + q = -3(uv + p/3)(u + v) = 0$ if $uv + p/3 = 0$ and $u^3 + u^{-3}(p/3)^3 + q = 0$.

37. $\omega/2 \pm z$ and their conjugates, where $\omega = \frac{-1+\sqrt{3}i}{2}$, $z^2 = (\omega/2)^2 + 1$.

Section 8.6

3. $I = \langle x^2 - 2\rangle$, maximal ideal in $\mathbb{Q}[x]$ **7.** $I = \langle x^2 - 2x - 1\rangle$, maximal ideal in $\mathbb{Q}[x]$

13. Do first Exercise 25 in Section 8.4. **19.** $\phi(f(x)) = f(\sqrt{3})$

27. *Hint:* Since $f(x) \neq 0$, there exists a least positive integer n such that $a_n \neq 0$ in F.

Section 8.7

1. $\mathbb{Z}_3[x]/\langle x^2 + 1\rangle$ **9.** Yes, since $x^2 + x + 1$ is irreducible over $\mathbb{Z}_2$

15. Do first Problem 29 in Section 8.4. **21.** $4x^2 - 9x - 13$

29. $x^3 - 6x^2 + 11x - 6$ **35.** $2x^2 + x + 2$

Section 8.8

3. $\mathbb{R} \times \mathbb{R}$ **7.** $I + J$ is an ideal containing 1.

9. Let $\phi_i: \mathbb{R} \to \mathbb{R}/I_i$ be the canonical homomorphisms and for $a \in \mathbb{R}$ let $\phi(a) = (\phi_1(a), \phi_2(a), \ldots, \phi_n(a))$. Then Kern $\phi = \cap I_i$ and ϕ is onto by Exercise 8.

Chapter 9 Euclidean domains

Section 9.1

3. First show that if $|u'| \leq 1/2$ and $|v'| \leq 1/2$, then $|u'^2 - 3v'^2| \leq 3/4$. Then imitate the proof of Proposition 9.1.4.

7. $q = 1$, $r = 2 + \sqrt{2}$ **9.** $d = 7 - \sqrt{3}$ **13.** $\langle 3 - i\rangle$

17. (b) $a|b$ implies $\langle b\rangle \subseteq \langle a\rangle$, and since $v(a) = v(b)$, we obtain $\langle a\rangle = \langle b\rangle$. (See the proof of Theorem 9.1.6.) Hence $a = ub = uva$.

19. See Exercise 17.

21. (c) $\Rightarrow$ (a) *Hint:* Let $0 \neq a \in D$ and consider the ideal $I = \langle a, x\rangle$.

23. *Hint:* If $\gcd(a, b) = d$, then $a = a'd$ and $b = b'd$, where $gcd(a', b') = 1$.

25. (a), (b) and (d) Since R is a Euclidean domain, it is a PID and $I = \langle a\rangle \cap \langle b\rangle$ is an ideal in R, hence principal. Let $I = \langle m\rangle$.

27. (a) and (b) Use Exercise 26.

(c) Let $0 < u$ be a unit. If u is not equal to $(1 + \sqrt{2})^n$ for any n, show that for some $k \in \mathbb{Z}$, $(1 + \sqrt{2})^k < u < (1 + \sqrt{2})^{k+1}$. This implies $1 < u < 1 + \sqrt{2}$. Show that this is impossible since u is a unit.

Section 9.2

3. No **5.** Yes

9. $11 = (1 + 2\sqrt{3})(-1 + 2\sqrt{3})$ in $\mathbb{Z}[\sqrt{3}]$

13. Yes **15.** Yes

Section 9.3

3. $1^2 + 6^2 = (1 + 6i)(1 - 6i)$ **7.** $(1 + i)(2 + 3i)$

13. Use the fact that $\mathbb{Z}[i]$ is a PID.

15. order 49, char 7 **17.** order 5, char 5

19. *Hint:* $p = (a + bi)(a - bi)$, where $a \pm bi$ are primes in $\mathbb{Z}[i]$. Show that $\mathbb{Z}[i]/\langle a + bi\rangle \cong \mathbb{Z}_p$.

21. (d) $\mathbb{Z}_5$ and $\mathbb{Z}_{13}$

Chapter 10 Field Theory

Section 10.1

5. Yes **7.** *Hint:* Use the subgroup test (Theorem 1.2.10).

11. Yes **13.** No **15.** No

19. $\{1, i\}$ **21.** Use induction on n **25.** (c) Yes

Section 10.2

3. $1 + i$ is a zero of $x^2 - 2x + 2$ **9.** 4

11. (d) 1 **15.** $\{1, \sqrt{2}i\}$ **19.** Theorem 10.2.11.

23. Use Corollary 10.2.24. **27.** Show that $\mathbb{Q}(\sqrt{p}) \subseteq \mathbb{Q}(\sqrt{p} + \sqrt{q})$.

Section 10.3

5. $K = \mathbb{Q}(5^{1/3}, \sqrt{3}\, i)$, $[K:\mathbb{Q}] = 6$ **11.** $K = \mathbb{Q}(\sqrt{3}\, i)$, $[K:\mathbb{Q}] = 2$
13. Use induction on n.
17. Show first that if $\gcd(n, m) = 1$, then $\phi(nm) = \phi(n)\phi(m)$.
21. Recall Theorem 1.3.26. **27.** Use Theorem 10.2.29.

Section 10.4

7. Let $F = \mathbb{Z}_2[x]/\langle x^3 + x + 1\rangle$ and let $\alpha = x + \langle x^3 + x + 1\rangle$. Then $F^* = \langle\alpha\rangle$, a cyclic group of order 7.
11. $|F^*| = 26$ and $\phi(26) = 12$.
17. Use Theorem 10.4.17 and imitate Example 10.4.16.
21. Do Exercise 20 first. **25.** Use Corollary 10.4.10.
27. Use Theorem 10.4.17. **29.** Do Exercise 28 first.

Chapter 11 Geometric Constructions

Section 11.1

3. No **7.** Yes **17.** Use Propositions 11.1.12 and 11.1.14.

Section 11.2

5. Yes **9.** No **13.** Yes
17. Name the pentagon $ABCDE$. Draw the diagonals AC, AD, and BD. Call F the point of intersection of the diagonals AC and BD. Show that $AFDE$ is a parallelogram.
19. Show that if n contains an odd factor k, $n = ak$, then $2^a + 1$ divides $(2^a)^k + 1$.

Section 11.3

3. Do first Exercise 2.
5. Do first Exercise 4, then construct an angle of measure $\theta/3$.
7. Show that $\angle OAD = 2\alpha$.

Section 11.4

5. Yes **7.** angle trisections **9.** cube roots
11. Use Theorem 11.4.9 and the primitive 11th root of unity.
13. First do Exercise 12 above and Exercise 12 in Section 8.5.

Chapter 12 Galois Theory

Section 12.1

3. $\mathbb{Z}_2$

7. S_3

13. $\sqrt{p} + \sqrt{q}$

17. Use Corollary 12.1.17.

19. Use Theorem 12.1.6.

Section 12.2

1. Yes

5. Use Example 12.2.18.

9. Find the minimal polynomial and all its zeros. Then compare with Examples 12.2.17 and 12.2.19.

11. Find all zeros, then compare with Examples 12.2.17 and 12.2.19.

13. Imitate the lattices in Example 12.2.17 or Example 12.2.19.

21. Use Theorem 1.3.23 and Theorem 12.2.16.

23. $N(2^{1/3}) = 1, \text{Tr}(2^{1/3}) = 0$.

27. Express $N(\alpha)$ and $\text{Tr}(\alpha)$ in terms of the coefficients of the minimal polynomial.

Section 12.3

3. $s_2s_1 - s_3$

5. 108

9. 512

13. $\mathbb{Z}_2$

19. (c) $\mathbb{Z}_2$ (d) D_4

23. (c) $\mathbb{Z}_2$ (d) $\mathbb{Z}_4$

25. (d) V_4

27. $p_0 = p_5 = 4, p_i = -1$ for $i \neq 0, i \neq 5$.

31. Calculate V^TV and use the definition of p_i in Exercise 26.

Section 12.4

5. Yes

9. Yes

15. Yes

21. Do Exercise 20 first.

Section 12.5

5. $\mathbb{Q}(\sqrt{2}, i)$

7. $\mathbb{Q}(i)$

13. Yes

15. No

17. Yes

19. Use Corollary 4.4.8.

21. Use the appropriate exercise from Section 5.3.

Part C Selected Topics

Chapter 14 Symmetries

1. $(E^1)^{-1} = I_n - ce_{ij}$, $(E^2)^{-1} = E^2$,
$(E^3)^{-1} = I_n$ if $c = 1$, $(E^3)^{-1} = I_n - (c-1)^{-1}e_{ii}$ if $c \neq 1$

3. $\det(E^1) = \det(I_n) = 1$ $\det(E^2) = -1$ $\det(E^3) = c$

5. ($\Leftarrow$) Show first that for some elementary matrices E_i, $E_sE_{s-1}\ldots E_1A$ is the identity matrix or a matrix with a zero nth row.

7. (a) σ is even if and only if σ^{-1} is even, and $a_{1\sigma(1)}\cdot a_{2\sigma(2)}\cdot\ldots\cdot a_{i\sigma(i)}\cdot\ldots\cdot a_{n\sigma(n)} = b_{1\sigma^{-1}(1)}\cdot b_{2\sigma^{-1}(2)}\cdot\ldots\cdot b_{i\sigma^{-1}(i)}\cdot\ldots\cdot b_{n\sigma^{-1}(n)}$, where $A = \{a_{ij}\}$ and $A^{\mathrm{T}} = \{b_{ij}\}$.

9. $\begin{bmatrix} c & 0 \\ 0 & -c \end{bmatrix}$ for $0 \neq c \in \mathbb{R}$.

13. Do Exercise 12 first.

15. (1) Let ρ_θ be given by the matrix $B(\theta)$ as in Example 14.13, and let $a = (x_1, x_2, x_3)$. Then $b = \rho_\theta(a) = (x_1 \cos\theta - x_2 \sin\theta,\ x_1 \sin\theta - x_2 \cos\theta,\ x_3)$, $t_b\rho_\theta((1,0,0)) = ((1 + x_1)\cos\theta - x_2 \sin\theta,\ (1 + x_1)\sin\theta - x_2 \cos\theta,\ x_3) = \rho_\theta t_a((1,0,0))$.

17. $\{P = (x, y, z) \in \mathbb{R}^3 \mid x + y + z = 1 \text{ and } (x - {}^1/_3)^2 + (y - {}^1/_3)^2 + (z - {}^1/_3)^2 = {}^{24}/_9\}$

19. $G = \langle A({}^{2\pi}/_5), r\rangle$ **21.** $\mathbb{Z}_3$ **23.** 48

Chapter 15 Gröbner Bases

3. $3x^2z - 2xy^4z^3 - 5x + 2y - 5x + 5$

7. $(2, 0, 1)$, $\mathrm{LM}(f) = x^2z$, $\mathrm{LT}(f) = 3x^2z$ **13.** $-z^6 + z^4 + z^3$

19. $\gamma = (2, 2, 0)$, $x^\gamma = x^2y^2$, $S(f, g) = xz^4 + yz^2$ **25.** $\{x, y, z\}$

Chapter 16 Coding Theory

3. 7 **7.** 5 **11.** 0110

15. $[x_1x_2x_3]G = [x_1x_2x_3x_4x_5x_6]$, where $x_4 = x_1 + x_3$, $x_5 = x_2 + x_3$, $x_6 = x_1 + x_2$

19. 3 **23.** detect 2, correct 1

25. $|C| = 4 \leq {}^{2^5}/_{1+5}$ and $t = 1$. **27.** Use Theorem 16.13.

31. No **35.** Use Theorem 10.1.10.

39. $c_0 + c_1\alpha + \ldots + c_{n-1}\alpha^{n-1} \in A_n$ if and only if $c_0c_1\ldots c_{n-1} \in A^n$.

41. ($\Rightarrow$) If C is a cyclic code, then for any $c_0 + c_1\alpha + \ldots + c_{n-1}\alpha^{n-1} \in C$ we have $\alpha(c_0 + c_1\alpha + \ldots + c_{n-1}\alpha^{n-1}) = c_{n-1} + c_0\alpha + \ldots + c_{n-2}\alpha^{n-1} \in C$. Hence C is an ideal in A_n.

43. (a) Consider the canonical ring homomorphism ϕ: $\mathbb{Z}_2[x] \to \mathbb{Z}_2[x]/I$. $\phi^{-1}(C)$ is an ideal in $\mathbb{Z}_2[x]$ that contains $I = \langle f(x)\rangle$.

Chapter 17 Boolean Algebras

3. No

5. Yes

11. $a \wedge (b \vee c) = a \neq (a \wedge b) \vee (a \wedge c) = 0$

15. For $|B| = 8$ see Example 17.23.

17. Use Lemma 17.29.

19. (a) For $y \in C$, since ϕ is onto, there exists $x \in B$ such that $\phi(x) = y$, $x \leq 1_B$, and $y = \phi(x) \leq \phi(1_B)$.

21. $a + 0 = (a \wedge 1) \vee (a' \wedge 0) = a$ $\quad a + a = (a \wedge a') \vee (a' \wedge a) = 0$

$a \cdot 1 = a \wedge 1 = a$

25. $(a \wedge b') \vee (a' \wedge b)$

29. $(((a \wedge b) \vee (a' \wedge b')) \wedge ((c \wedge d) \vee (c' \wedge d'))) \vee$
$(((a \wedge b') \vee (a' \wedge b)) \wedge ((c \wedge d') \vee (c' \wedge d)))$

Bibliography

Adams, W. W. and P. Loustaunau *An Introduction to Gröbner Bases* (Graduate Studies in Mathematics, vol. 3) Providence, RI: American Mathematical Society, 1994.

Armstrong, M. A. *Groups and Symmetry* New York: Springer-Verlag, 1988.

Artin, E. *Galois Theory* Notre Dame, IN: University of Notre Dame Press, 1944.

Artin, M. *Algebra* Englewood Cliffs, NJ: Prentice Hall, 1991.

Becker, T. and V. Weispfenning *Gröbner Bases: A Computational Approach to Commutative Algebra* New York: Springer-Verlag, 1993.

Birkhoff, G. and S. MacLane *A Survey of Modern Algebra*, 4th ed. New York: Macmillan, 1977.

Burn, R. P. *Groups: A Path to Geometry* Cambridge: Cambridge University Press, 1985.

Burnside, W. *The Theory of Groups of Finite Order* Cambridge: Cambridge University Press, 1911.

Childs, L. *A Concrete Introduction to Higher Algebra* New York: Springer-Verlag, 1979.

Cox, David, John Little, and Donald O'Shea *Ideals, Varieties and Algorithms* New York: Springer-Verlag, 1992.

Coxeter, H. M. S. *Introduction to Geometry* New York: Wiley, 1961.

——— and W. O. Moser. *Generators and Relations for Discrete Groups*, 2nd ed. Berlin: Springer-Verlag, 1965.

Dean, R. A. *Elements of Abstract Algebra* New York: Wiley, 1966.

Edwards, H. *Galois Theory* New York: Springer-Verlag, 1984.

Fraleigh, J. B. *A First Course in Abstract Algebra*, 6th ed. Reading, MA: Addison-Wesley, 1999.

Gardiner, C. F. *A First Course in Group Theory* New York: Springer-Verlag, 1980.

Gilbert, W. J. *Modern Algebra with Applications* New York: Wiley, 1976.

Grossman, I. and W. Magnus *Groups and Their Graphs*, MAA#14 Washington, DC: Mathematical Society of America, 1964.

Hall, M., Jr. *The Theory of Groups* New York: Macmillan, 1959.

Halmos, Paul *Boolean Algebras* Princeton: Van Nostrand, 1967.

Hartshorne, R. *Geometry: Euclid and Beyond* New York: Springer-Verlag, 2000.

Herstein, J. N *Topics in Algebra*, 2nd ed. New York: Wiley, 1975.

Hungerford, T. W. *Algebra* New York: Wiley, 1974.

Jacobson, N. *Basic Algebra*, vol. I San Francisco: Freeman, 1974.

——— *Basic Algebra*, vol. II San Francisco: Freeman, 1980.

Jones, A., S. A. Morris, and K. R. Pearson *Abstract Algebra and Famous Impossibilities* New York: Springer-Verlag, 1991.

Kaplansky, I. *Fields and Rings*, 2nd ed. Chicago: University of Chicago Press, 1972.

Lidl, R. and G. Pilz *Applied Abstract Algebra*, 2nd ed. New York: Springer-Verlag, 1998.

Lockwood, E. H. and R. H. Macmillan *Geometric Symmetry* Cambridge: Cambridge University Press, 1978.

Macdonald, I. D. *The Theory of Groups* Oxford: Clarendon Press, 1968.

Martin, G. E. *Transformation Geometry* New York: Springer-Verlag, 1982.

Niven, I., H. S. Zuckerman, and H. L. Montgomery *An Introduction to the Theory of Numbers*, 5th ed. New York: Wiley, 1991.

Rotman, J. J. *A First Course in Abstract Algebra*, 2nd ed. Upper Saddle River, NJ: Prentice Hall, 2000.

——— *An Introduction to the Theory of Groups*, 4th ed. New York: Springer-Verlag, 1995.

——— *Galois Theory*, 2nd ed. New York: Springer-Verlag, 1998.

Stewart, J. *Galois Theory*, 2nd ed. London: Chapman and Hall, 1989.

Stillwell, J. *Elements of Algebra* New York: Springer-Verlag, 1994.

Toth, G. *Glimpses of Algebra and Geometry* New York: Springer-Verlag, 1998.

van der Waerden, B. L. *Modern Algebra*, 4th ed. New York: Ungar, 1966.

——— *A history of Algebra: From al-Khwarizmi to Emmy Noether* Berlin: Springer-Verlag, 1985.

Weyl, H. *The Classical Groups* Princeton: Princeton University Press, 1946.

——— *Symmetry* Princeton: Princeton University Press, 1952.

Zassenhaus, H. J. *The Theory of Groups*, 2nd ed. New York: Chelsea, 1958.

Index